Handbook of Research on User Interface Design and Evaluation for Mobile Technology

Volume I

Joanna Lumsden
National Research Council of Canada
Institute for Information Technology – e-Business, Canada

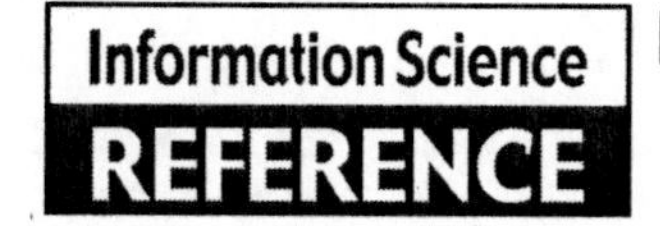

INFORMATION SCIENCE REFERENCE
Hershey · New York

Acquisitions Editor:	Kristin Klinger
Development Editor:	Kristin Roth
Senior Managing Editor:	Jennifer Neidig
Managing Editor:	Sara Reed
Copy Editor:	Joy Langel, Katie Smalley, and Angela Thor
Typesetter:	Jeff Ash
Cover Design:	Lisa Tosheff
Printed at:	Yurchak Printing Inc.

Published in the United States of America by
Information Science Reference (an imprint of IGI Global)
701 E. Chocolate Avenue, Suite 200
Hershey PA 17033
Tel: 717-533-8845
Fax: 717-533-8661
E-mail: cust@igi-global.com
Web site: http://www.igi-global.com

and in the United Kingdom by
Information Science Reference (an imprint of IGI Global)
3 Henrietta Street
Covent Garden
London WC2E 8LU
Tel: 44 20 7240 0856
Fax: 44 20 7379 0609
Web site: http://www.eurospanonline.com

Library of Congress Cataloging-in-Publication Data

Handbook of research on user interface design and evaluation for mobile technology / Joanna Lumsden, editor.

p. cm.

Summary: "This book provides students, researchers, educators, and practitioners with a compendium of research on the key issues surrounding the design and evaluation of mobile user interfaces, such as the physical environment and social context in which a device is being used and the impact of multitasking behavior typically exhibited by mobile-device users"--Provided by publisher.

Includes bibliographical references and index.

ISBN 978-1-59904-871-0 (hardcover) -- ISBN 978-1-59904-872-7 (ebook)

1. Mobile computing--Handbooks, manuals, etc. 2. Human-computer interaction--Handbooks, manuals, etc. 3. User interfaces (Computer systems)--Handbooks, manuals, etc. I. Lumsden, Joanna.

QA76.59.H36 2008

004.165--dc22

2007024493

British Cataloguing in Publication Data
A Cataloguing in Publication record for this book is available from the British Library.

Editorial Advisory Board

Table of Contents

Foreword xxviii

Preface xxx

Acknowledgment xxxiii

Section I
User Interface Design for Mobile Technologies

Chapter I
From Ethnography to Interface Design / *Jeni Paay* 1

Chapter II
Use of Experimental Ethno-Methods to Evaluate the User Experience with Mobile Interactive Multimedia Systems / *Anxo Cereijo Roibás and Stephen Johnson* 16

Chapter III
Problems Rendezvousing: A Diary Study / *Martin Colbert* 35

Chapter IV
User Experience of Camera Phones in Social Contexts / *Hanna Stelmaszewska, Bob Fields, and Ann Blandford* 55

Chapter V
Interaction Design for Personal Photo Management on a Mobile Device / *Hyowon Lee, Cathal Gurrin, Gareth J.F. Jones, and Alan F. Smeaton* 69

Chapter VI
Understanding One-Handed Use of Mobile Devices / *Amy K. Karlson, Benjamin B. Bederson, and Jose L. Contreras-Vidal* 86

Chapter VII
User Acceptance of Mobile Services / *Eija Kaasinen* 102

Chapter VIII
Transgenerational Designs in Mobile Technology / *Martina Ziefle and Susanne Bay* 122

Chapter IX
Learning-Disabled Children: A Disregarded User Group / *Susanne Bay and Martina Ziefle* 142

Chapter X
Human Factors Problems of Wearable Computers / *Chris Baber and James Knight* 158

Chapter XI
The Garment As Interface / *Sabine Seymour* 176

Chapter XII
Context as a Necessity in Mobile Applications / *Eleni Christopoulou* 187

Chapter XIII
Context-Awareness and Mobile Devices / *Anind K. Dey and Jonna Häkkilä* 205

Chapter XIV
Designing and Evaluating In-Car User-Interfaces / *Gary Burnett* 218

Chapter XV
Speech-Based UI Design for the Automobile / *Bent Schmidt-Nielsen, Bret Harsham,*
Bhiksha Raj, and Clifton Forlines 237

Chapter XVI
Design for Mobile Learning in Museums / *Nikolaos Tselios, Ioanna Papadimitriou,*
Dimitrios Raptis, Nikoletta Yiannoutsou, Vassilis Komis, and Nikolaos Avouris 253

Chapter XVII
Collaborative Learning in a Mobile Technology Supported Classroom / *Siu Cheung Kong* 270

Chapter XVIII
Design of an Adaptive Mobile Learning Management System / *Hyungsung Park,*
Young Kyun Baek, and David Gibson 286

Chapter XIX
Adaptive Interfaces in Mobile Environments: An Approach Based on Mobile Agents /
Nikola Mitrovic, Eduardo Mena, and Jose Alberto Royo 302

Chapter XX
Intelligent User Interfaces for Mobile Computing / *Michael J. O'Grady and*
Gregory M.P. O'Hare 318

Chapter XXI
Tools for Rapidly Prototyping Mobile Interactions / *Yang Li, Scott Klemmer, and*
James A. Landay 330

Chapter XXII
Modelling and Simulation of Mobile Mixed Systems / *Emmanuel Dubois, Wafaa Abou Moussa, Cédric Bach, and Nelly de Bonnefoy* 346

Chapter XXIII
Engineering Emergent Ecologies of Interacting Artefacts / *Ioannis D. Zaharakis and Achilles D. Kameas* 364

Section II
Novel Interaction Techniques for Mobile Technologies

Chapter XXIV
The Design Space of Ubiquitous Mobile Input / *Rafael Ballagas, Michael Rohs, Jennifer G. Sheridan, and Jan Borchers* 386

Chapter XXV
Text Entry/ *Mark David Dunlop and Michelle Montgomery Masters* 408

Chapter XXVI
Improving Stroke-Based Input of Chinese Characters / *Min Lin, Andrew Sears, Steven Herbst, and Yanfang Liu* 426

Chapter XXVII
Voice-Enabled User Interfaces for Mobile Devices / *Louise E. Moser and P. M. Melliar-Smith* 446

Chapter XXVIII
Speech-Centric Multimodal User Interface Design in Mobile Technology / *Dong Yu and Li Deng* 461

Chapter XXIX
Model-Based Target Sonification in Small Screen Devices: Perception and Action / *Parisa Eslambolchilar, Andrew Crossan, Roderick Murray-Smith, Sara Dalzel-Job, and Frank Pollick* 478

Chapter XXX
Unobtrusive Movement Interaction for Mobile Devices / *Panu Korpipää, Jukka Linjama, Juha Kela, and Tapani Rantakokko* 507

Chapter XXXI
EMG for Subtle, Intimate Interfaces / *Enrico Costanza, Samuel A. Inverso, Rebecca Allen, and Pattie Maes* 524

Chapter XXXII
Mobile Camera-Based User Interaction / *Tolga Capin and Antonio Haro* 543

Chapter XXXIII
3-D Visualization on Mobile Devices / *Andrea Sanna and Fabrizio Lamberti* 558

Chapter XXXIV
Navigation Support for Exploring Starfield Displays on Personal Digital Assistants /
Thorsten Büring 576

Chapter XXXV
Projected Displays of Mobile Devices for Collaboration / *Masanori Sugimoto* 594

Section III
Assistive Mobile Technologies

Chapter XXXVI
Designing Mobile Technologies for Individuals with Disabilities / *Rock Leung and*
Joanna Lumsden 609

Chapter XXXVII
Mobile Design for Older Adults / *Katie A. Siek* 624

Chapter XXXVIII
Designing Mobile Applications to Support Mental Health Interventions / *Mark Matthews,*
Gavin Doherty, David Coyle, and John Sharry 635

Chapter XXXIX
Widely Usable User Interfaces on Mobile Devices with RFID / *Francesco Bellotti,*
Riccardo Berta, Alessandro De Gloria, and Massimiliano Margarone 657

Chapter XL
Toward a Novel Human Interface for Conceptualizing Spatial Information in Non-speech Audio /
Shigueo Nomura, Takayuki Shiose, Hiroshi Kawakami, Osamu Katai, and Keiji Yamanaka 673

Chapter XLI
A Navigational Aid for Blind Pedestrians Designed with User- and Activity-Centered
Approaches / *Florence Gaunet and Xavier Briffault* 693

Chapter XLII
Trends in Adaptive Interface Design for Smart Wheelchairs / *Julio Abascal, Borja Bonail,*
Daniel Cagigas, Nestor Garay, and Luis Gardeazabal 711

Section IV
Evaluation Techniques for Mobile Technologies

Chapter XLIII
Evaluating Mobile Human-Computer Interaction / *Chris Baber* 731

Chapter XLIV
Usability Evaluation Methods for Mobile Applications / *Regina Bernhaupt, Kristijan Mihalic, and Marianna Obrist* 745

Chapter XLV
Evaluating Context-Aware Mobile Interfaces for Professionals / *Jan Willem Streefkerk, Myra P. van Esch-Bussemakers, Mark A. Neerincx, and Rosemarijn Looije* 759

Chapter XLVI
Appropriating Heuristic Evaluation Methods for Mobile Computing / *Enrico Bertini, Tiziana Catarci, Alan Dix, Silvia Gabrielli, Stephen Kimani, and Giuseppe Santucci* 780

Chapter XLVII
Using Wizard of Oz to Evaluate Mobile Applications / *Janet C. Read* 802

Chapter XLVIII
Cognitive Models as Usability Testing Tools / *Vanja Kljajevic* 814

Chapter XLIX
Assessing Human Mobile Computing Performance by Fitt's Law / *Thomas Alexander, Christopher Schlick, Alexander Sievert, and Dieter Leyk* 830

Chapter L
Multilayered Approach to Evaluate Mobile User Interfaces / *Maria de Fátima Queiroz Vieira Turnell, José Eustáquio Rangel de Queiroz, and Danilo de Sousa Ferreira* 847

Chapter LI
Theory and Application of the Privacy Regulation Model / *Jaakko T. Lehikoinen* 863

Chapter LII
Framework and Model of Usability Factors of Mobile Phones / *Dong-Han Ham, Jeongyun Heo, Peter Fossick, William Wong, Sanghyun Park, Chiwon Song, and Mike Bradley* 877

Chapter LIII
Will Laboratory Test Results be Valid in Mobile Contexts? / *Anne Kaikkonen, Aki Kekäläinen, Mikael Cankar, Titti Kallio, and Anu Kankainen* 897

Chapter LIV
Mobile Evaluations in a Lab Environment / *Murray Crease and Robert Longworth* 910

Chapter LV
Instrumented Usability Analysis for Mobile Devices / *Andrew Crossan, Roderick Murray-Smith, Stephen Brewster, and Bojan Musizza* 927

Chapter LVI
Three Eye Movement Studies of Mobile Readability / *Gustav Öquist* 945

Chapter LVII
Did You See That? / *Murray Crease and Joanna Lumsden* 972

Chapter LVIII
A Field Laboratory for Evaluating in Situ / *Rune T. Høegh, Jesper Kjeldskov, Mikael B. Skov, and Jan Stage* 982

Chapter LIX
Field Evaluation of Collaborative Mobile Applications / *Adrian Stoica, Georgios Fiotakis, Dimitrios Raptis, Ioanna Papadimitriou, Vassilis Komis, and Nikolaos Avouris* 997

Section V
Case Studies

Chapter LX
UI Design for Mobile Technology in a Closed Environment / *Kate Oakley, Gitte Lindgaard, Peter Kroeger, John Miller, Earl Bryenton, and Paul Hébert,* 1015

Chapter LXI
Designing a Ubiquitous Audio-Based Memory Aid / *Shwetak N. Patel, Khai N. Truong, Gillian R. Hayes, Giovanni Iachello, Julie A. Kientz, and Gregory D. Abowd* 1031

Chapter LXII
Visualisation of Meeting Records on Mobile Devices / *Saturnino Luz and Masood Masoodian* 1049

Chapter LXIII
A Proposed Tool for Mobile Collaborative Reading / *Jason T. Black and Lois Wright Hawkes* 1068

Chapter LXIV
Evaluating Learner Satisfaction in a Multiplatform E-Learning System / *Tiong T. Goh, Kinshuk, and Nian-Shing Chen* 1079

Detailed Table of Contents

Foreword xxviii

Preface xxx

Acknowledgment xxxiii

Section I
User Interface Design for Mobile Technologies

Chapter I
From Ethnography to Interface Design / *Jeni Paay*........ 1

Traditional design methods are often ill suited to the unique challenges inherent in interface design for mobile technology. This chapter looks at the way in which ethnography can inform the design of such technologies, and proposes a means of effectively transferring knowledge gained via ethnographic methods (such as rich understanding of mobile use contexts) to interface design. The author proposes an approach that bridges the gap between ethnography and interface design: the outcomes of field data-informed design sketching and iterative development of paper-based mock-ups can be used as a starting point for iterative prototype development. This chapter presents a design case study of a context-aware mobile information system to illustrate the proposed approach.

Chapter II
Use of Experimental Ethno-Methods to Evaluate the User Experience with Mobile Interactive Multimedia Systems / *Anxo Cereijo Roibás and Stephen Johnson* 16

Underpinning the work presented in this chapter is the assumption that commuters are a particularly relevant and interesting demographic in which to investigate novel interaction with mobile multimedia content. The chapter discusses the use of both validated and experimental ethnographic data gathering techniques to understand how mobile users (specifically nomadic users such as commuters) interact in real contexts. Through extensive use of scenarios, the authors demonstrate how handhelds are appropriate for the creation of new forms of multimedia content by users, and how novel forms of interaction could be implemented.

Chapter III
Problems Rendezvousing: A Diary Study / *Martin Colbert* 35

As reported in this chapter, one third of current mobile phone use is rendezvous-related, making rendezvousing support an important facet of mobile technology. This chapter discusses a diary study of university students' use of mobile telephones for rendezvousing—that is, arranging, and traveling to, informal meetings with friends and family—which highlighted a number of deficits in user performance. Based on the observed performance issues, this chapter discusses design implications (goals) associated with addressing these deficits. In particular, the author illustrates the kind of mobile technology that might address each identified deficit.

Chapter IV
User Experience of Camera Phones in Social Contexts / *Hanna Stelmaszewska, Bob Fields, and Ann Blandford* 55

This chapter reports on a study of people's use of camera phones for social interaction. The study, which looked at people's behaviour and positive experiences when camera phones were used in public and private spaces, discovered that camera phones influence social practices. The authors identified three distinct practices in copresent settings: sharing a moment now, sharing a moment later, and using photos to initiate social interaction with strangers, knowledge of which, they suggest, deepens our conceptual understanding of how camera phones are becoming incorporated into users' leisure-related practices, and informs the design of future camera phones to facilitate social interaction.

Chapter V
Interaction Design for Personal Photo Management on a Mobile Device / *Hyowon Lee, Cathal Gurrin, Gareth J.F. Jones, and Alan F. Smeaton* 69

This chapter constitutes a review of mechanisms that have the potential to enhance user interaction with personal photos on mobile devices. It reflects on recent technological innovations that influence personal photo management behaviour and needs, and discusses the design issues in supporting such behavioural patterns and needs on mobile devices. The authors introduce the concepts of content-based image analysis and context-awareness that, as mechanisms for supporting automatic annotation and organisation of photos, they suggest are becoming an important factor in helping to design efficient and effective mobile interfaces for personal photo management systems.

Chapter VI
Understanding One-Handed Use of Mobile Devices / *Amy K. Karlson, Benjamin B. Bederson, and Jose L. Contreras-Vidal* 86

The research presented in this chapter focuses on situations in which users engage in one-handed operation of their mobile devices. The chapter reports on three studies that were conducted to understand different aspects of one-handed mobile design requirements. The first of these studies, which was designed to capture an impression of the extent of current one-handed mobile phone use in the real world, comprised a field-based, naturalistic, anonymous observation of mobile phone users. This study showed that when walking while using their mobile phone, users were more likely to engage in one-handed use than when sitting, for example, often as a consequence of the other hand being otherwise occupied.

The results of this study suggest that one-handed use should be a serious consideration when designing mobile phones. Prompted by their initial study, the authors conducted a more extensive survey of mobile device use to highlight one-handed usage patterns relative to task and device type. The results of this study show both current usage patterns and preferred usages patterns, and from this, the authors conclude that designers should make one-handed usability a priority in order to bridge the gap between current and desired usage patterns. Finally, the chapter reports on the results of a thumb movement study designed to investigate the extent to which user performance was influenced by device size, task region, and movement direction. On the basis of this final study, the authors present guidelines concerning the use of thumb movement in mobile interaction design.

Chapter VII
User Acceptance of Mobile Services / *Eija Kaasinen* 102

Usability alone does not determine user acceptance of mobile services. This chapter introduces the technology acceptance model for mobile services (TAMM), which models the way in which a complex set of user acceptance factors affect user acceptance of mobile services. Based on field trials of several mobile services with more than 200 test users, the TAMM is designed to be used as a design and evaluation framework when creating new mobile services. Supporting the identification of issues that should be focused on during the design of new mobile services to ensure user acceptance, the motivation of the TAMM is different than that of the original TAM, which was built to *explain* user acceptance. The TAMM defines four main user acceptance factor (perceived value, perceived ease of use, trust, and perceived ease of adoption) and this chapter discusses design implications associated with each.

Chapter VIII
Transgenerational Designs in Mobile Technology / *Martina Ziefle and Susanne Bay*...................... 122

This chapter reflects on an examination of the complex relationship between user characteristics or diversity and menu navigation performance on mobile devices. The reported study considered the effect of age, gender, cognitive factors, motivational factors, and prior technological experience in terms of people's ability to effectively navigate menu structures on mobile phones. On the basis of a detailed analysis of individual interaction patterns to determine the effectiveness and efficiency of individuals' menu navigation, the authors discuss implications for the design of mobile phones. Most noticeably, the results show that although the use of small screen devices poses certain difficulties for *all* users, children and middle-aged adults were seen to be very sensitive to the cognitive demands imposed by current mobile phone designs.

Chapter IX
Learning-Disabled Children: A Disregarded User Group / *Susanne Bay and Martina Ziefle*............ 142

Usability studies often recruit participants from university student populations. This chapter reports on the benefits or results to be gained from including users with cognitive difficulties, in this case, learning-disabled children, a disregarded user group with respect to mobile technology design. By comparing learning-disabled children's interaction with mobile phones to that of "average" children and university students, the authors are able to demonstrate the qualitatively and quantitatively different insights into the impact of specific design decisions brought about by including this disregarded user group. The authors suggest that, when participants are restricted to university students, mobile technology evaluations typically fail to observe the full extent to which characteristics of the user interface impact on ease of use.

This chapter indicates that when designing a technology that is intended for general use, the "ergonomic worse case" should be included in the set of evaluation participants.

Chapter X
Human Factors Problems of Wearable Computers / *Chris Baber and James Knight* 158

This chapter introduces the concept of wearable computers as a form of cognitive prosthesis. On the basis that a prosthesis can be considered to fulfill a replacement, correction, or enhancement role, the authors argue that wearable computers could be considered as a cognitive prosthesis based on their potential to enhance cognitive performance. This chapter presents an overview of development in the field of wearable computing, stressing the importance of considering the physical and cognitive characteristics of the user when designing such technology.

Chapter XI
The Garment As Interface / *Sabine Seymour* 176

This chapter considers the concept of the surface of a smart garment as a dynamic interface (or interactive display), exposing a plethora of new applications. Recent developments in wearable technologies mean that textile surfaces can display data from input devices such as sensors and cell phones. Given the extremely personal nature of intelligent garments, their design and construction must consider the wearer's needs, the context of use, and other relevant factors of the wearing experience. This chapter outlines a list of issues regarding human interaction with smart garments and dynamic visual interfaces, which the author suggests it is essential to consider in order to design usable, smart garments.

Chapter XII
Context as a Necessity in Mobile Applications / *Eleni Christopoulou* 187

This chapter aims to demonstrate that the use of context information in mobile applications is a necessity. It discusses how context information can be used to support user interaction, arguing that its use is not restricted to locating users and providing them with suitable information, but that it can also be used to support automatic selection of appropriate interaction techniques. The chapter focuses on modeling devices, services, and context in a formal way, like ontologies, and presents an ontology-based context model that allows users to set up their own context-aware applications and define the way that artifacts react to changes.

Chapter XIII
Context-Awareness and Mobile Devices / *Anind K. Dey and Jonna Häkkilä* 205

This chapter introduces and defines the concepts of "context" and "context awareness" as they apply to mobile and ubiquitous computing. The chapter describes the challenges and complexities inherent in building context-aware mobile applications, and highlights some of the toolkits that have been built to address these challenges. The authors present a series of validated and evaluated design guidelines that can aid the designers of mobile context-aware applications in producing applications with both novel and useful functionality. The chapter highlights areas where user interface design effort needs to be focused in order to address the usability issues that are commonly found with mobile context-aware applications.

Chapter XIV
Designing and Evaluating In-Car User-Interfaces / *Gary Burnett*.. 218

Accompanying the introduction of computing and communications technologies within cars is a critical need to understand how user-interfaces to such technologies can best be designed to appropriately accommodate the constraints placed on users by the driving context. This chapter outlines the driving context, and highlights the range of computing systems (and associated user-interfaces) being introduced into this context. The author describes the factors that designers of user-interfaces to in-car technologies must consider, and compares the various facilities available the support the design and evaluation (e.g., simulators, instrumented vehicles) of such systems. The chapter illustrates the issues raised via discussion of a vehicle navigation system case study, and highlights continuing research challenges in this field.

Chapter XV
Speech-Based UI Design for the Automobile / *Bent Schmidt-Nielsen, Bret Harsham, Bhiksha Raj, and Clifton Forlines* 237

This chapter discusses issues regarding speech-based user interfaces for use in an automotive environment. The authors present a series of design principles or recommendations for such interfaces, illustrating their discussion with three case studies of current automotive navigation interfaces. The authors propose a new model for speech-based user interfaces in automotive environments that centers around selection of a desired command from a short list, and discuss experimental results that show this style of speech-based interface, compared to conventional user interfaces, has the potential to significantly reduce the cognitive load imposed on drivers.

Chapter XVI
Design for Mobile Learning in Museums / *Nikolaos Tselios, Ioanna Papadimitriou, Dimitrios Raptis, Nikoletta Yiannoutsou, Vassilis Komis, and Nikolaos Avouris*.............................. 253

This chapter discusses the design challenges associated with mobile learning applications for museums. The authors present a review of existing systems, from which they highlight design approaches and guidelines. The authors argue that the design of mobile learning applications for museums must consider an appropriate theoretical cognitive framework, as well as the context in which the application is to be used (including the device itself and the characteristics of the museum setting). To illustrate the concepts discussed, the authors describe a case study of designing a collaborative learning activity for a cultural history museum, reflecting on the experiences gained during the design process.

Chapter XVII
Collaborative Learning in a Mobile Technology Supported Classroom / *Siu Cheung Kong*............. 270

Based on a case study, this chapter discusses the process of migrating a Web-based cognitive tool from a desktop application to a mobile application in order to investigate the potential to increase the effectiveness of the cognitive tool by taking advantage of collaborative and mobile learning. This chapter outlines the theoretical design approach and empirical design methodology underpinning the migration, and then discusses the architectural and pedagogical design elements in order to illustrate aspects of the development and application of mobile technology in a classroom learning environment.

Chapter XVIII
Design of an Adaptive Mobile Learning Management System / *Hyungsung Park, Young Kyun Baek, and David Gibson* 286

This chapter introduces the adaptive mobile learning management system (AM-LMS) platform, which represents an adaptive environment that continually matches the needs and requirements of individual learners in a mobile learning environment to a mobile device's use of remote learning resources. The chapter discusses the concept of, and research trends associated with, mobile learnin,g as well as classifications of learning styles and how these relate to the AM-LMS platform, before describing the structure of the AM-LMS itself. The AM-LMS platform presents an ability to take advantage of the unique features of mobile devices combined with the ability to support individualized learning and learner-centered education.

Chapter XIX
Adaptive Interfaces in Mobile Environments: An Approach Based on Mobile Agents / *Nikola Mitrovic, Eduardo Mena, and Jose Alberto Royo* 302

Designing graphical user interfaces for effective use on mobile devices is challenging given the extent to which device capabilities differ and contexts, combined with user preferences, continually change. The authors of this chapter propose that the solution to this problem lies in using a single, abstract, user interface description that is then used to automatically generate user interfaces for different devices. Although there are a number of techniques possible for translating an abstract specification to a concrete interface, this chapter suggests that, by using indirect generation, it is possible to perform run-time analysis of computer-human interaction and application of artificial intelligence techniques, and thereby increase the resulting graphical user interface's performance and usability. The authors present their proposal for an indirect generation approach.

Chapter XX
Intelligent User Interfaces for Mobile Computing / *Michael J. O'Grady and Gregory M.P. O'Hare* 318

Meeting the demand for increasingly sophisticated applications and services poses new challenges for designers of mobile applications. The authors of this chapter suggest that careful adoption of intelligent user interfaces may offer a practical approach for evolving mobile services. To this end, using the intelligent agent paradigm for illustration purposes, this chapter discusses in detail the issues associated with adopting intelligent techniques in mobile applications.

Chapter XXI
Tools for Rapidly Prototyping Mobile Interactions / *Yang Li, Scott Klemmer, and James A. Landay* 330

This chapter introduces the potential for informal prototyping tools to speed up the early stage design of mobile interactions. The authors present two tools that address the early stage design of speech-based and location-enhanced interactions as proofs of concept for informal prototyping tools for mobile interactions. The chapter highlights the use of storyboarding and Wizard of Oz (WOz) testing, and discusses how these can be applied. The authors report on a case study to illustrate the iterative design of a location-aware application.

Chapter XXII
Modelling and Simulation of Mobile Mixed Systems / *Emmanuel Dubois, Wafaa Abou Moussa, Cédric Bach, and Nelly de Bonnefoy* 346

This chapter introduces the concept of a "mixed system" – that is, an interactive system that merges the physical and digital worlds. Where such systems support users' mobility, they are known as "mobile mixed systems". This chapter defines and classifies mixed systems, and presents an overview of existing support for their implementation. It then presents the collaboration between an existing design model (ASUR) for mixed systems and a 3-D environment (SIMBA) for simulating modeled mobile mixed systems as a first step toward an iterative method for designing such systems. With this combination, the authors aim to support investigation of mobile mixed system design and a better appreciation of the limits of modeled solutions through their simulation.

Chapter XXIII
Engineering Emergent Ecologies of Interacting Artefacts / *Ioannis D. Zaharakis and Achilles D. Kameas* 364

Everyday artifacts (including mobile devices) are being enhanced with sensing, processing, and communication abilities, with the result that we are surrounded by an increasingly complex environment of machine-machine and human-machine interaction. This chapter introduces a model that draws features from natural systems and applies them into ecologies inhabited by both humans and artifacts. It also introduces a high-level framework of Ambient Intelligent spaces that encapsulates the fundamental elements of bio-inspired self-aware emergent symbiotic ecologies. This chapter links the use of mobile devices to truly ubiquitous computing.

Section II
Novel Interaction Techniques for Mobile Technologies

Chapter XXIV
The Design Space of Ubiquitous Mobile Input / *Rafael Ballagas, Michael Rohs, Jennifer G. Sheridan, and Jan Borchers* 386

In addition to their core communications function, mobile phones are increasingly used for interaction with the physical world. This chapter introduces the notion of a "design space" and uses this concept as the basis for an in-depth discussion of existing interaction techniques, relating desktop to mobile phone techniques. The authors present a new five-part spatial classification for ubiquitous mobile phone interaction tasks, which covers supported subtasks, dimensionality, relative vs. absolute, interaction style, and feedback from the environment. The chapter identifies key design considerations in terms of real-world deployment of applications using these interaction techniques.

Chapter XXV
Text Entry/ *Mark David Dunlop and Michelle Montgomery Masters* 408

Today's mobile devices are too small to accommodate a full-size keyboard, making text entry a challenge. This chapter reviews text entry techniques for smaller keyboards and stylus input (including different hardware keyboard designs, different on-screen keyboard layouts, handwriting-based approaches, and more novel approaches such as gestures) and reflects on the nature of the evaluations that have been conducted to assess their validity. The authors discuss criteria for acceptance of new text entry techniques, and comment on how market perceptions can overrule laboratory successes. The chapter concludes with some interesting yet intentionally controversial statements to encourage the reader to consider the future of text entry on mobile devices.

Chapter XXVI
Improving Stroke-Based Input of Chinese Characters / *Min Lin, Andrew Sears, Steven Herbst, and Yanfang Liu* 426

Entering English text into a mobile phone can be challenging, but entering Chinese characters (of which there are thousands) using a mobile phone keypad is much more difficult. This chapter presents work, detailed on a step-by-step basis, that was undertaken to redesign the keypad graphics (that is, the symbols printed on the keys as the legends for Chinese strokes) for the Motorola iTap™ stroke-based input solution. Presented as a case study, this chapter introduces and compares the original iTap™ solution and the Pinyin method. It describes an alternative design, and details two longitudinal studies of the proposed design conducted in the USA and China. The chapter also discusses the simplification of the proposed design to allow it to fit on smaller keypads, and outlines its subsequent evaluation.

Chapter XXVII
Voice-Enabled User Interfaces for Mobile Devices / *Louise E. Moser and P. M. Melliar-Smith* 446

The authors of this chapter posit that the use of speech-based interaction, when combined with other modes of interaction, can enhance user experience with mobile technology. This chapter describes a prototype system that supports client-side, voice-enabled applications on mobile devices. The authors reflect on the design issues they faced, and evaluation methods they employed, during the development of a voice-enabled user interface for a mobile device. Interestingly, they discuss the need to evaluate the user interface and speech-recognizer independently to avoid mixing of data leading to inconclusive results, and describe how they achieved this separation when evaluating their system.

Chapter XXVIII
Speech-Centric Multimodal User Interface Design in Mobile Technology / *Dong Yu and Li Deng* 461

Multimodal user interfaces support interaction with a system via multiple human-computer communication channels (or modalities) and, as such, are particularly beneficial for mobile devices in terms of their ability to circumvent the complexities introduced by the limited form-factor of mobile devices and the varied contexts in which mobile devices are used. Taking a speech-centric view, this chapter surveys multimodal user interface design for mobile technology. Based on a selection of carefully chosen case studies, the authors discuss the main issues related to speech-centric multimodal user interfaces for mobile devices.

Chapter XXIX
Model-Based Target Sonification in Small Screen Devices: Perception and Action / *Parisa Eslambolchilar, Andrew Crossan, Roderick Murray-Smith, Sara Dalzel-Job, and Frank Pollick* 478

Despite the fact that our two primary senses are hearing and vision, the majority of interfaces to technology focus exclusively on the visual. Due to limited screen real estate, coupled with the need for users to attend to their physical environment and/or simultaneous task(s) rather than the interface, designing interfaces for mobile technology is problematic. This chapter reports on an investigation into the use of audio and haptic feedback to augment the display of a mobile device controlled by tilt input. The authors present the results of their investigation as a useful starting point for further investigation into appropriate feedback for users of a tilt-controlled mobile device with multimodal feedback. The chapter highlights the difficulty of designing experiments to test aspects of low-level perception of multimodal displays while avoiding the influence of prior knowledge.

Chapter XXX
Unobtrusive Movement Interaction for Mobile Devices / *Panu Korpipää, Jukka Linjama, Juha Kela, and Tapani Rantakokko* 507

Gesture-based interaction is evolving as a feasible modality for interaction with mobile technology. As an interaction paradigm, however, it is not without design challenges: for example, it is hard to detect gestural input reliably, it is hard to distinguish intended gestural movement from other user movement, the social acceptance of gestural input is as yet undetermined, and it can be hard to design meaningful feedback regarding the status of a gesture. In an attempt to address some of these challenges, this chapter presents an event-based movement interaction modality, tapping, which requires minimal user effort in interacting with a mobile device, and discusses the results of its evaluation.

Chapter XXXI
EMG for Subtle, Intimate Interfaces / *Enrico Costanza, Samuel A. Inverso, Rebecca Allen, and Pattie Maes* 524

This chapter asserts that mobile interfaces should be designed to enable subtle, discreet, and unobtrusive interaction. The authors discuss the ability of the electromyographic (EMG) signal, which is generated by muscle contraction, to act as a subtle input modality for mobile interfaces. To illustrate this concept, this chapter presents the Intimate Communication Armband, an EMG-based wearable input device that detects subtle, isometric (motionless) gestures from the upper arm. The authors discuss experimental results that attest to its reliability, effectiveness, and subtlety as a hands-free input device.

Chapter XXXII
Mobile Camera-Based User Interaction / *Tolga Capin and Antonio Haro* 543

This chapter introduces a camera-based approach for user interaction on mobile devices whereby a user interacts with an application by moving their device, and the captured video is used to determine interaction. The chapter reviews computer vision technologies and different uses to which such technology has been put in terms of mobile human-computer interaction. The authors present a camera-based toolkit prototype, reflect on design issues faced in its development, and illustrate their approach in several applications using 2-D and 3-D interaction.

Chapter XXXIII
3-D Visualization on Mobile Devices / *Andrea Sanna and Fabrizio Lamberti* 558

Mobile devices are now able to display 3-D graphic content, with the result that demand for visualization applications is rapidly increasing. This chapter reflects on the fact that the development of 3-D visualization environments on mobile devices demands careful consideration of performance constraints and user interaction requirements. After presenting a review of the main solutions for developing 3-D visualization applications, the authors introduce a complete framework for remote visualization in mobile environments based on distributed rendering and video streaming techniques and discuss the results of an evaluation of its effectiveness.

Chapter XXXIV
Navigation Support for Exploring Starfield Displays on Personal Digital Assistants /
Thorsten Büring 576

Mobile devices are now able to handle large-scale data sets, but effectively displaying and supporting navigation of such high information loads on limited screen real estate is challenging. This chapter suggests that starfield displays may offer a solution to this problem, but that user orientation is often problematic due to the clipping of orientation cues. This chapter provides an overview of recent research that has looked at improving the navigation and orientation features of starfield displays on small screens. Specifically, it focuses on smooth zooming, overview+detail, and focus+context approaches, and discusses their adaptation to small screens as well as the results of user testing of these adapted techniques.

Chapter XXXV
Projected Displays of Mobile Devices for Collaboration / *Masanori Sugimoto* 594

As their functionality continues to increase, mobile devices are moving from being personal tools to being used in a shared or collaborative fashion. This chapter focuses on shared mobile device use by means of display projection. It provides an overview of systems and technologies related to location-aware projection before introducing a system that implements intuitive manipulation techniques on projected displays of multiple mobile devices. This chapter describes the intuitive manipulation techniques employed within this system and discusses how user studies suggest that the manipulation techniques have the potential to support collaborative tasks in co-located situations. The chapter highlights the research issues associated with shared projected displays of mobile devices.

Section III
Assistive Mobile Technologies

Chapter XXXVI
Designing Mobile Technologies for Individuals with Disabilities / *Rock Leung and*
Joanna Lumsden 609

Mobile devices offer many innovative possibilities to help increase the standard of living for individuals with disabilities and other special needs, but the process of developing assistive technology can be extremely challenging. This chapter discusses key issues and trends related to designing and evaluating mobile assistive technology, and presents an overview of general design process issues. The authors suggest that individuals with disabilities and domain experts be involved throughout the development process. Although this presents its own set of challenges, many strategies have successfully been used to overcome the difficulties and maximize the contributions of users and experts alike. Guidelines based on these strategies are discussed and are illustrated with real examples from active research projects.

Chapter XXXVII
Mobile Design for Older Adults / *Katie A. Siek*.. 624

Faced with an aging population, there is a challenging need to design technologies to help older adults remain independent and preserve their quality of life. Current research is attempting to address this challenge by means of assistive technologies based on mobile devices such as personal digital assistants and cell phones, but the question remains as to whether or not older people can use such technologies effectively because of age-related problems. This chapter discusses issues surrounding the design, implementation, and evaluation of mobile applications for older adults, highlighting the unique challenges posed by, and best practices for working with and designing for, this population.

Chapter XXXVIII
Designing Mobile Applications to Support Mental Health Interventions / *Mark Matthews, Gavin Doherty, David Coyle, and John Sharry*.. 635

This chapter discusses issues that arise when designing and evaluating mobile software for sensitive situations, where access to end-users is extremely restricted and traditional design and evaluation methods are inappropriate and/or not viable. Specifically, this chapter focuses on the use of technology to support adolescents in mental health care settings, and highlights the corresponding contraints (both practical and ethical) that affect approaches to design and evaluation. The authors present some design recommendations for technological interventions of this nature, and suggest methods to maximise the value of evaluations conducted under such restricted situations. The issues outlined in this chapter are illustrated, or made tangible, by means of a case study relating the design and evaluation of a mobile phone-based "mood diary" application for use in clinical situations by adolescents undergoing mental health interventions.

Chapter XXXIX
Widely Usable User Interfaces on Mobile Devices with RFID / *Francesco Bellotti, Riccardo Berta, Alessandro De Gloria, and Massimiliano Margarone*... 657

Radio frequency identification (RFID) has the potential to enhance assistive technologies for mobile users; for example, RFID-based applications may support users with visual impairment by providing information on their current location and physical surroundings. This chapter discusses an extension to a mobile application development tool to include support for designing RFID-based applications. The authors describe an RFID-enabled location-aware tour-guide they developed using the development environment, highlighting the main concepts of the interaction modalities they designed to support visually impaired users. The authors describe a field evaluation of this application, and discuss the results which illustrate the costs, limits, strengths, and benefits of the new technology.

Chapter XL
Toward a Novel Human Interface for Conceptualizing Spatial Information in Non-speech Audio / *Shigueo Nomura, Takayuki Shiose, Hiroshi Kawakami, Osamu Katai, and Keiji Yamanaka* 673

This chapter looks at the concept of using non-speech audio for building interfaces to support visually impaired users. Specifically, it focuses on "unconventional" mechanisms by which to use non-speech audio to enable users with visual impairments to conceptualize spatial information. The authors report on studies they conducted towards meeting this goal, from which they were able to observe that sound effects, such as reverberation and reflection, enhance users' ability to localize pattern-associated sounds, and that "natural" sounds (as opposed to "artificial' sounds) better supported users' conceptualization of spatial information.

Chapter XLI
A Navigational Aid for Blind Pedestrians Designed with User- and Activity-Centered Approaches / *Florence Gaunet and Xavier Briffault* 693

This chapter reports on the process undertaken to design an interface for a mobile navigational aid for blind pedestrians. It presents a set of rules for producing route descriptions for blind users, and outlines the method (based on user- and activity-centred approaches) by which the rules were established. The authors reflect on the state of the art of wearable navigational aids, and present their approach to providing improved functional specifications for designing rules for producing verbal instructions and information for blind pedestrians.

Chapter XLII
Trends in Adaptive Interface Design for Smart Wheelchairs / *Julio Abascal, Borja Bonail, Daniel Cagigas, Nestor Garay, and Luis Gardeazabal* 711

This chapter reflects on design trends for interfaces to smart wheelchairs, and highlights the need to take into account the similarity between smart wheelchairs and autonomous mobile robots as well as the specific restrictions imposed by the users and task. The authors discuss the main aspects of the user-wheelchair interface, including the need for an adaptive design approach, and a case study is used to illustrate the design of user, context, and task models to support an intelligent adaptable interface. The chapter also includes discussion of the influence of new navigation models in the design of the user interface.

Section IV
Evaluation Techniques for Mobile Technologies

Chapter XLIII
Evaluating Mobile Human-Computer Interaction / *Chris Baber* 731

This chapter discusses, from a theoretical perspective, the concepts and issues involved in evaluating mobile human-computer interaction. The chapter assumes that "usability" is not a feature of a product, but rather it is the consequence of a given user employing a given product to perform a given activity in a given environment. The author argues that mobile context-of-use must be considered in order to assess usability as a concept of "fitness-for-purpose", and proposes a usability evaluation process specifically focused on mobile technology evaluation.

Chapter XLIV
Usability Evaluation Methods for Mobile Applications / *Regina Bernhaupt, Kristijan Mihalic, and Marianna Obrist* 745

The variability of users, uses, and contexts makes evaluating mobile applications challenging. This chapter outlines various traditional usability evaluation methods and discusses methodological variations to these approaches which make them better suited for evaluating usability aspects of mobile devices and applications. The authors suggest that a combination of both field evaluation methods and traditional laboratory testing should be used to address different phases in the user-centered design and development process for mobile technologies and introduce their "real world lab" concept as means by which to achieve both.

Chapter XLV
Evaluating Context-Aware Mobile Interfaces for Professionals / *Jan Willem Streefkerk, Myra P. van Esch-Bussemakers, Mark A. Neerincx, and Rosemarijn Looije* 759

Mobile user interfaces are often used in dynamic environments, with the result that user experiences can vary substantially. Consequently, this increases the complexity associated with effective evaluation of such interfaces. In response to this challenge, this chapter presents a framework for the systematic selection, combination, and tailoring of evaluation methods for context-aware applications based on seven evaluation constraints (the stage of development, the design complexity, and the purpose, participants, setting, duration, and cost of evaluation). This chapter describes how the framework was applied to the evaluation of a context-aware mobile interface for use by the police and how, as a result of this case study, the authors were able to derive specific guidelines for selecting evaluation methods and were able to reflect on the relationship between the mobile context and the user experience.

Chapter XLVI
Appropriating Heuristic Evaluation Methods for Mobile Computing / *Enrico Bertini, Tiziana Catarci, Alan Dix, Silvia Gabrielli, Stephen Kimani, and Giuseppe Santucci* 780

For desktop-based applications, heuristic evaluation is recognized as a cost-effective mechanism by which to identify a large proportion of usability flaws with limited resource investment. On the other hand, limited screen real estate, divided user attention, and elaborate contextual factors pose complex problems for usability assessment of mobile applications. This chapter describes a modified collection of usability heuristics, systematically derived from extensive literature and empirically validated, that are designed to be appropriate for evaluation of mobile technologies.

Chapter XLVII
Using Wizard of Oz to Evaluate Mobile Applications / *Janet C. Read* 802

This chapter introduces and describes the concept of Wizard of Oz studies. It discusses the use of such studies to evaluate mobile technologies, illustrating the issues raised with a case study. The author presents both a taxonomy for Wizard of Oz studies and a set of guidelines regarding the considerations that are essential when planning Wizard of Oz studies for mobile applications. The author argues that well planned Wizard of Oz studies can provide valuable information about user behaviour and experience that might otherwise be difficult to establish. Furthermore, this chapter suggests that the extent of use of Wizard of Oz studies will likely increase as the complexity of mobile systems increases and establishes greater demands for low-cost methods for early investigation and evaluation.

Chapter XLVIII
Cognitive Models as Usability Testing Tools / *Vanja Kljajevic* 814

This chapter introduces the concepts of computational cognitive models and cognitive architecture. The author asserts that given the complexity of mobile technologies, it is impossible to empirically assess all the possibilities of a mobile phone user interface design using traditional usability testing techniques. This chapter claims that computational cognitive models may prove to be a better alternative to theoretically unsupported, time-consuming, and often expensive traditional usability testing. Furthermore, the author argues that lack of solid theoretical underpinnings (common to many current mobile usability evaluation techniques) results in inconsistent and unreliable testing methods, and that quantitative testing is preferable to qualitative evaluation.

Chapter XLIX
Assessing Human Mobile Computing Performance by Fitt's Law / *Thomas Alexander, Christopher Schlick, Alexander Sievert, and Dieter Leyk* 830

This chapter reports on an investigation into the relationship between motion caused by walking and user input accuracy for mobile technology. The authors describe appropriate performance measures to support analysis of this interdependence, and explain how Fitt's Law can be used to support quantitative analysis. The chapter discusses, in detail, the investigative experimental protocol and outlines the results, which include an observation that error rates rise and performance levels drop significantly with increased walking speed. Quantitative estimation of these effects highlight the influence of input task difficulty, and as a result the authors are able to suggest threshold values for accuracy of user input which can be used to inform future mobile user interface design.

Chapter L
Multilayered Approach to Evaluate Mobile User Interfaces /
Maria de Fátima Queiroz Vieira Turnell, José Eustáquio Rangel de Queiroz, and Danilo de Sousa Ferreira 847

This chapter suggests that experience gained from evaluating conventional user interfaces can be applied to mobile interfaces and presents a multilayered approach or method (based on a combination of user opinion, standard conformity assessment, and user performance measurement) for evaluating user interfaces to mobile applications. The approach is illustrated by means of a case study which considered the influence of context (field vs. laboratory and mobile vs. stationary interaction) on the evaluation of mobile devices and applications.

Chapter LI
Theory and Application of the Privacy Regulation Model / *Jaakko T. Lehikoinen* 863

The issue of privacy protection is typically critical to user acceptance of any applications and services that require disclosure of personal information. This chapter argues that delivery of such applications and services in a mobile context heightens the need to consider privacy issues during application design and development. In response to this need, this chapter presents a privacy management model that facilitates evaluation of privacy aspects of communication technology. The model's applicability was evaluated by means of a field trial that was carried out to assess user acceptance of a mobile social awareness system; this case study is reported here as an example of how to apply the privacy regulation model in evaluation of a mobile communication solution.

Chapter LII
Framework and Model of Usability Factors of Mobile Phones / *Dong-Han Ham, Jeongyun Heo, Peter Fossick, William Wong, Sanghyun Park, Chiwon Song, and Mike Bradley*.. 877

This chapter proposes a framework and model for identifying, organizing, and classifying usability factors of mobile phones. The conceptual framework incorporates multiple views (including user, product, interaction, dynamic, and execution views) to explain different aspects of the interaction between users and mobile phones, and then describes usability factors in terms of these views. The authors describe a hierarchical model for the classification of usability factors in terms of goal-means relationships, which they developed based on the framework. The chapter outlines two case studies used to verify the usefulness of the framework and model, and presents a set of checklists designed to enhance their practicality for use.

Chapter LIII
Will Laboratory Test Results be Valid in Mobile Contexts? / *Anne Kaikkonen, Aki Kekäläinen, Mikael Cankar, Titti Kallio, and Anu Kankainen*.. 897

This chapter raises the question of whether or not lab-based usability tests of mobile technologies can return results that have ecological validity relative to real-world use. The chapter introduces the complexities inherent in the mobile usage context and provides an overview of studies conducted to compare lab and field studies. The authors describe a study they conducted to compare the results obtained via lab and field testing. They recommend that, for most testing, it is best to perform several quick laboratory tests iteratively during the design process, rather than concentrate efforts on a single field test. They do, however, acknowledge that in some instances lab-based testing is insufficient—for example, due to technical limitations such as testing a GPS-based system, where it is difficult to simulate the use context with sufficient realism in the lab, or where it is the intention to observe user behaviour in a natural environment. In these situations the authors suggest field trials may be beneficial and, on the basis of their own experience, suggest some guidelines for conducting field tests of mobile technologies.

Chapter LIV
Mobile Evaluations in a Lab Environment / *Murray Crease and Robert Longworth*........................ 910

This chapter argues that, while mobile application evaluation protocols increasingly reflect user mobility, they fail to place realistic demands on users' visual attention (e.g., to reflect the real-life need for users to be cognizant of hazards as they move through their physical environment). In this chapter, the authors present a simple classification for describing the kind of distractions which might typically surround a user, and report on two evaluations designed to determine the effect visual distractions have on users of a mobile application. The results, which showed that users' requirement to monitor their environment affected both task performance and measures of workload, indicated that it is important to include such distractions along with mobility in evaluations of mobile technology.

Chapter LV
Instrumented Usability Analysis for Mobile Devices / *Andrew Crossan, Roderick Murray-Smith, Stephen Brewster, and Bojan Musizza*.. 927

This chapter introduces the concept of instrumented usability analysis for mobile devices – that is, the use of sensors (such as accelerometers) to elicit quantitative, objective information about the "moment to moment" actions of users as they interact with mobile technology. Illustrated by a detailed case study of tapping while walking, this chapter demonstrates the benefits to be gained from fine-grained analysis of user actions and disturbances during a mobile usability study. The authors were able to show, for example, the significant effect of gait phase angle on tapping time and accuracy that would not have been possible without the introduction of sensors to the usability study. The work presented here highlights new directions for both design and evaluation of mobile technologies.

Chapter LVI
Three Eye Movement Studies of Mobile Readability / *Gustav Öquist* 945

Making text easy to read on mobile devices has proven to be a challenge, primarily because the way we are used to presenting textual information is incompatible with the limited screen space available on mobile devices. This chapter notes the importance of finding ways to present text on small screens in such a way that facilitates the level of readability we are used to and expect. The author argues that to achieve this requires the availability of methods for evaluating novel text presentation formats on mobile devices in an efficient yet reliable manner. This chapter reports on three readability studies which employed eye movement tracking to learn more about how to improve readability on mobile devices.

Chapter LVII
Did You See That? / *Murray Crease and Joanna Lumsden* 972

Experimental design for mobile technology evaluation needs to account for the environmental context in which such technologies will be used. In part, this requires the incorporation of relevant environmental distractions. This chapter reflects on different lab-based techniques for presenting visual distractions to participants and measuring the participants' cognizance of the distractions while mobile.

Chapter LVIII
A Field Laboratory for Evaluating in Situ / *Rune T. Høegh, Jesper Kjeldskov, Mikael B. Skov, and Jan Stage* 982

This chapter describes the evolution, and final version, of a field laboratory that was developed in response to recognised challenges faced when evaluating mobile technology in the field. The field laboratory was developed over a 4-year period as a result of the authors' direct experience evaluating a number of mobile systems in field settings. This chapter describes this evolution (including lessons learned along the way), and highlights rationale for technological and other design decisions. The current system—which is based on a system of small wireless cameras and wireless microphones—is outlined and its use is explained. The authors' posit that, using their field laboratory, it is possible to collect data which is of a quality equal to lab-based studies.

Chapter LIX
Field Evaluation of Collaborative Mobile Applications / *Adrian Stoica, Georgios Fiotakis, Dimitrios Raptis, Ioanna Papadimitriou, Vassilis Komis, and Nikolaos Avouris* 997

Based on a review of accepted techniques for data collection and evaluation relative to mobile applications, this chapter presents a method (based on a combination of techniques) for conducting usability evaluations of context-aware mobile applications that are to be deployed in semi-public spaces and that involve collaboration among groups of users. To illustrate their proposed method, the authors describe a case study of its application.

Section V
Case Studies

Chapter LX
UI Design for Mobile Technology in a Closed Environment / *Kate Oakley, Gitte Lindgaard, Peter Kroeger, John Miller, Earl Bryenton, and Paul Hébert,* 1015

This chapter introduces the notion of a "closed environment", such as a hospital or military context, for which designers of technology often have extremely limited access to end users, both for design and testing purposes. During the design and development of such systems, many of the typical protocols employed in user-centered design are inapplicable; requirements gathering becomes an indirect process and quasi-lab studies are used in place of real contexts. This chapter reports, in detail, on a case study of the design of a mobile application to monitor vital signs of hospital patients. It discusses the analysis and design challenges faced, as well as the alternative evaluation methods that had to be devised to ensure ecological validity of evaluation results despite lack of direct access to users.

Chapter LXI
Designing a Ubiquitous Audio-Based Memory Aid / *Shwetak N. Patel, Khai N. Truong, Gillian R. Hayes, Giovanni Iachello, Julie A. Kientz, and Gregory D. Abowd*................................ 1031

This chapter introduces the personal audio loop (PAL), an application designed to recover audio content from the recent past using the mobile phone platform. The authors discuss an evaluation of its potential usefulness in everyday life, the level of ubiquity and usability demanded of the service, and the social and legal considerations for long-term adoption. A detailed discussion of the various evaluation methods used (ranging from a controlled lab study to deployment of the system over a period of several weeks) is presented, as are analyses of the results obtained leading to an identification of issues critical to the use of PAL. This chapter raises and discusses interesting issues regarding the legality of a system such as PAL, highlighting that traditional privacy guidelines and policies may not adequately address personal ubicomp applications of this nature.

Chapter LXII
Visualisation of Meeting Records on Mobile Devices / *Saturnino Luz and Masood Masoodian*.. 1049

Mobile technology has the potential to provide convenient access to meeting records for users on the move. This chapter discusses issues surrounding the design, implementation, and evaluation of such interfaces, and proposes a general paradigm for meeting browsing which addresses the core information access requirements of the task within the constraints imposed by mobile technology. The authors illustrate their discussion with a case study and lessons learned developing a handheld meeting browser application.

Chapter LXIII
A Proposed Tool for Mobile Collaborative Reading / *Jason T. Black and Lois Wright Hawkes*.. 1068

This chapter describes the design of a collaborative m-Learning application that uses pair communication based on speech and text I/O. The authors present the process they undertook to develop their tool as a model for interface design, communication strategies, and data manipulation across mobile platforms. The chapter describes an evaluation of the system that was conducted based on a paper prototype, and highlights how this helped identify optimum interface layout, as well as confirm that children preferred the speech input. The authors outline the creative strategies for interface layout and data manipulation they adopted to design and develop their system, and reflect on the lessons learned throughout the process.

Chapter LXIV
Evaluating Learner Satisfaction in a Multiplatform E-Learning System / *Tiong T. Goh, Kinshuk, and Nian-Shing Chen*.. 1079

This chapter reports on a comparative evaluation between two e-learning systems from the perspective of the end user. The evaluation compared the difference in overall learner satisfaction between a blackboard e-learning system and a multiplatform e-learning system with three different accessing devices, and explored the factors that influenced learner satisfaction while engaged in a multiplatform e-learning system as well as the gain in learner satisfaction achieved with respect to three different accessing devices. The authors suggest that their findings are valuable in terms of improving the content adaptation process for multiplatform e-learning systems.

Foreword

Today there are over two billion mobile phones in use. Add this vast number to the growing portfolio of mobile devices—from music and video players to portable and wearable medical and health monitors, from tiny tags to intelligent garments—and its easy to appreciate the importance and timeliness of a book on effective interface design and evaluation.

Poor interface design of these devices is at best a cause of frustration—*how do I zoom out so I can view that web page?* —at worse, life-threatening—*did the nurse enter that drug dosage correctly in my portable medical pump?* Meanwhile, with many billions of interactions occurring daily with mobiles, great design can transform the world for good: simply reducing the number of key presses by one or two on a popular service could save a lifetime of human effort. While good design benefits all of us, for users with physical and cognitive impairments, as chapters in this book illustrate, the potential positive impact is huge. The Handbook is an important tool to help us all as we strive for even better mobile interactions in the future.

There are, of course, many other books on human-computer interaction design and evaluation. Should you read this one? Answer: absolutely. Other texts can help you understand broad issues and approaches but mobiles are very different from desktop-bound conventional computers. Consider traditional HCI as being a visit to a metropolitan zoo; in contrast, mobile HCI is like doing an adventurous animal safari, deep in the bush.

Take just two aspects: contexts and tasks. Mobiles are used in highly dynamic and demanding environments: people want to look up directions while walking; check flight changes while encumbered with luggage, children and jet-lag; tell each other stories in cafes using photos and video stored on their gadget. The office where this is written is a much more predictable, calmer context! Then, while a lot of HCI research has tackled work-based systems and tasks, mobiles are forcing the community to tackle other user 'goals' like fun, curiosity, and connecting.

The book comprehensively probes the unique problems and opportunities facing mobile designers and researchers, giving a much richer picture than the other general HCI volumes. You will learn not only of technologies and designs but, importantly, of methods and tools you can use when you are engaged in building your own systems.

The mobile research and practice community has grown considerably over the past 10 years and the full spectrum of approach and focus is represented in the book. So, there are chapters concerned with ethnographic methods and accounts of mobile use; and, others which detail engineering innovations and experimentation. You'll be able to learn about the state-of-the-art in interface and interaction technologies, from touch-based devices to RFID tagging; and be challenged by articles touching on issues like privacy, swarm intelligence, and technology acceptance models.

Whatever your area of mobile interest, you should be able to find material here that will inspire and inform design and evaluation processes. If you are a researcher, perhaps starting out on a particular topic such as mobile text entry, the book will provide you with good reviews of existing approaches and pointers to future challenges. If you are one of the many new developers working on designing mobile devices and services, the book also contains practical guidelines and case-study experience reports to help you make good choices. The book is not just for technologists, though: marketing and business strategists will be interested in the consumer and social analyses.

Mobiles are changing the world. People like you—researchers and developers—have an incredible opportunity to shape the future. This book will be a resource you should return to again and again to check and challenge your methods, tools, and approaches so that this future is enriched and not impoverished.

Matt Jones
Future Interaction Technology Lab
University of Wales Swansea
www.fitlab.eu

Matt Jones is a Senior Lecturer and is helping to set up the Future Interaction Technology Lab at Swansea University. He has worked on mobile interaction issues for the past ten years and has published a large number of articles in this area. He is the co-author of "Mobile Interaction Design", John Wiley & Sons (2006). He has had many collaborations and interactions with handset and service developers including Orange, Reuters, BT Cellnet, Nokia and Adaptive Info; and has one mobile patent pending. He is an editor of the International Journal of Personal and Ubiquitous Computing and on the steering committee for the Mobile Human Computer Interaction conference series. Married with three mobile, small children; when he's not working he enjoys moving quickly on a bike whilst listening to music and the occasional podcast.

Preface

In recent years, mobile technology has been one of the major growth areas in computing. Mobile devices are becoming increasingly diverse, and are continuing to shrink in size and weight. Although this increases the portability of such devices, their usability tends to suffer. Ultimately, the usability of mobile technologies will determine their future success in terms of end-user acceptance and, thereafter, adoption. Widespread acceptance will not, however, be achieved if users' interaction with mobile technology amounts to a negative experience. Mobile user interfaces need to be designed to meet the functional and sensory needs of users. In recognition of this need, a growing research area focusing on mobile human-computer interaction has emerged, and will likely continue to grow exponentially in the future.

The resource disparity between mobile and desktop technologies means that successful desktop user interface design does not automatically equate to successful mobile user interface design. Desktop user interface design originates from the fact that users are stationary (that is, seated at a desk) and can devote all or most of their attentional resources to the application with which they are interacting. As a result, the interfaces to desktop-based applications are typically very graphical (often very detailed) and use the standard keyboard and mouse to facilitate interaction. This has proven to be a very successful paradigm that has been enhanced by the availability of ever more sophisticated and increasingly larger displays. In contrast, users of mobile devices are typically in motion when using their device, which means that they cannot devote all of their attentional resources, especially visual resources, to the application with which they are interacting; such resources must remain with their primary task, often for safety reasons. Additionally, the form factor of mobile devices typically limits the applicability of standard input and output techniques, making mobile human-computer interaction design ineffective if we insist on adhering to the tried-and-tested desktop paradigm.

The design and evaluation of mobile human-computer interaction, unlike desktop-based interaction, needs to be cognizant of the implications brought to bear by complex contextual factors affecting both users and technology. Such contextual influences include, but are not limited to, the physical environment in which a mobile device is being used, the impact of multitasking behavior typically exhibited by users of mobile devices (e.g., using a device whilst driving), and the social context in which a device is used (e.g., consider social acceptability of interaction). All in all, designing the user interface for mobile applications is a very complex undertaking that is made even more challenging by the rapid technological developments in mobile hardware.

Not only is the design of human-computer interaction for mobile technologies difficult, so too is the evaluation of such designs. In fact, the most appropriate means by which to effectively evaluate mobile applications is currently a hotly debated topic in the field of mobile human-computer interaction. Evaluation techniques for mobile technology require as much consideration as the design of the user interfaces themselves; for the results of evaluations of mobile applications to be meaningful, the manner in which the evaluations are conducted needs to be, and is, the focus of considerable research in itself.

The purpose of the *Handbook of Research on User Interface Design and Evaluation for Mobile*

Technology is to offer a compendium of current research knowledge concerning the key issues surrounding the design and evaluation of mobile user interfaces such that students, researchers, educators, and practitioners alike may all derive benefit from the experience of leading experts working in this field. Its aim is to expose readers to, and heighten their awareness of, the complexity of issues concerning mobile human-computer interaction. Amongst the chapters included in the handbook, alternative points of view are included for some of the field's hotly debated topics in order to encourage readers to think out of the box and embrace the challenge of new paradigms both for interaction design and evaluation. Reliance on the tried-and-tested desktop design and evaluation paradigms has not worked; the mission of this handbook is to encourage people to *think out of the box* to ensure that novel, effective user interface design and evaluation strategies continue to emerge and, in turn, the true potential of mobile technology is realized.

To elicit the best and most balanced coverage of issues critical to the design and evaluation of mobile technologies, researchers from around the world were invited to submit proposals describing their intended contribution to the handbook. All proposals were carefully reviewed by the editor, with a view to assembling the finest contributions from leading experts in the field. Upon receipt of full chapter submissions, each submission was subjected to double-blind peer review, and only the best were then selected for final inclusion in the handbook. In many instances, the chapters were subjected to multiple revisions before final acceptance. The result of this rigorous process is a comprehensive collection of current research articles of high scholarly value written by distinguished researchers from many prominent research institutions and groups around the world.

ORGANIZATION OF THIS HANDBOOK

The goal of the *Handbook of Research on User Interface Design and Evaluation for Mobile Technology* is to improve our appreciation of the current and future challenges associated with the design and evaluation of user interfaces to mobile technologies. To achieve this goal, the handbook includes a comprehensive collection of 64 quality research contributions from leading experts around the world. It covers issues ranging from the use of ethnographic methods for design of mobile applications to instrumented lab-based methods for their evaluation. Additionally, each chapter includes a collection of related key terms and their definitions, contributing to a comprehensive compendium of terms, definitions, and concepts central to the field of mobile human-computer interaction.

Although most chapters touch on a number of the issues critical to user interface development for mobile technologies, and many include discussion of case studies for illustrative purposes, to assist you when searching for specific information, the 64 chapters have been organized according to their primary contribution. Hence, the handbook is organized into five sections that examine the following topics:

- **Section I: User Interface Design for Mobile Technologies**
  - Use of ethnography to inform mobile user interface design
  - Use of the technology acceptance mobile for mobile services to guide the design of mobile technologies
  - The impact of user characteristics on the design of mobile user interfaces
  - Wearable technologies and their design implications
  - Contextual information and awareness in mobile application design
  - Design of in-car user interfaces
  - Design of mobile learning applications
  - Adaptive and intelligent user interfaces in mobile computing

    - Rapid prototyping, modeling, and simulation tools for mobile applications
    - Ecologies of interacting artifacts for ubiquitous technologies
- **Section II: Novel Interaction Techniques for Mobile Technologies**
    - Classification of mobile interaction techniques
    - Novel interaction paradigms
    - Unobtrusive interaction
    - Visual interaction
- **Section III: Assistive Mobile Technologies**
    - Overview of key issues and trends for designing and evaluating mobile assistive technologies
    - Design for various special needs groups, including seniors, mental health interventions, and visually impaired users
    - Implications for designing the interface to smart wheelchairs
- **Section IV: Evaluation Techniques for Mobile Technologies**
    - Theoretical overview
    - Adaptation of traditional methods to suit mobile human-computer interaction
    - Method selection and combination strategies
    - Novel evaluation methods
    - Classification of usability factors for mobile technologies
    - Lab v. field evaluations
- **Section V: Case Studies**

The handbook provides literally thousands of references to existing literature and research efforts in the field of mobile human-computer interaction, and it includes a comprehensive index to support quick and convenient look up of topics and concepts. This handbook is an ideal reference for veteran and novice educators, researchers, students, and practitioners in the field of mobile human-computer interaction who require access to current information in this emerging field. The complementary combination of theoretical and practical content will enable readers to draw parallels with their own research or work, and apply and/or further the research efforts of others in their own projects.

Jo Lumsden
National Research Council of Canada
Institute for Information Technology – e-Business
Fredericton, Canada
2007

Joanna Lumsden is a Research Officer with the National Research Council of Canada's (NRC) Institute for Information Technology. Prior to joining the NRC, Lumsden worked as a research assistant in the Computing Science Department at the University of Glasgow, U.K. where she attained both her undergraduate software engineering Honours Degree and her Ph.D. in human computer interaction. Lumsden is also an Adjunct Professor at the University of New Brunswick in Fredericton, where she teaches graduate courses and supervises a number of graduate students. Lumsden is the lab manager for the NRC's Mobile Human Computer Interaction Lab – a facility dedicated to investigating mobile interaction design and evaluation.

Acknowledgment

Editing this handbook has been a unique, enlightening, and ultimately rewarding experience. I have learned a lot both personally and professionally, and have been amazed by the encouragement and enthusiasm I have received from the authors of this publication. This handbook has been 2 years in the making, and I would like to take a moment to acknowledge the efforts of those people who have made it possible.

My deepest gratitude goes to all the authors for their excellent contributions. I learned a lot from working with such knowledgeable and expert individuals around the world, and greatly appreciated their cooperation and kind words of encouragement throughout the process. Thank you! I hope the handbook does you proud!

Each chapter in this handbook was peer reviewed. I would like to acknowledge the assistance of *everyone* involved in the review process; their constructive reviews were invaluable to the quality of the handbook. In particular, I would like to thank two of my knowledgeable and insightful colleagues who stepped in and shouldered a heavier review load to help out when others let me down; Murray and Danny, thank you *so* much! Whilst I hope I have included all reviewers in the list, I appreciate there may be individuals of whom I am unaware who helped their colleagues review submissions "behind the scenes"; to any and all to whom this applies, my thanks.

I would like to extend my deep gratitude to the members of my Editorial Advisory Board: Dr. Gary Burnett from the University of Nottingham (UK), Dr. Murray Crease from the National Research Council of Canada (Canada), Dr. Mark Dunlop from the University of Strathclyde (UK), Mr. Philip Gray from the University of Glasgow (UK), Dr. Jesper Kjeldskov from Aalborg University (Denmark), Dr. Irina Kondratova from the National Research Council of Canada (Canada), Dr. Rod Murray-Smith from the University of Glasgow (UK), Dr. Antti Pirhonen from the University of Jyväskylä (Finland), and Dr. Janet Read from the University of Central Lancashire (UK). It was an honour to be able to turn to such a distinguished and wise group of people for support and sound advice during the collation of the handbook.

I am deeply indebted to the National Research Council of Canada's Institute for Information Technology for supporting me in this endeavour and sharing my belief in the value of this publication.

I owe a debt of gratitude to IGI Global for giving me the opportunity to realise my vision. In particular, thank you to Kristin Roth, my development editor, who guided me through the lengthy and complex process of publishing a book of this nature, and answered my myriad of questions and gave sound advice throughout.

I would like to thank Dr. Matt Jones from the Future Interaction Technology Lab at Swansea University (UK) for so willingly and enthusiastically writing the foreword to this handbook. Finally, my warmest thanks go to my husband, Keith, for his support, encouragement, patience, and love.

In closing, I am delighted to present this handbook and its comprehensive range of topics. I am certain that you will find it a useful reference, and source of better understanding, for all pertinent issues related to current research on the design and evaluation of mobile technologies.

Jo Lumsden
National Research Council of Canada
Institute for Information Technology – e-Business
Fredericton, Canada
2007

Section I
User Interface Design for Mobile Technologies

This section looks at many of the critical aspects concerned with effective design of mobile applications. The section begins with a series of chapters that discuss the adoption of ethnographic methods to inform the design of such technologies, including a selection of chapters that report on observed mobile device use and subsequent implications for design. This section covers issues such as how factors of user acceptance of mobile services can be used to guide the design of such technologies, as well as the impact of age and cognitive capacity on design. Chapters consider wearable technologies, the importance of contextual information in mobile application design, the design of in-car user interfaces, and issues surrounding the design and implementation of mobile learning applications. The section takes a look at adaptive and intelligent user interfaces for mobile computing, as well as tools for rapid prototyping, modeling, and simulation of mobile systems. The section concludes with a look to the future in terms of ecologies of interacting artifacts, reflecting an evolution from strictly mobile to more ubiquitous technologies.

Chapter I
From Ethnography to Interface Design

Jeni Paay
Aalborg University, Denmark

ABSTRACT

This chapter proposes a way of informing creative design of mobile information systems by acknowledging the value of ethnography in HCI and tackling the challenge of transferring that knowledge to interface design. The proposed approach bridges the gap between ethnography and interface design by introducing the activities of field-data informed design sketching, on a high level of abstraction, followed by iterative development of paper-based mock-ups. The outcomes of these two activities can then be used as a starting point for iterative prototype development—in paper or in code. This is particularly useful in situations where mobile HCI designers are faced with challenges of innovation rather than solving well-defined problems and where design must facilitate future rather than current practice. The use of this approach is illustrated through a design case study of a context-aware mobile information system facilitating people socialising in the city.

INTRODUCTION

This chapter looks at the mobile technology design problem of taking an ethnographic-based approach to gathering field data and making this data available to the design process in a form that is easily assimilated by designers to inform user-centred design of mobile technology. Interface design for mobile technologies presents unique and difficult challenges that sometimes render traditional systems design methods inadequate. Ethnography is particularly well-suited to design for mobile technology. Mobile usability is often highly contextual and ethnographic approaches can facilitate richer understandings of mobile use contexts providing insight into the user's perspective of the world. Exploring the huge potential of mobile devices presents designers with a unique opportunity for creativity. In thinking about mobile technology design for *future,* rather than *current* practice, the challenge becomes even greater.

Before this discussion proceeds further it is worth clarifying the use of the term *ethnography.* Traditionally, ethnographic studies within sociology are conducted from a particular theoretical viewpoint and for the purpose of contributing to theory. However, ethnography, as it is understood in HCI research, generally refers to a collection of techniques used for gathering and organizing field materials from observational studies (Dourish, 2006). By its very definition, ethnography is primarily a form of reportage. It provides both empirical observational data, and makes an analytical contribution in the organization of that data. The virtue of ethnography is that it takes place in real-world settings and provides access to the ways people perceive, understand, and do things (Hughes et al., 1997). Ethnographically-oriented field methods can be used in HCI to provide a deeper understanding of an application domain, a holistic understanding of users, their work, and their context, which can then be drawn into the design process at the earliest stages (Millen, 2000). Ethnographic studies involve detailed observations of activities within their natural setting, providing rich descriptions of people, environments and interactions, and acknowledging the situated character of technology use (Millen, 2000). These observations can provide valuable insights into the processes needed for systems requirements specifications (Sommerville et al., 1993).

In the literature, the terms ethnography and ethnomethodology are both used to refer to field studies using ethnographic methods to understand how people perceive their social worlds. Other terms such as technomethodology (Button & Dourish, 1996), rapid ethnography (Millen, 2000) and design ethnography (Diggins & Tolmie, 2003) are also used to distinguish different aspects of the use of ethnography in the design of technology. For the sake of simplicity, this chapter uses the term ethnography to encompass these understandings as being important to the discussion of the relationship between their outputs and the inputs they provide to the design process.

For ethnography to make a worthwhile contribution to the design of mobile technologies, we need to find ways for translating ethnographic findings into forms that are suitable for informing design processes. In the following sections, the historical relationship between ethnography and HCI is discussed, including how it has been incorporated into the process of interface design. The theoretical and methodological background for how to gather and interpret ethnographic data and use this for informing design is described. A design case study is then presented in which an ethnographic approach has been applied to mobile technology design in a real world research project through a structured series of activities. The overall process is described, and the two steps of developing *design sketches* and *paper-based mock-ups* are introduced as a way of bridging the gap between ethnography and interface design. Finally, lessons learned from using design sketches and paper-based mock-ups in the development process are outlined.

BACKGROUND

Ethnography and HCI

The issue of bridging the gap between ethnography and interface design has been a topic of discussion in HCI research for over a decade. Ethnography is now regarded as a common approach to HCI research and design (Dourish, 2006). Yet there is still no overall consensus on how best to incorporate the results of ethnographic fieldwork into the design processes (Diggins & Tolmie, 2003). In the early 90s seminal work by sociologists, such as Suchman, Hughes, Harper, Heath and Luff, inspired the use of ethnography for understanding the social aspects of work processes and informing user interface design (Hughes et al., 1995). However, researchers struggled with the challenge of utilizing insights provided by ethnography into the activity of designing. By the mid 90s, ethnography was hailed as a new approach to requirements elicitation for interactive system design, particularly through its application in the development of computer-supported cooperative work (CSCW) systems (Hughes et al., 1995). Even so, some researchers still held reservations about the ability of ethnographic methods to inform design (Hughes et al., 1997) and ethnography was regarded as a relatively untried approach to systems development, despite the fact that it was increasingly being used to inform and critique actual systems (Button & Dourish, 1996). Toward the end of the 90s, researchers were beginning to develop systematic approaches to social analyses for the purpose of influencing design (e.g., Viller & Sommerville, 1999). However, despite many research efforts, bridging the gap between ethnography and design still remains a matter of concern to HCI researchers today (Diggins & Tolmie, 2003).

The turn towards ethnography within HCI was motivated by a growing need to design for complex real world situations. This began with the belief that methods from the social sciences, such as ethnography, could provide means for understanding these contextual issues of technology use better. In the light of today's ubiquitous and mobile networked computing environments, the need to understand contexts of technology use, such as peoples' dynamic work and social practices, is challenging HCI researchers and designers more than ever. Supporting innovation in a world of emerging technologies can be done by submerging designers, who understand emerging technical possibilities, into rich ethnographic field data about potential users' lives and current practices (Holtzblatt, 2005). In this way technology design drives an understanding of the user's situation, which in turn, propels innovation.

Ethnography and Interface Design

The process of transition from field data to prototype design is a difficult one (Cheverst et al., 2005; Ciolfi & Bannon, 2003). A design process involving ethnography generally starts with observations and interviews collected through ethnographic methods. Key findings are then summarized and design ideas are drawn out with a set of features that can be tied back to the findings. The next step involves, "design suggestions" or "design implications," which may evolve into requirements through the development of a low-fidelity prototype. This prototype is then iterated with feedback from users and evolves into the operational system. The data collected by ethnographic methods reflects the richness of the user's situation in a way that is difficult to derive from a limited set of questions or measures as employed in traditional analysis methods (Wixon, 1995). In contrast to traditional systems analysis that looks at data, structures, and processing, ethnography is concerned with participants and interactions (Sommerville et al., 1993). This provides the designer with a rich understanding of the context of use for the artifacts that are being designed (Millen, 2000). In looking at a situation through the user's eyes rather than the designers, ethnography provides a view of the situation that is independent of design preconceptions (Hughes et al., 1997).

Ethnography has much to contribute to interface design—particularly in mobile device design due to the highly contextual nature of mobile usability and use. However, one of the main problems is finding a suitable mechanism for the transference of knowledge between these two fundamentally different disciplines. Ethnographic findings need to be understood and communicated to designers (Hughes et al., 1995). And yet, current mechanisms for incorporating ethnographic findings into the design process still fail to capture the value of these investigations (Dourish, 2006).

Ethnography deals in "the particular," and software design in "the abstract" (Viller & Sommerville, 1999). While willing to listen to each other, both disciplines speak different languages and use different methodologies. Ethnographers deal in text, notes, reports, and transcriptions, and produce detailed results giving a rich and concrete portrayal of the particulars of everyday practical action in context, presented in a discursive form; software designers and engineers deal in the creation and manipulation of more formal graphical abstractions, notations and description techniques to simplify the complexity of the situation and extract critical features. Ethnographers avoid judgements; designers make them. Where ethnographers take an analytic role, including gathering and interpreting data, software designers have a synthesis role, designing from abstract models of situations (Button & Dourish, 1996; Hughes et al., 1995). In addition to the problems of communication there are also problems of timing. Ethnography is generally conducted over a long period of time; in fact, it is difficult to define an end point for gathering understanding. On the other hand, software designers are often under restricted time pressure to deliver a product.

The problem has been in finding a timely method and a suitable form to present field findings that can be assimilated by and are readily usable for designers (Hughes et al., 1995; Viller & Sommerville, 1999). The needs of the software designer have to be aligned with a representation of the essential "real world" practices of users in context. Simply describing the social events being observed is not sufficient, designers need to be able to model and use this understanding in design.

USING ETHNOGRAPHY IN THE DESIGN PROCESS

Gathering Data

From HCI research it can be seen that using ethnography as a data gathering method requires the development of more structured approaches to conducting and reporting from ethnographic studies that better support the development of design requirements.

One approach is to conduct ethnography concurrently with design and bring ethnographic results into the design process in a more systematic way throughout the development process. This can, for example, be achieved through meetings between ethnographers and the design team (Hughes et al., 1995). This approach results in a change in the way that ethnography is conducted. Rather than extended periods in the field, ethnographers working in cooperation with software designers to create a system design, making short and focused field studies, reporting back to designers, and often taking design questions back into the field to focus their observations and questions to users. To structure the process, the communication of fieldwork to designers can be supported by dedicated software packages (Diggins & Tolmie, 2003; Sommerville et al., 1993). In this situation, the ethnographic record becomes a joint resource with ethnographers regularly reporting their findings in an electronic form, and designers using this content to develop structured design requirements. Constructing these records in a connected manner preserves backward and forward traceability between ethnographic findings and evolving system requirements.

Another approach is to lead into the design process through *rapid ethnography* (Millen, 2000). Rapid ethnography provides the field worker with a broad understanding of the situation which can then be used to sensitize designers to the use situation rather than identifying specific design issues. It is aimed at gaining a reasonable understanding of users and their activities in the short time available for this in a software development process. Rapid ethnography provides a more structured approach to ethnographic field studies by limiting the scope of the research focus before entering the field. It focuses time spent in the field by using key informants in the real situation and interactive observation techniques. Rapid ethnography also uses multiple observers in the field to ensure several views of the same events and to create a richer representation and understanding of the situation (Millen, 2000).

Interpreting Data

Ethnography is not simply about the collection of data in the field, it is also about reflection on and interpretation of that field data. Effective communication between ethnography and design is at the heart of the matter of bridging the gap between the two disciplines (Hughes et al., 1997). By recognizing the different natures and input and output requirements of ethnography and interface design, integration between the two disciplines can be achieved through enhancing and structuring the communication between them during the interpretation phase.

One approach to interpreting the data collected is to have a cross-discipline team participating in the fieldwork. In this situation designers go into the field with ethnographers to experience themselves how users work. They also contribute to the representation of the gathered data, shaping it into a form that is easier for designers to use (Diggins & Tolmie, 2003). Representing ethnographic findings through pictorial stories, drawings, data models, analogies and metaphors are ways to communicate field learning to cross-discipline teams (Millen, 2000). Videotapes of field observations and design documentaries play a similar role using a more designer-accessible communication mode than a written report (Raijmakers et al., 2006).

Another approach to interpretation is to have both ethnographers and designers involved in the conceptual design process. In this situation, the ethnographer is an ongoing member of the design team, providing grounded insights and interpretations into the abstracted requirements as they evolve and the design emerges. The ethnographer acts as a substitute user during the design process (Viller & Sommerville, 1999). Through their knowledge of the actual situation, they can participate in discussions with the designers, providing insights and access to instances of specific relevant situations.

A third approach is for the designer to play the part of a pseudo-ethnographer. This involves designers going "into the wild" and being exposed to users by watching real work while it is being done, and hence truly experiencing the richness of work (Wixon, 1995). Structured methods such as rapid ethnography and *contextual design* (Beyer & Holtzblatt, 1998) make this possible. In contextual design, the user and the designer explore the design space together using *contextual interview* or *facilitated enactment* of their practices in context (Holtzblatt, 2005). *Affinity diagramming*, from the contextual design method, provides a synthesis of the data into hierarchical classifications where the meaning contained in the data elements can be reflected on in relation to the design question, facilitating understanding and innovation for designers.

Informing Design

After the ethnographically gathered field data has been interpreted, abstracted findings are used to derive design opportunities and design requirements. The designer uses the outputs from the interpretation of the field data as input into the design process. Sometimes the ethnographers are involved in this design process bringing their intimate knowledge of the users and the situation of use, and their deep relationship to the data, to the team (Cheverst et al., 2005). They participate in the identification of design incentives by drawing attention to general design opportunities, and relevant topics and concerns. Otherwise, the designers must draw understanding entirely from the reports, discussions, diagrams and models, which represent the ethnographic record.

Design is a matter of making, and is used to create and give form to new ideas and new things (Fallman, 2003). A recent approach to informing design and achieving a close connection between the design team and the field data is the use of field observation videos or design documentaries. These videos mediate between ethnographic and design perspectives. As the design team watches them they incorporate interpretation of data into the design process on the fly through discussions drawing design sensitivities and identifying design concerns. Designers become sensitized to relevant issues visible in the real world interactions depicted in the video (e.g., Ciolfi & Bannon, 2003; Raijmakers et al., 2006). This method requires a high level of design experience, and in bridging the gap between ethnography and design, these designers work in an inspirational, ephemeral and creative way. For others this creative leap across the divide is very difficult, and more structured methods are needed to guide the process of envisioning design from ethnographic outputs. In response to new interface design challenges, including mobile technology, HCI researchers are investigating new techniques for guiding designers through this difficult transition – of particular interest to this chapter are the techniques of *design sketching* (Buxton, 2007), *paper-based mock-ups* (Ehn & Kyng, 1991) and *paper prototyping* (Snyder, 2003).

Design sketching is fundamental to the process of design, and can be used by information system designers to bring about the realization of an idea in the way designers think (Fallman, 2003). Sketching is the art of giving form to the unknown; it makes it possible to "see" ideas or envision whole new systems, and is especially critical in the early ideation phase of design (Buxton, 2007). According to Buxton, sketches should be rapid, timely, inexpensive, disposable, plentiful, clear, un-detailed, light, informal representations that practitioners can produce and interact with to suggest and explore ideas. Sketching is not only a way to visualize existing ideas, but it is about shaping new ideas. In making a sketch of something, the visualization talks back to the designer with a new perspective on that idea, providing a link between vision and realization of new ideas.

Paper-based mock-ups are closely related to the notion of design sketching. In this technique from the participatory design tradition, representational artifacts are constructed from paper, cardboard and materials at hand. Informed by studies of practice, mock-ups can play an important mediating role in connecting use requirements and design possibilities in a form recognizable to multi-disciplinary design teams (Ehn & Kyng, 1991). These mock-ups can be used to incorporate materials from the ethnographic study, embody envisioned new technological possibilities, convey design ideas in relation to existing practices and reveal requirements for new practices (Blomberg & Burrell, 2003).

Paper prototyping is a widely used technique for designing, testing and refining user interfaces (Snyder, 2003). This technique helps with the development of interfaces that are useful, intuitive, and efficient, by initiating testing of the interface at a stage when the design is in its formative stages and therefore still open to the input of new ideas. Paper prototyping can be used to reflect on field study findings while developing and refining the design (Holtzblatt, 2005). A collection of interface designs, drawn from ideas generated through design sketching and paper-based mock-ups are given functional and navigational connections through the process of paper prototyping. A paper prototype is a useful vehicle for giving visual form to identified design requirements. It forms the focus for design refinement discussions and cognitive walkthroughs by the design team, and is in itself part of the design specification for implementation of the system.

A DESIGN CASE STUDY

The project used as a design case study in this chapter involved the development of a context-aware mobile information system, *Just-for-Us,* designed to facilitate people socialising in the city by providing information about people, places, and activities in the user's immediate surroundings. The case study location was a specific city precinct covering an entire city block, Federation Square, Melbourne, Australia. This location was chosen because it is a new, award-winning architectural space providing a variety of activities through restaurants, cafes, bars, a museum, art galleries, cinemas, retail shops, and several public forums spanning an entire city block. The design intention for the civic space was to incorporate digital technologies into the building fabric creating a combination of virtual information space and physical building space for people to experience. Thus, this particular place provided a unique setting for studying people's situated social interactions in a "hybrid" space and for inquiring into the user experience of mobile technology designed to augment such a physical space with a digital layer.

Process

The Just-for-Us mobile information system was designed specifically for Federation Square on the basis of an ethnographic study of people socialising there. The development process involved seven major activities:

- Ethnographic field studies
- Field data interpretation
- Design sketching on a high level of abstraction

Figure 1. The overall process of designing the Just-for-Us mobile information system

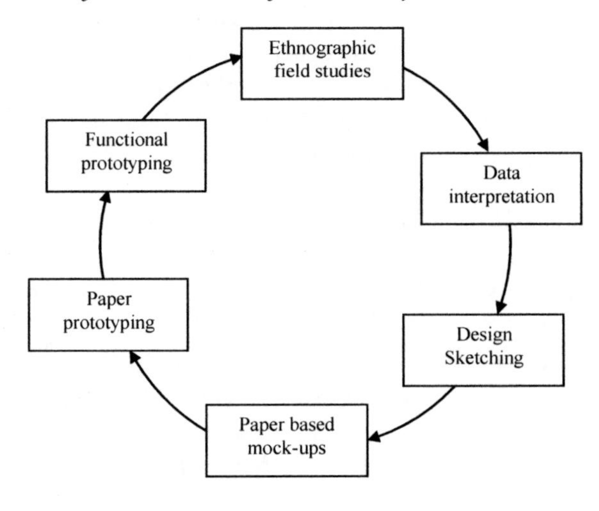

- Paper-based mock-up development
- Iterative paper prototyping
- Implementation of a functional prototype
- Field studies of prototype use in-situ

The specific content and outcome of these activities are described in the following subsections. Details of the implemented system and findings from the field study of its use are not covered here, but can be found in Kjeldskov and Paay (2006).

As illustrated in Figure 1, data from ethnographic field studies of situated social interactions in public were subjected to data interpretation, using the *grounded theory* approach (Strauss & Corbin, 1990) and affinity diagramming (Beyer & Holtzblatt, 1998). In trying to bridge the gap between our ethnographic data and actual mobile device interface design, outcomes from the interpretation of field data were used to inform a systematic activity of design sketching (Buxton, 2007). The purpose of this activity was to generate design ideas on a high level of abstraction inspired by ethnographic findings but without getting into too much detail about specific look, feel and functionality. On the basis of selected design sketches, we developed a number of paper-based mock-ups (Ehn & Kyng, 2991) of potential design solutions. This forced us to become more specific, but still allowed us to focus on overall functionality and interaction rather than on technical details. After this, we engaged in a number of paper prototyping (Snyder, 2003) iterations with the purpose of developing a detailed set of system requirements and a coherent interface concept prior to writing any program code. Finally, these specifications were implemented in a functional prototype allowing us to introduce new technology into the field and revisit peoples' socialising behavior in the city while using the operational Just-for-Us context-aware mobile information system.

Gathering and Interpreting Data

The aim of our ethnographic field study was to inquire into peoples' social interactions at Federation Square. The field study was guided by a subset of McCullough's typology of everyday situations (McCullough, 2004) for classifying peoples' social activities when out on the town: eating, drinking, talking, gathering, cruising, belonging, shopping, and attending. The study applied a rapid ethnography approach and consisted of a series of contextual interviews (Beyer & Holtzblatt, 1998) and ethnographic field observations (Blomberg & Burrell, 2003) with the designers acting as pseudo-ethnographers and gathering the field data (Figure 2). Three different established social groups participated in the study. Each group consisted of three young urban people, mixed gender, between the ages of 20 and 35, with a shared history of socialising at Federation Square. The groups determined the activities undertaken and the social interactions that they engaged in. Prior to the field visits, each group received a 10-minute introduction to the study followed by a 20-minute interview about their socialising experiences and preferences. This introduction occurred at a place familiar to the group, where they might meet before socialising in the city. This encouraged them to reflect on past social interactions, to relax about the visit, and gave the interviewer insight into the situated interactions that the group typically participated in. One of the members of the group was then taken to Federation Square and asked to arrange to meet up with the other members of the group. The group was then asked to do what

Figure 2. Ethnographic observations and contextual interviews at Federation Square

they would usually do as a group when socialising out on the town—while "thinking aloud" as they moved around the space, and responding to questions from the interviewer. Two researchers were present in the field, providing multiple views on the data collected.

Each field visits lasted approximately three hours and allowed the groups to engage in a number of social activities. The outcome of the ethnographic field studies amounted to eight hours of video and approximately 30 pages of written notes.

In addition to the observational studies of people socialising at Federation Square an architecturally trained observer carried out a single *expert audit* (Lynch, 1960) focusing on the physical space of Federation Square. The expert audit documented architectural elements and their relationships to surrounding context, including the people inhabiting the space through 124 digital photographs and corresponding field notes.

Interpreting data gathered from the ethnographic study involved two phases. Firstly, photographic data and written notes from the expert audit were analyzed using *content analysis* (Millen, 2000) and affinity diagramming (Beyer & Holtzblatt, 1998). Concepts and themes describing the physical space of Federation Square were overlaid onto a map of the precinct to produce a color-coded multi-layered abstraction of the space (Figure 3). This provided an overview of the spatial properties of Federation Square highlighting constraints and enablers for situated social interactions there with traceable links back to specific observations.

Secondly, video data from the contextual interview and observational field study of people socialising at Federation Square was transcribed and then analyzed using open and axial coding adapted from *grounded theory* analysis (Strauss & Corbin, 1990). Identifying key words or events in the transcript, and analyzing the underlying phenomenon created the initial open codes. Analysis of these codes resulted in a collection of categories relating to actions and interactions. After the codes were grouped into categories, higher-level themes were extracted using axial coding. Affinity diagramming was then used to draw successively higher levels of abstraction from the data by grouping and sorting the themes until a set of high-level concepts, representing the essence of the data and encompassing all lower level themes, had been formed. The process of affinity diagramming produced a hierarchical conceptual framework containing three overall clusters of themes abstracted from the transcripts (Figure 4). This provided a rich story about how people interact with each other while socialising in public, with traceable links back to specific observations in the field study sessions.

As illustrated in Figures 3 and 4, outcomes from the interpretation of our ethnographic field data were primarily on an abstract level, providing a deeper understanding of peoples' situated social interactions in the physical space of Federation Square. While this is an important part of the foundation for good design, in their current form these outputs did not point towards any particular design ideas. As an example, the analytical outcomes from interpreting the field data included a series of qualitative statements similar to those in the following list (For a detailed account of findings from the ethnographic field studies see Paay and Kjeldskov (2005)).

- Federation Square has four key districts with distinctly different characteristics, each with an associated landmark.

Figure 3. Graphical image of inhabited social context at Federation Square

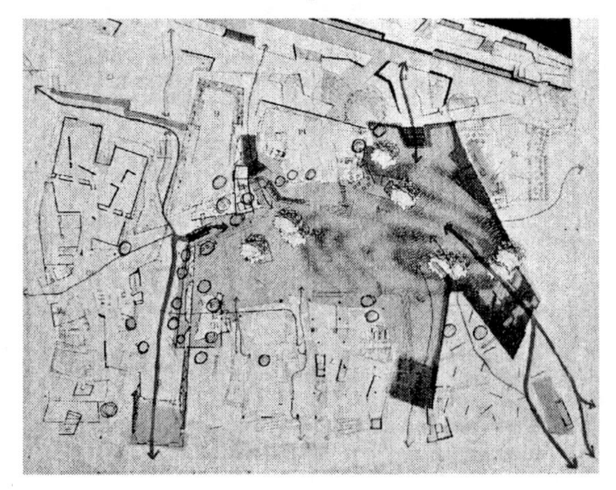

Figure 4. Affinity diagram of situated social interactions at Federation Square

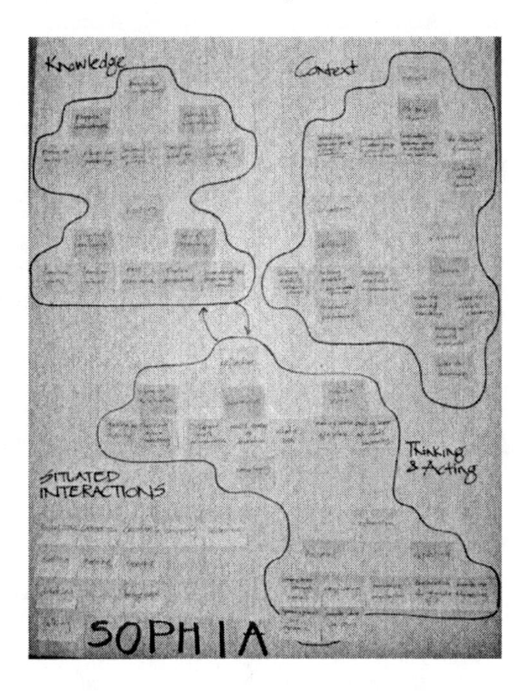

- Federation Square has visible surroundings, general paths, general entrances, focal structures and no clear paths, so people need to use the structures and surrounds in finding their way around the space.
- People socialising at Federation Square like getting an overview of what is happening around them, and want to know about the presence and activities of other people.
- People's past experience with places and people at Federation Square play an important factor in choosing places and activities for socialising.
- People give directions at Federation Square by referring to shared experiences and visible elements, and use their history and physical familiarity with a place to find their way around using familiar paths.

In order to move forward from data interpretation toward an overall design concept as well as actual interface design and system requirements for a context-aware mobile information system for people socialising at Federation Square, the design team engaged in two steps of developing design sketches and paper-based mock-ups (as described earlier). Each of these techniques produced interface design artifacts on different levels of detail and abstraction. These two "bridging" steps between ethnography and interface design are described in the following sections.

Design Sketching

The first step in the design of the Just-for-Us mobile information system was to develop a series of conceptual design ideas based on the insight from our data analysis. For this purpose, the design team spent two days generating, discussing, sketching, and refining design ideas on the basis of the abstract models of the architectural space of Federation Square and the clustering of themes in the affinity diagram from the analysis of people socialising there.

The design sketching activity was done in a dedicated design workspace with sheets of A1 paper lining the walls on which we could sketch and refine design ideas. Each sketch took its origin in a specific finding or observation from the interpreted field data. This field finding would firstly be discussed in more detail to ensure shared understanding among the design team. Secondly, we would start sketching possible design ideas, for example, how to facilitate an observed practice.

Figure 5. Design sketching informed by interpreted ethnographic field data. The delineated area corresponds to the paper-based mock-up produced later and highlighted on Figure 6.

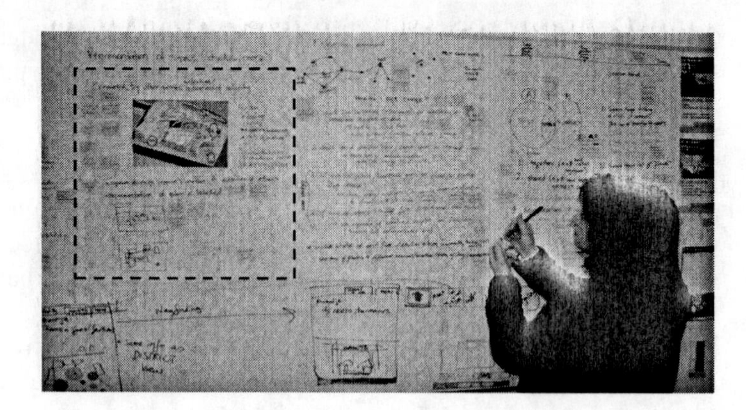

Hence, we were, in a sense, using collaborative data analysis, as described in the rapid ethnography method, to drive the generation of design ideas.

During the process of sketching, the conceptual outcomes from the data interpretation phase were continually revisited and, in turn, the sketches were continuously annotated with post-it notes referring to the data. For example, a section of the affinity diagram included the themes of "social experience," encompassing "past experience" and "shared experience." A diagram was then sketched to explore the intersections between past and shared experiences in groups of friends. In this way, we ensured a strong link between data and design, and maintained clear traceability between the two. This activity was about sketching the social concepts that came out of the data models, not about generating solutions. In doing this, we were able to explore the field data findings in a graphical form, and to explore derivations from these concepts by generating multiple understandings of them. Design sketching was used as a mechanism to understand the field outcomes, to generate graphical overviews of the design space, and create graphical representations of design opportunities within that space.

The outcome from the two-day design workshop was a collection of design sketches on A1 paper (Figure 5), each describing conceptually a potential design idea or design opportunity, for parts of the Just-for-Us mobile information system, including envisioned general functionality, general ideas for graphical design and user interaction, with clear references back to the empirical data.

The design sketches provided a new visual abstraction to the ethnographically interpreted field data, translating understanding encapsulated in the abstract findings into design parlance. Engaging in the process of design sketching rather than jumping straight to specifying system requirements, enabled us to see the ethnographic findings from a new perspective and to play with design ideas on a high level of abstraction. This allowed us to distance ourselves from the role of "problem solvers" and to explore instead, on a conceptual level, design ideas facilitating potential future practice in technology use.

Paper-Based Mock-Ups

While useful for generating and working with overall design ideas, conceptual design sketches are far too abstract for informing specific system requirements. Hence, moving directly on to detailed prototype design and implementation is likely to commit designers to specific solutions too early and impede their flexibility to try out new ideas. In an attempt to overcome this problem, the next step of our process from ethnography to interface design was to produce a series of

Figure 6. One of the paper-based mock-ups of possible mobile device screens

paper-based mock-ups of possible specific design solutions (Figure 6).

The production of paper-based mock-ups took place over several days and facilitated a series of long discussions within the design team leading to an overall concept for the Just-for-Us mobile information system providing functionality such as: an augmentation of the user's physical surroundings; chat capability with friends out on the town; content indexed to the user's physical and social context and history of interactions in the city; a graphical representation of places, people and activities within the user's vicinity; and way-finding information based on indexes to landmarks and familiar places. These design ideas were screen-based solutions to design opportunities identified during design sketching.

Working with each of these ideas in more detail, the paper-based mock-ups gave the design team a medium for trying out and modifying specific design ideas for what the system should be able to do and what it should look like—long before any actual coding was done. Consequently, the mock-ups coming out of this activity had already undergone several iterations of redesign and refinements.

Discussions during the mock-up phase took place on different levels of abstraction: from screen design, system functionality, privacy issues, problems designing for small screens, what aspects of the user's context to capture in the system, and how to do this. We also had several discussions about whether or not the implementation of the produced mock-ups would be feasible within current mobile technologies, and if not, which enabling technologies would have to be developed. Through these discussions and continued refinements and redesigns, a set of specific design requirements slowly began to take shape—gradually taking us into the "safer ground" of interface design.

Prototyping

Having completed the paper-based mock-up phase, the final steps of our development process were much more straightforward. On the basis of the mock-ups, more detailed paper prototypes were produced using Adobe Photoshop (Figure 7 left). This forced the design team to work within the graphical limitations of the target device and to use the specific graphical user interface elements available in the target browser, for this web based application. Also, the detailed paper prototypes allowed the designers to discuss some of the more dynamic interaction issues such as navigation structure and handling of pushed information. While most design changes were done in Adobe Photoshop at this time, some of the more serious issues, such as how to fit the Internet chat screen(s) into the limited design space, forced the design team back to working with paper-based mock-ups for a short time. After several cognitive walkthroughs, a full paper prototype with a detailed set of requirements was agreed upon and implemented as an operational mobile web site providing context-aware information to users, with very few modifications (Figure 7 right).

The design specified by the paper prototype was implemented as a functional Web-based system accessible through the Web browser of a PDA (personal digital assistant) providing context-related information, dynamic maps and location specific annotated graphics to the user. It also keeps a history of the user's visits to places around the city. The functional prototype uses WLAN or GPRS for wireless Internet access and resolves the user's location and the presence of friends in vicinity by

Figure 7. Detailed paper prototype screen (left) and the corresponding final functional prototype screen (right), designed from the paper-based mock-up highlighted in Figure 6

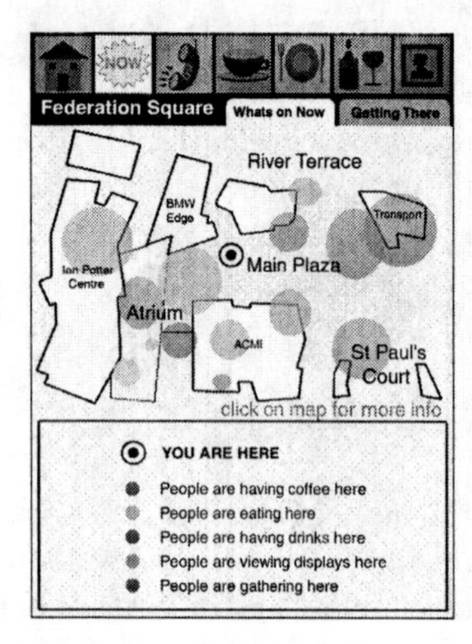

means of Bluetooth beacons potentially embedded into the environment. The implementation of a functional prototype allowed us to close the circle depicted in Figure 1 by returning to Federation Square to do an ethnographic field study of people socialising there—this time facilitated by the Just-for-Us system. For details on this use study see Kjeldskov and Paay (2006).

FUTURE TRENDS

The future trends for bridging between ethnography and interface design for mobile technologies are many. As a part of a drive toward more user centered innovative design for both current and future practice, new techniques are emerging, which respond to the specific challenges of mobile technology design and use. These include, for example, cultural probes, digital ethnography, video diaries, film documentaries, facilitated enactment, acting-out in context, role-playing and body storming. Through these new techniques, the roles of ethnographers, designers, and future users are becoming more interwoven, facilitating a smoother and more effortless transition from ethnography to interface design. Techniques such as these reflect the fact that mobile technology design is not only about designing for existing work practices but also about designing for future practices in peoples' private and social lives and responding to the challenge of innovating for non-work in as yet non-existing use situations. They also respond to issues raised by many researchers that mobile technologies are often used in dynamic and continually changing contexts, offering information directly related to those contexts, and that it can be very difficult to predict what future user-adaptations of mobile technology might evolve.

The techniques of sketching and mocking-up introduced in this chapter are not new. Both have a long tradition in other design disciplines. However, like many of the above emerging approaches, we have combined existing techniques in a new way that provides designers with a more structured path to follow when making the difficult transition of transferring knowledge from the field into the design process.

CONCLUSION

This chapter addresses the issue of ethnography informing interface design for mobile technologies. It has described how ethnographic studies can be used in HCI design and how such studies can be useful for understanding current practice as well as providing a backdrop for envisioning potential future practice. However, as confirmed in the literature, bridging between ethnography and design is difficult, and techniques are needed that enable designers to better use ethnographic findings in the design process. In response to this, the two steps of conceptual design sketching and creating paper-based mock-ups have been proposed as bridging activities between ethnographic data interpretation and iterative prototype development.

Illustrating how this can be done in practice, this chapter has described a recent project involving the design of a context-aware mobile information system on the basis of a rapid ethnographic field study. In this project, the process of design sketching from analytical data made a useful link between interpretation and design. It provided a means of communicating a conceptual understanding of current practice into the early stages of interface design, and helped "translate" findings from the field data into design parlance. Working with sketches allowed the design team to play with design ideas on a conceptual level rather than moving straight to specifying system requirements. It also allowed them to distance themselves from the role of "problem solvers" and to explore instead potential future practice of technology use.

The process of creating and refining paper-based mock-ups on the basis of selected design sketches gave the design team a medium for being a bit more specific while still maintaining a high level of flexibility. It allowed for drilling down into some specific design ideas and the exploration and modification of ideas for interface design and functionality before doing any coding. It also allowed the team to engage in discussions about possible screen designs, different functionality, privacy, small screens, etc., and to rapidly implement, evaluate, and refine design ideas. By working with paper-based mock-ups, it was possible to generate a strong set of specific design requirements, which provided a solid foundation for subsequent activities of paper and functional prototyping.

Innovative interface design for mobile technologies is both an art and a science. It requires us to be creative and inspired as well as structured and focused. Facilitating creativity and inspiration provides the art. Grounding interface design in empirically informed understanding of people and current practice provides the science. The challenge we are faced with is not just how to perform the art and science of design better individually, but more so how to support a fruitful interplay between the two. For this purpose, techniques such as conceptual design sketching and creation of paper-based mock-ups are valuable tools for researchers and designers on their journey from ethnography to interface design.

REFERENCES

Beyer, H., & Holtzblatt, K. (1998). *Contextual design—Defining customer centred systems*. San Francisco: Morgan Kaufmann.

Blomberg, J., & Burrell, M. (2003). An ethnographic approach to design. In J. Jacko & A. Sears (Eds.), *Handbook of human-computer interaction* (pp. 964-986). Mahwah, NJ, USA: Lawrence Erlbaum Associates Inc.

Button, G., & Dourish, P. (1996). Technomethodology: Paradoxes and possibilities. In *Proceedings of CHI 96*, (pp. 19-26). Vancouver, Canada: ACM.

Buxton, B. (2007). *Sketching user experiences: Getting the design right and the right design*. San Francisco, Morgan Kaufman Publishers.

Cheverst, K., Gibbs, M., Graham, C., Randall, D., & Rouncefield, M. (2005). Fieldwork and interdisciplinary design. *Notes for tutorial at OZCHI 2005*. Retrieved October 24, 2007, from http://www.comp.lancs.ac.uk/rouncefi/Tutout.html

Ciolfi, L. & Bannon, L. (2003). Learning from museum visits: Shaping design sensitivities. In *Proceedings of HCI International 2003* (pp. 63-67). Crete, Greece: Lawrence Erlbaum.

Diggins, T., & Tolmie, P. (2003). The 'adequate' design of ethnographic outputs for practice: some explorations of the characteristics of design resources. *Personal and Ubiquitous Computing, 7*, 147-158.

Dourish, P. (2006). Implications for Design. In *Proceedings of CHI 2006* (pp. 541-550). Montreal, Canada: ACM.

Ehn, P., & Kyng, M. (1991). Cardboard computers: Mocking-it-up or hands-on the future. In J. Greenbaum & M. Kyng (Eds.), *Design at work: Cooperative design of computer systems* (pp. 167-195). Hillsdale, NJ, USA: Lawrence Erlbaum Associates, Publishers.

Fallman, D. (2003). Design-oriented human-computer interaction. In *Proceedings of CHI 2003* (pp. 225-232). Florida, USA: ACM.

Holtzblatt, K. (2005). Customer-centred design for mobile applications. *Personal and Ubiquitous Computing, 9*, 227-237.

Hughes, J., King, V., Rodden, T., & Andersen, H. (1995). The role of ethnography in interactive systems design. *Interactions, 2*(2), 56-65.

Hughes, J., O'Brien, J., Rodden, T., & Rouncefield, M. (1997). Designing with ethnography: A presentation framework for design. *In Proceedings of DIS '97* (pp. 147-158). Amsterdam, Holland: ACM.

Kjeldskov, J., & Paay, J. (2006). Public pervasive computing in the city: Making the invisible visible. *IEEE Computer, 39*(9), 30-35.

Lynch, K. (1960). *The image of the city.* Cambridge, MA, USA: The MIT Press.

McCullough, M. (2004). *Digital ground—Architecture, pervasive computing and environmental knowing.* Cambridge, MA, USA: The MIT Press.

Millen, D. R. (2000). Rapid ethnography: Time deepening strategies for HCI field research. In *Proceedings of DIS '00* (pp. 280-286). Brooklyn, NY: ACM.

Paay, J., & Kjeldskov, J. (2005). Understanding situated social interactions in public places. In *Proceedings of Interact 2005* (pp. 496-509). Rome, Italy: Springer-Verlag.

Raijmakers, B., Gaver, W., & Bishay, J. (2006). Design documentaries: Inspiring design research through documentary film. In *Proceedings of DIS 2006* (pp. 229-238). Pennsylvania, USA: ACM.

Snyder, C. (2003). *Paper prototyping.* San Francisco: Morgan Kaufmann Publishers.

Sommerville, I., Rodden, T., Sawyer, P., Bentley, R., & Twidale, M. (1993). Integrating ethnography into the requirements engineering process. In *Proceedings of IEEE International Symposium on Requirements Engineering* (pp. 165-181). San Diego, CA, USA: IEEE Computer Society Press.

Strauss, A. L., & Corbin, J. (1990). *Basics of qualitative research.* Newbury Park, CA, USA: Sage Publications.

Viller, S., & Sommerville, I. (1999). Coherence: An approach to representing ethnographic analyses in systems design. *Human-Computer Interaction, 14*, 9-41.

Wixon, D. (1995). Qualitative research methods in design and development. *Interactions, 2*(4), 19-24.

KEY TERMS

Affinity Diagramming: One of the techniques of the contextual design process, used during data interpretation sessions to group related individual points together, creating a hierarchical diagram showing the scope of issues in the work domain being studied.

Content Analysis*:* A qualitative research technique for gathering and analyzing the content of text, where content can be words, meanings, pictures, symbols, ideas, themes, or any message that can be communicated, to reveal messages in the text that are difficult to see through casual observation.

Contextual Design: A collection of techniques supporting a customer-centered design process, created by Beyer and Holtzblatt (1998), for finding out how people work to guide designers to find the optimal redesign for work practices.

Design Sketch: A graphical representation of a concept or design idea on a high level of abstraction. It should be quick, timely, open, disposable, un-detailed, and informal, and is usually hand-drawn on paper.

Expert Audit: A field reconnaissance done by an architecturally trained observer maping the presence of various elements of the physical environment and making subjective categorizations based on the immediate appearance of these elements in the field and their visible contribution to the image of the city.

Ethnography: A collection of techniques used for gathering and organizing field materials from observational studies, involving detailed observations of activities within their natural setting, to providing rich descriptions of people, environments and interactions.

Grounded Theory: A theory based analytical approach, which takes a set of data collected using ethnographic methods and provides a set of specific procedures for generating theory from this data.

Paper Prototype: A paper representation of a system design, able to simulate operation of that system, which is independent of platform and implementation, and can be used for brainstorming, designing, testing and communication of user interface designs and for identifying usability problems at an early stage of the design process.

Paper-Based Mock-Up: A representation of a specific design idea that is built from simple materials such as paper and cardboard, keeping it cheap and understandable, but making it a physical representation of a design idea for a final system, good for envisioning future products in the very early stages of the design process.

Rapid Ethnography: A collection of field methods to provide designers with a reasonable understanding of users and their activities given a limited amount of time spent in the field gathering data.

Chapter II
Use of Experimental Ethno-Methods to Evaluate the User Experience with Mobile Interactive Multimedia Systems

Anxo Cereijo Roibás
University of Brighton, UK

Stephen Johnson
BT Mobility Research Centre, UK

ABSTRACT

This chapter discusses research initially supported by the Vodafone Group Foundation and the British Royal Academic of Engineering, and subsequently by the BT Mobility Research Centre. It aims to unfold the user experience in future scenarios of mobile interactive multimedia systems, such as mobile iTV with plausible significance in entertainment, work, and government environments. Consolidated and experimental ethnographic data gathering techniques have been used to understand how peripatetic and nomadic users such as commuters and travelers interact in real contexts, taking into account their physical and social environment together with their emotions and feelings during interaction with the system. This approach potentially enhances the consistency and relevance of the results. This chapter also envisages how mobile users could become a sort of 'DIY producers' of digital content, prompting the emergence of mobile communities that collaborate to create their own 'movies' and exchanging them not only with other users but also places (real and virtual environments) and objects (intelligent objects and other digital-physical hybrids). This work illustrates that mobile and pervasive TV would go further than merely broadcasting TV content on handhelds; it will be a platform that will support collaboration and enhancement of creative skills among users.

INTRODUCTION

Interactive TV demands active participation by viewers, and as a result, it considerably affects people's experience with television and their TV-related social behavior. Users' adoption of powerful handhelds with multimedia features, together with an increasing interoperability between platforms, results in the expansion of the iTV consumption beyond the domestic context. We can define this 'almost everywhere TV' as 'pervasive TV.'

The presented research explores realistic and relevant future scenarios for pervasive iTV and for pervasive interactive multimedia systems that address the demands, needs, and desires of a specific category of users: commuters. Likewise, novel processes and structures for content creation, sharing, and consumption that match the nomadic lifestyles of commuters, and embody their values, are investigated.

This research shows that there is, in fact, a growing interest by users in mobile interactive multimedia systems. However, these systems are different from the conventional concept of TV broadcasting on mobile phones. The scenarios that arise in this research are more related to non-professional users co-producing and sharing media content in applications for mobile devices, the internet and iTV for small network communities. According to a recent research by Deloitte Touche Tohmatsu, "Companies have invested significant sums in developing mobile television services so far, but mobile television has had muted commercial impact. Its disappointing performance is likely to continue in 2007. A key reason for this will be weak consumer demand" (The 2007 edition of DTT TMT Industry Group's Telecommunications Predictions, 2007).

There are several research projects addressing different aspects of pervasive interactive multimedia systems and distributed multimedia systems and services (Arreymbi, 2006; Butscher, 2006). Many of them explore either the area of mobile and pervasive games (Barrenho 2005; Capra et al., 2005) or the experimental interactive arts (Frisk 2005).

Furthermore, much of the current research focuses on specific interactive aspects such as the screen (Pham, 2000; Zheng, 2005), the sound (Scheible, 2005), or the digital content in general (Goularte, 2004), but disregards the influence of the context. However, the scope of this project is to analyze the user experience (UX) in a holistic way in order to understand which elements and applications of interactive multimedia systems are suitable in specific contexts, providing the user with a high quality experience.

BACKGROUND

Industry has often failed to understand and forecast users' needs and expectations in sectors that are normally characterized by innovation-driven approaches (such as telecommunications and iTV). Many companies developed applications for handhelds or iTV using inappropriate ICT resources that require massive modifications in users' habits resulting in perceptive or cognitive overload. Consequently, the market's response to investments in developing new products (e.g., mobile TV broadcasting) has not been positive to date. Rapid changes in users' habits and technological advances have generated enormous uncertainties and call for innovative research and development methodologies. As the aspects that need to be considered here have a diverse nature, a cross-disciplinary approach that includes human factor studies, behavioral theories, socio-cultural and economic trends, technological developments and emerging technologies markets, interactive arts, product design, and so forth is necessary. Moreover, several techniques such as collaborative and user-centered approaches that focus on users' cultural, social, behavioral and ergonomic backgrounds must be combined.

Many network operators in Europe, the USA, Japan, Korea and Canada are starting to broadcast TV on handhelds (see Table 1)[1]. This is commonly defined as mobile TV.

There are several reasons that might undermine the success of such operations. The first one is related to the intrinsic physical diversity between

Table 1. Commercial and trial mobile TV launches worldwide

Operator	Country	Platform	Channels	Trial/ Commercial
Bell Mobility	Canada	MobiTV	8	Commercial
Rogers	Canada	MobiTV	9	Commercial
TELUS Mobility	Canada	MobiTV	7	Commercial
Sonera & Elisa	Finland	DVB-H	9	Trial
SK Telecom	South Korea	S-DMB	9	Commercial
O2	UK	DVB-H	16	Trial
Orange	UK	MobiTV	9	Commercial
Virgin Mobile	UK	DAB	3	Trial
Cingular	US	MobiTV	23	Commercial
Midwest Wireless	US	MobiTV	23	Commercial
Sprint	US	MobiTV	23	Commercial

both interfaces (TV and handhelds) making them unsuitable for the same way of delivering of content. The second regards the context of use: TV is traditionally used in a domestic private environment (Spigel, 1992) and usually involves social sharing (Morley, 1986), while mobile phones are mainly used in public environments and entail an individual experience (Perry et al., 2001). Moreover, users are becoming more and more nomadic spending less time at home and in the office. This implies an increasing need for performing our daily tasks while on the move (Leed, 1991). Therefore, unlike TV, handhelds are regularly used in different situations and with different purposes (they are likely to be used as an auxiliary tool to assist users' in a main activity (Harper, 2003). In addition mobile services can be related to the specific context of the user (context awareness). Finally, there are operability differences: TV (including interactive TV) is considered a passive or low interactive medium while handhelds typically demand high interactivity and connectivity. These dissimilarities influence the way with which users interact with the medium and therefore necessitate distinct interaction patterns and content as well as different service formats and features.

MAIN FOCUS OF THE CHAPTER

One of the main assumptions of this work is that commuters are a particularly relevant and interesting population segment for investigating novel processes and forms of interaction with mobile multimedia content. The growing interest and dedication to mobility and mobile life among commuters is strikingly apparent through the rapidly increasing share of resources used for this lifestyle. Use of public transportation, bicycles, and walking in urban mobility[2] has been increasing steadily, as has the amount of money spent for mobility and telecommunications (Pooley et al., 2005). These target users have been further divided into two categories (with a balanced representation of different cultural backgrounds and professional areas and roles): 18-35 year olds and 35-60 year olds.

Furthermore, this work has been influenced and guided by the following premises: small mobile devices can provide both a functional and an effective interactive experience, being able to recreate an enjoyable immersive environment for the user; they are also appropriate for the creation of (and interaction with) new forms of multimedia content and finally, they are suitable tools for context awareness applications.

METHODOLOGY

This work integrates a variety of approaches to evaluate and understand the user experience. These methods include time studies of user panels, observation, mapping of movements and other ethnographic techniques in order to answer the factual questions about the UX in future scenarios of pervasive iTV, interpret the meaning of the findings and describe the relations between more levels of empirical experience and analytical outcome.

In-situ evaluation techniques have been used in several projects to assess the design of interactive systems in public or semi-public environments such as the evaluation of ambient displays at work and in a university (Mankoff et al., 2003); the evaluation of ambient displays for the deaf that visualize peripheral sounds in an office environment (Ho-Ching et al., 2003); the evaluation of a sound system to provide awareness to office staff about events taking place at their desks (Mynatt et al., 1998) and the evaluation of a system of interactive office door displays that had the function of electronic post-it notes to leave messages to the office occupant when they are not there (Cheverst et al., 2003).

Simulations and enactments are very useful when the usage contexts make the mediated data collection particularly difficult due to high privacy, technical, or legal issues (e.g., military environments) or when the system is at a very experimental level. Simulations using proof-of-concept mock-ups or explorative prototypes in labs have been largely used to evaluate the usability and accessibility of interactive systems. Although they might provide valuable information about the UX with a certain interface, they tend to disregard the contextual and emotional aspect of the interaction. Additionally, they can only be used when the conceptual model of the system reaches an adequate level of maturity as they presume the use of a functional prototype.

Such research needs to combine experience, data, analysis, and evaluations from many perspectives in order to achieve a multi-disciplinarily built platform for understanding how and why specific concrete needs, the demand for specific services and technological and aesthetic solutions are integrated in users' social, cultural and aesthetic practices; in short how these shifting trends among commuters evolve and shape. The work has been divided into three main phases.

The first phase is devoted to the analysis of the UX in future scenarios of mobile and ubiquitous i-TV and the elaboration of the usage scenarios consisting in the creation of the scripts for the storyboards. The second and the third phase involved enactments and simulations instead of mediated data collection (as in the previous stage). Creation of usage scenarios is a diffused ethnographic technique used to identify requirements and concept assessment, often combined with laboratory evaluation (Carroll, 2000). There are different typologies of scenarios each one being appropriate for a specific scope: activity scenarios (e.g., based on experiential narratives) are useful during preparatory fieldwork early in the design process (created in the first phase and validated in the second one); mock-up scenarios aim to understand how the designed system suits users' activities (used in the third phase); prototype evaluation scenarios aim to evaluate the interface models of the system; integration scenarios simulate the effect of the finished design. The last two categories have not been used in this project, as they require a rather sophisticated working prototype of the system (which was out of the scope of this work).

The first phase consisted of two initial focus group sessions with each of the target groups identified earlier. Each workshop involved 12 participants and aimed to get the user's view about trends on multimedia mobile applications, TV at home and on the move, new forms of content for mobile TV, advanced interaction possibilities and finally, potential interconnections between handhelds and other devices. This activity has been combined with a theoretical investigation of existing technologies together with successful interactive user experiences in other areas (e.g., games, HCI in Space, etc.). This phase also included ethnographic research using cultural probes, questionnaires and naturalistic observation (photo and video recording in-the-field and data analysis). While focus groups and analysis of study-cases were good sources of functional and data requirements; cultural probes and questionnaires provided good information about users' requirements and finally in-the-field observation has been a very valuable technique to identify environmental and usability requirements. Furthermore, the information collected here provided the basis for the scenario scripts that were evaluated in the following stage.

The second phase aimed to validate some significant usage scenarios and subsequently to identify and classify innovative related applications exploring, at the same time, radically new forms of 'smart' and 'malleable' content. This process consisted of two workshops that used role-playing as the basis for a collaborative design approach. It involved twelve representatives of both target groups of users to represent and discuss the scenario scripts (that were elaborated in the previous phase) in order to confirm the legitimacy of the scenarios and experience models proposed in terms of relevance, effectiveness and soundness.

The third phase involved the creation of proof of concept mock-ups and development of user experiments in order to bring to light the feasibility and usability of the scenarios, applications and forms of content previously identified. In this phase some experimental low-fi prototypes of applications were developed (and empirically evaluated in the field) that operate across an integrated system of interfaces and form factors that connote pervasive iTV (typically mobile phones, PCs and iTV). Thirty users aged between 18 and 60, with a peripatetic lifestyle and mixed cultural and professional backgrounds, took part in this evaluation through two sessions in a public plaza.

Focus groups in the first phase of the project provided a framework for discussion about the future use of multimedia content in handhelds in contexts of pervasive communication. During the focus group sessions, the facilitator stimulated brainstorming around the following topics: trends for nomadic users for work, leisure and government ('time spent at home/work/on the move,' 'what to do on the move and with whom'); TV versus mobiles ('What do we enjoy in TV when and why, what we don't enjoy and why,' 'TV on mobile'); novel smart multimedia content ('beyond images, video and sound,' 'poor content vs. rich and smart content,' 'contextualized content'); advanced interaction ('content malleability, gestures, haptics, multisensorial devices, holograms, voice navigation'); mobiles in connection with other devices, things and places ('mobile and TV, mobile and the office, mobile and the house, mobile and the street—buildings, objects, events, people'). Participants also provided short 'stories' (experiential narratives) regarding their view on the topic. Some stories were real, some other were fiction.

Cultural Probes (Figure 1) aimed to get inspirational responses to understand beliefs, desires, aesthetic preferences and cultural concerns of users without observing them directly. This technique was initially used by Gaver in industrial design (Gaver et al., 1999) and has recently been exported to HCI (Hulkko et al., 2004). Six selected users were given a cultural probes pack under the condition of completing and returning them after two weeks.

Each pack included four main items with the following instructions:

- **Maps:** World ('where would you imagine having a daydream?'); City ('Where would you like to go now but you can't?'); House ('Where would you like to be alone?', 'where would you like to meet people?'); Family, friends and colleagues relationships ('show frequency and nature of contacts').
- **Camera:** 'take a picture of an image/video you'd like to take with your mobile.'
- **Media Diary:** 'record TV, cinema and radio use (what, when, where, with whom).'
- **Photo album and colour pencils:** 'collect things, images and stories of your week; make sketches.'

Questionnaires (Figure 2) were designed as a set of 11 postcards in order to provide a very informal and open approach, encouraging instinctive and casual replies about the users' vision on the topic and were distributed to twelve target users[3].

Each card has an image on the front, and one question on the back such as: 'which device/s would you take with you if exiled to a desert island?' 'When and where can you feel over-loaded with information?' 'When & where would you like to connect it with other devices, things or places?' 'When and where the use of a mobile phone can be a collective experience (several users sharing the same application using their own devices)? And a public experience (several users operating the same device simultaneously)?' 'How would you feel about having awareness of other devices, people, places and things in your handheld?' 'Name a book or movie with a future scenario you liked and another one you disliked (or tell us about your own idea).'

Figure 1. Cultural probes pack and detail of the maps

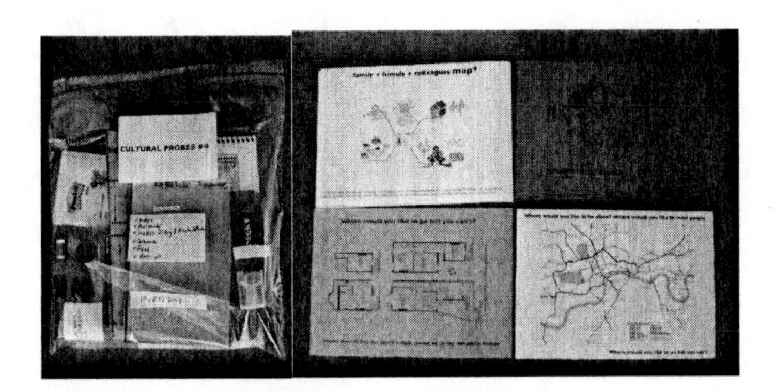

User-centered design researchers have historically favored direct observation because it is a method that places the researcher in the context in which technology use occurs (Hagen et al., 2005). However, mobile devices are designed for individual use within a personal body space. Therefore, observing the interface actions of the user can be physically arduous (Kjeldskov et al., 2005; Mark et al., 2003; Newcomb et al., 2003). Naturalistic observation (included note taking, photography and video recording) by following commuters for three months in public places in London such as theatres, parks, stores, stations and squares as well as on public transport between London and Brighton provided a useful insight into the use of mobile phones as multimedia communication tools (mainly taking video and pictures and sending, TV streaming, picture editing, etc.). During this activity, it was also possible to determine and map the behavior of users in common public meeting spaces such as stations and parks. The typical duration over which these maps were determined was in the order of a couple of hours and aimed to be a source of information regarding the sociability processes in public spaces as well as the effect of technology (especially mobile phones) in this phenomenon.

These techniques were useful not only to identify users' main requirements, but also to write scripts for the scenarios. The scenarios were then validated in two sessions, each one with 12 representatives of one of the user categories previously identified. Each session consisted of in-situ theatre performances carried out by the participants themselves following the scripts of the scenarios.

The last part of the project used horizontal proof of concept mock-ups to assess relevant and plausible applications that were identified during the research. This phase consisted of two evaluation sessions, again using role-playing, but this time in the field.

The plays were performed in public environments (mainly squares in the city) by some of the users while the others could comment on what they were watching. The advantage of using in-situ enactments was that they provided precious information about some contextual factors that had not been identified in the research process. Examples of these valuable outcomes included the users' embarrassment of watching a movie in a crowded train and concerns about the phone being stolen when using multimedia applications in public places in the city, and so on.

These very early prototypes incite experimentation, are easy to use and adopt, encourage discussion between users and designers and have a very low cost. However, due to their low-fi appearance they might appear unconvincing, raising criticism by the users. Moreover, they focus excessively on functionality not tested in the real usage contexts. Figure 3 shows one of the experimental prototypes—a mobile 'memory box.'

Figure 2. The questionnaire

Theatre Workshops: Personas and Scenarios (Phase Two)

Theatrical performances are a valuable technique that can be used to collect data and identify requirements and other crucial information from users such as feelings and emotions. This is because the user's self-esteem is not directly under scrutiny and so inhibiting factors like embarrassment or shyness are less likely to be elicited.

The use of drama (Iacucci et al., 2000; Sato, 1999) can be an effective tool in participatory design as it facilitates dialogue between designers and users. According to Newell, it can cross boundaries of technical language and knowledge, allowing elderly potential users to be involved effectively in the process of design at the pre-prototyping stage (Newell et al., 2006).

Live play gives the audience the possibility of interacting directly with the actors, providing feedback about the feasibility and realism of the situation played, but in the case of budget restrictions the use of video can be an effective alternative.

Three personas (Table 1) have been created to represent the three main typologies of users of these pervasive multimedia systems. Paul and Marina embody the two categories of users directly involved in this research. The third one, Alex, can be included in the younger typology, and represents a driver category for the diffusion of these systems: techno-fun users. Further users' profiles could have been considered, however according to the authors' experience in these sorts of projects, the benefit of increasing the number of personas in such systems is uncertain.

The scripts for the three following scenarios have been created from the stories[4] that participants brought to the focus groups (Howard et al., 2002). Each scenario corresponds to a prospected UX with pervasive multimedia systems for each user profile. The first scenario provides a more conservative approach to the problem, while the third one gives a more futuristic view of the prospective system. Although some elements that have been included in the scenarios are not novel individually, the intention was to recreate a complete view of how pervasive interactive multimedia systems could be totally embedded in the personas' daily lives. The following Summarizes the scripts for each one of the scenarios:

First Scenario: "Paul"

Morning routine: Paul has breakfast with his family whilst watching the morning news on the digital tablecloth. He then drives his son to school after which his wife gets an automatic notification (voice message on her mobile or digital tablecloth) that the boy has arrived at school. While driving to his office, Paul receives personalized local video news on his in-car-navigator.

Work routine: All of Paul's mobile communications go across his desktop when he is sitting in his chair at work. He uses the local satellite system to find a picnic area, the location of which he forwards to his wife in the form of a multimedia (video and map) message inviting her to lunch. After Paul leaves the office to meet her, he gets a video message on his in-car-navigator from a partner asking urgently for an important file. Paul retrieves the file in one of his company's folders using the voice interaction system, and replies to the message. He then sets the work message system to very-urgent-only mode since he has the afternoon off. He then sees by his wife's position on his navigator map that she has arrived at the arranged the meeting point for lunch.

Figure 3. Proof-of-concept mock-up for the 'memory-box'

Social lunch: As Paul joins his wife he receives an info alert message, which he accepts to get a short video clip about a nearby landmark. Using voice activation, Paul uses his mobile to provide him with related information about the area and other subjects of interest. Finding digital messages from a relative that has been to this landmark before, Paul and his wife view some of them and then leave one of their own for others to discover later. Following lunch, he notices in his phone's EDG (electronic diary guide) that his aged uncle is engaged in a global warming activist's discussion through videoconferencing. He decides to join and support him by providing some video documentary evidence of a dry lake nearby.

Afternoon leisure: Paul checks the state of the river through the rowing club live-cam and finds the water is too rough for rowing. He sends a link with the live video of the river to his pals together with a message suggesting they reschedule their plans to row together. Having changed his schedule, he visits a nearby market mentioned in the morning news and finds an interesting but pricey item. He starts a videoconference with a knowledgeable friend who examines the piece via Paul's mobile cam and offers an opinion on the value of the item. Unconvinced he should buy it, Paul uses a matchmaking system to locate similar collectors who are visiting the market and he finds two people. A quick look at their profiles tells him they are worth a short meeting so he sends them a message inviting them for a coffee. Meanwhile, Paul's wife checks her virtual map, spots a nice gallery nearby, and makes a short virtual visit. When she notices on the map that one of her friends is nearby, she sends an invitation for her to visit the gallery.

Evening routine: On the way back home, his wife uses her mobile shared-whiteboard with her best friend to help organize a party by choosing the menu, guests, and the decoration. After picking up their son from school, she receives a notification that her favorite TV show is about to begin so she remotely checks that the home video recorder has been set properly. She then accepts to watch a summary of today's episode on the in-car-navigator, which she also forwards to her friends.

Second Scenario: "Marina"

Morning routine: Waking up in a Shanghai hotel, Marina uses her mobile to activate personalized BBC news displayed on the interactive portrait near her bed. Although Mum is not feeling well these days, Marina notices through her EDG that her Mum is video-sharing a recipe with her cooking online community; this relieves her. A second message pops up to remind Marina of breakfast plans with a friend near the Yuyuan Garden.

On the move: On the bus heading to the café, Marina follows an electronic-paper map of the route in order to familiarize herself with the city. The live interactive local map includes links to helpful information. The map notifies her of digital messages attached to particular landmarks (when passing by), and provides the opportunity for Marina to leave her own comments for others to see. Passing a concert hall, Marina accepts to download a video clip informing her of an upcoming performance by her favorite band. Immediately she books using her mobile and sends the video message to two friends, inviting them to join her. The map shows her that there is one of her jogging partners in a café nearby, but she decides to disregard the notice as she is in a rush to her appointment.

Table 2. Summarized description of the three main personas

Paul is a married 52 years old mechanical engineer who has just moved to the countryside near Edinburgh and has 2 sons: one still in school while the other has just started at the university. Paul enjoys fiction movies, rowing, gardening, and collecting.
Marina, a single 28 years old lawyer living in east London and frequently travelling abroad for work, is interested in archaeology, travelling and jogging.
Alex, a 21 years old sociology student living in Cardiff, has a girlfriend in Copenhagen so he travels there very often. He is passionate about music, surfing and clubbing.

Social interaction: Still on the bus, Marina notices an attractive man reading a book on a nearby seat. Pointing her mobile toward the book, it reveals the book's title which she finds fascinating and increases her interest in meeting the person reading it. She approaches the man, introduces herself, and informs him that she is about to get off the bus but would like to talk with him another time. Shaking their mobiles, they exchange their contact information with each other. After leaving the bus, Marina checks the guy's data on her mobile and discovers it includes personal information about his interests, including video galleries, stories, and more, which reveal they have many points in common.

Context interaction: As Marina walks across the Yuyuan garden, a work of art grabs her attention. A quick query using her mobile (thanks to a pattern recognition search engine system) informs her of the artwork's significance and of a projected interactive discussion board related to it. Marina chooses to leave her own digital 'waymark' and to create a digital video card for her friends. Realizing she is now late for breakfast, Marina notices in the interactive map that her friend is already there, so she hurries.

Social interaction: While sitting in the café, Marina and her friend decide to go on holiday together. Using the electronic map they check travelers' advice and recommendations for prospective locations which includes travelers' self-authored audio-visual content about their experiences. Joining their two mobiles they double their screen size and access enhanced navigation features.

Third Scenario: "Alex"

Morning routine: Alex wakes up in his girlfriend's apartment in Copenhagen to find Linda has already left for work. Pointing his finger toward the electronic wallpaper, he voice-activates a holographic 3D videoconference with Linda. Using his mobile he activates a holographic projection of MTV-TRL to view whilst he starts his day. As he would like to go surfing, Alex requests activation of live-cams of other surfers' handhelds so he can check weather conditions. The only surfer on the beach accepts to tele-activate his cam, revealing a calm sea unsuitable for surfing, so Alex cancels his plans and heads out for breakfast. In a quiet café he catches up with news from his favorite mobloggers and participates in a video debate about a recent sailing race. He receives a live-cam activation request from a viewer of a reality TV show to which Alex is subscribed. Viewers of the show can access his mobile or home cam when they want.

Digital geo-caching: An alert on his mobile prompts Alex to receive a video message that turns out to have mysterious and cryptic significance.

With some research, Alex solves the riddle and gets the prize of a virtual ark on his mobile. An experienced player of this game, Alex knows he now has 24 hours to find a new hiding place for the ark and create a new riddle for the next person to solve. He hops on a bus heading across the Øresund Bridge to Malmö, Sweden, and streams a short video of the sky, sea, and coastline to the other players. While he searches for a good location to hide the ark, he uses his mobile to deposit multimedia messages on different landmarks on his way giving hints for other players to find the ark.

Romantic gesture: Back in Copenhagen and thinking of Linda stuck in a meeting at work, Alex records a video of himself with a romantic quote. He then buys a bunch of daisies and digitally attaches the video to the bouquet which he leaves with the receptionist at Linda's office. Later, when Linda receives the flowers, the registered video message appears on her digital book.

Virtual memory: Alex senses he is forgetting an obligation. He checks his electronic agenda but nothing is scheduled. When he rewinds his personal memory box system he recalls the previous night he had promised to make a video of a local band and narrowcast it to the interested people of the MTV TRL network. He adds this to his electronic agenda. Then Alex uses his mobile to check how many clips from his personal surf video gallery have been sold recently. Disappointed in what he finds, he records a note to himself to do more interactive advertising.

These scenarios have subsequently been assessed in two workshops through dramatizations. Some of the workshop participants were asked to act out the scripts to represent the scenarios. In this way, users and designers were able to discuss the feasibility of the different elements of the system proposed. This methodology has proved to be useful not only to confirm their reliability and relevance but also, to achieve a high engagement of the users during the design process (Newell et al., 2006). Some elements of the scenarios have not been intentionally described in detail (e.g., the EDG), because it wasn't enough information in the previous phase to understand how the users have envisaged these applications. Only if during the scenario dramatizations the participants found these applications relevant, designers generated discussion with the users in order to define them better.

During the role-play in the assessment workshops some applications presented here (e.g., the digital table cloth in the first scenario) were strongly criticized by the users and therefore they have not been considered in the development of the proof of concept mock-ups. The EDG produced general excitement among the participants, as they saw it as an effective tool for obtaining information about the wellbeing of older relatives and friends without letting them feel observed, thus supporting independent living and encouraging mobility. Other examples of applications described in the scenarios have been amended or refined in order to better satisfy the users' requirements (e.g., the interactive map and the in-car communication system).

Experimental Prototypes (Phase Three)

The scenarios provided the basis for identification of possible implementation settings and verification of system requirements. Once potential interfaces and applications were determined, assessing their quality in-the-field required the development of proof of concept mock-ups. Since this work focused on the nomadic and peripatetic behavior of users, ambient and home interfaces have not been considered. A description of the different applications that have been assessed in the form of low-fi experimental prototypes are detailed in the following paragraphs. Many of the devices and applications proposed are not original, however collectively they combine to provide a highly innovative pervasive interactive multimedia system. In fact, if any of the identified interfaces or applications had already been developed (e.g., a multimedia mobile phone), we would have considered it appropriate to incorporate it in the system instead of creating a new one. The proof of concept mock-ups was made by combining paper, cardboard, and real mobile-phones. The envisaged pervasive interactive multimedia system comprised of five interfaces; a mobile phone, an interactive map, public interactive displays, an in-car multimedia system and a memory pin.

Handheld

This mobile device has a traditional clamshell design with a pivotable color display, photo and video camera, and keypad-based standard interaction as well as voice-based interaction.

In addition, a small transmitter inside the device enables it to serve as a pointer and allows interaction with TV screens, public digital displays (much like how a mouse is used to point to a computer display) and with intelligent objects such as bus stops (to get information from them).

Also envisaged is the possibility combining two devices to double the size screen and permit enhanced navigation, which would increase the sociability potential of this application.

Applications for this device include:

- Context aware infotainment such as local video news, visualization of user's position on a map, reception of in-situ multimedia alert-messages from things, places (landmarks, building, etc.) and events.
- Distance vision such as remote cam activation and control (zoom, positioning). The remote can be an autonomous device or can be embedded in other users' handhelds.
- Customized multimedia content such as a embedded mobile live-encyclopedia with a pattern recognition search engine (linked to TV-video content).
- Self-authoring system enabling recording, editing and sharing, broadcasting or narrowcasting of personal videos, life TV video debate in videoconferencing, mob-blogging and co-production of reality-TV channels, notification of a live event or TV series and video clip summary of a TV show and possibility of storing, editing or sharing it to other users.
- 'Memory box' that enables users to register self-authored multimedia content (memories) that would be delivered in specified future occasions such as birthdays or graduation-day to their younger relatives or friends, to keep their presence alive even when they have passed away.
- Socialization and social awareness system such as a matchmaking system. It can be used to find users with desirable profiles, shared whiteboards or locator of buddies. New interaction models for this application include the possibility of pointing the mobile towards a person to get info about them, and exchange of personal information by shaking hands between users.
- Electronic diary guide where users can manage their daily appointments as well as checking whether their friends or relatives are currently engaged in any sort of interaction with the system (e.g., discussion group, etc.).

Public Interactive Display

Interactive digital billboard that displays customized info based on profiles of passers-by with a pointing-based interaction system and voice-based interaction capabilities. Although similar interfaces have been already developed in other projects, the scenarios showed that it was crucial in terms of completeness to integrate it in the proposed system (McCarthy et al., 200; Russell et al., 2002).

Applications for this device include:

- Storage and display of digital messages from people that have been there before, public interactive discussion board (using one's mobile as both pointing device and content editing tool), display of incoming personal messages (video, text or voice through sonic cones for direct sound to the user).

Personal Interactive Map

This device is best described as electronic foldable paper with full touch-screen display, voice based interaction and GPS location based system (see Figure 4).

Applications for this device include:

- Mapping and routing services such as local maps and interactive ads
- Social and context awareness such as location of the user, other people, things and places, routing system with multimedia information about nearby people, things and places including an instant sessaging system (IM).
- Micro-payment e-commerce applications such as the possibility of making bookings or purchases related to the above (e.g., concert) and forwarding the info to someone else.
- Display of self-authored content such as users edited travel guides and maps.

In-Car Multimedia Communication System

When the user sits in the driver's seat of a car with this built-in system, all of their mobile communications automatically route through it. The system consists of an adjustable monitor and semitransparent projection (for the driver's use) in the front glass (Figure 5), touch-screen capabilities, and voice-based interaction.

Applications for this device include:

- Context aware infotainment such as local video news, local satellite maps (and possibility of forwarding the location to someone with a voice or text message).
- Social awareness such as location of other people.
- *Busitainment* applications such as video message and videoconference system, retrieval of personal or work files, automatic addressing of messages, different 'screening' modes: family, personal, work and very-urgent-only.

Memory Pin

This device is simply a small, low-cost storage container capable of interacting with a user's mobile on request in order to store a wireless download of self-authored multimedia content. The 'pin' is then attached to a desired object (as shown in Figure 6) and ready to upload its multimedia content to nearby devices (PID, e-paper).

Applications for this device include:

- Download of text, sounds and movies from a mobile.
- Upload of the registered video-quote on a user's device (digital book, PC, PID, electronic paper) by simple touch.

The low-fi prototypes representing these interfaces and applications have been tested in-the-field. However, to overcome safety concerns in the case of the in-car system, the assessment was performed in a motionless vehicle. This unrealistic testing context has made questionable the validity of the results. However, the semitransparent GUI projection on to the front windscreen did elicit users' worries about safety and security.

In the other cases, when the evaluation considered the real context of use, they provided crucial information about how the physical and social environment can influence the use of the system. For example, the interactive map raised concerns about the management of the privacy, and the mobile multimedia phone about embarrassment in crowded areas. During the in-the-field assessment of the proof of concept mock-ups, the experience with some applications (like the 'memory box') was highly praised.

RESULTS ANALYSIS

In order to address complex issues such as understanding, emotion, security, trust and privacy, the data gathering techniques presented in this chapter focus on users rather than on their tasks or objectives with the analyzed interfaces. This research shows how the physical and social contexts have a strong impact on the users' attitudes towards mobile interactive multimedia applications: the context influences the users' emotions and feelings towards the interaction process, persuading or discouraging its use (Kjelskov et al., 2004). For example, during the in-the-field assessment of the proof of concept mock-ups some users felt unsafe recording video with their mobile phones in a crowded street, as they were very concerned about theft.

The questionnaires, observations and the focus groups revealed two main users' categories when considering the creation and sharing of self-authored multimedia content:

- Spontaneous or impulsive user (e.g., when travelling, during an exciting night out, when observing an interesting thing, place, or performance or just to update about domestic issues such as children, new partner, etc.). The addressees are the members of the user's restricted social personal circle: family, friends and colleagues.
- Reiterative or structured user (e.g., mob-blogs). The addressees belong to a broader social circle such as enlarged communities.

The cultural probes showed a clear desire by users for using their handhelds to create self-authored video content for two main purposes: as an enhanced democratic tool (e.g., voting on public issues or having 'five minutes of glory in TV') and to leave their 'signature' along their way (e.g., by putting down personal-digital content on public digital board at monuments or other places). Applications regarding exchange of multimedia content with objects and places have been explored in many different contexts: visiting a city (Brown et al., 2005; Cheverst et al., 2003), playing pervasive games (Benford et al., 2004), leaving signs and building communities (Burrell & Gay, 2002; Giles & Thelwall, 2005; Persson et al., 2002).

Figure 4. Interactive map from the Vodafone Futures

The probes also exposed users' preferences when receiving multimedia content on their handset from people, places or things: 'If on the move, it's better if related to my context.' Context awareness provides customized information that can be defined as the right information in the right place and in the right time. In this sense it is interesting to note the work of Abowd and Mynatt (2000) who apply a set of five questions to obtain what they call a good minimal set of necessary context.

Observations and mapping of movements provided qualitative information about how the social context influences the use of mobile phones in public spaces. For example, it revealed how mobile phones encouraged their owners to temporarily disconnect from a social group (such as a group of friends) during the period of use (for example, whilst reading a message).

The experiential narratives that participants presented during the focus groups, raised the following issues for mobile interactive multimedia systems: socialibility (e.g., to allow users traveling together to share the experience of viewing a video with their mobile phones) and collaboration (enable users who are in different places to exchange moods, share information and even work together) (Lull, 1980), context awareness (both services and content should be customized and related to the specific users' context), creativity (enable nomadic and peripatetic users to produce self-authored multimedia content), interactivity (interfaces need to support a high level of interactivity, by using new modalities such as gestures) (Palen et al., 2000), convergence (enable users to use the most plausible and appropriate interface in each context: iTV, mobile phones, in-car-navigators and

Figure 5. Semitransparent projection in the car windshield

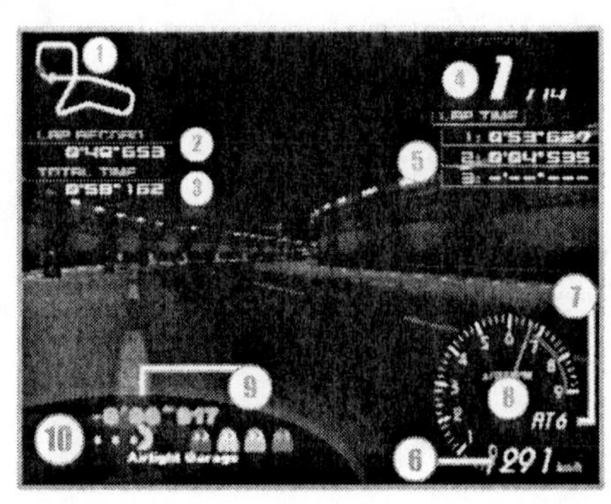

the Internet) and connectivity (enable different ways of communication among users: one-to-one and one-to-many).

Surprisingly, this research highlighted very little appeal by users in receiving broadcasts of traditional TV formats on their mobile phones (except some exceptions such as brief live updates of a decisive football match or extraordinary news). Therefore the concept of mobile or pervasive iTV is more likely to be related to the emergence of mobile communities that support 'DIY producers' of multimedia content: they will create multimedia content in specific contexts, with precise purposes and share it with others.

The interfaces and applications show how such open, diffuse, and pervasive interactive multimedia systems provide an exceptional virtual platform that might foster and enhance the development of new communities of creative users that can share moods, content and collaborate with different purposes such as work, entertainment or government. In the specific case of entertainment it is worth mentioning Davenport's view of the topic: 'Since the earliest days of cinema, artists and technologists have dreamt of a future in which everyone could create and share their vision of the world. With the evolution of ubiquitous mobile networks and the enhanced mobile handset as creative device, we are on the cusp of realizing improvisational media fabrics as an active expression in our daily lives'.[5]

CONCLUSION

Traditional data gathering and evaluation techniques based on cognitive psychology focus on the human machine interaction and disregard a crucial aspect in the process: the context of the user. The physical and social context might have a strong impact in the use of the analyzed interfaces: it influences in a positive or negative way the users' emotions and feelings towards the interaction process, persuading or discouraging its use.

This research tries to recognize the mutual influence between technology and society. Just as technology shapes society, we also need to investigate how society shapes technology. This particularly holds true with a social technology that needs to be integrated with household routines. In making predictions about new technology we need to explore the critical disconnections between the ways in which such technologies are produced and the ways in which they are consumed, naturalized and rejected (Fischer, 1992; Lee & Lee, 1995).

Handsets are becoming tools for creation, editing and diffusion of personalized and personal multimedia content and this attribute allow users to become 'DIY producers' of digital content (Cereijo Roibas & Sala, 2004). Users will be able to create their own multimedia content to share with others. Therefore new communities of nomad and peripatetic users will find themselves in original communication contexts and in novel expressive situations: they will be able to create their own 'movies' and share them with other users, places (real and virtual environments) and objects (intelligent objects and other digital-physical hybrids). This expression of users' creativity in pervasive interactive multimedia systems needs to be corroborated by interfaces that support some form of users' creativity, collaboration social interaction (Ducheneaut & Moore, 2004; Nardi et al., 2004; Preece & Maloney-Krichmar, 2003).

The research methodology adopted helped to confirm the validity of the three premises that guided this work. As shown in the second and third scenarios, small mobile devices can provide both a functional and an effective interactive experience. As it has been recently corroborated by the

Figure 6. Memory pin: Interfaces as containers of information

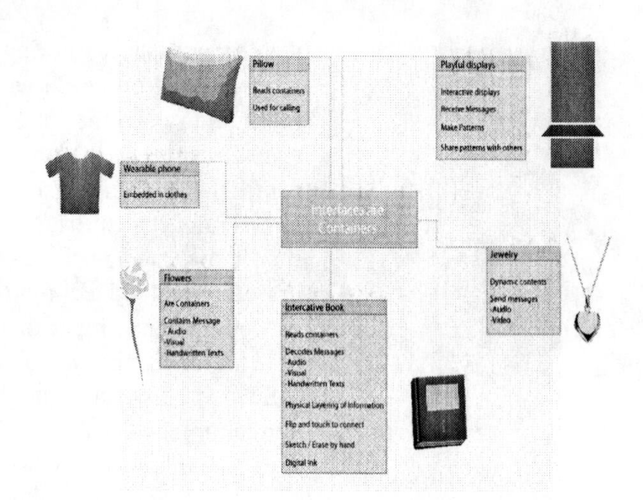

success of the Wii, the fifth video game console released by Nintendo (European consumers snap up 325,000 Wii consoles in two days. 2006), they are also able to recreate an enjoyable immersive environment for the user. The distinctive feature of the new game console is its small wireless controller, the Wii Remote, which can be used as a handheld pointing device and can detect motion and rotation in three dimensions providing the gamer with a high immersive experience.

While the first and second scenarios make it particularly clear how handhelds are appropriate for the creation of new forms of multimedia content by the user, the third, however, provides evidence of how novel forms of interaction could be implemented (see Romantic Gesture). The recently unveiled Apple iPhone promises deliver a fine example of novel interaction with content through its multi-touch screen (that uses users' fingers as the ultimate pointing device) and multi-touch sensing system. One of its innovative interactive features is the ability to zoom objects in and out such as photos, by placing two fingers on the screen and moving them farther apart or closer together as if stretching or squeezing the image (Apple's 'magical' iPhone unveiled. 2007).

All of the three scenarios showed how mobile phones are suitable tools for context related content and context awareness applications. The enhanced social interaction shown in the second scenario is particularly interesting.

Our future work will look at the exploration of how new technological paradigms will affect the perceived quality of experience in pervasive interactive multimedia systems. These paradigms include hybrid artifacts, use of biotechnology, advanced interaction modalities, new forms of content and novel intelligent environments, immersive environments such as collaborative virtual environments and multi-user environments. In this sense, an examination of the contributions that disciplines such as the interactive arts, space technology, medicine and games could give to this area might offer significant insights. At the same time, to achieve an adequate outcome in these new territories, we will also look at new immersive field research instruments that reduce, as much as possible, bias and subjectivity in ethnographic research (LeCompte, 1987).

REFERENCES

Abowd, G. D., & Mynatt, E. D. (2000). Charting past, present and future research in ubiquitous computing. ACM transactions on computer-human interaction, *ACM Press, 7*(1), 29-58.

Apple's 'magical' iPhone unveiled. Retrieved on January 9, 2007 from http://news.bbc.co.uk/1/hi/technology/6246063.stm

Arreymbi J., Gachanga, E. (2006). Interactive design and delivery challenges for wireless handheld multimedia systems. *International Conference on Internet Computing* (pp. 377-386).

Barrenho, F., Romão, T., Martins, T., & Correia, N. In Authoring environment: Interfaces for creating spatial stories and gaming activities. In *Proceedings of ACM SIGCHI International Conference on Advances in Computer Entertainment Technology (ACM ACE 2006).* New York: ACM Press.

Benford, S., Flintham, M., Drozd, A., Anastasi, R., Rowland, D., Tandavanitj, N. et al. (2004). Uncle Roy all around you: Implicating the city in a location-based performance. In *Proceedings Of ACM Advanced Computer Entertainment (ACE 2004*). Singapore: ACM Press

Brown B., Chalmers M., Bell M., MacColl I., Hall M., MacColl I. et al. (2005) Sharing the square: collaborative leisure in the city streets. I*n Proceedings of ECSCW 2005* (pp. 427-429). Paris: Springer

Burrell J., & Gay, G.K. (2002). E-graffiti: evaluating realworld use of a context-aware system. *Interacting with Computers, 14.*

Butscher B., Moeller E., & Pusch H. (1996, March 4-6). *Interactive Distributed Multimedia Systems and Services.* European Workshop IDMS '96. Berlin, Germany.

Capra M., Radenkovic, M., & Benford, S. (2005). Multimedia challenges raised by pervasive games. In *Proceedings in ACM Multimedia 2005.* New York: ACM Press.

Carroll, J. M. (2000). *Making use of scenario-based design of human computer interactions.* Cambridge, MA: MIT Press.

Cereijo Roibas, A., Sala, R. (2004). Main HCI issues for the design of interfaces for ubiquitous interactive multimedia broadcast. *Interactions Magazine,* 51-53.

Cheverst, K., Dix, A., Fitton, D., & Rouncefield, M. (2003). 'Out to lunch': Exploring the sharing of personal context through office door displays. In *Proceedings of the 2003 Australasian Computer-Human Conference, OzCHI 2003* (pp. 74-83), S. Viller and P. Wyeth (Eds.), Canberra: CHISIG.

Ducheneaut, N., & Moore, R. J. (2004). The social side of gaming: A study of interaction patterns in a massively multiplayer online game. In *Proceedings of CSCW 2004* (pp. 360-369). Chicago, IL: ACM Press.

European consumers snap up 325,000 Wii consoles in two days. *GamesIndustry.biz.* Retrieved December 13, 2006 from http://www.gamesindustry.biz/content_page.php?aid=21691

Fischer, C. S. (1992). *America calling: A social history of the telephone.* Berkeley, CA: University of California Press.

Frisk, H., & Yoshida, M. (2005). New communications technology in the context of interactive sound art: An empirical analysis. *Organised Sound, 10*(2), 121-127.

Gaver, W.W., Dunne, A., & Pacenti, E. (1999). Cultural probes. *Interactions, 6*(1), 21–29.

Giles L., & Thelwall, S. (2005). *Urban tapestries: public authoring, place and mobility.* London: Proboscis.

Goularte, R., Cattelan, R. G., Camacho-Guerrero, J. A., Valter, J., Inacio, R., & Pimentel, M. G. C. (2004). Interactive multimedia annotations: enriching and extending content. In *ACM DocEng '04* (pp. 84–86).

Hagen, P., Robertson, T., Kan, M., & Sadler, K. (2005). Emerging research methods for understanding mobile technology use. In *proceedings of OZCHI 2005* (pp. 1-10). Canberra, Australia.

Harper, R. (2003). People versus information: The evolution of mobile technology. In L. Chittaro (Ed.), *Human computer interaction with mobile devices* (pp. 1-15). Berlin: Springer.

Ho-Ching, F.W., Mankoff, J., & Landay, J.A. (2003). Can you see what I hear?: The design and evaluation of a peripheral sound display for the deaf. In *Proceedings of ACM CHI 2003* (pp. 161-168).

Howard, S., Carroll, J., Murphy, J, & Peck, J. (2002). Using endowed props in scenario-based design. In *NordiCHI 2002.*

Hulkko, S., Mattelmäki, T., Virtanen, K., & Keinonen, T. (2004). Mobile probes. In *proceedings of the 3rd Nordic Conference on Human-Computer Interaction* (pp. 43-51). Tampere, Finland: ACM Press.

Hutchinson, S. (2001). Urban moving 2030: transport typologies for the future city. In J. Myerson (Eds.), *The Helen Hamlyn research associates catalogue 2001* (p. 22). London: Helen Hamlyn Research Centre.

Iacucci, G., Kuutti, K., & Ranta, M. (2000). On the move with a magic thing: Role playing in concept design of mobile services and devices. At DIS 2000.

Kjeldskov, J., & Stage, J. (2004). New techniques for usability evaluation of mobile systems. *International Journal of Human-Computer Studies, 60*, 599-620

Kjeldskov, J., Graham, C., Pedell, S., Vetere, F., Howard, S., Balbo, S. et al. (2005). Usability of a mobile guide: The influence of location, participants and resource. *Behaviour and Information Technology, 24*, 51-65.

LeCompte, M.D. (1987). Bias in the biography: Bias and subjectivity in ethnographic research. *Anthropology and Education Quarterly, 18*(2), 43-52.

Lee, B., &Lee, R.S. (1995). How and why people watch tv: Implications for the future of interactive television. *Journal of Advertising Research, 35*(6).

Leed, E.J. (1991). *The mind of the traveller.* New York: Basic Book.

Lull, J. (1980). The social uses of television. *Human Communication Research, 6*(3).

Mankoff, J., Dey, A. K., Hsieh G., Kientz, J., Lederer, S., & Ames, M. (2003). Heuristic evaluation of ambient displays. In *Proceedings of CHI 2003.*

Mark, G., Christensen, U., & Shafae, M.(2001). A methodology using a microcamera for studying mobile IT usage and person mobility. In *Proceedings of CHI 2001.* Seattle, WA.

McCarthy, J.F.,Costa, T.J., & Liongosari, E.S. (2001). UniCast, OutCast & GroupCast: Three steps toward ubiquitous peripheral displays. *UBICOMP 2001*, Atlanta.

Morley, D. (1986). *Family television. Cultural power and domestic leisure.* London: Comedia.

Mynatt, E.D., Back, M., Want, R., Baer, M., & Ellis, J. B. (1998). Designing audio aura. In *Proceedings of CHI '98 (Los Angeles, California)* (pp. 566-573). ACM.

Nardi, B., Schiano, D., & Gumbrecht, M. (2004). Blogging as social activity, or, would you let 900 million people read your diary? In *Proceedings of CSCW 2004* (pp. 222-231). Chicago, IL: ACM Press.

Newcomb, E., Pashley, T., & Stasko, J. (2003). Mobile computing in the retail arena. In *proceedings CHI 2003.* Ft. Lauderdale, FL: ACM Press.

Newell, A. F., Gregor, P., & Alm, N. (2006). Theatre as an intermediary between users and CHI designers. *CHI 2006.* Montreal, Quebec, Canada.

Palen, L., Salzman, M., & Youngs, E. (2000). Going wireless: Behavior of practices of new mobile phone users. In *Proceedings CSCW 2000* (pp. 201-210).

Perry, M., O'Hara, K., Sellen, A., Harper, R., & Brown, B.A.T. (2001). Dealing with mobility: understanding access anytime, anywhere. *ACM Transactions on Computer-Human Interaction (ToCHI), 4*(8), 1-25.

Persson, P., Fagerberg, P. Geonotes: *A real-use study of a public location-aware community system* (Tech. Rep. T2002:27, ISSN 1100-3154).

Pham, T-L., Schneider, G., & Goose, S. (2000). *A situated computing framework for mobile and ubiquitous multimedia access using small screen and composite devices.* New York: ACM Press.

Pooley, C. G., Turnbull, J., & Adams, M. (2005). *A mobile century? Changes in everyday mobility in Britain in the twentieth century.* Ashgate, Aldershot, Hampshire.

Preece, J., Maloney-Krichmar, D. (2003). Online communities: Focusing on sociability and usability. In J.A. Jacko & A. Sears (Eds.), *2003: Handbook of human-computer interaction.* London: Lawrence Erlbaum Associates Inc.

Russell, D. M., Drews, C., Sue, A. (2002). Social aspects of using large public interactive displays for collaboration. Ubicomp 2002.

Sato, & Salvador. (1999). Playacting and focus troupes: Theatre techniques for creating quick, intense, immersive, and engaging focus groups. *Interactions, sept/oct.*

Scheible J., & Ojala, T. MobiLenin—Combining a multi-track music video, personal mobile phones and a public display into multi-user interactive entertainment. In *Proceedings ACM Multimedia 2005* (pp. 199-208). New York: ACM Press.

Spigel, L. (1992). *Make room for TV: Television and the family ideal in postwar america.* Chicago, Uuniversity of Chicago Press.

The 2007 edition of DTT TMT Industry Group's Telecommunications Predictions. Retrieved on January 16, 2007, from http://www.mondaq.com/article.asp?articleid=45596

Zheng, J Y., Wang, X. Applications 4: interactive multimedia systems: Pervasive views: area exploration and guidance using extended image media. In *Proceedings of the 13th annual ACM international conference on Multimedia MULTIMEDIA '05.* New York: ACM Press.

KEY TERMS

Context Awareness: Is a term from computer science that is used for devices that have information about the circumstances under which they operate and can react accordingly. Context aware devices may also try to make assumptions about the user's current situation.

Convergence of Technology: The coming together of two or more disparate technologies. For example, the so-called fax revolution was produced by a convergence of telecommunications technology, optical scanning technology, and printing technology.

ICT: Information technology as defined by the Information Technology Association of America (ITAA) is: "the study, design, development, implementation, support or management of computer-based information systems, particularly software applications and computer hardware." In short, IT deals with the use of electronic computers and computer software to convert, store, protect, process, transmit and retrieve information. Nowadays it has become popular to broaden the term to explicitly include the field of electronic communication so that people tend to use the abbreviation ICT (information and communication technology). Strictly speaking, this name contains some redundancy.

iTV: Interactive TV (iTV) is an umbrella term. Interactive TV is the content and services (in addition to linear TV and radio channels) which are available for digital viewers to navigate through on their TV screen.

Mobile TV: Watching TV on a mobile phone. There are several mobile TV air interfaces competing for prime time. Digital multimedia broadcasting (DMB) is based on the digital audio broadcasting radio standard; digital video broadcast-Handheld (DVB-H) is the mobile version of the international digital TV standard, and forward link only (FLO) is based on QUALCOMM's popular CDMA technology.

Pervasive iTV: An amalgamation between the concepts of iTV and pervasive TV. However this term goes beyond the concept of traditional TV programs data stream and focuses on content personalization and users' creativity, sociability, context awareness, advanced interactivity, immersive environments, convergence (iTV, mobile phones, in-car-navigators and Internet) and connectivity (one to one and one to many).

Pervasive TV: It is an adaptation of the term pervasive computing and it reflects the concept of accessing TV in different contexts such as home, the office, the auto, outdoors thanks to the convergence of technology.

Sociability: Regards the social character of the usage of TV and it involves the identification of suitable applications and interfaces that support social use.

User-Centered Design: UCD is a design philosophy and a process in which the needs, wants, and limitations of the end user of an interface are given extensive attention at each stage of the design process. User-centered design can be characterized as a multi-stage problem solving process that not only requires designers to analyze and foresee how users are likely to use an interface, but to test the validity of their assumptions with regards to user behavior in real world tests with actual users. Such testing is necessary as it is often very difficult for the designers of an interface to understand intuitively what a first-time user of their design experiences, and what each user's learning curve may look like. The chief difference from other interface design philosophies is that user-centered design tries to optimize the user interface around how people can, want, or need to work, rather than forcing the users to change how they work to accommodate the system or function.

UX: User experience (UX) is a term used to describe the overall experience and satisfaction a user has when using a product or system. It most commonly refers to a combination of software and business topics, such as selling over the web, but it applies to any result of interaction design. Interactive voice response systems, for instance, are a frequently mentioned design that can lead to a poor user experience.

ENDNOTES

1. The Virgin Mobile TB trial became a commercial offering in October 2006, but due to lack of interest the service will close in January 2008.
2. Mobility in city centres involves tasks such as commuting, entertainment-seeking, area visiting and dwelling (trips related to shopping or socialising for example) (Hutchinson, 2001).
3. Although Gaver included the postcards within the probes packet, it has been decided to treat them as an autonomous tool in order to extend the number of users involved in the data gathering process.
4. Experiential narratives
5. Glorianna Davenport, Principal Research Associate at the MIT Media Lab in Cereijo Roibas (2003), Ubiquitous media at the intersection: iTV meets Mobile Communications, Panel at the Proceedings of HCI 2003 Conference. Bath.

Chapter III
Problems Rendezvousing: A Diary Study

Martin Colbert
Kingston University, UK

ABSTRACT

This chapter seeks opportunities to use mobile technology to improve human mobility. To this end, the chapter reports a diary study of university students' use of mobile telephones for rendezvousing—arranging, and traveling to, informal meetings with friends and family. This diary study reveals, and suggests explanations for, a number of deficits in user performance: (1) rendezvousers occasionally become highly stressed and lose valuable opportunities; (2) outcomes are worse when rendezvousing at unfamiliar locations; (3) 31 to 45 year olds report more personal sacrifices than 18 to 30 year olds; and (4) when mobile phones are used on the move, the experience of communication is slightly worse than when phones are used prior to departure. Ways of using mobile technology to make good these deficits are suggested.

INTRODUCTION

Mobile Technology and Human Mobility

Between 1997 and 2001, ownership of GSM mobile telephones rose from 27 percent of the UK public to 73 percent (Butcher, 2003). One important reason for this rapid adoption of mobile technology was anytime, anywhere access to voice telephony. Talking on mobile telephones gave users the freedom to roam away from fixed access points and remain contactable, even when life took them to diverse, unpredictable locations (Palen et al., 2000). Mobile telephony also acted as a flexible, 'proxy' for resources elsewhere. Rather than endure unproductive, 'dead time,' mobile workers could use their mobile phone to have faxes read or sent, or to learn about developments on other projects (Perry et al., 2001). Mobile telephony was also useful for fine-grain, moment-to-moment awareness and co-ordination.

For example, friends out shopping together could split up to visit different shops, and then use their mobile phones to discuss interesting sale items, and arrange how to meet up again (Ling & Yttri, 2001). Mobile technology was adopted, it appears, and in addition to other reasons[1], because it made 'being on the move' less unproductive and smoother-flowing. In this sense, mobile phones improved everyday mobility.

Such improvements are of interest, because everyday mobility is an important activity (Pooley et al., 2005). Mobility has practical value as an activity that enables individuals to function—to eat, work, sleep—and, as such, it is fundamental to society. It also has social value—the movement that makes life possible also enables interactions that support personal relationships, social networks, and local communities. It adds meaning to life and contributes to society, for good or bad. Mobility also acquires meaning itself and so contributes to our definitions of self ("we are how we travel"), and mobility is part of the process by which individuals learn about, and give meaning to, place and space. Finally, mobility has psychological consequences. It encourages individuals to feel a certain way, and to hold certain attitudes.

Future Technology for Mobility and Rendezvousing

Subsequent generations of mobile technology, however, will not necessarily be adopted as widely or as rapidly as GSM phones. To be adopted, broader-band wireless networks, multimedia input and output capabilities, integrated cameras, positioning mechanisms, context sensors, and so on need to be combined into 'packages' of device, service and network that actually improve mobility for many segments of the general public. But what kind of improvement will bring measurable benefits in mobility to users?

Sometimes, potential improvements are relatively easy to identify and confirm. For example, commuting and long distance travel is often "boring." Consequently, the public may want to download music and video files, play computer games, send picture messages, and consume 'live' streams of audio and video data to escape the tedium of waiting rooms, train carriages and other kinds of transit locale (Antilla & Jung, 2006, p. 222). However, other potential improvements are less obvious and less certain. For example, consider rendezvousing, that is, the informal coordination of a face-to-face meeting between friends and family[2]. The shopping rendezvous described in Figure 1 appears "poor" at first glance, because one party arrived late. However, the delay was caused by a traffic jam (and we can not expect mobile IT to free the roads of congestion), and the rendezvousers used existing GSM telephony to adjust their plans and maintain their convenience and comfort. So where is 'the problem'?

One approach to identifying user problems begins by identifying deficits in user performance, that is, in this case, respects in which human mobility is observably 'worse' under some conditions than others. The identification of any deficit suggests the design goal of removing or "making good" the deficit—a deficit provides a starting point for discussing the improvements that technology might achieve[3].

AIM: TO IDENTIFY OPPORTUNITIES TO IMPROVE RENDEZVOUSING

The work reported here, then, seeks opportunities for mobile technology to improve rendezvousing. To this end, it reports a diary study of university students' use of mobile telephones for rendezvousing. The study reveals a number of deficits in rendezvousing performance, and describes the interactive behaviour that brought about these deficits. Design suggestions that illustrate how the deficits might be made good are then presented[5].

The study extends the literature about user performance with communication systems to cover mobile technology and 'mobile' contexts of use[6]. Previous work concerned table-top communication systems in stationary contexts. This work includes, for example, an investigation of the effects of prepared scripts upon the consensus reached during chat sessions (Farnham et

Figure 1. A rendezvous: Sonja is late meeting Kate to go shopping

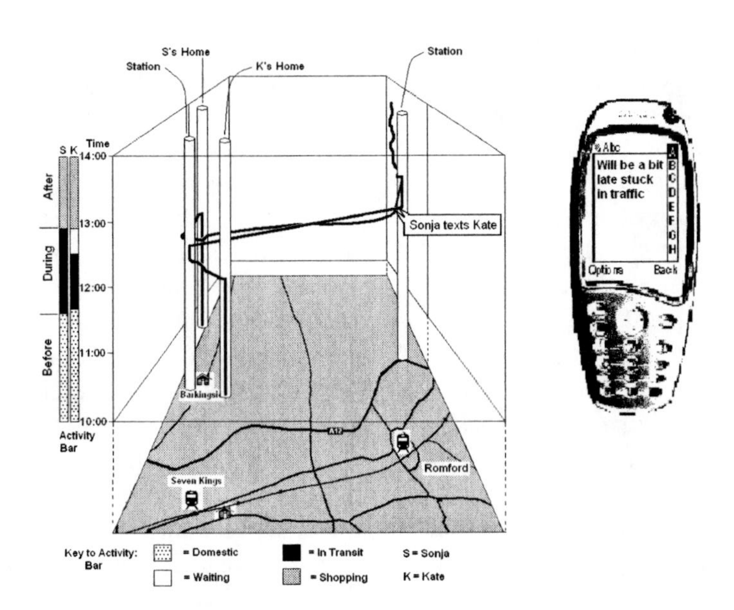

al., 2000), a study of the impact of a simplified 'communication computer' upon social interaction, mental stimulation, and attitude of elderly users (Czaja et al., 1993), and a comparison of the effect of video link quality upon overall system ratings and task completion time (Matarazzo & Sellen, 2000).

Participant 32 Entry No. 5 Traffic delays a rendezvous to go shopping. Sonja had arranged to meet an old friend at Romford railway station on Saturday at 12:30 p.m. Then, they would go shopping together. Sonja's friend Kate took the train from her home in Seven Kings, and arrived a few minutes early. Sonja drove from her home in Barkingside, but became stuck in a traffic jam in Romford town centre. At this point, Sonja phoned Kate (using her boyfriend's mobile phone) to say that she would be a little late. Sonja only knew a few landmarks in Romford, but conveyed her position accurately enough. Kate was half expecting traffic delays anyway, and went for a coffee rather than stand in the cold (It was February.). They met up 20 minutes later than planned. This rendezvous is represented in Figure 1[4]

Based upon data collected in 2000 (see Colbert, 2001). All places are in London, UK)

A DIARY STUDY OF RENDEZVOUSING

Study Design and Methodology Issues

Previous studies of communication systems design encourage methods that capture data in the field under naturalistic conditions. 'Practice' studies of communication—first, of landline telephone users (Frohlich et al., 1997; Lacohee & Anderson, 2001) and, more recently, of mobile phone use during urban journeys Tamminen et al., (2003) —and studies of media choice (Kraut et al., 1994; Preece, 2001; Whittaker et al, 1996) —report and illustrate the situated nature of communication. They show how, for example, the answering of a shared, household phone, or the choice to use e-mail rather than have a face-to-face meeting, reflects the situation in which communication occurs. Consequently, studies conducted under controlled conditions may simply distort the communication behavior being studied—participants may change their selection and use of media to better fit the 'experimental' situation. These studies also report and illustrate to the wide range of behavioral and performance parameters that are relevant to communication systems design. For example, the alternative communication services available, the richness of expression desired, the pace of exchange required, and the development of critical mass and social norms all influence usage and preference. These studies indicate, then, that it will be difficult to measure *all* relevant constructs within a single study. Also, any study needs to monitor the use of all the communications media at a users disposal, not just a single medium in isolation.

The work reported here satisfied the requirements for a naturalistic, field study that makes a large number of observations about all communication services on mobile phones[7] by asking participants to keep a diary about actual rendezvous they participated in, and the related communications. Diaries have long been used in user-centred development (Rieman, 1993), particularly to capture data in situ, with minimal observer effects (Carter & Mankoff, 2005). The diary used in this study collected both qualitative and quantitative data about each rendezvous, by having each diary entry comprise of free text, narrative descriptions as well as answers to preset, closed questions, and rating scales.

Method

Participants

The participants in the diary study were 22 male and 22 female students from the School of Computing and Information Systems, Kingston University. The aim of selecting participants was to obtain a sample that was large and gender-balanced from a source that mostly comprised of males. Between January, 2001, and April, 2002, students who took a module in human-computer interaction completed a diary as a minor part of coursework exercises. The diaries concern the participant's rendezvous for one week in January or February. Twenty-two female students completed a diary and consented to its anonymous use here. Then 22 male students were selected from the appropriate module year, to match the female participants as closely as possible in terms of age, ethnic background, marital status, number of children, and mobile phone ownership.

The diary keepers had a mean age of 25 years 11 months. Of the 44 participants, five did not state their age, 30 were in the 18 to 30 year age group, and nine were in the 31to 45 year group. Sixty-six percent were single, 23 percent had been with the same partner for more than one year, and 11 percent were married. Fourteen percent had children. All diary keepers were registered as full-time students, but 33 percent did more than 10 hours per week paid work in addition to their university studies. 89 percent owned a mobile telephone, 89 percent had access to a fixed line telephone, and 98 percent had a private e-mail account in addition to their university account. If they owned a mobile phone, 51 percent used it more than 10 times per week, and, if they had access to a fixed-line phone, 33 percent used that more than 10 times per week. Forty percent of participants were Asian in ethnic origin, 40 percent were European, nine percent were African-Caribbean, five percent were Middle Eastern and seven percent were 'Other'.

Materials

Each diary entry was comprised of: (i) an open-ended, narrative description in the participant's own words of what happened, and why; and (ii) the participant's responses to a questionnaire, which asked for specific details of each rendezvous and associated communication. This questionnaire comprised 37 questions as follows:

- **Questions 1–6:** The event (the who, when, where and why of the events)
- **Questions 7–11:** Outcomes (the additional stress and lost opportunity associated with attempts to meet at the time and place initially agreed)
- **Questions 12–24:** Usage and user experience of communication *prior to departure* for the rendezvous. User experience comprised satisfaction, convenience, social acceptability, disruption, frustration, and mental effort

- **Questions 25–37:** Usage and user experience of communication *whilst en route* to the rendezvous.

Two sets of answers about usage and user experience were returned per rendezvous—one for all communication that occurred prior to departure, and one for all communication that occurred en route. For example, a rendezvouser who spoke once on the phone and sent one text message before departing for the rendezvous point, and then spoke once on the phone and listened to one voice mail en route, provided one combined experience rating for the phone call and text message, and one combined experience rating for the phone call and voice mail. The prior to departure phase ends, and the en route phase begins when the first rendezvouser to do so departs for the rendezvous point.

Procedure

At the outset of the study, all participants were given an overview of future position-aware, computing and communications for mobile devices, and were introduced to the aims of the study and the obligations of diary keeping. To illustrate the kind of services that could be developed, participants examined fixed-access Web sites that provided map, transport, and venue information, such as www.multimap.com, and londontransport.co.uk. To encourage complete, and relevant free text descriptions of events, a possible future service was described, in which each member of a small group were able to display the positions of other group members on their mobile telephone. Participants made one diary entry for each rendezvous event. Participants were encouraged to complete their diary as soon after the event as possible, but were free to choose a time and place that was safe and suitable for thinking and writing. At the end of the diary keeping period, participants summarised their diary and its completeness. Questionnaire responses were processed automatically by an cccular reading Mmachine, which generated a text file that was checked and then read into statistical analysis software.

RESULTS

Rendezvous Reports

The following sections will be illustrated with reference to actual rendezvous reports and quotations from diaries.

Type and Frequency of Rendezvous and Communication

Diary keepers took part in a total of 248 rendezvous—a rate of approximately 5.6 rendezvous per week, or just under one per day. The rendezvous enabled a wide range of subsequent activity—from coffee in the student lounge, and shopping trips, to airport collections and wedding receptions (see Table 1). The rendezvous were very often in locations at which diary keepers had rendezvoused before (65 percent), and included people with whom they had close relationships (close friends 63 percent, immediate family 22 percent, acquaintances 22 percent, extended family 12 percent, and strangers 10 percent). The mean size of rendezvous was 3.6 people, including the diary keeper.

The plan for the rendezvous changed on average 0.56 times prior to departure, and 0.35 times en route—taken together, almost once per event.

Each rendezvous involved, on average, 4.08 communications. About half were telephone calls, and about a quarter were text messages.

The sample of participant activities reflects the relative freedom of student life. Failing to meet as agreed often led to stress and lost opportunity (e.g., P1 #3307), but undesired consequences were not necessarily entailed by being late or changing plans (e.g. lateness can be anticipated and accommodated P42 #3862). Nor was failing to meet as agreed necessary for stress and lost opportunity to arise (e.g. a loose plan was problematic in P33 #4381). Further details of this sample of rendezvous and communications are available in Colbert, 2005a.

Table 1. Example rendezvous

Participant Diary Sheet	Description of Rendezvous (paraphrased for clarity)
P1 #3307	I had arranged for a friend to drive by my house at 3:30 p.m. We would then continue into town and go shopping together. When my friend did not show up, I repeatedly attempted to contact her on the telephone, but each time my friend's phone was engaged. My friend did not reach my house until 4 p.m. Her departure from her house had been delayed because she had been talking to her boyfriend on her mobile phone. By 4 p.m., rush hour had begun, so by the time we got to town, the shop I specifically wanted to visit was closed. I was frustrated, because I wanted to talk to my friend urgently, but her line was busy. If her phone had a call waiting facility (which beeps when another person attempts to telephone the recipient), I may have been able to.
P21 #4340	I had arranged to meet my cousin at the theater. I decided to drive there. However, I was unable to locate the theater. I wanted to telephone my cousin to get information from her, but my mobile phone was not to hand—it was in the bottom of my bag. I was thinking about where it might be rather than the correct location and route to the theater. In the event, I was 10 minutes late.
P25 #1481	Myself and two friends, D and T, arranged to meet in a pub at Waterloo. The afternoon of the meeting, D phoned me saying that now he would probably not be able to make it. I said that he was to call T's mobile phone should he in fact be able to make it, and want to check that we were still in the pub. T kept his phone on the table in the pub, where it could have been stolen.
P33 #4381	I was to meet some friends for a pub crawl in Clapham to celebrate my birthday. No exact time was agreed. Everyone was to arrive at different times, depending upon when they finished work. We selected the pub at which to meet at the last minute, so I contacted my friends to let them know. Some of them had difficulty finding it. I made long phone calls explaining my location. It was very noisy in the pub, so I gave some people directions as text messages, which was very laborious. One person was very late due to "unforeseen complications." Had I been able to let my friends know my current location, it would have spared me the telephone calls and text messages, and we could have moved around more freely.
P39 #4040	I had arranged to meet 25 friends and relatives at my cousin's house, where a convoy of cars would then leave for an engagement in Gloucester. One person was late, so I sent him/her a text message, to discover their whereabouts. This message received no reply, so I sent another message. This message received no reply either. Not getting a reply was frustrating, but I was able to load my car whilst texting. It transpired that the latecomer had received all four messages, but had not replied because he/she was driving. The latecomer reasoned that if he/she stopped the car to answer the phone, it would just make him/her even later.
P41 #1922	I was at work (I serve in a shop) when a friend called me, inviting me over for dinner that evening. When the phone rang, I was serving a customer, so I was speaking on the phone and serving the customer at the same time. I don't see any best way to plan a meeting [other] than talking to the person you supposed to meet. (The rating for social acceptability was only 2 out of 5)
P42 #3862	I was driving to visit my parents in a suburb of North East London when I got stuck in a traffic jam near Kings Cross. I took the opportunity of stationary traffic to warn them I would be late.
P43 #1965	I received a text message inviting me to some drinks that night. They had already started and I was to join them after I finished work. The message read, "Were going to leic sq 4 drinks from 7 onwards, let us know when u there, c ya!" Unfortunately, when I finished work and tried to telephone them, I could get not get a response from them (so frustrating!!), so I left a text message "call me when u get this!" When I received a delivery confirmation for my message, I inferred that they now had reception, so I telephoned them to find out exactly where they were. All that effort for a few words!

Table 2. Frequency of occurrence and mean level of stress and lost opportunity x 'success' of rendezvous

Rendezvous Outcome	Stress Reported (percent)	Level of Stress (mean rating)	Lost Opportunity Reported (percent)	Level of Lost Opportunity (mean rating)
Met as Agreed	16 percent	2.0	12 percent	1.5
Did Not Meet as Agreed	64 percent	1.6	60 percent	1.5

Performance Deficits: Stress and Lost Opportunity

'Severe' Events

When stress or lost opportunity was reported, it was rated as 'medium/high' (4/5) or 'high' (5/5) for 13 percent and 11 percent of rendezvous respectively. This works out as a rate of around one 'severely problematic' rendezvous per month[8]. The most stressful rendezvous' were: an unconfirmed, last minute change of plan for meeting children after school; picking up friends one by one in a van on the way to catch a ferry; a dinner party host who rushed out to look for their dog just as their guests were due to arrive; forgetting to collect a brother from work; and trying to find a relative at the airport when the phone battery was running out. The most valuable opportunities lost arose when: multiple participants arrived late to discuss coursework; the rendezvousers arrived late and lost their reserved table at a restaurant; and, at a music venue, someone who had arrived on time was constantly interrupted from a latecomer who had got lost, and then they had to leave the venue to find him.

Lost opportunities frequently took the form of delay (27 percent). Other types of lost opportunity were reported less frequently—re-structuring—the activity went ahead, but in a different order, with different roles. (14 percent), less participation—the activity went ahead and someone joined in late (12 percent), individual sacrifices—an opportunity to do something for one's own sake was forgone (10 percent), non-participation—the activity went ahead without some individuals)(9 percent), and non-occurrence—the activity was cancelled/aborted (2 percent).

Diary-keepers attributed problems rendezvousing to two causes in particular (see Figure 2—the mode of travel (trains did not run on time, traffic was heavier than expected, etc.), and the over-run of previous activities. Other reasons for rendezvousing problems were cited less frequently—a poor plan (the plan was incomplete, inaccurate, never agreed or forgotten), the failure to value success (someone thought that arriving as agreed was not important), lack of information about other rendezvousers (rendezvousers were not aware that some others were delayed, lost or not coming), and lack of geographic information (rendezvousers become disorientated, or could not find the meeting place). Occasionally, problems were attributed to the performance of additional tasks, and lack of travel information (rendezvousers were unaware of routes, schedules, etc.).

Individual Sacrifices Amongst 31-45 year olds

A comparison of the outcomes reported by participants aged 31-45 years and those aged 18–30 years revealed a number of differences (see Table 3, and Colbert, 2005b):

1. 31-45 year olds more frequently attributed problems to the overrunning of previous-activities, and to taking the opportunity to perform additional, spontaneous tasks ('side-stepping')
2. 31-45 year olds more frequently report that the lost opportunities arising from problematic rendezvous take the form of individual sacrifices

Free text entries in the diaries suggested that these differences arise, because 31-45 year olds have commitments to spouses and children[9], and so pack their program of daily life with planned activities more tightly than 18-30 year olds (Carlstein et al., 1978). If one activity overruns, then it has a knock-on effect upon later rendezvous. There is no slack in the system. 'Side-stepping' is seen as a more frequent cause of problems by 31-45 year olds, because side-stepping is a useful technique for "getting everything done." It increases the proportion of time spent being "productive" relative to time spent travelling (which is "unproductive"). 31-45 year olds are more likely to perceive lost opportunities in the form of personal sacrifices, because 31-45 year olds are more aware of the activities they could have packed in to the time they actually 'wasted' failing to meet as initially agreed. It is as if 31-45 year olds have lengthy 'to do lists' (Taylor & Swan, 2004) continually at the back of their minds—tasks which they would like to perform, if only they could find the time.

Stress and Lost Opportunity at Unfamiliar Rendezvous Points

A comparison of rendezvous at familiar and unfamiliar meeting points revealed a number of differences. When meeting at unfamiliar places (places at which rendezvousers had *not* met before), the diary keeper:

- Reported stress more frequently
- Reported higher levels of stress and lost opportunity (see Table 4)

The reason for these performance deficits is as expected. When meeting at unfamiliar locations, rendezvousers more frequently attributed problems to lack of geographic and travel information ($p<0.001$ and $p=0.011$ respectively)(see Figure 3, and Participant 21 event #4340, and P33 #4381 in Table 1). The lack of this kind of information underlay the apparent, but not statistically significant, increase in related reasons for problems—the mode of travel, and poor planning (P43 #1965). It is possible, however, that the deficits noted may also be due to the fact that meetings at unfamiliar locations more often included strangers (Colbert, 2004, $p=0.013$). For example, a dinner party may occur at an unfamiliar location, because the diary keeper knows other guests, not the host. Meetings with strangers are slightly more formal, because "first impressions count," so stress and lost opportunity ratings increase, as rendezvousers are more sensitive to "being late."

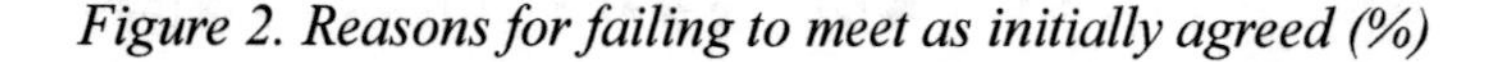

Figure 2. Reasons for failing to meet as initially agreed (%)

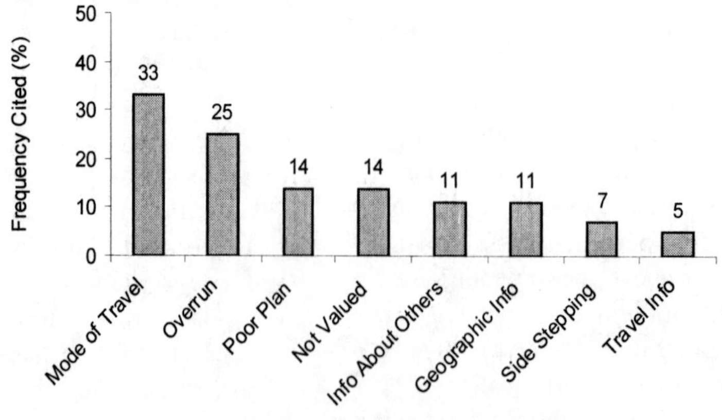

Table 3. Rendezvousing outcomes: Age differences

Measure	% 18-30s	% 31-45s	Sig.
lost opportunities take the form of personal sacrifices	7%	22%	p = 0.008
problems attributed to overrunning of previous activity	21%	39%	p = 0.050
problems attributed to 'side stepping'	5%	17%	p = 0.002

Performance Deficits: User Experience

A comparison of user experience of communication (phone use) in different contexts—prior to departure and en route to the rendezvous point—reveals that, when used en route, many aspects of user experience of communication are impaired by a small amount (Colbert, 2005b). Communication en route is significantly more frustrating, less convenient, more disruptive and less socially acceptable than communication prior to departure (see Table 5). These findings are strongly significant. At least the 0.001 level, using a paired-samples two-tailed T-test. Communication is also less satisfying, although this finding is only significant at the 0.05 level. Note, however, that the size of the impairment on each scale is not great (around a third of a rating point on a five point scale).

Free text entries in diaries suggest that these impairments are due to the cumulative effect of various adverse factors that tend to be more common and more severe in a 'mobile' context of use. These factors were:

- **Lack of network coverage:** Rendezvousers underground (e.g., in car parks, or tunnels), in the signal 'shadows' cast by tall buildings, or suffering interference from other activity in the airspace, may suffer low quality connections, be unable to connect, or 'cut off' in mid conversation (see P43 #1965, P2 #3164, P6 #0290 in Table 6).
- **'Phone free' zones:** Rendezvousers are asked to turn off their mobile phones in many public places, either for the sake of bystanders (as in 'quiet' railway carriages, theaters and some restaurants) or for reasons of safety (e.g., in hospitals, teaching laboratories and aeroplanes). P26 #1504 and P39 #4040 (see Table 6) were in a phone free zone, because holding a phone whilst driving is illegal in the UK.
- **Environmental noise:** Busy streets, and noisy vehicles or train stations, or entertainment venues sometimes make use of the telephone unpleasant or impossible, and tend to keep the duration of calls to a minimum (see P 26 #1506, P34 #4284, and P36 #4003, Table 6).

Table 4. Rendezvousing outcomes: Familiar vs. unfamiliar locations

Measure	Unfamiliar	Familiar	Significance
% reported stress	61%	45%	p=0.023
Stress (mean rating 1= low; 5=high)	2.54	2.13	p=0.04
Lost Opportunity (mean rating 1= low; 5=high)	2.26	1.87	p=0.04

Figure 3. Reasons for not meeting as initially agreed: Unfamiliar vs. familiar rendezvous points

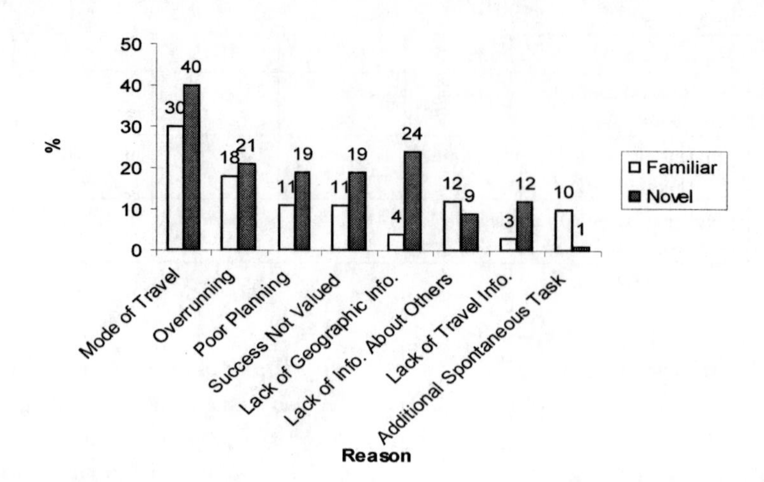

- **User availability:** The failure to be available for urgent communications sometimes reduces satisfaction and acceptability (see P1 #3307). Conversely, making oneself available to communicate whilst engaged in another activity sometimes feels disruptive (see P26 #1504 and P41 #1922).
- **Workspace constraints:** The space immediately around a rendezvouser is organised to ease the task of transiting to the rendezvous point, or to suit the locale, rather than communicating (see P21 #4340).
- **Security:** Using a mobile phone in some contexts sometimes feels 'unsafe,' because it might be lost or stolen (see P25 #1481).
- **Codes of conduct for public spaces:** The requirements of communication sometimes conflict with informal expectations for behaviour in transit locales (see P6 #0290).
- **Time pressure:** Time pressure sometimes impairs many aspects of user experience. For example, one diary keeper reported sending an e-mail to a group of friends and acquaintances, despite knowing that one recipient would not check her e-mail account soon enough for the message to be useful. However, the invitation had to be distributed rapidly, so he sent the e-mail anyway, albeit at the cost of excluding his friend. He felt this was not really acceptable, but he had to send the message immediately. Time pressure is also implicated in P1 #3307, P21 #4340, P39 #4040 and P43 #1965.
- **Lack of device power:** If a phone battery is running low, it is frustrating when the owner wishes to make a call, or inconvenient if he or she has to borrow a power adapter and recharge at university. One rendezvouser, whose batteries were on the verge of expiring, reported sending the briefest text message, when they would have preferred to telephone.
- **Conflict with preferred life-paths:** Rendezvousers have programmes of personal activities—plans that specify how they intend to use their limited resources of time and space. The need to communicate sometimes interferes with these plans, particularly if communication is constrained by lack of network coverage and phone free zones, limited user availability and time pressure. Taking detours, or delaying travel, in order to communicate is sometimes frustrating and inconvenient. The latecomer in P39 #4040 and P42 #3862 delayed communication until it did not interfere with the preferred life-path.

The impairment in user experience en route to a rendezvous, may also be due to rendezvousers'

Table 5. User experience of communication: Prior to departure vs. en route

User Experience	Prior to Departure (rating)	En Route (rating)	Significance
Satisfaction	3.92	3.75	p=0.036
Convenience	4.10	3.77	p=0.001
Social Acceptability	4.31	4.00	p=0.001
Frustration	1.42	1.69	p=0.001
Disruption	1.35	1.63	p=0.001

Table 6. Selected quotes from diary entries

Participant Diary Sheet	Quote
P2 #3164	"I also used text messaging when my cousin was on the train, as she did not have any reception to pick up the phone. . .bv.less agitated and frustrating. When in a distressing situation, or a situation that may cause anxiety, having a device that allows you to contact a person straight away can provide some relief."
P6 #0290	"During the rendezvous the mobile was slightly less satisfying to use as the reception was not very good on my travels. It was also slightly disruptive to the other passengers when I was on the train."
P26 #1504	"Driving whilst talking on the phone is not only illegal but hard! This forced me to pull over a couple of times and make a call to my girlfriend to get some more directions."
P26 #1506	"In the midst of a pub atmosphere i.e. where there is a lot of people talking, shouting and loud music, hearing your mobile phone ring is virtually impossible. There fore have a voice mail service that tracks your calls takes messages for you and then returns the messages to you is not only satisfying but reassuring."
P34 #4284	"My friend was unable to use his mobile phone because of the noise in side. He had to go outside to make the call."
P36 #4003	"It was annoying that I did not hear the phone beep to indicate a message. The message was from here friend who was late collecting me from my house en route to a dinner party"

tendancies to use the telephone more ($p=0.021$), and e-mail less ($p<0.001$), when en route (see Figure 4). En route to a rendezvous point, rendezvousers need to use the channel of communication that grounds information almost instantly (the telephone). Awkwardly, the telephone is the communication medium whose user experience is most impaired by the context of being en route.

DESIGN IMPLICATIONS

These performance deficits suggest the goals of developing applications of mobile technology that make good these deficits. The following paragraphs consider each deficit identified in turn, and illustrate the kind of mobile technology that could conceivably respond to it[10]. When deciding whether or not to respond to these deficits, and if so, then how, it is first worth noting the importance and general relevance of rendezvousing. Rendezvousing is important for mobile technology design, because around one third of current mobile phone use is rendezvous-related. One study estimated that 34 percent of telephone calls made from mobile phones by working parents were travel-related (Ling & Haddon, 2001). In Grinter and Eldridge's study, 36 percent of text messages sent by teenagers were related to coordination with friends or family, and coordination

Figure 4. Usage of communication services prior to departure vs. en route

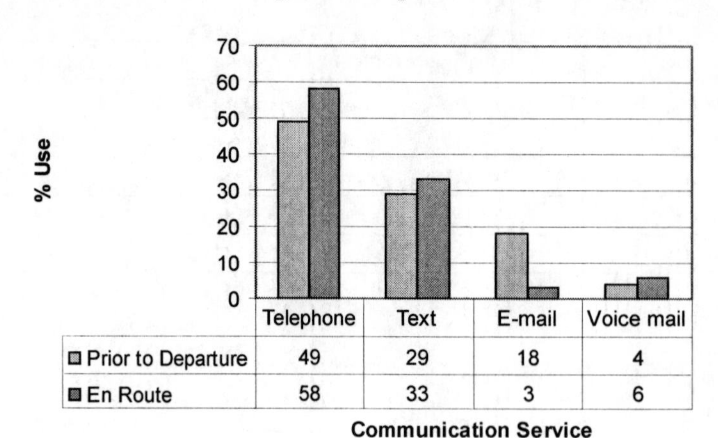

	Telephone	Text	E-mail	Voice mail
Prior to Departure	49	29	18	4
En Route	58	33	3	6

was also found to be the most common topic for text messages sent by Norwegian users (Telenor, 2002 cited in Ling, 2004). Rendezvousing is also relevant to a wide range of potential applications, including group-based communication and awareness systems (Milewksi & Smith, 2000; Nardi et al., 2000; Tang et al., 2001), electronic guides (Chincholle et al., 2002; Poposchil et al., 2002), position-aware reminder services (Marmasse & Schmandt, 2000), and diaries and event planners (Pousman et al., 2003).

Responding to Occasional High Stress and Lost Opportunity: Controlled Disclosure of Position Integrated with Group Communication

Assuming that one 'severely problematic' rendezvous per month is sufficient to limit individual mobility, position-aware, or location-enhanced communication for small groups appears potentially useful as a complement to telephony and text messaging. At face value, position and location[11] information is relevant to many of the frequently cited reasons for failing to meet as agreed (see Figure 2), for example, by letting rendezvousing 'A' decide whether that rendezvouser 'B' still intends to participate, and if so, then by how long is he or she delayed, or that 'B' has chosen to not come and do something else. The opportunity to obtain information about a number of other rendezvousers, without interrupting them, and without a lot of effort oneself, also appears to have some advantages (see the problems of scale P33 #4381 and 'no response' P39 #4040). However, the need to communicate via phone or text still remains, for example, to confirm inferences, in addition to negotiating changes of plan, and providing information.

A recent Wizard of Oz study asked participants to execute 'contrived' rendezvous with and without a mobile phone based 'friend finder' service in a local urban area. The study supports the potential efficacy of such systems for rendezvousing. Participants rapidly perceived the potential utility of the simulated service, and often used tracking in certain situations, for example, to discover context information before making a follow-up telephone call, if needed (Dearden et al., 2005).

The obvious limitation of tracking services, however, is that inappropriate use may invade other users' privacy. Preparatory studies suggest users will permit others to access the location details they require, provided they are satisfied with who wants to know, and why they need to know it (Barkhuus & Dey, 2003; Consolvo et al., 2005). Another study, partly in view of privacy issues, suggests an alternative to full integration of communication and tracking—simply broadcasting one's own location as an invitation to socialise

(Farnham, 2006). This study deployed SWARM, a service which broadcast short messages to a small group of extremely social, urban professionals. The broadcasts were frequently used for rendezvousing, notably for 'fishing' for company, or 'scouting' to find the best event. However, broadcasts could also make other contributions to rendezvousing—notably, announcing a delay or a decision not to attend. It will be interesting to see whether 'tracking' or 'broadcasting' position and location is adopted most widely, and whether either service reduces the frequency of 'severely problematic' rendezvous. Important questions for future work are, "Does the avoidance of occasional 'severely problematic' rendezvous constitute a 'need to know' (and so granting tracking rights)?" and "Is the avoidance of occasional 'severely problematic' rendezvous sufficient motivation to seek the permission to track?" Just encouraging others to broadcast their position, if they think it is necessary, might be enough.

Responding to More Individual Sacrifices by 31-45s: Reminder Systems

Reminder systems are relevant to rendezvousing because reminding prepares rendezvousers for side-stepping should the opportunity arise en route to a rendezvous. Reminders may also prove more useful than 'locator' services (e.g., "Where is the nearest ... <e.g., bank>?" because the majority of rendezvous occurred at familiar places, so rendezvousers probably know about the availability of relevant resources, they just fail to recall their need to make a visit when the opportunity to 'call in' arises.

The 'position-aware reminder' application is suggested in Figure 5, on the assumption that an effective reminder occurs when and where a user has the opportunity to act upon it, not where and when the user recalls the need to do something (see Colbert, 2004). This application seems most likely to be useful, when the rendezvous, and the route to it, are regular, say, collecting the kids from school, going to the gym, commuting home etc.

Interestingly, a recent Finnish field trial of 'De-De,' a context-enhanced phone, reports that users sometimes used a location-aware text messaging facility based upon cellID to deliver 'prompts' to friends to do something (Jung et al., 2005). Although De-De could have been used this way, participants did not send reminder messages to themselves to help them side-step, presumably because the participants were teenagers (so their life-situations did not warrant reminders to themselves), and because this usage was not prominently articulated in the user interface, as it would have been, for example, with a location enhanced to do list, or personal calendar.

An informal survey of another 20 students at Kingston University suggests that reminders near the shop, are not always the best possible reminders, because the minder comes too late. The time for the rendezvous is now so close that there is no time to take a detour. The best possible reminder, and particularly the widest acceptable reminder, was sometimes triggered by departure for the rendezvous point i.e. leaving campus, or leaving home. Kim recently conducted an exploratory user study with 'Gate Reminder' (Kim et al., 2004). Located at the front door of a family home, Gate Reminder detects RFID tags on participants and objects as they cross the threshold, and displays visual reminders on screen. Although the form of interaction was satisfactory to many participants, the additional infrastructure was felt to be too 'heavy weight' given the benefits. A more recent field study of PlaceMail (a system similar to De-De but augmented with GPS and with voice input) confirmed that users often found locational reminders to be useful (33 percent, Ludford et al., 2006, p896). Also, if the reminder is given at the "right place" according to the user, then the reminder is more likely to be seen as useful and acted upon.

From a rendezvousing point of view, future work needs to pay close attention the effect of such location-based reminders upon user performance and communication. In effect, their use in effect enables 31-45s to delay planning their day, and to pack their daily schedule even tighter with activities. This may create even greater demands for micro-coordination—and so greater use of the mobile phone for communication, more frequent changes of plan and more frequent 'failure' to meet as initially agreed.

Responding to Worse Outcomes when Meeting at Unfamiliar Locations: Personal Route Planning, Navigation and Information Seeking

It is not an original to suggest that future mobile technology could provide pedestrians, cyclists, and those on public transport with the same level of support for orientation, route planning and local information seeking that is already available to motorists. The distinctive requirements of pedestrians, for example, the usefulness of shop signs as landmarks for navigation, have already been investigated (May et al., 2003). It is probably reassuring to those already implementing such systems that there is, indeed, a problem to solve. However, many systems under development target tourists. This study reminds us that 'locals,' too, sometimes visit new parts of their own city, and perhaps locals have distinctive requirements. For example, they may be more likely to be listening to music whilst traveling, and so favor systems that provide navigational guidance through this audio channel (e.g., Warren, 2005). It also reminds us that rendezvousers want to work with spatial information at all phases of the activity—both prior to departure, and en route (and there may be many stages to their journey). In this light, integrated services that coordinate use of specialised desk-top, in-car, and personal devices (e.g., Baus et al., 2002) seem to have potential.

Responding to Impaired Experience en Route: Better Connectivity, Faster-Paced Messaging, Awareness and Negotiation, Systems Fit-For-Contexts

Better Connectivity

Given the wish the find an application for broader-band, 3G networks, it is a pity that this study suggests that better connectivity i.e., more complete coverage and constantly high quality of service, with existing 'second generation' (2G) networks will improve user experience of communication en route to a rendezvous, more than additional bandwidth (cf. the frustration and disruption caused by incomplete network coverage and poor quality of service.) The financial costs of 3G networks are such that they may never extend far beyond densely populated areas. A more promising development for rendezvousers is the provision of wireless local area network (LAN) access in public places (shopping centres, railway companies, etc.). Rendezvousers needing to communicate, but beyond the reach of a 2G network, may connect to a wireless LAN and, from there, to the world at large. Alas, it is not until the so-called Fourth Generation (4G) of wireless network that the

Figure 5. A location-enhanced to do list for mobile phones

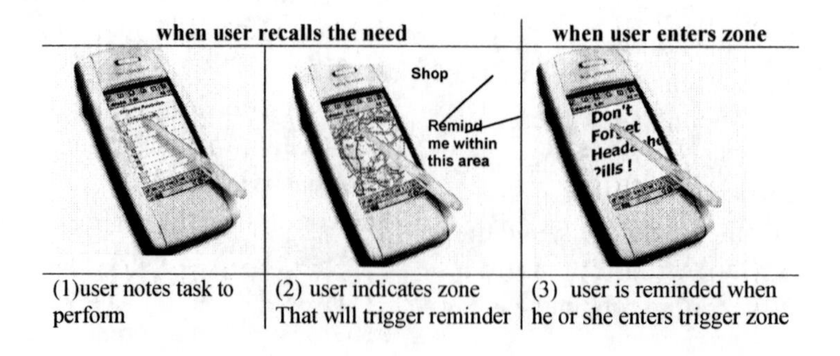

seamless integration of personal, local and wide area networks and "multi-mode" mobile phones are envisaged. Even then, the level of connectivity actually achieved with 4G remains to be seen.

Faster-Paced Messaging

This study also suggests that 'faster-paced' text messaging may also improve user experience of communication during rendezvous. Currently, rendezvousers tend to use the telephone during rendezvous rather than text messaging, because text messages are exchanged and grounded too slowly. Telephone calls are made despite the fact that the experience of telephony appears to be more impaired by the context of 'being en route' than texting. With a 'recorded delivery express text' service, selected messages receive priority transmission when networks are congested, so senders can be guaranteed a time of delivery, say, within 30 seconds of sending, provided that the recipient's device is on, and within coverage. This service would also acknowledge: (1) transmission from the sender (many networks/phones already provide this); (2) transmission to the recipient (some networks/phones provide this); and (3) reading by the recipient (not currently provided for text, but provided for e-mail). Another alternative resembles a 'response paid text' service. With this service, recipients are asked to select one of a limited number of responses to a message, and the sender pays for the response. For example, the message 'running 15 minutes late' may require the response 'OK,' 'Not OK,'or 'Ask again later.'

Awareness and Negotiation

Experience ratings may also be improved by providing users with means to negotiate appropriate contexts in which to communicate, and appropriate communication media to use in this context. Recent work has focused upon 'awareness systems' (Tang et al., 2001; Milewski and Smith, 2000), which inform users initiating communication about the other party's current, or previous activity. Relevant activity information may include a user's location, the status of a device (on/off), service usage, the other party's speed and direction, mode of transport and personal calendar. Devices and services could also provide users with greater control and pre-programming of their 'reachability.' For example, some users in this study turned their phone off to prevent the receipt of telephone calls whilst driving—a somewhat drastic action, as it also prevents receiving text messages. User experience may be improved if users could temporarily, or locally, 'bar' selected media (say, the telephone), but accept others (say, text messages, e-mail or voice messages). These preferences may be controlled by setting the device's 'mode'. However, they could also be pre-programmed in advance. Some users in this study took regular routes and stated, for example, that they preferred not to use the phone in certain contexts, such as crossing a busy road junction, or boarding their bus. On fixed networks, instant messaging is often used to negotiate good times to talk or to e-mail. If provided on mobile devices, it could serve a similar purpose for rendezvousers.

Devices Fit-for-Context

It is particularly important to improve the experience of telephone calls en route to a rendezvous point—these calls accounted for almost 60 percent of communications. Recent innovations, such as caller ID, which makes it easier for rendezvousers to ensure that they accept urgent calls, but leave non-urgent calls for a more suitable context, may already provide some help. Other innovations, such as mechanisms to cope with environmental noise (pre-recorded "Go on. I'm still listening messages," whisper mode, Nelson et al., 2001), and wearable devices (Sawhney & Schmandt, 2000) may also increase satisfaction and convenience. A possible limitation of such approaches, however, is that synchronous, real-time services such as the telephone require both participants to be in a suitable context at the same time to achieve good user experience. Such a situation may be difficult to achieve. Also, even if devices are more resilient to some adverse factors, such as environmental noise, synchronous audio services may still be impaired by other factors (phone-free zones, user non-availability, parallel task performance, etc.).

FUTURE WORK

This chapter has sought out opportunities to improve one kind of human mobility by identifying some deficits in current performance. Having identified these opportunities, it is logical for future work to make good those deficits by implementing and testing systems of the kind outlined in section 5. Quite independently of this work, many such systems, and other kinds too, are already under development.

Of course, various activities and information influence the design goals that are set for specific development projects—not just performance deficits from user studies, but user requirements from focus groups, feature lists from market research, and so on. The advantage of statements of current user performance obtained by this study, however, is that they provide a baseline against which levels of performance achieved by future users and systems can be compared. For example, one could imagine repeating the study reported here in say, 2012 (by which time future technology will have been widely adopted by student participants). How will context-aware communication for small groups, personal navigation, and reminding have affected rendezvousing performance? And will certain kinds of rendezvous become more popular, as users realise technology enables them to execute more fluid arrangements. It depends on the quality of the design and implementation. So let us hope this goes well.

It will be interesting to see how future devices and services benefit the occasional, highly problematic rendezvous that is the target of redesign. It will also be interesting or perhaps increase the popularity of certain kinds of rendezvous—flash mobs, en route, delay planning in detail until the last minute.

It is also worth recalling that this study concerned only one kind of everyday mobility. Mobile technology also has the potential to benefit, say, commuting, or long distance business travel. If we identified performance deficits for these kinds of mobility too, then we could similarly inform design goals for other kinds of mobile application (see Colbert & Livingstone, 2006 for an initial attempt).

REFERENCES

Anttila, A., & Jung, Y. (2006). Discovering design drivers for mobile media solutions. In *CHI 2006 Extended Abstracts on Human Factors in Computing Systems* (pp. 219-224). New York: ACM Press.

Barkhuus, L., & Dey, A. (2003). Location-based services for mobile telephony: A study of users' privacy concerns. In *Proceedings of Interact 2003* (pp. 709-712). Berlin: Springer Verlag.

Baus, J., Kruger, J., & Wahlster, W. (2002). A Resource-adaptive mobile navigation system. In *Proceedings of Intelligent User Interfaces 2002* (pp. 15-22). New York: ACM Press.

Butcher, M. (2003). Getting the cold shoulder. *The Guardian newspaper,* April 23rd, 2003. Retrieved January 19, 2007 from http://shopping.guardian.co.uk/gadgets/story/0,,928162,00.html

Cadiz, J. J., Czerwinski, M., McCrickard, S., & Stasko, J. (2003). Providing elegant peripheral awareness. In *CHI 2003 extended abstracts on human factors in computer systems* (pp. 1066-1067). New York: ACM Press.

Carlstein, T., Parkes, D., & Thrift, N. (1978). *Human activity and time geography: Timing space and spacing time (Vol 2).* London: Edward Arnold.

Carter, S., & Mankoff, J. (2005). When participants do the capturing: the role of media in diary studies. *Proceedings of CHI 2005 Human Factors in Computing Systems.* (pp. 899-908). New York: ACM Press.

Chincholle, D., Goldstein, M., Nyberg, M., & Eriksson, M. (2002). Lost or found? A usability evaluation of a mobile navigation and location-based service, in human-computer interaction with mobile devices. In *Proceedings of Mobile HCI 2002* (pp. 211 – 224). Berlin: Springer-Verlag.

Colbert, M. (2001). A diary study of rendezvousing: implications for position-aware computing and communications for the general public. *Pro-*

ceedings of ACM GROUP'01 (pp. 15-23). New York: ACM Press.

Colbert, M. (2004). Rendezvousing at familiar and unfamiliar places. *Journal of Navigation, 57*(3), 327-338.

Colbert, M. (2005a). User experience of communication before and during rendezvous. *Behaviour and Information Technology, 24*(6), 449-470.

Colbert, M. (2005b). Age differences rendezvousing: reminders for side-stepping. *Personal and Ubiquitous Computing, 9*(6), 404-412.

Colbert, M., & Livingstone, D. (2006). Context changes important for talking and text messaging during homeward commutes. *Behaviour and Information Technology, 25(5)*, 433-442.

Consolvo, S., Smith, I. E., Matthews, T., LaMarca, A., Tabert, J., & Powledge, P. (2005). Location disclosure to social relations: why, when, & what people want to share. In *Proceedings of CHI 2005 Human Factors in Computing Systems* (pp. 81-90). New York: ACM Press).

Czaja, S J., Guerrier, J. H., Nair, S. N., & Landauer, T. K.(1993). Computer communication as an aid to independence for older adults. *Behaviour and Information Technology, 12*(4), 197-207.

Dearden, D., Hawkey, K., & Inkpen, K. M. (2005). Rendezvousing with location-aware devices: enhancing social coordination. *Interacting with Computers, 17*, 542–566.

Farnham, S.D., & Keyani, P. (2006). Party on: Hyper awareness, hyper coordination, and smart convergence through mobile group text messaging. In *Proceedings of HICSS-39*, London: IEEE Computer Society.

Farnham, S., Chesley, H. R., McGhee, D. E., Kawal, R., & Landau, J. (2000). Structured onlineinteractions: improving the decision-making of small discussion Groups. In *Proceedings of CSCW 2000* (pp. 299-308). New York: ACM Press.

Frohlich, D.M., Chilton, K. & Drew, P. (1997). Remote homeplace communication: what is it like and how might we support it? In *People and Computers XII, Proceedings of HCI'97* (pp. 133-153) London: Springer-Verlag.

Gray W, D., John B. E., & Atwood M. E. (1993). Project Ernestine: Validating a GOMS analysis for predicting and explaining real-world task performance. *Human Computer Interaction*, 8, 237-309.

Grinter, R. E., & Eldridge, M. A. (2001). Y do tngrs luv 2 txt msg? *Proceedings of ECSCW 2001.* Bonn, Germany: Kluwer.

Jung, Y., Persson, P., & Blom, J. (2005). DeDe: Design and evaluation of a context-enhanced mobile messaging system. In *Proceedings of CHI 2005 Human Factors in Computing Systems* (pp. 351-360). New York: ACM Press.

Kim, S. W., Kim, M. C., Park, S. H., Jin, Y. K., & Choi, W.S. (2004). Gate reminder: A design case of a smart reminder. In *Proceedings of Designing Interactive Systems 2004* (pp. 81-90). New York: ACM Press.

Kraut, R. E., Cool, C., Rice, R. E., & Fish, R. S. (1994). Life and death of new technology: task, utility and social influences on the use of a communication medium from video phoning to video interacting. In *Proceedings of CSCW'94 Conference on Computer-Supported Cooperative Work* (pp.13-21). New York: ACM Press.

Lacohee, H., & Anderson, B. (2001). Interacting with the telephone. *International Journal of Human-Computer Studies, 54*(5), 665-699.

Ling, R. (2004). *The mobile connection: The cell phone's impact on society.* New York: Morgan Kaufmann.

Ling, R., & Haddon, L. (2001). Mobile telephony, mobility and the coordination of everyday life. paper presented at 'Machines that become us' conference, Rutgers

University, April 18-19, 2001. Retrived December 8, 2006 from http://www.richardling.com/papers/2001_mobility_and_the_coordination_of_everyday_life.pdf

Ling, R., & Yttri, B. (2001). Hyper-coordination via mobile phones in Norway. In J. Katz and M. Aakhus (Eds.) *Perpetual contact: mobile communication, private talk, public performance*. (pp. 139–169). Cambridge, UK: Cambridge.

Ludford, P. J., Frankowski, D., Reily, K., Wilms, K., & Terveen, L. (2006). Because I carry my cell phone anyway: functional location-based reminder applications. In *Proceedings of CHI2006, Human Factors in Computing Systems* (pp. 889-898). New York: ACM Press.

May, A. J., Ross, T., Bayer, S. H. & Tarkiainen, M. J. (2003). Pedestrian navigation aids: Information requirements and design implications, *Personal and Ubiquitous Computing, 7*(6), 331-338.

Marmasse, N., & Schmandt, C. (2000). Location-aware information delivery with comMotion. *Intl. Symposium on Handheld and Ubiquitous Computing* (pp. 64-73).

Matarazzo, G., & Sellen, A, (2000). The value of video in work at a distance: Addition or distraction? *Behaviour and Information Technology, 5*(19), 339-348.

Milewski, A. E., & Smith, T. M. (2000). Providing presence cues to telephone users. In *Proceedings of CSCW 2000 Conference on Computer-Supported Cooperative Work* (pp. 89–96). New York: ACM Press.

Nardi, B., Whittaker, S., & Bradner, E., (2000). Interaction and outeraction: Instant messaging in action. In *Proceedings of CSCW 2000 Conference on Computer-Supported Cooperative Work* (pp. 79–88). New York: ACM Press.

Nelson, L., Bly, S., & Skoler, T. (2001). Quiet calls: talking silently on mobile phones speech studies. In *Proceedings of CHI 2001 Conference on Human Factors in Computing Systems* (pp. 174-181). New York: ACM Press.

Newman, W.M., & Taylor, A. (1999). Towards a methodology employing critical parameters to deliver performance improvements in interactive systems. In *Proceedings of INTERACT'99* (pp. 605–612). Amsterdam: IOS Press.

Newman, W. M., Taylor, A. S., Dance, C. R. & Taylor, S. A. (2000). Performance targets, models and innovation in interactive system design. In *Proceedings of Designing Interactive Systems* (pp. 381-387). New York: ACM Press.

Palen L., Salzman, M., & Youngs, E. (2000). Going wireless: behaviour and practice of new mobile phone users. In *Proceedings of CSCW 2000 Conference on Computer-Supported Cooperative Work* (pp. 201–210). New York: ACM Press.

Perry, M., O'Hara, K., Sellen, A., Brown, B., & Harper, R. (2001). Dealing with mobility: understanding access anytime, anywhere. *ACM Transactions on Computer -Human Interaction, 8*(4), 323–347.

Pooley, C., Turnbull, J., & Adams, M. (2005). "... everywhere she went I had to tag along beside her": Family, life course and everyday mobility in England since the 1940s. *The History of the Family: An International Quarterly, 10*(2), 119-136

Poposchil, G., Umlauf, M. & Michlmayr, E. (2002). Designing Lol@, a mobile tourist guide for UMTS. In *Proceedings of Mobile HCI 2002* (pp. 155–169). Berlin: Springer-Verlag.

Pousman, Z., Iachello, G., Fithian, R., Moghazy, J., & Stasko, J. (2004). Design iterations for a location-aware event planner. *Personal Ubiquitous Computing, 8*(2), 117-125.

Preece, J. (2001). *Online communities: Designing usability, supporting sociability.* New York: Wiley.

Rieman, J. (1993). The diary study: a workplace-oriented research tool to guide laboratory efforts. In *Proceedings of INTERCHI'93* (pp. 321-326). Amsterdam: IOS Press.

Sawhney, N., & Schmandt, C. (2000). Nomadic radio: speech and audio interaction for contextual messaging in nomadic environments. *ACM Transactions on Computer-Human Interaction, 7*(3), 353-383.

Tamminen, S., Oulasvirta, A., Toiskalio, K., & Kankainen, A. (2003). Understanding Mobile

Contexts. In *Proceedings of Mobile HCI 2003* (pp. 135-143). Berlin: Springer-Verlag.

Tang, T., Yankelovich, N., Begole, J., Van Kleek, M., Li, F., & Bhalodia, J. (2001). ConNexus to Awarenex: cxtending awareness to mobile users. In *Proceedings of CHI2001 Conference on Human Factors in Computing Systems* (pp. 221-228). New York: ACM Press.

Taylor, A. S., & Swan, L. (2004). List making in the home. In *Proceedings of CSCW04 Computer Supported Collaborative Work* (pp. 542-545). New York: ACM Press.

Warren, N., Jones, M., Jones, S., & Bainbridge, B. (2005). Navigation via continously adapted music. In *Extended Abstracts on CHI 2005 Conference on Human Factors in Computing Systems* (pp.1849-1852). New York: ACM Press.

Whittaker, T. (1996). Talking to strangers: an evaluation of the factors affecting electronic collaboration. In *Proceedings of CSCW'96, Computer Supported Collaborative Work* (pp. 409-418). New York: ACM.

KEY TERMS

After the Rendezvous: From the time at which the last rendezvouser arrives at the rendezvous point.

En Route: From the time at which the first rendezvouser to depart does so, until the last rendezvouser arrives at the rendezvous point.

Life-Path Diagram: A continuous representation of human activity in time and space, and with respect to other entities and features of the environment.

Lost Opportunity: What the rendezvouser would have done, had the rendezvous occurred as intended, but which is now impossible.

Prior to Departure: Until the time at which the first rendezvouser to depart does so.

Rendezvousing: Everyday coordination, 'meeting up' of friends and family. The process of arranging, and traveling to a rendezvous point, in order to pursue some non-work, group activity, for example, to watch a movie, or to have lunch. Rendezvousers have personal relationships with each other—they are not impersonal embodiments of organisational roles. So rendezvousing does not include formal or anonymous attendance at institutions, such as 'reporting to the tax office for interview,' 'going to my electronics lecture,' or 'going to the annual general meeting.' It also does not include receipts of service, such as 'having a pizza delivered.'

Time Geography: A school of human geography that emphasises the development and use of continuous models of human activity with respect to time and space, and argues that these models are a basic component of the understanding of spatial behaviour. Classical, 'spatial' geography orients towards space over time, and concerns, for example, changes in settlement *size* and *layout*. Time geography, in contrast, orients towards the activities of human individuals in the context of time space, and concerns, for example, patterns of *commuting* and *migration* into and out of a settlement.

User Performance: Effectiveness of human-computer interaction. The quality of task outcomes achieved, for the costs that users incur achieving these outcomes.

ENDNOTES

1 Mobile technology was adopted for many reasons. See, for example, the communicative opportunities that text messaging provided teenagers (Grinter & Eldridge, 2001).

2 Rendezvousing is an important form of mobility, because it enables actual presence and participation in different social groups. The more and better we rendezvous, the less we have to 'tag along' with, or 'drag around' a group that temporarily does not interest us (Pooley et al., 2005). We are free to disperse,

because we are confident we can successfully get back together again, and that we will stay safe whilst apart.

3 User performance, here, refers to the quality of task outcomes achieved, for the costs that users incur achieving these outcomes. A performance-oriented approach is characteristic of an engineering approach in many domains (Newman & Taylor, 1999). Performance parameters have been productively applied to the design of telephone operator workstations (Gray et al., 1993), and video-based, document capture tools (Newman et al., 2000). They also seem applicable to peripheral notifications (Cadiz et al., 2003), so a performance-oriented approach to mobile technology and everyday mobility would be expected to be productive also.

4 Figure 1 is a life-path diagram—a kind of diagram devised by time geographers to represent human mobility. In life-path diagrams, two dimensions of geographic space (longitude and latitude) are represented 'horizontally' as an apparent surface (in this case, the map of East London suburbs at the foot of Figure 1). Time is represented along a vertical axis, and so successive snapshots of geographic space combine to form an apparent column of time-space. Locales within geographic space, such as Kate and Sonja's homes, and the railway stations, are represented as many-sided pillars within the main column of time-space. Individuals, such as the Kate and Sonja, are represented as continuous lines, or 'life-paths' running through time space (the thick, black line, snaking its way across Figure 1). A stationary entity, for example, Kate at home, has a vertical life-path. A moving entity, such as Kate on the train, has a life-path that projects through time-space, its slope indicating speed of movement, and its horizontal deflection indicating change in position. An activity bar adjacent to the time axis indicates the activity of each individual, such as transiting and shopping, and the times at which these activities start and stop.

5 The main contribution of the chapter, then, is the performance deficits, and so the direction, goals and priories of future work. Details of possible solutions are included only to illustrate what attempting to make good a deficit may involve. To prescribe details of particular solutions, other studies are required.

6 For practical purposes, at the time of this study, GSM mobile telephones essentially provided only communication services (telephony and text messaging).

7 At the time the study was conducted, these services were voice telephony and text messaging.

8 Stress or lost opportunity was only reported for 55 percent and 40 percent of rendezvous respectively, and participants reported 5.6 rendezvous per week, giving an overall frequency of severely problematic rendezvous of around one per month.

9 In this study, about 50 percent of 31-45 year olds were married with children, compared to only 3 percent of 18-30 year olds.

10 Of course, when setting design goals, actual projects will use various sources of information, not just the performance deficits identified here. The chapter is contributing to requirements analysis, not replacing it. Similarly, the chapter is setting a direction for and prioritising design, not replacing design. These details of user interfaces in the following paragraphs make the illustration concrete, and do not derive from user performance data.

11 Position information refers to spatial coordinates, such as map reference points. Location information refers to higher level information, such as proximity to a home place, or on board a certain train.

Chapter IV
User Experience of Camera Phones in Social Contexts

Hanna Stelmaszewska
Middlesex University, UK

Bob Fields
Middlesex University, UK

Ann Blandford
University College London, UK

ABSTRACT

This chapter reports on a qualitative study into people's use of camera phones for social interaction in co-present settings. The study examined people's behaviour and positive experiences (e.g., fun, enjoyment, or excitement) when camera phones were used in different spaces (public and private). It was found that camera phones influence social practices. Three distinct practices were observed: sharing a moment now, sharing a moment later, and using photos to initiate social interaction with strangers. The knowledge obtained through the study will offer a conceptual contribution that deepens our understanding of how this emerging and evolving technology is coming to be accommodated into the leisure-related practices of its users.

INTRODUCTION

What do we know about photography? Photography has been a part of our life for a long time. We document family celebrations, important events in our lives and those of our family and friends; we take pictures when visiting museums or if we want to illustrate everyday items and people in a funny way and when we want to create stories (Mäkelä et al., 2000). It seems that photography and photos bring either smiles when reminiscing about something pleasant or tears when emotions take over. They preserve memories, capture feelings, and provide a means to communicate with others. One of the most common and enjoyable experiences is to share photos with others through story telling (Balanovic et al., 2000; Chalfen, 1987). Photos can be shared using technology and then they can be used as means for interaction with others.

Recent technological developments not only support new ways of working but also provide new mechanisms for social interaction. Mobile phones and camera phones, in particular, are examples of such technology. In the past decade, mobile phones have allowed profound changes to take place in people's behavior and practices in relation to communication (Ling, 2004), from being extensively used as a medium of verbal and text communication to one that uses pictures to facilitate people's social life. Mobile phones with integrated camera and video features have changed forever the way people communicate and interact, and have shaped both their individual and their social lives (Ito, 2005; Kato, 2005; Kindberg et al., 2005a, 2005b;Okabe, 2004; Scifo, 2004).

Although there is a vast body of literature focussing on the use of camera phones (Kindberg, et al., 2005a, 2005b; Okabe, 2004; Scifo, 2004) the issues relating to how camera phones are used to mediate social interaction between co-located users have been neglected. In this chapter, we report on the study of the collaborative use of camera phones by co-located users in various spaces.

BACKGROUND TO THE RESEARCH

In recent years, there has been substantial interest in digital photography, with a particular interest on how the digital medium facilitates sharing of images (Balanovic et al., 2000; Frohlich et al., 2002, Van House et al., 2005). Studies of sharing digital photographs include the use of Web-based systems, mobile applications, and multimedia messaging. Most of the studies focus on personal applications for sharing images remotely (Kato, 2005; Kindberg et al., 2005a; Van House et al., 2005) work on sharing images in co-present settings is in its infancy.

The issues of what people capture on mobile phones and what they do with these images were extensively investigated by Kindberg et al (2005a). They proposed a six-part taxonomy to describe the intentions behind the use of camera phone images. Intentions were grouped along two dimensions. The first intention defines whether people captured the images for affective (e.g., sentimental) or functional reasons. The second one defines social or individual intensions.

Others, such as Licoppe & Heurtin (2001) and Taylor and Harper (2003), focused on teenagers using their phones for social practices. The latter claim teenagers' practices are similar to 'gift-giving' rituals, which shape the way teenagers understand and use their mobile phones. The 'gift-giving' practices included sharing certain text messages, call-credits and even the mobile phones themselves. All these practices establish and cement allegiances and sustain rivalries (Taylor & Harper, 2003).

A field study conducted by Kato (2005) explored how the use of mobile phones/camera phones changes people's daily activities in Japan. He argues that the new ways of pervasive photo taking through camera phones allows people to document their lives on a daily basis, which can be preserved and shared as a life of a local community.

A different approach to studying mobile phone users was taken by Okabe (2004). He studied practices of Japanese camera phone users, which included personal archiving, intimate sharing, and peer-to-peer news sharing. Okabe (ibid) argues that capturing and sharing visual information cannot be understood without also understanding the social relationships and contexts within which those activities take place. Scifo (2004) provides similar views on this matter, arguing that taking photographs on camera phones and using MMS communication allows users (particularly youngsters) to identify themselves within social groups, and will intensify communication within that community.

The relevance of social relations to the uses of photographs was also identified by Van House et al. (2005). They discovered five distinct social uses of personal photos. These are: creating and maintaining social relationships, constructing personal and group memory, self-expression, and self-presentation and functional communication with self and others.

Photos could also be used for social discourse. For example, a mobile picture system (MobShare) developed by Sarvas et al. (2005) supports that by transferring photos from the phone to different devices. These include transfers (1) to another phone over the network (e.g., MMS), (2) to a PC, (3) to a network server over the network, and (4) to a printer using a cable connection or Bluetooth.

Many methods have been used to study people's uses of mobile phones, including diaries, interviews and field studies (Kato, 2005; Kindberg et al., 2005a, 2005b; Okabe, 2004; Sarvas et al., 2005). The approach employed by Sarvas et al. (2005) involved asking people to fill out a diary including all activities their performed using their camera phones. This was followed by a set of interviews focusing on photographic habits and social networking involving photography. The same methods were employed by Okabe (2004) when investigating social practices, situations and relations of the use of camera phone.

Kato (2005) applied a fieldwork study to observed and record the practices of camera phone users encouraging them not only to take pictures but also to collect and store them as visual field notes on a specially designated web site. When conducting an in-depth study of camera phone use Kindberg et al. (2005a, 2005b) applied a set of interviews asking the subject to show images that were not private from their camera phones and talk about them.

Taking inspiration from such research, semi-structured in-depth interviews and field observational studies were employed in the study reported here, which will be discussed later in this chapter.

METHODOLOGY

This study is specifically concerned with peoples' experiences when using camera phones for social interaction in a co-present setting (i.e., when participants are present at the same location at the same time). The chapter builds on an earlier more general study into peoples' experience and emotions using personal technologies such as PDAs, digital cameras and mobile phones (Stelmaszewska et al., 2005).

Because we wanted to obtain the insights of the ways people use, their camera phones as a medium for social practices we adopted Kindberg et al.'s (2005a) method of asking participants about circumstances and reasons for taking these images and their life cycle. A series of observational field studies was conducted to develop a better understanding of peoples' practices using camera phones. The use of dual methods strengthened the results obtained and provided a means of triangulation between the interviews and observations to confirm that the reported practices really did occur when the observations took place. In addition, field observations of the phenomena provided richer insights into the circumstances and contexts in which practices described in interviews actually take place.

Five students were interviewed including two PhD students, two undergraduates, and one college student, all aged between 18 and 27; all participants had been camera phone users for at least a year. Each interview took between 25 and 45 minutes and was recorded and later transcribed. The participants were asked to describe how and for what reasons they used their camera phones. The participants were also asked to show a few of the images (pictures or video) stored on their phones and encouraged to discuss where the images were taken, in what circumstances, by whom and for what reason. Also of interest was whether pictures were taken by the participant or received from another person, the means of storage and transfer employed (e.g., infrared, Bluetooth, MMS, e-mail), how long these pictures were stored, and whether they were shared with others, or retained for a private use.

The data from the field studies was gathered in a variety of public spaces, including pubs, restaurants, leisure and entertainment places, museums, and public transport (tube and buses).

The first author spent around 35 hours in public spaces observing camera phone usage. In this time, 18 individual instances of individuals and groups interacting with photos on cameras were observed and noted.

As the data gathered from interviews and field observations was of a qualitative nature, data collection and analysis was carried out iteratively. This allows for 'theoretical sampling' on the basis of concepts and themes that emerge from the analysis and allows concepts to be explored and hypotheses to be tested as they are developed from the data (Strauss & Corbin, 1998, p.46). Data from both studies was transcribed and then analyzed by first, coding it using qualitative methods to identify emerging themes, and then the themes were merged to extract the high level concepts that gave the outline of the use and practices of camera phones.

SITUATED USE OF CAMERA PHONES

The field observation study revealed many instances of people being engaged in social interaction using camera phones in different co-present settings. The in-depth interviews provided extended information to support these phenomena. The data shows the relationships between space and place as well as the photo/video sharing practices, which will be discussed in the following sections.

The concepts of place and space have been researched by many like Casey (1997), Ciolfi (2004), Dourish (2001), and Salovaara et al. (2005) just to name a few. Casey (1997) discusses this phenomena as 'space refers to abstract geometrical extension and location' whereas 'place describes our experience of being in the world and investigating a physical location or setting with meaning, memories and feelings' (cited in Ciolfi, 2004, p.1). A similar view has been taken by Dourish (2001) who gives an example of a space like a shopping street being a different kind of a place depending on the time of a day. According to Salovaara et al. (2005) the 'concepts of space and place are mutually dependent and co-occur in the context' (p. 1).

Camera Phone Use in Different Spaces

Camera phones have become a part of our lives. People carry them to work, to social events, to leisure activities, even when going shopping. Every time we use camera phones, we experience something. The experience, however, does not exist in a vacuum, but rather in a dynamic relationship with other people, places and objects (Mulder & Steen, 2005). What we experience and how camera phones are used is also determined by place and space, which will be explored in the consecutive sections.

Public Space

It appeared in the data that people use their camera phones differently depending on where they are. It was observed that when using public spaces like a tube or a bus people tend to use their camera phones for individual purposes; that includes reading and answering text messages, playing games, viewing and sorting out images, playing music or ring tones, or examining different functions on their camera phones. Interview data indicated that people do these things to overcome the feeling of boredom or simply to 'kill time' while waiting for a bus, as one of the participants (Steve) commented:

I listen to the radio ... when I'm on the tube, when walking around or waiting for a bus and I don't have anything to amuse me. To amuse me, I use the calendar and the diary quite a bit. Otherwise I'd forget everyone's birthday.

Similarly, another participant (Luisa), on using camera phone on a bus, commented:

...the setting itself is boring not much inspiration to take pictures and things ... you have to be with someone to do it.

It was reported in the literature that some public spaces are regulated by different means: signage, announcements and by more informal peer-base regulations (Ito, 2003, 2004; Okabe & Ito, 2005).

The former claims that these regulations are mostly exercised in public transport. Posters and signage exhort passengers from putting their feet on the seats or not smoking. The study by Okabe & Ito (2005) reported that people use email rather then voice calls when on trains and subways following 'sharing the same public space' regulations. Although, this kind of behavior was observed amongst Japanese youth population similar findings were reported by Klamer et al. (2000) who conducted a European survey investigating if the mobile phones used in public spaces disturb people.

A different kind of behavior was observed in museums (Science Museum and Natural History Museum in London). Camera phones were rarely used and only for individual purposes: receiving calls or messages, making phone calls, or texting. People treat museums as places to go on outings with friends and family, which they plan for and therefore they take a digital camera with them to capture something specific that they would like to keep as a reminder. In this case, the quality of pictures is of high importance. The comments of Maria confirm this:

...I like to take pictures of a nice scenery or ... er... flowers or trees or just a really nice views or things... then I use my digital camera because of the quality of the picture.

Other public spaces like pubs, restaurants, clubs, places of entertainment and leisure provide a different social context for camera phone activities, which is in line with our previous research reported elsewhere (Stelmaszewska et al., 2005, 2006). The data illustrates that people more often engage themselves in social interaction using camera phones during gatherings with friends and family, when going out with friends or during trips or excursions with friends (see Figures 1 and 2). Most of the participants claimed that the important issue for using camera phones is to be with other people. It is people who create experiences that people enjoy, as Adam noted:

When you have other people around you then you have a different kind of experience. ... you are more likely to do silly things. So then you take pictures and when you view them you can laugh and have fun. When you are on your own ... no, you don't do these things. You need to have people around you to have fun.

Private Space

A similar behavior was reported when groups of participants use their camera phones in private spaces (e.g., homes or cars); that is people took pictures or videos of friends, members of family or even themselves behaving funny or silly and

Figure 1. A girl sitting with her family and taking picture of the artist playing

Figure 2. People taking photos of the pantomime artist

then shared them with others co-present or they viewed pictures and videos taken previously. The comment from Adam supports this view:

...so what we did was just running through clips and passing them from one group of people to another ... [laughing] this was funny... I like to take pictures of funny situations and when my friends are drunk they do funny things so we go back and try to remember what happen and we always have a good laugh. Sometimes we like to compare who managed to take the most funny shots ... it is really funny seeing people doing crazy things.

Since the camera screens are small and do not support easy and clear viewing for a group of people when sharing pictures in the home environment, people often made use of external display technology, such as TV or computer. This issue will be explored further in the next section, *'Sharing a moment later.'*

As discussed in this section peoples' use of camera phones changes in relation to the space they are in; private vs. public. It was found that people's practices when using camera phones differs in different spaces. The next section will discuss this phenomenon in more detail.

Social Uses of Camera Phones

Camera phones have been used for individual as well as group purposes. Consistent with other studies (Kindberg et al., 2005a, 2005b) we found that people take photos for individual purposes that include creating memories and evocations of special events, trips, holidays, or beautiful landscapes. A common practice is to share images with friends and family, in a way that is deeply embedded in social interaction (Stelmaszewska et al., 2005, 2006). Sharing digital photos is often done remotely via email or by posting them on the web (Counts & Fellheimer, 2004; Stelmaszewska et al., 2005). Despite the growing popularity of using web-based applications and services (e.g., Flickr, YouTube, or Mobido) that allow their users to share photos there were no accounts reported using these services by the participants involved in this study.

However, we observed other practices that occur in co-present social contexts. These include 'sharing a moment now,' 'sharing a moment later,' or using photos to initiate social interaction with strangers.

'Sharing a Moment Now'

This study shows a different way people share photos taken on a camera phone that appears to be less about evoking or recreating an event or scene after the fact, and more about augmenting that event as it happens. It was observed that people take a 'spur of the moment' photo or video and share it with people who are present at the same location at the same time. People reported having fun when taking photos or videos of their friends behaving funnily and then viewing them collectively at the location. This kind of behavior seems to motivate and shape social interaction, as Adam reported:

...she was happy and funny (referring to a friend) ... far too engaged with dancing to notice what was happening around her ... and I just thought that I'll just take that picture. ... there were few of us friends so then I showed them and then other friends were taking more pictures of her dancing and we were waiting for her to realize what was going on ... we were all taking pictures of her ... we shared all the pictures and picked out the funniest ones. It was so funny because she couldn't believe that we did that and she didn't even notice it.

Whereas Lucy said:

When I'm out with my friends then I'll definitely use it (referring to a camera phone). ... Sometimes I take pictures of my friends and then we'll sit down and go through them selecting the best once.

Data shows that photos were used for functional purposes as well, which is consistent with the findings of other research (e.g., Kindberg et al., 2005a; Van House et al., 2005). It was observed that when on a trip, people took a picture of a map displayed by a leader and then pursued his instructions using a display on their camera phones. This kind of activity allowed every person within the group to see clearly the map and use it for further reference.

Another common practice observed and reported by participants was to transfer photos between phones using the Bluetooth technology so that everybody concerned could store and use them when needed. The following observed episode is a typical example:

Episode 1: Pub, evening

Ten people are sitting at the table (three females and 7 males). Jim takes the camera phone out of his pocket and plays with it.

Jim: 'I have something really cool to show you.' He does something with his phone. After a while Jim said: 'OK, I've got it.' He plays the video and passes his phone over to a neighbor, Roy.

Jim: 'Just press the button.' Roy plays the video and moves the phone towards another male, Paul. Another male, Martin moves from his seat and stands behind Roy and Paul watching the video clip.

Martin: 'I want this clip. Can you Bluetooth it?'

Jim: 'Yeah' Jim takes his phone back from Roy and sets up the Bluetooth. Martin does the same on his phone. After a short while Jim transfers the clip over to Martin's phone.

However, it appeared that some people found it difficult to use it and either abandon the transfer or asked for help. When discussing issues related to managing pictures on the phone Maria said:

I Bluetooth them ... I can do it now but I had to ask my friend to show me how to do it so I'm OK now.

'Sharing a Moment Later'

When people who you want to share photos with are around, it creates opportunities for social interaction to take place so that people can enjoy the moment of sharing pictures together. What happens when they are not around? Other studies reported this kind of practice; that is to view the photos when the occasion arises, and not immediately after they have been taken. For example, Okabe (2004) described situations where people show their friends the photos from their archives (photo gallery) on occasions that they get together.

A co-present social interaction was reported to be associated with participants' experience when viewing pictures or videos stored on individual's phones but taken previously (not at the time of gathering). The intentions behind it were reported to include sharing memories of special events, reporting on events to those who were absent at the time of events, or creating and sharing a documentary of a friendship or family life as Maria remarked:

with the cam_phone I can capture the moment ... and being able to view them later will bring all the memories and the fact that those pictures can be shared ... so people can have fun.

People were more inclined to use photos for storytelling, which is in line with (Balanovic et al., 2000; Kindberg et al., 2005a) and, as suggested by Fox (2001) and Vincent & Harper (2003), mobile phones have been used to maintain personal relationships between friends and family. Since camera phones are becoming a part of our everyday lives, it is not surprising that the same behavior was observed in the context of camera phone use when photos or videos were shared during social gatherings.

However, given that phone screens were claimed to be very small it was common amongst participants to use other media like computer or

TV to display photos in order to improve their visibility and enhance the experience of people participating. Adam reported:

I transferred them onto my computer ... I'm quite organized with my pictures so I categorize them and put them in kind of albums and sometimes when I'm with friends we like to go through pictures and have fun.

Maria commented:

...sometimes what we do is we Bluetooth to transfer our pictures to one of our computers and then have a slide show so everybody can see it ...you see the phone screens are very small and if we all want to have fun we need to see those pictures simultaneously. With camera phones we can't see it clearly if there are more then two or three people looking. It's just not enough space ...

Sharing photos at co-present settings proved to be a way of social interaction that brings fun and joy to people's lives. The remarks of an interviewee, Steven, appear to confirm this point:

I'll show them (referring to family) what I managed to capture and then we have a good laugh.

Supporting the view, Lucy commented:

...you take pictures and when you view them you can laugh and have fun.

Ito & Okabe (2003, p.6) claim that: "Mobile phones ... define new technosocial situations and new boundaries of identity and place ... create new kinds of bounded places." We argue that camera phones go beyond that. When people view pictures together and tell the story behind them, they are transported to the place and space where those pictures were taken. Pictures conjure memories, feelings, and emotions and evoke sensations associated with the events that were photographed. Lee, another study participant, remarked when showing pictures from a group trip:

... The first dive was really s.... it was sooo cold, remember, ... and we didn't see much... The vis was absolutely s.... yeah and then we had to get warmer ha, ha, ha ...

Comments from other participants suggest the same:

Adam: *... when you are having a good time you don't always know what's happening around you. ... I don't always know what everybody is doing so I miss a lot of stuff but when we view all the pictures taken during a particular party or we go for a short trip together ... so only then you really can see what happened. We really like doing that.*

Maria: *... you can not only see the pictures but there are always some stories behind every picture. ... so later when you show the pictures everybody gets involve and just add a story to it and that's great. I like it. And others who were not there can feel like they were there err...kind of.*

Social Interaction with Strangers

Studies reported by Weilenmann and Larson (2002) explored the collaborative nature of mobile phones use in local social interaction amongst teenagers. They suggest that mobile phones are often shared in different forms including: minimal form of sharing (SMS messages), taking turns (several people handling a phone), borrowing and lending of phones, and sharing with unknown others. The latter involves the phones being handled by teenagers who are unacquainted until one of them makes the initial contact. Weilenamm and Larson (2002) describe practices of teenagers (boys giving girls their mobile phone) to enter their phone numbers. This kind of social interaction is similar to the one that emerged from our studies.

Social interaction can coalesce around different media, from text and graphics, to interactive games (Stelmaszewska et al., 2005, 2006). Such interactions often occur between friends or family members sharing the same technology (i.e., computer, digital camera or mobile/camera phone).

However, a striking finding was that camera phones were used as a new channel and medium for initiating social interaction with strangers. It was reported that people take photos of others (whom they like) in order to show their interest, introduce themselves, or simply start a new social relationship.

The comment from Luisa supports this claim:

I was at the Harvester, a restaurant/pub thing, ... and there was a small window with glass between it looking like a fake door and the guys were looking through that doing (mimicking facial expressions) and then I saw one holding his camera phone against one of the window things and there was a picture of me going (shows facial expression) and I didn't know that they were taking it ... I didn't really mind. It's a good humor... it was kind of friendly, sort of vague flirting without talking ... just taking pictures.

So does another comment by Maria:

We were in the bar ... having fun and there was this guy dancing [laughing] kind of a very funny dance ... almost like an American Indian kind of dance ... and one of the girls from our group took a photo of him because she liked him and she was showing it to us so instead of looking at him we could see his picture ... and when he saw her taking pictures of him he did the same to her... the whole situation was funny ... at least we had fun watching them two taking pictures of each other instead of talking ...

This kind of behavior typically occurred in public spaces such as pubs, bars, or clubs where people usually gather for social events, and interaction with others is a part of the entertainment. In our study, the focus was on social interaction that took place through and around digital photos. Such interaction is not always appreciated by those involved. Some participants felt offended and annoyed with those taking photos without obtaining agreement. For example, Lucy noted:

I don't know if I would be offended so much. I think it depends what for ... sometimes you get photographers going like around pubs and clubs ... and I never said yes to the photo. The other night when I was there with my friend and this group of guys we met before errr ... this guy said: 'Oh yeah, let's get a picture' but we went like: 'no, we really don't want to.' And they had one done anyway and this kind of annoyed me a bit because ... it's fair they wanted the picture of us but we didn't really want to be in it. ... I think it depends how much choice you are given as whether or not you want your photo taken.

It appeared that pictures are not the only phone-related way people try to 'chat up' others. Phone features like Bluetooth can be used to connect to strangers and initiate communication. This kind of behavior was observed in public places (pubs, restaurants, bars). The practice was to switch on the Bluetooth and ask others (whoever is picked up by the Bluetooth) to activate the connection. However, this kind of interaction often raised some suspicions, as people did not know who wants to 'chat up' to them. Here is an extract from one of the participants expressing his concerns:

... someone wants me to activate the connection ... but what do I do ... I don't want any 'Boss' [the name of the Bluetooth connection] connecting to my phone. What if they do something to my phone?

The fact that people do not see the 'talker' and they do not have the full control of who they interact with seems to be a barrier to engaging in interaction with a stranger.

It seems that communication takes place not only through technology but also alongside it, a finding that is consistent with our earlier studies (Stelmaszewska et al., 2005). Moreover, Van House et al. (2005) argue that technology (e.g., online photo blogs) is used to create new social relationships. Although this study is at an early stage and further evidence is required, we suggest that camera phones provide new channels and foci for social interaction within co-present settings.

Barriers to Sharing

Although camera phones appear to be a new medium for social interaction that is enjoyable and fun, they are not without problems that limit the extent to which they are used. The data illustrates that people experience different kinds of trouble that hinder their experience or make it impossible for sharing to happen.

Firstly, the lack of compatibility between different camera phones stops people from sending photos. Several participants reported not using MMS features because it was difficult to use. In addition, people often know (not always) that those who they want to send pictures to will not be able to retrieve them as was commented by Luisa:

... none of mine friends really do this ... you have to have the same phone or something to be able to send it and for them not to just say: 'message not being able to deliver or whatever.' Some people tried to send pictures on my phone but I never got them.

Secondly, for many camera phone users it is difficult to send pictures either via MMS or Bluetooth. People reported having difficulties to find the functions to do so or they could not set them up (in case of the Bluetooth—see comments in the section on 'Sharing a moment now').

Another barrier to sharing photos was the lack of a quick and easy way to find archived pictures. People spent time, sometimes a long time, trying to find the pictures they wanted to share with their friends. This caused frustration and dissatisfaction as Jim said:

Where is it?!!! S... Hrrrrrrrrr

Quick access to camera functionality and photo image features is an important issue in a context of sharing and it raised concerns amongst participants as Maria noted:

... one of my friends helped me to set it up so I can use it by pressing just a couple of buttons instead of going through menus and stuff. It was horrible. I missed so many great pictures because of that and I was very upset about it. ... it's very important. I could have so many great pictures but couldn't find the camera function on my phone ... it was very frustrating.

All these barriers affect not only experience of camera phone users but also their engagement in social interaction. So providing functionality that is transparent and supports users sharing activities is of a paramount importance when designing systems. It might also enhance the use of camera phones by creating pleasurable and fun experiences instead of satisfying only functional purposes.

DISCUSSION AND CONCLUSION

It seems that phone technology is moving from facilitating its original primary goal, supporting distance communication, to supporting new ways of social interaction that happens through sharing activities (photos and videos) as well as providing bridges between contexts. When people share photos or videos, they are transported from the context of a present space (pub, restaurant, or home) to the one that a specific photo or video clip conjures up.

In addition to providing resources for communication and interaction, camera phones have been used as a kind of archive of a personal life, a viewpoint on the world, or a collection of fragments and stories of everyday life. Okabe (2004) suggests that photos are often taken for purely personal consumption, whereas text messages are generally created with the intent to share with others. However, the findings from this study contradict Okabe's claim; people often take photos with the intention to share them with others, which is a more selective and intimate activity than sharing text.

When technologies are used in different places and spaces they become part of a specific environment and this often shapes the use of technology and experiences connected to it. As a consequence of this, technologies are often used in unexpected ways (Taylor & Harper, 2003). In the case of this study, these ways are 'sharing the moment now,' 'sharing the moment later' and using camera phones for 'social interaction with strangers.'

This chapter has described distinctive practices of camera phone users occurring in co-present settings, and how these practices change in relation to the place and space in which they were used. It has been argued that camera phones provide a new medium through which people can sustain and enrich they social interaction through taking and sharing photo images or videos. However, these activities are inseparable from social relations and context, which is in line with Okabe's (2004) and Scifo's (2004) findings. Moreover, we argue that this study provides a better understanding of how this emerging and evolving technology facilitates social interaction in the leisure-related practices of its users.

We agree with Rettie's (2005) view that mobile phone communication affects the role of space and we have shown that camera phones go beyond this: they bring people together, creating experiences through social interaction. No other technology has supported this to such an extent, and to so many people. The multi-functionality of camera phones provides a different means of social interaction, which is unique to a place and space.

More generally, when designing camera phones that facilitate social interaction, understanding of emerging uses, practices and social activities is essential for the effective design of camera phones and related systems. Moreover, identifying problems within existing systems might be a good starting point for discussing user requirements, helping designers to develop systems that fulfill utilitarian as well as user experience needs.

Although the notions of 'sharing' might be a new phenomenon it is a manifestation and reflection of needs that relate to social identity (Scifo, 2004; Taylor & Harper, 2002) and are shaped by social context (Okabe, 2004; Stelmaszewska et al., 2005, 2006). This study is part of an ongoing effort to explore issues related to the use of camera phones for social interaction within co-present settings, and further studies will be required to investigate what affects such interaction, how camera phones' design, usability and context of use influence the nature of users' experience.

Furthermore, more work is needed to identify and understand problems when camera phones are used for social interaction, and how we can improve the design of camera phones so that they can evoke experiences such as pleasure, excitement, or fun.

ACKNOWLEDGMENT

We would like to thank all anonymous participants who took part in this study.

REFERENCES

Balanovic, M., Chu, L., & Wolff, G. J. (2000). Storytelling with digital photographs. In *Proceedings of CHI 2000: Conference on Human Factors in Computing Systems* (pp. 564-571). ACM Press.

Casey, E. S. (1997). *The fate of place: A philosophical history.* Berkeley: University of California Press.

Chalfen, R. (1987). *Snapshot version of life.* Bowling Green, OH: Bowling Green State University Press.

Ciolfi, L. (2004). *Digitally making places: An observational study of people's experiences of an interactive museum exhibition.* Paper presented at the Proceedings of the 2nd workshop on 'Space, Satiality and Technologies.' Edinburgh, UK.

Counts, S., & Fellheimer, E. (2004, April 24-29). Supporting social presence through lightweight photo sharing on and off the desktop. In *Proceedings of CHI 2004: Conference on Human Factors in Computing Systems* (pp. 599-606). Vienna: ACM Press.

Dourish, P. (2001). *Where the action is: The foundation of embodied interaction.* Cambridge, MA: MIT-Press.

Fox, K. (2001). Evolution, alienation and gossip: The role of mobile telecommunications in 21st century. *Social Issues Research Centre Report.* Retrieved from http://www.sirc.org

Frohlich, D., Kuchinsky, A., Pering, C., Don, A., & Ariss, S. (2002). Requirements for photoware. *Proceedings of the 2002 ACM conference on Computer Supported Cooperative Work* (pp. 166-175). ACM Press.

Ito, M. (2003). *A new set of social rules for a newly wireless society. Japan Media Review.* Retrieved July 28, 2006 from http://www.ojr.org/japan/wireless/1043770650.php

Ito, M. (2004). *Personal Portable Pedestrian: Lesson from Japanese Mobile Phone Use.* Paper presented at The 24 International Conference on Mobile Communication Social Change. Seoul, Korea.

Ito, M., & Okabe, D. (2003, June 22-24). Mobile phones, Japanese youth, and the re-placement of social contact. In R. Ling (Ed.), *Front stage—back stage: Mobile communication and the renegotiation of the public sphere.* Grimstad, Norway.

Ito, M., & Okabe, D. (2005). Technosocial situations: Emergent structurings of mobile email use. In M. Ito, Okabe, D. & Matsuda, M. (Ed.), *Personal, portable, pedestrian: Mobile phones in japanese life.*

Kato, F. (2005, April 28-30). *Seeing the "seeing" of others: Conducting a field study with mobile phones/mobile cameras.* Paper presented at the T-Mobile conference, Budapest, Hungary.

Kindberg, T., Spasojevic, M., Fleck, R., & Sellen, A. (2005a). An in-depth study of camera phone use. *Pervasive Computing, 4*(2),42-50.

Kindberg, T., Spasojevic, M., Fleck, R., & Sellen, A. (2005b). I saw this and thought of you: Some social uses of camera phones. *In CHI '05 extended abstracts on Human factors in computing systems* (pp. 1545-1548). ACM Press.

Klamer, L., Haddon, L., & Ling, R. (2000). The qualitative analysis of ICTs and mobility, time stress and social networking. (No. P-903): EURESCOM.

Licoppe, C., & Heurtin, J.P. (2001). Managing one's availability to telephone communication through mobile phones: A French case study of the development dynamics of mobile phone use. *Personal and Ubiquitous Computing, 5*(2), 99-108.

Ling, R. (2004). *The mobile connection: The cell phone's impact on society.* Morgan Kaufmann.

Mäkelä, A., Giller, V., Tscheligi, M., & Sefelin, R. (2000). Joking, storytelling, artsharing, expressing affection: A field trial of how children and their social network communicate with digital images in leisure time. In *Proceedings of CHI 2000: Conference on Human Factors in Computing Systems* (pp. 548-555). ACM Press.

Mulder, I., & Steen, M. (2005, May 11). *Mixed emotions, mixed methods: Conceptualising experience of we-centric context-aware adaptive mobile services.* Paper presented at the Pervasive 2005. Presented at workshop on User Experience Design for Pervasive Computing, Munich, Germany.

Okabe, D. (2004, October 18-19). *Emergent Social practices, situations and relations through everyday camera phone use.* Paper presented at the International Conference on Mobile Communication, Seoul, Korea.

Okabe, D., & Ito, M. (2005). Ketai and public transportation. In M. Ito, Okabe, D., & Matsuda, M. (Eds.), *Personal, Portable, Pedestrian: Mobile Phones in Japanese Life.* Cambridge: MIT.

Rettie, R. M. (2005). Presence and embodiment in mobile phone communication. *PsychNology Journal, 3*(1), 16-34.

Salovaara, A., Kurvinen, E., & Jacucci, G. (2005, September 12-16). *On space and place in mobile settings.* Paper presented at the Interact '05, Rome.

Sarvas, R., Oulasvirta, A., & Jacucci, G. (2005). Building social discourse around mobile photos—A systematic perspective. *Proceedings of the 7th international conference on Human computer interaction with mobile devices & services* (pp. 31-38). ACM Press.

Scifo, B. (2004, June 10-12). *The domestication of camera-phone and MMS communication: The early experience of young Italians.* Paper presented at the T-Mobile Conference, Hungary.

Stelmaszewska, H., Fields. B., & Blandford, A. (2005, September 5-9). *Emotion and technology: An empirical study.* Paper presented at the Emotions in HCI design workshop at HCI '05.

Stelmaszewska, H., Fields, B., & Blandford, A. (2006, September 11-15). Camera phone use in social context. In *proceedings of HCI 2006* (Vol. 2, pp. 88-92), Queen Mary, University of London, UK.

Strauss, A., & Corbin, J. (1998). *Basics of qualitative research. Techniques and procedures for developing grounded theory.* Newbury Park: Sage.

Taylor, A., & Harper, R. (2002, April 20-25). *Age-old practices in the 'new world': A study of gift-giving between teenage mobile phone users.* Paper presented at the CHI, Minneapolis, MN, USA.

Taylor, A., & Harper, R. (2003). The gift of the grab: A design oriented sociology of young people's use of mobiles. *Journal of Computer-Supported Cooperative Work, 12*(3), 267-296.

Van House, N., Davis, M., Ames, M., Finn, M., & Viswanathan, V. (2005, 2-7 April). The uses of personal networked digital imaging: An empirical study of cameraphone photos and sharing. *CHI '05 extended abstracts on human factors in computing systems* (pp.1853-1856). ACM Press.

Vincent, J., & Harper, R. (2003). *Social Shaping of UMTS: Preparing the 3G Customer.* University of Surrey, Digital World Research Centre. Retrieved July 26, 2006, from http://www.dwrc.surrey.ac.uk

Weilenamann, A., Larsson, C. (2002). Local use and sharing of mobile phones. In B. Brown, Green, N., and Harper, R. (Ed.), *Wireless world: Social and interactional aspects of the mobile age* (pp. 92-107): Springer.

KEY TERMS

Bluetooth: A wireless protocol that is used to connect compliant devices that are in close proximity with each other in order to transfer information between them. Bluetooth is commonly used with phones, hand-held computing devices, laptops, PCs, printers, digital cameras.

Camera Phone: A mobile phone with a camera built-in that allows the user to take pictures and share them instantly and automatically via integrated infrastructure provided by the network carrier. Camera phones can transfer pictures via Bluetooth, Infrared, or MMS messaging system.

Co-Present Interaction: Interaction that happens between two or more people that are physically present at the same time and location.

Digital Photo Sharing: An activity of two or more people, who share images by showing pictures to others. Sharing digital photos can occur at the co-present location or remotely. The former happens using different devices like camera phone screen, digital cameras, TV screen, or computer screen. The latter is often done via email or by posting them on the web.

Digital Photography: A type of photography where pictures are taken on digital cameras or camera phones. Images can be viewed, edited, stored, or shared with others using different means of communication medium such as email, Web-based applications and services, Bluetooth, Infra-red, MMS, computers or TV screens.

Field Observation Studies: A qualitative data collection method, which is used to observed naturally occurring behavior of people in their natural settings. The data can be gathered in a form of: film or video recording, still camera, audio type (to record spoken observation), or hand-written note taking.

Qualitative Data Analysis: A collection of methods for analyzing qualitative data, such as interviews or field notes. One example of such method is Grounded Theory, which is used to generate theory through the data gathering and analysis. Data is sorted to produce categories and themes of concepts emerging from the data.

Social Interaction: Interaction that happens between individuals typically mediated by, or in the presence of technological artifacts.

Theoretical Sampling: The process of data collection for generating theory where the researcher collects, codes and analyses data and makes decisions about what data to collect next. Researchers consciously select additional cases to be studied according to the potential for developing new insights or expanding and refining those already gained. Sampling decisions depend on analysis of data obtained, which relate to the developing theory.

Triangulation: The application and combination of at least two research methods or data gathering exercises to research the same phenomena in order to cross-checking one result against another, and increasing the reliability of the results.

User Experience: A term that is used to describe the overall experience and satisfaction of a user while using a product or system.

Chapter V
Interaction Design for Personal Photo Management on a Mobile Device

Hyowon Lee
Dublin City University, Ireland

Cathal Gurrin
Dublin City University, Ireland

Gareth J.F. Jones
Dublin City University, Ireland

Alan F. Smeaton
Dublin City University, Ireland

ABSTRACT

This chapter explores some of the technological elements that will greatly enhance user interaction with personal photos on mobile devices in the near future. It reviews major technological innovations that have taken place in recent years which are contributing to re-shaping people's personal photo management behavior and thus their needs, and presents an overview of the major design issues in supporting these for mobile access. It then introduces the currently very active research area of content-based image analysis and context-awareness. These technologies are becoming an important factor in improving mobile interaction by assisting automatic annotation and organization of photos, thus reducing the chore of manual input on mobile devices. Considering the pace of the rapid increases in the number of digital photos stored on our digital cameras, camera phones and online photoware sites, the authors believe that the subsequent benefits from this line of research will become a crucial factor in helping to design efficient and satisfying mobile interfaces for personal photo management systems.

INTRODUCTION

Long before digital technology came into everyday use, people have been managing personal photos with varying degrees of effort. Individuals' photo management strategies ranged from stacking photos in shoe boxes to carefully placing them into a series of photo albums with detailed notes of where and when each photo was taken or a witty caption beside it. Reminiscing and story-telling past events that have been visually recorded in personal photos is a highly-valued activity for many people. This gives meaning to the person's past events and also works as a socially-binding and relationship-enhancing device at gatherings of family or friends. With the Internet revolution, and the arrival of inexpensive digital cameras, people's photo organizing and sharing behavior has been evolving as new technologies allow different ways of managing photo collections. This is exemplified with online photoware applications such as Flickr[1], with which people can now upload personal photos taken from their digital cameras onto a shared web space on which collaborative annotation, browsing and sharing photos with other people is possible.

Another aspect of the development of digital photography is that people's behavior in capture of photos is changing as well. In particular, due to the low cost and ease of capture nowadays people are taking many more photos than in the past. This is possibly best illustrated by the ubiquity of camera phones, mobile devices that can be used as digital camera as well as a phone. Many people carry their phone with them at all times meaning that they can capture their everyday lives and holiday scenes whenever they want. This change in capture behavior can also have a significant impact on people's personal photo management activity. Once captured, the phone can be used to send photos to a friend's mobile phone or to upload them to a public Website for instant sharing and receiving comments back. This means that when designing personal photo management tools, we should consider the implications of the changes in user photo capture behavior arising from the emergence of the ubiquitous availability of the means of photo capture. For example, there is a need to design specific user-interfaces for photo management on a camera phone itself. A camera phone may be used merely as a capture device that takes photos and stores them, to be copied later to a PC for further photo management. However, the quality of screens now commonly available on mobile phones means that it is quite reasonable to look to design tools that enable users to organize, annotate and browse photos *on* the mobile phone itself. Between these two extreme cases, there is a spectrum of varying degrees to which a camera phone or other mobile device can be integrated into overall photo management functions and tasks, effectively a continuum of trade-off among technological resources and the user's effort and time. For example, due to the difficulty of text input on a camera phone arising from physical constraints, it may be easier for the user to fully annotate photos after copying them on to a desktop PC at home. Even so, a user in some situations might still want to make the effort to annotate their photos using the mobile and to send them to a friend for the benefit of its immediacy and not having to do the extra work of copying photos to a PC at home in the evening before performing the annotation. On occasion some users will want to bulk-upload a large number of photos taken at a party directly to a website without any annotation, e.g., to share with close friends. Depending on the design decisions on the allocation of photo management tasks for different devices, the optimal user-interface for such tasks on the mobile device will vary. Currently available interfaces on camera phones and digital cameras for photo management illustrate this possible diversity of user task requirements.

Particular challenges faced in designing and evaluating mobile interfaces for personal photo management arise due to, among other things, the following:

- New technology regularly emerges and applications constantly evolve.
- Mobile users are difficult to observe.

Consequently, it is difficult to rely on the traditional system development cycle of user study, user needs, and requirements establishment, followed by prototyping and evaluation. By the time this established process has been completed for an application, a new technological innovation may have appeared bringing in a new possible line of products to be developed. Conducting a user study on mobile devices can pose significant problems compared to desktop-based systems because a proper observation of users using a mobile device is technically difficult. When the user is on the move, we are simply unable to hook up an observation camera and recorder, and keep following the user while he or she is out and about using the device.

However, recently study methodologies have been developing to cope with these difficulties. For example, quicker and cheaper prototyping techniques are starting to be adopted to keep pace with changing technology (e.g., PC-based simulated prototypes or using a general mobile platform (Jones & Marsden, 2006, p. 179)). In another example, user studies are emerging which adopt light ethnographic and indirect observations using diaries and self-reporting to cope with test users on the move (Palen & Salzman, 2002; Pascoe et al., 2000; Perry et al., 2001). More specifically on the use of *camera phones*, some early longitudinal user studies have appeared (Kindberg et al., 2005; Sarvas et al., 2005; Van House et al., 2005) that aim to better understand the different motivations and current practices in camera phone use. These studies use diaries and interviews to capture usage data from 30-60 users within a period ranging from one to two months.

In understanding the current status of mobile interaction design for personal photo management and in setting the right direction for future applications, we need to look at the way such applications have been developed so far. As will be described in more detail in the following sections, applications for personal photo management have been incrementally shaped by the major technology innovations that have appeared during the last two decades or so. By looking at other new technologies, which are likely to be available in the near future, we can roughly determine what kind of personal photo applications or requirements will emerge in which mobile devices are an important component. The primary technology that we anticipate will appear is the *automatic organization* of the photos. The need for and adoption of this technology will be driven by the very rapid increase in the number of digital photos that users will accumulate on camera phones, desktop PCs, and online photoware. The sheer volume of photos means that it is increasingly difficult and frankly becomes unrealistic to manually annotate these photos.

Fortunately, by leveraging context data such as the time and location of photo capture, the bulk of the organization task can be done automatically. In addition, content-based image analysis techniques, although considered still not mature enough for many other applications, are also proving a very promising element that can further contribute to effective automatic organization and annotation of large photo collections. Use of these organization automation tools reduces the user's annotation burden to such an extent that a user working with large digital photo collections can focus on enjoyable browsing, searching and sharing tasks, rather than the nuisance of file extension and ongoing manual annotation. In this chapter we explore how mobile interaction design for personal photo management can take advantage of these emerging technological factors to overcome potential interaction design problems for photo management applications on a mobile device.

BACKGROUND: TECHNOLOGY TREND AND PHOTO MANAGEMENT ON MOBILE

Starting from physically printed photos, which are manually organized in an album, moving to current camera phones to capture, annotate, and browse digital pictures, the way we manage personal photos is evolving very rapidly. Current emerging applications have largely been geared to the major technological innovations that have occurred. These include the wide uptake of PCs

Figure 1. Supporting personal photo management at different stages of technological development

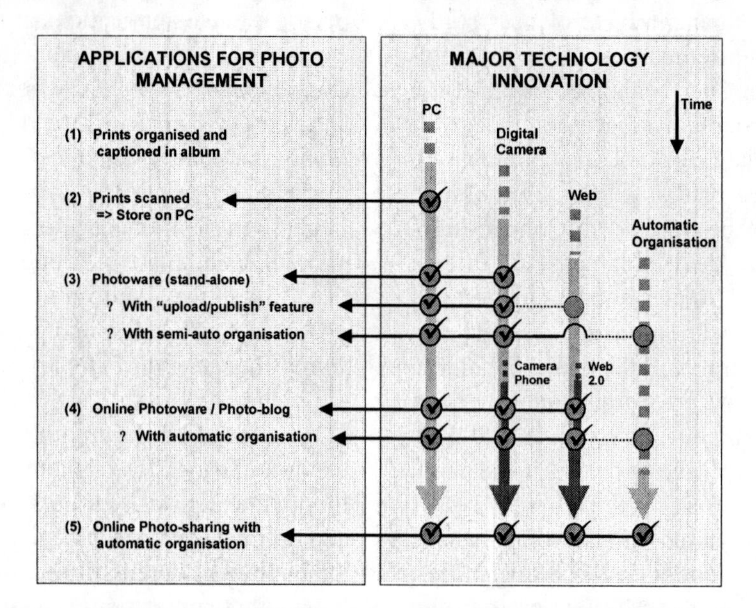

to the general public since the mid-80s, the World Wide Web and inexpensive digital cameras that have become commonplace since the mid- and late-90s. Each of these has brought a set of new ways of managing personal photos and the applications (with their associated features) have been quickly developed and used. This is illustrated in Figure 1.

On the right side of the figure, four vertical arrows indicate the major technologies: PCs, digital cameras and camera phones, Web and Web 2.0 technology, and finally a newly appearing technology, Automatic Organization. The first three of these have already shaped the way people manage their personal photos, in effect the *enabling technology* for personal photo management. These technologies were not mature when first emerging (thus the dotted line at the start of each of the vertical lines), but in time have become reliable enough to be taken up as full applications. The rows on the left side of the diagram show the way people manage their photos taking advantage of these technologies. For each row the cross point with the vertical lines indicates which technology has been used (circle with a tick mark). Circles without a tick mark indicate cases where the uptake of the technology has only been partial or done in a work-around way due to insufficient development of this particular technology, or of our understanding of its potential at that time. The last row is the envisaged future photo applications that take full advantage of the four technologies including Automatic Organization.

In this section we briefly go through each of the rows as numbered in the figure, highlighting how the enabling technologies influenced its development, what specific features became possible due to such technologies, and where mobile device and associated user-interface design issues arise from this.

Prints Organized and Captioned in an Album

Since the camera became a common household gadget, the most traditional way to manage personal photos was to get film developed and then select well-taken prints to insert into an album. Because the user selected only good photos for inclusion in a photo album, it was most likely to contain high quality or useful photos (Rodden, 1999). The major feature of an album is the grouping or organization into pages and the *captioning*, often appearing beside the chosen photos where the user adds a short description of the photo, often with a humorous comment. When people meet at a family event or party, they might gather around the album and flip through pages discussing and appreciating photos and their captions; good photos were also ordered for reprints to be given or posted to families and friends for sharing. These social activities have been popular as the main ways of browsing and sharing personal photos (Frohlich et al., 2002) and have often been adapted in photoware as a metaphor.

Prints Scanned and Stored on a PC

Before digital cameras became inexpensive and commonplace, those who owned PCs could scan prints of physical photos to store them digitally on their PCs. Preservation, digital enhancement or novelty were more likely motives rather than as a true replacement of chemical photos. Sharing was possible by copying files onto a disk and passing it to somebody who also owned a PC or more recently sending the photos as an email attachment (Frohlich et al., 2002; Rodden & Wood, 2003), but the PC did not come to be used as a tool for photo management until people started taking photos with digital cameras.

Stand-Alone Photoware

The availability of inexpensive digital cameras meant people could now capture their interesting events directly into digital format and copy them onto their PC. As the cost of capture is virtually zero and unwanted photos are easily deleted, the average number of photos taken increased dramatically. However, digital photos stored in PC directories often have cryptic file names such as "1430XX23-02.jpg" which have been generated automatically by the camera. The large quantity of such photos means that users often do not attempt to make the effort to rename them with more meaningful titles. It is thus not possible for users to find individual photos based on filenames.

As the Web started to become more common, some users manually created Web pages containing their photos to share with family and friends. However, this required much time and effort to find and select photos from within their collections and then to generate the Web pages, and only carefully selected photos were made available for sharing by users interested in both web and digital photography technologies.

This situation improved with the introduction of *photoware* software that imports photos from a digital camera and supports easy management of photos by allowing the grouping, sorting by date, annotation, and subsequently allowing searching and browsing. As people started accumulating large numbers of photos the utility of photoware grew. Examples of popular photoware include Photoshop Album[2], ACDSee[3] and Picasa[4].

Some experimental photoware systems support automatic grouping of photos based on time/date as the photos are imported from the camera (O'Hare et al., 2006), removing some of the organizing burden from the user while at the same time motivating higher quality annotation (Kustanowitz & Shneiderman, 2004). Other systems feature a convenient "upload" or "publish" feature whereby after organizing photos on a PC, the system generates attractive Web pages with the selected photos and captions.

Currently a large number of similar photoware tools are available each providing some combination of features including photo manipulation (rotation, brightness/contrast change, sharpening, red-eye removal, adding visual effects, etc.), organization (adding titles and descriptions, fast photo tagging, searching for duplicate photos, synchronising directories, etc.) and browsing and searching (thumbnail and full-screen views, slideshows, easy zoom-in/out, search by annotation, view colour histogram, etc.).

Online Photoware for Sharing and Photo-Blogging

With the advent of Web 2.0 in which the web is itself a platform, a highly interactive user-interface can be realized directly within the Web browser. Conventional stand-alone software applications such as word processors, time schedulers and e-mail clients are now available online, and photoware applications can be deployed in this way as well. Organization, annotation and other useful photo manipulation such as rotation and cropping, as well as various photo collection visualizations are now featured in online photoware in which the user directly interacts with a Web-based interface.

In addition, the proliferation of *camera phones* has resulted in ubiquitous use of capturing and sharing of digital photos, spurring ever higher quantities of digital photos to be taken. In parallel, some mobile phone services have started allowing digital photos taken on a camera phone to be directly uploaded to online photoware applications and shared with other users, bypassing the step of copying photos onto a PC. Use of a camera phone for saving, organizing, annotating, browsing, searching, and sharing photos raises multitudes of user interaction issues, as we will see in more detail in the next section.

Finally, some users have started regular online posting of their daily photos with annotations allowing a community of online citizens to comment on their photos and annotations, referred to as *photo-blogging*, merging the uploading of digital photos with the features of text blogging.

Emerging from the combination of these digital and online technologies, we have popular online photoware such as Flickr[5] and Yahoo! 360[6] providing highly interactive photoware features, while focusing on the online community and sharing of photos with such features as searching over community photos, popular photos of the day, commenting and photo grouping across users. Although, not yet a mature genre, the early design guidelines for online photoware can already be found drawn from general design principles and developers' experiences (Frohlich et al., 2002).

Online Photo-Sharing with Automatic Organization

Manually organizing photos into groups and subgroups and annotating them one by one is at times pleasant, but as our photo-taking habits change from being selective to taking as many photos as possible, creating meaningful captions for individual photos becomes more and more time consuming. However, to be able to subsequently access the photos, appropriate organization and annotation of each photo is crucial. While the subjective nature of indexing a visual medium is a problem and in itself an important research area (Enser, 1995), to help a user index hundreds or thousands of personal photos for efficient searching and browsing is a challenging problem both in terms of design and technique.

Photoware which *automatically* organizes and annotates photos for the user is an attractive possibility, and as with any other digital library project where a system automatically indexes the documents in the database, is becoming more and more feasible with technical advancements that are happening today. As will be described in detail in the following sections, much of the automatic organization and annotation for personal photos can be achieved by recording context information at the time of photo capture (such as time and location) in conjunction with content-based image analysis techniques to detect, for example, the existence of faces and buildings in the photos. These features, although promising, require further research and development to be able to deliver robust performance in real-life photo applications. At this point other semi-automatic or work-around schemes to leverage a user's manual annotation input have been proposed. For example, an initial annotation by the user at the time of photo capture can then be used later by the system to suggest more annotation options to the user (Wilhelm et al., 2004); a user does manual bulk annotation of faces appearing in a group of photos, after which the system automatically assigns the annotation to the faces in each photo (Zhang et al., 2004); and how a system could motivate its users to do enjoyable annotation is also considered (Kustanowitz & Shneiderman, 2004).

As seen in this section, photo management applications have been developed incrementally adding feature after feature whenever a new technology allowed it. In drawing up the last row in Figure 1 for a truly online automatic photo-sharing application for personal photo management, the role of mobile devices should be considered carefully especially in the light of the upsurge in the use of camera phones.

From this progression, we can now envisage a scenario in which a user takes photos at a party with her camera phone, these are instantly uploaded to an online photo server where they are fully and automatically processed and annotated, then a few days later, the user visits her uncle's house and searches on her phone for those photos taken at the party, and shares them by passing her phone around or by playing a slide show on the TV screen or an interactive wall in the house.

DESIGNING MOBILE INTERFACE FOR PHOTO MANAGEMENT

From the foregoing discussions we can see that personal photo management with a mobile device could involve activities such as:

- Capturing (taking photos)
- Storing and/or uploading
- Organizing and annotating
- Browsing and searching
- Sending and sharing of personal photos

However, not all of these activities need to be conducted on the mobile device itself. We need to consider ways in which a mobile device is best used in conjunction with other technology in varying degrees of division and overlap of task. In considering *mobile information ecologies* (Jones & Marsden, 2006, pp. 280-286) such as how a mobile device's usage should fit with other devices, physical resources, network availability and other context sources should be considered. For example, where processing power is not sufficient on the mobile device which captured a photo, uploading to a server which can index the photo more quickly and then send the result back to the mobile device could be a better solution. It is often more convenient to enter long textual descriptions for a photo using a desktop PC when the user returns home, while a short annotation at the time of capture might be still useful for facts that could have otherwise been forgotten by the time the user returns home. Viewing slide shows of photos at a gathering of relatives may be best served with a TV screen rather than viewing on the mobile device.

Depending on how a particular service or product has been designed in its use of resources in the chain of activities from capturing to storing to searching to sharing with friends, the user may need to interact with different types of interfaces, or this could be to some extent transparent and hidden from the user. For example, photos stored on a remote server or on somebody else's mobile device (in the case of peer-to-peer architecture in resource allocation) could be downloaded to the user's mobile device in the background while the user is browsing photos without them having to use a separate interface to download or browse photos stored on the server. In either case, the often quoted problems of user-interface design for mobile devices seem to remain true for personal photo management. These include details such as;

- Limited screen space
- Limited input mechanism especially awkward text input
- Potentially distracting usage environments

Many interaction and visualization related issues are raised due to these limitations. Unfortunately, we have to live with them because these natural limitations arise from the fact that most mobile devices need to be, by definition, small and mobile. Studies on Web page searching on mobile devices (Jones et al., 1999; Jones et al., 2003), though not specifically on photo searching, form a useful starting point for searching within a mobile photo management context; ideas for visualization on mobile devices have been proposed, especially for displaying interactive maps for location-based navigation (Chittaro, 2006).

The minimal attention user interface (MAUI) (Pascoe et al., 2000) tries to design a mobile interface that requires minimal user attention. This is especially intended to assist field workers who use a PDA during their physically demanding tasks. In a similar vein, use of "push" technology has been proposed to reduce the amount of user interaction on a mobile device by shifting the user's interaction burden to background processing by the device (or the server the device is connected to). For example, if the device can predict which photos the user wants to see on the mobile screen at this moment, the system can display those photos without the user needing to make the effort to enter the search query, possibly delivering the relevant photos as soon as the user turns on the device. How to accurately predict which photos the user will want to view at a given time is of course the main challenge of such an approach. Examples of such technologies include the use of collaborative filtering with the data collected from explicit preference indications (Gurrin et al., 2003), and use of attention data collected from the user's daily web browser log data (Gurrin et al., 2006). The key point of this approach is to reduce the frequency and amount of the user's interactions with the device. We will see an example of how a small number of selective personal photos are displayed before a user's query input on the first browsing screen in the next section.

The following are some of the photo browsing techniques that could be suitable for a mobile interface:

- **Thumbnail browsing:** Spatially presenting multiple miniaturized photos allows easy browsing that leverages the efficient human visual system, and has been used in almost all desktop photo management systems. Although the screen is much smaller on a mobile device, it is still a useful technique and widely used in mobile interfaces for photo browsing.
- **Smart thumbnail view** (Wang et al., 2003): A photo usually contains a main focus of interest (for example, the face of a friend) as well as unnecessary visual elements (for example, strangers in the background or a large background area). By automatically determining the "regions of interest" within a photo, the interface can crop the photo to show only the area that is pertinent to the viewer. By identifying multiple regions of interest in a photo, the interface can guide the user, automatically moving the view window over a photo from one region to another in the order of importance of the regions.
- **Rapid, serial, visual presentation (RSVP)** (De Bruijn & Spence, 2000)**:** Temporally presenting multiple photos one by one as in a slide show seems particularly suitable for a small screen (De Bruijn, Spence & Chong, 2002), although having to keep focusing on the flipping-through of the images requires continuous user attention, and is thus a disadvantage when using a device with such an interface if the user needs to check their environment frequently (for example, while walking or waiting for a bus).
- **Speed Dependent automatic zooming (SDAZ):** When scrolling a page, the photos become smaller (zoomed-out) showing more of them, while it is zoomed-in when scrolling speed is reduced or ceases. This technique attempts to use the context of screen browsing. Some variations of this idea have been evaluated on a mobile device with promising results (Patel et al., 2004).
- **Key photo selection:** When multiple photos need to be displayed on a small screen, the system can determine one of those photos that is most representative of the photos and simply show the one chosen photo (while indicating that there are more to be viewed if desired). In this way, the screen space is saved and the user can browse more photos one by one if they wish to. Of course, selecting one representative photo from a group of photos taken at a particular event is an interesting research question in itself.

Some of these techniques were originally developed for desktop interfaces while others originated in PDA interfaces. These techniques and variants of them are currently being investigated for mobile interfaces. We can expect to see some of these techniques appearing in mobile photo management applications in the near future. Two strong features for enhancing photo management on mobile devices come from the technology camp, and are the subject of the remainder of this chapter. These are:

- **Content-based image analysis:** Computer vision techniques can be used to analyze the image content to classify, label, or identify something meaningful in photos for searching and browsing.
- **Context-awareness:** Context such as time, location (from GPS) and people present at the time of capture can be recorded and used to enhance metadata for searching and browsing.

As we will see in the following section, leveraging these technical elements can significantly enhance mobile interaction for photo management by enriching metadata. They can also be used subsequently to derive other useful metadata, which can in turn reduce the user's photo organization and annotation effort, and possibly enable the use of simple yet powerful time- and map-based interfaces suitable for mobile devices.

ENHANCING INTERACTION FOR MOBILE PHOTO MANAGEMENT WITH CONTENT AND CONTEXT

Content-Based Image Analysis

The use of content-based image analysis techniques for indexing and retrieving images has been an active area of research in the field of computer vision and information retrieval for many years and is neatly summarized in Smeulders et al., (2000), although even in the intervening years there have been further developments. Current approaches to content-based image retrieval can broadly be divided into three different approaches, namely using low-level features, using high-level semantic features and using segmented objects, which we now describe in turn.

Analyzing visual features of an image into low-level features such as color, shape, and texture has been the major building block for indexing image databases in order to classify and retrieve images in terms of their visual characteristics. This approach can be characterized as computationally efficient and undemanding, since these image features can be identified directly from the encoded (compressed) form of the images. Similarity between the low-level features of images can be computed simply based on intersecting histograms representing color or texture bands, where these histograms are derived for the entire image or for regions within the image. While they are computationally efficient and scalable to replicate, low-level representations of images most often do not correspond to high-level, semantic concepts that humans use when we see images. We can say that color, texture and shape are a crude first approximation to semantic image content, but very often they do not satisfy our requirement for recognizing, understanding, searching and browsing images. This difference between what low-level features offer, and what users require, is known in the literature as the "semantic gap," and has been a difficult research problem to tackle (Har et al., 2006; Smeulders et al., 2000).

The second general approach to image retrieval addresses the semantic gap head-on by trying to automatically detect semantic units directly from image content. Such semantic concepts can include almost anything, but generic concepts such as faces, buildings, indoor/outdoor, and landscape/cityscape are often used because they give general applicability. The set of possible semantic features we could detect is influenced by the use that individual detected features can offer, and as this is mostly in image classification and image retrieval, the set of possible features which we *could* calculate is enormous. In order to provide some structure and to limit the set of semantic features to use in image retrieval, we usually arrange features into an ontology which is a hierarchical arrangement of semantic topics, like the LSCOM ontology (Naphade et al., 2006). The LSCOM ontology has just under 1,000 concepts taken from the domain of broadcast TV news, but most of these concepts could be applied to any visual media, including personal photos.

The main challenge with using semantic features in applications such as personal photo management is in building classifiers to automatically detect the features. Semantic features are usually detected based on an analysis of low-level features like color, texture and shape. They are usually constructed by using a machine learning algorithm to *learn* the presence and absence of features associated with individual semantic concepts based on some training set. In the early days of using semantic features where the number of features was of the order of dozens, this was a scalable approach, but as we move towards detecting several hundred features or more, then the approach of building and training individual feature detectors does not scale, and this is one of the main challenges facing the field currently.

A second major challenge in automatically detecting features is improving the accuracy and reliability of the feature detection. Performance assessment of feature detection is carried out as part of the annual TRECVid evaluation benchmarking campaign[7] in which many (70+) participating research groups from around the world benchmark the performance of their systems for automatically detecting high-level concepts appearing in video sequences. In particular, and what makes this activity relevant to content-based image retrieval tasks such as photo managements, is that TRECVid feature detection is mostly based on shot keyframes, which are still images taken from within video shots[8]. At TRECVid 2006 benchmarking the performance of only 39 feature detectors including the presence of buildings, desert, roads, faces, animals, airplanes, cars, explosions and so on, was a significant activity for the participating groups. It is believed that building semantic feature detectors which depend upon each other, in the same way that concepts in the LSCOM ontology are arranged in a hierarchical dependency, will lead to improved feature detection accuracy (Naphade et al., 2002; Wu et al., 2004).

The final approach to content-based image retrieval that we will mention is to detect, and then use, *objects* that appear in an image as the basis for retrieval. In the approaches described so far, the processing is done on the entire image whereas in this approach we seek to identify and segment the major objects that appear within an image and to use them, rather than the whole frame, for retrieval. For example if we seek to find photos of boats then we can use a segmented image of a boat object taken from an image, independent of the background, and retrieve other objects from a photo collection based on their color, texture and/or shape. As an example of this, Sav et al. (2006) describe a system to allow manual segmentation of semantic objects from query images which are then matched against segmented objects in database images. A similar approach, albeit applied to video rather than to image retrieval, is reported by Sivic et al. (2006) used for the Google Video Search Engine[9]. What these, and a number of object-based retrieval applications which are experimental in nature, have in common is that they use segmented objects as the basis for retrieval, yet the task of automatic or semi-automatic image segmentation remains one of the most challenging image processing tasks and represents a significant hurdle towards making object-based image retrieval more widespread.

Applying content-based analysis methods of these types can enable the use of photos as queries to search for similar ones or objects within one photo to find similar objects in other photos. However, as is made clear from the above, these technologies are either rather unreliable, since they lack the power to capture perceived semantic features because they are too low-level, such as the color-based features, or they are rather specialized, in the form of learned features of particular objects. These methods thus have considerable possible utility, but are not, at present at any rate, suitable for robust and reliable photo management, but they do offer interesting potential when used in combination with context features associated with photo capture as explored in the next section.

Context-Awareness

Context information recorded at the time of photo capture can be used to assist with photo management on a mobile device in a number of ways. Particularly important given the high volume of photos often taken with devices such as camera phones and the interaction issues reviewed earlier in this chapter, a key feature in the use of the context of photo capture is that it can generally be used entirely automatically. In this section we examine the following easily captured and used context features: time, location, lighting levels and weather.

Time

Chronology is one of the most important clues when a user is looking for photos (Rodden, 2002). We know that users often remember at least the rough date/time of photo capture, even if they cannot remember exact details. Digital cameras routinely record the time of photo capture in the EXIF header of photos captured. This data allows photos to be indexed using fields such as year, month, day of the month, day of the week, hour of the day. In addition it is possible to derive more descriptive fields, such as season, weekday, or weekend, which will further aid user interactions. Indexing by time along multiple dimensions like this is useful when the user only remembers certain facets of the temporal context surrounding photo capture. For example, they may remember only that a photo was taken in the summer, in the evening, on a certain day of the week, or on the weekend.

Location

The integration of a location capture device and a camera provides the ideal scenario for location stamping of digital photo collections. However, at this point consumer digital cameras do not have integrated location stamping capabilities. While awaiting the arrival to the market of cameras which incorporate this capability, it is possible to utilize a separate GPS device to record the location at which photos are taken, and then via a timestamp matching process, incorporate the location of photo capture into the EXIF header. In our own work, we capture the locations using a small portable GPS device tracklog stored every 10 seconds, and utilize this tracklog in the location stamping process. In order to map raw GPS coordinates to real world locations, we utilize a gazetteer which typically allows the indexing of each photo at three separate levels: country, city and state, and town. The level and accuracy of location stamping depends on the granularity of the available gazetteers.

The key benefits of labeling digital photos with their location are that it enables us to support a number of access methodologies: search by actual location (country, city,town, even street), search by proximity to a location, or by proximity to other photos. By using such information the browsing space (number of photos that a user has to browse through) when seeking a particular photo can be drastically reduced.

In addition, it is possible to present a user with a map-based interface to their photo collection, with photos, or icons, plotted on a map. For example the Microsoft WWMX system (Toyama et al., 2003) takes this approach, while Google Maps[10] allows its map-specific APIs to be easily incorporated into a Web-based personal photo application thus saving development effort.

Previous research (Gurrin et al., 2005) shows that the integration of location context into a time-context based system reduces mean time to locate a given photo within an experimental collection of 8,000 photos from 32 seconds to 18 seconds, and reduces the mean number of query iterations required to locate the given photo from 3.7 to 2.8.

Other Context Issues

However powerful time and location are individually at supporting user search of digital photo collections, by combining these two contextual features, one can derive additional contextual features, such as *lighting levels* and *weather.* Standard astronomical algorithms (Meeus, 1999) allow us to calculate the environmental lighting level at the time and location of photo capture. A photo taken at 10 a.m. will be in daylight in most parts of the world, but this is not always the case, for example, in parts of Scandinavia and similar high-latitude locations this time could signify dawn, or even darkness, depending on the time of year. We use astronomical algorithms to calculate sunrise and sunset times for any location on any date, and using these algorithms we can associate a daylight status (daylight, darkness, dawn or dusk) with each photo based on its time and GPS location of capture, and thereby automatically annotate each photo with this information. When searching for a photo it is probably more likely that a user will remember that it was dark when a particular photo was being taken, than the exact time that they took the picture.

Another feature that can be used to annotate each photo is the prevailing weather conditions. There are 10,500 international weather stations dotted all across the globe which log weather data a number of times each day. Given this information, and readily available access to the weather data logs via the Web, one can annotate each photo with the weather data (clear, cloudy, rainy, or snowy) from the closest international weather station at the time the photo was taken.

Finally, people present at the time of photo capture could be yet another potentially useful context that can be captured. By using a Bluetooth device, people nearby who have Bluetooth-enabled devices can be picked up and recorded, and this information can complement other methods such as face recognition (Davis et al., 2005) effectively combining context with content-based techniques.

Content-based analysis and context-awareness as discussed so far can be applied to user access to photo collections via a mobile device to significantly enhance the user interaction on such a device. The next section introduces a prototype of such a system under development in our laboratory.

Mobile Photo Access: An Example

The MediAssist mobile interface (Gurrin et al., 2005) to personal digital photo libraries has been designed to minimize user input and proactively recommend photos to the user. Consequently, it supports the following three access methodologies from a mobile device:

- **'My Favorites':** The first screen a user sees when accessing their archive using a mobile device (see Figure 2a) is a personalized thumbnail listing of the top 10 most popular photos based on a user's history of viewing full-size photos, where this history data is gathered both from mobile devices and conventional desktop device access.
- **Search functionality:** Primarily based on location and the derived annotations. The aim is to reduce the level of user interaction required to quickly locate relevant content. In order to maximize screen real-estate available for browsing the photo archives, search options are hidden in a panel that slides into view when a user wants to search (see Figure 2b) and then disappears afterwards until required again (Figure 2a).
- **Browsing the collection by events:** Even by supporting the two access methods on a mobile device a user may still end up having to spend time scrolling through screens of photos if many were taken at the same time and place. To address this issue, the interface presents results to the user clustered into *events* and ordered by date and time. Events are logical combinations of photos taken in close proximity of location and time. Event clustering of photos can either be a rule-based process (e.g., no photos taken for a period of 90 minutes signifies the end of an event), or a clustering process where photos are grouped together based on location and/or time and the unique clusters extracted to comprise events in a personal photo collection.

A user accessing the photo archive is immediately presented with the 'My Favorites' screen, of their most accessed photos, helping to reduce user interaction. If the required photo is not in the favorites, the user engages in a process of searching, followed by browsing of the search results, so in effect it is a two-phase search. The search options are: three level location (country, state, and city or town), season, weather, and lighting status, as shown on the sliding panel in Figure 2b. The contents of the location drop-down boxes are personalized to the user's collection to minimize user input. Season, weather, and lighting status are included to filter the search results thereby reducing the amount of browsing effort required to locate the desired photos.

Figure 2. MediAssist mobile interface takes advantage of context information to automatically organize personal photos

(a) 'My favorite photos' with search panel down

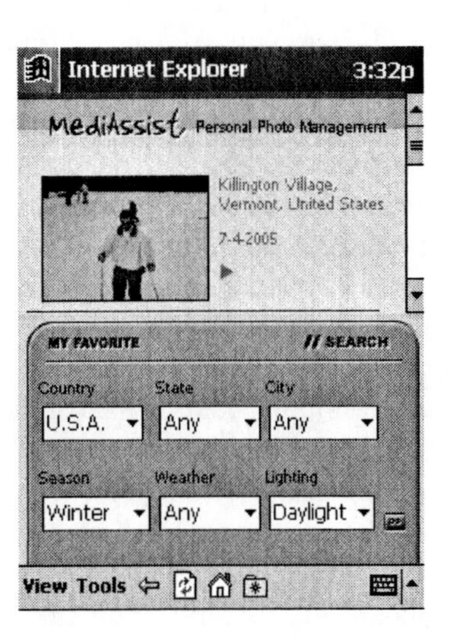

(b) Searching the archive with a representative photo results as event summary

Simply presenting a (potentially long) list of photo thumbnails in response to a query is not an ideal interaction scenario for the user of a mobile device. A more 'mobile friendly' technique is to group photos into events and using a single 'key' thumbnail which represents an event, as shown in Figure 2b (one event displayed). This is done by automatically grouping photos together into logical sets by examining when and where clusters of photos co-occur and choosing a single representative photo to represent the whole cluster. Typically the photo chosen to represent the cluster is the middle photo from a temporally organized listing of the cluster photos. In future work we will focus on judiciously choosing the most representative photo in a query-biased manner taking account of context and content data associated with photos in the event. These clusters are then presented to the user, ordered by time and date. Tapping on a thumbnail photo on screen presents the user with a full-screen photo, and it is this detailed viewing of a photo that is used to support the 'My Favorites' access method. Associated with each thumbnail is a small arrow button on the right side of the thumbnail. Tapping on this arrow brings the user to a screen showing all photos from that particular event, once again organized by date and time. In an experiment, the broad context searching capabilities of the mobile MediAssist system were shown to clearly outperform a more conventional time-only based system (Gurrin et al., 2005).

CONCLUSION

Interesting avenues for application scenarios are already appearing in literature which leverage context and/or content analysis for mobile photo annotation and searching. A mobile photo management system (Sarvas et al., 2004) records location, time and user data at the time of photo capture and then compares this with other already annotated metadata from other users and presents an inferred annotation for the new photo to the user. The Photo-to-Search system (Fan et al., 2005) allows a user to take a photo with a camera phone, the system then searches for visually similar images from the web and returns the result on the mobile device. While these make interesting applications and their evaluation with users will be highly important for developing future mobile systems, we need further application ideas and testing in other tasks of photo management to be able to explore more diverse application possibilities and new kinds of functions.

As we have seen in this chapter, as the number of digital photos each person needs to manage in their collection continues to grow, it will be inevitable that some form of automatic management is used as an integral module of photo management systems to help the user cope with the number of photos, even if running only at the background. Leveraging context data at the time of photo capture and use of steadily improving content-based image analysis will form a crucial part in making automatic organization of personal photos feasible. The role of mobile devices such as camera phones in this application area is also growing very rapidly as such devices become more powerful technically and more ubiquitous and more accepted socially. However, a mobile device will not, in itself, be designed to do every task of photo management. They will be designed to be optimally used in conjunction with other devices such as desktop PCs and laptops, TVs, and other information appliances, depending on technical and social situations. Automatic organization, then, is a vital back-end technology in the chain of personal photo management tasks of which mobile interaction is a part.

ACKNOWLEDGMENT

We gratefully acknowledge support from the aceMedia Project under contract FP6-001765, Science Foundation Ireland under grant number 03/IN.3/I361, and Enterprise Ireland under grant number CFTD-03-216.

REFERENCES

Chittaro, L. (2006). Visualizing information on mobile devices. *IEEE Computer, 39*(3), 40–45.

Davis, M., Smith, M., Canny, J., Good, N., King, S., & Janakiraman, R. (2005). Towards context-aware face recognition. In *Proceedings of the 13th Annual ACM International Conference on Multimedia '05* (pp. 483-486). Singapore

De Bruijn, O., & Spence, R. (2000). Rapid Serial Visual Presentation: a Space-Time Trade-off in Information Presentation. In *Proceedings of the Working Conference on Advanced Visual Interfaces AVI '00* (pp.189-192). Palermo, Italy.

De Bruijn, O. Spence, R., & Chong, M. Y. (2002). RSVP browser: Web browsing on small screen devices. *Personal and Ubiquitous Computing, 6*(4), 245-252.

Enser, P. (1995). Pictorial information retrieval. *Journal of Documentation, 51*(2), 126-170.

Fan, X., Xie, X., Li, Z., Li, M., & Ma, W.-Y. (2005). Photo-to-search: Using multimodal queries to search the web from mobile devices. In *MIR '05 - Proceedings of the 7th ACM SIGMM International Workshop on Multimedia Information Retrieval* (pp. 143–150). Singapore.

Frohlich, D., Kuchinsky, A., Pering, C., Don, A., & Ariss, S. (2002). Requirements for photoware. In *CSCW'02 - Proceedings of the 2002 ACM Conference on Computer Supported Cooperative Work* (pp. 166–175). New Orleans, LA.

Gurrin, C., Brenna, L., Zagorodnov, D., Lee, H., Smeaton, A. F., & Jahansen, D. (2006). Supporting mobile access to digital video archives without requiring user queries. In *MobileHCI '06 - Proceedings of the 8th International Conference on Human-Computer Interaction with Mobile Devices and Services* (pp. 165-168). Espoo, Finland.

Gurrin, C., Jones, G., Lee, H., O'Hare, N., Smeaton, A. F., & Murphy., N. (2005). Mobile Access to Personal Digital Photograph Archives. In *MobileHCI'05 - Proceedings of the 7th International Conference on Human Computer Interaction with Mobile Devices and Services* (pp. 311-314). Salzburg, Austria.

Gurrin, C., Smeaton, A. F., Lee, H., Donald, K. M., Murphy, N., O'Connor, N., & Marlow, S. (2003). Mobile access to the Físchlár-news archive. In F. Crestani, M. Dunlop, & S. Mizzaro (Eds.), In *Mo-*

bileHCI'03 – Proceedings of the 5th International Symposium on Human Computer Interaction with Mobile Devices and Services, Workshop on Mobile and Ubiquitous Information Access (LNCS Vol. 2954) (pp. 124-142), Berlin/Heidelberg, Germany: Springer.

Har, J. S., Lewis, P. H., Enser, P., & Sandom, C. J. (2006, January15-19). Mind the Gap: Another Look at the Problem of the Semantic Gap in Image Retrieval. In *Proceedings of the SPIE Multimedia Content Analysis, Management, and Retrieval, 6073(1).* San Jose, CA.

Jones, M., Buchanan, G., & Thimbleby, H. (2003). Improving web search on small screen devices. *Interacting with Computers, 15*(4), 479-495.

Jones, M., & Marsden, G. (2006). *Mobile interaction design.* Wiley.

Jones, M., Marsden, G., Mohd-Nasir, N., & Boone, K. (1999). Improving Web Interaction on Small Displays. In *Proceedings of the 8th World Wide Web Conference* (pp. 1129-1137). Toronto, Canada.

Kindberg, T., Spasojevic, M., Fleck, R., & Sellen, A. (2005). The ubiquitous camera: An in-depth study of camera phone use. *IEEE Pervasive Computing, 4*(2), 42-50.

Kustanowitz, J., & Shneiderman, B. (2004). Motivating Annotation for Digital Photographs: Lowering Barriers While Raising Incentives. In *Technical Report, HCIL-2004-18, CS-TR-4656, ISR-TR-2005-55, HCIL University of Maryland, 2004.*

Meeus, J. (1999). *Astronomical Algorithms, 2nd ed.* Willmann-Bell.

Naphade, M., Kozintsev, I., & Huang, T. (2002). A factor graph framework for semantic video indexing. *IEEE Transactions on Circuits and Systems for Video Technology, 12*(1), 40-52.

Naphade, M., Smith, J. R., Tesic, J., Chang, S.-F., Hsu, W., Kennedy, L. et al.(2006). Large-scale concept ontology for multimedia. *IEEE Multimedia, 13*(2), 86-91.

O'Hare, N., Lee, H., Cooray, S., Gurrin, C., Jones, G., & Malobabic, J. (2006). MediAssist: Using content-based analysis and context to manage personal photo collections. In H. Sundaram, M. R. Naphade, J. R. Smith, & Y. Rui: (Eds.), *CIVR2006 - 5th International Conference on Image and Video Retrieval* (Vol. LNCS 4071) (pp. 529-532). Tempe, AZ.

Palen, L., & Salzman, M. (2002). Beyond the handset: Designing for wireless communications usability. *ACM Transactions on Computer-Human Interaction, 9*(2), 125–151.

Pascoe, J., Ryan, N., & Morse, D. (2000). Using while moving: HCI issues in fieldwork environments. *ACM Transactions on Computer-Human Interaction, 7*(3), 417–437.

Patel, D., Marsden, G., Jones, S., & Jones, M. (2004). An Evaluation of Techniques for Browsing Photograph Collections on Small Displays. Lecture Notes in Computer Science (LNCS). In *MobileHCI'04 - Proceedings of the 6th International Conference on Human-Computer Interaction with Mobile Devices and Services* (pp. 132-143). Glasgow, Scotland.

Perry, M., O'hara, K., Sellen, A., Brown, B., & Harper, R. (2001). Dealing with mobility: Understanding access anytime, anywhere. *ACM Transactions on Computer-Human Interaction, 8*(4), 323–347.

Rodden, K. (1999). How do people organise their photographs? In *IRSG'99 - Proceedings of the BCS IRSG 21st Annual Colloquium on Information Retrieval Research,* Glasgow, Scotland.

Rodden, K. (2002). Evaluating Similarity-based Visualisations as Interfaces for Image Browsing. In *University of Cambridge Technical Report, UCAM-CL-TR-543.*

Rodden, K., & Wood, K. (2003). How do people manage their digital photographs? In *CHI'03 - Proceedings of CHI '03: Proceedings of the SIGCHI conference on Human factors in computing systems,* (pp. 409-416). Fort Lauderdale.

Sarvas, R., Herrarte, E., Wilhelm, A., & Davis, M. (2004). Metadata Creation System for Mobile Images. In *MobiSys '04 - Proceedings of the 2nd International Conference on Mobile Systems, Applications, and Services* (pp. 36–48). Boston, MA.

Sarvas, R., Oulasvirta, A. and Jacucci, G. (2005). Building social discourse around mobile photos: a systematic perspective. In *MobileHCI' 05 – Proceedings of the 7th International Conference on Human Computer Interaction with Mobile Devices and Services* (pp. 31-38). Salzburg, Austria

Sav, S., J. F. Jones, G., Lee, H., O'Connor, N. E., & Smeaton, A. F. (2006). Interactive experiments in object-based retrieval. In H. Sundaram, M. R. Naphade, J. R. Smith, & Y. Rui: (Eds.), *CIVR 2006 – Proceedings of the 5th International Conference on Image and Video Retrieval* (LNCS 4071). Tempe, AZ.

Sivic, J., Schaffalitzky, F., & Zisserman, A. (2006). Object level grouping for video shots. *International Journal of Computer Vision, 67*(2), 189–210.

Smeulders, A. W. M., Worring, M., Santini, S., Gupta, A., & Jain, R. (2000). Content-based image retrieval at the end of the early years. *IEEE Transactions on Pattern Analysis and Machine Intelligence, 22*(12), 1349–1380.

Toyama, K., Logan, R., Roseway, A., & Anandan, P. (2003). Geographic location tags on digital images. In *MULTIMEDIA'03 - Proceedings of the 11th ACM International Conference on Multimedia* (pp. 156-166). Berkeley, CA.

Van House, N., Davis, M., Ames, M., Finn, M., & Viswanathan, V. (2005). The uses of personal networked digital imaging: an empirical study of cameraphone photos and sharing. In *CHI'05 - Proceedings of the CHI 2005 Extended Abstracts on Human Factors in Computing Systems* (pp. 1853-1856). Portland, OR.

Wang, M.-Y., Xie, X., Ma, W.-Y., & Zhang, H.-J. (2003). MobiPicture: Browsing Pictures on Mobile Devices. In *MULTIMEDIA'03 - Proceedings of the 11th ACM International Conference on Multimedia* (pp.106–107). Berkeley, CA.

Wilhelm, A., Takhteyev, Y., Sarvas, R., House, N. V., & Davis, M. (2004). Photo annotation on a camera phone. In *CHI'04 Extended Abstracts on Human Factors in Computing Systems* (pp.1403-1406). Vienna, Austria.

Wu, Y., Tseng, B., & Smith, J.R. (2004). Ontology-based multi-classification learning for video concept detection. In *ICME 2004 - Proceedings of the IEEE International Conference on Multimedia and Expo* (pp. 1003-1006). Taipei, Taiwan.

Zhang, L., Hu, Y., Li, M., Ma, W., & Zhang, H. (2004). Efficient propagation for face annotation in family albums. In *MULTIMEDIA'04 - Proceedings of the 12th annual ACM international conference on Multimedia* (pp. 716-723). New York, NY.

KEY TERMS

Blog: A type of website in which the user adds regular written contributions on his/her own life or thoughts, as in a journal or diary. Contracted from *weblog,* usually the entries are in reverse chronological order, and readers are allowed to add their own comments.

Context-Awareness: A system that can use information about the circumstances under which it is being used. For example a context-aware device will use the current time and location where it is being used to infer what would be the most beneficial piece of information to display for the user.

Content-Based Image Retrieval (CBIR): An application of computer vision to image retrieval, in which an image's content (its color, texture, shapes, objects or faces in it, etc.) is automatically analyzed to index the image for subsequent retrieval.

Information Retrieval (IR): An interdisciplinary field of study that deals with searching for information in documents (papers, books, pictures, video clips, or any other item that contain useful information). IR systems seek to return to users documents which satisfy their current information need as expressed through some form of search request which may comprise components in one or more media.

Global Positioning System (GPS): A satellite navigation system in which more than two dozen satellites broadcast precise timing signals by radio, allowing any GPS receiver device to accurately determine its location.

Photoware: A software application used for personal photo management. Although the term emerged when online sharing of photos became common in personal photo management software, in this chapter we use the term in a more general sense.

Web 2.0: The second generation of Internet-based services in which the Web itself is a platform for users to directly use and share information on the Web, often characterized by its highly-dynamic and highly-interactive Web interfaces and pulling together the distributed resources from independent developers of contents.

ENDNOTES

1. Flickr. Available at http://www.flickr.com/ (Retrieved January 2007)
2. Adobe Photoshop Album. Available at http://www.adobe.com/products/photoshopalbum/starter.html (Retrieved January 2007)
3. ACDSee. Available at http://www.acdsee-guide.com/ (Retrieved January 2007)
4. Google Picasa. Available at http://picasa.google.com/ (Retrieved January 2007)
5. Flickr. Available at http://www.flickr.com/ (Retrieved January 2007)
6. Yahoo! 360. Available at http://360.yahoo.com/ (Retrieved January 2007)
7. TRECVid 2006 Guideline. Available at: http://www-nlpir.nist.gov/projects/tv2006/tv2006.html (Retrieved January 2007)
8. A "video shot" is an unbroken sequence of frames taken by a single camera. Shot boundaries occur at camera changes.
9. The Google Video Search Engine. Available at: http://video.google.com/ (Retrieved January 2007)
10. Google Maps. Available at http://maps.google.com/ (Retrieved January 2007)

Chapter VI
Understanding One-Handed Use of Mobile Devices

Amy K. Karlson
University of Maryland, USA

Benjamin B. Bederson
University of Maryland, USA

Jose L. Contreras-Vidal
University of Maryland, USA

ABSTRACT

Mobile phones are poised to be the world's most pervasive technology, already outnumbering land lines, personal computers, and even people in some counties. Unfortunately, solutions to address the usability challenges of using devices on the move have not progressed as quickly as the technology or user distribution. Our work specifically considers situations in which a mobile user may have only one hand available to operate a device. To both motivate and offer recommendations for one-handed mobile design, we have conducted three foundational studies: a field study to capture how users currently operate devices; a survey to record user preference for the number of hands used for a variety of mobile tasks, and an empirical evaluation to understand how device size, interaction location, and movement direction influence thumb agility. In this chapter we describe these studies, their results, and implications for mobile device design.

INTRODUCTION

The handheld market is growing at a tremendous rate; the technology is advancing rapidly and experts project that over one billion mobile phones will be sold in this year (2007) alone (Milanesi et al., 2007). To meet customer demand for portability and style, device manufacturers continually introduce smaller, sleeker profiles to the market. Yet advances in battery power, processing speed, and memory allow these devices to come equipped with increasing numbers of functions, features, and applications. Unfortunately these divergent trends are at direct odds with usability: richer content accessed through shrinking input and output channels simply makes devices harder to use. The unique requirements for mobile computing only compound the problem, since mobile use scenarios can involve unstable environments, eyes-free interaction, competition for users' attention, and varying hand availability (Pascoe, Ryan, & Mores, 2000). While each of these constraints requires attention in design, we are currently interested in issues of usability when a user only has only one hand available to operate a mobile device.

Devices that accommodate single-handed interaction can offer a significant benefit to users by freeing a hand for the host of physical and mental demands common to mobile activities. But there is little evidence that current devices are designed with this goal in mind. Small, light mobile phones that are easy to control with one hand are unfriendly to thumbs due to small buttons and crowded keypads. Larger devices, such as personal digital assistants (PDAs) are not only harder to manage with a single hand, they tend to feature more (rather than larger) buttons, as well as stylus-based touchscreens whose rich interface designs emphasize rich information content, but often offer targets too small, and/or too distant, for effective thumb interaction.

While it may seem obvious which features inhibit single-handed use, there has been relatively little systematic study of enabling technologies and interaction techniques. Most commercial and research efforts in one-handed device interaction have focused primarily on either a specific technology or task. For example, accelerometers have been explored to support tilt as a general input channel for handheld devices (Dong, Watters, & Duffy, 2005; Hinckley, Pierce, Sinclair et al., 2000; Rekimoto, 1996), while media control (Apple, 2006; Pirhonen, Brewster, & Holguin, 2002) and text entry (Wigdor & Balakrishnan, 2003) have been popular tasks to consider for one-handed device operation. But in the varied landscape of mobile devices and applications, one-handed design solutions must ultimately extend to a wide range of forms and functions. We began our investigation of this problem by looking at the fundamental human factors involved in operating a device with a single hand.

In this chapter, we report on three studies conducted to understand different aspects of one-handed mobile design requirements. We first ran a field study to capture the extent to which single-handed use is currently showing up "in the wild." Second, we polled users directly to record personal accounts of current and preferred device usage patterns. The results from these studies help motivate one-handed interface research, and offer insight into the devices and tasks for which one-handed techniques would be most welcomed. Finally, we performed an empirical evaluation of thumb tap speed to understand how device size, target location, and movement direction influence performance. From these results we suggest hardware-independent design guidelines for the placement of interaction objects. Together our findings offer foundational knowledge in user behavior, preference, and motor movement for future research in single-handed mobile design.

BACKGROUND

The physical and attention demands of mobile device use were reported early on for fieldworkers (Kristoffersen & Ljungberg, 1999; Pascoe et al., 2000), from which design recommendations for minimal-attention and one-handed touchscreen interface designs emerged (Pascoe et al., 2000). Though well suited to the directed tasks of fieldwork, the guidelines do not generalize to the varied and complex personal information management tasks of today's average user. Research of the effects that mobility has on attention and user performance continues (Oulasvirta, Tamminen, Roto et al., 2005), as well as how these factors can be replicated for laboratory study (Barnard, Yi, Jacko, & Sears, 2005).

Several approaches for one-handed device interaction have been proposed. Limited gestures sets have been explored for mobile application control with both the thumb (Apple, 2006; Karlson, Bederson, & SanGiovanni, 2005; Pascoe et al., 2000) and index finger (Pirhonen et al., 2002), but none have specifically considered ergonomic factors. Since text entry remains the input bottleneck for mobile devices, many are working on improvements, and some targeting one-handed use. Peripheral keyboards for one-handed text entry are available, such as the Twiddler (Lyons et al., 2004), but the mobile device itself must be supported by another hand, desk or lap, which violates our definition of one-handed device control. Text entry on phone keypads is generally performed with a single thumb, but methods to improve input efficiency have focused on reducing the number of key presses required, such as T9 word prediction, rather than by improving ergonomics by optimizing button sizes, locations, or movement trajectories. Accelerometer-augmented devices allow for the device's spatial orientation to serve as an input channel, and have been shown to support one-handed panning (Dong, Watters, & Duffy, 2005), scrolling (Rekimoto, 1996), and text entry (Wigdor & Balakrishnan, 2003). However, the coarse level of control tilt offers, and the potential for confusion with the normal movements of mobile computing necessarily limit the viability of tilt for generalized input.

Scientists in the medical community have studied the biomechanics of the thumb extensively for the purposes of both reconstruction and rehabilitation. The structure of the thumb is well understood (Barmakian, 1992), but only now are scientists beginning to reliably quantify the functional capabilities of the thumb. Strength has been the traditional parameter used to assess biomechanical capabilities, and recent research has established the effect movement direction has on thumb strength (Li & Harkness, 2004). Unfortunately, only standard anatomical planes have been considered, which excludes movements toward the palm that are typical of mobile device interaction. As a complement to force capabilities, others have looked at range as a characteristic of thumb movement. Kuo, Cooley, Kaufman et al. (2004) have developed a model for the maximal 3D workspace of the thumb and Hirotaka (2003) has quantified an average angle for thumb rotation. The experimental conditions for these studies, however, do not account for constraints imposed by holding objects of varying size, such as alternative models of handheld device.

FIELD STUDY

One motivation for our research in single-handed mobile designs was our assumption that people already use devices in this manner. Since current interaction patterns, whether by preference or necessity, are predictive of future behavior, they are likely to be transferred to new devices. This suggests that designs should become more accommodating to single-handed use, rather than less, as the tradition has been. To capture current behavior, we conducted an in situ study of user interaction with mobile devices. The study targeted an airport environment for the high potential of finding mobile device users and ease of access for unobtrusive, anonymous observation.

Field Study Method

We observed 50 travelers (27 male) at Baltimore Washington International Airport's main ticketing terminal during a six hour period during peak holiday travel. Because observation was limited to areas accessible to non-ticketed passengers, seating options were scarce. We expected to observe the use of both PDAs and cell phones since travelers are likely to be coordinating transportation, catching up on work, and using mobile devices for entertainment purposes. Since most users talk on the phone with one hand, we recorded only the cell phone interactions that included keypad interaction as well. All observations were performed anonymously without any interaction with the observed.

Note that while any subject observation without consent presents a legitimate question for ethical debate, in our research we follow the federal policy on the protection of human research subjects (Department of Health and Human Services, 2005) as a guideline. The policy states that the observation of public behavior is not regulated if the anonymity of the subjects is maintained and that disclosure of the observations would not put the subjects at risk in terms of civil liability, financial standing, employability, or reputation. Since we were interested in capturing natural behavior, did not record identifying characteristics, and consider phone use while standing, walking and sitting relatively safe activities, we did not obtain subject consent.

Field Study Measures

For each user observed, we recorded sex, approximate age, and device type used: candy bar phone, flip phone, Blackberry, or PDA. A "candy bar" phone is the industry term for a traditional-style cellular phone with a rigid rectangular form, typically about 3 times longer than wide. For phone use, we recorded the hand(s) used to dial (left, right or both) and the hand(s) used to speak (left, right or both). We also noted whether users were carrying additional items, and their current activity (selected from the mutually exclusive categories: walking, standing, or sitting).

Field Study Results

Only two users were observed operating devices other than mobile phones—one used a PDA and the other a Blackberry. Both were seated and using two hands. The remainder of the discussion focuses on the 48 phone users (62.5 percent flip, 37.5 percent candy bar). Overall, 74 percent used one hand for keypad interaction. By activity, 65 percent of one handed users had a hand occupied, 54 percent were walking, 35 percent were standing, and 11 percent were sitting. Figure 1 presents the distribution of subjects who used one vs. two hands for keypad interaction, categorized by the activity they were engaged in (walking, standing, or sitting). The distribution of users engaged in the three activities reflects the airport scenario where many more people were walking or standing than sitting. It is plain from Figure 1 that the relative proportion of one handed to two handed phone users varied by activity; the vast majority of walkers used one hand, about two-thirds of standers used one hand, but seated participants tended to use two hands. However, we also recorded whether one hand was occupied during the activity, and found walkers were more likely to have one hand occupied (60 percent), followed by standers (50 percent), and finally sitters (25 percent), which may be the true reason walkers were more likely than standers to use one hand, as well as why standers were more likely than sitters to use one hand. Regardless of activity, when both hands were available for use, the percentage of one vs. two handed phone users was equal.

Analysis of Field Study

Although Figure 1 suggests a relationship between user activity and keypad interaction behavior, it is unclear whether activity influences the number of hands used, or vice versa. Furthermore, since the percentage of users with one hand occupied correlates with the distribution of one-handed use across activities, hand availability, rather than preference, may be the more influential factor

Figure 1. Airport field study: Number of hands used for keypad interaction by activity.

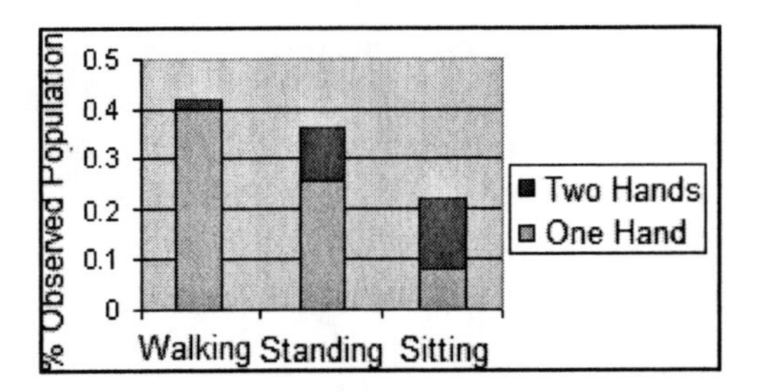

in the number of hands used to interact with the keypad. While use scenario certainly impacts usage patterns, the fact that users were as likely to use one hand as two hands when both hands were available suggests that preference, habit and personal comfort also play a role. Regardless of scenario, we can safely conclude that one-handed phone use is quite common, and thus is an essential consideration in mobile phone design.

Generalizability. The choice of observation location may have biased our results from those found in the general population since travelers may be more likely to be: (1) carrying additional items; (2) standing or walking; and (3) using a phone vs. PDA. Different environments, information domains, populations, and scenarios will yield unique usage patterns. Our goal was not to catalogue each possible combination, but to learn what we could from a typical in-transit scenario.

WEB SURVEY

While informative for a preliminary exploration, shortcomings of the field study were (a) a lack of knowledge about motivation for usage style; (b) the limited types of devices observed (phones); and (c) the limited tasks types observed (assumed dialing). To broaden our understanding of device use over these dimensions, we designed a Web survey to capture user perceptions of, preferences for and motivations surrounding their own device usage patterns.

Survey Method

The survey consisted of 18 questions presented on a single Web page which was accessed via an encrypted connection (SSL) from a computer science department server. An introductory message informed potential participants of the goals of the survey and assured anonymity. Notification that results would be posted for public access after the survey period was over provided the only incentive for participation. Participants were solicited from a voluntary subscription mailing list about the activities of our laboratory. In addition the solicitation was propagated to one recipient's personal mailing list, a medical informatics mailing list, and a link to the survey was posted on two undergraduate CS course Web pages.

Survey Measures

For each participant, we collected age, sex and occupation demographics. Users recorded all styles of phones and/or PDAs owned, but were asked to complete the survey with only one device in mind—the one used for the majority of information management tasks. We collected general information about the primary device, including usage frequency, input hardware, and method of text entry. We then asked a variety of questions to understand when and why people use one vs. two hands to operate a device. We asked users to record the number of hands used (one and/or two) for 18 typical mobile tasks, and then to specify the number of hands (one *or* two) they would *prefer* to use for each task. Three pairs of activities were designed to distinguish between usage patterns for different tasks within the same application, which we differentiated as "read" (e-mail reading, calendar lookup, and contact lookup) vs. "write" (e-mail writing, calendar entry, and contact entry) tasks. Users then recorded the number of hands used for the majority of device interaction and under what circumstances they chose one option over the other. Finally, users were asked how many hands they would prefer to use for the majority of interactions (including no preference), and were also asked to record additional comments.

Survey Results

Two hundred twenty-nine participants (135 male) responded to the survey solicitation. One male participant was eliminated from the remaining analysis because his handheld device was specialized for audio play only, leaving 228. The median participant age was 38.5 years. Participant occupations reflected the channels for solicitation, with 25 percent in CS, IT or engineering, 23 percent students of unstated discipline, 20 percent in the medical field, 10 percent in education, and the remainder (22 percent) from other professional disciplines.

Devices owned. The three most common devices owned were flip phones (52 percent), small candy bar phones (23 percent) and Palm devices without a Qwerty keyboard (20 percent). Palm devices with an integrated Qwerty keyboard were as common as Pocket PCs without a keyboard (14 percent). Since interaction behavior may depend on device input capabilities, we reclassified each user's primary device into one of four general categories based on the device's input channels: (1) *keypad-only (51 percent)* are devices with a 12-key numeric keypad but no touchscreen, (2) *TS-no-qwerty (23 percent)* are devices with a touchscreen but no Qwerty keyboard, (3) *TS-with-qwerty (21 percent)* are devices with a touchscreen as well as an integrated Qwerty keyboard, and finally (4) q*werty-only (5 percent)* are devices with an integrated Qwerty keyboard but no touchscreen.

Figure 2. Web survey: Number of hands (a) currently used and (b) preferred (1 hand is shown as solid, 2 hands is shown as striped) for 18 mobile tasks as a percentage of the observed population. Hand usage for each task is broken down by device type (TS = touchscreen)

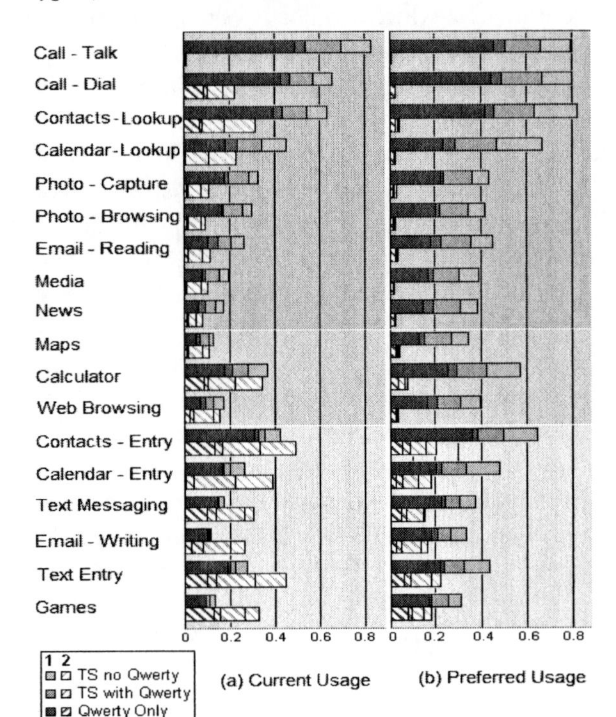

For users with multiple devices, we derived their primary device type from the text entry method reported.

Current usage patterns. Of the 18 activities users typically perform with devices, 9 were performed more often with one hand, 6 more often with two hands, and 3 were performed nearly as often with one vs. two hands. Figure 2a displays these results, with the shaded backgrounds grouping the activities by those used with one, either or two hands. Upon inspection, all of the "reading" activities were performed more often with one hand (top) and all "writing" activities with two hands (bottom). Considering users' device types, we notice that with the exception of gaming, owners of *keypad-only* devices were more likely to use one hand regardless of activity, owners of *TS-no-qwerty* were more likely to use two-hands for most activities, and those owning Qwerty based devices were more likely to use two hands when performing writing tasks, but not reading tasks.

Overall, 45 percent of participants stated they use one hand for nearly all device interactions, as opposed to only 19 percent who responded similarly for two hands. Considering device ownership, however, users of touchscreen-based devices were more likely to use two hands "always" than they were one hand (Figure 3). When participants use one hand, the majority (61 percent) perceive they do so whenever the interface supports it, the reason cited by only 10 percent of those who use two hands. Device form dictated usage behavior when the device was too small for two hands, too large for one hand, or when large devices could be supported by a surface and used with one hand. Participants cited task type as a reason for hand choice, primarily as a trade off between efficiency and resources usage: 14 percent of users selected one hand only for simple tasks (conserving resources), while 5 percent selected two hands for entering text, gaming, or otherwise for improving the speed of interaction (favoring efficiency). Finally, according to respondents, the majority of two-handed use occurs when it is the only way to accomplish the task given the interface (63 percent).

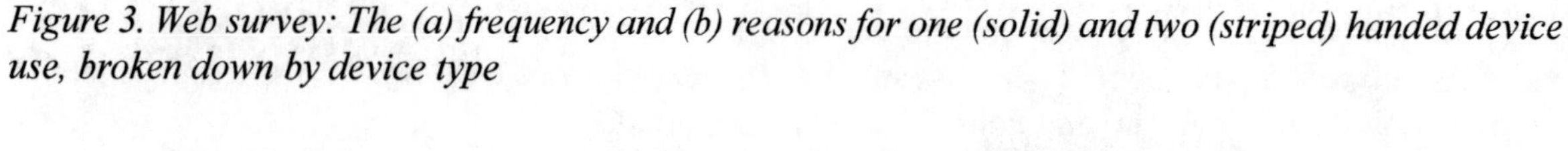

Figure 3. Web survey: The (a) frequency and (b) reasons for one (solid) and two (striped) handed device use, broken down by device type

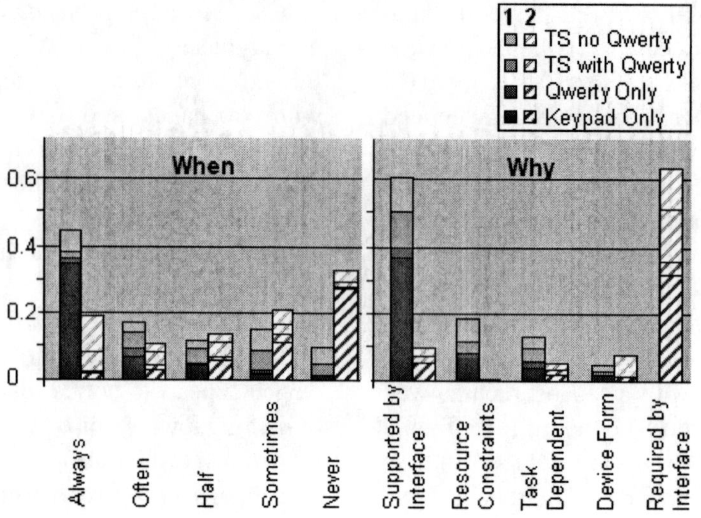

Hand preferences. When asked how many hands users preferred to use while performing the same 18 tasks, one hand was preferred overwhelmingly to two hands for all tasks (Figure 2b). The activities with the closest margin between the number of participants who preferred one vs. two hands were playing games (13 percent) and composing e-mail (16 percent). With one exception (gaming), the activities for which more than 14 percent of users stated a preference for two hands were "writing" tasks (e.g., those that required text entry): text entry, contact entry, calendar entry, e-mail writing, and text messaging, in decreasing order. Even so, except for users of *TS-with-qwerty* devices, the majority of users stated a preference for using one hand, regardless of task or device owned. Users of *TS-with-qwerty* devices preferred two hands for text messaging, email composition, and text entry. Based on these data, it is consistent that 66 percent of participants stated they would prefer to use one hand for the majority of device interaction, versus nine percent who would prefer two hands for all interaction. Twenty-three percent did not have a preference and six users did not respond.

Survey Summary

Considering current usage patterns only, there is no obvious winner between one and two handed device use. Excluding phone calls, the number of activities for which a majority of respondents use one (seven) vs. two hands (six) is nearly balanced. However, device type certainly influences user behavior; users of keypad-only devices nearly always use one hand, while users of touchscreen devices more often favor two hands, especially for tasks involving text entry. But user justifications for hand choice indicate that the hardware/software interface is to blame for much two-handed use occurring today. Most use one hand if at all possible and only use two hands when the interface makes a task impossible to do otherwise. Other than gaming, tasks involving text entry are the only ones for which users may be willing to use two hands, especially when the device used provides an integrated Qwerty keyboard. It seems, therefore, that the efficiency gained by using two hands for such tasks is often worth the dedication of physical resources, which is also true of the immersive gaming experience.

While most users can imagine the ideal of single-handed text entry, enabling single-handed input may not be enough—*throughput* is also important. Ultimately, it is clear that interface designers of all device types should make one-handed usability a priority, and strive to bridge the gap between current and desired usage patterns.

THUMB MOVEMENT STUDY

The third component of our exploration was an examination of thumb movement in the context of mobile device interaction. As input technologies and device forms come and go, biomechanical limitations of the thumb will remain. Although the thumb is a highly versatile appendage with an impressive range of motion, it is most adapted for grasping tasks, playing opposite the other four fingers (Bourbonnais, Forget, Carrier et al., 1993). Hence thumb interaction on the surface of today's mobile devices introduces novel movement and exertion requirements for the thumb—repetitive pressing tasks issued on a plane parallel to the palm. We believe a fundamental understanding of thumb capabilities when holding a device can help guide the placement of interaction targets for both hardware and software interfaces designed for one-handed use. Although we can make reasonable guesses about thumb capabilities, empirical evidence is a better guide. Since no strictly relevant studies have yet been conducted, we developed a study to help us understand how device form and task influences thumb mobility.

Since thumb tapping is the predominant means of interaction for keypad-based devices, and has also proven promising for one-handed touchscreen use (Karlson et al., 2005), we focused our investigation on surface tapping tasks. We hypothesized that the difficulty of a tapping task would depend on device size, movement direction, and surface location of the interaction. We captured the impact of these factors on user performance by using movement speed as a proxy for task difficulty, under the assumption that harder tasks would be performed more slowly than easier tasks.

Equipment

Device models. For real devices, design elements such as buttons and screens communicate to the user the "valid" input areas of the device. We instead wanted outcomes of task performance to *suggest* appropriate surface areas for thumb interaction. We identified four common handheld devices to represent the range of sizes and shapes found in the market today: (1) a Siemens S56 candy bar phone measuring 4.0 x 1.7 x 0.6 in (10.2 x 4.3 x 1.5 cm); (2) a Samsung SCH-i600 flip phone measuring 3.5 x 2.1 x 0.9 in (9 x 5.4 x 2.3 cm); (3) an iMate smartphone measuring 4 x 2.0 x 0.9 in (10.2 x 5.1 x 2.3 cm) and (4) an HP iPAQ h4155 Pocket PC measuring 4.5 x 2.8 x 0.5 in (11.4 x 7.1 x 1.3 cm). These devices are shown in the top row of Figure 4. We refer to these as simply SMALL, FLIP, LARGE, and PDA. To remove the bias inherent in existing devices, we created a 3D model of each device, removing all superficial design features. The models were developed using Z Corp.'s (http://www.zcorp.com/) ZPrinter 310 3D rapid prototyping system. Device models were hollow, but we reintroduced weight to provide a realistic feel. Once "printed"

Figure 4. Thumb movement study: Devices we chose to represent a range of sizes and forms (top row) together with their study-ready models (bottom row): (a) SMALL, (b) FLIP, (c) LARGE, and (d) PDA.

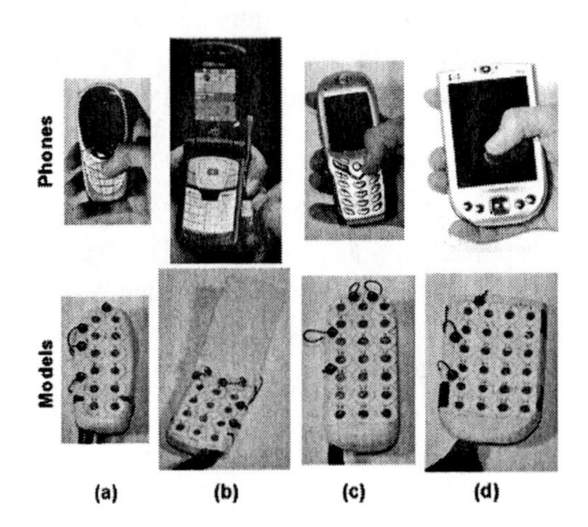

and cured, the models were sanded and sealed to achieve a smooth finish.

Target design. A grid of circular targets 1.5 cm in diameter was affixed to the surface of each device. Circles were used for targets so that the sizes would not vary with direction of movement (MacKenzie & Buxton, 1992). The target size was selected to be large enough for the average-sized thumb, while also providing adequate surface coverage for each device. The grid dimensions for each device were: SMALL (2x5), FLIP (3x4), LARGE (3x7) and PDA (4x6), as shown in the bottom row of Figure 4.

Measurement. A typical measurement strategy for tapping tasks would involve a surface-based sensor to detect finger contact. Unfortunately, due to the number and variety of device sizes investigated, no technical solution was found to be as versatile, accurate or affordable as required. Instead we used Northern Digital Inc's OPTOTRAK 3020 motion analysis system designed for fine-grained tracking of motor movement. The OPTOTRAK uses 3 cameras to determine the precise 3D coordinates of infrared emitting diodes (IREDs). Three planar IREDs attached to the surface of each device defined a local coordinate system, and a fourth IRED provided redundancy (see Figure 4, bottom row). The spatial positions of two markers affixed to each participant's right thumb were then translated with respect to the coordinate system of the device to establish relative movement trajectories. Diode positions were sampled at 100Hz, and data were post processed to derive taps from thumb minima.

Software. Data collection and experiment software was run on a Gateway 2000 Pentium II with 256 MB of RAM running Windows 98.

Participants

Twenty participants were recruited via fliers posted in our department of computer science, with the only restriction that participants be right-handed. Participants (15 male) ranged in age from 18 to 35 years with a median age of 25 years. Participants received $20 for their time.

Design

For each target on each device (SMALL, FLIP, LARGE, and PDA), users performed all combinations of *distance* (1 or 2 circles) x *direction* (↕, ↔, ↙↗ , ↖↘) tasks that could be supported by the geometry of the device. For example, SMALL could not accommodate trials of distance 2 circles in the directions (↔, ↙↗, ↖↘). Note that the grid layout results in *actual* distances that differ between *orthogonal* trials (↕, ↔) and *diagonal* trials (↙↗, ↖↘), which we consider explicitly in our analysis. For LARGE and PDA, trials of distance 4 circles were included as the geometry permitted. Finally each device included a ↙↗ and ↖↘ trial to opposite corners of the target grid. For each device, a small number of trials (1 for SMALL, LARGE and PDA, 3 for FLIP), selected at random, were repeated so as to make the total trial count divisible by four. The resulting number of trials for each device were: SMALL (32), FLIP (48), LARGE (108), and PDA (128). Since the larger devices had more surface targets to test, they required more trials.

Tasks

Users performed reciprocal tapping tasks in blocks as follows. For SMALL and FLIP, trials were divided equally into two blocks. For the LARGE and PDA, trials were divided equally into four blocks. Trials were assigned to blocks to achieve roughly equal numbers of *distance* x *direction* trials, distributed evenly over the device. Trials were announced by audio recording so that users could focus attention fully on the device. Users were presented with the name of two targets by number. For example, a voice recording would say "1 and 3." After one second, a voice-recorded "start" was played. Users tapped as quickly as possible between the two targets, and after five seconds, a "stop" was played. After a 1.5 second delay the next trial began. Trials continued in succession to the end of the block, at which point the user was allowed to rest as desired, with no user resting more than two minutes. Device and block orders were assigned to subjects using a Latin Square, but the presentation of within-block trials was randomized for each user.

Procedure

Each session began with a brief description of the tasks to be performed and the equipment involved. Two IRED markers were then attached to the right thumb with two-sided tape. One diode was placed on the leftmost edge of the thumb nail, and a second on the left side of the thumb. The orthogonal placement was intended to maximize visibility of at least one of the diodes to the cameras at all times. The two marker wires were tethered loosely to the participant's right wrist with medical tape.

The participant was seated in an armless chair, with the device held in the right hand, and the OPTOTRAK cameras positioned over the right shoulder. At this point the participant was given more detailed instruction about the tasks, and informed of the error conditions that might occur during the study: if at any point fewer than three of the device-affixed IREDs or none of the thumb IREDs were visible to the cameras, an out-of-sight error sound would be emitted, at which point he or she should continue the trial as naturally as possible while attempting to make adjustments to improve diode visibility. Next, the participant was given the first device and performed a practice session of 24 trials, selected to represent a variety of distances, directions, and surface locations. During the practice trials, the administrator intentionally occluded the diodes to give the participant familiarity with the out-of-sight error sound and proper remedies. After completion of the practice trials and indication that the participant was ready, the study proper was begun.

During trials, participants were allowed to hold the devices in whatever manner supported their best performance. Since the instructions were presented audibly and with a short pause before the trial began, users could prepare their grip if desired. We chose not to control for grip in our study under the reasoning that it resembled real world settings, in which users have the freedom to adjust their grips to best suit the environment and task.

After all trials for a device were completed, users were allowed to rest while the next device was readied, typically three to five minutes; together with the rest period users were offered between trial blocks and counterbalanced device order, we hoped to minimize as much as possible the impact of fatigue on the results. After completing all trials for the last device, the participant completed a questionnaire, recording demographics and subjective ratings. Total session time was two hours, approximately an hour of which was devoted to data collection.

Measures

Raw 3D thumb movement data for each five second trial were truncated to the middle three seconds to eliminate artifacts resulting from initiation lag and anticipated trial completion, phenomena routinely observed by the administrator. In a post processing phase, taps were identified within the remaining three second interval and a single average tap time was computed from the difference in time between the onset of the first tap to the onset of the last tap, divided by one fewer than the total number of taps detected. In a post experiment questionnaire, participants assigned an overall rating of difficulty to each device (1-7, where 1 = easy, 7 = difficult), and indicated the device regions that were both easiest and hardest to interact with.

Data post processing. Since the 3D thumb position (x,y,z) was recorded relative to the device surface, the z-value represented the thumb height above the device. While one might assume that taps were those thumb positions for which the z-distance was 0, the IREDs were mounted on participants' thumbnails, and so never actually

Figure 5. Example MATLAB output of the thumb's distance from the surface of the device. Stars depict the auto-detected peaks and valleys.

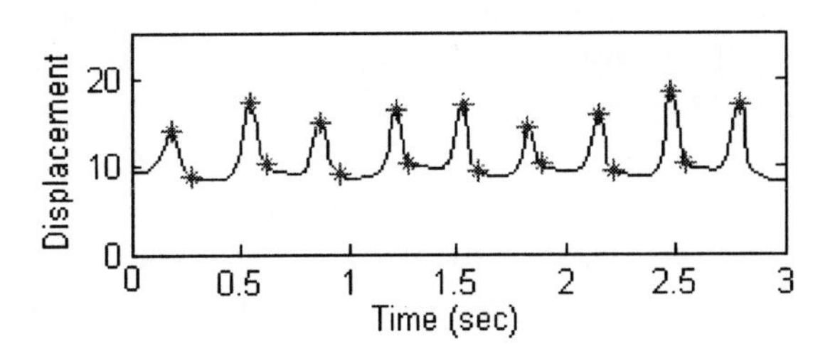

reached the surface of the device. Taps were instead defined as points when both the *z*-value and change in *z*-value (velocity) were minimal. For example, plotting *z*-values over time reveals a wave pattern whose valleys indicate taps (Figure 5).

Raw data was first preprocessed to extract the middle three seconds of each trial as well as to select the thumb diode with the most complete data set (e.g., the fewest number of missing frames, or if equal, the one with the most compact windows of missing frames). Linear interpolation was performed on missing frames if the gap was less than 100 ms. Missing frames included those lost due to out-of-sight errors, as well as occasional frames dropped by the collection hardware.

The data was then analyzed by the PICKEXTR MATLAB function to identify extrema in a signal. This function is provided with the RelPhase.Box Matlab toolbox for relative phase analysis of oscillatory systems (Dijkstra, Giese, & Schöner, 1997). The accuracy of the tap classifier was verified by inspecting a visual representation (Figure 5) of each trial. When required, corrections were made as follows: (1) valid endpoints were preserved, (2) if intermediate taps were missing, they were added, (3) if intermediate taps were incorrect, they were recoded by hand, and (4) if endpoints were invalid, the entire signal was coded by hand. Since average tap time was calculated as the number, not placement, of intervening taps, this method minimized as much as possible the bias of human annotation. Of the trials included for statistical analysis, 1.3 percent were discarded because they could not be encoded by machine or human, or had less than 1.5 seconds of encodable signal.

Results

The goal of our analysis was to understand whether user performance was influenced by device size, interaction location, and movement direction. To allow for comparison among the devices, we limited the analysis to trials with distances of 1 or 2 circles since the geometries of all but the smallest device (SMALL) supported these trials in all four movement directions. To address the fact that actual movement distance differed between orthogonal and diagonal trials, we analyzed these groups separately. For all analyses, Huynh-Felt corrections were used when the sphericity assumption was violated, and Bonferroni corrections were used for post hoc comparisons.

Movement Direction. A 2 (*distance*) x 2 (*direction*) repeated measures analysis of variance (RM-ANOVA) was performed on mean task time data for both orthogonal trials (distances: 1, 2; directions: ↕, ↔) and diagonal trials (distances: 1.4, 2.8; directions: ⤢, ⤡) for the three largest devices. Since SMALL did not support distance 2 trials in all four directions, a one-way RM-ANOVA was performed on mean task time for trials of distance 1 and 1.4.

SMALL: A main effect of direction was observed for diagonal trials ($F(1,19) = 65.1$, $p < .001$). Post hoc analyses showed that trials in the ⤢ direction were performed significantly faster than those in the ⤡ direction (0.26 v. 0.28 ms, $p < .001$).

FLIP, LARGE, and PDA: Results were similar across the analyses of the three largest devices. Unsurprisingly, a main effect of distance was observed for both orthogonal and diagonal trials, with shorter trials significantly faster than longer trials. There were no further effects of direction or interaction between direction and distance for orthogonal trials (↕, ↔). However, for diagonal trials, a main effect of direction was observed, with trials in the ⤢ direction significantly faster than those in the in ⤡ direction for all devices. In addition, a *distance* x *direction* interaction showed performance differences between the diagonal trials were more pronounced for longer trials than shorter trials (Table 1).

Device Size

To determine if device size impacted comparable tasks across devices, we analyzed all trials performed in the lower right 3x4 region of the three largest devices using a 3 (*devices*) x 43 (*trials*) RM-ANOVA. While a main effect of trial was observed, this was expected, as trials of every distance and direction were included for analysis. Yet no effects of *device* or *device* x *trial* were found.

Table 1. Mean time for movement direction and distance x direction for FLIP, LARGE, and PDA

	Direction (⤢ v. ⤡)	$F_{1,19}$	p	Dist. x Dir. (⤢ v. ⤡)	$F_{1,19}$	p
FLIP	.31 v. .35 ms	50.5	<.001	(1.4) .27 v. .30 ms (2.8) .34 v. .41 ms	14.6 34.5	<.001 <.001
LARGE	.31 v. .36 ms	46.1	<.001	(1.4) .27 v. .28 ms (2.8) .35 v. .42 ms	28.0 44.2	<.001 <.001
PDA	.32 v. .36 ms	46.5	<.001	(1.4) .28 v. .30 ms (2.8) .36 v. .43 ms	23.0 38.0	<.001 <.001

Target Location

To determine if target location affected performance, we analyzed task time for the shortest tasks for each device. We chose short tasks because they provide high granularity for discriminating among device locations. Since direction was shown to affect task time for diagonal trials, only orthogonal tasks could be considered. For each device, a one-way RM-ANOVA was performed on mean trial time, with the number of trials varying by device.

A main effect of target location was observed for SMALL ($F_{(8.6, 163.3)} = 2.1$, $p = .032$), FLIP ($F_{(11.5, 218.4)} = 3.5$, $p < .001$) and PDA ($F_{(9.8, 188.1)} = 3.9$, $p < .001$), but not for LARGE. However, in post hoc analyses, only PDA had a reasonable number of trials that differed significantly from one another. Since it is difficult to draw helpful conclusions from specific pairs of trials, we explored two aggregation techniques.

Subject-derived regions. Based on subjective opinion of which regions were easiest to reach for each device, we divided tasks into three groups (E)asy, (M)edium, and (H)ard. Tasks for SMALL and FLIP were assigned to only E and M groups. A one-way RM-ANOVA on mean group task time was performed for each device. A main effect of group was found for FLIP ($F_{(5.5, 105.1)} = 11.3$, $p < .001$), LARGE ($F_{(3.1, 58.7)} = 8.4$, $p < .001$), and PDA ($F_{(4.8, 91.0)} = 22.0$, $p < .001$). Post hoc analyses showed all groups differed significantly from each other for FLIP and PDA. For LARGE, E and M were significantly faster than H, but were indistinguishable otherwise, so we collapsed them to E (Table 2, third column).

Data-derived regions. For each device we ordered tasks by mean tap time, and then segmented them into seven groups. If the number of trials was not divisible by seven, the remainder trials were included in the middle group. A one-way RM-ANOVA on mean group task time was performed for each device. A main effect of group was found for FLIP ($F_{(5.5, 105.1)} = 11.3$, $p < .001$), LARGE ($F_{(3.1, 58.7)} = 8.4$, $p < .001$), and PDA ($F_{(4.8, 91.0)} = 22.0$, $p < .001$). From these results, groups were labeled *fastest* and *slowest* such that all groups in *fastest* were significantly faster than all groups in *slowest*, according to post hoc analyses. Trials in these groups are shown visually in the rightmost column of Table 2. Mean task time for *fastest* v. *slowest* trials for each device were FLIP (0.26 v. 0.28 ms), LARGE (0.25 v. 0.28 ms), and PDA (0.26 v. 0.29 ms).

Subjective Preferences

After completing all trials, users were presented with diagrams of each device similar to those in the first two columns of Table 2 and asked to identify the targets they found most easy and most difficult to interact with. Aggregating results across users yielded a preference "map" for the least and most accessible targets of each device (columns 1 and 2 of Table 2), with darker regions indicating more agreement among participants. We see that for each device the two representations are roughly inverses of one another.

In addition to region marking, we asked users to rate the overall difficulty of managing each device with one hand on a 7-point scale (7 = most

Table 2. Preference and movement time maps for each device. Depth of color in columns 1 and 2 indicate stronger user agreement.

	Subjective Difficult Regions	Subjective Easy Regions	Preference-Derived Regions Easy / Medium / Hard	Data-Derived Regions Fastest / Slowest
SMALL	8 5 1 9 10	10 11 15 13 16 15 13 11 4 4		
FLIP	8 3 1 1 2 5 4 10	11 12 10 12 16 14 14 16 10 7 8 4		
LARGE	12 6 5 5 2 1 3 3 4 1 2 6 5 7 13 12 13	4 5 4 7 10 9 14 15 12 16 16 13 13 13 9 6 7 7 1 2 1		
PDA	13 8 7 9 10 4 2 4 9 3 1 4 9 3 1 5 11 4 2 8 14 9 9 14	2 5 5 3 3 11 11 7 4 15 16 8 6 15 16 5 4 11 8 4 1 2 2		

comfortable). Average ratings from most to least comfortable were as follows: SMALL (6.4), FLIP (5.4), LARGE (4.1) and PDA (3.0).

Thumb Movement Summary

The findings from our analysis of thumb movement suggest the following guidelines. First, thumb movement in the ↖↘ direction is difficult for right-handed users regardless of device size. Presumably the difficulty arises from the considerable flexion required to perform these types of tasks. Under this reasoning, the opposite movement ↙↗ would be difficult for left-handed users, so conservative designs should constrain repetitive movement to ↕ and ↔ directions to accomodate all users, and especially for repetitive tasks such as text or data entry.

Second, device region affects both task performance and perceived difficulty. Not only did the slowest trials correspond to those regions users found most difficult, but fastest trials also matched those regions users found most easy (Table 2). In general, regions within reach of the thumb were fastest and most comfortable, favoring those toward the midline of the device—a "sweet spot" that required movement primarily from the base of the thumb. The lower right corners of the devices present an exception in that they are biomechanically awkward to reach because they are "too close" rather than "too far."

Because the absolute time differences between the fastest and slowest regions of the devices were quite small (at most 30 ms), we do not think performance speed is the main concern in forming design recommendations from these data. Rather, it is the fact that the speed differences between the regions were statistically significant (7%-12% slowdown between the fastest and slowest regions) that suggests a mechanical and/or physical encumbrance was to blame. The data, therefore, are concerning primarily from an ergonomics perspective. In fact, we believe that the slowdowns we found should be thought of as optimistic, since they capture only localized movement and required substantial changes in user grip between tasks; subjective opinion, user observations and practical experience indicate that designers should be cautioned against using the entire device surface for thumb interaction, especially for larger devices. We instead recommend placing interaction objects centrally to accommodate both left and right handed users, or offering configurable displays. Since hand size and thumb length will differ by individual, designs should strive to support a range of users.

Finally, the result that users performed trials in the lower right 3x4 sub-grid of the three largest devices equally well suggests that large devices do not inherently impede thumb movement. Rather, larger devices simply have more areas that are out of thumb reach, and so have more regions that are inappropriate for object placement in one-handed designs. Together with user opinion that larger devices were more difficult to manage suggests that the current trend toward smaller device forms benefits one handed device control.

FUTURE TRENDS

Market forces are driving the sizes of mobile devices down and at the same time driving feature sets up. These concurrent trends *alone* compromise the ease and expressiveness of information interaction and presentation, as shrinking keypads and screens are being used to access ever increasing data sets. Yet at the same time, mobile users often are carrying personal effects, opening doors, holding handrails, or otherwise needing the use of a hand to manage the environment, which then leaves only a single hand available for device operation. Indeed, results from our field study and survey of mobile device use have confirmed that one-handed device operation is widespread—not limited to a niche user segment—and that users would prefer to use one hand more often than current designs allow.

Commercial trends for supporting single handed device operation have focused on thumb operation of touchscreen based devices. Several thumb-based virtual keypads are available from third party vendors, such as the Phraze-It® keypad from Prevalent Devices, www.prevalentdevices.com. One of the few handheld *systems* dedicated to single-handed operation is the touchscreen-based N1m phone by Neonode, www.neonode.com, which supports application navigation and interaction using only thumb taps and sweeps. However, the N1m's primary use is as a phone, camera and media player, rather than a personal data manager. As such, the N1m is not designed to support rich graphical interfaces or data interactions.

As the mobile user base expands, so do device storage capacities and wireless services. Not only are mobile devices accumulating more resident data, but they are increasingly used as front-end interfaces to external data sets. Given the broad range of data sets and tasks users expect of today's devices, single-handed device support will need to generalize beyond specific applications and technologies seen today. Our work seeks to understand some of the basic human factors involved in single handed device use, in order that this knowledge can be applied generally to the variety of tasks and device forms users demand for portable computing.

CONCLUSION

In an effort to understand the one-handed interaction needs of mobile device users, we looked at a broad range of device use. Our field study showed that for at least one class of user (travelers), mobile phones are most often used with one hand, and that this behavior seems to correlate with activity, such as walking or holding items in the other hand. Our survey revealed that the vast majority of users want to use one hand for interacting with mobile devices, but that current interfaces, especially for touchscreens, are not designed to support dedicated single handed use. Finally, an empirical evaluation of thumb interaction on varying-sized devices suggests that (1) mid-device regions are easiest to access; (2) the position of a target with respect to the thumb impacts performance more than device size, and finally (3) ↖↘ movement is difficult for right-handed users and degrades with movement distance.

REFERENCES

Apple. (2006). *iPod.* Retrieved January 16, 2007, from http://www.apple.com/ipod/ipod.html.

Barmakian, J. T. (1992). Anatomy of the joints of the thumb. *Hand Clinics, 8*(4), 681-691.

Barnard, L., Yi, J. S., Jacko, J. A., & Sears, A. (2005). An empirical comparison of use-in-motion evaluation scenarios for mobile computing. *International Journal of Human-Computer Studies, 62*, 487-520.

Bourbonnais, D., Forget, R., Carrier, L., & Lepage, Y. (1993). Multidirectional analysis of maximal voluntary contractions of the thumb. *Journal of Hand Therapy, 6*(4), 313-318.

Department of Health and Human Services. (2005). 45 CFR Part 46, Protection of Human Subjects. In (Section 46.101(b)). Retrieved January 16, 2007 from http://www.hhs.gov/ohrp/humansubjects/guidance/45cfr46.htm

Dijkstra, T. M., Giese, M. A., & Schöner, G. (1997). RelPhase.box: A Matlab toolbox for relative phase analysis of oscillatory systems. Retrieved January 17, 2007, from http://www.psy.ohio-state.edu/visionlab/dijkstra/Software/RelPhase.sit

Dong, L., Watters, C., & Duffy, J. (2005). Comparing two one-handed access methods on a PDA. *International Conference on Human Computer Interaction with Mobile Devices and Services,* (pp. 235-238). ACM Press.

Hinckley, K., Pierce, J., Sinclair, M., & Horvitz, E. (2000). Sensing techniques for mobile interaction. *ACM Symposium on User Interface Software and Technology*, (pp. 91-100). ACM Press.

Hirotaka, N. (2003). Reassessing current cell phone designs: using thumb input effectively. *Extended Abstracts of the ACM Conference on Human Factors in Computing,* (pp. 938-939). ACM Press.

Karlson, A. K., Bederson, B. B., & SanGiovanni, J. (2005). AppLens and LaunchTile: two designs for one-handed thumb use on small devices. *ACM Conference on Human Factors in Computing*, (pp. 201-210). ACM Press.

Kristoffersen, S., & Ljungberg, F. (1999). Making place to make it work: empirical exploration of HCI for mobile CSCW. *ACM Conference on Supporting Group Work,* (pp. 276-285). ACM Press.

Kuo, L., Cooley, W., Kaufman, K., Su, F., & An, K. (2004). A kinematic method to calculate the workspace of the TMC joint. In *Proceedings of the Institute of Mechanical Engineering H, 218*(2), 143-149.

Li, Z.-M., & Harkness, D. A. (2004). Circumferential force production of the thumb. *Medical Engineering and Physics, 26*, 663-670.

Lyons, K., Starner, T., Plaisted, D., Fusia, J., Lyons, A., Drew, A. et al. (2004). Twiddler typing: one-handed chording text entry for mobile phones. *ACM Conference on Human Factors in Computing,* (pp. 671-678). ACM Press.

MacKenzie, I. S., & Buxton, W. (1992). Extending Fitts' Law to two-dimensional tasks. *ACM*

Conference on Human Factors in Computing, (pp. 219-226). ACM Press.

Milanesi, C., Liang, A., Vergne, H.J.D.L., Mitsuyama, N., & Nguyen, T.H. (2007). *Data insight: Market share for mobile devices, 1*(7). Gartner, Inc.

Oulasvirta, A., Tamminen, S., Roto, V., & Kuorelahti, J. (2005). Interaction in 4-second bursts: the fragmented nature of attentional resources in mobile HCI. *ACM Conference on Human Factors in Computing,* (pp. 919-928). ACM Press.

Pascoe, J., Ryan, N., & Mores, D. (2000). Using while moving: HCI issues in fieldwork environments. *Transactions on Computer-Human Interaction, 7*(3), 417-437.

Pirhonen, P., Brewster, S. A., & Holguin, C. (2002). Gestural and audio metaphors as a means of control in mobile devices. *ACM Conference on Human Factors in Computing,* (pp. 291-298). ACM Press.

Rekimoto, J. (1996). Tilting operations for small screen interfaces. *ACM Symposium on User Interface Software and Technology,* (pp. 167-168). ACM Press.

Wigdor, D., & Balakrishnan, R. (2003). TiltText: using tilt for text input to mobile phones. *ACM Symposium on User Interface Software and Technology,* (pp. 81-90). ACM Press.

KEY TERMS

Biomechanics: The study of muscular mechanics and activity.

Candybar Phone: A traditional-style mobile phone with a rigid rectangular form, typically about 3 times longer than wide.

Ergonomics: Often used synonymously with "human factors," ergonomics is the scientific discipline concerned with the understanding of interactions among humans and other elements of a system, and the profession that applies theory, principles, data and methods to design in order to optimize human well-being and overall system performance (definition adopted by the International Ergonomics Association in August 2000).

Flip Phone: A popular form factor for a mobile phone that structures the phone into two halves that are hinged like a clamshell: the numeric keypad is placed on the lower half, and the display is placed on the upper half. In the closed position the screen and the keypad are protected from inadvertent damage or activation, while in the open position users can angle the screen for optimal visibility.

Mobile or Cellular Phone: A portable electronic device which, at a minimum, supports long-range telecommunications. Today's mobile phones support a much broader range of functions for personal data and data management activities, including data exchange with a personal desktop computer, music play, and photo capturing.

PDA: Personal digital assistant. The original handheld device designed expressly for supporting personal data management on the go. PDAs typically have touch sensitive screens and are operated with a stylus. Recent models may also include a miniaturized Qwerty keyboard for text entry.

Chapter VII
User Acceptance of Mobile Services

Eija Kaasinen
VTT Technical Research Centre of Finland, Finland

ABSTRACT

Personal mobile devices are increasingly being used as platforms for interactive services. User acceptance of mobile services is not just based on usability but includes also other interrelated issues. Ease of use is important, but the services should also provide clear value to the user and they should be trustworthy and easy to adopt. These user acceptance factors form the core of the Technology Acceptance Model for Mobile Services introduced in this chapter. The model has been set up based on field trials of several mobile services with altogether more than 200 test users. The model can be used as a design and evaluation framework when designing new mobile services.

INTRODUCTION

Research on mobile services has thus far mainly concentrated on the usability of alternative user interface implementations. Small mobile devices pose significant usability challenges and the usability of the services is still worth studying. However, more attention should be paid to user acceptance of the planned services. The reason for many commercial failures can be traced back to the wrongly assessed value of the services to the users (Kaasinen, 2005b).

User evaluations of mobile services often have to be taken into the field as the service would not function properly otherwise, or it would not make sense to evaluate it in laboratory conditions. This would be the case, for instance, with GPS systems and route guidance systems. In long-term field trials with users, it is possible to gather feedback on the adoption of the service in the users' everyday lives. Such studies gather usage data beyond mere usability and pre-defined test tasks (Figure 1). Field trials help in studying which features the users start using, how they use them and how often, and which factors affect user acceptance of the service.

Figure 1. Taking user evaluations from the laboratory to the field makes it possible to evaluate user acceptance on new services

Business and marketing research already have approaches whereby new technology is studied on a wider scale. The Technology Acceptance Model by Davis (1989) defines a framework to study user acceptance of a new technology based on perceived utility and perceived ease of use. Each user perceives the characteristics of the technology in his or her own way, based for instance on his or her personal characteristics, his or her attitudes, his or her previous experiences and his or her social environment. The Technology Acceptance Model has been evolved and applied widely, but mainly in the context of introducing ready-made products rather than in designing new technologies.

In this chapter an extension to the Technology Acceptance Model will be introduced. The model is based on a series of field trials and other evaluation activities with different mobile Internet and personal navigation services and over 200 test users (Kaasinen, 2005b). The Technology Acceptance Model for Mobile Services constitutes a framework for the design and evaluation of mobile services.

BACKGROUND

Technology acceptance models aim at studying how individual perceptions affect the intentions to use information technology as well as actual usage (Figure 2).

In 1989, Fred Davis presented the initial technology acceptance model (TAM) to explain the determinants of user acceptance of a wide range of end-user computing technologies (Davis 1989). The model is based on the Theory of Reasoned Action by Ajzen and Fishbein (1980). TAM points out that perceived ease of use and perceived usefulness affect the intention to use. Davis (1989) defines perceived ease of use as "*the degree to which a person believes that using a particular system would be free from effort*" and perceived usefulness as "*the degree to which a person believes that using a particular system would enhance his or her job performance.*" Perceived ease of use also affects the perceived usefulness (Figure 3). The intention to use affects the real usage behavior. TAM was designed to study information systems at work to predict if the users will actually take a certain system into use in their jobs. The model provides a tool to study the impact of external variables on internal beliefs, attitudes and intentions.

TAM deals with perceptions; it is not based on observing real usage but on users reporting their conceptions. The instruments used in connection with TAM are surveys, where the questions are constructed in such a way that they reflect the different aspects of TAM. The survey questions related to usefulness can be, for instance: "Using this system improves the quality of the work I do" or "Using this system saves my time." The survey questions related to ease of use can be, for instance:

Figure 2. The basic concept underlying technology acceptance models (Venkatesh et al., 2003)

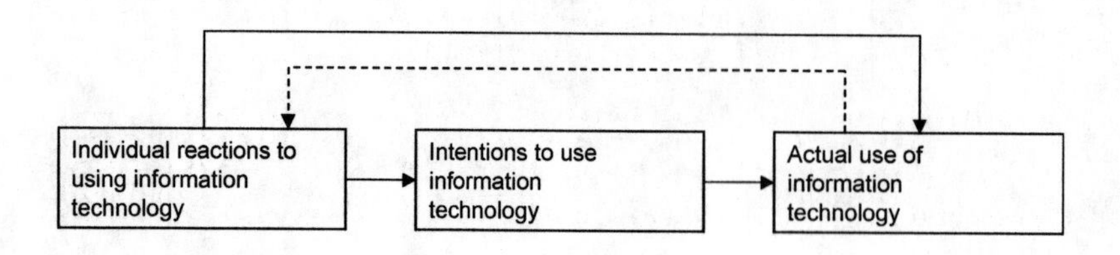

Figure 3. Technology acceptance model (Davis, 1989)

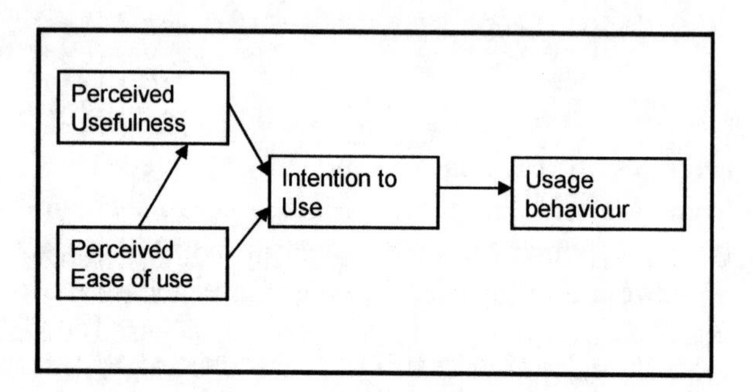

"The system often behaves in unexpected ways" or "It is easy for me to remember how to perform tasks using this system."

TAM has been tested and extended by many researchers, including Davis himself. Venkatesh and Davis (2000) have enhanced the model to TAM2 (Figure 4), which provides a detailed account of the key forces underlying judgments of perceived usefulness, explaining up to 60 percent of the variance in this driver of usage intentions. TAM2 showed that both social influence processes (subjective norm, voluntariness and image) and cognitive instrumental processes (job relevance, output quality, result demonstrability, and perceived ease of use) significantly influenced user acceptance.

Mathieson, Peacock and Chin, (2001) have extended TAM by analyzing the influence of perceived user resources. They claim that there may be many situations in which an individual wants to use an information system, but is prevented by lack of time, money, expertise and so on (Mathieson et al., 2001) classify resource-related attributes into four categories: user attributes, support from others, system attributes and general control-related attributes that concern an individual's overall beliefs about his or her control over system use. In their extended model, external variables affect perceived resources that further affect perceived ease of use and the intention to use.

TAM was originally developed for studying technology at work, but it has often been used to study user acceptance of Internet services as well (Barnes & Huff, 2003; Chen, Gillenson & Sherell, 2004; Gefen, 2000; Gefen & Devine, 2001; Gefen, Karahanna & Straub, 2003). Gefen et al. (2003) have studied TAM in connection with e-commerce. They have extended TAM for this application area and propose that trust should be included in the research model to predict the purchase intentions of online customers.

The Technology Acceptance Model constitutes a solid framework to identify issues that may affect user acceptance of technical solutions. Davis and Venkatesh (2004) proved that the model can be enhanced from the original purpose of

Figure 4. Enhanced technology acceptance model (TAM2) by Venkatesh and Davis (2000)

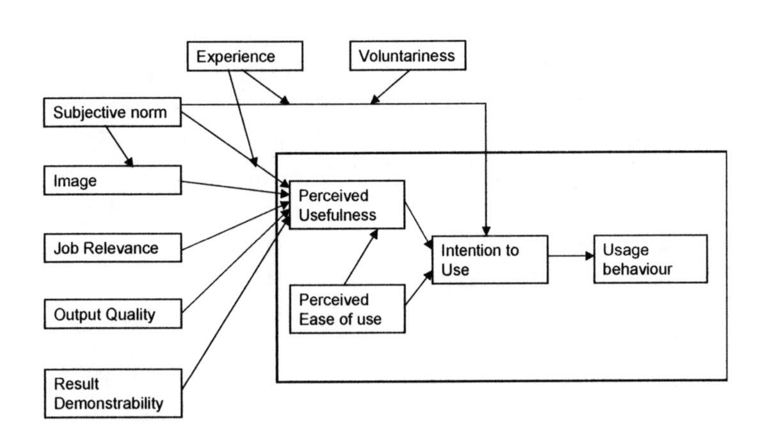

studying user acceptance of existing products to study planned product concepts, for example, in the form of mock-ups. This indicates that TAM could also be used in connection with technology development projects and processes to assess the usefulness of proposed solutions.

APPLICABILITY OF EARLIER APPROACHES FOR MOBILE SERVICES

The focus of traditional usability studies is on specified users performing specified tasks in specified contexts of use (ISO13407, 1999). In field trials the users can use prototype services as part of their everyday life. The research framework can then be enhanced to identify the actual tasks that users want to perform and the actual contexts of use. Technology acceptance models provide a framework for such studies.

Mobile services targeted at consumers have several specific characteristics that may mean that their user acceptance cannot be studied using the same models as with information systems in the workplace. When dealing with consumer services, individuals make voluntary adoption decisions and thus the acceptance includes assessing the benefits provided compared with either competing solutions or the non-acquisition of the service in question. As pointed out by Funk (2004), mobile services are disruptive technology that may find their innovation adopters elsewhere than expected, as highlighted by the experiences with the Japanese i-mode. Focusing too early on only limited user groups may miss possible early adopters. With the Japanese i-mode, other services were boosted through e-mail and personal home pages (Funk, 2004). This suggests that the focus of user acceptance studies of mobile services should be extended to interrelated innovations, as proposed by Rogers (1995).

Perceived usefulness included in TAM may not indicate an adequate purchase intention in a market situation. Product value has been proposed as a wider design target both in software engineering and HCI approaches. A value-centered software engineering approach was proposed by Boehm (2003) to define more clearly what the design process is targeted at, and identifying the values that different stakeholders—including end-users—expect of the product. Although not using the actual term "value," Norman (1998) emphasizes the importance of identifying big phenomena related to user needs and communicating them early on to the design. Cockton (2004b) points out that in value-centered HCI existing HCI research components, design guidance, quality in use and fit to context need to be reshaped to subordinate them to the delivery of product value to end-users and other stakeholders.

Mobile services are increasingly handling personal information of the user, for instance due to the personalization and context-awareness of the services. The functionalities of the increasingly complex systems are not always easy for the users to comprehend. Context-aware services may include uncertainty factors that the users should be able to assess. Mobile service networks are getting quite complex and the users may not know with whom they are transacting. Technical infrastructures as well as the rapidly developed services are prone to errors. All these issues raise trust as a user acceptance factor, similar to TAM applied in e-commerce (Chen et al., 2004; Gefen et al., 2003). Trust has been proposed as an additional acceptance criterion for mobile services by Kindberg, Stellen and Geelhoed, (2004) and Barnes and Huff (2003). Trust has also been included in studies of personalization in mobile services (Billsus et al., 2002) and studies of context-aware services (Antifakos, Schwaninger & Schiele, 2004).

Ease of adoption is included in the studies by Sarker and Wells (2003) and Barnes and Huff (2003). Sarker and Wells (2003) propose a totally new acceptance model that is based on user adoption. Barnes and Huff (2003) cover adoption in their model within the wider themes of compatibility and trialability. *Perceived user resources* in the extension of TAM by Mathieson et al. (2001) and *Facilitating conditions*, in the Unified Theory of Acceptance and Use (Venkatesh et al., 2003) also include elements related to ease of adoption.

In the following, the technology acceptance model for mobile services (Kaasinen, 2005b) is described in detail. The model aims at taking into account the aforementioned special characteristics of mobile consumer services, and previous studies on user acceptance described in this chapter. The model can be utilized when designing new services and assessing them to ensure that key user acceptance factors are considered in the design.

TECHNOLOGY ACCEPTANCE MODEL FOR MOBILE SERVICES (TAMM)

The technology acceptance model for mobile services (TAMM) was constituted based on a series of field trials and other user evaluation activities involving over 200 users. The studies were carried out as parts of technology development projects in 1999-2002 by project usability teams comprising altogether 13 researchers from VTT and three researchers from other research organizations. The focus of the studies was in particular on mobile Internet services and location-based services targeted at consumers (Kaasinen, 2005b). Mobile Internet studies were carried out in connection with the development of mobile browsers and the first WAP (wireless application protocol) services for mobile phones. In addition to commercial services, the test users could access many Web services because our project developed a Web-WAP conversion proxy server. Based on identified user needs, our research team also developed specific WAP services, for instance, for group communication. The services were evaluated in long-term field trials with users. The studies of location-based services were carried out within a horizontal usability support project, part of the Personal Navigation (NAVI) research and development program in Finland. The aim of the program was to facilitate co-operation between different actors who were developing personal navigation products and services. Our research group supported individual projects in usability and ethical issues and, beyond this, identified general guidelines for acceptable personal navigation services. We studied user attitudes and preliminary acceptance by evaluating different service scenarios in focus groups. In addition we evaluated some of the first commercial location-based services and carried out user evaluation activities in co-operation with the NAVI projects that were developing location-based services. Table 1 gives an overview of the user evaluation activities that the technology acceptance model for mobile services is based on.

The original technology acceptance model was chosen as the starting point for the new model because it provided a framework for connecting field study findings of ease of use and usefulness. The user acceptance framework is especially suitable for field trials where the focus is to study how different users start using the mobile services

Table 1. The user evaluation activities that the technology acceptance model for mobile services was based on.

Service, application or device	Research methods	Users	Original results published in
WAP services	Laboratory evaluation with phone simulator	6	Kaasinen et al., 2000
WAP-converted Web services	Laboratory evaluation with phone simulator	4	
WAP services WAP-converted Web services	Field trial 2 months	40	Kaasinen et al., 2001
	Interviews with service providers	25	
WAP services WAP-converted Web services Web/WAP Message board for group communication	Field trial 2 months	40	
	Interviews with service providers	11	
Scenarios of personal navigation services	Group interviews	55	Kaasinen, 2003
Benefon GPS phone and services	Field evaluation	6	
Sonera Pointer location-aware WAP services	Laboratory evaluation	5	
Garmin GPS device	Field evaluation	5	
Magellan GPS device	Field evaluation	5	
Location-aware SMS services	Field evaluation	6	Kaasinen, 2005a
Weather and road conditions by SMS	Field trial, 1 month	10	
Location-aware integrated service directory	Field trial, 3 weeks	7	
Mobile topographic maps	Field evaluation	6	
Mobile 3D maps	Laboratory evaluation	6	
	Field evaluation	4	
Scenarios of context-aware consumer services	Interviews in anticipated contexts of use	28	

in their everyday lives and which features make the services acceptable in actual usage. As not all the field study findings could be fit to the original TAM model, it was necessary to update the model according to the repeated field study findings and themes identified in related research. The new model extends the original core model by Davis (1989) by identifying two new perceived product characteristics that affect the intention to use, that is trust and ease of adoption, and by redefining the theme of usefulness as value to the user.

The framework (Figure 5) suggests that perceived ease of use, perceived value, and trust affect the intention to use a mobile service. To get from an intention to use to real usage, the user has to take the service into use. This transition is affected by the perceived ease of adoption. Perceived value, perceived ease of use, trust and perceived ease of adoption need to be studied in order to assess user acceptance of mobile services.

Figure 5. Technology acceptance model for mobile services (Kaasinen, 2005b) as an extension and modification of TAM by Davis (1989)

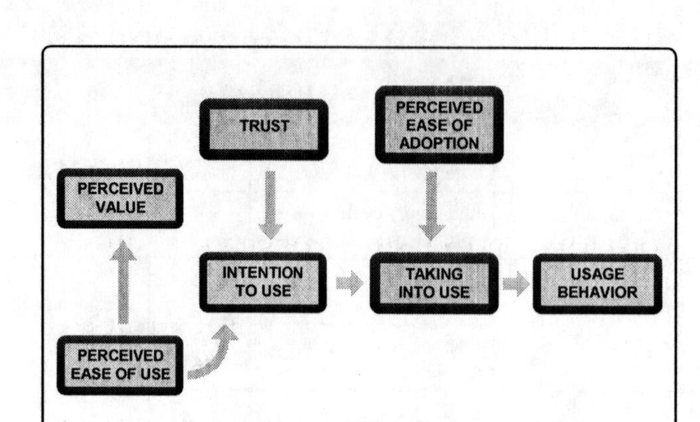

The technology acceptance model for mobile services (Kaasinen, 2005b) constitutes a framework that helps designers of mobile services to identify key issues that should be focused on in the design to ensure user acceptance. Thus the motivation of the model is different than the motivation of the original TAM, which was built to explain user acceptance and underlying forces for existing technical solutions.

Perceived ease of use was included in the original TAM and it is also included in the TAMM model. Davis (1989) defined perceived ease of use as "*the degree to which a person believes that using a particular system would be free from effort.*" At first, perceived ease of use is based on external factors such as the user's attitude towards technology in general, experiences of using similar services and information from other people. In actual use and sustained use, perceived ease of use is increasingly affected by the user's own experiences of using the system in different contexts of use.

In the case of mobile services that are used on small devices such as mobile phones or PDAs, the limitations of the device have a major influence on perceived ease of use. The limitations include the small screen, small and limited keyboard, the absence or limited functionality of pointing devices, limited amount of memory, limited battery power, and slow connections. As new devices and mobile networks are being introduced to the market, these limitations have somewhat diminished but still mobile networks are slower than fixed ones and the requirements for ease of carrying and holding the device do not allow very large screens or large keyboards. Designing mobile services for ease of use is to a large extent about coping with the limitations of the device. In addition, the design should adapt to the variety of client devices and available networks and other infrastructures.

The ease of use of mobile services has been studied quite a lot and different usability guidelines are available. It is a pleasure to note that many of the usability problems identified in early mobile Internet studies have already been corrected in current mobile devices, browsers and services. However, location-aware services pose even more challenges for ease of use. Location-aware services are not just mobile in the sense that they can be easily carried around but, typically, they are used while the user is moving. These kinds of usage situations require extreme ease of use. Personalization and context-awareness are expected to improve ease of use, but they may also introduce new usability problems, for example in the form of personalization dialogues.

Perceived value replaces perceived usefulness in the TAMM model because in our field trials with consumers it became evident that in the consumer market, perceived usefulness may not indicate adequate motivation to acquire the mobile service. As the focus group studies by

Järvenpää et al. (2003) point out, consumers may lack a compelling motivation to adopt new mobile services unless those services create new choices where mobility really matters and manage to affect people's lives positively. In a value-neutral setting each requirement is treated as equally important in the design (Boehm, 2003). This easily leads to featurism—the product becomes a collection of useful features but as a whole it may not provide enough value to the user. Value not only includes rational utility but also defines the key features of the product that are appreciated by the users and other stakeholders, that is the main reasons why the users are interested in the new product. As Roto (2006) points out, costs of using the service also affect the perceived value as user expectations tend to be higher for more expensive products. Values are made explicit by the identification of objectives, which are statements about what the user wants to achieve. Fundamental objectives are directly related to the user's current problem or situation at hand, whereas means objectives help to achieve the fundamental objectives (Nah et al., 2005).

Defining the targeted values and concentrating on them in design and evaluation helps to focus the design on the most essential issues. This is in line with the concept of value-centered software engineering proposed by Boehm (2003) and value-centered HCI proposed by Cockton (2004a, b). Focusing on perceived value in user acceptance studies supports the wider scope of value-centered design, where user value can be studied in parallel with business value and strategic value as proposed by Henderson (2005).

Trust is added as a new element of user acceptance in the TAMM model. The original TAM (Davis, 1989) was defined for information systems at work, and in those usage environments the end-users could rely on the information and services provided and the ways their personal data was used. When assessing user acceptance of e-commerce applications, Gefen et al. (2003) proposed to enhance TAM with trust in the service provider, as in their studies trust-related issues turned out to have a considerable effect on user acceptance. In our studies with mobile Internet, consumers were using mobile services that were provided to them via complex mobile service networks. In this environment trust in the service providers turned out to be an issue. As location-based services collect and use more and more information about the usage environment and the user, ethical issues arise. Especially ensuring the privacy of the user was a common concern of our test users. As the users get increasingly dependent on mobile services, reliability of the technology and conveying information about reliability to the user becomes more important.

In the technology acceptance model for Mobile Services, trust is defined according to Fogg and Tseng (1999). Trust is an indicator of a positive belief about the perceived reliability of, dependability of, and confidence in a person, object or process. User trust in mobile services includes perceived reliability of the technology and the service provider, reliance on the service in planned usage situations, and the user's confidence that he or she can keep the service under control and that the service will not misuse his or her personal data.

Perceived ease of adoption is related to taking the services into use. In the original TAM settings with information systems at work, this certainly was not an issue as users typically got their applications ready installed. In our field trials it turned out that a major obstacle in adopting commercial mobile services was the users' unawareness of available services, as well as problems anticipated in taking services into use (Kaasinen, 2005b). Furthermore, as usage needs were typically quite occasional, people often did not have enough motivation to find out about these issues. And finally, configuration and personalization seemed to require almost overwhelming efforts (Kaasinen, 2005b). Introducing the services to users would definitely require more attention in service design (Kaasinen et al., 2002).

As mobile services are typically used occasionally and some services may be available only locally in certain usage environments, ease of taking the services into use becomes even more important. The user should easily get information about available services and should be able to install and start to use the services easily. Finally, he or she should be able to get rid of unnecessary services.

Compared with the original TAM (Davis, 1989), the technology acceptance model for mobile services includes an additional phase between the intention to use and the actual usage behavior. Taking a service into use may constitute a major gap that may hinder the transfer from usage intention to actual usage (Kaasinen, 2005b). Perceived ease of adoption is added to the model at the stage when the user's attention shifts from intention to use to actually taking the service into use.

The characteristics of the user and his or her social environment affect how the user perceives the service. These issues are not included in the core TAMM model that aims to identify key characteristics of mobile services that generally affect user acceptance of mobile services. Further research is needed to fit previous TAM enhancements such as TAM2 (Venkatesh & Davis, 2000) and UTAUT (Venkatesh et al., 2003) to the model to identify external factors such as characteristics of the users and their social environment that affect the user acceptance factors in the model.

In the following section the technology acceptance model for mobile services is analyzed further and design implications are presented for each user acceptance factor, based on the synthesized results of the original case studies (Kaasinen, 2005b). The technology acceptance model for mobile services, together with the design implications, communicates previous user acceptance findings to the design of future mobile services.

DESIGN IMPLICATIONS

The technology acceptance model for mobile services defines four main user acceptance factors: perceived value, perceived ease of use, trust and perceived ease of adoption. How these factors should be taken into account in the design of individual mobile services depends on the service in question. However, there are many attributes of the acceptance factors that repeat from one service to another. These attributes form a set of design implications that can be used in the design of mobile services. The design implications can additionally be used in designing user acceptance evaluations to define the issues to be studied in the evaluation. In the following, design implications for each user acceptance factor are presented by combining results from the original studies (Kaasinen et al., 2000; Kaasinen et al., 2001; Kaasinen, 2003; Kaasinen, 2005a; Kaasinen 2005b) and results from related research.

Because of the quality of the case study material, the design principles cover best mobile information services targeted at consumers. For other kinds of services, the technology acceptance model for mobile services as well as the design implications can certainly be used as a starting point but they may need to be revised.

Perceived Value

Values define the key features of the services that are appreciated by the users and other stakeholders, that is the main reasons why the users are interested in the new services. Defining the targeted values helps in focusing the design on the most essential issues. Value is also related to the costs of using the service, and for commercial products the relationship of these two attributes should be studied, as proposed by Roto (2006). The following list gives some ideas about where in our studies the value was found.

Successful Service Content is Comprehensive, Topical, and Familiar

In the early days of mobile Internet, service providers often thought that small devices would require just a small amount of contents. Our studies showed that mobile users need access to all relevant information, as deep as they are ready to go, but the information has to be structured in such a way that the user can choose to get the information in small portions. Users appreciate comprehensive services in terms of geographic coverage, breadth (number of services included) and depth (enough information in each individual service).

Topical information is likely such that the mobile service is the best way to keep up to date with what is going on. In our field trials examples of successful topical content included weather forecasts, traffic information, news topics and event information. Topical travel information, for instance, does not just give timetables but informs about delays and traffic jams and recommends alternative routes.

The user may perceive the service as being familiar because it resembles other mobile services that he or she has been using or because it resembles the same service or brand in a different environment such as Web, TV or newspaper. For instance, in our mobile Internet studies teletext services converted from the Web were well accepted because of their familiarity. Familiarity was also related to the provider of the service, as test users pointed out that they preferred using, for instance, news services from a familiar and trusted service provider.

The Service Should Provide Personal and User-Generated Content

Personalization is not just about selecting services and contents within services but also about making the user's own personal items available, as illustrated in Figure 6 by the setup of the personal mobile Internet pages in our trials. Also with location-based services the users appreciated the possibilities to complement, for example map data with their own information such as important places, favorite routes, and self-written notes.

In the mobile Internet trials many users were keen to use services such as discussion groups where they could contribute as content providers. Letting the mobile users contribute to content creation could enhance many services. Such content may enrich the service, bring in additional users and encourage a sense of community among users. For instance, information generated by users at a particular location may be of interest to the next visitors. With the growing trend of social media services, the role of users' own content generation is expected to become increasingly important. Mobile users have key roles in many social media services as they can contribute by bringing in topical information from the field, such as mobile video of important occasions.

The Users Appreciate Seamless Service Entities Rather than Separate Services

In the mobile Internet trials it turned out that usage needs for many individual services were quite occasional, even if the users would have assessed the services as being very useful in those occasional situations. The value of mobile Internet to the user was based on the wide selection of services rather than any individual service.

The studies with location-based services pointed out the need for seamless service entities, whereby the user is supported throughout the whole usage situation, for example while looking for nearby services, getting information on the services, contacting the services, and getting

Figure 6. The shift from common to personal increases the appeal of the services

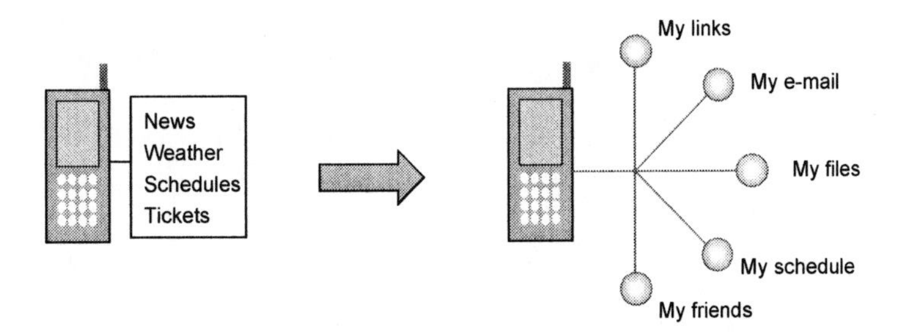

route guidance to find those services. The usage may even extend from one terminal device to another.

The Services Need to Provide Utility, Communication or Fun

In addition to personally selected content, interactive services also take mobile services to a more personal level, providing the users with new ways of communicating and participating. A mobile phone is basically a communication device and thus it is no wonder that services that enhanced or enriched communication were well accepted in our field trials.

Location-awareness can provide the users with services that are really intended for mobile use, not just secondary access points to Web services. Examples of such services include traffic information, weather forecasts, route guidance, travel information, event information and help services in emergency situations. Those services turned out to be popular as location-awareness made them both easier to use and more personal.

Perceived Ease of Use

Many ease of use attributes are already well known but as mobile services are getting increasingly complex and enhanced with new characteristics such as personalization and context-awareness, new usability challenges are raised. Key design principles that in our trials turned out to affect the perceived ease of use of mobile services are described in the following.

Clear Overview of the Service Entity

The most common usability problem with both mobile Internet services and location-based services was that when accessing the services, the users did not know what to expect from the service. The users would need a clear and intelligible overview of the whole range of available information, services and functions. The first impression may encourage and motivate the user or frighten him/her away. Enough design efforts and user evaluations should be invested in designing the main structure and the front page of the service. There are already efficient solutions available such as Minimap introduced by Roto (2006). Minimap gives the overview by showing a miniaturized version of the original Web page layout on the mobile device, and Roto's (2006) studies showed that this approach clearly improved the usability of Web browsing with mobile phones.

The information and functions that the user will most probably need should be the easiest to access. By proceeding further, the user should be able to access any information available within the service. Occasional usage typical of mobile services emphasizes the need for a clear overview of available services, including information on how the service should be used, where the content comes from, how often it is updated, and how comprehensive it is.

Fluent Navigation on a Small Screen

The mobile Internet trials showed that a single scrollable page (Figure 7) is good for browsing through information, whereas separate pages are better for navigation. The users need ways to browse quickly through less interesting information: for instance, an adaptive scroll speed and an illustrative scroll bar are useful.

The user needs clear feedback on which service and where in it he or she currently is. In our evaluations, this was facilitated by descriptive and consistent link/page header pairs for back, forward, exit, home and other safe heavens within the service.

The usability of the sites can be further improved by making the structure adaptive according to each user. A novice user may want to get instructions first, whereas more experienced users may want to go straight into the service. For frequent users, the structure could be adaptive so that the most recently or most often used items are easily available.

Figure 7. On a small screen, there is a lot to scroll, even when accessing a simple Web page

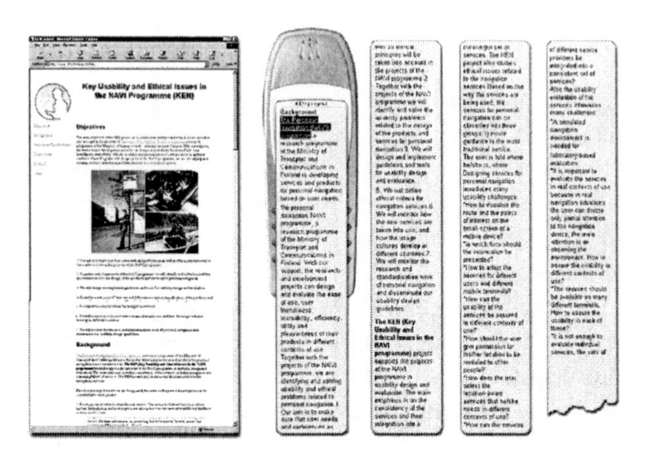

Smooth User Interaction with the Service

User interface restrictions of mobile devices and the implementation of the user interface elements may hinder smooth user interaction. In our trials text input was often a major effort for the users, especially if the usage took place while moving. Still the users needed and wanted to give input to the services. In the mobile Internet evaluations ready-made selection lists turned out to be useful when the user was getting acquainted with the service, whereas experienced users often preferred text input. Preferably, both alternatives should be available. Text input should be predicted and interpreted to suggest corrections to possible misspellings. Location-awareness as such could also be utilized in text input, as suggested by Ancona et al. (2001). For instance, nearby streets or the most popular search terms at a certain location could be suggested to the user. Even though user input may be tedious, it should not be avoided in the services. However, the user should not be obliged to input information that is not absolutely necessary.

Personally Relevant Services and Information without Expending Effort on Personalization Set-up

Our trials repeatedly showed that users were not willing to do much regarding personalization, although they would have appreciated the personalized solution. Personalization should be voluntary, and strongly supported in the beginning.

Users could be provided with ready-made service package alternatives, as we did in our mobile Internet trials, or they could be guided through personalization services. New service offerings could be sent automatically based on user profiles if the user accepts that. The user should be able to see and refine the personalization with his or her mobile device on the fly, even though the personalization could also be done with a desktop PC. New approaches such as group profiles, profiles shared by several services and learning profiles may ease personalization.

Easy Access to Situationally Relevant Information and Services

Mobile contexts vary a lot and may even change in the middle of a usage session. Our trials with location-based services showed that in services targeted at a limited area, such as travel guides, service catalogues and event guides, the context of use can be predicted quite well according to user location and time. This gives possibilities for different context-aware features in the services, easing their use and giving the users personalized access to the services.

Location-awareness can be utilized to provide the users with local services such as tourist guides, event information and shopping guides. Context-awareness can be complemented with personalization to adapt to user preferences that in different contexts may vary from one individual to another. This may, however, indicate lots of personalization efforts.

Facilitating Momentary Usage Sessions on the Move

On the move the users can devote only part of their attention to using the service while their main attention is on their main task of moving. In our trials with location-based services, on-the-move use was typically non-continuous. A user could, for instance, activate a route guidance service and start using it but occasionally he or she had to put the device aside and do something else. Later on he or she returned to the service. For these kinds of usage sessions task resumability should be supported both in the terminal device and in the services. Pousman et al. (2004) point out that resumability can be supported, for example by atomic interaction sessions, by appropriate timeouts on unfinished operations, and by a stateless interaction model. The users should be able to use the services both on and offline.

Design for Device and Network Variety

One of the main challenges in designing mobile services is the growing variety of mobile devices, networks and other infrastructures. The Design for All approach (EDeAN, 2007) with regard to mobile services requires taking into account all kinds of devices, not just the most advanced ones. In our development work on mobile Internet services we found that a good starting point is a simple service, suitable for any device. The usability and the attractiveness of the service can then be improved by utilizing the unique features of each device in separate implementations. Our experiences from mobile Internet trials show that in mobile environments there may be needs for adaptive search services that would not only look for particular content, but also take into account the current client device. The search results could be prioritized according to how suitable the content is for the device and network that the user is currently using.

Trust

In the TAMM model, user trust in mobile services is quite a wide concept that includes perceived reliability of both the technology and the information and functions provided, reliance on the service in planned usage situations, and the user's confidence that he or she can keep the service under control and that the service will not misuse his or her personal data. The design principles that in our evaluations turned out to affect user trust in mobile services are described in the following.

The User should be able to Rely on the Service in Intended Contexts of Use

In our user trials errors with mobile services were often difficult to cope with for the users as they did not know whether the problems were in the mobile device, in the network or in the services. Repeated malfunctions that the user could not understand or solve were a major source of bad usage experiences and often made the user stop using the service in question. To avoid these kinds of situations, the user should get easy-to-understand information to help him or her to understand and recover from the error situation. User errors should be prevented by all means, for example by trying to interpret, correct or complete user input. In the event of the user losing the connection to the service, it should be assured that no harm will be done.

With location-based services the users often would have liked to get feedback on the power still available and estimates of the sufficiency of batteries with different combinations of add-on devices and functions (Kaasinen, 2003). A user on the move may need to make decisions regarding which combination of functions he or she can afford to keep on in order to avoid exhausting the battery power totally.

Evaluations of personal navigation scenarios and prototype services revealed that users may get quite dependent on mobile services such as navigation services. That is why the users should be made aware of the possible risks of using the product and they should be provided with information about the reliability of the service so that they can assess whether they can rely on the service in the planned usage situations.

Measurement without Estimated Accuracy is of no Use

The accuracy of the location information was often questioned in our trials. In addition to location, future mobile services will be using and providing the user with increasing amounts of different measurement data (Kaasinen et al., 2006). Accuracy requirements for the data need to be considered in the design. The accuracy should be sufficient for the kinds of tasks for which the user will be using the service. The users should get feedback on the freshness of the data and its accuracy, especially if these vary according to the usage situation. Both actual reliability and perceived reliability need to be ensured in the design as these may be only loosely mapped, as found out by Kindberg et al. (2004).

Context-aware systems have several error possibilities: the system may offer the user wrong things either because it predicted the context wrongly or because it predicted the context correctly but predicted the user's needs in that context wrongly. Displaying uncertainty to the user may improve the acceptability of the services by making them more intelligible, as pointed out by Antifakos et al. (2004).

The Privacy of the User must be Protected Even if the User would not require it

User data should be protected even if—like in some of our trials—the users themselves would be trusting enough not to require it. The user should be provided with easy mechanisms for giving permission to use the data for a predefined purpose. Histories of user data should not be stored purposelessly and without user consent. When location data is conveyed to others, it is worth considering whether they will need the exact location coordinates or a more descriptive but less intrusive description. It should also be considered whether it is necessary to connect personal data to the user identity.

The legislation in most countries requires the user's permission before he or she can be located. Also social regulation can create rules and norms for different situations in which location-aware services are used (Ackerman et al., 2001). In practice, trade-offs between privacy protection and effortless use need to be resolved.

In future services, it can be expected that in addition to user location, a lot of other personal data may be collected. This may include health-related measurements, shopping behavior; services used and so on (Kaasinen et al., 2006). The same principles as with location are to great extent valid also with this data.

The User Needs to Feel and Really be in Control

The more complicated the mobile services and the service networks behind them get, the less possibilities the user has to understand what is happening in the service. The services need to be somewhat seamless to ensure effortless use. On the other hand, some issues need to be clearly differentiated so as to ensure that the user understands what is going on. Seamless services may hide details from users when aiming to provide ease of use. This may prevent the user from understanding what is happening "behind the scenes" (Höök, 2004).

Based on the findings of the trials with location-aware services, the main user requirement is that the user needs to feel and really be in control. For instance, the users more easily accepted context-aware behavior of the services if they could understand the reason for the behavior. To be able to be in control, the user needs to understand enough about the system's capabilities and rules of reasoning. The user needs to get feedback on what is going on and why, even if it is unnecessary to understand all the details. As automated functions may take control away from the user, the user should be able to control the degree of automation and intrusiveness. The user should be able to override the recommendations of the system, as suggested by Cheverst et al. (2000).

Similar to the findings by Cheverst et al. (2002), also in our trials the users tended to accept push services because of the effortless use. However, as the amount of push features grows, the attitude of the users may soon change. That is why the user should be able to fine-tune or cancel the push feature easily—ideally as he or she receives a push message.

Perceived Ease of Adoption

As mobile services will increasingly be available from different sources and in complex service networks, it becomes important to ensure that the users get reliable information about available services and the necessary guidance when taking the services into use. Based on user feedback in our trials, key design principles regarding ease of adoption of mobile services are described in the following.

Real Values of the Services Need to be Emphasized in Marketing

Users often have a poor understanding of mobile devices and services (Kolari et al., 2002). The users may have misconceptions about the services behind acronyms or different technologies. In our

Table 2. Trade description model for personal navigation products and services (Kaasinen et al., 2002)

Classification	*Trade description*
User	**Is this product/service suitable for me?** • Targeted specially at a certain user group • Targeted only at a certain group • Accessibility for disabled users
User goal	**What can I do with this product / service?** • Locate myself • Be located by other people • Locate other people • Track my property • Get route guidance • Find and use nearby services • Get help in emergency situations • Have fun
Environment	**Where can/cannot I use this product/service?**
Equipment	**What do I need to know about the technology?** • What kind of technology do I need to be able to use the service? • How compatible is this product/service with other products/services? • How accurate is the positioning? • To what extent can I rely on this product?
Service characteristics	**What specific features does this service include, what is the added value of this product compared with competing products or current ways to act?**

trials the users were often unaware of the features and services available on their personal phones.

As a part of our research work a Trade Description Model (Kaasinen et al., 2002) was set up to help consumers to compare different products and, on the other hand, to help service providers to describe their products in a consistent way (Table 2). Although the model was designed for personal navigation services, it is general enough to be adopted for the description of other mobile services as well. The trade description model can also be used as a checklist of issues to be covered when writing "Getting started" manuals.

Disposable Services for Occasional Needs

IBM has issued guidelines on how to design out-of-box experiences that are productive and satisfying for users (IBM, 2005). Ideally, the services should be installed on the user device at the point of sale, and the user should at the same time get personal usage guidance, but presumably this will be possible with only a few services.

In our trials with location-based services, the users often said that they wanted to have the services easily available when a spontaneous need for a certain service arose. Context-aware services pose additional challenges for taking new services into use. The services may be available only locally or in certain contexts. The user should be able to identify, understand and take into use these services easily while on the move. As the selection of available services grows, it will also become increasingly important to get rid of unnecessary services easily.

The Service has to Support Existing and Evolving Usage Cultures

Personal mobile devices should be designed to be both intuitive for first-time use and efficient in long-term use (Kiljander, 2004). This is true also with mobile services, which should be designed for gradual learning. New services shape the usage, but the usage should also shape the services (Norros et al., 2003). Existing and evolving usage cultures should be studied in parallel with the technology development to identify and support natural usage patterns. The design should fit in with the social, technical and environmental contexts of use, and it should support existing usage cultures. Ideally, the technology should provide the users with possibilities that they can utilize in their own way, rather than forcing certain usage models fixed in the design (Norros et al., 2003). Although the users will benefit from clear usage guidance, they should also be encouraged to discover and innovate their own ways to utilize new services.

FUTURE TRENDS

The current technology acceptance model for mobile services (TAMM) is based on studies with mobile Internet services and location-based information services targeted for consumer use. The identified user acceptance factors can be utilized in designing these kinds of services, but they can also be applied when designing other kinds of mobile services. In future visions, mobile devices are increasingly interacting with their environment and are transforming into tools with which the user can orient in and interact with the environment. As the user moves from one environment to another, the available services will change accordingly (Kaasinen et al., 2006). These kinds of services will require extreme ease of adoption, and, as the services will increasingly deal with personal data, the user's trust in the services will become an even more important user acceptance factor.

Further studies will be needed to study the mutual relations of the four user acceptance factors. As with the original TAM, the model can be enhanced by studying key forces underlying the judgments of perceived value, perceived ease of use, trust and perceived ease of adoption.

The technology acceptance model for mobile services was set up by analyzing and combining the results of several individual evaluation activities of different mobile services. When developing future mobile technologies and infrastructures, human-centered design can be expanded similarly. By synthesizing and generalizing the results of parallel research activities, key user acceptance factors and design implications for future service development can be identified.

The technology acceptance model for mobile services seems to have potential as a framework for ubiquitous computing applications as well. The model has already been successfully applied in connection with a project that aims to develop a mobile platform for ubiquitous computing applications that utilize wireless connections to sensors and tags (Kaasinen et al., 2006).

CONCLUSION

In this chapter, the technology acceptance model for mobile services has been introduced. According to the model, user acceptance of mobile services is built on three factors: perceived value of the service, perceived ease of use, and trust. A fourth user acceptance factor: perceived ease of adoption is required to get the users from intention-to-use to actual usage. Based on the technology acceptance model for mobile services, design implications for each user acceptance factor have been proposed.

Instead of implementing collections of useful features, the design of mobile services should be focused on key values provided to the user. The value of mobile services can be built on utility, communication or fun. Successful service content is comprehensive, topical and familiar, and it includes personal and user-generated content. The users appreciate seamless service entities rather than separate services. Ease of use requires a clear overview of the service entity, fluent navigation on a small display, and smooth user interaction with the service. The users should get personally and relevant services and information without needing to expend effort on personalization. The services should be designed to be adaptive to a wide variety of devices and networks. As the services increasingly support individual users in their daily tasks and increasingly deal with personal data, user trust in the services is becoming more and more important. The user should be able to assess whether he or she can rely on the service in the intended contexts of use. The user needs to feel and really be in control, and the privacy of the user must be protected.

Occasional usage and momentary usage sessions on the move are typical of mobile services. In addition, services are increasingly available only locally or in certain contexts of use. This indicates the need for disposable services: services that are easy to find, take into use, use and get rid of when no longer needed. The user needs realistic information about the actual values of the services, so that he or she can realize how to utilize the service in his or her everyday life and discover new usage possibilities.

The technology acceptance model for mobile services provides a tool to communicate key user acceptance factors and their implications to the design. The model can be used in all design and evaluation activities throughout the design process, but it is especially useful in identifying issues that should be examined in field studies.

REFERENCES

Ackerman, M., Darrel, T., & Weitzner, D. J. (2001). Privacy in context. *Human-Computer Interaction, 16*, 167–176.

Ajzen, I., & Fishbein, M. (1980). *Understanding attitudes and predicting social behaviour.* Prentice Hall.

Ancona, M., Locati, S., & Romagnoli, A. (2001). Context and location aware textual data input. *SAC 2001* (pp. 425-428). Las Vegas: ACM.

Antifakos, S., Schwaninger, A., & Schiele, B. (2004). Evaluating the effects of displaying uncertainty in context-aware applications. In Davies, N., Mynatt, E., & Siio, I. (Eds.), in *Proceedings of Ubicomp 2004: Ubiquitous Computing 6th International Conference* (pp. 54-69). Springer-Verlag.

Barnes, S. J., & Huff, S. L. (2003). Rising Sun: imode and the wireless internet. *Communications of the ACM, 46*(11), 79–84.

Billsus, D., Brunk, C. A., Evans, C., Gladish, B., & Pazzani, M. (2002). Adaptive interfaces for ubiquitous Web access. *Communications of the ACM, 45*(5), 34–38.

Boehm, B. (2003). Value-based software engineering. *Software Engineering Notes, 28*(2), 1–12.

Chen, L., Gillenson, M. L., & Sherell, D. (2004). Consumer acceptance of virtual stores: A theoretical model and critical success factors for virtual stores. *ACM SIGMIS Database archive, 35*(2), 8–31.

Cheverst, K., Davies, N., Mitchell, K., Friday, A., & Efstratiou, C. (2000). Developing a context-aware electronic tourist guide: some issues and experiences. *CHI 2000 Conference Proceedings* (pp. 17–24). ACM.

Cheverst, K., Mitchell, K., & Davies, N. (2002). Exploring context-aware information push. *Personal and Ubiquitous Computing, 6*, 276–281.

Cockton, G. (2004a). From quality in use to value in the world. In *Proceedings of CHI2004* (pp. 1287-1290). ACM.

Cockton, G. (2004b). Value-centred HCI. In *Proceedings of the Third Nordic Conference on Human-Computer Interaction* (pp. 149–160).

Davis, F. D. (1989). Perceived usefulness, perceived ease of use, and user acceptance of information technology. *MIS Quartely, 13*, 319–339.

Davis, F. D., & Venkatesh, V. (2004). Toward preprototype user acceptance testing of new information systems: implications for software project management. *IEEE Transactions on Engineering Management, 51*(1).

EDeAN. (2007). European design for all e-accessibility network. *Homepage*. Retrieved February 13, 2007, from www.e-accessibility.org

Fogg, B. J., & Tseng, H. (1999). The elements of computer credibility. In *Proceedings of CHI 99 Conference* (pp. 80–87).

Funk, J. L. (2004). *Mobile disruption. The technologies and applications driving the mobile Internet.* Wiley-Interscience.

Gefen, D. (2000). E-commerce: The role of familiarity and trust. *Omega: International Journal of Management Science, 28*, 725–737.

Gefen, D., & Devine, P. (2001). Customer loyalty to an online store: The meaning of online service quality. *Proceedings of the 22nd International Conference on Information Systems* (pp. 613–617).

Gefen, D., Karahanna, E., & Straub, D. W. (2003). Inexperience and experience with online stores: The importance of TAM and Trust. *IEEE Transactions on Engineering Management, 50*(3), 307–321.

Henderson, A. (2005). Design: The innovation pipeline: Design collaborations between design and development. *ACM Interactions, 12*(1), 24–29.

Höök, K. (2004). Active co-construction of meaningful experiences: but what is the designer's role? *Proceedings of the Third Nordic Conference on Human-Computer Interaction* (pp. 1-2). ACM Press.

IBM. (2005). Out-of-box experience. Retrieved January 10, 2005, from www-3.ibm.com/ibm/easy/eou_ext.nsf/publish/577

ISO 13407. (1999). *Human-centred design processes for interactive systems.* International standard. International Standardization Organization. Geneve.

Järvenpää, S. L., Lang, K. R., Takeda, Y., & Tuunanen V. K. (2003). Mobile commerce at crossroads. *Communications of the ACM, 46*(12), 41–44.

Kaasinen, E. (2003). User needs for location-aware mobile services. *Personal and Ubiquitous Computing, 6*, 70–79.

Kaasinen, E. (2005a). User acceptance of location-aware mobile guides based on seven field studies. *Behaviour & Information Technology, 24*(1), 37–49.

Kaasinen, E. (2005b). *User acceptance of mobile services—Value, ease of use, trust and ease of adoption*. Doctoral dissertation. VTT Publications 566. Espoo: VTT Information Technology.

Kaasinen, E., Aaltonen, M., Kolari, J., Melakoski, S., & Laakko, T. (2000). Two approaches to bringing internet services to WAP devices. *Computer Networks, 33*, 231–246.

Kaasinen, E., Ermolov, V., Niemelä, M., Tuomisto, T., & Välkkynen, P. (2006). *Identifying user requirements for a mobile terminal centric ubiquitous computing architecture*. FUMCA 2006: System Support for Future Mobile Computing Applications. Workshop at Ubicomp 2006.

Kaasinen, E., Ikonen, V., Ahonen, A., Anttila, V., Kulju, M., Luoma, J., & Södergård, R. (2002). *Products and services for personal navigation—Classification from the user's point of view*. Publications of the NAVI programme. Retrieved January 4, 2005, from www.vtt.fi/virtual/navi

Kaasinen, E., Kasesniemi, E.-L, Kolari J., Suihkonen, R., & Laakko, T. (2001). Mobile-transparent access to web services—Acceptance of users and service providers. In *Proceedings of International Symposium on Human Factors in Telecommunication*. Bergen, Norway

Kiljander, H. (2004). *Evolution and usability of mobile phone interaction styles*. Doctoral thesis. Helsinki University of Technology. Publications in Telecommunications Software and Multimedia. TML-A8. Espoo: Otamedia.

Kindberg, T., Sellen, A., & Geelhoed, E. (2004). Security and trust in mobile interactions—A study of users' perceptions and reasoning. In Davies, N., Mynatt, E., & Siio, I. (Eds.), *Proceedings of Ubicomp 2004: Ubiquitous Computing 6th International Conference* (pp. 196-213). Springer-Verlag.

Kolari J., Laakko T., Kaasinen E., Aaltonen M., Hiltunen T., Kasesniemi, E.-L., Kulji, M., & Suihkonen, R. (2002). *Net in pocket? Personal mobile access to web services*. VTT Publications 464. Espoo: Technical Research Centre of Finland.

Mathieson, K., Peacock, E., & Chin, W. W. (2001). Extending the technology acceptance model: The influence of perceived user resources. *The DATA BASE for Advances in Information Systems, 32*(3), 86–112.

Nah, F. F.-H., Siau, K., & Sheng, H. (2005). The value of mobile applications: A utility company study. *Communications of the ACM, 48*(2), 85–90.

Norman, D. A. (1998). *The invisible computer*. Cambridge, MA: MIT Press.

Norros, L., Kaasinen, E., Plomp, J., & Rämä, P. (2003). *Human-technology interaction. Research and design. VTT Roadmap*. VTT Research Notes 2220. Espoo: Technical Research Centre of Finland.

Pousman, Z., Iachello, G., Fithian, R., Moghazy, J., & Stasko, J. (2004). Design iterations for a location-aware event planner. *Personal and Ubiquitous Computing, 8*, 117–125.

Rogers, E. M. (1995). *The diffusion of innovations*.(Fourth Ed.). New York: Free Press.

Roto, V. (2006). *Web browsing on mobile phones—Characteristics of user experience*. Doctoral dissertation. Espoo: Helsinki University of Technology.

Sarker, S., & Wells, J. D. (2003). Understanding mobile handheld device use and adoption. *Communications of the ACM, 46*(12), 35–40.

Venkatesh, V., & Davis, F. D. (2000). Theoretical extension of the technology acceptance model: Four longitudinal field studies. *Management Science, 46*(2), 186–204.

Venkatesh, V., Morris, M. G., Davis, G. B., & Davis, F. D. (2003). User acceptance of information technology: Toward a unified view. *MIS Quarterly, 27*(3), 425–478.

KEY TERMS

Ease of Adoption (TAMM): Perceived ease of identifying, understanding and taking into use new products.

Innovation Diffusion: User adoption of different innovations in target populations

Location-Aware Service: A special case of location-based service: a mobile service that adapts according to the location.

Location-Based Service: A mobile service that utilizes location data.

Perceived Ease of Use (TAM and TAMM): The degree to which a person believes that using a particular system would be free from effort (Davis, 1989).

Perceived Usefulness (TAM): The degree to which a person believes that using a particular system would enhance his or her performance in a certain task (Modified from Davis, 1989).

Technology Acceptance: User's intention to use and continue using a certain information technology product (Davis, 1989).

Technology Acceptance Model (TAM): Technology acceptance models aim at studying how individual perceptions affect the intentions to use information technology as well as the actual usage. The Technology Acceptance Model was originally defined by Davis (1989), but it has subsequently been modified and augmented by other researchers.

Technology Acceptance Model for Mobile Services (TAMM): Extension of the original Technology Acceptance Model to take into account the specific characteristics of mobile services (Kaasinen, 2005b)

Trust (TAMM): An indicator of a positive belief about the perceived reliability of, dependability of, and confidence in a product (modified from Fogg & Tseng, 1999).

Value (TAMM): The key features of the product that are appreciated by the users and other stakeholders, i.e. the main reasons why the users are interested in the new product (Kaasinen, 2005b).

Chapter VIII
Transgenerational Designs in Mobile Technology

Martina Ziefle
RWTH Aachen University, Germany

Susanne Bay
RWTH Aachen University, Germany

ABSTRACT

Mobile devices have proliferated into most working and private areas and broad user groups have access to mobile technology. This has considerable impact on demands for usable designs. As users differ widely regarding age, upbringing, experience and abilities, it is a basic question whether there are user interface designs feasible that meet the demands of user diversity and trans-generational designs. The aim of the present research was to uncover effects of user diversity on menu navigation. Users of a wide age range were examined when interacting with mobile phones. In a detailed way, individual navigation routes were analyzed and effectiveness and efficiency of menu navigation was determined. In addition, effects of individual variables were considered. The results show that the usage of small-screen devices imposes considerable difficulties for all users, but in particular for children and middle-aged adults, who were very sensitive for cognitive demands imposed by current mobile phone designs.

INTRODUCTION

The distribution of mobile devices represents one of the fastest growing technological fields ever. Especially, small interface devices are omnipresent and can be characterized as important technical devices in today's societies. Mobile devices promise to be ubiquitously applicable and cover basic communication as well as office functionalities and allow Internet access. Moreover, the devices are used for route and traffic information, but provide also fun and entertainment applications.

The ubiquity and penetration of mobile devices raise new usability concerns. Many users show considerable problems with respect to the handling, learning, and understanding of these devices, which in turn reduce the ease of use and the perceived usefulness (e.g., Arning & Ziefle, 2006, 2007; Jakobs, 2005; Tuomainen & Haapanen; 2003; Ziefle & Bay, 2004; 2005). Yet, commonly agreed rules, which complexity of functions and which interface design is appropriate, have not been defined, and perhaps due to this fact, usability is not an issue that manufacturers are primarily investing in.

Several factors can be referred to that contribute to these difficulties. While formerly the usage of information technology was mainly restricted to technology-prone users, today, all user groups are addressed by technology. The diversity of the target groups, however, requires a basic understanding of the human factor and should be adequately addressed by device design. Users differ considerably with regard to their needs, motivation, competencies, and aptitudes, which is reflected in users' age, gender, and experience with technical devices.

In addition, more and more transactions include the utilization of technical devices and demand the acceptance and the competence of using technical devices. Thus, technical device usage is increasingly less optional, but represents more and more an indispensable qualification for many working settings. Furthermore, the nature and number of the devices' functionalities is elementarily changing. The traditional functionality of mobile phones, making calls, is only one among many other functions and the devices have an increasing complexity. Aggravating, numerous different device types within and across brands can be found on the market. While the applications and functions are increasingly merging across device types, though, devices differ considerably with respect to their basic structure and interface design. Within cross-platform-designs, it is thus difficult to understand, which operation modes and "device logic" is specific for a certain device and which is valid across devices (e.g., Ziefle, Arning & Bay, 2006). Finally, the miniaturization of the devices also contributes to cognitive difficulty when using technology. The tiny devices have small keys and miniature displays, thus the key handling and the visibility of the displayed information is considerably complicated. Furthermore, due to the restricted display, only few functions can be seen at a time. This increases memory load, as users have to remember function names and their menu location. Also, spatial orientation in the menu is problematical. Users do not experience how the menu is structured and how many functions are in the menu. As a consequence, users often lose their way in the menu.

BACKGROUND

The development of mobile technology and the device interface design still seems mainly to concentrate on what young and experienced users want (Maguire & Osman, 2003). However, children (mobileyouth.org, 2005) and older adults (Arning & Ziefle, 2007; Ziefle & Bay, 2006) are now also major user groups and, though, have not been considered adequately so far. This may be due to the fact that there is only little knowledge whether these groups have specific difficulties when using small-screen devices, and also regarding the factors, which might hamper or benefit a purposeful interaction with these devices (e.g., Arning & Ziefle, 2006, Tuomainen & Haapanen; 2003). Instead, a lot of preconceptions are prevailing. According to casual comments of many participants in our lab, we experienced that there is a "common knowledge" about aptitudes and abilities of age groups interacting with technical devices. Older adults are assumed to be the taillight regarding technical competence (and interestingly, they characterize themselves the same way), and quite low interest in technical developments is ascribed to them. As they have a different upbringing and were educated in times when technical devices were far less complex, they are thought to be considerably penalized. Conversely, children are supposed to easily master the interaction with technical devices. They are believed to understand the mode of operation of those devices much faster by virtue of their contact with interactive technology (e.g., computers, video games) from early on.

Additionally, children's fascination for explorative and inquisitive activities is well known, therefore they are assumed to be especially qualified for the interacting with technical devices.

Contrary to these statements, it was found that adults and children show a similar performance when using technical devices. Both were very sensitive to the demands imposed by the devices and showed considerable performance losses in sub-optimally designed interfaces (e.g., Bay & Ziefle, 2005; Ziefle & Bay, 2004; 2005; 2006; Ziefle, Bay & Schwade, 2006). But the exclusive focusing on users' age for technical performance is not sufficient. Rather, age must be characterized as the carrier of individual characteristics that are known to affect technical performance: cognitive abilities, attitudes, gender, or computer experience. Therefore, we need to understand the interrelation of these factors. If we want to learn if there are designs feasible, that are suited for all user groups or if we want to identify shortcomings, we also need to understand the specific impacts of individual variables, and their interaction with age and gender. The knowledge of the factors, which might underlie the aging and gender impact, though, is mostly limited to the examination of adults. Moreover, the interplay of different factors and performance has not been investigated satisfactorily so far.

Among the individual variables, which are known to play a role for adults' menu navigation performance, spatial ability is very prominent. Persons with high spatial abilities outperformed those with lower levels of spatial ability (e.g., Arning & Ziefle, 2007; Egan, 1988; Goodman et al., 2004; Kim & Hirtle, 1995; Vicente, Hayes & Williges, 1987; Westerman, 1997; Ziefle & Bay, 2006). Also, (verbal) memory is essential for the performance in technical menus. Users with high memory abilities had a better orientation in the menu, because they better memorized the functions and menu locations (e.g., Arning & Ziefle, 2007; Bay & Ziefle, 2003; Hasher & Zack, 1988; Ziefle & Bay, 2006). Moreover, the gender factor is crucial, especially in combination with computer self-efficacy. Female users often show lower self-efficacy and higher computer anxiety (e.g., Busch, 1995 Davies, 1994; Downing, Moore & Brown,2005). Rodger and Pendharkar (2004) referred performance differences between women and men to differences in computer experience levels, which are often lower in women. The interrelation of gender effects and computer experience is corroborated by studies showing that playful and active exploring of technical menus are forming an incidental knowledge of the system, which in turn contributes to computer experience (e.g., Bay & Ziefle, in press; Beckwith et al., 2006). Interestingly, playful interacting with computer systems is a behavior that is more often observed in male than female users (e.g., Van den Heuvel-Panheizen, 1999).

MAIN FOCUS OF THE CHAPTER

Comprising, the usability of small-screen devices, is an important but sophisticated demand, especially when taking user diversity into account. The interplay of user characteristics for the performance in technical menus is complex and needs a detailed examination. This is what the present paper wants to contribute to. The main focus was directed to a detailed analysis of users' navigation performance when using mobile phones. To understand the impact of user diversity, children, younger, and middle-aged adults were examined. Furthermore, gender effects were explored. Also, the influence of previous experience with technology and cognitive factors, as well as motivational factors, were taken into account. On the basis of a detailed analysis of individual interaction patterns, some implications for the design of mobile phones are discussed.

Method

Independent and Dependent Variables

Two independent variables were examined. The first independent variable was users' age, comparing the navigation performance of children (9-10 years), younger (20-30 years) and middle-aged adults (40-61 years). The second independent variable was gender, comparing female and male participants. Furthermore, participants' experience with technical devices, their spatial ability and short-term memory capacity were determined and treated as between subject variables, possibly affecting performance when using mobile phones.

As dependent variables, the effectiveness and efficiency of navigation were determined. In order to get a detailed insight in navigation behavior and to identify individual navigation patterns, six different measures were surveyed.

For the task *effectiveness*, the percentage of successfully solved tasks (within the time limit of five minutes per task) was measured. A maximum of eight tasks (four tasks solved twice) were to be completed.

Efficiency: (1) The *time* needed to process the tasks was surveyed. However, the time is a rather unspecific measure (as it does not tell us what users actually do in the menu). Therefore, more specific measures were determined in addition. (2) The number of *detour steps* (steps executed in the menu that were not necessary when solving the task on the shortest way possible (3) The number of *hierarchical returns* to higher levels in menu hierarchy, indicating that users in the belief of having taken the wrong path go back to a known menu position, consequently re-orientating themselves. (4) The number of *returns to the top*. This measure was assumed to reflect utter disorientation, as users had to re-orientate by returning to the top menu level, beginning from scratch.

Experimental Tasks

Four typical and frequently used mobile phone tasks were selected. In total, a minimum of 47 steps was necessary to solve the four tasks. Participants had to:

1. Call a number (11 steps)
2. Hide one's own number when calling someone (14 steps)
3. Send a text message (11 steps)
4. Make a call divert to the mailbox (11 steps)

In order to determine learnability effects, the tasks had to be solved twice consecutively. The order of tasks in the two trials was held constant over participants.

Apparatus and Materials

For the mobile phone, a well-known mass model, the Siemens S45, was chosen. In order to experimentally examine the quality of users' menu navigation performance, it was of methodological importance to analyze individual navigation routes in detail, and controlling for confounding factors at the same time. Therefore, the phone was simulated as software solution, run on a PC and displayed on a touch screen (Iiyama TXA3841, with a touch logic by ELO RS232C). Figure 2 shows a snapshot of the emulated phone. The display size corresponded to the original size, but the chassis of the phone and the keys were enlarged in order to enable easy operation of them with the finger on the touch screen. Moreover, a logging software tool was developed, which enabled us to log any user interaction with the system. By this, the number and type of keys used, the functions selected, and the individual navigation routes taken through the menu could be reconstructed in detail.

Figure 1. Snapshot of the emulated mobile phone (Siemens S45)

Participants

In total, 108 participants (58 females, 50 males) volunteered for the study. They were divided into three age groups: In the children group, 22 girls and 14 boys participated (M = 9.4; SD = 0.7). From the 36 young adults, 18 were females and 18 males (M = 24.1; SD = 2.8). Finally, 36 middle-aged adults (18 females, 18 males) took part (M = 47.1; SD = 7.6). The children were in their fourth school year. In the younger group, students of different academic fields volunteered. Participants of the middle-aged group were reached by an advertisement in a local newspaper and had a wide educational range. It was instructed that the study aimed at an evaluation of the usability of mobile phones. The motivation to join the study was high.

Assessing Users' Characteristics Interacting with Navigation Performance

As it was a major aim of the study to learn how the three age groups were interacting with the mobile phone and, moreover, which user characteristics might be crucial for navigation performance, the participants were surveyed regarding their spatial ability, verbal memory and the experience using technical devices. Here, the frequency and the reported ease of using these devices were assessed. Moreover, participants' interest in technology was determined.

Assessing Spatial Abilities and Verbal Memory

Assessing spatial and memory abilities in the children group, two subtests of the HAWIK-R were carried out. In the test on spatial ability ("Mosaic Test") the experimenter showed the child a picture (Figure 2) and the child's task was to reproduce the picture using cubes having different patterns on each of the sides. A maximum of 26 points could be attained in this test.

The test on short-term memory required the children to verbally repeat a row of numbers read aloud by the experimenter. The test consisted of seven rows of between three and nine numbers, which had to be reproduced directly after. The children were given two trials to correctly reproduce each row. A maximum score of 14 points could be reached.

For the adult group, spatial abilities were assessed with the paper-folding test (Ekstrom et al., 1976) in an online version (http://www.lap.umd.edu/vz2). Each of the 20 items includes successive drawings of two or three folds made in a square sheet of paper. The final drawing shows a hole punched in the folded paper. Participants had to mentally rotate the paper from the folded into the fully opened form and to indicate which of a number of possibilities shows the correct drawing. The 20 items had to be solved within 180 seconds. In Figure 3, an example item of the paper-folding test is given.

To assess memory ability in the adult groups, the verbal memory test adapted (Bay & Ziefle, 2003) from the learning and memory test (Bäumler, 1974) was used. Fifteen Turkish words (unknown to German participants) were presented in succession for three seconds each. Directly after the presentation, participants had to recognize the target items among three distractors, each being phonologically or visually similar. The maximum score to be reached was 15. An example from this test is given in Figure 4.

Figure 2. Mosaic test (Hawik-R): The upper row represents the single cube sides. The lower row represents one of the spatial tasks. The spatial demand for the children was to mentally deconstruct the figure into single cubes and to mentally rotate and arrange the cubes according to the figure.

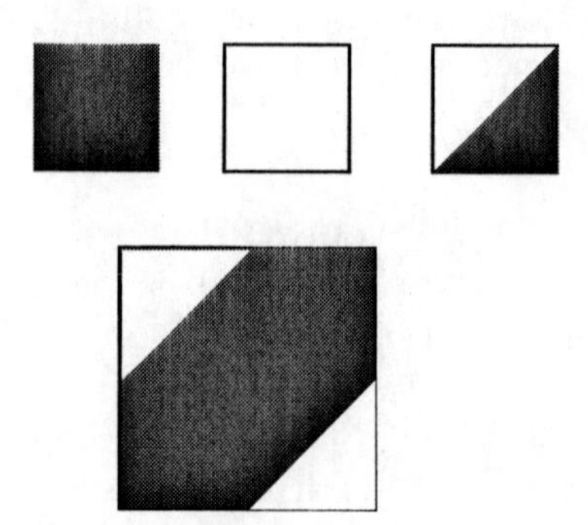

Figure 3. Item example of the paperfolding test (Ekstrom et al., 1976)

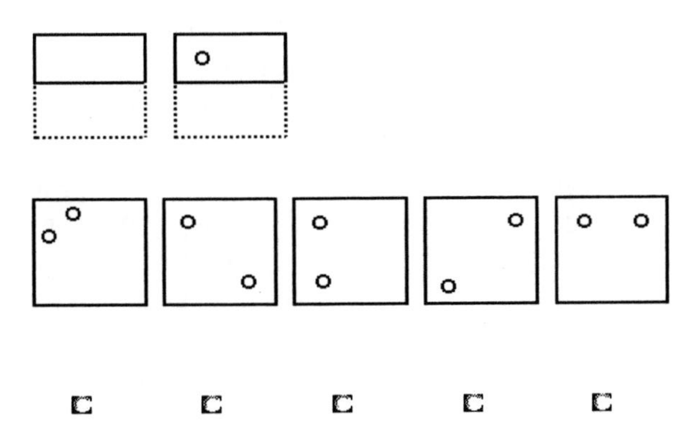

Assessing Previous Experience with Technical Devices

A detailed pre-screening of participants' experience with mobile phones and other technical devices as well as the reported ease of using them was carried out. Participants reported if and how often they use technical products (mobile phone, PC, video cassette recorder (VCR) and DVD), using a 5-point scale (1 = several times per day, 2 = once per day, 3 = once or twice a week, 4 = once or twice per month and 5 = less than once or twice a month). Furthermore, the estimated ease of using different technical devices had to be stated on a scale with four answering modes (1= the usage is easy, 2 = the usage is rather easy, 3 = the usage is rather difficult and 4 = the usage is difficult). Finally, participants indicated their interest in technology, using a 4-point scale (1 = low interest; 2 = rather low interest, 3 = rather high interest, 4 = high interest).

Results

The results were analyzed by multivariate analyses of variance assessing effects of age and gender on navigation performance in terms of effectiveness (number of tasks solved) and efficiency measures (time, detour steps, hierarchical returns, and returns to the top. Furthermore, the relationship between the users' age, gender and user characteristics was determined by correlation analyses. The level of significance was set at $p < 0.1$.

The result section is organized as follows. First the outcomes regarding user characteristics in the three age groups are described and how characteristics interrelate. Second, navigation performance is looked at, differentiating performance outcomes regarding age groups and also regarding gender effects. Then, learnability effects are focused upon, comparing the first contrasted to the second trial, determining if and to what extent performance improved for the three ages and for gender groups. A final analysis is concerned with the impact of user characteristics for performance outcomes.

Figure 4. Item example from the verbal memory test (Bay & Ziefle, 2003)

TATIL

☐ TAFIL ☐ TARAK ☐ TITAL ☐ TATIL

User Characteristics of Participants

In this section, the experience with technical devices, and the reported ease of using them are focused on. Also, the rated interest in technology is illustrated. Furthermore, the users' verbal memory and spatial abilities are described. It is of interest, if these variables are modulated by age or the gender of participants.

Previous Experience with Technical Devices

First, all participants reported to have a high experience using mobile phones. While younger and middle-aged adults possessed an own mobile phone, only 14 of the 36 kids did so. The children, who did not own a phone, though, reported to have frequent access to mobile phones (friends, siblings, or parents). Even though devices offer an increasing number of functionalities, astoundingly, only a very small fraction of these functionalities were used, across all age groups. However, there were age differences in the type of functionalities, which were commonly used. In the children group, the phones were mainly used for games (first choice) followed by calling and sending text messages. The young adult group indicated to use the phones mainly for text messaging and calling, but they also reported to play games and to use the phone as an alarm clock. Contrary, the middle-aged adult group reported to use the phones mainly for calling purposes, and emphasized that the majority of functions are "quite unnecessary." In Table 1, key results (means, standard deviations) are given for the experience measures and the interest in technology.

The frequency of mobile phone usage was significantly different for the three age groups (χ^2= 28.6; p = .000). The children and the middle-aged adult group used it 1-2 times a week (not differing from each other), while younger adults reported to use it once daily. With respect to PC usage, another significant age difference was found (χ^2 = 31.2; p = .000). The PC was used least in the child group (1-2 times a week), while both adult groups indicated to use it at least once a day. Finally, the frequency of using VCR/DVD is low, with the middle-aged adults using it about 1-2 times per month, while children and young adults use VCR/DVD about 1-2 times a week (χ^2 = 17.1; p = .000). When focusing on the ease of using these devices, all participants reported the usage as easy or at least rather easy. Age differences were

Table 1. Means (standard deviations) in user characteristics in all age groups

	Children	Young adults	Middle-aged adults
Gender	22 girls, 14 boys	18 women, 18 men	18 women, 18 men
Age	9.4 (0.7)	24.1 (2.8)	47.1 (7.6)
Frequency using a...	1= several times a day; 2 = Once daily; 3 = 1-2 times a week; 4 = 1-2 times a month; 5 = less than once a month		
Mobile phone	3 (1)	1.5 (0.8)	2.8 (1.8)
PC	3.2 (1.3)	1.5 (0.8)	1.9 (1.2)
VCR/DVD	2.9 (1.4)	3.2 (1)	4.2 (1.2)
Ease of using is...	1= easy; 2 = rather easy; 3 = rather difficult; 4 = difficult		
Mobile phone	1.4 (0.6)	1.2 (0.5)	2 (1.4)
PC	1.6 (0.7)	1.5 (0.6)	1.8 (1)
VCR/DVD	1.3 (0.5)	1.5 (0.7)	2.1 (1.4)
Interest in technology	1= low; 2 = rather low; 3 = rather high; 4 = high		
	3.4 (0.8)	2.8 (1.1)	2.4 (1.1)

found only with respect to VCR/DVD usage (χ^2= 8.8; p = .000). Finally, the interest in technology revealed another significant age difference (χ^2= 15.9; p = .000). The highest interest was present in the children group (M = 3.4). The lowest interest in technology was reported by the middle-aged adult group (M = 2.8), however, the young adults' interest was also comparably low (M = 2.4).

Across all age groups, female and male users did not differ regarding to the frequency of using mobile phones, VCR/DVD and PC, but introduced themselves as frequent users of common technical products. Also, no correlation of gender and the ease of using mobile phones and VCR/DVD devices were revealed. However, the interest in technology (r = -.26; p = 0.008) and the ease of using the computer (r= -.38; p = 0.000) showed significant correlations to gender: Female users reported the PC to be more difficult to use than the males and their interest in technology was lower compared to male users' interest in technology. Though interrelations were present in all age groups, interestingly, they were most pronounced for children.

Verbal Memory and Spatial Ability

First, outcomes in verbal memory in all age groups are addressed. On the left side of Figure 5, the scores of the children are illustrated. From the 14 points, the children reached, on average, "only" 5.4 points (SD = 1.5), and none of the children was able to reach the maximum score. Young adults (Figure 5, center) reached, on average, 12.4 points (out of 15). The middle-aged adult group (Figure 5, right) showed also a solid memory performance (M = 10.3; SD = 2.6), even though their memory score differed significantly from the younger adults' score (F (1,71) = 15.3; p=0.000).

In Figure 6, the outcomes in spatial abilities are pictured. The children (Figure 6, left) differed considerably with respect to spatial abilities. The inter-individual variance among children was high (range 4-26 points; M = 15.6; SD = 5.5), showing big developmental differences among 9-10 years old kids. For the younger group (Figure 6, center), the spectrum of correct answers ranged between 8 and 19 points (out of 20), reaching a mean performance of 13.2 (SD = 3.1). Finally, the middle-aged group reached an average score of 12.9 out of 20 points (SD= 3.7). The range of answers (4 points minimum and 20 points maximum) also represents a high variance, showing that spatial abilities do not follow a systematic decrease with increasing age. Statistical testing revealed no significant differences between spatial abilities of younger and middle-aged adults. Also, no gender differences were present neither with respect to verbal memory, nor spatial ability. However, for the children, there was a significant correlation of gender and the level of spatial ability (r = 0.6; p=0.03), with boys having higher spatial abilities (M = 18/26 points) than girls (M = 14/26 points).

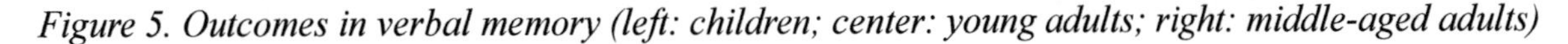

Figure 5. Outcomes in verbal memory (left: children; center: young adults; right: middle-aged adults)

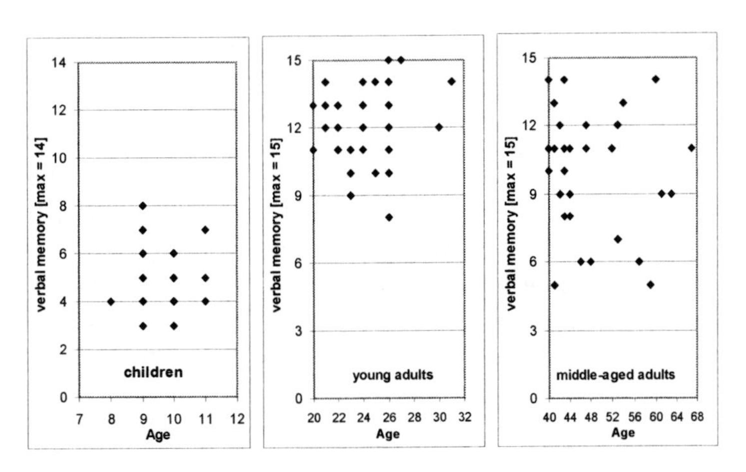

Figure 6. Outcomes in spatial abilities (left: children; center: young adults; right: middle-aged adults)

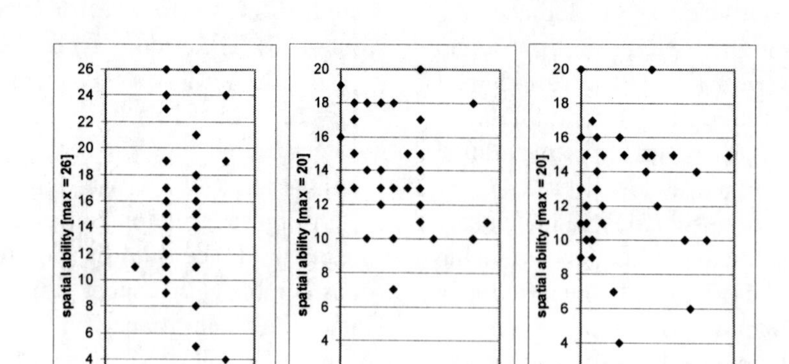

Menu Navigation Performance

When comprising the navigation performance in the eight tasks, a significant omnibus effect of age was found (F (2,102) = 18,1; p=0.00). Moreover, a significant omnibus effect of gender was revealed (F (1,102) = 3.1; p=0.01) as well as a significant interaction effect of age and gender (F (2,102) = 2.9; p=0.002). The age effect (single F-test; F (2,102) = 59.9; p<0.000) was based on significant differences between all age groups.

In Figure 7, the outcomes in task effectiveness are illustrated (left side: effectiveness for all age groups; right side; effectiveness for age and gender groups).

As can be seen there, the children showed the lowest effectiveness, reaching a mean task effectiveness of 51 percent. The best performance was present in the younger adult group, yielding a task effectiveness of 94 percent, while the middle-aged adults' effectiveness ranked in between (79 percent). Even though adult participants showed a considerably better performance than the children, it is quite astounding that not even the students were able to solve the eight tasks completely successfully. Also the effects of gender (F (1,102) = 11.3; p=0.001) as well as the interaction of age and gender (F (2,102) = 3.1; p=0.05) become obvious. The significant gender effect was originated by the girls' lower task success compared to the boys (girls: 43 percent; boys: 64 percent), while gender differences in the adults groups were not found to yield significant effects.

Furthermore, task efficiency is considered. Effects of age were significant for each of the single measures (time: F (2,102) = 50.5; p=0.000; detour steps: F (2,102) = 41; p<0.000); hierarchical returns: (F (2,102) = 22.6; p<0.000; returns to the top: (F (2,102) = 12.3; p=0.000). Independently of the measure, the children showed the lowest task efficiency, followed by the middle-aged adult group. The best task efficiency was present for young adults.

In Figure 8, the key results in task efficiency are illustrated for age groups and gender. As can be seen there, the children needed 20 minutes and 36 seconds and made 672.7 detour steps, when processing the phone tasks. The detouring of the children is considerable when taking into account that, overall, a minimum of 94 steps were needed to solve the tasks (47 steps per trial). Furthermore, the children made about 81.1 returns in menu hierarchy, and left the menu 8.1 times, to begin from scratch. Compared to the children, both adult groups were much more efficient even when the young adults significantly outperformed the middle-aged adult group. For the student group, it took 6 minutes 39 seconds to complete

Figure 7. Task effectiveness (%) in the three age groups (left) and both gender groups (right)

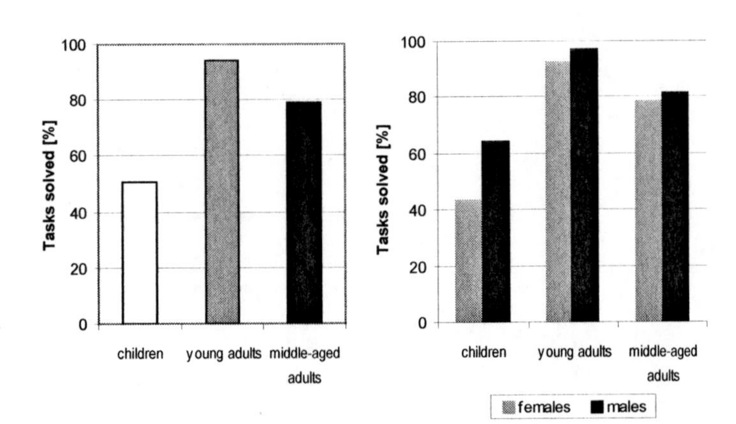

the tasks; compared to 14 minutes 19 seconds in the middle-aged group- this equals a benefit of 57 percent. Also, the students carried out only—but still—213.7 detour steps, returned 15.5 times to higher levels in menu hierarchy and began 0.9 times from the very beginning. In contrast, the middle-aged group made 414 detour steps, and returned, on average, 52.5 times to higher levels in menu hierarchy. Also, they made 5.4 complete returns to the top menu level.

Looking on gender effects, male participants showed a generally more effective, and also a more efficient navigation style. However, gender effects were not symmetrical across ages and measures, respectively. The most pronounced gender differences were present in the children group, while gender effects decrease with increasing age. A quite interesting effect was revealed for the navigation strategy, which was different for girls and boys. The boys were more successful solving the tasks compared to the girls. Also, they needed less time and showed overall a smaller amount of returns to the top. So far, their navigation style is similar to adult male participants when compared to female participants. Yet, the boys' higher effectiveness was reached by a higher amount of exploration behavior in the menu. This can be taken from the high number of detour steps and hierarchical returns. Actually, their detouring was larger than that of all participants. Apparently, the boys capitalize the additional detouring on an overall better navigation performance, in contrast to the girls but also in contrast to both adult groups. Interestingly, the boys returned to a lesser amount to the top menu level compared to the girls, which shows from another side that the boys' detouring is more probably reflecting an active menu exploration rather than disorientation—otherwise they should have started from scratch more often.

Learnability Effects: Comparison of Navigation Performance in the First vs. Second Trial

Learnability effects, especially their interaction with age and gender effects might give additional insights in the difficulties users experience when interacting with small screen devices.

Again, first the task effectiveness is looked at. A significant learnability effect ($F(1,102)=17$; $p=0.000$) was revealed, showing that in the second trial more tasks were solved successfully compared to the first trial (children: first trial: 47 percent, second trial 55 percent; young adults: first trial: 93 percent, second trial 95 percent; middle-aged adults: first trial: 75 percent, second trial 84 percent). No interaction effects of learnability and gender and learnability and age were revealed. Thus the higher task success in the second run was equally large for all participants.

Figure 8. Task efficiency (time on task, number of detour steps, hierarchical returns and returns to the top)

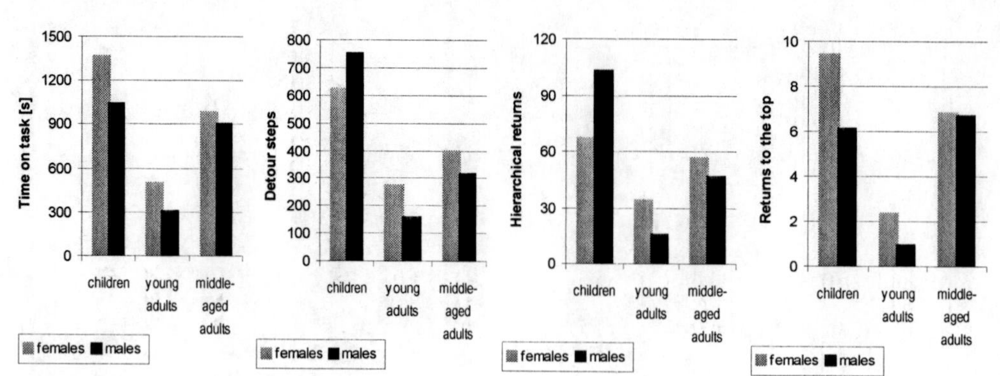

Next, efficiency is taken into account. In Figure 9 efficiency measures (time on task, detour steps, hierarchical returns and returns to the top) are depicted, differentiating the three age groups. Independently of the measure, there is a clear age pattern. Young adults showed the best performance and children the lowest performance.

According to the learnability effects regarding effectiveness, one would expect hat efficiency should also be higher in the second trial, therefore participants should spent less time on tasks, accompanied by fewer detour steps. In addition, also the number of returns in menu hierarchy and returns to the top should considerably decrease in the second run. Even though significant learnability effects for efficiency measures were present (time: $F(1,102) = 9.8$; $p=0.002$; detour steps: $F(1,102)=6,4$; $p = 0.01$), it becomes obvious that within navigation efficiency learnability effects are not equally high for all groups, but interact with the age of participants (time: $F(1,102) = 22.3$;$p=0.000$) as well as with gender (time: $F(1,102) = 7.8$; $p=0.006$; hierarchical returns: $F(1,102) = 5.5$; $p=0.002$; returns to the top: $F(1,102) = 2,5$; $p<0,1$). Moreover, there were also three-fold interactions between learnability, age and gender (time: $F(1,102) = 2, 3$; $p<0.1$; hierarchical returns: $F(1,102) = 2, 5$; $p<0.1$; returns to the top: $F(1,102) = 2, 4$; $p<0.1$). In order to disentangle the complex interrelation, first the nature of the interacting effect between learnability and age is addressed (Figure 9).

From Figure 9, it can be seen that actually only the young adult group profit from executing the tasks a second time. They were faster, executed less detour steps, and carried out fewer returns in menu hierarchy and also fewer returns to the top in the second trial compared to the first. The children and the older group, however, showed a different pattern. As the young adults, they were also faster in the second trial, however, in contrast to students, their detouring behavior did not improve in the second trial, as taken from the number of detours steps, the hierarchical returns and the returns to the top level. In short, one could characterize children's and older adults' navigation style as less cautious in the second compared to the first run (as they were faster), but still inefficient.

However, it is a basic question, whether learnability effects are similar across female and male users. This is analyzed in Figure 10. Here, tasks' efficiency is pictured for all participants, males (gray lines) and females (black lines) as well as for all age groups. The upper row of Figure 10 represents task efficiency of the children, the middle row navigation efficiency of the older group and the lower row shows efficiency measures of the young adults.

Again, we see the better overall performance of male compared to female users as well as the clear performance superiority of younger adults. Furthermore, it becomes obvious that the children group is considerably different compared to

Figure 9. Task efficiency (time on task, number of detour steps, hierarchical returns and returns to the top) in the first compared to the second trial for all age groups

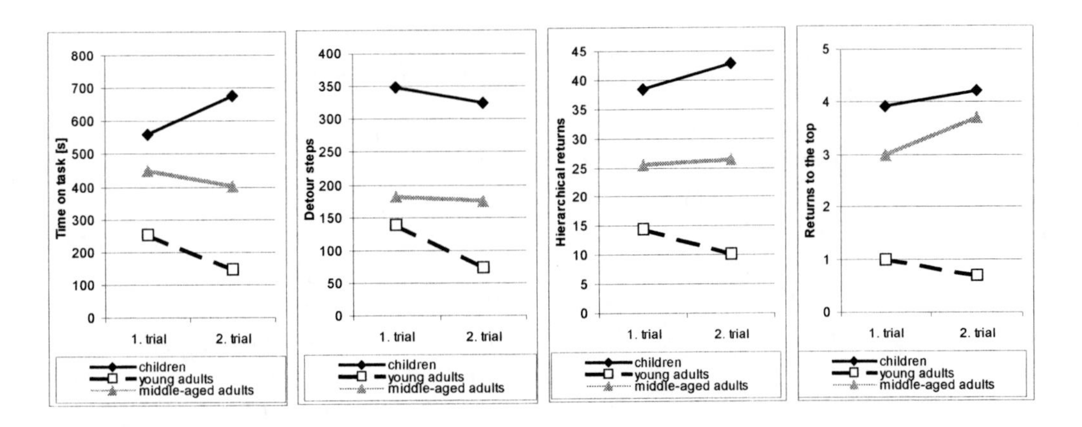

Figure 10. Task efficiency (time on task, number of detour steps, hierarchical returns and returns to the top) in the first compared to the second trial for all age groups (upper: children, middle: younger adults and lower row: middle-aged adults) as well as for female (black lines) and male (gray lines) participants

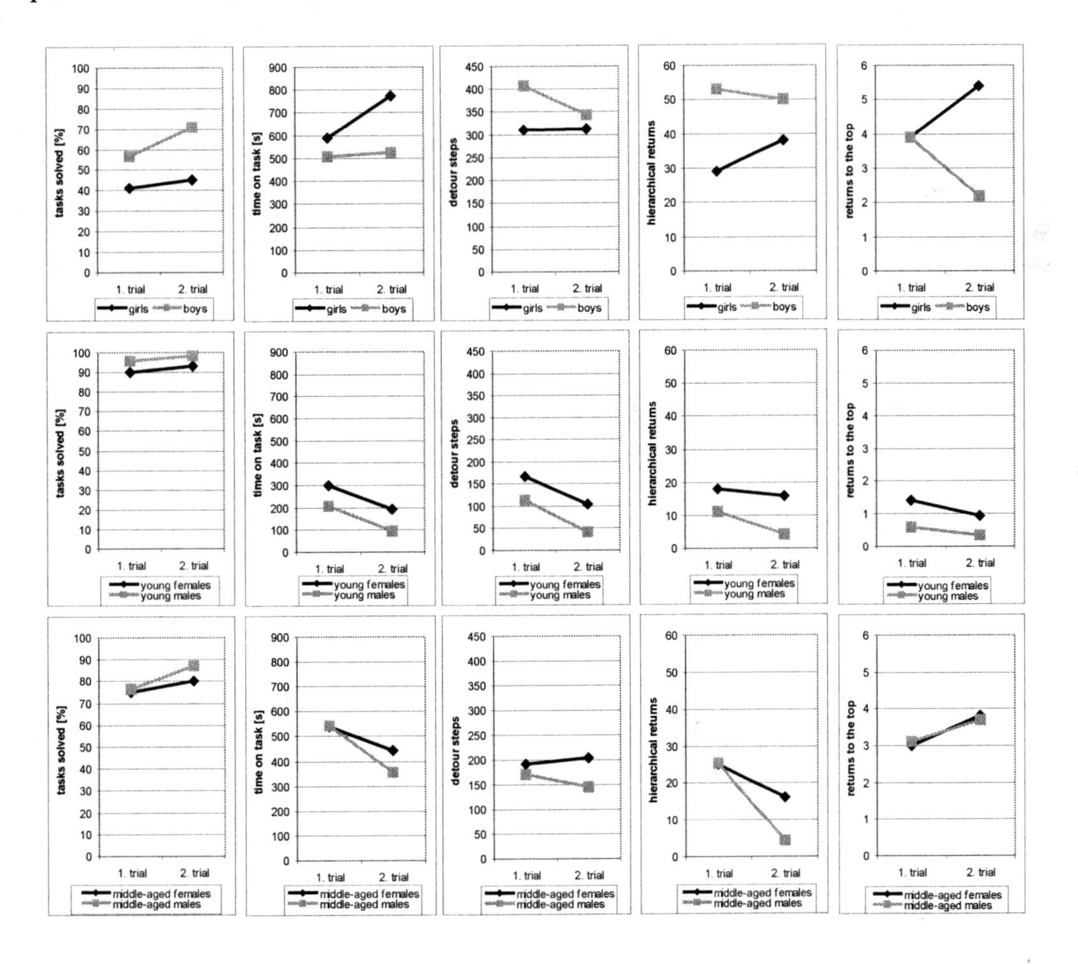

both adult groups: First, children had overall the lowest efficiency. Second, children did not profit from learnability, at least not the whole children group. Third, gender effects are most pronounced at this age, with boys having not only an overall better performance, but also revealing a different navigation style than the girls. The boys showed learnability effects, and profit from executing the tasks a second time, while, quite contrary, the girls did not show learnability effects and even had a lower performance in the second compared to the first run. Fourth, again, the boys' different navigation strategy becomes evident, which is characterized by a higher tasks success, a faster navigation with lower returns to the top. This effective strategy is reached by a more active and explorative menu navigation pattern—the boys carried out even more detour steps and hierarchical returns in the second trial than the girls in the first. With respect to the returns to the top, the measure for menu disorientation, the boys left the menu only twice in the second trial, compared to four times in the first.

User Characteristics and their Effects on Menu Navigation Performance

So far, we found performance differences between age and gender respecting effectiveness and efficiency when using a mobile phone. In this section it is analyzed, which of the user characteristics may account for these differences. Is users' experience with technical devices a substantial source of performance or is their interest in technology the source of the differences? Which role are the differences in spatial ability and verbal memory playing for menu navigation performance? First, the interrelations of the experience with technical device and the interest in technology are focused (Table 2).

Neither the reported interest in technology was interrelated with performance outcomes, nor the reported ease of using the devices showed significant correlations. Thus, motivational factors as the individual interest in technology were not decisive for performance. Also the perceived ease of using the devices did not reflect the actual navigation performance. However, participants' previous experience in terms of the self-reported

*Table 2. Correlations between user characteristics and navigation performance (***p = 0.000; ** p =0.05; * p = 0.1)*

N = 108	Tasks solved	Time on task	Detour steps	Hierarchical returns	Returns to the top
Frequency using a...					
Mobile phone	**r = -0.46*****	**r = 0.5*****	**r = 0.44*****	**r = 0.45*****	**r = 0.45*****
PC	**r = -0.48*****	**r = 0.41*****	**r = 0.42*****	**r = 0.36*****	**r = 0.26****
DVD/VCR	**n.s**	**n.s**	**n.s**	**n.s**	**n.s**
Ease of using a...					
Mobile phone	**n.s.**	**r = 0.27****	**n.s.**	**n.s.**	**n.s.**
PC	**n.s.**	**r = 0.23****	**n.s**	**n.s**	**r = 0.26****
DVD/VCR	**n.s.**	**n.s.**	**n.s**	**n.s**	**n.s**

frequency of using technical devices was strongly interrelated with performance outcomes, showing that frequent usage and activities with the devices lead to a elaborated knowledge that is basically benefiting performance for all age groups (even though it should be considered that the experience level was not sufficient to reach an "optimal" performance (100 percent task success) in neither of the groups).

With respect to the impact of cognitive factors (spatial ability and verbal memory), correlation analyses were run for each age group, separately. The outcomes are summarized in Table 3.

For the children group, neither spatial abilities nor verbal memory had a significant impact for children's navigation performance. Thus, even those kids, who had—relatively to other children of this group—possessed a high verbal memory capacity and spatial ability were not specifically advantaged in menu navigation over those kids, who have only lower cognitive abilities. In other words: navigating through mobile phone menus was a high cognitive demand for all children. In contrast, both adult groups showed interrelations with spatial abilities, even though the relation was much more pronounced in the young adult group compared to the older adult group. Users with a high spatial score solved more tasks, were faster, and also made fewer detour steps and hierarchical returns. Also they did not return to the start level as often than adults with lower spatial abilities.

The impact of verbal memory capacity for navigation performance was comparably low. For the young adults, none of the performance measures showed significant interrelations with verbal memory. In the middle-aged adult group, however, effects of variability in memory capacity were revealed: users with high memory abilities solved significantly more tasks and were faster in comparison to those with a lower memory capacity.

Implications for Design

Even though the central focus of this work was directed to a detailed analysis of user behavior interacting with mobile phones and, also, to the impact of user diversity for performance, the present findings may also give some insights for design concerns to be considered for mobile devices. Also some training and tutoring issues may be derived.

The results reported here uncover both, similarities as well as differences in the navigation behavior of kids, young and older adults. Thus, we learn that there are design implications, which favor "a design for all approach" as well as differential aspects, which should be pursued to support specific user groups.

Across age groups, considerable difficulties were revealed in completing these common and easy phone tasks on a standard mobile phone. Not

*Table 3. Correlations between user characteristics and navigation performance (***p = 0.000; ** p =0.05;* p = 0.1)*

	Tasks solved	Time on task	Detour steps	Hierarchical returns	Returns to the top
Spatial ability					
Children	**r = 0.29***	n.s	n.s	n.s	n.s
Young adults	**r = 0.29***	**r = -0.36****	**r = -0.29***	**r = -0.48*****	**r = -0.29***
Middle-aged adults	**r = 0.37****	**r = -0.40****	**n.s**	**n.s**	**r = -0.45*****
Verbal Memory					
Children	n.s	n.s	n.s	n.s	n.s
Young adults	n.s	n.s	n.s	n.s	n.s
Middle-aged adults	**r = 0.40****	**r = -0.36****	n.s	n.s	n.s

even the students, bright and technology prone were able to solve the four tasks completely successfully. Nevertheless students showed the best performance compared to kids and older adults. All participants carried out a lot of detouring in the menu, as taken form the high number of detour steps, the returns in menu hierarchy and the returns to the top menu level, beginning from scratch. Thus, it must be concluded that current small screen devices are—cognitively—challenging to use. This is valid for kids and older adults, which—due to their developmental status—can be categorized as "weaker" users, but it is also valid for the "best case" student user group. According to statements of participants after the experiment, the difficulties they experienced in the menu had mainly three sources. The first difficulty referred to the complexity of the menu. Even though the task complexity was relatively low, with three menu levels at the most, participants (especially kids and older adults) had considerable problems to orientate, often not knowing were they were in the menu and where they had to go next. A second point refers to keys' complexity. Ambiguous functionality and design of keys lead to difficulties and provoked many unnecessary key actions. This was especially the case for navigation keys, which had a high complexity (keys, with several functions on different menu levels). Third, the naming of menus, sub-menus, and functions is also of crucial importance for good usability of a mobile phone.

However, there were also findings that hint at specificities of user groups, which should be considered. The first refers to different coping styles of children and older adults, when confronted with suboptimal and not very intuitive interfaces. Older users were nearly annoyed and insistently emphasized that they want to have devices that meet their demands of low complexity and all necessary functions within easy reach. Otherwise, they are not willing to use these devices. For the children, mobile phones still represent a high status and attractive gadget. Nevertheless, the children reacted highly sensitive to their failure and tended to attribute the failure to their own incompetence, but they also criticized that the mobile phone was "pretty hard" to use. This is of specific pedagogic impact. The success and the ease with which devices can be used contribute considerably to users' self-efficacy, the perceived competency and usability of technical devices (e.g., Arning & Ziefle, 2007; Ziefle, Bay & Schwade, 2006). Sub-optimally designed interfaces might lead to a lower frequency of using technical devices, in order to avoid negative feelings. However, frequent interaction with technical devices is an essential precondition for the formation of technical expertise, which in turn benefits navigation performance. Gender effects were also identified. Female users reported a significant lower interest in technology, and also, rated the ease of using technical devices as lower. While gender differences in performance did not reach significance level in both adult groups, for the children, they were most pronounced. The boys reached overall a higher performance, which was supported by a specific and successful navigation strategy. The boys were highly explorative and active in the phones' menu, pursuing a trial and error style. In contrast, the girls showed a much lower activity when interacting with the phone. Playful experimentation is assumed to yield educational benefits because the users may incidentally gain knowledge of the system by exploring its structure.

As practical implications from the current research, the following recommendations can be given:

Interface-design:

- Keep menu structures as flat as possible and avoid high complexity, this helps to reduce menu disorientation.
- Keys complexity should be held as low as possible. Avoid the allocation of many functions to single keys (multimode). Whenever keys with more than one function have to be used, those functions should be grouped to one key, which have a (semantically) similar meaning ("correct," and "step back"). Avoid mode keys with semantically dissimilar functions ("step back," "hang up"). Also, the spatial position of the keys is important. Very frequent functions should be allocated

to centrally located keys. Do not change position of frequently used keys.

- Respecting the naming of menu functions, basically, very similar terms ("phone setting" vs. "call setting") should be avoided. Also, do not use abstract terms ("incognito"), unfamiliar abbreviations ("GBG," "GSM") or technical terms ("D2 services," "SIM-activity"). For many users, they are not easy to understand and therefore, are not easily to be learned. Furthermore, very generic terms ("options," "settings") are highly misleading terms. Even though these terms are not difficult to understand, it is nevertheless not easy to deduce which functions are summarized under these category labels. Apparently, it is difficult to know which functions are and which are not present within these categories, as the general terms basically allow—semantically spoken—a targeted function to be housed, however, in many cases they actually do not. Categorizing the functions in intuitively understandable menus and sub-menus is also of great importance for a good design.

Training and pedagogic issues:

- It is essential to encourage active and playful interaction with the device quite early in the learning process, especially for children and middle-aged users, in order to enable the development of perceptions of achievement and competence. By playful interacting with technical devices, users get casually to know which features and functions are available, where these functions are located within the menu, and also how to activate the functions. This may especially benefit users, who cannot rely on a high spatial ability, a high verbal memory capacity or a high technical self-competence, to develop a solid expertise respecting using small screen devices. Thus, it is of high importance to motivate and encourage users to actively interact with mobile devices.

CONCLUSION

This study aimed at a critical actual inventory of user characteristics and the competence using a mobile phone. On the user side, effects of age and gender were analyzed, as well as technical experience and perceived ease of device usage. Moreover, interest in technology was taken into account. Further, spatial ability and verbal memory capacity were psychometrically determined and related to menu navigation performance. On the performance side, an elaborate analysis of navigation patterns was undertaken. Beyond the task success and the processing time, the individual extent of detouring behavior was analyzed. In this context we determined the number of detour steps, but also, how often participants returned to higher levels in menu hierarchy. As a measure for utter disorientation, we analyzed how often participants, after they delved into distraction, returned to the first menu level, beginning from scratch. Methodologically, this detailed analyzing procedure can be strongly recommended, as it mirrors exactly what users actually do and enables the determination of the relation between performance and user judgments, which are often biased. Considering that the majority of manufacturers evaluate mobile phones primarily operating with user ratings for evaluation purposes, the validity of user ratings is questionable. Of course, preference ratings can be obtained much more easily, but they possibly do not reflect the actual difficulties of users in the system. If a device is supposed to be accepted in the long run and also acknowledged by a diverse user group, the impact of a detailed analysis of navigation patterns seem essential.

Our participants had a solid experience with different state-of-the-art technical devices. It could be shown that the experience with technical devices considerably advantaged menu navigation performance. Interestingly, and perhaps contrary to expectations, the children had, relatively, the smallest computer experience, but nevertheless a high interest in technology—higher than the interest reported by both adult groups. As shown, the common prejudice of children to easily master the handling of technical devices—due to their early contact with technology—is not true, at least not in the most stringent form. From a pedagogic point of view, it is important to motivate especially kids to frequently use and handle technical devices, in order to support the formation of technical experience.

With respect to cognitive abilities, which underlie the performance in technical systems, spatial abilities turned out to be important for the menu performance. This shows that the navigating in small screen devices imposes considerable demands on users' ability to "spatially" orientate within the menu. Spatial abilities are assumed to provide a specific advantage. Persons with high spatial abilities are able to construct a mental representation of the systems' structure during navigation (Sein, Olfman, Bostrom et al., 1993), and therefore have a better orientation in the system. However, the impact of spatial abilities for performance turned out to be age-related. While younger and middle-aged adults were able to profit from high spatial abilities, this was not the case for the 9-10 year old kids in our experiment. This confirms earlier findings (e.g., Bay & Ziefle, in press; Shemakin, 1962), according to which the ability to cognitively process spatial hints and to mentally represent structural knowledge is fully developed not until children are 12-13 years. It would be insightful to examine if there are specific trainings or software tutors feasible, which can support younger children and help them to achieve a good performance.

A final remark is concerned with some limitations regarding the methodology used. Our results are based on laboratory experiments and on the interaction with a simulated mobile phone. This was accomplished in order to provide experimental control and to rule out confounding effects. However, we acknowledge that the results presented here might represent a solid underestimation of the real situation. In our experiment, the cognitive workload to use mobile devices was much lower than they usually are in the interaction with mobile devices in real environments. In a mobile context users have to manage different and complex demands, simultaneously, and in the laboratory setting a quiet setting was present and users were able to concentrate on the tasks. Also key handling and visibility problems may occur in real contexts, which were controlled for in the experiment. Another limitation refers to the selection of the middle-aged adult group, which is definitively not representative for the whole group of adult users, especially older users (65+). Thus, overall, we have to concede that the performance levels reached in our setting might be higher compared to more realistic settings.

FUTURE TRENDS

Due to the fast cycles of technical innovations and the development of novel and still more complex technical devices, usability demands will still increase. This is of vital interest facing the demographic change and the increasing prominence of mobile devices. Therefore, research activities should address user diversity more strongly than hitherto. Many topics in this context should be pursued in greater detail. One is to examine the nature and benefit of the exploratory behavior of users when interacting with technical devices. It is a central question if there are specific interaction strategies, which should be encouraged or supported by trainings, and if there are differential aspects, which should be applied in specific user groups. Another interesting research question is the question whether the findings reported here are limited to devices with an exclusively hierarchical menu structure (as the mobile phone) or if the navigation patterns found here can be transferred to devices with a network data structure, which might provoke a completely different interaction pattern.

REFERENCES

Arning, K., & Ziefle, M. (2006). What older users expect from mobile devices: An empirical survey. In *Proceedings of the IEA 2006*. Amsterdam: Elsevier.

Arning, K., & Ziefle, M. (2007). Barriers of information access in small screen device applications: The relevance of user characteristics for a transgenerational design. In C. Stephanidis and M. Pieper (Eds.), *User interfaces for all: Universal access in ambient intelligence environments* (pp. 117-136). Berlin, Germany: Springer.

Arning, K., & Ziefle, M. (2007). Understanding Age differences in PDA acceptance and performance. *Computers in Human Behavior, 23*(6), 2904-2927.

Bäumler, G. (1974). *Learning and memory test.* Göttingen, Germany: Hogrefe.

Bay, S., & Ziefle, M., (2003). Design for all: User characteristics to be considered for the design of phones with hierarchical menu structures. In H. Luczak, K. J. Zink (Eds.), *Human factors in organizational design and management,* (pp. 503-508). Santa Monica: IEA.

Bay, S., & Ziefle, M. (2005). Children using cellular phones. The Effects of shortcomings in user interface design. *Human Factors, 47*(1), 158-168.

Bay, S., & Ziefle, M. (in press). Landmarks or surveys? The impact of different instructions on children's performance in hierarchical menu structures. *Computers in Human Behavior,* (2007), doi:10.1016/j.chb.2007.05.003

Beckwith, L., Kyssinger, C., Burnett, M., Wiedenbeck, S., Lawrence, J., Blackwell, A., & Cook, C. (2006). Tinkering and gender in end-user programmers debbuging. In *Proceedings of the ACM Conference on Human Factors in Computing Systems 2006* (pp. 231-240).

Busch, T. (1995). Gender differences in self-efficacy and attitudes toward computers. *Journal of Educational Computing Research, 12*, 147-158.

Davies, S. (1994). Knowledge restructuring and the acquisition of programming expertise. *International Journal of Human-Computer Studies, 40*, 703-726.

Downing, R.W., Moore, J.L., & Brown, S.W. (2005). The effects and interaction of spatial visualization and domain expertise on information seeking. *Computers in Human Behavior, 21*, 195-209.

Egan, D. (1988). Individual differences in human-computer-interaction. In M. Helander (Ed.), *Handbook of human-computer-interaction* (pp. 543-568). Amsterdam: Elsevier.

Ekstrom, R. B., French, J. W., Harman, H. H., & Dermen, D. (1976). *Manual for the kit of factor-referenced cognitive tests.* Princeton, NJ: Educational Testing Service.

Goodman, J., Gray, P., Khammampad, K., & Brewster, S. (2004) Using landmarks to support older people in navigation. In S. Brewster & M. Dunlop (Eds.), *Mobile human computer interaction* (pp. 38-48). Berlin, Germany: Springer.

Hasher, L., & Zacks, R. (1988). Working memory, comprehension, and aging: A review and a new view. In G.H. Bower (Ed.), *The psychology of learning and motivation* (pp. 193-225). San Diego: Academic.

Kim, H., & Hirtle, S. (1995). Spatial metaphors and disorientation in hypertext browsing. *Behaviour & Information Technology, 14*, 239-250.

Jakobs, E.-M. (2005). Technikakzeptanz und Technikteilhabe [acceptance and usage of Technology]. *Technikfolgenabschätzung, 14*(3), 68-75.

Maguire, M., & Osman, Z. (2003). Designing for older and inexperienced mobile phone users. In C. Stephanidis (Ed.), *Universal access in HCI* (pp. 439-443). Mahwah, NJ: LEA.

Mobileyouth.org, 5 Million School Children Own A Mobile Phone in UK. Available online at: http://www.mobileyouth.org/my_item.php?mid=1471 (accessed 2005).

Rodger, J. A., & Pendharkar, P. C. (2004). A field study of the impact of gender and user's technical experience on the performance of voice-activated medical tracking application. *International Journal of Human-Computer Studies, 60*, 529-544.

Sein, M. K., Olfman, L., Bostrom, R. P., & Davies, S.A. (1993). Visualisation ability as a prediction of user learning success. *International Journal of Man-Machine-Studies, 39*, 599-620.

Shemakin, F. N. (1962). Orientation in space. In B.G. Anan'yev et al. (Eds.). *Psychological sciences in the USSR*. (Vol. 1, Part 1) (Rep. No. 11466) (pp. 379-384). Washington: U.S. Office of Technical Reports.

Tuomainen, K., & Haapanen, S. (2003). Needs of the active elderly for mobile phones. In C. Stephanidis (Ed.), *Universal access in HCI: Inclusive design in the information society* (pp. 494-498). Mahwah, NJ: LEA.

Van-den Heuvel-Panheizen, M. (1999). Girls' and boys' problems: gender differences in solving problems in primary school mathematics in the Netherlands. In T. Nunes and P. Bryant (Eds.), *Learning and teaching mathematics* (pp. 223-253). Psychology Press.

Vicente, K.J., Hayes, B.C., & Williges, R.C. (1987). Assaying and isolating individual differences in searching a hierarchical files system. *Human Factors, 29(3)*, 349-359.

Westerman, S.J. (1997). Individual differences in the use of command line and menu computer interfaces. *International Journal of Human Computer Interaction, 9(2)*, 183-198.

Ziefle, M., & Bay, S. (2004). Mental models of a cellular phone menu. Comparing older and younger novice. In S. Brewster & M. Dunlop (Eds.), *Mobile human computer interaction* (pp. 25-37). Berlin, Germany: Springer.

Ziefle, M., & Bay, S. (2005). How older adults meet complexity: Ageing effects on the usability of different mobile phones. *Behaviour & Information Technology, 24*(5), 375-389.

Ziefle, M., Arning, K., & Bay, S. (2006). Cross platform consistency and cognitive compatibility: the importance of users' mental model for the interaction with mobile devices. In K. Richter, J. Nichols, K. Gajos & A. Seffah (Eds.), *MAFOC' 06. The Many Faces of Consistency in Cross-Platform Design* (pp. 75-81). Retrieved in 2007 from http://CEUR-WS.org/Vol.198

Ziefle, M., Bay, S., & Schwade, A. (2006). On keys' meanings and modes: The impact of navigation key solutions on children's efficiency using a mobile phone. *Behaviour & Information Technology, 25*(5), 413-434.

KEY TERMS

Ease of Use: The ease of use describes the extent to which users believe a technical system to be free from effort and easy to handle.

Effectiveness: The term reflects the degree to which system objectives (e.g., tasks) are being achieved.

Efficiency: The term describes the degree to which a certain performance is achieved in terms of productivity. For example, it can be analyzed how many detouring routes are carried out until users reach their targeted goal in the menu, and how often they return to higher levels in menu hierarchy to re-orientate.

Navigation Performance: The term navigation in this context describes the process of moving through a menu structure in order to retrieve information or choose functions. The individual navigation routes the users take while searching for a specific target function may give valuable insights into shortcomings of menu design.

Trans-Generational Designs: Trans-generational designs are interface designs, which are usable and understandable by a broad user group, thus meeting the needs and demands of user diversity. They aim at coming up with developmental (cognitive, physical and sensory) specificities, which are present in users of different ages.

Usability: The term describes users' effectiveness, efficiency, and satisfaction with which users achieve specified goals in a technical system.

Spatial Ability: It is conceptualized as the ability to mentally manipulate and integrate visual stimuli consisting of more than one part. This includes the ability to imagine of rotations of objects or their parts.

Verbal Memory: Verbal memory is the basic ability to store and retrieve verbal or semantic information without additional processing.

Chapter IX
Learning-Disabled Children: A Disregarded User Group

Susanne Bay
RWTH Aachen University, Germany

Martina Ziefle
RWTH Aachen University, Germany

ABSTRACT

In usability research it is a common practice to take young and healthy university students as participants for usability evaluations. This chapter focuses on the "weaker" mobile phone users, which have been mostly disregarded in this field: Learning-disabled children. Their interaction with mobile phones is compared to that of average children and students. Results show that the consideration of the "ergonomic worst case," which means a user group with cognitive deficiencies, leads to qualitatively and quantitatively different insights into the impact of specific design decisions. In contrast, when only students are involved as participants in the evaluation of technical devices, the impact of characteristics of the user interface on the ease of use is dramatically underestimated. One factor hampering the ability of learning-disabled children to interact meaningfully with a technical device may be their big difficulty building a correct mental representation of it. Therefore, this process should be especially supported.

INTRODUCTION

In most research projects, focusing on the usability of technical devices students serve as participants for the experimental evaluations. As students are bright and technically skilled, highly performance-motivated, have high cognitive and verbal abilities and no fear of being tested, the examination of this user group can be regarded as benchmark. Of course, this may give an insight into the effects of a specific design on users' performance interacting with the device because results can be interpreted as mainly caused by the design of the technical device, and no shortcomings have to be considered from the users' side. Furthermore, there are practical reasons for this procedure, as students can be recruited very easily by research institutions. On the other hand, the fact that some devices such as the mobile phone can be found in all age groups and levels of society give reason to seriously doubt whether students as participants in usability tests will be able to identify the real impact of specific user interface alternatives on the ease of use of the device. Taking only students' performance as basis for design decisions seems to be risky.

The purpose of the present study was to learn if and to which extent the performance achieved by students in usability studies may be generalized to a broader (or weaker) population. The motivation was to assess with a common technical device and typical tasks whether not only the quantitative performance level but also specific difficulties of the special user group could be identified. If the same difficulties may be found this means that the user interface design should be aligned with the "weakest" user's needs. If specific problems are encountered, a "design for all" approach would not be feasible but special design recommendations for different user groups would be needed.

BACKGROUND

Considering the variance in all factors characterizing the users it is highly debatable if the benchmark procedure for usability evaluations meets the demands of easily usable devices for all target users. There are differences, for example in expertise, experience with technology in general, domain knowledge, cultural factors and upbringing, but also developmental aspects with respect to the huge field of cognitive abilities, ascending in children and descending in older adults. As shown in earlier studies (e.g., Bay & Ziefle, 2003a; Liben, Patterson & Newcombe, 1981; Vicente, Hayes & Williges, 1987; Westermann, 1997; Ziefle & Bay, 2005a, Ziefle & Bay, 2006) a number of cognitive abilities, for example spatial cognition or verbal memory, show a considerable change over the lifespan.

Given that diversity, it may be problematic to focus only on best case conditions and to neglect weaker users. Rather, it might be more advisable to pursue an inverse proceeding in usability research in order to reach what usable designs promise.

Everyday, products as the mobile phone should be conceptualized bearing in mind the "weaker" user, that is, for example, a user with cognitive abilities below average. These users are the ones who need to be supported much more than those who are well trained with technical devices and office software because otherwise they may not be able to handle a device even after a substantial time of exposure. Also, more and more children possess mobile phones, which have not been specifically designed for this user group. Probably children would not even want to use a "kid's phone" because of "image" issues. In the recent past a number of studies have been concerned with enlightening children as a special user group of technical devices or technology in general (Berg, Taylor & Harper, 2003; Carusi & Mont'Alvao, 2006; Hanna et al., 1998; Jones & Liu, 1997; Ketola & Kohonen, 2001; Lieberman, 1998). While some knowledge was collected on children's attitudes (e.g., Vincent, 2004) and general usage criteria (e.g., Crenzel & Nojima, 2006), only few studies have investigated *how* children actually interact with different mobile phones in terms of efficiency and effectiveness (e.g., Bay & Ziefle, 2003b; Bay & Ziefle, 2005; Ziefle, Bay & Schwade, 2006). And in even fewer studies a direct comparison of the children with the performance of other user groups (i.e., young and older adults) was undertaken (Ziefle & Bay, 2004; 2005b).

Similar to the small HCI research output regarding children there is even less knowledge about mentally impaired users' interaction with technology (e.g., Oliver et al., 2001; Petrie et al., 2006; Mátrai, Kosztyán & Sik-Lányi, in press). Especially for these people the importance of usable mobile devices is high. Given the fact that mere calling is not longer the most frequent interaction but impaired users could be supported by memory functionality (e.g., medical monitoring) or navigation aids of mobiles, the mobile device could be a supportive aid enabling more independency and higher mobility of this special group.

Learning-disabled children are said to show the same developmental process as "normal" children, their developmental speed is only slower (on average about 1 to 2 years behind average children). Usually, they do not reach the highest stadium of cognitive development, which is characterized by abstract reasoning about problems (Schröder, 2000). Learning-disabled show a permanently constricted learning field, which means they are only susceptive to concrete and needs-related material. They have a reduced ability for abstractions, limited capacity to structure tasks and are generally slow, shallow and time-limited in their learning process.

In Germany, around 2.2 percent of all pupils are characterized as learning-disabled and need therefore special education. They are a marginal group but they deserve some attention from ergonomists and designers, because they are currently totally disregarded in usability research. Furthermore, a detailed look at their difficulties interacting with a mobile phone can give interesting insights into problems an average user will also very likely experience, for example when his attention is not entirely focused on the phone because he is on the move. When pursuing a "design for all" approach, learning-disabled are certainly a prototypical user group that should be considered for participation in usability studies.

MAIN FOCUS OF THE CHAPTER

In the present study the performance of different user groups interacting with mobile phones is compared: A student group and two groups of children—one being average school kids between 9 and 12 years of age, the second consisting of learning-disabled children and teenagers between 11 and 15 years of age. Their performance when solving typical tasks on a widespread phone model, the Siemens C35i, is evaluated. To assess the impact of specific user interfaces of a mobile phone in different user groups, only one aspect of the phone is experimentally varied: the keys that are used to interact with the mobile phone's menu. For reasons of ecological validity the navigation keys of a second widespread mobile phone model, the Nokia 3210, were chosen. Only few adaptations of the menu (such as changing the position of the soft key labels on the display) were necessary to operate the menu of the Siemens C35i with the Nokia 3210 keys.

Method

Participants

In the experiment, three different user groups with a total of 80 participants took part. Thirty students, 20 children with normal intelligence and 30 learning-disabled children and teenagers. The 20 children with normal intelligence were between 9 and 12 years of age, the 30 learning-disabled between 11 and 15 years. (The differences between those two groups regarding number of participants and age were due to difficulties recruiting these special users, yet the age difference is of minor importance since the "cognitive" age of learning-disabled is reduced compared to their "real" age. Half of the participants in each user group processed tasks using the phone with menu and keys both stemming from the Siemens C35i, the other half using the phone consisting of the Siemens C35i menu and the Nokia 3210 navigation keys.

Apparatus

The two mobile phones were simulated on a touch screen connected to a PC where user actions were logged on the keystroke level. To ensure good visibility and avoid difficulties hitting the keys because if the missing tactile feedback the display and the keys of the mobile phones were enlarged compared to the original devices. The appearance of the simulated phones was also modified to exclude effects of preferences for specific brands. The touch screen was fixated on a table in an angle of 35° which enabled an interaction in approximately the same posture as when using a real mobile phone.

Key Description

The two variants of navigation keys differed with regard to the number of keys (more specifically, the number of different options to be pressed) and the number of different functions each of these keys can exert. Moreover, among those keys which exert different functions at different points within the menu, there are such that have a similar meaning (e.g., confirm and save) and such that are quite dissimilar (e.g., end calls and return to a higher menu level).

The two original navigation key solutions which were simulated in the present experiment are shown in Figure 1 and 2.

The C35i keys consist in total of seven key options. Each of the two rocker switches contains two options (marked by a dot on each side of the rocker switch) were the key may be pressed, thus resulting in two "stroke options" per key. Sometimes (depending on the menu level) however, there are not two different options to be selected, but the same function is exerted, independently which side of the rocker switch that was pressed. The label displayed on the display above the key indicates which functions can be exerted. The left rocker switch has six different functions. The functions are scrolling (left: up, right: down), selecting the mailbox, changing, saving entries, and sometimes it has no function at all, depending on the point of the menu. All those actions are semantically very different from the scrolling function. The right rocker switch serves to enter the menu, to select, to correct (left part) and confirm (right part), or to correct (left) and save (right), and to send a message (eight functions/combinations of functions, where six are semantically dissimilar from selecting/confirming). Additionally, there is an extra key with an icon (open book) to open the phone directory. This function is most of the time not active at all. Furthermore, there is a big, centrally positioned key with a green receiver icon on it is used to make calls, which also exerts a function in specific cases (e.g., when a number is displayed), otherwise having no function. Finally, there is a smaller key with a red receiver sign to end calls as well as for hierarchical steps back in the menu.

Figure 1. Navigation keys of the Siemens C35i

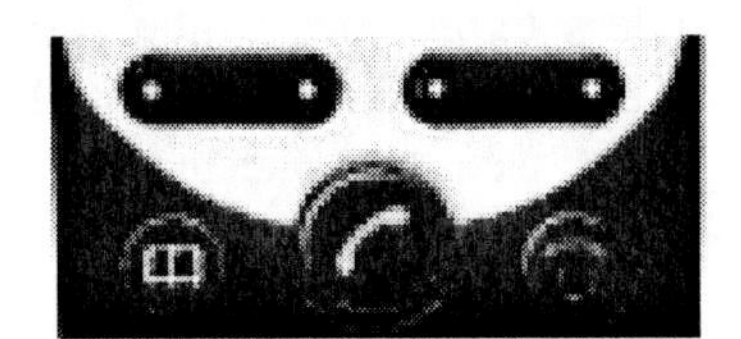

Figure 2. Navigation keys of the Nokia 3210

The Nokia 3210 navigation keys exhibit four key options. Two of them have several functions: The c-key is used for corrections of letters and digits as well as for returns to higher menu levels. These two functions can be regarded as similar, as they both mean "undo." The centrally positioned key is a softkey used to enter the menu, to select highlighted menu entries, to confirm and to effect calls (four functions, three of them semantically similar representing confirmation actions, but entering the menu is not a confirmation action and can therefore be regarded as semantically dissimilar). The scrolling-key is used for movements up and down within any level of the menu.

Overall, the Nokia keys can be judged as simple with respect to both the number of keys and the number of keys with different functions compared to the Siemens keys.

Procedure

To assess the participants` previous experience using different kinds of technical devices, including the mobile phone, they were asked to complete a questionnaire before processing tasks on the mobile phone. The questionnaire was shown on the touch screen and required participants to activate fields by touching the screen. Thus, they were able to get used to the reaction of the touch logic. On a five-point scale the frequency of using different devices had to be answered (1= "several times a day," 2 = "once a day," 3 = "once a week," 4 = "once a month" and 5 = "less than once a month"). The estimated ease of use of those devices was to be judged on a four-point scale (1 = "very easy," 2 = "rather easy," 3 = "rather difficult," and 4 = "very difficult"). Also, the general interest in technology had to be rated on a four-point scale (1 = "low," 2 = "rather low," 3 = "rather strong," 4 = "strong").

Afterwards, participants had to complete four tasks on the simulated mobile phone:

1. Enter a telephone number and make a call
2. Send a text message (SMS) to a specific phone number (to compensate differences in the speed of typing the text, the message was provided and only had to be sent)
3. Hide your own phone number when calling
4. Redirect all phone calls to the mailbox

The participants were given a period of five minutes to solve each of the tasks. When a task was solved correctly, a "congratulations" message was shown on the screen. When a participant did not succeed in solving a task within the period of five minutes the experimenter told the participant that the specific tasks was very hard to solve (in order to prevent user's frustration) and that he should go on with the next.

Independent and Dependent Variables

The first independent variable was the key solution (3210 vs. C35i). The second independent variable was the user group (students vs. average children vs. learning-disabled children).

As dependent variable the participants' performance was assessed by counting the number of ineffective keystrokes carried out. That is, each key stroke that did not lead to any task related effect on the display. This includes:

- Hash (#) and asterisk (*) at any point within the menu
- Numbers when not task related
- Soft keys, function keys and scroll-buttons when not exerting a function

Additionally, the number of steps executed, the time needed to process tasks as well as the number of tasks solved were measured.

Results

The Participants' Experience with Technology

In a pre-experimental questionnaire the participants' experience with different technological devices was surveyed.

The students reported to use a mobile phone between daily and once a week (M = 2.5; SD = 1.6), the wireless phone once a week (M = 3.0; SD = 1.8), the PC between several times and once a day (M = 1.4; SD = 0, 7) and a DVD or VCR once a month (M = 4.1; SD = 0.9). The perceived ease of use of all devices was between "very easy" and "rather easy" (mobile phone: M = 1.7, SD = 0.9; wireless phone: M = 1.4, SD = 0.7; PC: M = 1.8, SD = 0.8; DVD: M = 1.7, SD = 0.7). On average, the students' interest in technology was rated between "rather strong" and "rather low" (M = 2.4, SD = 0.9).

The learning-disabled children had quite some experience using technical devices: The mobile phone (M = 2.1; SD = 1.6) and a PC (M = 2.2; SD = 0, 9) were used daily, the wireless phone (M = 3, 0; SD = 1.6) and DVD player (M = 2.9; SD = 1.3) once a week. The estimated ease of use of all devices was between "very easy" and "rather easy" (mobile phone: M = 1.5; SD = 0.8; PC: M = 1.7, SD = 0.9; wireless phone: M = 1.3, SD = 0.8; DVD: M = 1.6, SD = 1.1). The general interest of the learning-disabled in technology was rated as "rather high" (M = 3.3; SD = 1.1).

Looking at the average children's answers in the questionnaire it may be said that they were somewhat less experienced than the learning-disabled. They reported to use a mobile phone between once a week and once a month (M = 3.5; SD = 1.4), the PC between once a day and once a month (M = 2.4; SD = 1.1), the wireless phone

Figure 3. Frequency of using a mobile phone, a PC and a DVD player in the three user groups (1 = "several times a day," 2 = "once a day," 3 = "once a week," 4 = "once a month," 5 = "less often")

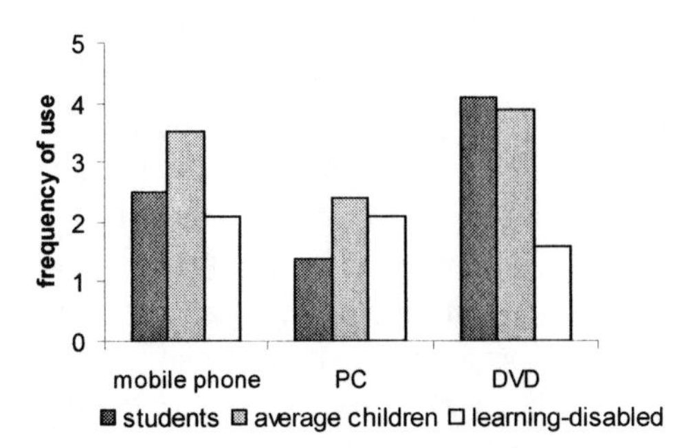

once a day (M = 2.0; SD = 0.9) and the DVD once a month (M = 3.9; SD = 1.0). The reported ease of use was between "very easy" and "rather easy" as in the two other user groups (mobile phone: M = 1.8, SD = 0.9; wireless phone: M = 1.1, SD = 0.3; PC: M = 1.8, SD = 0.8; DVD: M = 1.8, SD = 0.9). Children's general interest in technology was rather low (M = 2.1; SD = 1.0).

As visualized in Figure 3, it has to be stated that the learning-disabled are not less experienced with technology. Quite the contrary, their reported frequency with which they use a mobile phone and a DVD-player is even higher than that reported by the other two groups. Therefore, from this perspective no big performance differences should be expected.

In order to draw back performance differences between the participants using the different key solutions to the experimental manipulation, we had to make sure that the groups did not differ regarding their experience with technology. Therefore non-parametric Mann-Whitney tests were carried out for the variables surveyed in the questionnaire. No significant differences could be detected.

Performance Using the Two Phones

For each of the three groups of participants the number of ineffective keystrokes carried out with the two navigation key solutions is assessed.

First the total number of ineffective keystrokes carried out by the students, the average children and the learning-disabled using the two phones is analyzed. Figure 4 shows the outcomes.

The students carried out 5.1 (SD = 6.3) ineffective keystrokes when using the C35i keys and 0.2 (SD = 0.6) with the 3210 key solution. Children in the C35i group made 22.8 (SD = 31.5) ineffective keystrokes, those of the 3210 group only 7.7 (SD = 7.7). Learning-disabled made 10 times as many ineffective keystrokes than the students and more than twice as many as average children when using the C35i (M = 55.3, SD = 71.2). Learning-disabled using the 3210 key solution made 14.9 (SD = 16.1) ineffective keystrokes, which also represents a considerably higher number compared to students and children. However, the performance difference between the two groups using different key solutions becomes more obvious in learning-disabled children (55.3 vs. 14.9 ineffective keystrokes) than in other user groups. Thus, the huge impact of different key solutions on user's performance becomes only apparent when participants other than students are taken into consideration. For a deeper insight into the performance of the different groups, the number of keystrokes carried out in each of the tasks is looked at in detail.

Effects of Key Solutions on Students

Figure 5 shows the performance outcomes for the first user group, the students, where the number of ineffective key strokes in each of the four tasks

Figure 4. Ineffective keystrokes carried out in the four tasks by the three user groups when using the Siemens C35i keys and the Nokia 3210 keys

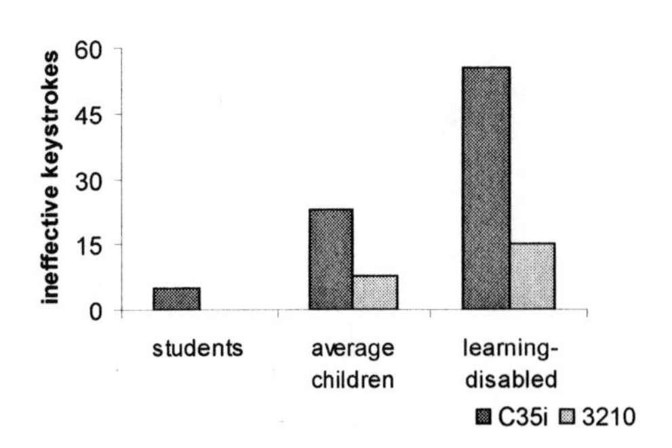

is visualized. Students using the mobile phone with the Nokia 3210 keys carried out almost no ineffective keystroke (on average only 0.1 in tasks 2 (SD = 0.5) and 3 (SD = 0.3)) and also using the Siemens C35i keys the number of ineffective keystrokes may be regarded as negligible: When effecting a call no ineffective keystroke was carried out. When sending an SMS 2.9 (SD = 5.6), when hiding their own number 1.9 (SD = 3.8) and when making a call divert to the mailbox only 0.3 (SD = 0.7) times keys were stroked ineffectively. For the tasks of sending a short message ($t(28) = 1.87; p < .1$) and of hiding their own number ($t(28) = 1.91; p < .1$) marginally significant differences, depending on the key solution used, could be detected. Nevertheless, due to the small total number of ineffective keystrokes affected, it could be concluded, that the different key solutions do not significantly affect the performance interacting with the mobile phone.

But, before drawing this conclusion, a closer look should be taken at the performance of the other user groups.

Effects of Key Solutions on Average Children

Figure 6 visualizes performance outcomes for the average children carrying out the four tasks with different phones. It becomes evident that this user group undertakes considerably more keystrokes and also the difference between the two key solutions is somewhat clearer.

When effecting a call, children using the 3210 keys make 0.3 (SD = 0.9) ineffective keystrokes. With the C35i keys 0.6 (SD =2.0) ineffective keystrokes are undertaken. Thus, this task seems not to impose high demands on the children. In task two, where a text message was to be sent, with the 3210 keys 3.7 (SD = 6.8) and with the C35i 8.9 (SD = 10.0) ineffective keystrokes are undertaken. In the task of hiding their own number, participants press 2.8 (SD = 4.6) keys ineffectively when using the 3210, and 5.8 (SD =8.7) keys with the C35i solution. The last task, in which a call divert had to be carried out, led to the greatest performance difference between the two key solutions. Using the 3210, keys were stroked without exerting an effect only 0.7 (SD = 1.1) times, whereas in case of the C35i keys this happened on average 7.8 (SD = 15.2) times. In spite of the big numerical differences in performance between the users of the two phones, t-tests did not reveal any significant effect for any task, which may be due to the big variance in the data.

From these results obtained with children possessing average intelligence, it may be assumed that the navigation key solution exerts some effect on the users' performance interacting with a mobile phone, but the impact does not seem to be dramatic. After all, when solving tasks using the Siemens C35i on average only 9 times a key is stroked without exerting any effect. This should not unsettle a user. However, the results give a

Figure 5. Number of ineffective key strokes carried out by students using the two navigation key solutions

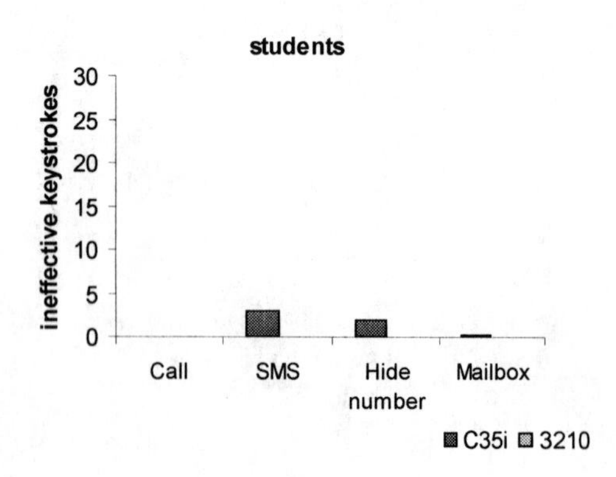

Figure 6. Number of ineffective keystrokes carried out by average children using the two navigation key solutions

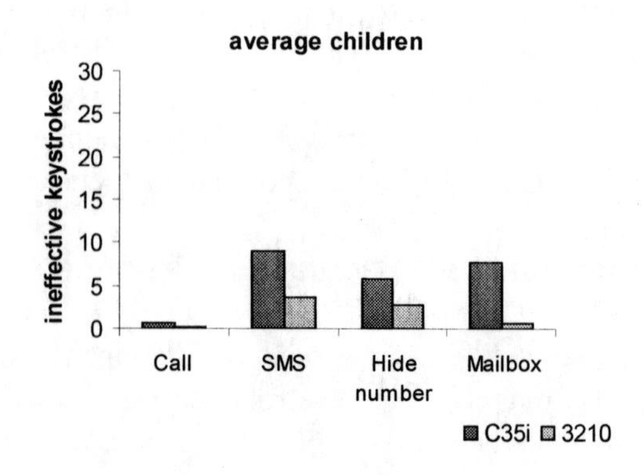

first hint that different navigation key solutions used to control the same menu can lead to some interesting differences in performance.

For a deeper insight into the effects, a user group, which should be more sensitive towards complex rules of interaction, is surveyed.

Effects of Key Solutions on Learning-Disabled Children

Figure 7 shows the number of ineffective keystrokes carried out by the group of learning-disabled children when using the Nokia 3210 keys and when using the original keys of the Siemens C35i.

Huge performance differences become evident. When using the 3210 keys between one and six ineffective keystrokes were carried out by the learning-disabled in the four tasks (call: M = 1.1, SD = 1.7; SMS: M = 6.1, SD = 7.4; hide own number: M = 3.8, SD = 5.7 and call divert to mailbox: M = 3.9, SD = 5.0). In contrast, when the learning-disabled children used the original keys of the C35i phone, averages of up to 30 ineffective keystrokes—when sending a text message, SD = 48.5—were reached. And also when hiding their own number and diverting calls to the mailbox with 14.8 (SD = 16.4) and 10.1 (SD = 15.3) a substantial number of ineffective keystrokes was undertaken. In the task of calling a number only 0.4 (SD = 1.1) ineffective keystrokes were made (Figure 7). T-test show significant differences between the two key solutions in the task of hiding their own number ($t(28) = 2.46$; $p < 0.05$) and a marginally significant difference in sending a text message ($t(28) = 1.89$; $p < 0.1$). This impressive performance differences between the two key solutions when they are used by learning-disabled should not be ignored. Interestingly, it is not the original key solution that leads to the best performance, but the solutions originating from an alternative phone, the Nokia 3210.

Which Keys Lead to the Difficulties?

The question, why the Siemens C35i keys led to many more ineffective keystrokes, may be answered by looking at the specific keys, which were stroked very often without exerting a function. The key that led to the biggest number of ineffective strokes was the receiver key, which is used to effect calls in the C35i solution. This key was used on average 28.7 times by the learning-disabled, 4.9 times by average children and on average 0.6 times by the students without exerting any effect. The key is comparably large, green and centrally positioned on the mobile phone, and was therefore presumably mistaken as a confirmation key.

Figure 7. Number of ineffective keystrokes carried out by learning-disabled children using the two navigation key solutions

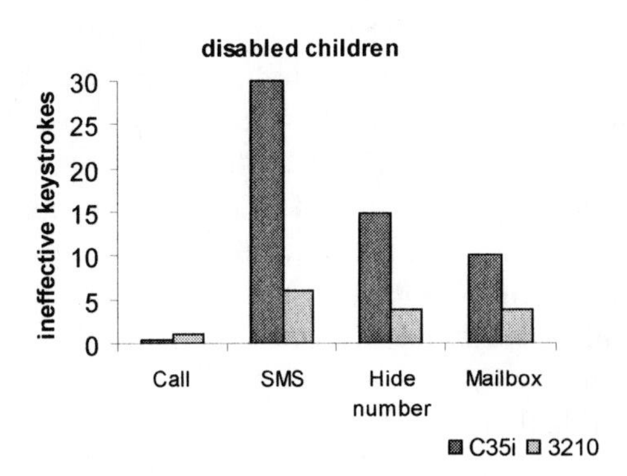

The second type of keys, which were often used without exerting a function at the current point within the menu, were the soft keys of the Siemens C35i. Learning-disabled pressed these keys 17.7 times ineffectively, average children 9.2 times and students 4.1 times. The two soft keys of the C35i model exert many different functions at different points within the menu, and can sometimes exert different functions when stroked left or right. Sometimes, they have only one or even no functionality at all. This changing assignment of modes-of-operation has probably confused the users and especially the learning-disabled children, which led to many ineffective keystrokes.

Performance Differences between the User Groups in other Variables

For an insight into the performance differences between the three user groups of the present study two other variables, processing time and detour steps, are looked at. These variables are not directly related to the usability of the key solution, but are caused mainly by the difficulty imposed by the menu of the mobile phone. Therefore no differentiation between the two navigation key solutions used is made.

When effecting a call, both groups of children needed more than double the time (average children M = 44.2s; learning-disabled: M = 41.3s) than the students (M = 19.1s). In the task of sending a text message the differences between the groups increased: Students needed 83.1s, children 161.8 and learning-disabled 212.3s. Similar patterns of results were found for the tasks of hiding their own number and making a call divert to the mailbox. To hide their own number students needed 116.8 s, average children 221.4s and learning-disabled 242.7s. To make a call divert to the mailbox students needed 109.1s, average children 204.3s and learning-disabled 225.7s (Figure 8).

Thus, when considering only student participants, the difficulty imposed by the phone with a maximum of two minutes for completing the task of hiding their own number is not irrelevant, but still limited. However, when considering the performance of learning-disabled, who needed on average four minutes for this task, many were not able to actually solve it. The need for an improvement of the mobile phone's user interface becomes obvious.

A look at the number of steps executed while trying to solve the tasks confirms the argument. In the first task, where a number had to be entered and a call effected, performance between the groups does not differ meaningfully. Students need 12.9 steps, average children 16.9 and learning-disabled 17.4 steps. In the other tasks, learning-disabled children need mostly more than twice as many steps as the students. To send an SMS students need 51.2 steps, average children 68.6 and learning-disabled 120 steps. In the task of hiding their own number students execute 86.7 steps, average children 129.9 and learning-disabled 163.4 steps. In the last task students made 76.9 steps during their attempts to solve the task, average children made 141.4 and learning-disabled made 155.7 steps.

It becomes evident that the difficulty to perform different tasks on a widespread mobile phone is underestimated when only students are selected as participants for usability tests.

Figure 8. Time needed to process the four tasks by the three user groups

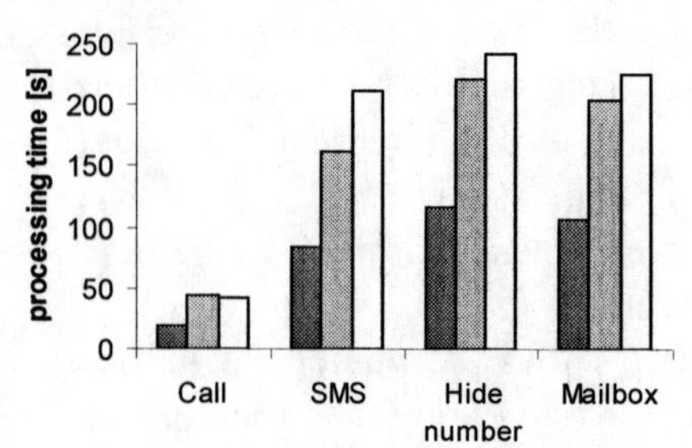

Figure 9. Steps executed while processing the four tasks by the three user groups

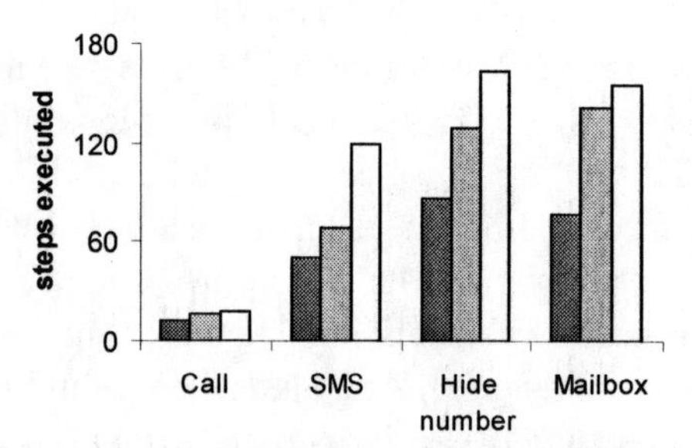

What is so Special About Learning-Disabled?

The results outlined above give reason to argue that designers of interfaces for mobile phones (and presumably of other electronic devices, too) should focus on weaker mobile phone users, such as learning-disabled, if they want to make sure that their device is really usable for a broad range of users. However, to understand what makes the learning-disabled a special group, it is worthwhile to examine their specific characteristics that may be of importance for the interaction with mobile devices.

Memory

The first important aspect, which differentiates learning-disabled from other users, is their memory capacity. The high correlation between memory capacity and the performance using a mobile phone was shown in a study by Bay and Ziefle (2003a). Participants of this study who were between 27 and 61 years of age had reached 12.6 points (SD = 2.2) in a test of figural memory. The learning-disabled of the present study were asked to solve the same memory test (LGT-3, Bäumler, 1974) and it revealed that the memory ability of the learning-disabled was somewhat lower with a mean of 9.8 (SD = 2.2) out of 20 points. This may be one reason for the big difficulties experienced by the learning-disabled. They may have had troubles remembering the functions, each of the different keys exerted, and which of the menu functions they had already selected. However, correlations between the scores in the memory test and performance measures, when using the mobile phone, did not reach the significance level in the present study.

Locus of Control

In earlier studies (Bay & Ziefle, 2003a; Ziefle, Bay & Schwade, 2006) it was found, that users' "experienced competency" with respect to the use of technical devices in general can also affect performance outcomes when using mobile phones. The locus of control regarding the use of technology (LOC) as measured through a standardized test (Beier, 1999) showed significant correlations with the number of tasks successfully solved on a mobile phone in a group of younger and older adults. Thus, users having high values in LOC showed a better performance than those with lower values. Even if it is not clear, whether the LOC is the antecedent of a good performance or if the successful interaction with technology is the antecedent of the high LOC value, the correlation shows that the felt competency and the real competency of using technical devices go in parallel in adults.

It is a basic question whether the learning-disabled have equally high self-assessment. On the one hand, it may be assumed that learning-disabled are intimidated by technical devices. This may be due to their negative experience interacting with them. Such an attitude could have negative impact on the way they approach technical problems and is therefore worth looking at. On the other hand, it is also equally plausible that the learning-disabled have no valid or realistic self-estimation with respect to their own competencies when using technical devices.

The learning-disabled children's locus of control interacting with technology was therefore surveyed with a standardized instrument by Beier (1999). It consists of eight statements, such as "I like cracking technical problems" or "Whenever I solve a technical problem this happens mostly by chance." These statements have to be affirmed or denied by the participants on a six point scale.

The learning-disabled reached an average score of 73.4 (SD = 15.7) of a maximum of 100 points. The participants in an earlier study (Bay & Ziefle, 2003a) had reached 66.1 points. The average children, in comparison, revealed a similar level with, on average, 67.4 points (SD = 14.1). The students showed a somewhat higher LOC level with 76.6 points (SD = 10.5), however, surprisingly, they did not reach values close to the upper end of the scale, as one could have expected.

Does the LOC really affect performance and can this be found in all user groups?

To assess interrelations between performance using the mobile phone and locus of control regarding technical devices, Spearman rank correlations were carried out. To begin with the benchmark, students showed a weak, but marginally significant correlation between the number of tasks solved and LOC values ($r = -0.3$; $p < 0.1$). Also, for the average children group LOC values were found to considerably affect performance (number of keys used ineffectively: $r = -0.47$; $p < 0.05$; tasks solved: $r = -0.50$; $p < 0.05$; time on task: $r = -0.36$; $p < 0.05$). Thus, for younger and older adults as well as for average children the LOC was interrelated with performance, even though to a different extent.

It characterizes the specificity of the disabled children that a significant correlation between LOC values and performance using the mobile phone could not be found in this group, for none of the dependent variables. This shows that this user group was not able to self-assess themselves realistically with respect to their technical competency. On the one hand, their absolute level in LOC was not very different from the other user groups, however, their performance level was distinctly lower: Their efficiency was 10 times lower with respect to the number of ineffectively used keys and the children needed double the time and twice as many detour steps when compared to the benchmark, the students' performance.

Mental Models

The importance of mental models for a purposeful interaction with technical devices has been emphasized by a number of studies (e.g., Norman, 1983), also more specifically for the interaction with mobile phones (Bay & Ziefle, 2003b; Ziefle & Bay, 2005). It was found that the better the mental representation of the spatial structure of the device (that is, the hierarchical nature in the case of mobile phones), the better was the performance of a user.

The difficulties of the learning-disabled children interacting with the mobile phones may therefore be due to a deficiency in building an appropriate mental representation of the menu structure. Therefore the users' mental model of the menu was assessed by showing the children a number of drawings that are supposed to visualize different kinds of mental models. The children had to process the phone tasks first, and were then asked to choose the one of the shown alternatives that was most appropriate, according to them. The different drawings of the mental models are shown in Figure 10.

The menu of a mobile phone has a hierarchical structure. However, only seven of the 30 learning-disabled children chose this drawing.

It is of interest, whether users who chose the correct drawing also performed better than those without a correct mental representation of the mobile phone menu. And indeed the analyses revealed a somewhat superior performance of the learning-disabled who were aware of the hierarchical nature of the menu. They executed 394.3

Figure 10. Drawings used to assess the users' mental representation of the menu structure

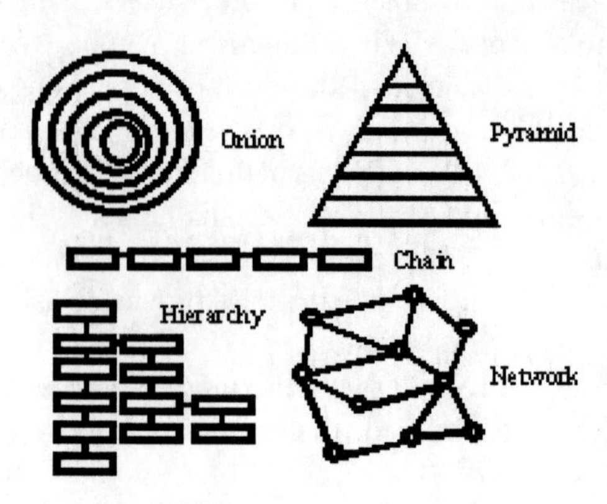

steps (SD = 148.2) in contrast to 569.6 steps (SD = 255.6) needed by the rest (Figure 11). This difference was marginally significant ($t(28) = 1.73$; $p < .1$). Users with a hierarchical mental map also solved somewhat more tasks (M = 2.9; SD = 1.1) and executed less ineffective keystrokes (M = 21.6; SD = 18.1) compared to the majority without a correct mental map who solved only 2.5 tasks (SD = 1.1) and made nearly twice as many ineffective keystrokes (M = 39.2; SD = 61.5). With regard to the processing time, differences were much smaller, but still showing a benefit of the correct mental representation (M = 829.4, SD = 251 versus M = 853.1, SD = 251.5 of the majority without a correct mental map).

Thus, the impact of a correct mental representation of the mobile phone menu on learning-disabled children's performance could be shown.

In a study by Bay and Ziefle (2003b), the benefit of having a correct model on the performance interacting with mobile phones was also found for average children. It was shown that nearly all (80 percent) of the examined children aged nine to 16 had a correct mental representation of the hierarchical nature of the menu. The results of the present study confirm the importance of a hierarchical mental representation of the menu and suggest that this is a crucial factor, which may explain the huge inferiority of the learning-disabled compared to other user groups including other children.

Figure 11. Performance difference between users with a hierarchical and a non-hierarchical mental representation of the menu

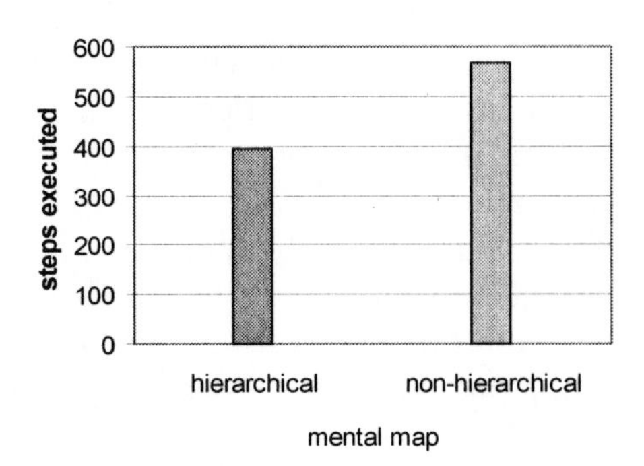

CONCLUSION

A study was conducted with the aim to explore the differences in performance between learning-disabled children, average children and students when interacting with a mobile phone. This study was undertaken out of two reasons: Firstly, learning-disabled have been completely disregarded as users of everyday technical products as the mobile phone. Secondly, the consideration of participants in usability research who do not possess high cognitive abilities can provide meaningful insights into the real implications of specific user interface design decisions. As one important example for a design decision the navigation key solution was varied. Participants processed four typical phone applications on two mobile phones simulated on a touch screen. One phone corresponded to the Siemens C35i regarding menu and navigation keys, the second phone had the same menu, which was to be operated with navigation keys stemming from the Nokia 3210 model.

Thirty students, 20 average children aged between 9 and 12 years of age and 30 learning-disabled children and teenagers between 11 and 15 years took part in the study. As expected, results revealed similarities between the three user groups as well as differences. The similarity between groups is the basic pattern of performance outcomes when using the C35i keys in contrast to the 3210 keys: All user groups showed a considerably better performance when using the simple and easy to understand 3210 keys compared to the rather complex C35i key solution. This means, that there are design solutions that benefit every user independent of age and cognitive abilities. This result makes the idea of the "design for all" approach feasible.

The difference that showed up between the user groups is the huge inferiority of the learning-disabled compared to the student group as well as to the average children. Learning-disabled students needed more than double the time and steps to process the tasks than the students and about one third more time and steps than average children for example when sending a text message. Also, the effect of the different key solutions becomes more evident in the learning-disabled compared to the other groups, which can be taken from the total amount of ineffectively used keys. In total, using the C35i keys the learning-disabled made on average 55 ineffective keystrokes while solving the four tasks, and this is actually ten times more than students and more than twice as many as the average children carried out when using the C35i keys. However, learning-disabled children made only 15 ineffective keystrokes with the 3210 keys, which shows how much this user group benefits from good design. With student users, this difference between the two key solutions did not become as obvious, since with the more complex phone only five ineffective keystrokes were carried out on average—an number that apparently may be neglected. Thus, the real difficulties that can be caused by a navigation key solution and the importance of creating less complex solutions only becomes evident when a user group like the learning-disabled is considered. If only students are taken as participants in usability tests—as it is very often the case—important aspects are completely ignored. For example, the present study could show how dramatic the impact of inconsistent assignment of functions to keys can be, which only reached a meaningful effect on performance in the group of learning-disabled users. Average children and students experienced the same difficulties but were able to overcome them after some practice.

It may be concluded that whenever a "design for all" approach is pursued, the ergonomic "worst case" has to be taken into consideration. Even if the recruitment of those users requires more time, it is worth the effort and represents a very insightful experience. But what is so special about the learning–disabled as user group?

As a first hypothesis it may be assumed that learning-disabled children and teenagers have lower expertise and experience using different technical devices. Results show, however, that the contrary is the case. The learning-disabled of the present study showed a higher frequency of using a mobile phone, a DVD-player and a PC than average children and even partly outreached the students. Thus, this hypothesis has to be rejected.

When memory ability, which was found to influence the ability to interact with technical devices, was assessed some, but again, no meaningful differences between learning-disabled and other user groups were detected. However, for this user group, correlations between memory ability and performance measures did not lead to significant interrelations.

Furthermore, the locus of control of the learning-disabled children did not show to differ meaningfully from that of other users. In contrast, the learning-disabled children and teenagers surveyed in the present study seemed to be pretty convinced of their ability to handle technology well (nearly as convinced as the average children and the students were). Not only their locus of control was high but also the estimated ease using a mobile phone, a PC or a DVD player was high: They were rated between "very easy" and "rather easy" to use by the participants. Thus, the reported expertise has also to be ruled out as explanatory variable.

A last aspect that may account for the inferior performance of learning-disabled children and teenagers is the lack of an appropriate mental model of the functioning of the mobile phone, more specifically of the menu structure. And indeed only seven out of 30 learning-disabled were able to identify the hierarchy as the correct model of the menu structure. For a comparison, in Bay & Ziefle (2005) it was found that 80 percent of 9 to 16 years old children were aware of the hierarchical nature of the menu. Also, the present study showed that learning-disabled children possessing a correct mental map performed better in solving tasks on the mobile phone than participants whose mental map was incorrect.

It is often assumed that people automatically build-up cognitive representations of the functioning of technical devices while interacting with them (e.g., Norman, 1983). Thus, frequent exploration and active handling of the technical device is believed to support the development of adequate representations with respect to how the mental room, which has to be navigated, is structured. Furthermore, the exploration of the menu structure is assumed to pre-structure the interconnections and relations between functions and sub-categories present in the menu. However, learning-disabled apparently have big difficulties building a correct representation of the menu structure, considering that they reported to use a mobile phone on a daily basis. At least, the mental representation they have built does not correspond to the real information structure of the phone. Therefore, it may be concluded that the process of building a mental representation needs to be actively supported. This may be done through training or better visual cues on the spatial structure to be incorporated on the mobile phone's display (Ziefle & Bay, 2006).

Some final remarks are concerned with potential methodological limitations of the presented research study. One could critically argue that the phones under study were simulations on a touch screen rather than real mobile phones. This criticism can be met with three arguments: First, as only the keys were under study and have been experimentally varied independently of differences with regard to the menu, there were no real phones available that met these requirements. Second, different mobile phones differ in so many attributes and aspects (size, form, color, key shape, labels, haptics, etc.) that any comparison would be misleading for the question at issue, because we actually do not know which of the aspects leads to performance differences. Third, with the special user group under study—learning- disabled children—it was a deliberate aim to design the experimental situation as easy and comfortable as possible and with the simulation visual as well as psychomotor difficulties could be ruled out. It is clear that our results therefore represent an underestimation of the real performance but performance differences can be unequivocally traced back to the experimental variation. The same argument can serve to explain the feedback provided when a task was solved successfully. Whenever the children had solved a task they were presented a "Congratulations!" message—something that is not given in real life situations and definitively makes the experimental situation easier than in reality. However, as average as well as handicapped children use to quickly lose their motivation it was of central importance to sustain children's enthusiasm taking part in the experiment. Even though an underestimation of performance differences between the phones may have taken place, the key result that handicapped children show the very same structural problems with phones but react more sensitive on specific user interface design decisions in not affected by these methodological issues.

FUTURE TRENDS

Future studies should focus on supporting the process of cognitive mapping in participants. Even if the performance outcomes were qualitatively very similar between the different user groups, it may not necessarily be deduced that the development of a proper mental model is also similar. In the contrary, it is very likely that different user groups need different types of support for this purpose. Different types of training, instructions in manuals or information visualization on the display of the mobile phone itself should be evaluated with respect to their helpfulness for different user groups.

ACKNOWLEDGMENT

Thanks to Lisa Ansorge and Alexander Schwade for their help collecting and analyzing the data as well as Hans-Jürgen Bay for helpful comments on an earlier version of this chapter.

REFERENCES

Bäumler, G. (1974). *Lern- und Gedächtnistest 3 (LGT-3)* [Learning and Memory Test 3]. Göttingen, Germany: Hogrefe

Bay, S., & Ziefle, M. (2003a). Design for all: User characteristics to be considered for the design of devices with hierarchical menu structures. In H. Luczak & K.J. Zink (Eds.), *Human factors in organizational design and management* (pp. 503-508). Santa Monica: IEA.

Bay, S., & Ziefle, M. (2003b). Performance in mobile phones: Does it depend on proper cognitive mapping? In D. Harris; V. Duffy; M. Smith & C. Stephanidis (Eds), *Human centred computing. Cognitive, social and ergonomic aspects* (pp. 170-174). Mahwah, NJ: LEA.

Bay, S., & Ziefle, M. (2005). Children using cellular phones. The effects of shortcomings in user interface design. *Human Factors, 47*(1), 158-168.

Beier, G. (1999) Kontrollüberzeugung im Umgang mit Technik [Locus of control interacting with technology]. *Report Psychologie, 9*, 684-693.

Berg, S., Taylor, A., & Harper, R. (2003). Mobile phones for the next generation. Devices for Teenagers. *CHI, 5*(1), 433-440.

Carusi, A., & Mont'Alvao, C. (2006). Navigation in children's educational software: The influence of multimedia elements. In R. N. Pikaar; E.A. Konigsveld & P.J. Settels (Eds.), *Meeting diversity in ergonomics. Proceedings of the International Ergonomics Society 2006 Congress.* Amsterdam: Elsevier.

Crenzel, S.R., & Nojima. V.L. (2006). Children and instant messaging. In R. N. Pikaar; E.A. Konigsveld & P.J. Settels (Eds.), *Meeting diversity in ergonomics. Proceedings of the International Ergonomics Society 2006 Congress.* Amsterdam: Elsevier.

Hanna, L., Risden, K., Czerwinski, M., & Alexander, K.J. (1998). The role of usability research in designing children's computer products. In A. Druin (Ed.), *The design of children's technology.* San Francisco: Morgan Kaufmann Publishers.

Jones, M., & Liu, M. (1997). Introducing interactive multimedia to young children: A case study of how two-year-olds interact with the technology. *Journal of Computing in Childhood Education, 8*(4), 313-343.

Ketola, P., & Korhonen, H. (2001). ToyMobile: Image-based telecommunication and small children. In: A. Blanford, J. Vanderdonckt & P. Gray (Eds.), *People and computers XV — interaction without frontiers, Proceedings of HCI 2001,* (pp. 415-426).

Liben, L. S., Patterson, A. H. & Newcombe, N. (Eds.), *Spatial representation and behavior across the life span* (pp. 195-233). New York: Academic Press.

Lieberman, D.A. (1998). The researcher's role in the design of children's media and technology. In A. Druin (Ed.), *The design of children's technology* (pp.73-97). San Francisco: Morgan Kaufmann Publishers.

Mátrai, R., Kosztyán, Z. T., & Sik-Lányi, C. (in press). *Navigation methods of special needs users in multimedia systems.* Computers in Human Behavior (2007), doi10.1016/j.chb.2007.07.015.

Norman, D. A. (1983). Some observations on mental models. In D. Gentner and A.L. Stevens (Eds.), *Mental models* (pp.7-14). Hillsdale, NJ: LEA.

Oliver, R., Gyi, D., Porter, M., Marshall, R. & Case, K. (2001). A survey of the design needs of older and disabled people. In M.A. Hanson (Ed.), *Contemporary ergonomics* (pp. 365-270). London, New York: Taylor & Francis.

Petrie, H., Hamilton, F., Fraser, N. & Pavan, P. (2006). Remote usability evaluations with disabled people. *In Proceedings of CHI* (pp. 1133-1141). Montreal: ACM.

Schröder, U. (2000) *Lernbehinderten Pädagogik, Grundlagen und Perspektiven sonderpädagogischer Lernhilfe.* [Pedagogics for learning-disabled, basics and perspectives of special education learning aid]. Stuttgart, Germany: Kohlhammer.

Vicente, K.J., Hayes, B.C., & Williges, R.C. (1987). Assaying and isolating individual differences in searching a hierarchical files system. *Human Factors, 29(3)*, 349-359.

Vincent, J. (2004). *'11 16 Mobile.' Examining mobile phone and ICT use amongst children aged 11 to 16.* Digital World Research Centre. Retrieved January 2007, from http://www.dwrc.surrey.ac.uk/Portals/0/11-16Mobiles.pdf

Westerman, S.J. (1997). Individual differences in the use of command line and menu computer interfaces. *International Journal of Human Computer Interaction, 9(2)*, 183-198.

Ziefle, M., & Bay, S. (2004). Mental models of a cellular phone menu. Comparing older and younger novice users. *Proceedings of Mobile HCI 2004, 6th International Symposium on Mobile Human-Computer Interaction* (pp. 25-37). Heidelberg, Germany: Springer.

Ziefle, M., & Bay, S. (2005a). How older adults meet cognitive complexity: Aging effects on the usability of different cellular phones. *Behaviour & Information Technology, 24*(5), 375-389.

Ziefle, M., & Bay, S. (2005b). The complexity of navigation keys in cellular phones and their effects on performance of teenagers and older adult phone users. In: *Proceedings of the HCI International 2005.(Vol. 4)* Theories, Models and processes in Human Computer Interaction. Mira Digital Publishing.

Ziefle, M., & Bay, S. (2006). How to overcome disorientation in mobile phone menus: A comparison of two different types of navigation aids. *Human Computer Interaction, 21*(4), 393-432.

Ziefle, M., Bay, S., & Schwade, A. (2006). On keys' meanings and modes: The impact of navigation key solutions on children's efficiency using a mobile phone. *Behaviour and Information Technology, 25*(5), 413-431.

KEY TERMS

Ecological Validity: Degree to which results of experiments are transferable to behavior in real world situations. The higher the ecological validity of an experiment the higher is the probability that results found in the experiment can be found in the same fashion in the field.

Ineffective Keystrokes: Measure for performance evaluation. Counting each key stroke carried out by a user that does not lead to any task related effect on the display enables to measure the difficulty imposed by the navigation keys of a mobile phone independent of the difficulties caused by the menu. Ineffective keystrokes include Hash (#) and asterisk (*) at any point within the menu, number keys when not task related as well as soft keys, function keys and scroll-buttons when not exerting a function.

Learning-Disabled Children: Children with a developmental speed that is slower than that of average children (about 1 to 2 years behind). Usually learning-disabled also do not reach the highest stadium of cognitive development which is characterized by abstract reasoning about problems. Learning-disabled show a permanently constricted learning field, which means they are only susceptive to concrete and needs-related material, they have a reduced ability for abstractions, limited capacity to structure tasks and are generally slow, shallow and time-limited in their learning process.

Mental Models: concepts in the mind of users about the functioning of devices, metaphors, and ideas which lead the user while interacting with the device.

Mobile Phone Menu: Form of displaying mobile phone functions that go beyond effectuation of calls to the user. Mobile phone menus usually have a hierarchical tree structure, which the user needs to navigate through via keys in order to find and select the desired function.

Navigation Keys: Keys used to operate the menu of a mobile phone, usually consisting at least of two keys fro scrolling up and down within one level, one key for selection and one key for returning to higher menu levels.

Chapter X
Human Factors Problems of Wearable Computers

Chris Baber
The University of Birmingham, UK

James Knight
The University of Birmingham, UK

ABSTRACT

In this chapter wearable computers are considered from the perspective of human factors. The basic argument is that wearable computers can be considered as a form of prosthesis. In broad terms, a prosthesis could be considered in terms of replacement (i.e., for damaged limbs or organs), correction (i.e., correction to 'normal' vision or hearing with glasses or hearing aids), or enhancement of some capability. Wearable computers offer the potential to enhance cognitive performance and as such could act as cognitive prosthesis, rather than as a physical prosthesis. However, wearable computers research is still very much at the stage of determining how the device is to be added to the body and what capability we are enhancing.

INTRODUCTION

There is a wide range of technologies that have been developed to be fitted to the person. Depending on one's definition of "technology," this could range from clothing and textiles, through to spectacles, to cochlear implants. The use of these different technologies can be basically summarized as the supplementation or augmentation of human capability, for example the ability to regulate core temperature (clothing), to see (spectacles) or to hear (cochlear implant). One reason why such supplementation might be required is that the current capability does not fit with environmental demands, either because the environment exceeds the limits over which the human body can function or because the capability is impaired or limited. From this perspective, a question for wearable computers should be what are the current human capabilities that are exceeded by the environment and require supplementation by wearable computers? In the majority of cases for wearable computers, the answer to this question hinges on communicative, perceptual, or cognitive ability. As Clark (2006) notes, "...the use, reach, and transformative powers of these cognitive technologies is escalating"(p. 2). but the essential point to note is that "Cognitive technologies are best understood as deep and integral parts of the problem-solving systems that constitute human intelligence"(p.2). Thus, such technologies could represent a form of 'cognitive prosthesis' in that they are intended to support cognitive activities, for example, having a camera (performing face-recognition) to advise the wearer on the name of the person in front of them. The immediate challenge is not necessarily one of technology but of cognition. If the technology is 'doing the recognition,' the question is raised what is the human left to do? To draw on a commonly cited analogy, spectacles serve as a perceptual prosthesis, that is, to improve or correct a person's vision. For wearable computers (and the related field of augmented reality), the 'improvement' could be to reveal to the person objects that are not present by overlaying an artificial display onto the world. The display could simply take the form of labels or directional arrows, or could be more a sophisticated presentation of moving (virtual) objects. In both cases, the 'augmentation' could either enhance the person's understanding of the environment or could substitute this understanding.

In terms of communication, mobile telephones and MP3 devices contain significant computing power such that they can easily be considered as being computers, albeit with limited functionality. These devices can be worn, for example MP3 players can be worn on the upper arm, attached to belts or neck-straps, or placed in hats, and mobile telephones can be attached to belts. Furthermore, with the wireless (Bluetooth) headset, the user interface of a mobile telephone can be worn on the ear at all times. Thus, both can be always present and both can be considered part of the person. A definition of wearable computers ought to, at least, allow differentiation from devices that can slip into the user's pockets (if this technology is to be treated as a new area of research and development). Two early definitions of wearable computers, from Bass (1996) and Mann (1997), emphasize that wearable computers are designed to exist within the corporeal envelope of the user and that this makes them part of what the user considers himself or herself. In many respects this allows an analogy to be drawn between wearable computers and prosthetic devices. Having something added to the body, whether externally, such as spectacles, artificial limbs, hearing aids, or internally, such as pace-makers, or cochlear implants, changes the performance of the person and (for external prosthesis) the appearance of the body. This now becomes a very different concept from the mobile telephone, MP3 player and the computer that we traditionally encounter. This raises all manner of interesting questions relating to physical, perceptual, and cognitive aspects of human factors, as well as a whole host of emotional aspects of wearing devices (for the wearer and the people with whom they interact). At the moment, there remains a gap between what a wearable computer is intended to be and what mobile telephones and MP3 players currently are. This gap can best be considered as a form of perceptual and cognitive prosthesis, in which the wearer's ability to view the world, retrieve pertinent information, and respond to environmental demands are enhanced by the technology across all aspects of everyday life. At present mobile telephones and MP3 players are able to be tailored (by the user) and can deal with a limited set of situations (relating to communications or music playing) but do not fill the specification that one might have for a wearable computer. The basic difference between a wearable computer and these other technologies lies in the question of how well the user can interact with both the device and the environment simultaneously. Obviously, listening to an MP3 player or speaking on a mobile telephone can be performed while walking through an environment. However, performing control actions on the devices can be sufficiently demanding to draw the user's attention from the environment. The ideal wearable computer would allow the user to manage attention to both device and environment.

In other words, it should allow the user to manage both foreground and background interaction (Hinckley et al., 2005).

FORM-FACTOR AND PHYSICAL ATTACHMENT

Moore's law continues to guarantee that the processors will get smaller, and work in the field of micro-electrical-mechanical systems (MEMS) shows how it is possible to create application specific processors that are small enough to be incorporated into buttons on clothing or into jewelry (Figure 1a).

Furthermore, it is feasible to assume the widespread development of general purpose processors, such as the mote (and similar) concept, that combine low power with sufficient processing capability to deal with a number of different sensors (figure 1b).

One direction for wearable computers is that the miniaturization of technology will mean it is possible to implant processors under the skin (or embed them into clothing). This is not a particularly novel idea as the medical world has been experimenting with mechanical implants for many years and has, over the past decade or so, developed digital implants that can, for example, regulate drug administration or improve hearing ability. While the development of such technology is exciting and likely to lead to fascinating discoveries, it lies somewhat outside the remit of this chapter. The problem is how can these extremely small devices support *interaction* with the person wearing (or supporting) them? To a great extent, the human simply has these devices fitted into them. While this represents the logical extension of the argument that the wearable compute exists within the wearers 'corporeal envelope,' it does preclude the possibility of developing means of interacting with the device. Indeed, as technology continues to get smaller, ever more difficult the challenge of supporting human-computer interaction becomes, for example, how can one read displays with very small font or press buttons that are much smaller than the human finger? This could mean that we should redefine our concept of *interaction*, for example, one could argue that 'interaction' should be any activity that the person performs, that is sensed by the device and that allows the device to make a response. A potential problem with such a concept is that the person might not be able to selectively control what the device senses or to fully understand why the device is behaving in the manner that it is. This means that one either locks the human out of the interaction (in which case one has a sensor system that acts on the environment and which coincidentally affects people in that environment) or one must develop different ways in which the person can understand and manage the behavior of the device. This is not a trivial problem and one can point to many developments in the ubiquitous and pervasive computing domain in which the role of the human in device performance is merely one of passive recipient of device activity. From the point of view of human-computer interaction, this will not only lead to

Figure 1a. MEMS device

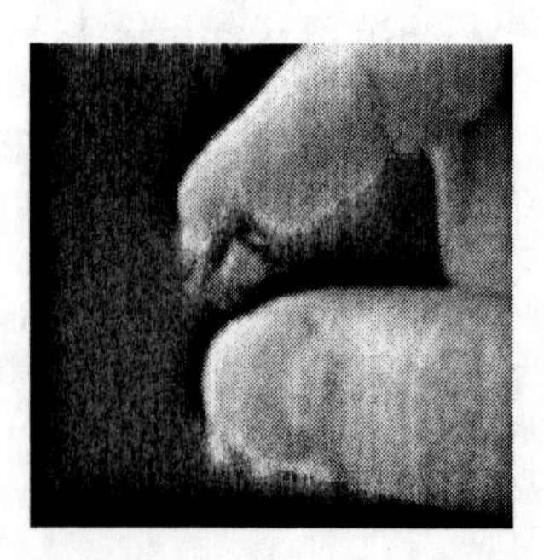

Figure 1b. Intel's Mote prototype[1]

the 'irony of automation'[2] but also to frustration, annoyance and a perceived lack of control. In other words, who wants to live in a world that is basically run by the descendants of the Microsoft Office Paperclip?

Physical Effects of Wearing Computers

We might expect a computer to consist of a processor and storage, some form of power, a display, and an interaction device. For wearable computers, each of these components can be separated and worn on different parts of the body. Thus, as the prototypes in Figure 2 illustrate, the processor and storage could be incorporated into a single unit and mounted on the waist or the back, the display could be mounted on the head (either near the eyes or the ears, depending on the feedback provided to the wearer), the power could be located near the processor unit, and the interaction device could be mounted within easy reach of the hand (if manual control) or the mouth (if speech input). These prototypes are typical of much contemporary work on wearable computers in terms of the relative size and placement of components.

Wearable computers represent a load on the person and consequently can affect the physical activity of the person (Zingale et al., 2005). Legg (1985) proposes that the human body can carry loads at the following points: head, shoulder, back, chest, trunk, upper arm, forearm, hands, thighs, feet, a combination of these, and aided (i.e., by pulling, pushing or sharing a load). A glance through the literature of wearable computers shows that most of these sites have been experimented with in various designs. Gemperle et al. (1998) offer the term 'wearability' to describe the use of the human body to physically support a given product and, by extension, the term 'dynamic wearability' to address the device being worn while the body is in motion. Given this notion of wearability, there is the question of where a device might be positioned on the body, and, once positioned, how it might affect the wearer in terms of balance, posture and musculoskeletal loading Given the notion of dynamic wearability, there are the questions of how the device will be carried or worn, how this might affect movement, and how this might lead to either perceptions of differences in movement patterns, physiological strain or psychological stress on the wearer. These changes can be assessed, using subjective self-report techniques (Bodine & Gemperle, 2003; Knight et al., 2002, 2006; Knight & Baber, 2005) and through objective analysis (Nigg & Herzog, 1994).

Figure 2. Wearing prototype wearable computers

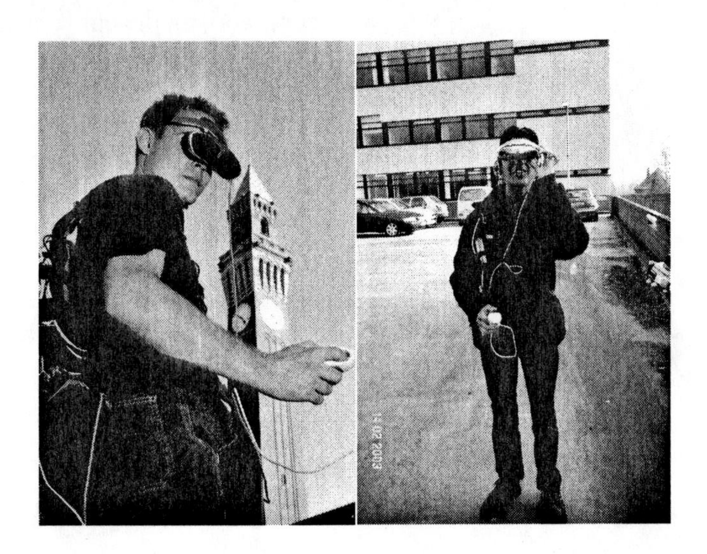

Energy Expenditure, Muscle Activity and Using Wearable Computers

An attached load to the body will have a direct affect on the energy expended by the body, as the muscles burn more energy to generate force to counteract the weight of the load. This situation sees a wearer of a wearable computer potentially increasing their overall energy expenditure to overcome inertial changes. Thus, *any* additional weight on the body can create demands on the musculoskeletal system to support and move the object. In addition, the kinematics of posture and movement can compound the loading effects of a wearable computer. Knight and Baber (2004) report the following about head-mounted loads, such as head mounted displays: (1) different head postures have a measurable effect on musculoskeletal loading; (2) the heavier the frontal load (as is representative of the loading of a HMD), the greater the muscle activity required to keep the head in a fixed position and; (3) in neutral or extended postures, the wearer can sustain larger loads than if the head is in a flexed or rotated head position Thus, it is not simply a matter of reducing load of head-mounted equipment, but one must determine the posture that the wearer of such equipment is likely to adopt. An alternative location for display technology is on the forearm. In their study of pointing devices, Thomas et al. (1998) demonstrate that for general purpose activity, the forearm would be an appropriate place to mount the device. This location is attractive as it allows the wearer to easily move the display into the field of vision. However, Knight and Baber (2007) have questioned this location or at least raised concerns. Recording shoulder and upper arm muscle activity and measuring perceptions of exertion while participants interacted with arm mounted computers of different weights they found that the mere act of holding the arm in an appropriate posture to interact with an arm mounted computer was sufficient to exceed recommended levels of muscle activity for sustained activity . In addition it induced symptoms of fatigue after only two minutes where the addition of weight in the form of the mounted technology compounded this physical effect.

Figure 3. SenseWear device from BodyMedia[3]

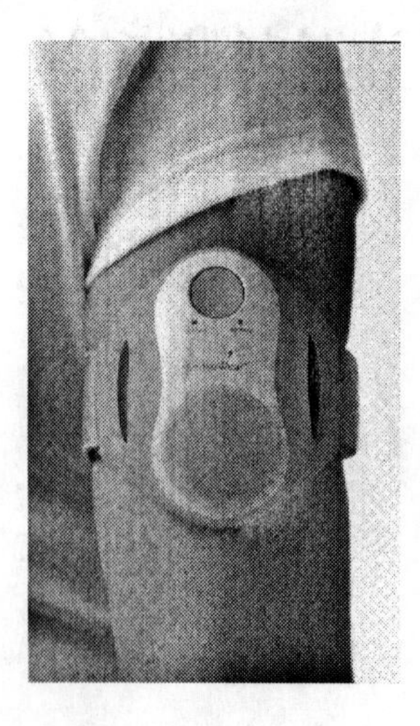

Reducing Size and Separating Components

The discussion so far has suggested ways in which weight and placement of loads on the body can lead to problems The implication is that most wearables take the form factor of the ‘brick-on-the-back’ (as illustrated by Figure 2). It is possible to reduce the size of components further by removing the interaction and display components. Thus, the Bluetooth headset for a mobile phone effectively reduces the user interface (for talking on the phone) to a unit that clips on the ear. Various MP3 players can be worn on the upper arm while jogging (see comment about usability). Alternatively, there has been much interest in the use of ‘active badges’ that signal a person’s location to a network (Want et al., 1992). These devices can be made small enough to be worn as badges or incorporated into items of clothing, for example Schmidt et al. (1999) report a device, mounted in the wearer’s tie, which detects changes in ambient sound levels or wearer movement. In these examples, the ‘on-body’ part of the computer system is basically a sensor and/or transmitter and/or receiver that link the wearer to a network of other technologies. BodyMedia have developed a device that places sensors on the person’s upper arm in order to record data relating to everyday activity (in much the same way that an MP3 player is a data display device) and needs to be connected to a computer in order to process the tunes on it (see Figure 3). However, it makes sense to ask whether reducing the size of the devices will reduce or even eliminate these problems. Physical discomfort may arise from pressure of the device on the body. The pinnar is very sensitive and many users complain of discomfort around the ear from using ear plugs and ill-fitting head phones. HMDs place pressure around the forehead and crown. Even the small displays produced by Microoptical (see Figure 4) can cause discomfort around the ears and nose; especially given that the glasses they come with are hard plastic, not specifically fitted for the individual wearer. Items attached around the arm, specifically during physical activity (i.e., for jogging), have to be attached tightly so that they do not bounce against the arm or slip down, as such they may result in discomfort and result in vasoconstriction leading to sensations of numbness and tingling in the lower arm and hand, not to mention that the display is not easily viewable when the device is in position which raises usability issues.

Perceptual Impacts of Wearing Computers

A wearable computer can be considered a 'perceptual prosthesis' in the ways that it can provide additional information to, or enhance perceptual capability of, the wearer. Information provision can simply mean allowing the wearer access to information that is not currently present in the world, for example through a visual or audio display. The underlying concept in many of these applications is that the wearer's perception of the world can be augmented or enhanced (Feiner et al., 1998). Before considering this argument, we can consider the MP3 player as a device that can present media that is added to the world (in the form of a personal soundtrack to the person's activity). To some extent this could be considered as form of augmenting perception. What makes augmented reality different from simply displaying information is that (in most systems) the information is presented in accordance with the context in which the person is behaving. Such information could be in the form of an opaque visual display or could be visual information overlaid onto the world. In basic systems, the information can be called up by the wearer, or pushed from another source, for example a radio or telephone link. In more sophisticated systems, the information is presented on the basis of the computer's interpretation of 'context.'

While augmented reality displays could be beneficial, there are potential problems associated with the merging of one source of information (the computer display) with another (the world). Contemporary wearable computers tend to combine monocular head-mounted displays (see Figure 4) with some form of interaction device, for example keyboard, pointing device or speech. NRC (1997) point out that monocular head-mounted displays could suffer from problems of binocular rivalry, that is, information presented to one eye competes for attention with information presented to the other, which results in one information source becoming dominant and for vision to be directed to that source. A consequence of this phenomenon is that the wearer of a monocular display might find it difficult to share attention between information presented to the 'display eye' and information seen through the 'free eye.' This problem is compounded by the field of view of such displays.

Figure 4. Monocular head-mounted displays can restrict field of view

The field of view for the commercial monocular HMDs ranges from 16°-60°, which is considerably less than that of normal vision (around 170° for each eye, Marieb 1992). Narrow field of view can degrade performance on spatial tasks such as navigation, object manipulation, spatial awareness, and visual search tasks. Restrictions on field of view will tend to disrupt eye-head coordination and to affect perception of size and space (Alfano & Michel, 1990). One implication of a restricted field of view is that the wearer of a see-through HMD will need to engage in a significant amount of head movement in order to scan the environment (McKnight & McKnight, 1993). Seagull and Gopher (1997) showed longer time-on-task in a flight simulator when using a head down visual display unit than with a head-mounted, monocular display. Thus, it appears that a monocular display might impair performance. Apache helicopter pilots currently wear monocular, head-mounted displays and a review of 37 accidents concluded that 28 of these accidents could be attributed to wearing of the display (Rash et al., 1990).

Computer Response to Physical Activity

While it is possible that wearing technology affects user performance, it is also possible that the physical activity of the person can affect the computer, for example through the use of sensors to recognize actions and use this recognition to respond appropriately.

Given the range of sensors that can be attached to wearable computers, there has been much interest in using data from these sensors to define and recognize human activity. This has included work using accelerometers (Amft et al., 2005; Junker et al., 2004; Knight et al., 2007; Ling & Intille, 2004, Van Laerhoven & Gellersen, 2004; Westeyn et al., 2003) or tracking of the hand (Ogris et al., 2005). The approach is to collect data on defined movements to train the recognition systems, and then use these models to interpret user activity. At a much simpler level, it is possible to define thresholds to indicate particular postures, such as sitting or standing, and then to use the postures to manage information delivery (Bristow et al., 2004). The use of data from sensors to recognize human activity represents the merging of the research domains of wearable computers with that of pervasive computing, and implies the recognition not only of the actions a person is performing but also the objects with which they are interacting (Philipose, et al., 2004; Schwirtz & Baber, 2006).

An area of current interest for the wearable or ubiquitous computing communities is the interpretation of human movement related to maintenance and related activity. The artifacts with which the person is interacting could be instrumented. For example, a simple approach is to fit switches on components (in a self-assembly furniture pack) and for the user to depress the switches when each components is handled (Antifakos et al., 2002). A less intrusive approach would be to fit radio frequency identification person is using (Scwhirtz et al., 2006) and to use the activation of these tags to infer user activity. This requires that either the components or the tools be adapted to the task. The decreasing costs of RfiD suggest that, within a few years, tags will be universally used in a wide range of consumer products. Alternatively, the activity of the user could be taken to develop predictive models in order to infer the activity that is being performed. For some activities it might be sufficient to use data from very generic sensors, such as microphones, to collect data to define actions (Ward et al., 2006), while in other it might be necessary to rely on more specific sensors, such as the use of accelerometers to define movements that are characteristic of assembly and maintenance tasks (Schwirtz & Baber, 2006; Westyn et al., 2003). Steifmeier et al. (2006) show how tracking the motion of the hand (e.g., using ultrasonic tracking and inertial sensors) can be used to define specific types of movement that relate to maintenance tasks, such as spinning a wheel or rotating the pedals, unscrewing a cap or using a bicycle pump. In a related domain, there is some interesting work on the collection of activity data relating to nursing, for example through a combine infra-red proximity sensing and accelerometers (Noma et al., 2004).

USING WEARABLE COMPUTERS

Given that a wearable computer provides a means of 'anytime, anywhere' access to information, most forms of cognitive activity have been mooted as possible areas that can be supported. In this section, we consider three areas: (i.) supporting memory; (ii.) supporting navigation; and (iii.) information search and retrieval.

Supporting Memory

A common example used to illustrate the benefits of a wearable computer is what can be termed the 'context-aware memory.' Imagine you are attending a conference (or any large social gathering) and having to remember someone's name. Systems using some form of badge or face-recognition have been proposed to help with such situations; the computer would register the person and provide you with name and some additional details about the person, for example when you last met, what research interests are listed on their web-page, where they work, and so forth. There has been little research on whether and how these systems improve memory, and this example points to a possible confusion between supporting processes involved in recalling information from memory and the provision of contextually-relevant information. An associated question is whether wearable computers (particularly having a head-mounted display) positioned on the eye can have an impact on recall. The analogy is with the tourist watching a parade through the lens of a video camera—does the act of recording something weaken the ability to process and recall information? Baber et al. (2001) use a search task coupled with surprise recall to show that, in comparison with not using any technology, participants using a digital camera and wearable computer conditions showed lower performance, and that overall the wearable computer showed the biggest impairment in recall. There are many reasons why interruption at initial encoding can limit the ability to remember something, and the question is whether the head-mounted display serves to interrupt encoding; either due to distraction (with a host of information appearing on the screen), or through limitations of field of view, or for some other reasons.

Navigation and Way-Finding

Wayfinding requires people to travel through the world in order to reach specific locations. Thus, there is a need to manage both the act of traveling (for wearable computers this usually consists of walking) and relating a view of the world to the defined location. Sampson (1993) investigated the use of monocular, head-mounted displays for use when walking. Participants were presented with either 'spatial' or alphanumeric information and required to traverse paths with or without obstacles. In general, participants performed equally well when standing or when traversing paths without obstacle, but were significantly worse when obstacles were present. Thus, the need to maintain visual attention on display and environment can be seen to impair performance which suggests that the amount and type of information which can usefully be presented on such displays needs to be very limited. As global positioning systems (GPS) become smaller, cheaper, and more accurate, there has been an increase in their application to wearable computers. Seager and Stanton (2004) found faster performance with a paper map than GPS, that is routes completed faster. This was due, in part, with the increase in time spent looking at the GPS display and the number of updates made on the digital view (presumably because the GPS was moving a marker along the map to show the participants location). Participants also were less likely to orientate the digital view in the direction of travel (possibly because they might have assumed that the display would orient to their direction of travel rather than North up). Studies into the effect of perspective views, that is aligning a map with direction of travel, have not shown significant performance advantage over 2D views to date (Suomela et al., 2003), although this approach does seem beneficial in systems that support wayfinding in moving vehicles (Aretz, 1991). Systems that overlay routes onto the head-mounted display (Figure 5) could also assist in simple wayfinding tasks.

The use of visual support for navigation requires the user to divide attention between the environment and a visual display. In terms of the background or foreground of activity (discussed by Hinckley et al., 2005), this essentially places all the tasks in the users foreground. It is possible that navigation prompts could be provided using auditory cues. In simple terms, the heading could be indicated by varying the parameters of simple 'beeps,' for example changing in pitch or in intensity as the person deviates from a path (rather like auditory glide-slope indicators in aircraft). More recent developments have replaced the simple 'beeps' with music. In this way, a more subtle (background) form of cueing can be achieved to useful effect. The music could be manipulated to vary quality with deviation from a path, for example through distortion (Strachan et al., 2007, 2005), or through modifying the panning of music in stereo presentation (Warren et al., 2005).

Finding and Retrieving Information

Having the ability to access information as you need it is a core concept of wearable computer research. Early examples had the user enter queries and view information on a monocular display. This approach of having the user ask for information has been superseded by having the computer push information, on the basis of its interpretation of context. In such examples, the benefit of wearing the computer comes from its permanent presence and state of readiness, allowing access either of data stored on the computer or via the World Wide Web. Rhodes and Starner (1997) describe the 'remembrance agent,' which monitors the information that a user types into a computer and makes associations between this information and data it has stored. Obviously this creates an overhead on the user, in terms of the need to type information into the computer. However, it is only a relatively small modification to replace the typed entry with speech recognition (see Pham et al., 2005). While the 'agents' in either of these examples can run on desktop computers, it is the fact that they are continuously running on the computer worn by the user that makes them interesting. The role of the wearable computer in these examples is to discretely run searches in the background and alert the wearer to interesting links and associations between the current topic of conversation (or typing) and information to which the computer has access. It might also be useful for the computer to track user activity and then to either record patterns of activity (in order to refine its model of context) or to offer information relevant to the user. Thus, a very common application domain for context-aware, wearable computer research is the museum visitor (Sarini & Strapparava, 1998). The idea is that when the visitor stands at a certain location, say in front of a painting, the computer offers information relating

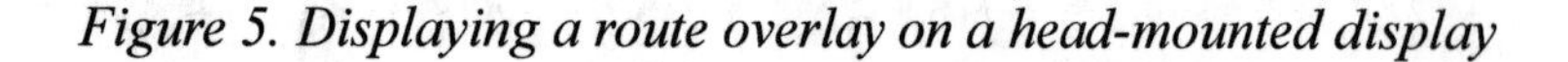

Figure 5. Displaying a route overlay on a head-mounted display

to the objects in that location. The manner in which the location (and other contextual factors) is used to select information and the manner in which the information is presented to the user might vary across applications. The examples considered thus far present the wearable computer as a device that is able to 'push' potentially relevant information to the wearer. However, rather than simply overloading the wearer with information, the role of the agents or context-awareness is to manage and tailor the information to the person.

Impact on User Performance

While there are many applications reported in the literature, there is a surprising lack of research into how effective these applications are in improving or otherwise changing user performance. One domain in which wearable computers have received both interest and support has been in maintenance (Mizell, 2003). A study exploring the use of head-mounted displays to support maintenance work show that performance could be improved, providing information was displayed in an appropriate format (Kancler et al., 1998). However, other studies have been more equivocal. Tasks requiring participants to follow instructions on a wearable computer or printed on paper have shown both the wearable computer (Baber et al., 1998) and the paper (Baber et al., 1999c; Ockerman et al., 1997; Siegel & Bauer, 1997) to lead to superior performance. One explanation of these differences lies in the design of the information that was presented, for example the manner in which information is presented can impact on overall performance times (Baber et al., 1999b; Sampson et al., 1993). One implication of these studies is that participants using the wearable computer tend to follow the same sequence of tests (as defined on the visual display), whereas the participants in the paper condition would order the tests as they saw fit (Baber et al., 1999b, Ockerman & Pritchett, 1998).

In terms of interacting with wearable computers and the appropriate devices to use, there has been very little work to date. While one might assume that the optimal interaction techniques would be ones that support hands-free interaction, such as speech recognition, studies suggest that walking has a negative impact on speech recognition performance (Oviatt, 2000; Price et al., 2004). In terms of entering data, Thomas et al. (1997) showed that a forearm mounted QWERTY keyboard led to superior performance over a five-button chording device or a virtual keyboard controlled using an isometric button. However, one might question the recommendation of a forearm mounted device, based on consideration of musculoskeletal strain. In terms of selecting objects on a display, Thomas et al. (1998) found that a touchpad mounted on the forearm was preferred by users, but that one mounted on the thigh lead to superior performance when sitting, kneeling or standing. Thus, the mounting of a pointing device can have a bearing on performance (although one might question whether pointing is an appropriate means of performing selection tasks on a wearable computer). Zucco et al. (2006) considered the performance of 'drag and drop' tasks while stationary and whilst walking, using different devices. They found that a gyroscopic mouse lead to best performance while stationary, but that touchpad or trackball were lead to better performance when the user was walking (and that all devices were superior to the Twiddler keypad).

Rather than seeing these studies as recommendations for specific interaction devices, I feel that they illustrate that the relationship between the activity that a person is performing and the demands of the ongoing activity in the environment interact in ways that require careful planning in the design of wearable computers. This brief discussion raises questions on one might develop standards for the deployment of wearable computers in these (and related) applications, and also what level of performance improvement one might expect from this technology.

SMART CLOTHING AND TEXTILES

It is worth mentioning the efforts to incorporate at least some aspects of technology into clothing and textiles (Berzowska, 2005). This could then take the form of clothing that has been adapted to incorporate some of the components (see Figure 6).

Alternatively, the textiles can exhibit some form of 'smartness.' Often this refers to "Textiles that are able to sense stimuli from the environment, to react to them and adapt to them..." (Van Langehoven & Hertleer, 2004). Taking these terms separately, we can ask how could textiles sense, react and adapt. In terms of sensing, there is a wide range of possible approaches, including thermal, chemical, mechanical, as well as biosensors. Buechley (2006) shows how simple off-the-shelf sensors and actuators can be incorporated into items of clothing. In particular, this work, following the earlier work of Post and Orth (1997) demonstrates how fabrics can be knitted or woven to provide some electrical conductivity, and then fitted with components. In contrast, other researchers report ways in which it is possible to use to electrical properties of prepared fabrics, such as change in resistance, inductance or capacitance (Wijesiriwardana et al., 2003, 2004), to incorporate specific sensing capabilities, such as stretching (Farringdon et al., 1999; Huang et al., 2006) or impact (Lind et al., 1997). What has yet to be fully realized from this smart textile work is that manner in which the user would interact with the 'computing' aspects in order to entry data or perceive displayed information. Thus, much of the research is still concerned with the development of the 'bus' onto which sensors, processors, displays, batteries, and so forth can be mounted.

EMOTIONAL IMPACT OF WEARING COMPUTERS

There are three broad categories of impact that will be considered in this section. The first concerns the emotional response to wearable computers by the people wearing these devices and the people with whom they interact. The second concerns the response of the computer to the emotions of the wearer, that is affective computing. The third concerns the manner in which collaboration can be supported within a group of wearable computer wearers.

While head-mounted displays have been used for many years in the domain of military aviation, they have yet to find widespread use on the ground. Most cinema goers will be familiar with concepts of wearable computers from science fiction films and might feel uncomfortable with 'cyborgs' in their midst. This discomfort could be particularly acute in stressful situations (as one of the paramedics in an earlier study pointed out to us, *If you're just coming round from a heart attack, the last thing you'd expect to see if some robot headed bloke trying to take your pulse*). As there are so few commercial applications of wearable computers, these devices still represent something of a novelty and there has been very little research into how people might respond to people wearing such devices. One collection of anecdotal evidence can be found in Thad Starner's Ph.D. Starner completed his Ph.D. at MIT, and with Brad Rhodes and Steve Mann developed a variety of wearable computers that they wore

Figure 6. Incorporating consumer electronics into clothing—the Philips-Levi Strauss jacket[4]

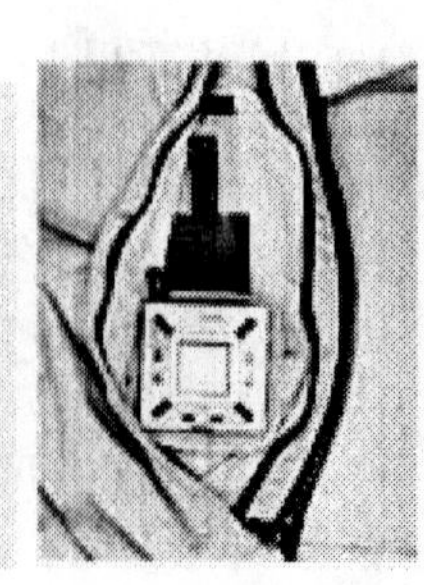
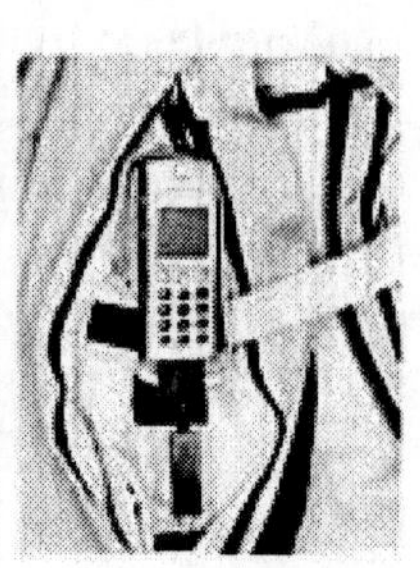

for prolonged periods of time. This meant that they would encounter members of the public on a regular basis and Starner recorded some of the reactions. A common response was to assume that the wearable computer was simply some form of familiar technology, perhaps a very sophisticated video camera or an advanced form of video game console or a medical device. The implication of this is that people might try to explain the technology using a familiar mental model but, as Starner points out, people have yet developed a mental model of a wearable computer on the basis of familiarity and exposure. While this is partly a matter of the unfamiliar appearance of the wearer it is also related to the manner in which the wearer interacts with the device. For example, Starner points out that when you ask someone the time they will raise their wrist in order to consult their watch, but with a head-mounted display one might simply glance up at the screen. This means that the physical behavioral cues might be far less obvious. This is, of course, similar to the way in which Bluetooth headsets allow people to speak on their mobile telephone in a manner that makes it looks as if they are speaking to themselves (indeed, some users of Bluetooth headsets make gestures that look as if they are holding a phone, for example holding the microphone or holding their hands near their faces). There will be a trade-off between the familiarity of wearable computers, their appearance, and the ability of people to explain their use. However, there is an additional factor at play here (which is also hinted at by the Bluetooth headset discussion), and that is that the wearable computer provides information to an individual in a highly individualized manner—it is not possible for other people to see what the wearer is looking at or guess what they are doing. An analogy can be drawn in this instance with the person on a train reading a humorous book at laughing out loud—the other people feel uncomfortable because they can not read the what the person is reading. The removal of the observer from the source of information can be disconcerting and can lead to such comments as (from Starner) "we can't tell if you're talking about us behind our backs" or "when you wear your display, how can I tell if you are paying attention to me or reading your e-mail?"

One implication of the physical appearance and individualized interaction of wearable computers is the sense that people who wear such technology are different from 'normal' people. As Sheridan et al. (2000) note, people wearing computers could be perceived (by themselves or by the people around them) as a 'cyborg community' that is different from other people. With the use of communication and networking capabilities, it is possible for a group of wearable computer users to be able to share information and maintain contact as a community (Wellman, 2001). However, we should be careful to distinguish between the ability to maintain contact with other people (which one can do easily with mobile telephones, even to the extent of setting up talk groups) and the ability to share the wide range of information that wearable computing can support. This could simply mean the sharing of the same documents or video, but could also allow new forms of collaboration, sharing and exchange of information.

Context could be defined by changes in the physiological state of the wearer (Picard, 1997). This requires a more intimate means of recording data from the wearer, perhaps through monitoring of pulse or heart activity. The use of 'context' to initiate image capture has been demonstrated by several projects; most notably in Healey and Picard's (1998) 'StartleCam,' in which changes in galvanic skin response (GSR) was used to trigger image capture. There has been surprisingly little attempt at extending this work in the years since it was reported, although over uses of 'context' in image capture have explored the use of ambient sound which is captured at the same time as the image (e.g., Frolich & Tallyn, 1999; Ljunblad et al., 2004). Bristow et al. (2005) used a set of context identifiers to take still images when 'context' changed, and showed that these were surprisingly consistent with photographs taken by humans. In addition to the computer 'sensing' the physiological responses of the wearer (and hence drawing some inference as to the affective state), it is also possible to infer the state of people with whom the wearer is interacting in order to develop 'emotionally intelligent interfaces.' By monitoring the changing facial expressions of ones conversational partner, it might be possible to provide support for people who find it difficult to judge the emotional response, for example people with Autism (El Kaliouby & Robinson, 2003).

DISCUSSION

Wearable computers continue to raise many significant challenges for human factors research. These challenges involve not only cognitive aspects of presenting information but also perceptual aspects of displaying the information against the backdrop of the everyday environment and physical aspects of mounting the devices on the person. This chapter has overviewed some of the developments in the field and offered some consideration of how these human factors can be considered. While the field is largely motivated by technological advances there is a need to carefully ground the developments in the physical and cognitive characteristics of the humans who are intended to wear them.

REFERENCES

Alfano, P.L., & Michel, G.F.(1990). Restricting the field of view: Perceptual and performance effects. *Perceptual and Motor Skills, 70,* 35-45.

Amft, O., Junker, H., & Troster, G. (2005). Detection of eating and drinking arm gestures using inertial body-worn sensors, *Ninth International Symposium on Wearable Computers ISWC 2005* (pp.160-163). Los Alamitos, CA: IEEE Computer Society.

Antifakos, S., Michahellis, F., & Schiele, B. (2002). Proactive instructions for furniture assembly. *UbiComp 2002 4th International Conference on Ubiquitous Computing* (pp. 351-360). Berlin: Springer.

Aretz, A.J. (1991). The design of electronic map displays. *Human Factors, 33*(1), 85-101.

Baber, C. (1997). *Beyond the desktop: designing and using interaction devices.* San Diego, CA: Academic Press.

Baber, C., Haniff, D.J., Knight, J., Cooper, L., & Mellor, B.A. (1998). Preliminary investigations into the use of wearable computers. In R. Winder (Ed.), *People and computers XIII* (pp. 313-326). London: Springer-Verlag.

Baber, C., Haniff, D.J., & Woolley, S.I. (1999a). Contrasting paradigms for the evelopment of wearable computers. *IBM Systems Journal, 38*(4), 551-565.

Baber, C., Knight, J., Haniff, D., & Cooper, L. (1999b). Ergonomics of wearable Computers. *Mobile Networks and Applications, 4,* 15-21.

Baber, C., Arvanitis, T.N., Haniff, D.J., & Buckley, R. (1999c). A wearable computer for paramedics: studies in model-based, user-centred and industrial design, In M.A. Sasse and C. Johnson (Eds.), *Interact'99* (pp. 126-132). Amsterdam: IOS Press.

Baber, C., Sutherland, A., Cross, J., & Woolley, S. (2001). A wearable surveillance system: Implications for human memory and performance. In D. Harris (Ed.), *Engineering psychology and cognitive ergonomics industrial ergonomics, HCI, and applied cognitive psychology, 6,* (pp. 285-292). Aldershot: Ashgate.

Bainbridge, L. (1987) Ironies of automation. In J. Rasmussen, J. Duncan & J. Leplat (Eds.), *New technology and human errors* (pp. 271-286). New York: Wiley

Bass, L. (1996). Is there a wearable computer in your future? In L.J. Bass and C. Unger (Eds.), *Human computer interaction* (pp. 1-16). London: Chapman and Hall.

Bass, L., Siewiorek, D., Smailagic, A,. & Stivoric, J. (1995). On site wearable computer system. *CHI'95* (pp. 83-84). New York: ACM.

Berzowska, J. (2005). Electronic textiles: wearable computers, reactive fashions and soft computation. *Textile, 3,* 2-19.

Bodine, K., & Gemperle, F. (2003). Effects of functionality on perceived comfort of wearables, *Digest of Papers of the 7th International Symposium on Wearable Computers,* (pp. 57-61). Los Alamitos, CA: IEEE Computer Society.

Buechley, L. (2006) A construction kit for electronic textiles. In *Digest of Papers of the 10th International Symposium on Wearable Computers* (pp. 83-90). Los Alamitos, CA: IEEE Computer Society.

Bristow, H., Baber, C., Cross, J., Knight, J., & Woolley, S. (2004). Defining and evaluating context for wearable computers *International Journal of Human Computer Studies,* 60, 798-819.

Clark, A. (2006). Natural born cyborgs? *Edge. Retrieved from* www.edge.org/3rd_culture/clark/clark_index.html

El Kaliouby, R., & Robinson, P. (2003). The emotional hearing aid—An assistive tool for autism, *HCI International*

Farringdon, J., Moore, A.J., Tilbury, N., Church, J., & Biemond, P.D. (1999). Wearable sensor badge and sensor jacket for contextual awareness. *Digest of Papers of the 3rd International Symposium on Wearable Computers* (pp. 107-113). Los Alamitos, CA: IEEE Computer Society.

Feiner, S., MacIntyre, B., Hollerer, T., & Webster, A. (1997). A touring machine: Prototyping 3D mobile augmented reality systems for exploring the urban environment. *Digest of Papers of the 1st International Symposium on Wearable Computers,* (pp. 74-81). Los Alamitos, CA: IEEE Computer Society.

Frolich, D., & Tallyn, E. (1999). AudioPhotography: Practice and prospects. *CHI'99 Extended Abstracts* (pp. 296-297). New York: ACM.

Gemperle, F., Kasabach, C., Stivoric, J., Bauer, M., & Martin, R. (1998). Design for Wearability. *Digest of Papers of the 2nd International Symposium on Wearable Computers* (pp. 116-123). Los Alamitos, CA: IEEE Computer Society.

Healey, J., & Picard, R.W. (1998). StartleCam: a cybernetic wearable computer, *Proceedings of the 2nd International Symposium on Wearable Computers* (pp. 42-49). LosAlamitos, CA: IEEE Computer Society.

Hinckley, K., Pierce, J., Horvitz, E., & Sinclair, M. (2005). Foreground and background interaction with sensor-enhanced mobile devices. *ACM Trans. Computer-Human Interaction, 12*(1), 31-52.

Huang, C-T., Tang, C-F., & Shen, C-L. (2006). A wearable textile for monitoring respiration, using a yarn-based sensor. *Digest of Papers of the 10th International Symposium on Wearable Computers,* Los Alamitos, CA: IEEE Computer Society, 141-142

Junker, H., Lukowicz, P., & Troster, G. (2004). Continuous recognition of arm activities with body-worn inertial sensors. *Eighth International Symposium on Wearable Computers ISWC'04* (pp.188-189). Los Alamitos, CA: IEEE Computer Society.

Kancler, D.E., Quill, L.L., Revels, A.R., Webb, R.R., & Masquelier, B.L. (1998). Reducing cannon plug connector pin selection time and errors through enhanced data presentation methods. *Proceedings of the Human Factors and Ergonomics Society 42nd Annual Meeting* (pp. 1283-1290). Santa Monica, CA: Human Factors and Ergonomics Society.

Knight, J. F., Baber, C., Schwirtz, A., & Bristow, H. W. (2002). The comfort assessment of wearable computers. *The Sixth International Symposium of Wearable Computers* (pp. 65-72). Los Alamitos, CA: IEEE Computer Society.

Knight J.F., & Baber, C. (2004). Neck muscle activity and perceived pain and discomfort due to variations of head load and posture. *Aviation, Space and Environmental Medicine, 75,* 123-131.

Knight, J. F., & Baber, C. (2005). A tool to assess the comfort of wearable Computers. *Human Factors, 47*(1), 77-91.

Knight, J. F., & Baber, C. (2007). Assessing the physical loading of wearable computers. *Applied Ergonomics, 38,* 237-247.

Knight, J.F., Williams, D.D., Arvanitis, T.N., Baber, C., Sotiriou, S., Anastopoulou, S. et al. (2006). Assessing the wearability of wearable computers. *Digest of Papers of the 10th International*

Symposium on Wearable Computers (pp. 75-82). Los Alamitos, CA: IEEE Computer Society.

Knight, J., Bristow, H.W., Anastopoulou, S., Baber, C., Schwirtz, A., &Arvanitis, T.N. (2007). Uses of accelerometer data collected from a wearable system. *Personal and Ubiquitous Computing 11*(2), 117-132.

Legg, S. J. (1985). Comparison of different methods of load carriage. *Ergonomics, 28*(1), 197-212.

Lin, R., & Kreifeldt, J. G. (2001) Ergonomics in wearable computer design. *International Journal of Industrial Ergonomics, 27*(4), 259-269.

Lind, E.J., Jayaraman, S., Park, S., Rajamanickam, R., Eisler, R., Burghart, G., et al. (1997). A sensate liner for personnel monitoring applications. *Digest of Papers of the 1st International Symposium on Wearable Computers* (pp. 98-107). Los Alamitos, CA: IEEE Computer Society.

Ling, B., & Intille, S. S. (2004). Activity recognition from user-annotated acceleration data. *Pervasive 2004* (pp. 1–17). Berlin: Springer.

Ljungblad, S., Hakansson, M., Gaye, L & Holmquist, L.E. (2004). Context photography: Modifying the digital camera into a new creative tool, *CHI'04 Short Papers*, (pp. 1191-1194). New York: ACM.

Mann, S. (1997). An historical account of 'WearComp' and 'WearCam' inventions developed for applications in 'Personal Imaging.' In *The First International Symposium on Wearable Computers* (pp. 66-73). Los Alamitos, CA: IEEE Computer Society.

Marieb, E. N. (1992). *Human anatomy and physiology* [2nd edition]. Redwood City, CA: Benjamin/Cummings.

McKnight, A.J., & McKnight, A.S. (1993). The effects of motor cycle helmets upon seeing and hearing. In *Proceedings of the 37th Annual conferences of the Association for the Advancement of Automotive Medicine* (pp. 87-89). Des Plaines, Ill.

Mizell, D. (2001). Boeing's wire bundle assembly project. In W. Barfield & T. Caudell (Eds.), *Fundamentals of wearable computers and augmented reality* (pp. 447-469). Mahwah, NJ: LEA.

Naya, F., Ohmura, R., Takayanagi, F., Noma, H., & Kogure, K. (2006). Workers' routine activity recognition using body movement and location information. *Digest of Papers of the 10th International Symposium on Wearable Computers* (pp. 105-108). Los Alamitos, CA: IEEE Computer Society.

Nigg, B.M., & Herzog, W. (1994). *Biomechanics of the musculoskeletal system*. England: Wiley

NRC. (1997). *Tactical display for soldiers*. Washington, DC: National Academy.

Noma, H., Ohmura, A., Kuwhara, N., & Kogure, K. (2004). Wearable sensors for auto-event recording on medical nursing—User study of ergonomic requirements. *Digest of Papers of the 8th International Symposium on Wearable Computers* (pp. 8-15). Los Alamitos, CA: IEEE Computer Society.

Ockerman, J.J., & Pritchett, A.R. (1998). Preliminary investigation of wearable computers for task guidance in aircraft inspection. *Digest of Papers of the 2nd International Symposium on Wearable Computers* (pp. 33-40). Los Alamitos, CA: IEEE Computer Society.

Ogris, G., Stiefmeier, T., Junker, H., Lukowics, P., & Troster, G. (2005). Using ultrasonic hand tracking to augment motion analysis based recognition of manipulative gestures. *ISWC 2005* (pp. 152-159). Los Alamitos, CA: IEEE Computer Society.

Oviatt, S. (2000). Taming recognition errors with a multimodal interface. *Communications of the ACM, 43*, 45-51.

Patterson, D. (2005). Fine grained activity recognition by aggregating abstract object usage. *Digest of Papers of the 9th International Symposium on Wearable Computers* (pp. 44-51). Los Alamitos, CA: IEEE Computer Society.

Pham, N.V., Terada, T., Tsukamato, M., & Nishio, S. (2005). An information retrieval system for supporting casual conversation in wearable computing environments. *Proceedings of the Fifth International Workshop on Smart Appliances and Wearable Computing* (pp. 477–483). Washington, DC: IEEE Computer Society.

Philipose, M., Fishkin, K. P., Perkowitz, M., Patterson, D. J., Fox, D., Kautz, H., & Hahnel, D. (2004). Inferring activities from interactions with objects. *Pervasive Computing 2004,* 50–57.

Picard, R. (1997). *Affective computing.* Cambridge, MA: MIT Press.

Post, E.R. & Orth, M. (1997). Smart fabric, or "wearable clothing." *Digest of Papers of the 1st International Symposium on Wearable Computers* (pp. 167-168). Los Alamitos, CA: IEEE Computer Society.

Price K. J., Min L., Jinjuan, F., Goldman, R., Sears, A., & Jacko, J.A. (2004). Data entry on the move: An examination of nomadic speech-based text entry. *User-centered interaction paradigms for universal access in the information society* (pp. 460-471). Berlin: Springer.

Rash, C.E., Verona, R.W., & Crowley, J.S. (1990). Human factors and safety considerations of night vision systems flight using thermal imaging systems. *Proceedings of SPIE–The International Society for Optical Engineering, 1290,* 142-164.

Rhodes, B.J., & Starner, T. (1997). Remembrance agent: A continuously running automated information retrieval system. *The Proceedings of The First International Conference on The Practical Application Of Intelligent Agents and Multi Agent Technology* (PAAM '96), pp. 487-495.

Rohaly, A. M., & Karsh, R. (1999). Helmet-mounted displays. In J. M. Noyes & M. Cook (Eds.), *Interface technology: The leading edge* (pp. 267-280). Baldock: Research Studies Press Ltd.

Sampson, J. B. (1993). Cognitive performance of individuals using head-mounted displays while walking. In *Proceedings of the Human Factors and Ergonomics Society 37th Annual Meeting* (pp. 338-342). Santa Monica, CA: Human Factors and Ergonomics Society.

Sarini, M., & Strapparava, C. (1998, June 20-24). *Building a user model for a museum exploration and information-providing adaptive system.* Proceedings of the Workshop on Adaptive Hypertext and Hypermedia. Pittsburgh, PA.

Sawhney, N., & Schmandt, C. (1998). Speaking and listening on the run: Design for wearable audio computing. *Digest of Papers of the 2nd International Symposium on Wearable Computers* (pp. 108-115). Los Alamitos, CA: IEEE Computer Society.

Schmidt, A., Gellerson, H-W., & Beigl, M. (1999). A wearable context-awareness component: Finally a good reason to wear a tie. *Digest of Papers of the 3rd International Symposium on Wearable Computers* (pp. 176-177). Los Alamitos, CA: IEEComputer Society.

Schwirtz, A., & Baber, C. (2006). Smart tools for smart maintenace. *IEE and MoD HFI-DTC Symposium on People in Systems* (pp. 145-153). London: IEE.

Seager, W., & Stanton, D. (2004, April). *Navigation in the city.* Ubiconf *2004*, Gresham College, London.

Seagull, F. J., & Gopher, D. (1997). Training head movement in visual scanning: An embedded approach to the development of piloting skills with helmet-mounted displays. *Journal of Experimental Psychology: Applied, 3,* 163-180

Siegel, J., & Bauer, M. (1997). On site maintenance using a wearable computer system, *CHI'97* (pp. 119-120). New York: ACM.

Sheridan, J. G., Lafond-Favieres, V., & Newstetter, W. C. (2000). Spectators at a Geek Show: An Ethnographic Inquiry into Wearable Computing, *Digest of Papers of the 4th International Symposium on Wearable Computers* (pp. 195-196). Los Alamitos, CA: IEEE Computer Society.

Starner, T. (1999). *Wearable computing and contextual awareness.* Unpublished Ph.D. thesis, Massachusetts Institute of Technology, USA.

Stiefmeier, T., Ogris, G., Junker, H., Lukowicz, P. & Tröster, G. (2006). Combining motion sensors and ultrasonic hands tracking for continuous activity recognition in a maintenance scenario. *Digest of Papers of the 10th International Symposium on Wearable Computers* (pp. 97-104). Los Alamitos, CA: IEEE Computer Society.

Stein, R., Ferrero, S., Hetfield, M., Quinn, A., & Krichever, M. (1998). Development of a commercially successful wearable data collection system. *Digest of Papers of the 2nd International Symposium on Wearable Computers* (pp. 18-24). Los Alamitos, CA: IEEE Computer Society.

Strachan, S., Williamson, J., & Murray-Smith, R. (2007). Show me the way to Monte Carlo: density-based trajectory navigation. In *Proceedings of ACM SIG CHI Conference*, San Jose.

Strachan, S., Eslambolchilar, P., Murray-Smith, R., Hughes, S., & O'Modhrain, S. (2005). Gps-Tunes—Controlling navigation via audio feedback. *Mobile HCI 2005.*

Suomela, R., Roimela, K., & Lehikoinen, J. (2003), The evolution of perspective view in WalkMap. *Personal and Ubiquitous Computing, 7,* 249-262.

Tan, H. Z., & Pentland, A. (1997). Tactual displays for wearable computing. *Digest of Papers of the 1st International Symposium on Wearable Computers* (pp. 84-89). Los Alamitos, CA: IEEE Computer Society.

Thomas, B., Tyerman, S., & Grimmer, K. (1997). Evaluation of three input mechanisms for wearable computers. In *Proceedings of the 1st International Symposium on Wearable Computers* (pp. 2-9). Los Alamitos, CA: IEEE Computer Society.

Thomas, B., Grimmer, K., Makovec, D., Zucco, J., & Gunther, B. (1998). Determination of placement of a body-attached mouse as a pointing input device for wearable computers. *Digest of Papers of the 3rd International Symposium on Wearable Computers* (pp. 193-194). Los Alamitos, CA: IEEE Computer Society.

Van Laerhoven, K., &Gellersen, H. W. (2004). .Spine versus porcupine: A Study in distributed wearable activity recognition. In *Proceedings of the 8th IEEE International Symposium on Wearable Computers (ISWC'04),* (pp. 142-149). Los Alamitos, CA: IEEE Computer Society.

Van Langehoven, L., & Hertleer, C. (2004). Smart clothing: a new life. *International Journal of Clothing Science and Technology, 13,* 63-72.

Want, R., Hopper, A., Falcao, V., & Gibbons, J. (1992). The active badge location system. *ACM Transactions on Information Systems, 10*(1), 91-102.

Ward, J.A., Lukowicz, P., & Tröster, G. (2006). Activity recognition of assembly tasks using using body-worn microphones and accelerometers. *IEEE Transaction on Pattern Analysis and Machine Intelligence.*

Warren, N., Jones, M., Jones S., & Bainbridge, D. (2005, April 2-7). *Navigation via continuously adapted music.* Extended Abstracts ACM CHI 2005, Portland, Oregon.

Wellman, B. (2001). Physical place and cyberspace: The rise of personalized networks. *International Journal of Urban and Regional Research, 25,* 227-252.

Westeyn, T., Brashear, H., Atrash, A. & Starner, T. (2003). Georgia Tech Gesture Toolkit: Supporting Experiments in Gesture Recognition, *ICMI'03,* ??

Wijesiriwardana, R., Dias, T. & Mukhopadhyay, S. (2003). Resisitivefiber meshed transducers. *Digest of Papers of the 7th International Symposium on Wearable Computers* (pp. 200-205). Los Alamitos, CA: IEEE Computer Society.

Wijesiriwardana, Mitcham, K. R., & Dias, T. (2004). Fibre-meshed transducers based on real-time wearable physiological information monitoring system. *Digest of Papers of the 8th International Symposium on Wearable Computers* (pp. 40-47). Los Alamitos, CA: IEEE Computer Society.

Zingale, C., Ahlstrom, V., & Kudrick, B. (2005). *Human factors guidance for the use of handheld, portable and wearable computing devices.* Atlantic City, NJ: Federal Aviation Administration, reprt DOT/FAA/CT-05/15

Zucco, J. E., Thomas, B. H. & Grimmer, K. (2006). Evaluation of four wearable computer pointing devices for drag and drop tasks when stationary and walking. *Digest of Papers of the 10th International Symposium on Wearable Computers* (pp. 29-36). Los Alamitos, CA: IEEE Computer Society.

KEY TERMS

Activity Models: Predictive models of human activity, based on sensor data

Augmentation Means: Devices that can augment human bevaior—a term coined by Doug Engelbart, and covering: *Tools & Artifacts:* the technologies that we use to work on the world which supplement, complement or extend our physical or cognitive abilities; *Praxis:* the accumulation and exploitation of skills relating to purposeful behavior in both work and everyday activity; *Language:* the manipulation and communication of concepts; *Adaptation:* the manner in which people could (or should) adapt their physical and cognitive activity to accommodate the demands of technology.

Comfort: Subjective response to wearing a wearable computer (ranging from physical loading to embarrassment)

Context-Awareness: The capability of a device to respond appropriately to changes in a person's activity, environment, and so forth.

Form-Factor: The overall size (and shape) of a device

Sensors: Devices that produce digital output in response to some change in a measured parameter, for example dependent on environmental change or on user activity

Wearable Computers: Devices worn on the person that provided personalized, context-relevant information

ENDNOTES

1. http://www.intel.com/research/exploratory/motes.htm
2. Bainbridge (1987) argued that full automation can lead to the ironic situation that, the role of the human operator is to intervene when something goes wrong. However, the automation is such that the human is locked out of the process and has little understanding as to what is happening. Consequently, the human will not be able to intervene in an informed and efficient manner. Ultimately, it means that, by designing the human out of the system, the potential for a flexible and intelligent response to unknown situations is lost.
3. http://www.bodymedia.com/main.jsp
4. http://www.extra.research.philips.com/pressmedia/pictures/wearelec.html

Chapter XI
The Garment as Interface

Sabine Seymour
Moondial Fashionable Technology, Austria

ABSTRACT

This chapter focuses on the surface of a smart garment as a dynamic interface. The use of the garment's surface as an interactive display opens up an array of new applications. Novel developments in interactive and wearable textile surfaces for garments display data from input sources like sensors and cell phones. The integration of these surfaces into the garments is evaluated regarding wearability and the wearer's interaction. This chapter provides a list of considerations for human interaction with smart garments and dynamic visual interfaces, which are an essential tool to design usable, smart garments.

ELECTRONIC TEXTILES

Electronic textiles or wearables result from the integration of technology into a textile, a garment, or a wearable object. Such objects or devices can either be embedded into the skin or into a textile or wearable material, or be portable. "An electronic textiles—or smart fabric/textile—refers to a textile substrate that incorporates capabilities for sensing (biometric or external), communication (usually wireless), power transmission, and interconnection technology to connect sensors and microprocessors to be networked together within the fabric" (Berzowska, 2005, p. 60). Wearable technologies are closely related to electronic textiles. The term wearable technologies covers in particular electrical engineering, physical computing, and wireless technologies. Electronic textiles and wearable technologies are literally interwoven. Intelligent garments have an enhanced functionality through embedded technologies. Integrated sensors monitor vital signs, built-in speech-recognition systems allow for an interface independent from a physical interaction, and embedded wireless systems enable hands-free communication. Intelligent garments is functional clothing constructed with textiles and materials that are considered smart. "Fibre sensors, which are capable of measuring temperature, strain/stress, gas, biological species and small, are typical smart fibres that can be directly applied to textiles" (Tao, 2001, p. 4). Smart textiles are capable of reacting to a stimulus with or without the use of competition. Outlast explains that PCMs—phase changing materials—can absorb, store, and release heat while the material changes from solid to liquid and back to solid.

H2: Dynamic Garment Interface

What is a dynamic interface of a garment? Technologies enrich the cognitive characteristics of our second skin—the surface of our garments. Currently, the surface of a garment is mostly used for static displays of information or for safety features. An example is the glow-in-the-dark function of running and biking gear to warn drivers on the road. The dynamic information, the output, can either be real-time or static. It can be controlled through a microprocessor or a simple non-computational input like light. The dynamic character of the surface, its colors, animation, lengths of appearance, subject, speed of movement, and so forth is influenced by the input. Though, the interface of the surface is not limited to a visual output. Its effectiveness can be achieved through an array of outputs for example motors or speakers. The focus of this paper is on the surface of the garment as a visual output devices, the garment as interface. The inputs can be many, and the interactions may vary, though the output is pre-defined through the surface. The objective is to describe the considerations for the design of an intelligent garment regarding the human interaction with the embedded electronic components, their placement, and the breath of functions that need to be considered in this novel field. It furthermore describes the importance of aesthetics in the design of a very personal yet public surface of the garment fueled with numerous preconceptions. Today's interface design in a ubiquitous environment is not restricted to two-dimensional displays stationary of computer monitors with the use of a mouse or mobile devices.

Multidisciplinary Character

The design of an intelligent garment is complex because of the breadth of disciplines needed for the development and because of the constraints the embedded technologies cause. A common vocabulary needs to be developed to allow for the many disciplines—like physical computing, fashion design, industrial design, wireless networking, software engineering, graphic design—to collaborate efficiently and fruitfully. Often the hands-on expertise of the craft—in particular garment construction—is not considered in the design process. A seamstress that understands the flow of electricity in a garment is rare and could soon be a know-how in high demand. "All too often projects covering this area fail fashion design—a flaw that often follows when engineers are dealing with the integration of technology in fashion. Conversely, where fashion designers who have no background in physical computing or programming work in the field, the actual technical integration is often flawed or absent" (Seymour, 2004, p. 13). A textile designer and an electrical engineer need to find a suitable common vocabulary in this novel field, as do all trades and disciplines involved.

The term wearable technologies covers in particular electrical engineering, physical computing, and wireless technologies.

EMBEDDED TECHNOLOGIES

Embedded technologies influence the wearability and comfort, the interaction system, and the aesthetic of the intelligent garment. If the functionality requires an active input by the wearer, the simplicity in understanding the technical features is key. Thus, the inclusion of wearable technologies know-how in the beginning stages of the design phase of a functional garment is central for its success. McCann, Hurford, and Martin (2005) describe the critical path during the design process of smart garments with end-user requirements on one hand and appropriate technology to fulfill such needs on the other hand. They identify the following items or processes: fiber/yarn, fabric, dying and finishing, coatings and laminates, body measurements/sizing and fitting, garment development, pattern development and grading, modeling, simulation and initial prototyping, integration of smart and wearable technologies, fitting, manufacture (cutting/bonding): mass and custom, distribution, display and point of sale: in-store and online, and end of life/recycling.

Inputs

Inputs, often sensors, answer where, who, what, when, how, and why. The huge variety of sensors makes it obvious how many possibilities there are in using them in the construction of intelligent clothing. For example, the inputs influence the user interaction extremely. Pressure is used for buttons and switches. The resistance varies with the application of force when using force-sensitive resistors. Pressure can also be produced through displacement. The resistance of bend sensors varies with the materials used. The firmer the material the harder it is to bend and as a result the interaction with the input varies. Other typical inputs are sensors such as temperature, proximity, magnetic hall sensors, and humidity sensors and do not require an active wearer interaction The environment delivers a number of inputs like macro-particles in smoke, smell, or optical scattering. The use of the environment as an input is particularly popular when referring to pollution. The wearer offers a dynamic display as the output—a human billboard—for current environmental data and is intertwined with the environment. This presents an interesting aspect of interaction. It seems to be passive but the act of choosing to display the information already requires the wearer to interact with the garment and to make conscious, active decision. As a result the wearer is consciously and actively involved. Sensors for an active interaction are for example acoustic, photodiodes, optical, accelerometers, touch, capacitive, compass, and orientation sensors. Many inputs can be both active and passive depending on the way they are used. The translation from input to output depends on the envisioned functionality.

The Visual (Dynamic) Surface as Output

A garment is seen, felt, heard, and touched. It stimulates the five senses. This chapter primarily focuses on the output of data on the surface of the garment, which is the display of information on a visual interface. All other outputs, for example motors, buzzers, speakers, fog or smoke will not be addressed. The use of the garment's surface as an interactive display is researched at various institutions and subject of interest for many artists. In 2003 France Telecom developed prototypes of flexible color screens using light-emitting diodes (LEDs) integrated in clothing calling it an optical fiber flexible display (OFFD). The prototype allows downloading, creating or exchanging visual data via the appropriate Internet gateway. (Konar, Deflin, & Weill, 2005). Barbara Layne, a researcher at Hexagram in Montreal works on a project called 'Animated Textiles,' which integrates light-emitting diodes (LEDs) and electronic circuitry into the structure of hand woven fabrics. 'Electronic Plaid,' by International Fashion Machine, layers thermochromic pigments using textile-printing technologies on top of an electronic textile with resistors woven into patterns. Laura and Lawrence MacCary, an artist and retired engineer based in Seattle create conductive fibers by weaving lead into 36 amplifier circuits, which illuminate LEDs depending on where you touch the fabric. The art-

Table 1. Classification of inputs

Origin	Input formats (examples)
Person	pressure, motion, vital signs, speech
Environment/Ambience	light, humidity, sound, temperature, smoke

work is entitled 'Dialectric: Connections.' In 2006 Philips launched Lumalive, a dynamic wearable textile surface. Lumalive fabrics feature arrays of colored light-emitting diodes (LEDs) fully integrated into the fabric that make it possible to create garments that can show dynamic messages, graphics or multicolored surfaces.

There are many factors that influence the dynamic textile-based display on the surface of the smart garment. For example the time the thermochromic needs to react to heat generated by conductive fibers woven into the material or the selection of the color of LEDs that are embedded into the fabric.

1. The variables captured from input sources (for example from sensors) are software-based data and consequently allow computation.
2. The data that goes through the microprocessor, the brain of the garment, and the computation programmed obviously determine the output.

The research into flexible organic light emitting diodes (FOLEDs) for garments seems hampered due to new developments in textile-based flexible displays. However, a visual dynamic textile output does not have to be dynamic. Very simple visual outputs, like one LED, are often sufficient to inform the wearer. Humans are already conditioned through their every day lives. Red means stop, green means go. Therefore a simple visual key through color might often be enough. If the heart rate of a heart patient is exceeding a certain beat an integrated red LED lights up indicating that the wearer of the intelligent garment needs to rest. This indication can be hidden in an area primarily visible for the wearer or obviously placed for care personnel to be able to monitor more easily individual patients without restraining them to a machine with a monitor.

Microprocessors

The microprocessor is necessary for the computation of the data derived from the input sources. Through computation the outputs are addressed on either the garment itself—referring to a BAN (body area network)—or the data is transferred wirelessly to a larger computer that processes the data and either gets back to the garment or keeps the data depending on the purpose of the application. Microprocessors or microcontrollers are the brains of the intelligent garment or wearable object. The single-chip computer can run and store a program but despite major research has still limited possibilities. Today the Arduino prototyping board seems to be the microprocessor of choice at universities and research settings where a quick and simple prototype is necessary. Such prototyping is very important to understand the functionalities and the interaction necessary by the user. Failure of an intelligent garment can thus be avoided. Microprocessors are becoming more and more relevant in the use for everyday garments due to their size, flexibility, and energy consumption, as are chips in everyday household appliances.

Issues Around Networks

The regional differences in wireless communication networks even just five years ago made it difficult to develop for a global audience. Today, however, the global collaborations in developing common standards make it easier to conceptualize and design embedded systems that endure geographical changes and standards. The need for flexibility leads to an obvious step in designing modular systems than can be removed and interchanged easily, thus influencing the architecture of the garment. The user needs to be able to switch off the network. An empowerment of the user is necessary for the success of such systems. The user might not be the actual wearer in particular when dealing with elderly care or neurologically sick patients like Alzheimer. However, the wearer needs to be in charge, not the machine.

Typical connectors in fashion design like buttons, zippers, grommets, snaps, or hooks and eye can easily be transformed into connectors and conductors of electricity, for example metal-based snaps. Other conductive connectors are conductive threads and fabric, conductive Velcro, conductive glue and solder. Such connectors act as electronic switches and instead of a wire a conductive thread is used in a smart garment; thus creating a fiber network. Humans are already conditioned using such connectors through their everyday experience with clothing. The interaction with a smart garment using such typical connectors makes the explanatory character of the functionality of a smart garment less difficult to understand for a wearer. A far more complicated issue deals with conductive yarns or thread. The + and – of a conductive thread needs to be separated when no conductive connection is necessary. Though, to allow this function the electronic fibers need to be insulated from each other. Their contact with the human skin is a main issue that needs to be addressed in the construction process. The skin is humid and can conduct electricity and as a result also create a short circuit. It can demolish the network the function is used for. The conductive thread is therefore best to be run on the surface of the textile or sandwiched between layers of textile or textile substrates. This constraint forms a new craft allowing to sustain the expertise of traditional textile design and to include nanotechnologies.

Power, Radiation, and the Environment

Besides the transport of energy throughout an item of clothing, the creation of energy is an important issue and strongly related to the way the wearer will interface and interact with the garment's functionalities. Energy for smart garments usually still comes from batteries, which are thrown away after their lifespan expires, causing huge environmental problems. Also, the life cycle of digitally enhanced devices in general is getting shorter and shorter. What happens with all the hazardous components and who is considering the energy consumption for the fabrication of such devices?

A battery's life is limited, thus the failure of a system depending on a battery is obvious. How can we avoid that a system, whose failure can be life threatening, collapses because of the lack of energy? If larger amounts of electricity are needed, research in solar power is probably the most advanced. Products like ScotteVest's (SeV) solar panels seem to be a first step towards using the surface of the garment that is constantly exposed to sunlight as a source of energy. ScotteVest's solar panels, trademarked PowerFLEX™, are flexible and can get wet, however, there are many 'do nots' in the user manual that make a widespread usage still questionable. An obvious and attractive energy source for intelligent clothing is the human body derived through movement or fluctuations in body temperature because of its characteristic of being less dependent on placement of the energy converter than solar or thermal energy. Today the energy that can be harvested through human kinetic energy based on movement or heat-exchange from the body can be measured in microwatts. It is, thus, too low to drive wearable technologies. Much research about sustainable energy consumption and the effects of wireless transmissions is still needed. Radiation and the effects of electro magnetic fields create a controversy amongst scientists and the industry, in particular when dealing with healthcare. Whether it is questionable to expose patients who are already weakened to additional pollution of electro magnetic frequency has yet to be understood. The effects of a monitoring system attached to a patient already wearing a pacemaker are one of the many issues that need to be tackled. As in the case of a network, the wearer needs to be able to switch off the power, to control the power consumption. Another option is to create a 'sleep' function like on a laptop; however, how can the wearer 'wake up' the system actively? What are the parameters for the automatic re-starting?

ERGONOMICS OF INTELLIGENT CLOTHING

To which extent do technical constraints impact the aesthetic and ability to wear the garments? An important difference makes the degree of body integration. The degree of intimacy is determined by personal preference or the functionality desired. The least integrated is a mobile device or handhelds like cell phones. These mobile devices can be attached to the garment, which serves as a container or a simple carrier. When such devices become accessories and are connected to the garment they become less mobile and more wearable. Tattoos are the most visible fully integrated visual display on the skin of the body. Less visible, but technically enhanced, are medical devices like implants. The degree of intimacy also depends on the context of use of the intelligent garment. Questions like how is the garment going to be used, when, by whom, for what, and why are asked to understand the use and therefore the human interaction with the garment.

Wearable technologies are essentially close to the skin, in particular through the use of sensors in the medical field. Textiles are more rigid than, for example, a hard cased microprocessor that is integrated into a garment. Gemperle, Kasabach, Stivoric et al. (1998) developed guidelines for wearability with their study 'Design of Wearability' and observed placement, form language, human movement, proximity, sizing, attachments, containment, weight, accessibility, sensory interaction, thermal, aesthetics, and long-term use. These considerations are a useful start for dealing with wearable technologies on our body and the construction of functional clothing. "A product that is wearable should have wearability. Wearability is defined as the interaction between the human body and the wearable object. Dynamic wearability extends that definition to include the human body in motion." (Gemperle, Kasabach, Stivoric et al., 1998, p. 1). Knight, Baber, Schwirtz et al. (2002) evaluated comfort across six dimensions being emotion, attachment, harm, perceived change, movement, and anxiety. Attachment, perceived change, and movement are sensed significantly stronger by the wearers, confirming the findings of Gemperle et al. (1998). A strong focus in the design of an intelligent garment needs to be placed on size, weight, weight distribution, placement, and attachment. The study also illustrates the importance of cognitive components like emotion and anxiety. In describing her research project 'whispers,' Schiphorst refers to these issues. "The research of whisper is based on wearable body architectures, extrapolated as small wearable devices, embedded within garments, worn close to skin: proximity creating resonance, contact and communication, body as carrier to device, device as devising the body" (Schiphorst, 2004, p. 1). Fashion designers, interaction designers, textile designers, and technologists are looking into innovative ways to integrate electronic components in the fashion design of the garments, while ensuring their wearability.

The Cut, Connectors, and Material

A wetsuit is closed with a zipper in the back of the suit with a long band to enable the wearer to reach it and use it. Thus the body ergonomics define the cut of the garment. Even though flexible antennas, batteries, and even microprocessors are developed the problem for their disappearance in a smart garment is an issue of the cut of the garment. Where can components such as these be hidden? How can a 'short-circuit be avoided? All this is an issue of the right cut, a major element in the design of the intelligent garment. The understanding of garment construction, the technical components, the network, the interaction, and the aesthetics is necessary for the success for the construction of an intelligent garment. Sizing looks at the volume of an embedded system; the dimensions of the microprocessor, the battery, and sensor, the antenna, etc determine the size of the garment. Layering can avoid short cuts and determines whether a separate layer for a specific sensor is necessary. The components and textiles have a specific weight and influence the garment's appearance. Draping creates the right shapes. The cut needs to allow for modular systems similar to Lego. Computer components need to be able to be easily exchanged due change of standards, failure, or simply the fact the garments needs to be washed. This requires modular systems including simple connection systems.

What are the current interactions with our garments? We need to button, zip, Velcro, pull over, slip into, and so much more. To understand our current interaction with garments it is necessary to develop a cut that integrates new, additional interactions. An example is the Burton Audex system with its textile switch interface on the sleeve to control an iPod in the jacket's interior pocket. The iPod is connected to the interface on the sleeve through a little microprocessor and conductive woven fabric in the sleeve of the jacket. The iPod can be removed and leaving a jacket that can easily be cleaned. Not only the technical functionality but also the extreme conditions were considered making it successful as an end-consumer product. In science and research, however, comfort and aesthetics are often not considered. Copper-coated conductive threads or fabrics might work ideally, conducting enough energy needed for the function the garment is promised to do. But, the textile is uncomfortable or even hazardous to wear or simply aesthetically not pleasing. The need for specialists in garment construction and textile design is apparent. Fashion designers often do not have a hands-on experience with the actual construction necessary for intelligent garments.

Aesthetics vs. Function

Building upon the considerations for wearability that define the relation between the human body and the wearable object, the question of aesthetics needs to be strongly examined. The cut is a major design element of architecture and construction of the garment. It defines the physical aesthetic of the garment—its appearance. The perceptions are many and range from cultural expectations to rules of human communication. The surface of a garment, "the cult of the body as an object of public display" (Warwick & Carvalho, 2004, p. 136), is once again noticeable in our culture. Thus a dynamic visual display on the garment calls for the wearer's control and need for confirmation of exposure if data. The functions of the visual display might change with its purpose and always need to be re-evaluated. Depending on the use of the garment either aesthetics or the functional components are more dominant. The wearer feels safe, protected from electromagnetic frequencies, which can also be considered a psychological function obvious. Social or cultural functions like ceremonial or religious strongly influence the look and feel of the garment. The functions of a specific intelligent garment are pre-defined in regards to their necessity. It is essential that technology and design work together to create garments that humans want to wear. And to understand the major differences in individual human needs. Another such cultural function of a garment is its association with a group. Groups of snowboarders sometimes tend to stencil their boards and their jackets. This is nothing new in our western culture but new technologies are used to create new applications by the users; not planned by those creating the technology.

THE FUTURE OF INTERACTION

"Not only did the garment impose a demeanor, it obliged me to live towards the exterior of the world" (Eco, 1976, p. 193). Commercially available successful interfaces are usable and intuitive. The use of durable technology is essential; wearable technology cannot break down. Users are conditioned that their PCs might fail but a cell phone never breaks down. Therefore the technology has to work; a wearer cannot just take off the garment and reboot it. Flexibility is a main characteristic of a textile interface and closely related to the comfort for the wearer and the cut the designer can choose. Washability is yet another major success factor; a garment needs to be able to be cleaned. Often it is enough to be able to hand-wash a garment that is not worn on the skin. Eleksen developed a smart fabric interface by pairing electronics with textile and calling it a fabric interface. Many of the claims Eleksen makes with its product are also the ones of other commercially successful textile interface manufacturers like Fibretronic. The interaction on the sleeve touchpads, however, is developed for right-handed people who are the majority of the population. Modular systems with plug-and-play connectors might seem to be a solution for many of the interaction problems. It also allows the quick removal of sensitive components that are essential for the washability of an intelligent garment. But, still a common standard needs to be developed to allow the use of components from different manufacturers in construction and use of an intelligent garment.

The human-garment interaction can either be passive when the wearer does not actively participate in the interaction or active, when a physical activation of the function by the wearer is necessary. Emotion-related physiological data from skin conductivity or skin temperature are an automatic, passive input. Such information needs to be translated in real-time and enable an immediate reaction to the current wearer-state. Active inputs can range from simple active mechanic input ranging from touch, pull, push, and so forth. to voice, eye tracking or facial recognition. The wearer needs to close the circuit through an active interaction. The Adidas_1, a running shoe by Adidas, first senses the runner's amount of compression in the heel, then the data is sent to a processor in the sole of the shoe, and the necessary adoption information is then sent to a motor-driving cable system in the shoe's muscle. Such a system requires the active usage of the shoe by the wearer but no actual interaction is required. The interaction, thus, is subtle and not obvious to the wearer. Picard's research in affective computing is particularly interesting when the user interaction is limited to inputs that are not using an input device or human hands. Picard (1998, p. 227) describes, "Wearables are computational devices that are worn as an article of clothing or jewelry. In particular, because of their potential for long-term intimate contact with you, wearables have a unique opportunity to become affective. An 'affective wearable' is a wearable system equipped with sensors and tools that enables recognition of its wearer's affective patterns." Very often the line blurs and passive inputs seem to become active when the wearer starts to influence the functionality of the intelligent garment.

The User Experience and its Considerations

The user experience with the intelligent garment is influenced by the ergonomics, the psychonomics, the interaction, and the context of use. The ergonomics describe the form of the wearable or an intelligent garment. The body ergonomics and the cut of the garment are the main influencers of wearability. The psychonomics refer to the perception from within and by the outside world. It is closely related to the aesthetics of the garment and social and cultural functions. The interaction, the feedback, is one of the most important factors to indicate that the communication took place and was successful or failed. The wearer always needs feedback in particular when the output, the dynamic visual element, is placed where the wearer cannot immediately see it. Depending on the use of the smart garment the interaction might change. The considerations for extreme sports situation with very low temperatures not only include the basic body ergonomics but also the human psychology. In a crisis situation, for example when hit by an avalanche, the wearer might react very differently to the same functionality the intelligent garment provides.

The comfort of an embedded system is experienced strongly by the wearer. A sensor that reads biometric data from the skin needs to be close to the body but also comfortable to wear. Bodymedia is promoting its 3.0 armband to be worn like a Band-Aid close to the skin without the need of an armband bulkier than a watch. The output is shown on a computer screen, the information transferred wirelessly. However, an immediate display of the data onto the wearers garment would allow for a self-enclosed body area network with and immediate ability for changes. The applications and the areas of display on the surface need to be considered wisely. Such a system not only requires comfort, it also needs to be safe. The wearer needs to feel in control and as a result safe; in particular in wellness, rehabilitation, and chronic diseases where the wearer is much aware of what is happening. Such a system requires special attention regarding the interaction possibilities. A successful intelligent garment interaction considers all factors: human, environment, and machine. The guideline for design considerations for an intelligent garment in Table 2 refers to the factors that influence the wearer's experience.

Table 2. Design considerations

Factors	Considerations
Body ergonomics	wearability and overall comfort (placement, form language, human movement, proximity, sizing, attachments, weight, accessibility, heat, material, cut)
Psychonomics	perceptions by the wearer and the environment, aesthetics, psychological function
Interaction	interface with the system (e.g. inputs), practicality of daily use (e.g., washing/ cleaning)
Context of use	functions (social, cultural, physical), environment, wearer

Applications

The applications and functions of conductive fibers are countless. Clothing with embedded technologies is evident in sport, work wear, healthcare and rehabilitation, rescue services, and elderly care and prevention. The consumer interest in functional wear in particular in sports is steadily increasing and the monetary threshold is rising. Smart fabrics like Goretex have set the price for 'extreme' sports wear with enhanced functionality high for the consumer market. It is easier now for Nike with its Nike+ or Burton's Audex system to reach the price levels needed to sustain the development of these products with enhanced embedded technology. Other industries will follow when solutions for the issues around usability, the interaction, and the aesthetics are honestly confronted.

The dynamic real-time output on the surface of the intelligent garment is limitless and depends on the applications and functions that are needed. A visual output is limiting and its suitability needs to be examined. Obvious commercial applications like wearable advertising, safety information or the visual output of body area network data in healthcare are relevant future applications.

CONCLUSION

"Technology has enabled a great deal of personalization in fashion" (Seymour, 2004, p. 534) in intelligent garments. Intelligent garments and wearable devices are very personal. The personal relation to a garment is very sound because of its many perceptions. They are stronger than those associated with mobile devices that can be hidden in a pocket. The human-garment relationship is therefore an important consideration. The multidisciplinary team that designs and constructs the intelligent garment needs to understand the wearer's needs, the context of use, and the relevant factors of the wearing experience. Besides the obvious trades like textiles, electronics, and fashion design the knowledge and expertise of psychologists, philosophers, representatives from the medical professions, anthropologists, and so on are essential to create successful and sustainable intelligent garments. 1) The user experience is influenced by the wearer's interaction with the intelligent garment. 2) Only the thoughtful construction of wearable systems allow for a great user experience and useful functionality. Consequently the design of user experience comprises all main design factors to enable successful use of an intelligent garment by an individual.

REFERENCES

Anttila, A., Ribak, A., Cereijo-Roibás, A., Seymour, S., Svanteson, S., Weiss, S. et al. (2004). Mobile communication versus pervasive communications: The role of handhelds. In Brewster, S. & Dunlop, M. (Eds.), *Mobile human-computer interaction, MobileHCI 2004* (pp. 531–535). Berlin, Germany: Springer-Verlag.

Baurley, S. (2005). Interaction design in smart textile clothing and applications. In Tao, X. (Ed.), *Wearable electronics and photonics* (pp. 223–243). Cambride, UK: Woodhead Publishing Ltd.

Berzowska, J. (2005). Electronic textiles: Wearable computers, reactive fashion, and soft computation. In Jeffries, J. (Ed.), *Textile, the journal of cloth & culture, digital dialogues 2: Textiles and technology, 3*(1), 2-19. Biggleswade, UK: Berg Publishers.

Braddock, S. E., & O'Mahony, M. (2006). *Techno textiles 2. Revolutionary fabrics for fashion and design.* New York, NY: Thames & Hudson.

Eco, U. (1990). *Travels in hyperreality.* Fort Washington, PA: Harvest Books.

Farren, A. & Hutchison, A. (2004). Digital clothes: Active, dynamic, and virtual textiles and garments. In Jeffries, J. (Ed.), *Textile, The Journal of Cloth & Culture, Digital Dialogues 2: Textiles and Technology, 2*(3), 290-306. Biggleswade, UK: Berg Publishers.

Gemperle, F., Kasabach, C., Stivoric, J., Bauer, M., & Martin, R. (1998). *Design for wearability.* Pittsburgh, PA: Carnegie Mellon University. Retrieved April 22, 2002, from http://www.ices.cmu.edu/design/wearability

Knight, J. F., Baber, C., Schwirtz, A., & Bristow H. W. (2002). The comfort assessment of wearable computers. In *6th International Symposium on Wearable Computers (ISWC 2002)* (pp. 65-74). Las Alamitos, CA: IEEE Computer Society.

Koncar, V., Deflin, E., & Weill, A. (2005). Communication apparel and optical fibre fabric display. In Tao, X. (Ed.), *Wearable Electronics and Photonics* (pp. 155-176). Cambride, UK: Woodhead Publishing Ltd.

McCann, J., Hurford, R., & Martin, A. (2005). A design process for the development of innovative smart clothing that addresses end-user needs from technical, functional, aesthetic and cultural view points. In *the 9th IEEE International Symposium on Wearable Computers* (pp. 70 – 77). Las Alamitos, CA: IEEE Computer Society.

McQuaid, M. (Ed.). (2005). *Extreme textiles. Designing for high performance.* New York, NY: Princeton Architectural Press.

Peter, C., Ebert, E, & Beikirch, H. (2005). A wearable multi-sensor system for mobile acquisition of emotion-related physiological data. In J. Tao, T. Tan, & R. W.Picard (Eds.), *Affective computing and intelligent interaction. First international conference, ACII 2005* (pp. 691-698). Berlin, Germany: Springer-Verlag.

Picard, R. W. (1997). *Affective computing.* Cambridge, MA: MIT Press.

Raffle, H., Ishi H., & Tichenor, J. (2004). Super cilia skin: A textural interface. In Jeffries, J. (Ed.), *Textile, the journal of cloth & culture, digital dialogues 2: Textiles and technology, 2*(3) (pp. 328-347). Biggleswade, UK: Berg Publishers.

Rosson, M. B., & Carroll, J. M. (2002). *Usability engineering. Scenario-based development of human-ccomputer interaction.* San Francisco: Morgan Kaufmann Publishers.

Schiphorst, T. (2004). *Soft, softer and sofly: [whispering] between the lines.* Surrey, Canada: Simon Fraser University. Retrieved August 16, 2006, from http://whisper.surrey.sfu.ca/PDF/whispering-between-lines.pdf

Seymour, S. (2004). The complexity of collaboration in wearable computing. In Tralla, M. (Ed.), *RAM3. Reclaiming cultural territory in new media.* (pp. 13-16). Tallinn, Estonia: RAM.

Warwick, A., & Cavallaro, D. (2001). *Fashioning the frame. Boundries, dress and the body.* Oxford, UK: Berg.

KEY TERMS

Body Area Network (BAN): Connects independent nodes dispersed in a smart garment. For example implanted medical devices and on-body sensors are connected wirelessly with monitoring tools to provide patient health data in real-time.

Conductive Threads, Electronic Fibers, or Conductive Yarns: Textile-based yarns that conduct electricity. The most rudimentary form is copper-coated yarn.

Connectors in Smart Garments: Are conductive fasteners such as metal-based snaps or zippers that function like a switch in an electronic circuit.

Electronics Textiles, Smart Fabrics, or Conductive Textiles: Result from the integration of technology into a textile. A textile-based circuit board is achieved through incorporating conductive yarns into the fabric.

Fiber Network: An electrically conductive network within the fabric and works with all technical components like inputs, outputs, microprocessors, and communication networks.

Intelligent Garments: Functional clothing constructed with electronic textiles or smart fabrics and wearable technologies to enable various functionalities.

Wearables: Objects with embedded technologies that are wearable and may have wireless communication capabilities.

Wearable Technologies: Technologies or systems that are embedded into intelligent garments or wearable objects. Components are inputs, outputs, microprocessors, and communication networks.

Chapter XII
Context as a Necessity in Mobile Applications

Eleni Christopoulou
University of Patras & Ionian University, Greece

ABSTRACT

This chapter presents how the use of context can support user interaction in mobile applications. It argues that context in mobile applications can be used not only for locating users and providing them with suitable information, but also for supporting the system's selection of appropriate interaction techniques and providing users with a tool necessary for composing and creating their own mobile applications. Thus, the target of this chapter is to demonstrate that the use of context in mobile applications is a necessity. It will focus on the current trend of modeling devices, services and context in a formal way, like ontologies, and will present an ontology-based context model.

INTRODUCTION

The future of computer science was marked by Weiser's vision (Weiser, 1991), who introduced the term ubiquitous computing (ubicomp) by defining a technology that can be seamlessly integrated into the everyday environment and aid people in their everyday activities. A few years later, the European Union, aiming to promote "human-centered computing," presented the concept of ambient intelligence (AmI) (ISTAG, 2001), which involves a seamless environment of computing, advanced networking technology and specific interfaces. So, technology becomes embedded in everyday objects such as furniture, clothes, vehicles, roads, and smart materials, providing people with the tools and processes that are necessary in order to achieve a more relaxing interaction with their environment.

Several industry leaders, like Philips and Microsoft, have turned to the design of ubicomp applications with a focus on smart home applications. However, people nowadays are constantly on the move, travel a lot, and choose to live in remote or mobile environments. In the near future, each person will be "continually interacting with hundreds of nearby wirelessly connected computers" (Weiser, 1993). Therefore, the need for mobile applications is now more evident than ever.

Recent years have seen a great breakthrough occur in the appearance of mobile phones. Initially they were used as simple telephone devices. Today, mobiles have evolved into much more than that. Although the majority of people still use mobile phones as communication devices, an increasing number of users have begun to appreciate their potential as information devices. People use their smart mobile phones to view their e-mails, watch the news, browse the Web, and so forth. Eventually, mobile phones and other mobile handheld devices became an integral part of our daily routine.

Both scientists and designers of ubicomp applications have realized that the mobile phone could be considered as one of the first AmI artefacts to appear. As mobile phones are becoming more powerful and smarter this fact is increasingly proven true. Thus, scientists wanting to take advantage of the emerging technology have implemented a great number of mobile applications that enable human-computer interaction through the use of handheld devices like mobile phones or personal digital assistants (PDAs). Such applications include visitor guides for cities and museums, car navigation systems, assistant systems for conference participants, shopping assistants and even wearable applications.

A closer examination of mobile applications shows that most of them are location-aware systems. Specifically, tourist guides are based on users' location in order to supply more information on the city attraction closer to them or the museum exhibit they are seeing. Nevertheless, recent years have seen many mobile applications trying to exploit information that characterizes the current situation of users, places and objects in order to improve the services provided. Thus, context-aware mobile applications have come to light.

Even though significant efforts have been devoted to research methods and models for capturing, representing, interpreting, and exploiting context information, we are still not close to enabling an implicit and intuitive awareness of context, nor efficient adaptation to behavior at the standards of human communication practice. Most of the current context-aware systems have been built in an ad-hoc approach, deeply affected by the underlying technology infrastructure utilized to capture the context (Dey, 2001). To ease the development of context-aware ubicomp and mobile applications it is necessary to provide universal models and mechanisms to manage context.

Designing interactions among users and devices, as well as among devices themselves, is critical in mobile applications. Multiplicity of devices and services calls for systems that can provide various interaction techniques and the ability to switch to the most suitable one according to the user's needs and desires. Context information can be a decisive factor in mobile applications in terms of selecting the appropriate interaction technique.

Another inadequacy of current mobile systems is that they are not efficiently adaptable to the user's needs. The majority of ubicomp and mobile applications try to incorporate the users' profile and desires into the system's infrastructure either manually or automatically observing their habits and history. According to our perspective, the key point is to give them the ability to create their own mobile applications instead of just customizing the ones provided.

The target of this chapter is to present the use of context in context-aware ubicomp and mobile applications and to focus on the current trend of modeling devices, services and context in a formal way (like ontologies). Our main objective is to show that context in mobile applications can be used not only for locating users and providing them with suitable information, but also for supporting the system's selection of appropriate interaction techniques and for providing them with a tool necessary for composing and creating their own mobile applications.

In the background section, which follows, we define the term context and present how context is modeled and used in various mobile applications focusing on ontology-based context models. In the subsequent sections we present our perspective of context, an ontology-based context model for mobile applications as well as the way in which human-computer interaction can be supported by the use of context. The Future section embraces our ideas of what the future of human-computer interaction in mobile applications can bring by taking context into account. Finally we conclude with some prominent remarks.

BACKGROUND

What is Context

The term "context-aware" was first introduced by Schilit and Theimer (1994), who defined context as "the location and identities of nearby people and objects, and changes to those objects." Schilit, Adams, and Want (1994) defined context as "the constantly changing execution environment" and they classified context into computing environment, user environment, and physical environment. Schmidt (2000) also considered situational context, such as the location or the state of a device, and defined context as knowledge about the state of the user and device, including surroundings, situation and tasks and pointing out the fact that context is more than location.

An interesting theoretical framework has been proposed by Dix et al. (2000), regarding the notions of space and location as constituent aspects of context. According to this framework context is decomposed into four dimensions, which complement and interact with each other. These dimensions are: system, infrastructure, domain, and physical context.

One of the most complete definitions for context was given by Dey and Abowd (2000); according to them context is "any information that can be used to characterize the situation of an entity. An entity should be treated as anything relevant to the interaction between a user and an application, such as a person, a place, or an object, including the user and the application themselves."

When studying the evolution of the term "context" one notices that the meaning of the term has changed following the advances in context-aware applications and the accumulation of experience in them. Initially the term "context" was equivalent to the location and identity of users and objects. Very soon, though, the term expanded to include a more refined view of the environment assuming either three major components; computing, user and physical environment, or four major dimensions; system, infrastructure, domain, and physical context. The term did not include the concept of interaction between a user and an application until Dey and Abowd (2000). This definition is probably at present the most dominant one in the area.

Context Modeling in Context-Aware Applications

A number of informal and formal context models have been proposed in various systems; the survey of context models presented in Strang and Linnhoff-Popien (2004) classifies them by the scheme of data structures. In Partridge, Begole and Bellotti (2005) the three types of contextual models, which are evaluated, are environmental, personal, and group contextual model.

Among systems with informal context models, Context Toolkit (Dey, Salber & Abowd, 2001) represents context in the form of attribute-value tuples, and Cooltown (Kindberg et al., 2002) proposed a Web-based model for context in which each object has a corresponding Web description. Both ER and UML models are used for the representation of formal context models in Henricksen, Indulska, and Rakotonirainy (2002).The context modeling language is used in Henricksen and Indulska (2006) in order to capture user activities, associations between users and communication channels and devices and locations of users and devices.

Truong, Abowd and Brotherton (2001) point out that the minimal set of issues required to be addressed when designing and using applications are: who the users are, what is captured and accessed, when and where it occurs, and how this is performed. Designers of mobile applications should also take these issues into account. Similar to this approach Jang, Ko and Woo (2005) proposed a unified model in XML that represents user-centric contextual information in terms of 5W1H (who, what, where, when, how, and why) and can enable sensor, user, and service to differently generate or exploit a defined 5W1H-semantic structure.

Given that ontologies are a promising instrument to specify concepts and their interrelations (Gruber, 1993; Uschold & Gruninger1996), they can provide a uniform way for specifying a context model's core concepts as well as an arbitrary amount of subconcepts and facts, altogether enabling contextual knowledge sharing and reuse in a Ubicomp system (De Bruijn, 2003). Ontologies are developed to provide a machine-processable semantics of information sources that can be communicated between different agents (software and humans). A commonly accepted definition of the term ontology was presented by Gruber (1993) and stated that "an ontology is a formal, explicit specification of a shared conceptualization." A "conceptualization" refers to an abstract model of some phenomenon in the world which identifies the relevant concepts of that phenomenon; "explicit" means that the type of concepts used and the constraints on their use are explicitly defined and "formal" refers to the fact that the ontology should be machine readable. Several research groups have presented ontology-based models of context and used them in ubicomp and mobile applications. We will proceed to briefly describe the most representative ones.

In the Smart Spaces framework GAIA (Ranganathan & Campbell, 2003) an infrastructure that supports the gathering of context information from different sensors and the delivery of appropriate context information to ubicomp applications is presented; context is represented as first-order predicates written in DAML+OIL. The context ontology language (Strang, Linnhoff-Popien & Frank, 2003) is based on the aspect-scale-context information model. Context information is attached to a particular aspect and scale and quality metadata are associated with information via quality properties. This contextual knowledge is evaluated using ontology reasoners, like F-Logic and OntoBroker.

Wang, Gu, Zhang et al. (2004) created an upper ontology, the CONON context ontology, which captures general features of basic contextual entities, a collection of domain specific ontologies and their features in each subdomain. An emerging and promising context modeling approach based on ontologies is the COBRA-ONT (Chen, Finin & Joshi, 2004). The CoBrA system provides a set of OWL ontologies developed for modeling physical locations, devices, temporal concepts, privacy requirements and several other kinds of objects within ubicomp environments.

Korpipää, Häkkilä, Kela et al. (2004) present a context ontology that consists of two parts: structures and vocabularies. Context ontology, with the enhanced vocabulary model, is utilized to offer scalable representation and easy navigation of context as well as action information in the user interface. A rule model is also used to allow systematic management and presentation of context-action rules in the user interface. The objective of this work is to achieve personalization in mobile device applications based on this context ontology.

Although each research group follows a different approach for using ontologies in modeling and managing context in ubicomp and mobile applications, it has been acknowledged by the majority of researchers (Biegel & Cahill, 2004; Dey et al., 2001; Ranganathan & Campbell, 2003) that it is a necessity to decouple the process of context acquisition and interpretation from its actual use, by introducing a consistent, reliable and secure context framework which can facilitate the development of context-aware applications.

Context Utilisation in Mobile Applications

In context-aware mobile applications location is the most commonly used variable in context recognition as it is relatively easy to detect. Thus, a lot of location-aware mobile systems have been designed, such as shopping assistants (Bohnengerger, Jameson, Kruger et al., 2002) and guides in a city (Davies, Cheverst, Mitchell et al., 2001) or campus area (Burrell, Gay, Kubo et al., 2002). Many location-aware mobile applications are used in museum environments; a survey is presented in (Raptis, Tselios & Avouris, 2005). In the survey of Chen and Kotz (2000) it is evident that most of the context-aware mobile systems are based on location, although some other variables of context like time, user's activity and proximity to other objects or users are taken into consideration.

User activity is much more difficult to identify than location, but some aspects of this activity can be detected by placing sensors in the environment. Advanced context-aware applications using activity context information have been put into practice for a specific smart environment (Abowd, Bobick, Essa et el., 2002). The concept of activity zones (Koile, Tollmar, Demirdjian et al., 2003) focuses on location, defines regions in which similar daily human activities take place, and attempts to extract users' activity information from their location.

Sensor data can be used to recognize the usage situation based on illumination, temperature, noise level, and device movements, as described for mobile phones in Gellersen, Schmidt and Beigl (2002) and PDA in Hinkley, Pierce, Sinclair et al.(2000), where it is suggested that contextual information can be used for ring tone settings and screen layout adaptation. The mobile device can observe the user's behavior and learn to adapt to a manner that is perceived to be useful at a certain location as was the case with the comMotion system (Marmasse & Schmandt, 2000).

Sadi and Maes (2005) propose a system that can make adaptive decisions based on the context of interaction in order to modulate the information presented to the user or to carry out semantic transformation on the data, like converting text to speech for an audio device. CASIS (Leong, Kobayashi, Koshizuka et al., 2005) is a natural language interface for controlling devices in intelligent environments that uses context in order to deal with ambiguity in speech recognition systems. In Häkkilä and Mäntyjärvi (2005) context information is used in order to improve collaboration in mobile communication by supplying relevant information to the cooperating parties, one being a mobile terminal user and the other either another person, group of people, or a mobile service provider.

Perils of Context-Awareness

The promise and purpose of context-awareness is to allow computing systems to take action autonomously; enable systems to sense the situation and act appropriately. Many researchers, though, are skeptical and concerned because of the problems that emerge from context-awareness.

A main issue regarding context-aware computing is the fear that control may be taken away from the user (Barkhuus & Dey 2003). Experience has shown that users are still hesitant to adopt context-aware systems, as their proactiveness is not always desired. Another aspect of this problem is that users often have difficulties when presented with adaptive interfaces.

Apart from control issues, privacy and security issues arise. The main parameters of context are user location and activity, which users consider as part of their privacy. Users are especially reluctant to exploit context-aware systems, when they know that private information may be disclosed to others (Christensen et al., 2006).

Even recent research projects suffer from difficulties in automated context fetching; in order to overcome this, the user is asked to provide context manually. Studies have shown that users are not willing to do much in order to provide context and context that depends on manual user actions is probably unreliable (Christensen et al., 2006). Additionally, systems that ask from users to supply context fail, as this affects the user's experience and diminishes his benefit from the system.

Practice has shown that there is a gap between how people understand context and what systems consider as context. The environment in which people live and work is very complex; the ability to recognize the context and determine the appropriate action requires considerable intelligence. Skeptics (Erickson, 2002) believe that a context-aware system is not possible to decide with certainty which actions the user may want to be executed; as the human context is inaccessible to sensors, we cannot model it with certainty. They, also, argue whether a context-aware system can be developed to be so robust that it will rarely fail, as ambiguous and uncertain scenarios will always occur and even for simple operations exceptions may exist. A commonly applied solution is to add more and more rules to support the decision making process; unfortunately this may lead to large and complex systems that are difficult to understand and use.

An issue that several researchers bring forward (Bardram, Hansen, Mogensen et al., 2006) is that context-aware applications are based on context information that may be imperfect. The ambiguity over the context soundness arises due to the speed at which the context information changes and the accuracy and reliability of the producers of the context, like sensors.

It is a challenge for context-aware systems to handle context, that may be non accurate or ambiguous, in an appropriate manner. As Moran and Dourish (2001) stated, more information is not necessarily more helpful; context information is useful only when it can be usefully interpreted.

WHAT IS CONTEXT FOR MOBILE APPLICATIONS?

Considering the use of context in the mobile applications discussed in the background section, we may conclude that, for these applications, context is almost synonymous to location and, specifically, to user location. However, context is quite more than just that. In this section, we will present our perspective on the parameters of context that are necessary for mobile applications. In order to figure out these parameters we have to identify the concepts that constitute the environment in which mobile applications exist. The primary concepts are indubitably people, places, time, objects and physical environment.

A mobile application is context-aware if it uses context to provide relevant information to users or to enable services for them; relevancy depends on a user's current task and profile. The user context issue has been addressed by many researchers of context-awareness (Crowley, Coutaz, Rey et al., 2002; Schimdt, 2002). However, the key for context-aware mobile applications is to capture user activity and preferences. Apart from knowing who the users are and where they are, we need to identify what they are doing, when they are doing it, and which object they focus on. In the background section we mentioned that, until now, most mobile applications determine user activity by their location; it is apparent, however, that a more elaborate model is necessary for representing this activity. Stahl (2006) proposes a model that represents a user's goals, activities and actions; he suggests that the distinction between an activity and an action lies in the fact that an activity takes a time span, while actions occur instantaneously. The system can define user activity by taking into account various sensed parameters like location, time, and the object that they use. For example, when a user opens the front door he is thought to be either entering or leaving the house, when the bed is occupied and the television is turned on he is watching a movie, but when the television is turned off he is probably sleeping. User preferences are also very important for context-aware mobile applications, but it is difficult for the system to define them. Users have to incorporate their preferences into the application on their own, although the system can also gather information from the interaction with them in order to acquire experience based on history. By exploiting system experience the application may also infer a user's mood, a factor that cannot be measured by any sensor.

In order to identify user location various technologies are being used. In outdoors applications, and depending on the mobile devices that are used, satellite supported technologies, like GPS, or network supported cell information, like GSM, IMTS, WLAN, are applied. Indoors applications use RFID, IrDA and Bluetooth technologies in order to estimate the users' position in space. Although location is the determining factor in identifying where users are, orientation is also a very important parameter; the system has to know what users are looking at or where they are going to. However, in order to efficiently exploit the information on user location and orientation, the mobile application needs to have a representation of the layout of the place in which users are. Spaces can be classified into the following types: public,

Figure 1. Context in mobile applications

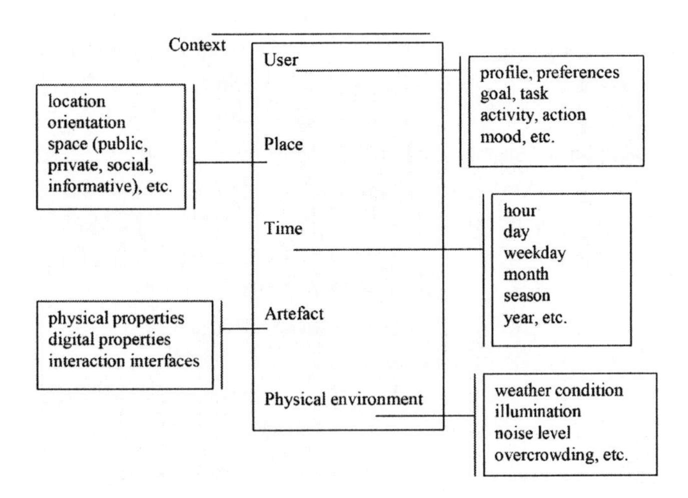

private, an area in which restrictions may apply, transient, places where people do not congregate easily or frequently, like hallways and corridors, social, public places where people arrange to meet, like coffee shops, informative, places that are used for public announcements (Mitchell, Race & Suggitt, 2006). Additionally a space can also be divided into districts, for example a home may have a living room, kitchen and bedroom, while a museum could have ancient Greek, paintings and modern art sections, as well into zones, such as lower left, upper left, and so forth.

Time is another significant parameter of context as it can play an important role in order to extract information on user activity; for example if it is early in the morning and the front door is opening the user is probably leaving the house, not entering it. Time can be used in various forms such as hour (daytime), night, day, weekday, week, month, season and year.

The objects that are used in mobile applications are the most crucial context sources. In mobile applications the user can use mobile devices, like mobile phones and PDAs and objects that are enhanced with computing and communication abilities (AmI artefacts). Sensors attached to artefacts provide applications with information about what the user is utilizing. However, this is not the most important parameter of context sensed by the artefacts. In order to present the user with the requested information in the best possible form, the system has to know the physical properties of the artefact that will be used, for example the display size of the artefact is determinant for the modulation of information. Additionally, the types of interaction interfaces that an artefact provides to the user need to be modeled; the system has to know if an artefact can be handled by both speech and touch techniques or if a mobile phone can vibrate. Apart from the physical properties of an artefact, the system must know how it is designed. A table with only one weight sensor in the centre cannot provide to the application information on whether an object is at its edge; thus the system has to know the number of each artefact's sensors and their position in order to gradate context information with a level of certainty. Based on information on the artefact's physical properties and capabilities, the system can extract information on the services that they can provide to the user; this is considered to be the most crucial context information related to artefacts. The application has to know if a printer can print both black-and-white and color text or if it can supply free maps and guidelines to a user that is close enough to a city's info center.

Finally, context from the physical environment may include current weather conditions, illumination, noise level, overcrowding. Taking into account the illumination of a room the application may decide to turn on an additional light when a user is reading a book or, if a user is in a noisy public space, the system may decide to vibrate his mobile phone when he has a call.

We selected to model the parameters of context illustrated in Figure 1 creating an ontology and taking into account the acknowledgement, shared by the majority of researchers (Biegel & Cahill, 2004; Dey et al., 2001; Ranganathan & Campbell, 2003), that it is a necessity to decouple the process of context acquisition and interpretation from its actual use. In the next section the details of this ontology-based context model are discussed.

AN ONTOLOGY-BASED CONTEXT MODEL FOR MOBILE APPLICATIONS

The key idea behind the proposed context model is that artefacts of AmI environments can be treated as components of a context-aware mobile application and users can compose such applications by creating associations between these components. In the proposed system, artefacts are considered as context providers. They allow users to access context in a high-level abstracted form and they inform other application's artefacts so that context can be used according to the application needs. Users are able to establish associations between the artefacts based on the context that they provide; keep in mind that services enabled by artefacts are provided as context. Thus defining the behavior of the application that they create, they can also denote their preferences, needs and desires to the system.

The set of sensors attached to an artefact measure various parameters such as location, time, temperature, proximity, motion, and so forth; the raw data given by its sensors is the artefact's low level context. As the output of different sensors that measure the same artefact parameter may differ, for example sensors may use different metric system, it is necessary to interpret the sensors' output into higher level context information. Aggregation of context is also possible meaning that semantically richer information may be derived based on the fusion of several measurements that come from different homogeneous or heterogeneous sensors. Thus, an artefact based on its own experience and use has two different levels of context; the low level which represents information acquired from its own sensors and the high level that is an interpretation of its low level context information. Additionally, an artefact can get context information from the other artefacts; this context can be considered as information from a "third-person experience."

When a user interacts and uses an artefact it affects its state; for example turning on the television sets it in a different state. An artefact may decide to activate a response based on both a user's desires and these states; for example when the user's PDA perceives that it is close to a specific painting in a museum, it will seek information about this painting. Such decisions may be based on the artefact's local context or may require context from other artefacts. The low and high level context, their interpretation and the local and global decision-making rules can be encoded in an ontology.

The ontology that we propose to represent the context of mobile applications is based on the GAS Ontology (Christopoulou & Kameas, 2005). This ontology is divided into two layers: a common one that contains the description of the basic concepts of context-aware applications and their inter-relations representing the common language among artefacts and a private one that represents an artefact's own description as well as the new "knowledge or experience" acquired from its use.

The common ontology, depicted in Figure 2, defines the basic concepts of a context-aware application; such an application consists of a number of artefacts and their associations. The concept of artefact is described by its physical properties and its communication and computational capabilities; the fact that an artefact has a number of sensors and actuators attached is also defined in our ontology. Through the sensors an artefact can perceive a set of parameters based on which the state of the

Figure 2. The common ontology

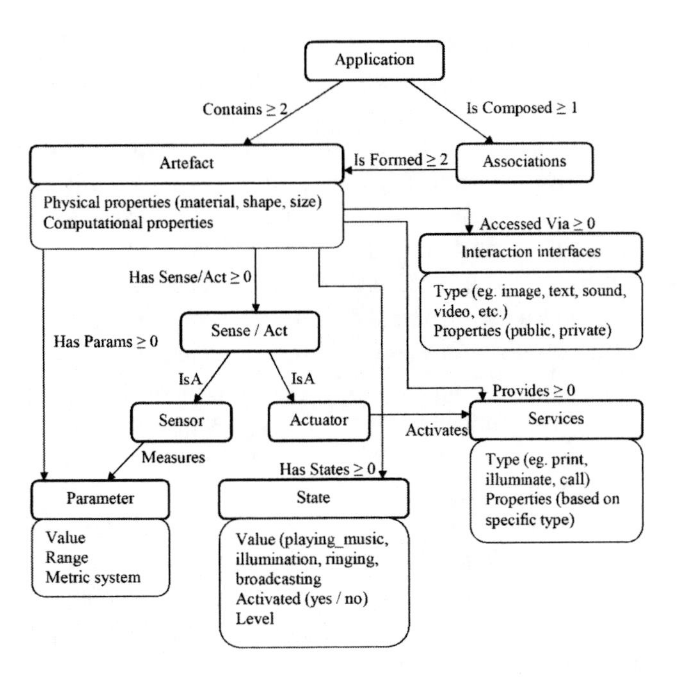

artefact is defined; an artefact may also need these parameters in order to sense its interactions with other artefacts as well as with the user. Artefacts may provide various services to the environment, for example a printer provides the print service, a lamp provides illumination and a phone the call service; these services are activated either by the user or by other artefacts using the actuators attached to artefacts. The interaction interfaces via which artefacts may be accessed are also defined in our ontology in order to enable the selection of the appropriate one.

We have decided that each parameter of context in our context-aware mobile applications, for example user, space, time and physical environment, is represented as an application's artefact. For instance, the notion of time is integrated into such applications only if a watch or a clock may provide this context as a service. The necessary information about the users that interact with such applications may be provided by the users' mobile phone or PDA. The services provided by such artefacts may be regarded as context; for instance the information that a thermometer provides is context related to the weather and we consider that the thermometer provides a temperature service. So, based on the concepts of context and their subcategories as presented in Figure 1, we have designed a service classification.

The common ontology represents an abstract form of the concepts represented, especially of the context parameters, as more detailed descriptions are stored into each artefact's private ontology. For instance, the private ontology of an artefact that represents a house contains a full description of the different areas in a house as well as their types and their relations.

The question that arises is where should these ontologies be stored? The system's infrastructure is responsible for answering this question. For a centralized system the common ontology as well as all the artefacts' ontologies can be stored in a central base. However, the majority of context-aware mobile applications are based on ad-hoc or p2p systems. Therefore, we propose that each artefact should store the common ontology and its private one itself; although when an artefact has limited memory resources its private ontology could be stored somewhere else. Another issue is where should place, time, environment and user ontologies be stored? The artefact that measures time, for example a clock, is responsible to store the time ontology; similarly there is an artefact for the environmental context. The place ontology can be stored either in a specific artefact that represents the space, for example an info kiosk in the entrance of a museum, or in the digital representation of the space managed by the application, for example the context from the sensors located in a room should be handled by the system through the use of the e-room and stored in an artefact with sufficient memory and computational capabilities. The user ontology in a similar way to the place ontology can be stored either in the user's mobile phone or in a digital self. These ontologies could also be stored in a web server in order to be accessible from artefacts.

The basic goal of the proposed ontology-based context model is to support a context management process, presented in Figure 3, based on a set of rules that determine the way in which a decision is made and are applied to existing knowledge represented by this ontology. The rules that can be applied during such a process belong to the following categories: rules for an artefact's state assessment that define the artefact's state based on its low and high level context, rules for local decisions which exploit an artefact's knowledge only in order to decide the artefact's reaction (like the request or the provision of a service) and finally rules for global decisions that take into account various artefacts' states and their possible reactions in order to preserve a global state defined by the user (Christopoulou, Goumopoulos & Kameas, 2005).

The ontology that is the core of the described context-management process was initially developed in the extrovert-Gadgets (eGadgets) project (http://www.extrovert-gadgets.net). In the e-Gadgets project our target was to design and develop an architectural framework (the Gadgetware Architectural Style—GAS) that would support the composition of ubicomp applications from everyday physical objects enhanced with sensing, acting, processing and communication abilities. In this project we implemented the GAS Ontology (Christopoulou & Kameas, 2005), which served the purpose of describing the semantics of the basic concepts of a ubicomp environment and defining their inter-relations. The basic goal of this ontology was to provide a common language for the communication and collaboration among the heterogeneous devices that constitute these environments; it also supported a service discovery mechanism necessary for that ubicomp environment. Already, at this early stage, we had decided on issues like how this ontology would be stored in each artefact, by dividing it into two layers, and a module had been implemented, which was responsible for managing and updating this ontology.

Figure 3. Context-management process

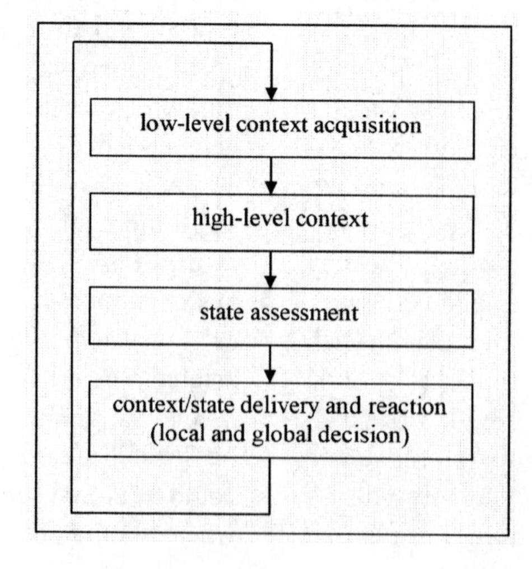

This work evolved in the PLANTS project (http://plants.edenproject.com) that aimed to enable the development of synergistic, scalable mixed communities of communicating artefacts and plants (Goumopoulos, Christopoulou, Drossos et al., 2004). In this project we extended the concept of "context" in order to allow for the inclusion of plants as components of our ubicomp applications, by attaching sensors to them that provided information regarding the plants' state. The ontology that was inherited from the e-Gadgets project was extended and refined in order to include all the parameters of context that were identified as necessary for our applications. The ontology-based context model and the context-management process, presented in Figure 3, were defined at that stage. Experience showed that our system managed to decouple the process of context acquisition and interpretation from its actual use. Our context-management process is based on a set of rules that define the state of each artefact or plant in an application. Based on these rules and their state, each artefact determines its local decisions; the set of rules on various artefacts determine global decisions made by the whole application. These rules are defined by the users themselves via a graphical user interface. Each artefact stores its ontology as well as its rule base as defined by the user; the decision-making process, part of the context-management process, is supported by an inference engine. Experience has shown that users could easily define their own applications, denoting the rules that govern both each artefact and the whole application; the fact that the reasoning process permits user-defined rules that can be dynamically updated was another positive point. A drawback of our system is that the inference engine, which was used required significant memory that was not always available; a workaround to this problem was to host the inference engine in an artefact with the required capabilities. Details on the design and implementation of this system as well as a case study of an application in the e-health domain and an evaluation of the outcome are presented in Christopoulou et al. (2005).

HOW CONTEXT CAN SUPPORT USER INTERACTION IN MOBILE APPLICATIONS

Recalling the use of context in mobile applications presented in the background section, we reach the conclusion that context has not been adequately exploited so far in order to support human-computer interaction. In this section we will present how our ontology-based context model enables the use of context in order to assist human-computer interaction in mobile applications and to achieve the selection of the appropriate interaction technique.

The goal of context in computing environments is to improve interaction between users and applications. This can be achieved by exploiting context, which works like implicit commands and enables applications to react to users or surroundings without the users' explicit commands (Schmidt, 2000). Context can also be used to interpret explicit acts, making interaction much more efficient. Thus, context-aware computing completely redefines the basic notions of interface and interaction.

The future of human computer interaction is going further than WIMP (Windows Icons Menus Pointing) interfaces. Jones and Marsden (2005) present various mobile interaction techniques that are trying to better exploit a user's capabilities like auditory (hearing) and haptic (touch and movement sensing) abilities as well as gestural skills, such as the expressive movements users can make with their hands or heads. More senses (vision, hearing, touch) and more means of expression (gestures, facial expression, eye movement and speech) are involved in human-computer interaction. A comparable analysis of mobile interaction techniques is presented in (Ballagas, Borchers, Rohs et al., 2006).

Rukzio et al. (2006) conclude from their experimental comparison of touching, pointing and scanning interaction techniques that users tend to switch to a specific physical mobile interaction technique dependent on location, activity and motivation; for example when a user is close enough to an artefact he prefers to touch it, otherwise he has no motivation for any physical effort. Thus, mobile systems have to provide multi-modal interfaces so that users can select the most suitable technique based on their context.

The ontology-based context model that we presented in the previous section captures the various interfaces provided by the application's artefacts in order to support and enable such selections. The application based on context can adapt to the information provided to the user; for example if a user tries to hear a message sent by his child on the mobile phone in a noisy environment the application may adjust the volume.

Similarly the context can determine the most appropriate interface when a service is enabled. Imagine that a user is in a meeting and an SMS is received by his mobile phone; even though he may have forgotten to enable the phone's silent profile, the application can select to enable the vibration interface instead of the auditory one based on the context about place and activity. Another example is the following: a user is with his children in a museum and he receives a high priority e-mail and the display of his PDA is too small for him to read the whole document that a colleague sent him; the application tries to identify a larger display to present the document based on proximate artefacts' context and taking into account environmental parameters, like whether there are other users close to it, and issues of privacy and security, like whether the document is confidential.

This infrastructure could also be useful for people with special needs. Consider how useful a museum guide application could be if it can provide more auditory information or even a model that the user can touch when it identifies a user with impaired vision entering a gallery.

Another aspect of mobile applications is that they are used simultaneously by several users. The mobile application has to consider the number of users and their preferences and attempt to form groups of people with similar profiles and interests. The application can base its decisions on place context when many users exploit it. In a museum guide, it is easier to form groups of people with similar interests than in city guides. People in social places are more willing to share artefacts and services than in private spaces. In a home application the system can give priority to a father to print his last version of a work instead of first printing a child's painting, whereas in a work environment application it is arguable whether the boss should have greater priority.

An important issue in mobile applications is system failure because of device unavailability; a mobile phone may run out of battery or be out of range. The service classification represented in the proposed context-ontology can handle such situations, as it merely needs to identify another artefact that provides the same or similar services, therefore is abstracting the user from such problems.

Ubiquitous and mobile interfaces must be proactive in anticipating needs, while at the same time working as a spatial and contextual filter for information so that the user is not inundated with requests for attention (Brumitt, Meyers, Krumm et al., 2000). At the same time, ubiquitous interfaces must allow the user control over the interface (Abowd & Mynatt, 2000). Barkhuus and Dey (2003) presented an interesting case study on some hypothetical mobile phone services and have shown that users prefer proactive services to personalized ones. Providing proactive context aware services based on perceived user context is one of the major focuses of mobile and ubiquitous computing. However, proactive systems involving multiple smart artefacts often create complex problems if their behavior is not inline with user preferences and implicit understandings.

The ontology-based context model that we propose empowers users to compose their own personal mobile applications. In order to compose their applications they first have to select the artefacts that will participate and establish their associations. They set their own preferences by associating artefacts, denoting the sources of context that artefacts can exploit and defining the interpretation of this context trough rules in order to enable various services. As the context acquisition process is decoupled from the context management process, users are able to create their own mobile applications avoiding the problems emerging from the adaptation and customisation of applications like disorientation and system failures. A similar approach is presented in Zhang and Bruegge (2004).

Finally context can also assist designers to develop mobile applications and manage various interfaces and interaction techniques. Easiness is an important requirement for mobile applications; by using context according to our approach, designers are abstracted from the difficult task of context acquisition and have merely to define how context is exploited from various artefacts by defining simple rules. Our approach presents an infrastructure capable of handling, substituting and combining complex interfaces when necessary. The rules applied to the application's context and the reasoning process support the application's adaptation. The presented ontology-based context model is easily extended; new devices, new interfaces as well as novel interaction techniques can be exploited into a mobile application by simply defining their descriptions in the ontology.

FUTURE TRENDS

A crucial question that emerges is what the future of user interaction techniques and interfaces in mobile and ubicomp applications is. Aarts (2004) presented that the ultimate goal of user interaction in such applications is realizing "magic." Watching the movie Matilda (DeVito, 1996), a number of interaction techniques that designers try to integrate into mobile applications are presented as magic; eyes blinking can lead to opening or closing of the blinds, simple gesture movements may open or close the windows and pointing at specific devices switches them on and off.

However, can ubiquitous and mobile computing enable forms of magic? The answer is yes. As Scott (2005) mentions "by embedding computing, sensing and actuation into everyday objects and environments, it becomes feasible to provide new abilities to users, allowing them to exert levels of control and sensing in the physical world that were not previously possible." All superhuman or magic powers related to mobile applications are closely connected with context as defined in the previous sections. When users establish associations among artefacts define how artefacts should react on various context changes; a form of telekinesis is implemented as devices are ubiquitously controlled. Teleathesia can also be implemented using context; having associated their mobile phone with their house, users can be informed via their phone if someone is entering or leaving house by merging place's and family members' context. When a user drives back to home, this context information about the user's activity can be presented via a toy's display to his child who is playing waiting to go to the zoo; thus telepresence is enabled by context. Precognition and postcognition abilities can also be supported by exploiting context; from system experience and artefacts's knowledge important results from the past can be concluded, whereas precognition is also feasible if users have particularly incorporated information into the applications about future meetings, appointments, and so forth.

Magic is not applicable only to user interaction and interfaces in mobile applications. The artefacts that will be created may embody forms of magic. Consider the Weasley's clock in the Harry Potter book series (Rowling), it presents information about each member to the family based on their current activity and state. Context could enable the design and development of such artefacts.

Ontologies will play an important role in context representation for mobile applications as well as rule-based infrastructures and inference engines will be exploited for context reasoning in such applications. However a number of critical questions arise. For example, the location where ontologies are stored is still in dispute. Various infrastructures propose general ontologies centrally stored, whereas others prefer smaller and application-specific ontologies stored in distributed locations. Concerning the context-reasoning based on rule-based infrastructures, the issue that emerges is whether existing inference engines are suitable for mobile applications or need we turn our focus on different, more light-weight systems.

A research opportunity within the domain of this topic is how various interaction techniques and interfaces can be classified and represented into the ontology-based context model in order to provide a more effective selection of interaction techniques. During the previous years a number of markup languages were created in order to represent and describe interfaces; we believe that ontologies are the most suitable formal model for representing interfaces for mobile applications. Additionally, a formal model of interfaces described by an ontology may also assist the evaluation of interfaces used in mobile applications.

It is evident that the progress made in the last decade in the field of context-awareness in mobile systems is significant; however, certain critical issues remain open. Proactive mobile applications need to be certain for the context information based on which they decide their reaction in order to be trusted by the users; furthermore, mobile applications are usually multi-user so privacy and security are crucial.

CONCLUSION

The objective of this chapter was to present how context can support user interaction in mobile applications. Context-aware applications exploit location information in order to deliver location-aware services; when a user is identified by the system, personalized and adaptive services are provided. Whenever the user activity can be determined, the infrastructure provides the user with a proactive system that transforms his environment to a smart one; when the environmental parameters can be exploited along with the activity the system can best adapt the conditions or select the most suitable interaction method and interface. More advanced scenarios of proactive systems can even accommodate for the failures of particular system components.

However, users are still hesitant to adopt context-aware systems. The major reason for this is the fear that control may be taken away from them (Barkhuus & Dey 2003). Also, the gap between human expectations and the abilities of context-aware systems is sometimes big, especially when systems must handle ambiguous and uncertain scenarios or when the context on which decisions are based is imperfect.

The ontology-based context model that we presented in a previous section offers the benefits that were described above. Additionally, it allows users to setup their own context-aware applications and define the way that artefacts react to changes, giving them at the same time the sense of retaining control over the system. The context-management process assesses the state of an artefact in a two step process; the low-level context may contain impure information that is refined in order to produce the high-level context. In our system the user is able to dynamically update the rules that define the environment; so he is capable of foreseeing possible exceptions.

ACKNOWLEDGMENT

I would like to deeply thank the various people who, during the several months in which this endeavor lasted, provided me with useful and helpful assistance.

As part of the research described in this chapter carried out in the e-Gadgets and PLANTS projects. I would like to thank all my fellow researchers in these projects; especially thank Achilles Kameas, Christos Goumopoulos, Irene Mavrommati, and all my colleagues in the DAISy team of the Research Unit 3 of the Research Academic Computer Technology Institute for their encouragement and patience throughout the duration of these projects.

I would like to thank the anonymous reviewers, who read an early (and rather preliminary) proposal of this chapter and provided me with helpful feedback and invaluable insights, as well as Joanna Lumsden, the editor of this book, for her personal invitation to me to contribute to this book and her support.

I would like to commend the interest and great job done by Dimitris Dadiotis and Ourania Stathopoulou, who reviewed and proofed this chapter.

Most important, to Dimitris, who put up with lost weekends and odd working hours.

REFERENCES

Aarts, E. (2004). Keynote speak. *Adaptive Hypermedia Conference 2004*, Eindhoven, Netherlands.

Abowd, G., & Mynatt, E. (2000). Charting past, present, and future research in ubiquitous computing. *ACM Transactions on Computer-Human Interaction, 7*(1), 29-58.

Abowd, G., Bobick, A., Essa, I., Mynatt, E., & Rogers, W. (2002). The aware home: Developing technologies for successful aging. *Workshop held in conjunction with American Association of Artificial Intelligence (AAAI) Conference, Alberta, Canada.*

Ballagas, R., Borchers, J., Rohs, M., & Sheridan, J. G. (2006). The smart phone: A ubiquitous input device. *IEEE Pervasive Computing, 5(1),* 70-77.

Bardram, J., Hansen, T., Mogensen, M., & Soegaard, M. (2006). Experiences from real-world deployment of context-aware technologies in a hospital environment. *In proceedings of Ubicomp 2006* (pp. 369-386). Orange County, CA.

Barkhuus, L., & Dey, A. K. (2003). Is context-aware computing taking control away from the user? Three levels of interactivity examined. In *Proceedings of UbiComp 2003* (pp. 150-156). Springer.

Biegel, G., & Cahill, V. (2004, March 14-17). *A framework for developing mobile, context aware applications.* In 2nd IEEE Conference on Pervasive Computing and Communications. Orlando, FL

Bohnengerger, T., Jameson, A., Kruger, A., & Butz, A. (2002). User acceptance of a decision-theoretic location-aware shopping guide. In *Proceedings of the Intelligent User Interface 2002* (pp. 178-179). San Francisco: ACM Press

Brumitt, B., Meyers, B., Krumm, J., Kern, A., & Shafer, S. A. (2000). EasyLiving: Technologies for intelligent environments. *In proceedings of the 2nd international symposium on Handheld and Ubiquitous Computing* (pp.12-29). Bristol, UK.

Burrell, J., Gay, G. K., Kubo, K., & Farina, N. (2002). Context-aware computing: A test case. In *Proceedings of Ubicomp 2002* (pp. 1-15).

Chen, G., & Kotz, D. (2000). *A survey of context-aware mobile computing research.* (Tech. Rep. TR2000-381). Department of Computer Science, Dartmouth College.

Chen, H., Finin, T., & Joshi, A. (2004). An ontology for context aware pervasive computing environments. *Knowledge Engineering Review—Special Issue on Ontologies for Distributed Systems.* Cambridge: Cambridge University Press.

Christensen, J., Sussman, J, Levy, S., Bennett, W. E., Wolf, T. V., & Kellogg, W. A. (2006). Too much information. *ACM Queue, 4*(6).

Christopoulou, E., & Kameas, A. (2005). GAS Ontology: An ontology for collaboration among ubiquitous computing devices. *International Journal of Human-Computer Studies, 62*(5), 664-685.

Christopoulou, E., Goumopoulos, C., & Kameas, A. (2005). An ontology-based context management and reasoning process for UbiComp applications. *In proceedings of the 2005 joint conference on Smart objects and ambient intelligence: innovative context-aware services: usages and technologies* (pp. 265-270). Grenoble, France.

Crowley, J. L., Coutaz, J., Rey, G., & Reignier, P. (2002). Perceptual Components for Context Aware Computing. In *the proceedings of UbiComp 2002.*

Davies, N., Cheverst, K., Mitchell, K., & Efrat, A. (2001). Using and determining location in a context-sensitive tour guide. *In IEEE Computer, 34*(8), 35-41.

De Bruijn, J. (2003). *Using ontologies—Enabling knowledge sharing and reuse on the semantic Web.* (Tech. Rep. DERI-2003-10-29). Digital Enterprise Research Institute (DERI), Austria.

DeVito, D. (Director). (1996). Matilda. *Sony and TriStar Pictures.*

Dey, A. K. (2001). Understanding and using context. *Personal and Ubiquitous Computing, Special issue on Situated Interaction and Ubiquitous Computing, 5*(1), 4-7.

Dey, A.K., & Abowd, G.D. (2000). Towards a better understanding of context and context-awareness. *CHI 2000, Workshop on The What, Who, Where, When, Why and How of Context-awareness* (pp.1-6). ACM Press

Dey, A. K., Salber, D., & Abowd, G. D. (2001). A conceptual framework and a toolkit for supporting the rapid prototyping of context-aware applications. *Human-Computer Interaction Journal, 16*(2-4), 97-166.

Dix, A., Rodden, T., Davies, N., Trevor, J., Friday, A., & Palfreyman, K. (2000). Exploiting space and location as a design framework for interactive mobile systems. *ACM Transactions on Computer-Human Interaction, 7*(3), 285-321.

Erickson, T. (2002). Some problems with the notion of context-aware computing. *Communications of the ACM, 45*(2), 102-104.

Gellersen, H.W., Schmidt, A., & Beigl, M. (2002). Multi-sensor context-awareness in mobile devices and smart artefacts. *Mobile Networks and Applications, 7,* 341-351.

Goumopoulos, C., Christopoulou, E., Drossos, N. & Kameas, A. (2004). The PLANTS System: Enabling Mixed Societies of Communicating Plants and Artefacts. In *proceedings of the 2nd European Symposium on Ambient Intelligence* (pp. 184-195). Eindhoven, the Netherlands.

Gruber, T. G. (1993). A translation approach to portable ontologies. *Knowledge Acquisition, 5*(2), 199–220.

Häkkilä, J., & Mäntyjärvi, J. (2005). Collaboration in context-aware mobile phone applications. In *Proceedings of the 38th International Conference on System Sciences.* Hawaii

Henricksen, K., & Indulska, J. (2006). Developing context-aware pervasive computing applications: Models and approach. *Journal of Pervasive and Mobile Computing,* 2(1), 37-64.

Henricksen, K., Indulska, J., & Rakotonirainy, A. (2002). Modeling context information in pervasive computing systems. In F. Mattern & M. Naghshineh (Eds.), *Pervasive 2002* (pp. 167–180). Berlin: Springer Verlag.

Hinkley, K., Pierce, J., Sinclair, M., & Horvitz, E. (2000). Sensing techniques for mobile interaction. *In CHI Letters, 2*(2), 91-100.

IST Advisory Group (ISTAG). (2001). *Scenarios for Ambient Intelligence in 2010-full.* http://www.cordis.lu/ist/istag-reports.htm

Jang, S., Ko, E. J., & Woo, W. (2005). Unified context representing user-centric context: Who, where, when, what, how and why. In *proceedings of International Workshop ubiPCMM05.* Tokyo, Japan.

Jones, M., & Marsden, G. (2005). *Mobile interaction design.* John Wiley & Sons.

Kindberg, T., Barton, J., Morgan, J., Becker, G., Caswell, D., Debaty, P., Gopal, G., Frid, M., Krishnan, V., Morris, H., Schettino, J., Serra, B., & Spasojevic M. (2002). People, places, things: Web presence for the real world. *Mobile Networks and Applications, 7*(5), 365–376.

Korpipää, P., Häkkilä, J., Kela, J., Ronkainen, S., & Känsälä, I. (2004). Utilising context ontology in mobile device application personalisation. In *proceedings of the 3rd international conference on Mobile and ubiquitous multimedia* (pp.133–140).

Koile, K., Tollmar, K., Demirdjian, D., Shrobe, H., & Darrell, T. (2003). Activity zones for context-aware computing. In *proceedings of UbiComp 2003 conference* (pp. 90-106). Seattle, WA.

Leong, L. H., Kobayashi, S., Koshizuka, N., & Sakamura, K. (2005). CASIS: A context-aware speech interface system. In *Proceedings of the 10th international conference on Intelligent user interfaces* (pp.231-238). San Diego, CA.

Marmasse, N., & Schmandt, C. (2000). Location-aware information delivering with comMotion. In *Proceedings of HUC 2000* (pp.157-171). Springer-Verlag.

Mitchell, K., Race, N. J.P., & Suggitt, M. (2006). iCapture: Facilitating spontaneous user-interaction with pervasive displays using smart sevices. In *PERMID workshop at the Pervasive 2006*. Dublin, Ireland.

Moran, T. P., & Dourish, P. (2001). Introduction to this special issue on Context-Aware Computing. *Human Computer Interaction 16*(2-4), 1-8.

Partridge, K., Begole, J., & Bellotti, V. (2005, September 11). Evaluation of contextual models. In *Proceedings of the First Internaltional Workshop on Personalized Context Modeling and Management for UbiComp Applications*. Tokyo, Japan.

Ranganathan, A., & Campbell, R. (2003). An infrastructure for context-awareness based on first order logic. *Personal and Ubiquitous Computing*, 7(6), 353–364.

Raptis, D., Tselios, N. & Avouris, N. (2005). Context-based design of mobile applications for museums: a survey of existing practices. In *Proceedings of the 7th international Conference on Human Computer interaction with Mobile Devices &Amp; Services. MobileHCI '05, 111* (pp. 153-160).. *ACM Press*.

Rowling, J. K. Harry Potter book series. *Bloomsbury Publishing Plc.*

Rukzio, E., Leichtenstern, K., Callaghan, V., Holleis, P., Schmidt, A., & Chin, J. (2006). An experimental comparison of physical mobile interaction techniques: Touching, pointing and scanning. In *proceedings of the 8th International Conference UbiComp 2006*, Orange County, CA.

Sadi, S. H., & Maes, P. (2005). xLink: Context management solution for commodity ubiquitous computing environments. In *proceedings of International Workshop ubiPCMM05*. Tokyo, Japan.

Schilit, B., Adams, N., & Want, R. (1994). Context-aware computing applications. In *proceedings of the IEEE Workshop on Mobile Computing Systems and Applications* (pp.85-90). Santa Cruz, CA.

Schilit, B., & Theimer, M. (1994). Disseminating active map information to mobile hosts. *IEEE Network*, *8*, 22-32.

Schmidt, A. (2000). Implicit human computer interaction through context. *Personal Technologies, 4*(2-3), 191-199.

Schmidt, A. (2002). *Ubiquitous computing—Computing in context*. Unpublished Ph.D. thesis, Department of Computer Science, Lancaster University, UK.

Scott, J. (2005). UbiComp: Becoming superhuman. *In the UbiPhysics 2005 workshop, Designing for physically integrated interaction*. Tokyo, Japan.

Stahl, C. (2006). Towards a notation for the modeling of user activities and interactions within intelligent environments. In *proceedings of the 3rd International Workshop on the Tangible Space Initiative (TSI 2006).* In Thomas Strang, Vinny Cahill, Aaron Quigley (Eds.), *Pervasive 2006 Workshop Proceedings* (pp. 441-452).

Strang, T., & Linnhoff-Popien, L. (2004). A context modeling survey. In *proceedings of the 1st International Workshop on Advanced Context Modelling, Reasoning And Management* (pp. 33-40). Nottingham, UK.

Strang, T., Linnhoff-Popien, L., & Frank, K. (2003). CoOL: A context ontology language to enable contextual interoperability. In *LNCS 2893 Proceedings of 4th IFIP WG 6.1 International Conference on Distributed Applications and Interoperable Systems* (pp. 236–247). Paris, France.

Truong, K. N., Abowd, G. D., & Brotherton, J. A. (2001). Who, what, when, where, how: Design issues of capture & access applications. In *proceedings of the International Conference: Ubiquitous Computing (UbiComp 2001)* (pp. 209-224). Atlanta, GA.

Uschold, M., & Gruninger, M. (1996). Ontologies: Principles, methods, and applications. *Knowledge Engineering Review, 11*(2), 93–155.

Wang, X. H., Gu, T., Zhang, D. Q., & Pung, H. K. (2004). Ontology based context modeling and reasoning using OWL. *Workshop on Context Modeling and Reasoning at IEEE International Conference on Pervasive Computing and Communication*. Orlando, FL.

Weiser, M. (1991). The computing for the 21st century. *Scientific American, 265*(3), 94-104.

Weiser, M. (1993). Some computer science issues in ubiquitous computing. *Communications of the ACM, 36*(7), 75-84.

Zhang, T., & Bruegge, B. (2004, August). *Empowering the user to build smart home applications.* Second International Conference on Smart homes and health Telematics. Singapore.

KEY TERMS

Ambient Intelligence (AmI): Implies that technology will become invisible, embedded in our natural surroundings, present whenever we need it, enabled by simple and effortless interactions, accessed through multimodal interfaces, adaptive to users and context and proactively acting.

Context: Any information that can be used to characterize the situation of entities (i.e., whether a person, place or object) that are considered relevant to the interaction between a user and an application, including the user and the application themselves.

Context-Aware Application: An application based on an infrastructure that captures context and on a set of rules that govern how the application should respond to context changes.

Mobile Computing: The ability to use technology in remote or mobile (non static) environments. This technology is based on the use of battery powered, portable, and wireless computing and communication devices, like smart mobile phones, wearable computers and personal digital assistants (PDAs).

Ontology: A formal, explicit specification of a shared conceptualisation. A tool that can conceptualise a world view by capturing general knowledge and providing basic notions and concepts for basic terms and their interrelations.

Ubiquitous Computing (Ubicomp): Technology that is seamlessly integrated into the environment and aids human in their everyday activities. The embedding computation into the environment and everyday objects will enable people to interact with information-processing devices more naturally and casually than they currently do, and in whatever locations or circumstances they find themselves.

Chapter XIII
Context-Awareness and Mobile Devices

Anind K. Dey
Carnegie Mellon University, USA

Jonna Häkkilä
Nokia Research Center, Finland

ABSTRACT

Context-awareness is a maturing area within the field of ubiquitous computing. It is particularly relevant to the growing sub-field of mobile computing as a user's context changes more rapidly when a user is mobile, and interacts with more devices and people in a greater number of locations. In this chapter, we present a definition of context and context-awareness and describe its importance to human-computer interaction and mobile computing. We describe some of the difficulties in building context-aware applications and the solutions that have arisen to address these. Despite these solutions, users have difficulties in using and adopting mobile context-aware applications. We discuss these difficulties and present a set of eight design guidelines that can aid application designers in producing more usable and useful mobile context-aware applications.

INTRODUCTION

Over the past decade, there has been a widespread adoption of mobile phones and personal digital assistants (PDAs) all over the world. Economies of scale both for the devices and the supporting infrastructure have enabled billions of mobile devices to become affordable and accessible to large groups of users. Mobile computing is a fully realized phenomenon of everyday life and is the first computing platform that is truly ubiquitous. Technical enhancements in mobile computing, such as component miniaturization, enhanced computing power, and improvements in supporting infrastructure have enabled the creation of more versatile, powerful, and sophisticated mobile devices. Both industrial organizations and academic researchers, recognizing the powerful combination of a vast user population and a sophisticated computing platform, have focused tremendous effort on improving and enhancing the experience of using a mobile device.

Since its introducion in the mid-1980s, the sophistication of mobile devices in terms of the numbers and types of services they can provide has increased many times over. However, at the same time, the support for accepting input from users and presenting output to users has remained relatively impoverished. This has resulted in slow interaction, with elongated navigation paths and key press sequences to input information. The use of predictive typing allowed for more fluid interaction, but mobile devices were still limited to using information provided by the user and the device's service provider. Over the past few years, improvements to mobile devices and back-end infrastructure has allowed for additional information to be used as input to mobile devices and services. In particular, context, or information about the user, the user's environment and the device's context of use, can be leveraged to expand the level of input to mobile devices and support more efficient interaction with a mobile device. More and more, researchers are looking to make devices and services *context-aware*, or adaptable in response to a user's changing context.

In this chapter, we will define context-awareness and describe its importance to human-computer interaction and mobile devices. We will describe some of the difficulties that researchers have had in building context-aware applications and solutions that have arisen to address these. We will also discuss some of the difficulties users have in using context-aware applications and will present a set of design guidelines that indicate how mobile context-aware applications can be designed to address or avoid these difficulties.

What is Context-Awareness

The concept of context-aware computing was introduced in Mark Weiser's seminal paper 'The Computer for the 21st Century' (Weiser, 1991). He describes ubiquitous computing as a phenomenon *'that takes into account the natural human environment and allows the computers themselves to vanish into the background.'* He also shapes the fundamental concepts of context-aware computing, with computers that are able to capture and retrieve context-based information and offer seamless interaction to support the user's current tasks, and with each computer being able to *'adapt its behavior in significant ways'* to the captured context.

Schilit and Theimer (1994a) first introduce the term *context-aware computing* in 1994 and define it as software that "adapts according to its location of use, the collection of nearby people and objects, as well as changes to those objects over time." We prefer a more general definition of context and context-awareness:

Context is any information that can be used to characterize the situation of an entity. An entity is a person, place or object that is considered relevant to the interaction between a user and an application, including the user and applications themselves, and by extension, the environment the user and applications are embedded in. A system is context-aware if it uses context to provide relevant information and/or services to the user, where relevancy depends on the user's task. (Dey, 2001)

Context-aware features include using context to:

- Present information and services to a user
- Automatically execute a service for a user and
- Tag information to support later retrieval

In supporting these features, context-aware applications can utilize numerous different kinds of information sources. Often, this information comes from sensors, whether they are software sensors detecting information about the networked, or virtual, world, or hardware sensors detecting information about the physical world. Sensor data can be used to recognize the usage situation for instance from illumination, temperature, noise level, and device movements (Gellersen, Schmidt & Beigl, 2002; Mäntyjärvi & Seppänen, 2002). Typically, sensors are attached to a device and an application on the device locally performs the data analysis, context-recognition, and context-aware service.

Location is the most commonly used piece of context information, and several different location detection techniques have been utilized in context-awareness research. Global positioning system (GPS) is a commonly used technology when outdoors, utilized, for example, in car navigation systems. Network cellular ID can be used to determine location with mobile phones. Measuring the relative signal strengths of Bluetooth and WLAN hotspots and using the hotspots as beacons are frequently used techniques for outdoors and indoors positioning (Aalto, Göthlin, Korhonen et al., 2004; Burrell & Gay, 2002; Persson et al., 2003). Other methods used indoors include ultrasonic or infrared-based location detection (Abowd et al., 1997; Borriello et al., 2005).

Other commonly used forms of context are time of day, day of week, identity of the user, proximity to other devices and people, and actions of the user (Dey, Salber & Abowd, 2001; Osbakk & Rydgren, 2005). Context-aware device behavior may not rely purely on the physical environment. While sensors have been used to directly provide this physical context information, sensor data often needs to be interpreted to aid in the understanding of the user's goals. Information about a user's goals, preferences, and social context can be used for determining context-aware device behavior as well. Knowledge about a user's goals helps prioritize the device actions and select the most relevant information sources. A user's personal preferences can offer useful information for profiling or personalizing services or refining information retrieval. The user may also have preferences about quality of service issues such as cost-efficiency, data connection speed, and reliability, which relate closely to mobile connectivity issues dealing with handovers and alternative data transfer mediums. Finally, social context forms an important type of context as mobile devices are commonly used to support communication between two people and used in the presence of other people.

Relevance to HCI

When people speak and interact with each other, they naturally leverage their knowledge about the context around them to improve and streamline the interaction. But, when people interact with computers, the computing devices are usually quite ignorant of the user's context of use. As the use of context essentially expands the conversational bandwidth between the user and her application, context is extremely relevant to human-computer interaction (HCI). Context is useful for making interaction more efficient by not forcing users to explicitly enter information about their context. It is useful for improving interactions as context-aware applications and devices can offer more customized and more appropriate services than those that do not use context. While there have been no studies of context-aware applications to validate that they have this ability, anecdotally, it is clear that having more information about users, their environments, what they have done and what they want to do, is valuable to applications. This is true in network file systems that cache most recently used files to speed up later retrieval of those files, as well as in tour guides that provide additional information about a place of interest the user is next to.

Relevance to Mobile HCI

Context is particularly relevant in mobile computing. When users are mobile, their context of use changes much more rapidly than when they are stationary and tied to a desktop computing platform. For example, as people move, their location changes, the devices and people they interact with changes more frequently, and their goals and needs change. Mobility provides additional opportunities for leveraging context but also requires additional context to try and understand how the user's goals are changing. This places extra burden on the mobile computing platform, as it needs to sense potentially rapidly changing context, synthesize it and act upon it. In the next section, we will discuss the difficulties that application builders have had with building context-aware applications and solutions that have arisen to address these difficulties.

BUILDING MOBILE CONTEXT-AWARE APPLICATIONS

The first context-aware applications were centered on mobility. The Active Badge location system used infrared-based badges and sensors to determine the location of workers in an indoor location (Want et al., 1992). A receptionist could use this information to route a phone call to the location of the person being called, rather than forwarding the phone call to an empty office. Similarly, individuals could locate others to arrange impromptu meetings. Schilit, Adams and Want,(1994b) also use an infrared-based cellular network to location people and devices, the PARCTAB, and describe 4 different types of applications built with it (Schilit et al., 1994b). This includes:

- **Proximate selection:** Nearby objects like printers are emphasized to be easier to select than other similar objects that are further away from the user;
- **Contextual information and commands:** Information presented to a user or commands parameterized and executed for a user depend on the user's context;
- **Automatic contextual reconfiguration:** Software is automatically reconfigured to support a user's context; and
- **Context-triggered actions:** If-then rules are used to specify what actions to take based on a user's context.

Since these initial context-aware applications, a number of common mobile context-aware applications have been built: tour guides (Abowd et al., 1997; Cheverst et al. 2000; Cheverst, Mitchell & Davies, 2001), reminder systems (Dey & Abowd, 2000; Lamming & Flynn, 1994) and environmental controllers (Elrod et al., 1993; Mozer et al., 1995). Despite the number of people building (and re-building) these applications, the design and implementation of a new context-aware application required significant effort, as there was no reusable support for building context-aware applications. In particular, the problems that developers faced are:

- Context often comes from non-traditional devices that developers have little experience with, unlike the mouse and keyboard.
- Raw sensor data is often not directly useful to an application, so the data must be abstracted to turn it into useful context.
- Context comes from multiple distributed and heterogeneous sources, and this context often needs to be combined (or fused) to be useful. This process often results in uncertainty that needs to be handled by the application.
- Context is, by its very nature, dynamic, and changes to it must be detected in real time and applications must adjust to these constant changes in order to provide a positive user experience to users.

These problems resulted in developers building every new application from scratch, with little reuse of code or design ideas between applications.

Over the past five years or so, there has been a large number of research projects aimed at addressing these issues, most often trying to produce a reusable toolkit or infrastructure that makes the design of context-aware applications easier and more efficient. Our work, the Context Toolkit, used a number of abstractions to ease the building of applications. One abstraction, the context widget is similar to a graphical user interface widget in that it abstracts the source of an input and only deals with the information the source produces. For example, a location widget could receive input from someone manually entering information, a GPS device, or an infrared positioning system, but an application using a location widget does not have to deal with the details of the underlying sensing technology, only with the information the sensor produces: identity of the object being located, its location and the time when the object was located. Context interpreters support the interpretation, inference and fusion of context. Context aggregators collect all context-related to a specific location, object or person for easy access. With these three abstractions, along with a discovery system to locate and use the abstractions, an application developer no longer needs to deal with common difficulties in acquiring context and making it useful for an application, and instead can focus on how the particular application she is building can leverage the available context. Other similar architectures include JCAF (Bardram, 2005), SOCAM (Gu, Pung & Zhang, 2004), and CoBRA (Chen et al., 2004).

While these architectures make mobile context-aware applications easier to build, they do not address all problems. Outstanding problems needing support in generalized toolkits include representing and querying context using a common ontology, algorithms for fusing heterogeneous context together, dealing with uncertainty, and inference techniques for deriving higher level forms of context such as human intent. Despite these issues, these toolkits have supported and continue to support the development of a great number of context-aware applications. So, now that we can more easily build context-aware applications, we still need to address how to design and build *usable* mobile context-aware applications. We discuss this issue in the following section.

USABILITY OF MOBILE CONTEXT-AWARE APPLICATIONS

With context information being provided as implicit input to applications and with those applications using this context to infer human intent, there are greater usability concerns than with standard applications that are not context-aware. Bellotti and Edwards discuss the need for context-aware applications to be *intelligible*, where the inferences made and actions being taken are made available to end-users (Bellotti & Edwards, 2001). Without this intelligibility, users of context-aware applications would not be able to decide what actions or responses to take themselves (Dourish, 1997).

To ground our understanding of these abstract concerns, we studied the usability and usefulness of a variety of context-aware applications (Barkhuus & Dey, 2003a; 2003b). We described a number of real and hypothetical context-aware applications and asked subjects to provide daily reports on how they would have used each application each day, whether they thought the applications would be useful, and what reservations they had about using each application. All users were given the same set of applications, but users were split into three groups with each group being given applications with a different level of proactivity. One group was given applications that they would personalize to determine what the application should do for them. Another group was provided with information about how their context was changing, and the users themselves decided how to change the application behavior. The final group was evaluating applications that autonomously changed their behavior based on changing context. Additional information was also gathered from exit interviews conducted with subjects.

Users indicated that they would use and prefer applications that had higher degrees of proactivity. However, as the level of proactivity increased, users had increasing feelings that they were losing control. While these findings might seem contradictory, it should be considered that owning a mobile phone constitutes some lack of control as the user can be contacted anywhere and at anytime; the user may have less control but is willing to bear this cost in exchange for a more interactive and smoother everyday experience. Beyond this issue of control, users had other concerns with regards to the usability of context-aware applications. They were concerned by the lack of feedback, or intelligibility, that the applications provided. Particularly for the more proactive versions of applications, users were unclear how they would know that the application was performing some action for them, what action was being performed, and why this action was being performed. A third concern was privacy. Users were quite concerned that the context data that was being used on mobile platforms could be used by service providers and other entities to track their location and behaviors. A final concern that users had was related to them evaluating multiple context-aware applications. With potentially multiple applications vying for a user's attention, users had concerns about information overload. Particularly when mobile and focusing on some other task, it could be quite annoying to have multiple applications on the mobile device interrupting and requesting the user's attention simultaneously or even serially.

In the remainder of this chapter, we will discuss issues for designing context-aware applications that address usability concerns such as these.

Support for Interaction Design

Despite all of the active research in the field of context-aware computing, much work needs to be done to make context-awareness applications an integral part of everyday life. As context-awareness is still a very young field, it does not have established design practices that take into account its special characteristics. The development of applications has so far been done primarily in research groups that focus more on proof-of-concept and short-term use rather than deployable, long-term systems. For most of these applications, the interaction design has rarely been refined to a level that is required for usable and deployable applications. Particularly for applications aimed at consumers and the marketplace, robustness, reliability and usability must be treated more critically than they are currently, as these factors will have a significant impact on their success.

Currently, the lack of existing high-quality, commercial, and publicly available applications limits our ability to assess and refine the best practices in interaction design of context-aware mobile applications. As there is very little experience with real-life use of these applications, the ability of developers to compare and iterate on different design solutions is very restricted. As user groups for a particular application mostly do not exist yet, much of the current research is based on hypothesized or simulated systems rather that actualized use situations. Knowledge of what device features people fancy and which they just tolerate, and when application features become insignificant or annoying, are issues that are hard to anticipate without studies of long-term real-life usage.

As with any other novel technology, bringing it to the marketplace will bring new challenges. Bringing context-awareness to mobile devices as an additional feature may lead to situations where the interaction design is performed by people with little experience in context-aware computing. Using well-established commercial platforms such as mobile phones or PDAs often means that user interface designers only have experience with conventional mobile user interfaces. On the other hand, the technical specifications of an application are often provided by people who have no expertise in human-computer interaction issues. When entering a field that involves interdisciplinary elements, such as mobile context-awareness, providing tools and appropriate background information for designers helps them to recognize the risks and special requirements of the technology.

Hence, there are several factors which make examining context-awareness from the usability and interaction design perspective relevant. Failures in these may lead not only to unprofitable products, but may result in an overall negative effect—they may slow down or prevent the underlying technology from penetrating into mass markets.

Usability Risks for Mobile Context-Aware Applications

A system and its functionality are often described with mental models that people form from using the system. According to Norman (1990), one can distinguish between the designer's mental model and the user's mental model. The designer's model represents the designer's understanding and idea of the artefact being constructed, whereas the user's model is the user's conceptual model of the same artefact, its features and functionality, which has developed through her interaction with the system. In order to respond to the user's needs, efficiently fulfil the user's goals and satisfy the user's expectations, the designer's and user's understanding of the device or application should be consistent with each other, in other words, the user's model and designer's model should be the same (Norman, 1990).

To ensure the best possible result, the mental models of different stakeholders in application development and use have to meet each other. First, the mental models of the application's technical designer and user interface designer should be consistent. This means that the user interface designer should have a basic understanding of the special characteristics of context-aware technology. Second, the designer's and user's mental models of the application should be the same. People's perception of context may differ significantly from each other, and both attributes and the measures used to describe context may vary greatly (Hiltunen, Häkkilä & Tuomela, 2005; Mäntyjärvi et al., 2003). The relationship between the designer's and the user's mental models should be checked with user tests several times during the design process. Without this careful design, there are two significant usability risks that may result: users will be unable to explain the behavior of the context-aware application, nor predict how the system will respond given some user action. While this is true of all interactive systems, it is especially important to consider for context-aware systems as the input to such systems is often implicit.

Context-awareness has several characteristics that can be problematic in interaction design. Figure 1 summarizes potential usability risks with context-aware applications.

A fundamental cause of potential usability risks is *uncertainty in context recognition*, which can be due to different reasons, such as detection accuracy, information fusion, or inferring logic. This is a key issue for designing the user interface for a mobile context-aware application, as it affects the selected features, their functionality and accuracy. In practice, features such as the proactivity level may be designed differently if the confidence level in context recognition can be estimated correctly. Uncertainty is a part of the nature of context-aware applications. Thus, it is important that the application and UI designers share a common understanding of the matter and take it into account when designing both the application and its user interface.

Application complexity has a tendency to grow when functions are added and it forms a potential risk for context-aware applications, as they use a greater number of information sources than traditional mobile applications. Hiding the complex nature of the technology while maintaining a sufficient level of feedback and transparency so that the user can still make sense of the actions the device is performing (i.e., intelligibility) is a challenging issue. Here, the involvement of user-

Figure 1. Sources of usability risks and their potential consequences related to context-aware mobile applications. Consequences that are unique to context-aware mobile applications are in the smaller rectangle on the right.

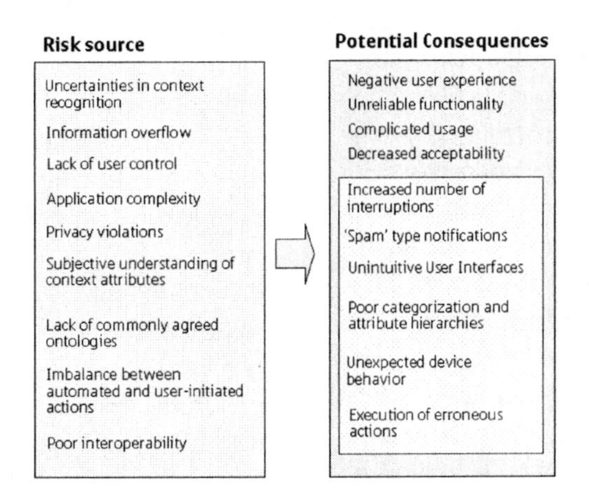

centric design principles is emphasized. Usability testing and user studies performed in an authentic environment combined with iterative design are key elements to producing well-performing user interface solutions.

Poor interoperability of services and applications relates to the absence of standardization in this maturing field and it limits the application design, available services, and seamless interaction desired across a wide selection of devices and users. Interoperability issues have gained much attention with the current trend of mobile convergence, where different mobile devices resemble each other more and more, yet providing services for them must be performed on a case-by-case basis.

Subjective understanding of context attributes creates a problem for user interface design, as the measures, such as the light intensity or noise level in everyday life are not commonly understood by end-users in terms of luxes or decibels but in relative terms such as 'dark,' 'bright,' 'silent' or 'loud.' This issue is connected to the *lack of commonly agreed ontologies,* which would guide the development of context-aware applications. The difficulties in categorizing context attributes and modeling context is evident from the literature (Hiltunen, Häkkilä & Tuomela, 2005; Mäntyjärvi et al., 2003).

As indicated earlier, *privacy violations* are possible with mobile context-aware systems collecting, sharing and using a tremendous amount of personal information about a user. When such information is shared with a number of different services, each of which will be contacting the user, *information overflow* often results. One can imagine a potential flow of incoming advertisements when entering a busy shopping street, if every shop within a radius of one hundred meters was to send an advertisement to the device. Information overflow is particularly a problem for the small screens that are typical with handheld devices.

As our earlier studies illustrated, the *lack of user control* can easily occur with mobile device automation, when context-triggered actions are executed proactively. However, the promise of context-awareness is that it provides "ease of use" by taking over actions that the user does not want to do or did not think to do for themselves. Any solution for correcting the *imbalance between the set of automated actions and user-initiated actions,* must take user control into account.

The consequences resulting from these usability risks are numerous. The general outcome can be a negative user experience. This may result from an increased number of interruptions, spam, and the execution of erroneous or otherwise unintuitive device behavior. Unreliable device functionality, and unintelligible user interfaces can lead to reduced acceptability of context-aware applications in the marketplace.

Design Guidelines for Mobile Context-Aware Applications

Context-awareness typically contains more risks than conventional, non-context-aware technology. At the same time, context-awareness can offer much added value to the user. In order to provide this value to end-users and avoid these negative design consequences and minimize usability risks, we have sought to provide a set of design guidelines that can offer practical help for designers who are involved in developing context-aware mobile applications (Häkkilä & Mäntyjärvi, 2006). These general guidelines have been validated in a series of user studies (Häkkilä & Mäntyjärvi, 2006) and should be taken into account when selecting the features of the application and during the overall design process.

GL1. Select appropriate level of automation. A fundamental factor with context-awareness is that it incorporates uncertainty. Uncertainty in context-recognition is caused by several different sources, such as detection accuracy, information fusion, or inferring logic. This is a key issue in designing user interfaces, as it affects the selected features, their functionality and accuracy. In practice, features such as the automation level or level of proactivity may be designed differently if the confidence level of context recognition can be estimated correctly. The relationship between uncertainty and selected application automation level is illustrated in Figure 2. As shown in Fig-

Figure 2. How uncertainty in context-recognition should affect the selected level of automation/proactivity

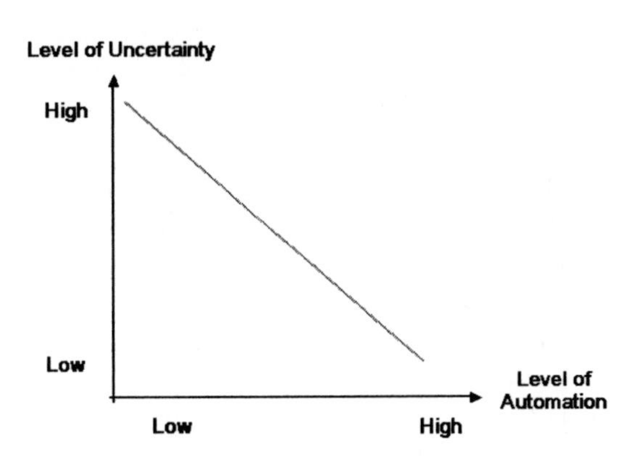

ure 1, *uncertainties in context recognition* create significant usability risks, however, by selecting an appropriate level of automation, an application designer can acknowledge this fact and address it appropriately. The greater the uncertainty is in the context-recognition, the more important it is not to automate actions. The automation level has also a direct relationship with user control, and its selection has a large impact on the number of expected interruptions the system creates for the user. The level of automation must be considered in relation to the overall application design, as it affects numerous issues in the user interface design.

GL 2. Ensure user control. The user has to maintain the feeling that he is in the control over the device. The user, who normally has full control over his mobile device, has voluntarily given some of it back to the device in order to increase the ease of use of the device. To address this *lack of user control*, an important usability risk, the user must be able to take control of the device and context-aware application at any time. The desire to take control can happen in two basic circumstances—either the device is performing erroneous actions and the user wants to take a correcting action, or the user just wishes to feel in control (a feeling that users often have). The user has to have enough knowledge of the context-aware application and the device functionality in order to recognize malfunctioning behavior, at least in the case where context-recognition errors lead to critical and potentially unexpected actions. The perception of user control is diminished if the device behaves in unexpected manner or if the user has a feeling that the device is performing actions without him knowing it. User control can be implemented, for example, with confirmation dialogues however, this must be balanced with the need to minimize unnecessary interruptions, our next guideline.

GL3. Avoid unnecessary interruptions. Every time the user is interrupted, she is distracted from the currently active task, impacting her performance and satisfaction with the system. In most cases, the interruption leads to negative consequences, however if the system thinks that the interruption will provide high value or benefit to the user, allowing the interruption is often seen as positive. Examples of this are reminders and alarm clocks. The user's interruptibility depends on her context and the user's threshold for putting up with intrusion varies with each individual and her situation. Some context-aware functionality is so important that the user may want the application to override all other ongoing tasks. This leads to a tension between avoiding unnecessary interruptions and supporting user control (GL2).

GL4. Avoid information overflow. The throughput of the information channel to each user is limited, and users can fully focus only on a small number of tasks at one time. In order to address the usability risk of *information overflow* where several different tasks or events compete for this channel, a priority ordering needs to be defined. Also, the threshold for determining the incoming event's relevancy in the context must be considered in order to avoid unnecessary interruptions (GL3). Systems should not present too much information at once, and should implement filtering techniques for to avoid messages that may appear to be spam to users. Also, information should be arranged in a meaningful manner to maintain and maximize the understandability of the system.

GL 5. Appropriate visibility level of system status. The visibility level of what the system is doing has to be sufficient for the user to be aware of the application's actions. While this guideline has been co-opted from Nielsen and Molich's user interface heuristics (1990), it has special meaning in context-aware computing. The implicit nature of context-awareness and natural *complexity of these types of applications* means that users may not be aware of changes in context, system reasoning or system action. When uncertainty in context-awareness is involved, there must be greater visibility of system state in order to allow the user to recognize the risk level and possible malfunctions. Important actions or changes in context should also be made visible and easily understandable for the user, despite the fact that users may have *subjective understandings of context attributes* and that there may be *no established ontology*. System status need not be overwhelming and interrupting to the user but can be provided in an ambient or peripheral fashion, where information is dynamically made more visible as the importance value grows, and may eventually lead to an interruption event to the user if its value is high enough.

GL 6. Personalization for individual needs. Context-awareness should allow a device or application to respond better to the individual user's personal needs. For instance, an application can implement filtering of interruptions according to the user's personal preferences. Personalization may also be used to improve the subjective understanding of context attributes. Allowing the user to name or change context attributes, such as location names or temperature limits, may contribute to better user satisfaction and ease of use. User preferences may change over time, and their representation in the application can be adjusted, for example implicitly with learning techniques or explicitly with user input settings.

GL 7. Secure user's privacy. *Privacy* is a central theme with personal devices, especially with devices focused on supporting personal communication, and impacts, for example trust, frequency of use, and application acceptability. Special care should be taken with applications that employ context sharing. Privacy requirements often vary between who is requesting the information, the perceived value of the information being requested and what information is being requested, so different levels of privacy should be supported. If necessary, users should have the ability to easily specify that they wish to remain anonymous with no context shared with other entities.

GL 8. Take into account the impact of social context. The social impact of a context-aware application taking an action must be part of the consideration in deciding whether to take the action or not. The application and its behavior reflects on users themselves. In some social contexts, certain device or user behavior may be considered awkward or even unacceptable. In such situations, there must be an appropriate *balance of user-initiated and system-initiated actions*. Social context has also has an effect on interruptibility. For example, an audible alert may be considered as inappropriate device behavior in some social contexts.

Once an application has been designed with these guidelines, the application must still be evaluated to ensure that the usability risks that have been identified for mobile context-aware systems have been addressed. This evaluation can take place in the lab, but is much more useful when conducted under real, *in situ*, conditions.

SUMMARY

Context-aware mobile applications, applications that can detect their users' situations and adapt their behavior in appropriate ways, are an important new form of mobile computing. Context-awareness has been used to overcome the deficit of the traditional problems of small screen sizes and limited input functionalities of mobile devices, to offer shortcuts to situationally-relevant device functions, and to provide location sensitive device actions and personalized mobile services.

Context-awareness as a research field has grown rapidly during recent years, concentrating on topics such as context-recognition, location-awareness, and novel application concepts. Several toolkits for enabling building context-aware research systems have been introduced. Despite their existence, there exist very few commercial or publicly available applications utilizing context-awareness. However, the multitude of research activities in mobile context-awareness allow us to make reasonable assumptions about tomorrow's potential applications. For example, navigation aids, tour guides, location-sensitive and context-sensitive notifications and reminders, automated annotation and sharing of photographs, use of metadata for file annotation, sharing or search are topics which frequently appear in the research literature and will likely be relevant in the future. In addition, using context-awareness to address the needs of special user groups, for example in the area of healthcare also appears to be a rich area to explore.

Despite the active research in context-awareness, there is much that remains to be addressed in interaction design and usability issues for context-aware mobile applications. Due the novelty of the field and lack of existing commercial applications, design practices for producing usable and useful user interfaces have not yet evolved, and end-users' experiences with the technology are not always positive. We have presented a set of 8 design guidelines which have been validated and evaluated in a series of user studies, which point to areas where user interface designers must focus efforts in order to address the usability issues that are commonly found with mobile context-aware applications.

While context-aware applications certainly have more usability risks than traditional mobile applications, the potential benefits they offer to end-users are great. It is important that application designers and user interface designers understand each other's perspectives and the unique opportunities and pitfalls that context-aware systems have to offer. With context-aware applications, careful application and interface design must be emphasized. The consequences resulting from usability risks include an overall negative user experience. Unsuccessful application design may result in diminished user control, increased number of interruptions, spam, and the execution of erroneous device actions or otherwise unintuitive behaviour. Unreliable device functionality and an unintuitive user interface can lead to decreased acceptability of the context-aware features in the marketplace.

In this chapter we have discussed the notion of context-awareness and its relevance to both mobile computing and interaction design in mobile computing. We have described technical issues involved in building context-aware applications and the toolkits that have been built to address these issues. Despite the existence of these toolkits in making context-aware applications easier to build, there are several additional issues that must be addressed in order to make mobile context-aware applications usable and acceptable to end-users. We have presented a number of design guidelines that can aid the designers of mobile context-aware applications in producing applications with both novel and useful functionality for these end-users.

REFERENCES

Aalto, L, Göthlin, N., Korhonen, J., & Ojala T. (2004). Bluetooth and WAP Push based location-aware mobile advertising system. In *Proceedings of the 2nd International Conference on Mobile Systems, Applications and Services* (pp. 49-58).

Abowd, G. D., Atkeson, C. G., Hong, J., Long, S., Kooper, R.. & Pinkerton, M. (1997). Cyberguide: a mobile context-aware tour guide. *ACM Wireless Networks, 3*, 421-433.

Bardram, J. (2005). The java context awareness framework (JCAF)—A service infrastructure and programming framework for context-aware applications. In *Proceedings of Pervasive 2005* (pp. 98-115).

Barkhuus, L., & Dey, A.K. (2003a). Is context-aware computing taking control away from the user? Three levels of interactivity examined. In *Proceedings of UBICOMP 2003* (pp. 149-156).

Barkhuus, L., & Dey, A.K. (2003b). Location-based services for mobile telephony: A study of users' privacy concerns. In *Proceedings of INTERACT 2003* (pp. 709-712).

Bellotti, V., & Edwards, K. (2001). Intelligibility and accountability: Human considerations in context-aware systems. *HCI Journal, 16*, 193-212.

Borriello, G., Liu, A., Offer, T., Palistrant, C., & Sharp, R. (2005). WALRUS: Wireless, Acoustic, Location with Room-Level Resolution using Ultrasound. In *Proceedings of the 3rd International Conference on Mobile systems, application and services (MobiSys'05)*, (pp. 191-203).

Burrell, J., & Gay, G. K. (2002). E-graffiti: Evaluating real-world use of a context-aware system. *Interacting with Computers, 14*, 301-312.

Chen, H., Finin, T. and Joshi, A. Chen, H., Finin, T., & Joshi, A. (2004). Semantic Web in the Context Broker Architecture. (2004). In *Proceedings of the Second IEEE international Conference on Pervasive Computing and Communications (Percom'04)*, (pp. 277-286).

Cheverst, K., Davies, N., Mitchell, K., & Friday, A. (2000). Experiences of developing and deploying a context-aware tourist guide: The GUIDE project. In *Proceedings of the 6th annual international conference on Mobile computing and networking (MobiCom)*, (pp. 20-31).

Cheverst, K., Mitchell, K., & Davies, N. (2001). Investigating Context-Aware Information Push vs. Information Pull to Tourists. In *Proceedings of MobileHCI'01*,

Dey, A.K., & Abowd, G.D. (2000). CybreMinder: A context-aware system for supporting reminders. In *Proceedings of the International Symposium on Handheld and Ubiquitous Computing* (pp. 172-186).

Dey, A.K., Salber, D., & Abowd, G.D. (2001). A conceptual framework and a toolkit for supporting the rapid prototyping of context-aware applications. *Human-Computer Interaction Journal 16*(2-4), (pp. 97-166).

Dourish, P. (1997). Accounting for system behaviour: Representation, reflection and resourceful action. In Kyng and Mathiassen (Eds.), *Computers and design in context* (pp. 145-170). Cambridge, MA: MIT Press.

Elrod, S., Hall, G., Costanza, R., Dixon, M., & des Rivieres, J. Responsive office environments. *Communications of the ACM* 36(7), 84-85.

Gellersen, H.W., Schmidt, A., & Beigl, M. (2002). Multi-sensor sontext-awareness in mobile devices and smart artefacts. *Mobile Networks and Applications, 7,* 341-351.

Gu, T., Pung, H.K., & Zhang, D.Q. (2004). A middleware for building context-aware mobile services. In *Proceedings of IEEE Vehicular Technology Conference* (pp. 2656-2660).

Häkkilä, J., & Mäntyjärvi, J. (2006). Developing design guidelines for context-aware mobile applications. In *Proceedings of the IEE International Conference on Mobile Technology, Applications and Systems.*

Hiltunen, K.-M., Häkkilä, J., & Tuomela, U. (2005). Subjective understanding of context attributes – a case study. In *Proceedings of Australasian Conference of Computer Human Interaction (OZCHI) 2005*, (pp. 1-4).

Lamming, M., & Flynn. M. (1994). Forget-me-note: Intimate computing in support of human memory. In *Proceedings of Friend21: International Symposium on Next Generation Human Interface* (pp. 125-128).

Mäntyjärvi, J., & Seppänen, T. (2002). Adapting applications in mobile terminals using fuzzy context information. In *Proceedings of Mobile HCI 2002* (pp. 95-107).

Mäntyjärvi, J., Tuomela, U., Känsälä, I., & Häkkilä, J. (2003). Context Studio—Tool for Personalizing Context-Aware Application in Mobile Terminals. In *Proceedings of Australasian Conference of Computer Human Interaction (OZCHI) 2003* (pp. 64-73).

Mozer, M.C., Dodier, R.H., Anderson, M., Vidmar, L., Cruickshank III, R.F., & Miller, D. The Neural Network House: An Overview. In L. Niklasson & M. Boden (Eds.), *Current trends in connectionism*, (pp. 371-380). Hillsdale, NJ: Erlbaum.

Nielsen, J., & Molich, R. (1990). Heuristic evaluation of user interfaces. In *Proceedings of CHI 1990* (pp. 249-256).

Norman, D. A. (1990). *The design of everyday things*. New York, NY: Doubleday.

Osbakk, P., & Rydgren, E. (2005). Ubiquitous computing for the public. In *Proceedings of Pervasive 2005 Workshop on Pervasive Mobile Interaction Devices (PERMID 2005)*, (pp. 56-59).

Persson, P., Espinoza, F., Fagerberg, P., Sandin, A., & Cöster, R. (2003). GeoNotes: A Location-based information System for Public Spaces. In K. Hook, D. Benyon & A. Munro (Eds.), *Readings in Social Navigation of Information Space* (pp. 151-173). London, UK: Springer-Verlag.

Schilit, B., & Theimer, M. (1994a). Disseminating active map information to mobile hosts. *IEEE Computer 8*(5), 22-32.

Schilit, B., Adams, N., & Want, R. (1994b). Context-aware computing applications. In *Proceedings of the IEEE Workshop on Mobile Computing Systems and Applications* (pp. 85-90).

Want, R., Hopper, A., Falcao, V., & Gibbons, J. (1992). The Active Badge Location System. *ACM Transactions on Information Systems, 10*(1), 91-102.

Weiser, M. (1991). The computer for 21st century. *Scientific American,* 265(3), 94-104.

KEY TERMS

Context: Any information that can be used to characterize the situation of an entity. An entity is a person, place or object that is considered relevant to the interaction between a user and an application, including the user and applications themselves, and by extension, the environment the user and applications are embedded in.

Context-Awareness: A system is context-aware if it uses context to provide relevant information and/or services to the user, where relevancy depends on the user's task.

Design Guidelines: Guidelines or principles that, when followed, can improve the design and usability of a system.

Interaction Design: The design of the user interface and other mechanism that support the user's interaction with a system, including providing input and receiving output.

Mobile Context-Awareness: Context-awareness for systems or situations where the user and her devices are mobile. Mobility is particularly relevant for context-awareness as the user's context changes more rapidly when mobile.

Usability Risks: Risks that result from the use of a particular technology (in this case, context-awareness) that impact the usability of a system.

Chapter XIV
Designing and Evaluating In-Car User-Interfaces

Gary Burnett
University of Nottingham, UK

ABSTRACT

The introduction of computing and communications technologies within cars raises a range of novel human-computer interaction (HCI) issues. In particular, it is critical to understand how user-interfaces within cars can best be designed to account for the severe physical, perceptual and cognitive constraints placed on users by the driving context. This chapter introduces the driving situation and explains the range of computing systems being introduced within cars and their associated user-interfaces. The overall human-focused factors that designers must consider for this technology are raised. Furthermore, the range of methods (e.g., use of simulators, instrumented vehicles) available to designers of in-car user-interfaces are compared and contrasted. Specific guidance for one key system, vehicle navigation, is provided in a case study discussion. To conclude, overall trends in the development of in-car user-interfaces are discussed and the research challenges are raised.

INTRODUCTION

The motor car is an integral part of modern society. These self-propelled driver-guided vehicles transport millions of people every day for a multitude of different purposes, for example as part of work, for visiting friends and family, or for leisure activities. Likewise, computers are essential to many peoples' regular lives. It is only relatively recently that these two products have begun to merge, as computing-related technology is increasingly implemented within road-going vehicles. The functions of an in-car computing system can be broad, supporting tasks as diverse as navigation, lane keeping, collision avoidance, and parking. Ultimately, by implementing such systems car manufacturers aim to improve the safety, efficiency, and comfort and entertainment of the driving experience (Bishop, 2005)

Designing the user-interface for in-car computing systems raises many novel challenges, quite unlike those traditionally associated with interface design. For instance, in many situations, the use of an in-car system is secondary to the complex and already demanding primary task of safely controlling a vehicle in 2D space, whilst simultaneously maintaining an awareness of hazards, largely using the visual sense. Consequently, the level of workload (physical, visual, and mental) when using displays and controls becomes a critical safety-related factor. As a further example, in-car computing systems have to be used by a driver (and possible also, a passenger) who is sat in a constrained posture and is unlikely to be able to undertake a two handed operation. Therefore, the design (location, type, size, etc.) of input devices has to be carefully considered, accounting in particular for comfort, as well as safety, requirements.

This chapter aims primarily to provide the reader with an overall awareness of novel in-car computing systems and the key HCI design and evaluation issues. The focus is on the user-interface, that is, "the means by which the system reveals itself to the users and behaves in relation to the users' needs" (Hackos & Redish, 1998, p.5). Topics of relevance to both researchers and practitioners are raised throughout. Given the complexity of the driving task and the wide range of computing systems of relevance, the chapter principally provides breadth in its consideration of the subject. Nevertheless, some depth is explored in a case study investigation on the design and evaluation of user-interfaces for vehicle navigation systems.

TYPES OF IN-CAR COMPUTING SYSTEMS

Technology is increasingly being seen to have a critical role to play in alleviating the negative aspects of road transport, such as congestion, pollution and road traffic accidents (Bishop, 2005). Many technological initiatives are considered under the umbrella term, intelligent transport systems (ITS), where "ITS provides the intelligent link between travelers, vehicles, and infrastructure" (www.itsa.org, September, 2006). In this respect, in-vehicle computing systems are an important facet of ITS. Specifically, there are two core types of computing and communications systems which are either being implemented or developed for use in vehicles:

- **Information-based systems:** These systems provide information relevant to components of the driving environment, the vehicle or the driver. Examples of systems include navigation (facilitating route planning and following), travel and traffic information (traffic conditions, car parking availability, etc.), vision enhancement (providing an enhanced view of the road ahead, when driving at night, in fog or in heavy rain), driver alertness monitoring (informing the incapacitated driver if they are unfit to drive) and collision warnings (presenting warnings or advice regarding hazards).
- **Control-based systems:** These systems affect the routine, operational elements of the driving task. Examples of systems include adaptive cruise control (where the car is kept at a set time gap from a lead vehicle), speed limiting (the car speed cannot exceed the current limit), lane keeping (the driver's vehicle is kept within a given lane), self parking (vehicle automatically steers in low speed operation to position itself within a selected parking space) and collision avoidance (the vehicle automatically responds to an emergency situation). Clearly, such systems fundamentally change the nature of what we consider to be 'driving.'

It is important to note that there is a third category of in-car computing system, those which do not provide any functionality to support the driving task. These systems are an important consideration though, as they can negatively influence safety, particularly through the potential for distraction (Young, Regan & Hammer, 2003). Such systems may aim to enhance work-oriented productivity whilst driving (e.g., mobile phones, e-mail and Internet access) or be primarily conceived for entertainment and comfort purposes (e.g., music and DVD players, games). Moreover, they may be designed for dedicated use in a vehicle or for operation in a range of different contexts (often termed nomadic devices).

OVERALL HUMAN FACTORS ISSUES

Driving is a complex task involving a large number of subtasks that can be conceptualised as existing within three levels of an overall hierarchical structure (Michon, 1985):

- Strategic tasks (highest level global travel decisions—e.g., which car to take, which route to take);
- Tactical tasks (making concrete maneuvers requiring interaction with other road users—e.g., changing lane, turning at a roundabout);
- Operational tasks (motor execution of tasks planned at higher levels—e.g., turning steering wheel, pressing brake).

Inevitably, the introduction of new technologies into the driving context will have a considerable impact across all three levels. As a result, there are many human-focused issues that must be considered in the design and evaluation process for in-car computing systems. To provide structure to a discussion of these issues, two overall scenarios are envisaged which may arise from poor design and/or implementation of the technology.

- **Overload:** Many of these systems (particularly those providing novel types of information and/or interactions) lead to situations in which a driver must divide their attention between core driving tasks (e.g., watching out for hazards) and secondary system tasks (e.g., inputting information). Furthermore, systems may provide excessive information in an inappropriate way leading to high levels of mental workload, stress and frustration. Such issues often manifest themselves as distraction to the driver (biomechanical, visual, auditory and/or cognitive).
- **Underload:** Control-based systems clearly automate certain aspects of driving, transferring certain responsibilities from operator to computer (e.g., staying in lane), whilst potentially providing new tasks for the driver (e.g., monitoring system performance). Automation is a fundamental human factors topic with a considerable research literature (see Wickens et al., 2004). Key concerns in this context relate to the potential for a driver exhibiting reduced situational awareness (e.g., for other road users), negative behavioral adaptation (e.g., by taking greater risks) and de-skilling (e.g., driver not able to resume control in the event of system failure).

THE HUMAN-CENTRED DESIGN PROCESS

The fundamental components of a human-focused approach hold true for in-car computing, as much as for any interactive product or system, that is, early focus on users and tasks, empirical measurement and iterative design (Gould & Lewis, 1985). A comprehensive understanding of the context in which in-car computing devices will be used is especially important early in the design process. Context of use refers to "the users, tasks and equipment (hardware, software, and materials), and the physical and social environments in which a product is used" (Maguire, 2001, p.457). A context of use analysis assists in developing the initial requirements for a design and also provides an early basis for testing scenarios. Moreover, context of use analysis provides a focused approach that helps to ensure a shared view among a design team. In the driving situation, there are several context of use issues which will have a significant effect on how an in-car computing system is subsequently designed. Accounting for these raises many unique challenges for in-car user-interface designers.

Users

As with many other consumer products, there will be a large variability in user characteristics (e.g., in perceptual and cognitive abilities, computer experience, anthropometry) to consider when designing in-car computing systems. Car manufacturers may have particular socio-economic groups in mind when designing a vehicle, but the user base may still be extremely large.

One fundamental individual difference factor often addressed in research is driver age—drivers can be as young as 16 (in certain countries) and as old as 90. In this respect, younger drivers may be particularly skilled in the use of computing technology, in comparison with the population at large, but are especially prone to risk taking (Green, 2003). Moreover, studies have shown a limited ability to divide attention and prioritize sources of information, largely due to lack of driving experience (Wickman, Nieminem & Summala, 1998). Subsequently, system block outs, which prevent the use of complex functions in inappropriate driving situations, are likely to be of particular benefit for these individuals.

In contrast, older drivers often suffer from a range of visual impairments that can lead to a range of problems with in-vehicle displays. For instance, presbyopia (loss of elasticity in the lens of the eye) is extremely common amongst older people, as is reduced contrast sensitivity. Studies consistently show that older drivers can take 1.5 to 2 times longer to read information from an in-vehicle display compared to younger drivers (Green, 2003). Given that drivers have a limited ability to change the distance between themselves and an in-vehicle display, the size, luminance and contrast of presented information are obviously critical design factors.

Tasks

A key task-related issue is that the use of an in-car computing system is likely to be discretionary. Drivers do not necessarily have to use the system to achieve their goals and alternatives will be available (e.g., a paper map, using the brake themselves). As a result, the perceived utility of the device is critical. Furthermore, drivers' affective requirements may be particularly important. In certain cases, this requirement may conflict with safety-related needs, for instance, for a simple, rather than flashy or overly engaging user-interface.

The factor that most differentiates the driving context from traditional user-interface design is the multiple-task nature of system use, and in this respect, there are two critical issues that designers must take into consideration. The first concerns the relationship between primary driving tasks and secondary system tasks, as drivers seek to divide their attention between competing sources of information. Driving is largely a performance and time-critical visual-manual task with significant spatial components (e.g., estimating distances). Consequently, secondary tasks must not be overly time-consuming to achieve or require attentional resources that are largely visual, manual, and spatial in nature, if they are to avoid having a significant impact on primary driving.

A second fundamental issue is the amount of information processing or decision making required for successful task performance, known as mental workload (Wickens et al., 2004). Novel in-car computing systems may provide functionality of utility to a driver or passengers, but interaction with the technology will inevitably increase (or in some cases decrease) overall workload. Context is very important here, as driving is a task in which workload varies considerably from one situation to another (compare driving in city traffic versus on the motorway). In this respect, certain authors (e.g., Green, 2004; Jones, 2002; Markkula, Kutila, & Engström, 2005) have taken the view that workload managers must be developed which make real-time predictions of the workload a driver is under and only present information or enable interactions to occur when overall workload is considered to be at an acceptable level. As an example, an incoming phone call may be sent straight to voice mail when the driver is considered to be particularly loaded (e.g., when driving in an unfamiliar city), but may be permitted in a lower workload scenario (e.g., driving along a dual carriageway and following a lead vehicle). Simple workload managers already exist in some vehicles (e.g., http://driving.timesonline.co.uk/article/0,,12929-2319048,00.html, September 2006), nevertheless, there are several complex research issues which must be addressed to fully realize the benefits of adaptive software in this context. For instance, workload managers need a comprehensive and accurate model of the driver, driving tasks and the driving environment. Given the vast range of variables of relevance to these categories, many of which do not lend to accurate and reliable measurement, extensive workload managers are likely to remain in the research domain for several years.

Equipment

The driving situation necessitates the use of input and output devices which are familiar to the majority of user-interface designers (pushbuttons, rockers, rotaries, LCDs, touchscreens, digitized or synthesized speech), together with equipment which is perhaps less known. For instance, there is a considerable research literature regarding the use of Head-Up Displays (HUDs) within vehicles. A HUD uses projection technology to provide virtual images which can be seen in the driver's line of sight through the front windscreen (see Figure 1). They are widely used within the aviation and military fields, and are now beginning to be implemented on a large-scale within road-based vehicles. HUDs will potentially allow drivers to continue attending to the road ahead whilst taking in secondary information more quickly (Ward & Parkes, 1994). As a consequence, they may be most applicable to situations in which the visual modality is highly loaded (e.g., urban driving), and for older drivers who experience difficulties in rapidly changing accommodation between near and far objects (Burns, 1999).

From a human-focused perspective, there are clear dangers in simply translating a technology from one context to another, given that vehicle-based HUDs will be used by people of varying perceptual and cognitive capabilities within an environment where there is a complex, continually changing visual scene. Specifically, researchers have established that poorly designed HUDs can mask critical road information, disrupt distance perception and visual scanning patterns, and negatively affect the ability of drivers to detect hazards in their peripheral vision (known as perceptual tunneling)—summarized by Tufano (1997) and Ward and Parkes (1994). Critical design factors that emerge from these findings include: display complexity; contrast and luminance; color choice; size of image; spatial location; and virtual image distance. Perhaps the most important design-related requirements are to consider carefully what and how much information is most appropriate to present on a HUD. There are temptations for designers to present ever-increasing amounts of information on HUDs. However, in contrast with traditional in-vehicle displays, a HUD image, by its very presence in the driver's line of sight, will demand focused attention (Burnett, 2003).

Figure 1. Example of a head-up display (HUD)

Environments

The physical environment is also a specific area that designers need to be aware of. In particular, the light, sound, thermal and vibration environment within a car can be highly variable. A range of design requirements will emerge from a consideration of these factors, for instance, potential for glare, problems with speech interfaces, use with gloves, and so on.

From anthropometric and biomechanical perspectives, the vehicle cabin environment provides many challenges for designers. This is an area in which designers make considerable use of CAD modeling to analyze different locations for displays and controls, ultimately aiming to ensure good fit for the design population. However, drivers sit in a constrained posture, often for several hours and have limited physical mobility (e.g., to comfortably view displays or reach controls). Consequently, there is limited space within a vehicle for the placement of a physical user-interface, a key problem for designers hoping to implement additional functionality within the vehicle.

To a large extent, this factor has fueled the development of multi-modal user-interfaces, where a small number of controls, together with menu-driven screens, provide access to many functions within the vehicle. Clearly, such visually-oriented user-interfaces are likely to promote a considerable amount of "eyes-off-road" time, and empirical studies have confirmed this prediction (Dewar, 2002). Moreover, users mistaking the current mode is a well-established problem in user-interface design, and clear feedback is an important design requirement (Preece, Rogers & Sharp, 2002). In many respects, there is a trade-off in design between the number of discrete controls that a user must scan within a vehicle and the number of levels within a menu-based system that must be explored and understood. This is a very similar problem to that considered by HCI researchers in the 1980s and 1990s interested in the breadth versus depth of menus in graphical user-interfaces (Shneiderman, 1998). An overall recommendation from such HCI research is that breadth should generally be favored over depth, as excessive depth can cause considerably more problems for the user than an equivalent breadth, largely due to the cognitive problems of navigation (Shneiderman, 1998). Whilst such guidance is considered to be of relevance to the design of in-car computing, research is still required which considers the trade-off existing in the multiple-task driving environment.

METHODS FOR USE IN DESIGN AND EVALUATION

In considering the range of methods that a designer can utilize when designing and evaluating in-car computing systems, the first difficulty is in establishing what is meant by a method. In this respect, a "human factors method for testing in-car systems" can be seen to be a combination of three factors:

1. Which environment is the method used in (road, test track, simulator, laboratory, etc.). As can be seen in Figure 2 (redrawn and adapted from Parkes, 1991), there is a fundamental trade off in choosing a method environment between the need for control and the validity of results. Choosing an environment will also be largely influenced by practical considerations, the knowledge/skills of the design and evaluation team and resource limitations.
2. Which task manipulations occur (multiple task, single task loading, no tasks given, etc.)? In certain methods, there is an attempt to replicate or simulate the multiple task nature of driving. For other methods, performance and/or behavior on a single task may be assessed and the potential impact on other tasks inferred from this. Most removed from actual driving, some methods do not involve users, but instead aim to predict impacts or issues, for instance through the use of expert ratings or modeling techniques.
3. Which dependent variables (operationalized as metrics) are of interest. In assessing an in-car computing user-interface, a

Figure 2. Environments for evaluation of in-car computing devices and the relationship between validity and control

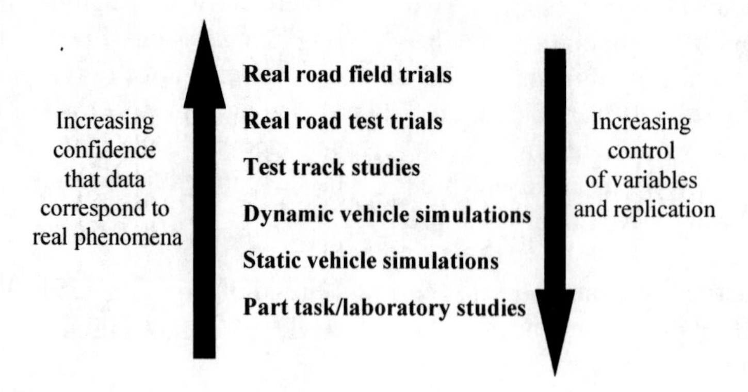

large range of possible metrics could be implemented. Some will relate to drivers' performance with primary driving tasks (e.g., lane position, hazard detection) or their use of primary vehicle controls (e.g., use of brake, steering wheel). Other metrics focus on driver performance and/or the demand of secondary tasks (e.g., task times, errors, display glances). As noted by Parkes (1991), usability evaluations of in-car computing devices should incorporate a wide range of measures relevant to the different levels of the driving task. For instance, at the strategic level, observation techniques and surveys are of relevance, whereas verbal protocols, interviews and questionnaires can capture the behavior of drivers at the tactical level. As noted by Parkes, such an approach provides "complete, rather than partial, pictures of product usability" (p.1445).

There is presently considerable research investigating specific methods for use in the design and evaluation of in-car user-interfaces. As noted by Preece et al. (2002), in deciding on any HCI method, the design team must consider the overall goals of the work, specific questions to be addressed, the practical and ethical issues and how data will need to be analyzed and reported. For in-car computing, these principles still hold, and many of the same global techniques used in the HCI area (for example, questionnaires, interviews, guidelines/checklists) will be used. However, by necessity, bespoke methods (or at least specific versions of generic methods) are required that account for the particular complex, safety-critical characteristics of the driving context. The following section summarizes key methods currently used and highlights some of the important research issues under investigation. Moreover, primary advantages and disadvantages are given. Table 1 summarizes some of the key issues.

Field Trials

Participants are given a car fitted with an operational system for several months for use in everyday activities. This method tends to look at broad issues relating to the long-term use of a system, for example, drivers' acceptance of the technology, and whether any behavioral adaptation effects arise. Objective data can be measured using on-board instrumentation (e.g., cameras, speed sensors) whereas subjective data is often captured using survey or interview-based approaches. Clearly, such a method provides an ecologically valid test of a system, and is particularly appropriate to the late stages of the design process where a robust prototype is available. Nevertheless, field trials can be extremely expensive and various ethical and liability considerations must be accounted for. An example of a field trial that was carried out in Sweden concerned drivers' use of intelligent speed adaptation systems (whereby a vehicle's speed is

Table 1. Overview of methods used to evaluate the user-interface for in-car computing systems

Method	Environment	Task manipulations	Overall Measures	Primary Advantages	Primary Disadvantages
Field trials	Real road (in everyday driving)	Multi-task (according to driver motivation)	Primary/ secondary task performance/ behavior, user opinions, etc.	Ecological validity, can assess behavioral adaptation	Resource intensive, ethical/liability issues to consider
Road trials	Real road (in pre-defined settings)	Multi-task (commonly, evaluator-manipulated)	Primary/ secondary task performance/ behavior, user opinions, etc.	Balance of ecological validity with control	Resource intensive, ethical/liability issues to consider
Simulator trials	Virtual driving environment (varying in fidelity)	Multi-task (commonly, evaluator-manipulated)	Primary/ secondary task performance/ behavior, user opinions, etc.	Control over variables, safe environment, cost-effective	Validity of driver behavior, simulator sickness
Occlusion	Laboratory/ statically in car	Secondary task achieved in controlled visual experience	Visual demand of user-interface	Standardized approach, control over variables	Limited scope, concern over validity of approach and metrics
Peripheral detection	Road/virtual driving environment	Multi-task (although commonly, evaluator-manipulated)	Visual/ cognitive workload	Assesses cognitive, as well as visual demand	Can be resource intensive, range of approaches
Lane change task	Specific lo-fidelity virtual driving environment	Multi-task motorway driving scenario	Primary lateral control of vehicle	Standardized approach, control over variables	Difficult to relate results to interface characteristics
15 second rule	Laboratory/ statically in car	Secondary task achieved without presence of driving task	Secondary task time (whilst stationary)	Simple approach	Only relates to certain aspects of visual demand
Keystroke-Level Model (KLM)	Modeling exercise	No user trials take place - models expert performance	Secondary task time (whilst stationary)	Quick/cheap, analysis explains results	Only relates to certain aspects of visual demand
Extended KLM	Modeling exercise	As for KLM, but with additional assumptions	Visual demand of user-interface	Quick/cheap, analysis explains results	Requires reliability assessments

automatically kept within the speed limit for the current area. Wallen, Warner & Aberg, 2005).

Road Trials

Drivers take part in a short-term (normally less than one day) focused study using a system in an instrumented car on public roads (occasionally on test tracks). For such trials, a wide range of variables may be measured and analyzed (e.g., visual behavior, workload, vehicle control, subjective preference) depending on the aims of the study. Road trials enable more experimental control than field trials, but are still potentially affected by a wide range of confounding variables (e.g., traffic conditions, weather). Furthermore, such a method remains costly to implement and requires robust protocols to ensure the safety of all concerned. Many road trials are reported in the literature, particularly concerning information and entertainment/productivity oriented systems. For instance, Burnett and Joyner (1997) describe a study which evaluated two different user-interfaces for vehicle navigation systems.

Simulator Trials

Drivers take part in a short-term (normally less than one day) focused study using a system fitted or mocked up within a driving simulator. The faithfulness that a simulator represents the driving task (known as its fidelity) can vary considerably, and configurations range from those with single computer screens and game controller configurations, through to real car cabins with multiple projections and motion systems. An example of a medium fidelity driving simulator is shown in Figure 3.

Driving simulators have become increasingly popular in recent years as a result of reduced hardware and software costs, and potentially offer an extremely cost-effective way of investigating many different design and evaluation issues in a safe and controlled environment (Reed & Green, 1999). Nevertheless, there are two key research issues concerning the use of driving simulators. Firstly, it is well known that individuals can experience symptoms of sickness in driving simulators, manifested as feelings of nausea, dizziness, and headaches. There has been considerable research regarding such sickness in virtual environments, and whilst there is still debate regarding the theoretical basis for the phenomenon (see for instance Nichols & Patel, 2002), there is practical guidance for those using driving simulators. For instance, screening questionnaires can be used to eliminate individuals who are most likely to experience sickness during a trial (Kennedy et al., 2001). Furthermore, various countermeasures can be used in the development of the simulator and its environment to reduce the prevalence of sickness (e.g., high, consistent frame rate, air-conditioning, natural background lighting. Nichols & Patel, 2002).

A second and more complex issue concerns validity, particularly behavioral (or construct) validity, that is, the extent to which drivers behave in the simulator as they would in the real world (Blaauw, 1982; Reed & Green, 1999). Driving simulator validity is problematic to study for several reasons. Running both road and simulator trials which are comparable (in terms of participants, tasks, measures, procedures, etc.) can be extremely difficult to achieve, and ultimately will be resource intensive. Furthermore, validity in this area is widely recognized to be a function of a large number of variables, including those relating to how the vehicle is represented (e.g., primary and secondary control design, the sense of enclosure, viewing angles, engine noise, vibration, motion, etc.) and those concerning the driving environment (e.g., visual field of view, screen resolution, graphical complexity, traffic representation, wind/road noise, etc. Kaptein et al.,1996; Peters & Peters, 2002). Most importantly, our understanding of validity must consider the driving task itself. Driving is a complex task, involving a substantial number of discrete physical, perceptual and cognitive behaviors, and a specific simulator configuration will only enable a subset of these to be investigated (e.g., speed control, headway maintenance).

As a consequence, despite the importance of the topic, there are few driving simulator validity studies in the open literature. Moreover, various limitations can be expressed for previous research in this area:

Figure 3. Example of a medium fidelity driving simulator

- It is difficult to generalize from existing validity studies, as they tend to be very specific to (a) the simulator configuration under investigation, and (b) the technology (hardware and software) available at that time (see for instance, Tornros (1998) compared with Blaauw (1982)).
- Studies inevitably only concern a small number of variables, for instance the effect of screen resolution and forward field of view on speed and headway choice (Jamson, 2001); or the effect of characteristics of torque feedback for steering on curve negotiation (Toffin et al., 2003).
- Studies often do not report critical data regarding the simulator configuration (e.g., steering sensitivity, max/min acceleration, driver eye height) which, for given types of study will be critical in results interpretation and cross-study comparison.

Occlusion

This is a laboratory-based method which focuses on the visual demand of in-vehicle systems. Participants carry out tasks with an in-vehicle system (stationary within a vehicle or vehicle mock up) whilst wearing computer-controlled goggles with LCDs as lenses which can open and shut in a precise manner (see Figure 4). Consequently, by stipulating a cycle of vision for a short period of time (e.g., 1.5 seconds), followed by an occlusion interval (e.g., 1.5 seconds), glancing behaviour is mimicked in a controlled fashion. Occlusion offers a relatively simple method of predicting visual demand, but is has been pointed out that its emphasis on user trials and performance data means that it requires a robust prototype and is therefore of limited use early in the design process (Pettitt et al., 2006).

Following considerable research, the occlusion method has recently been formalized as an international standard (ISO, 2005). In particular, guidance is given on how many participants are required, how much training to give, how many task variations to set, data analysis procedures, and so on. Moreover, two key metrics are stipulated: total shutter open time (the total time required to carry out tasks when vision is available); and resumability (the ratio of total shutter open time to task time when full vision is provided). For resumability, there is considerable debate regarding the merit of the measure. Advocates believe the metric provides an indication of the ease by which a task can be resumed following a period without vision (Baumann et al., 2004). Critics point out that the metric is also influenced by the degree to which participants are able to achieve tasks during occluded (non-vision) periods (Pettitt et al., 2006). Consequently, it can be difficult for a design team to interpret the results of an occlusion trial.

Peripheral Detection Task

This method requires drivers to carry out tasks with an in-car system (either on road or in a simulator)

Figure 4. The occlusion method with participant wearing occlusion goggles, with shutters open (left) and closed (right)

and to respond to the presence of lights within their periphery. The speed and accuracy of responses are considered to relate to the mental workload and distraction associated with secondary tasks (Young et al., 2003). The advantage of this method over occlusion is that it offers an assessment of cognitive, as well as visual demand (of relevance to the assessment of speech interfaces, for instance). The primary disadvantage is that the method still requires some form of driving task. Moreover, in contrast with occlusion, the method has not been fully standardized, and the ability to make cross study comparisons is severely limited by the specific choice of driving task scenarios (affecting task load and the conspicuity of the peripheral stimuli). It has also been noted that it is very difficult to discern between the level of cognitive demand and the visual demand for a given user-interface (Young et al., 2003).

An interesting recent development addresses some of these limitations. Engstrom, Aberg and Johansson (2005) considered the potential for the use of a haptic peripheral detection task, where drivers respond to vibro-tactile stimulation through the wrist whilst interacting with an in-vehicle system. Clearly, such a variation of peripheral detection is not affected by variations in lighting conditions. Furthermore, the authors argue on the basis of their validation work that this method provides "a 'pure' measure of cognitive load not mixed up with the effect of simply looking away" (p.233).

Lane Change Task

This method occurs in a basic PC simulated environment in which drivers are requested to make various lane change maneuvers whilst engaging with an in-vehicle system. The extent to which the profile of maneuver made by a driver varies from the optimum maneuver (the normative model) is considered to be a measure of the quality of their driving. Specifically, the method has the ability to assess the impact of an in-car computing system on a driver's awareness of the driving environment (perception, reaction), and, their ability to safely control the vehicle (maneuvering, lane keeping) Mattes (2003). Considerable research is ongoing with the lane change task in an attempt to develop an international standard (Transport Canada, 2006). Key research issues concern participant choice, training requirements and developing acceptable limits for performance.

15 Second Rule

Participants carry out tasks with an in-car computing system whilst stationary within a vehicle or mock up (i.e., with no driving task) and with full vision. The mean time to undertake a task is considered to be a basic measure of how demanding visually it is likely to be when driving (Green, 1999). A "cut-off" of 15 seconds has been set by the Society for Automotive Engineers (SAE). If the task on average takes longer than 15 seconds to achieve when stationary, it should not be allowed in a moving vehicle. The method is simple to implement and has the key advantage that it has been formalized in an SAE statement of best practice (SAE, 2000).

Research by Green (1999) and other research teams (e.g., Pettitt et al., 2006) has shown strong correlations between static task times and the total amount of time spent looking away from the road at displays/controls, both in simulator and road studies. However, the correlation between static task times and the duration of single glances towards an in-vehicle display is generally poor. This is important because a user-interface may promote a small number of very long glances (e.g., as a result of dynamically changing visual information) which can have a considerable negative effect on driving performance (Burnett & Joyner, 1997). It is for this primary reason that many authors advocate the use of the occlusion method as a better low-cost method for investigating the visual demand of an in-car user-interface (Pettitt et al., 2006; Stevens et al., 2004).

Keystroke Level Model (KLM)

The KLM method from the GOMs family of techniques is well known to HCI researchers and (to a lesser extent) practitioners (Preece et al., 2002;

Shneiderman, 1998). It is a form of task analysis in which system tasks with a given user-interface are broken down into their underlying physical and mental operators, e.g., pressing buttons, moving hand between controls, scanning for information. This is a method that is extremely cheap to implement, as there is no need for participants, and the method can be used with very basic prototypes early in the design process. Time values are associated with each operator and summed to give a prediction of task times. Researchers have developed new operator values relevant to the in-car situation (e.g., time to search a visual display, locate a control, move hand back to steering wheel) and have reported strong correlations between predicted task times and times based on user trials (Green, 2003; Pettitt et al., 2005). Task times can be related to certain measures of visual demand for in-car user-interfaces.

In an extension of the KLM method, Pettitt, Burnett and Stevens (2007) recently developed new rules that enable designers to develop predictions for a broader range of visual demand measures. In particular, the extended KLM considers a time-line view of an interaction in which a cycle of vision/non-vision occurs with a user-interface (similar to the occlusion protocol). The authors have found that their version of KLM can differentiate between tasks as effectively as does the occlusion technique, but recommend that further development is carried out to ensure that practitioners can utilize the method reliably.

CASE STUDY: VEHICLE NAVIGATION SYSTEMS

To ground many of the issues previously mentioned, a specific system type has been chosen for further discussion (vehicle navigation systems). Many of the individual points made for this system can be generalized and are applicable to other in-car computing technologies.

Vehicle navigation systems aim to support the strategic (e.g., route planning) and tactical (e.g., route following) components of the overall driving task. They have the greatest potential to assist drivers who undertake many unfamiliar journeys, for instance as part of work, or during leisure trips (e.g., when on holiday) and those who experience extreme difficulties with existing methods of navigation (particularly paper maps). When linked with reliable, real-time traffic information (thus providing dynamic guidance), the perceived utility of navigation systems to the everyday motorist is significantly enhanced (Bishop, 2005).

The market potential for vehicle navigation systems has already been demonstrated in Japan, where the technology has been available since the early 1990s. Approximately 40 percent of all vehicles on Japan's roads now have a navigation system installed (http://www.jetro.go.jp/en/market/trend/topic/2004_12_carnavi.html, September 2006). In many other countries, the popularity of navigation systems is currently reduced in relation to Japan, but is predicted to rise rapidly over the next few years (Bishop, 2005).

The majority of human factors issues relevant to this form of technology relate to overload, although as shall be seen, underload is increasingly being researched. With respect to overload, clearly, a key concern is the potential for driver distraction and there has been considerable research on this topic since the mid 1980s (see Young et al., 2003 and Srinivisan, 1999, for reviews). In using a vehicle navigation system, drivers must interact with controls (e.g., to enter a destination, chance map scale) and view/understand displays (e.g., to decide which turn to make, to examine options within a menu). In many cases, these interactions will arise when the vehicle is in motion. Consequently, to provide guidance for designers, researchers have aimed to understand how the user-interface design for a vehicle navigation system impacts on both navigating and primary driving performance. Specifically, research has aimed to answer the following three design-oriented questions:

What Information Should a Navigation System Provide?

To support route following, there are a wide range of different information types that a system could present, either referring to something real in the road environment (junction representations, street/road signs, landmarks, etc.) or indirectly referring to or pointing at aspects of the environment (distance to turn, directions, etc.). In this respect, researchers have established through a range of methodologies that the use of distinctive features of the environment (landmarks) within navigation instructions (e.g., "turn right at the church") offer considerable advantages over the use of distance to turn information (e.g., "turn right in 300 meters" Burnett, 2000; Ross, May & Grimsley, 2004). Moreover, research has identified the fundamental characteristics of landmarks which designers of vehicle navigation systems and providers of underlying map databases must consider in choosing appropriate landmarks for presentation by a navigation system (Burnett, Smith & May, 2001).

How Should Information be Presented?

Navigation and related information has to be presented to the driver in some way, and there has been considerable research on a range of topics. One key concern has been the impact of system modality (voice and/or visual) on driving and navigating performance. The general consensus here is the primary modality for presentation of navigation instructions should be auditory to reduce the conflict with the predominately visual driving task. However, information should also be presented visually, in particular, to support driver's understanding of more spatially complex maneuvers which cannot be represented easily in voice directions (Ross et al., 1995). Recently, Van Erp (2005) investigated empirically the potential for the use of passive touch as a novel modality for presentation of navigation instructions (specifically, vibro-tactile direction and distance to turn presented through the driver's seat). They concluded that haptic navigation displays offer various advantages over visual displays, for example, they provide a 'private' display to the driver appropriate for very simple maneuvers. Nevertheless, it must be noted that the authors did not make comparisons with the prevailing visual and auditory interfaces. Other research related to information presentation has considered a wide range of issues, such as the format of information (map-based vs. turn-by-turn based), the scheduling of information (when to present instructions), and the location of information (positioning of displays). On these topics, the reader is directed to Ross et al. (1995) and Srinivisan (1999).

How Should Drivers Interact with a Navigation System?

For drivers (or passengers) to interact with a vehicle navigation system, there must be a means by which they can enter data (e.g., postcode for an address), select from continuous/discrete options (e.g., voice volume levels, stored destinations), request/repeat information (e.g., voice directions), and move through the system (e.g., within and between menu screens). There is understandably a natural tendency for designers to utilise the familiar desktop computing paradigms, thus utilizing specific hardware devices (e.g., joysticks, touchscreens, buttons) and associated software approaches (e.g., use of menus, lists, scrolling). Historically, such paradigms were conceived as a means of overcoming the significant limitations of command-line user-interfaces and provided a what-you-see-is-what-you-get (WYSIWYG) experience for the user (Shneiderman, 1998). In the driving context, several studies have shown that such highly visual-manual user-interfaces can have a considerable impact on safety (Nowakowski, Utsui & Green, 2000; Tijerina, Palmer & Goodman, 1998).

As an alternative to such user-interfaces, speech shows promise as a largely non-visual/manual input method for navigation systems (Tsimhoni, Smith, & Green, 2002). Nevertheless, research has also shown that there is considerable potential for cognitive distraction with speech interfaces (Gärtner, König, & Wittig, 2001), and it is critical that recognition accuracy is very high. Moreover, designers must provide clear dialogue structures, familiar vocabulary, strong feedback and error recovery strategies. These issues are of particular importance given the potentially large number of terms (e.g., towns, street names) that might be uttered and the difficulties that a speech recognition system can experience with alphabet spelling (specifically, the 'e-set'—b, c, d, e, g etc.).

Recent research has also shown the potential for handwriting recognition in a driving context for inputting alphanumeric data (Burnett et al., 2005; Kamp et al., 2001). Whilst handwriting requires manual input, there is a reduced cognitive component and it is a more familiar method for users in contrast with speech interfaces. Nevertheless, issues relating to recognition accuracy remain and it is critical to place a handwriting touchpad in a location that facilitates the use of a driver's preferred hand (Burnett et al., 2005).

The difficulties for complex interactions with vehicle navigation systems are considered to be so significant that many authors believe that systems should disable "overly demanding" functionality when the vehicle is in motion (e.g., by "greying out" options when the vehicle is moving. Burnett, Summerskill & Porter, 2004; Green, 2003). This is currently a rich area for research, requiring an understanding of (a) what is meant by "overly demanding," (b) establishing valid/reliable metrics for the assessment of demand and finally, c) deciding where to put limits on acceptability (Burnett et al., 2004).

Underload for Vehicle Navigation Systems

In contrast with the overload perspective, over the last five years some researchers have viewed navigation systems as a form of automation, where underload issues become central. Vehicle navigation systems calculate a route for a driver according to pre-defined algorithms and then present filtered information, often via visual and auditory instructions. Two related concerns are emerging as important research questions, of particular relevance to user-interfaces which place a reliance on turn-by-turn guidance.

Firstly, it has been noted that there may be a poor calibration in the perceived versus objective reliability of in-car computing systems (Lee & See, 2004). This is of relevance as a navigation system (particularly the underlying digital map) that is unlikely ever to be 100 percent reliable. Nevertheless, drivers, largely based on their accumulated experience, may believe this to be the case. In certain situations, such overtrust in a system (commonly referred to as complacency) may lead to drivers following inappropriate routes and potentially making dangerous decisions, for instance, turning the wrong way down a one-way street. There is plenty of anecdotal evidence for such behavior in the popular press (e.g., http://www.timesonline.co.uk/article/0,,2-2142179,00.html, September 2006). Recently, research has replicated the effect in a simulated environment and indicated that there are considerable individual differences in the likelihood of a driver showing a complacency effect (Forbes & Burnett, 2007). Further research is considering what role the extended user-interface (training procedures, manuals, marketing information) can have in reducing complacency effects.

Secondly, drivers who use vehicle navigation systems may not develop a strong mental representation of the environments in which they travel, commonly referred to as a cognitive map. It has been stressed that traditional methods (e.g., using a paper map) require drivers to be active in the navigation task (route planning and following. Jackson, 1998; Burnett and Lee, 2005). Whilst the demands (particularly the cognitive demands) can initially be high, drivers who are engaged are able to develop landmark, then route knowledge, ultimately progressing to a map-like mental understanding (survey knowledge). Such a well-developed cognitive map means that drivers are able to navigate independent of any external source of information. Empirical research in this area has shown that drivers using current forms of user-interface for vehicle navigation system do indeed experience reduced environmental knowledge in relation to drivers using traditional methods (Burnett & Lee, 2005; Jackson, 1998). A key research question here is how user-interfaces can be developed which balance the need for low demands (workload) whilst simultaneously aiding drivers in developing a well formed cognitive map (Burnett & Lee, 2005).

FUTURE TRENDS AND CONCLUSION

The incessant growth in the use of cars and worries about road safety have led car manufacturers to offer more intelligent cars providing a range of novel functions to drivers. Moreover, existing mobile technologies such as PDAs, MP3 players, mobile phones, and so on, are increasingly being used within cars, as drivers seek to be more productive and to enjoy the time spent in their vehicles.

All of these computing-based systems offer potential benefits to drivers. This chapter has focused on some key design issues for user-interfaces from the perspective of the individual driver. However, as systems become commonplace within vehicles, there are fundamental conflicts to resolve between the requirements of an individual versus the overall traffic system. In this respect, the design of an in-car computing user-interface will be a critical consideration. As an example scenario, one can envisage many drivers using information systems providing the same information at the same time. Such a situation may lead to a range of problems, for instance the use of roads not designed for high volumes of traffic. Clearly, there is a need for overall management and an understanding of the impact that specific styles of user-interface will have on driver behavior.

A second broad issue for research concerns the interaction between multiple systems. This chapter has introduced the overload and underload concepts and discussed them in turn relating them to different individual systems. It is highly likely that in the short to medium term, overload will be given a prominent position in research and development work, whereas underload will emerge as an increasingly important topic in the medium to long term. However, this singular view neglects the fact that information and control-based systems are likely to be used together in a vehicle. Clearly, there will be various interaction effects for researchers to investigate. Moreover, there is a fundamental need to find the right balance between the two extremes of overload and underload. As noted by Dewar (2002, p. 330), "humans operate best at an optimal level of arousal, and either too much or too little workload can be detrimental to performance."

The development of suitable methods for designing and evaluating in-car computing user-interfaces will continue to be an important research topic. Reliable and valid methods are required which are accepted within industry. A key motivation will be to establish 'quick and dirty' methods (and associated metrics) enabling designers to understand the likely demands of their user-interfaces early in the design process when very rudimentary prototypes are available. A further critical requirement is for "benchmarking," that is, establishing a point of reference from which user-interfaces can be compared or assessed. Such benchmarks will be of particular benefit when identifying user-interface designs that are considered acceptable or unacceptable, particularly from a safety perspective.

REFERENCES

Baumann, M., Keinath, A., Krems, J.F., & Bengler, K (2004). Evaluation of in-vehicle HMI using occlusion techniques: Experimental results and practical implications. *Applied Ergonomics, 35*(3), 197-205.

Bishop, R. (2005). *Intelligent vehicle technology and trends.* London: Artech House Publishers.

Blaauw, G.J. (1982). Driving experience and task demands in simulator and instrumented car: A validation study. *Human Factors, 24*(4), 473-486.

Burnett, G.E. (2000). "Turn right at the traffic lights." The requirement for landmarks in vehicle navigation systems. *The Journal of Navigation, 53*(3), 499-510.

Burnett, G.E. (2003). A road-based evaluation of a head-up display for presenting navigation information. In *Proceedings of HCI International conference (Human-Centred Computing)*, Vol. 3, (pp. 180-184).

Burnett, G.E., & Joyner, S.M. (1997). An assessment of moving map and symbol-based route guidance systems. In Y.I. Noy (Ed.), *Ergonomics and safety of intelligent driver interfaces* (pp. 115-137). Lawrence Erlbaum.

Burnett, G.E., & Lee, K. (2005). The effect of vehicle navigation systems on the formation of cognitive maps, In G. Underwood (Ed.), *Traffic and transport psychology: Theory and application* (pp. 407-418). Elsevier.

Burnett, G.E., Lomas, S., Mason, B., Porter, J.M., & Summerskill, S.J. (2005), Writing and driving: An assessment of handwriting recognition as a means of alphanumeric data entry in a driving context. *Advances in Transportation Studies, 2005 Special Issue*, 59-72.

Burnett, G.E., Smith, D., & May, A.J. (2001). Supporting the navigation task: Characteristics of 'good' landmarks. In Hanson, M.A. (Ed.), *Proceedings of the Annual Conference of the Ergonomics Society* (pp. 441-446). Taylor & Francis.

Burnett, G.E., Summerskill, S.J., & Porter, J.M. (2004). On-the-move destination entry for vehicle navigation systems: Unsafe by any means? *Behaviour and Information Technology, 23*(4), 265-272.

Burns, P. (1999). Navigation and the mobility of older drivers. *Journal of Gerontology. Series B, Psychological sciences and social sciences, 54*(1), 49-55.

Dewar, R.E. (2002). Vehicle design. In R.E. Dewar & P.L.Olson (Eds.), *Human factors in traffic safety,* (pp. 303-339). USA: Lawyers & Judges Publishing Company, Inc.

Engstrom, J., Aberg, N., & Johansson, E. (2005). Comparison between visual and tactile signal detection tasks applied to the safety assessment of in-vehicle information systems. In *Proceedings of the third international driving symposium on Human Factors in driver assessment, training and vehicle design.* (University of Iowa public policy center, Iowa).

Forbes, N., & Burnett, G.E. (2007). *Can we trust in-vehicle navigation systems too much?* In preparation, School of Computer Science and IT, University of Nottingham, UK.

Gärtner, U., König, W., & Wittig, T. (2001). Evaluation of manual vs. speech input when using a driver information system in real traffic. In *Proceedings of International driving symposium on human factors in driver assessment, training and vehicle design. (University of Iowa public policy center, Iowa)*, Available at: http://ppc.uiowa.edu/driving-assessment//2001/index.html

Gould, J. D., & Lewis, C. H. (1985). Designing for usability: key principles and what designers think. *Communications of the ACM, 28*(3), 300-311.

Green, P. (1999). The 15 second rule for driver information systems. In *Proceedings of the ITS America 9th annual meeting*, Washington, D.C.

Green, P. (2003). Motor vehicle driver interfaces. In J. A. Jacko & A. Sears (Eds.), *The human computer interactions handbook* (pp. 844-860). UK: Lawrence-Earlbaum Associates.

Green, P. (2004). Driver Distraction, telematics design, and workload managers: Safety issues and solutions. (SAE paper 2004-21-0022). *In Proceedings of the 2004 International Congress on Transportation Electronics (Convergence 2004, SAE publication P-387)*, (pp. 165-180). Warrendale, PA: Society of Automotive Engineers.

Hackos, J. T., & Redish, J. C. (1998). *User and task analysis for interface design*. NY: John Wiley and Sons.

ISO. (2005). *Road vehicles–Ergonomic aspects of transport information and control systems–Occlusion method to assess visual distraction due to the use of in-vehicle information and communication systems.* Draft International Standard ISO/DIS 16673. ISO/TC 22/SC 13.

Jackson, P. G. (1998). In search of better route guidance instructions. *Ergonomics, 41*(7), 1000-1013.

Jamson, H. (2001). Image characteristics and their effect on driving simulator validity. In *Proceedings of International driving symposium on Human Factors in driver assessment, training and vehicle design. (University of Iowa public policy center, Iowa).*

Jones, W.D. (2002). Building safer cars. *IEEE Spectrum, 39*(1), 82-85.

Kamp, J.F., Marin-Lamellet, C., Forzy, J.F., & Causeur, D. (2001). HMI aspects of the usability of internet services with an in-car terminal on a driving simulator. *IATSS Research, 25*(2), 29-39.

Kaptein, N. A., Theeuwes, J., & Van der Horst, R. (1996). 'Driving simulator validity: Some considerations. *Transp. Res. Rec., 1550,* 30–36.

Kennedy, R. S., Lane, N. E., Stanney, K. M., Lanham, S., & Kingdon, K. (2001, May 8-10). Use of a motion experience questionnaire to predict simulator sickness. In *Proceedings of HCI International*, New Orelans, LA.

Lee, J.D., & See, K. (2004). Trust in automation: Designing for appropriate reliance. *Human Factors, 46*(1), 50-80.

Maguire, M. (2001). Context of use within usability activities. *International Journal of Human-Computer Studies, 55*, 453-483.

Markkula, G., Kutila, M., & Engström, J. (2005, November,14 - 16). *Online Detection of Driver Distraction - Preliminary Results from the AIDE Project.* Proceedings of International Truck & Bus Safety & Security Symposium Alexandria, VA.

Mattes, S. (2003). The lane change task as a tool for driver distraction evaluation. IHRA-ITS Workshop on Driving Simulator Scenarios, October 2003 - Dearborn, Michigan. Retrieved from www.nrd.nhtsa.dot.gov/IHRA/ITS/MATTES.pdf

Michon, J.A. (1985). A critical view of driver behaviour models. In L. Evans & R.S. Schwing (Eds.), *Human behaviour and traffic safety.* NY: Plenum Press.

Nichols, S., & Patel, H. (2002). Health and safety implications of virtual reality: A review of empirical evidence. *Applied Ergonomics, 33*(3), 251-271.

Nowakowski, C., Utsui, Y., & Green, P. (2000). *Navigation system evaluation: The effects of driver workload and input devices on destination entry time and driving performance and their implications to the SAE recommended practice.* (Tech. Rep. UMTRI-2000-20). USA: The University of Michigan Transportation Research Institute.

Parkes, A.M. (1991). Data capture techniques for RTI usability evaluation. In *Advanced telematics in road transport—The DRIVE conference*: Vol. 2. (pp. 1440-1456). Amsterdam: Elsevier Science Publishers B.V.

Peters, G.A., & Peters, B.J. (2002). *Automotive vehicle safety.* London: Taylor and Francis.

Pettitt, M.A., & Burnett, G.E. (2005, November). Defining driver distraction. In *Proceedings of World Congress on Intelligent Transport Systems.* San Francisco.

Pettitt, M.A., Burnett, G.E., Bayer, S., & Stevens, A. (2006). Assessment of the occlusion technique as a means for evaluating the distraction potential of driver support systems. *IEE Proceedings in Intelligent Transport Systems, 4*(1), 259-266.

Pettitt, M.A., Burnett, G.E., & Karbassioun, D. (2006). Applying the Keystroke Level Model in a driving context. In Bust, P.D. (Ed.), *Proceedings of the Ergonomics Society Annual conference, Taylor & Francis, Contemporary Ergonomics* 2006 (pp. 219-223).

Pettitt, M.A., Burnett, G.E., & Stevens, A. (2007). An extended Keystroke Level Model (KLM) for predicting the visual demand of in-vehicle information systems. In *Proceedings of the ACM Computer-Human Interaction (CHI) conference.* May 2007, San Jose, CA.

Preece, J., Rogers, Y., & Sharp. H. (2002). *Interaction design: Beyond human-computer interaction.* NY: John Wiley and Sons.

Reed, M.P., & Green, P.A. (1999). Comparison of driving performance on-road and in a low-cost simulator using a concurrent telephone dialling task. *Ergonomics, 42*(8). 1015-1037.

Ross, T., May, A.J., & Grimsley, P.J. (2004). Using traffic light information as navigation cues: implications for navigation system design. *Transportation research Part F, 7*(2), 119-134.

Ross, T., Vaughan, G., Engert, A., Peters, H., Burnett, G.E., & May, A.J. (1995). *Human factors guidelines for information presentation by route guidance and navigation systems (DRIVE II V2008 HARDIE, Deliverable 19).* Loughborough, UK: HUSAT Research Institute.

Shneiderman, B. (1998). *Designing the user-interface* (3rd edition). Addison Wesley.

Srinivisan, R. (1999). *Overview of some human factors design issues for in-vehicle navigation and route guidance systems.* Transportation Research Record 1694 (paper no. 99-0884). Washington, DC: National Academy Press.

Society of Automotive Engineers (SAE) (2000). *Navigation and Route Guidance Function Accessibility While Driving (SAE Recommended Practice J2364)*, version of January 20th, 2000. Warrendale, PA: Society of Automotive Engineers.

Stevens, A., Bygrave, S., Brook-Carter, N., & Luke, T. (2004). *Occlusion as a technique for measuring in-vehicle information system (IVIS) visual distraction: a research literature review.* Transport Research Laboratory (TRL) report no. TRL609. Crowthorne, Berkshire: TRL.

Tijerina, L., Palmer, E., & Goodman, M.J. (1998). *Driver workload assessment of route guidance system destination entry while driving.* (Tech. Rep. UMTRI-96-30). The University of Michigan Transportation Research Institute.

Toffin, D., Reymond, G., Kemeny, A., & Droulez, J. (2003, October 8-10). Influence of steering wheel torque feedback in a dynamic driving simulator. In *Proceedings of the Driving Simulation Conference North America.* Dearborn, USA.

Tornros, J. (1998). Driving behaviour in a real and a simulated road tunnel–A validation study. *Accident Analysis and Prevention, 30*(4), 497-503.

Transport Canada. (2006, June). *R Occlusion Research.* Presentation given to the AAM driver focus task force. Detroit.

Tsimhoni, O., Smith, D., & Green, P. (2002). *Destination entry while driving: Speech recognition versus a touch-screen keyboard.* (Tech. Rep. UMTRI-2001-24). The University of Michigan Transportation Research Institute.

Tufano, D.R. (1997). Automotive HUDs: The overlooked safety issues. *Human Factors, 39*(2), 303-311.

Van Erp, J.B.F. (2005). Presenting directions with a vibrotactile torso display. *Ergonomics, 48*(3), 302-313.

Wallen Warner, H.M., & Aberg, L. (2005). Why do drivers speed? In Underwood, G. (Ed.), *Traffic and transport psychology: theory and application* (pp. 505-511). Oxford, UK: Elsevier.

Ward, N. J., & Parkes, A.M.. (1994). Head-up displays and their automotive application: An overview of human factors issues affecting safety. *Accident Analysis and Prevention, 26*(6), 703-717.

Wickens, C.D., Lee, J.D., Liu, Y., & Becker, S.E.G., (2004). An introduction to human factors engineering (2nd Edition). New Jersey, USA: Pearson.

Wikman, A., Nieminen, T., & Summala, H. (1998). Driving experience and time-sharing during incar tasks on roads of different width. *Ergonomics, 41*(3), 358-372

Young, K., Regan, M., & Hammer, M. (2003). *Driver distraction: a review of the literature.* Monash University Accident Research Centre, Report No. 206, November 2003.

KEY TERMS

Driver Distraction: Occurs when there is a delay by the driver in the recognition of information necessary to safely maintain the lateral and longitudinal control of the vehicle. Distraction may arise due to some event, activity, object or person, within or outside the vehicle that compels or tends to induce the driver's shifting attention away from fundamental driving tasks. Distraction may compromise the driver's auditory, biomechanical, cognitive or visual faculties, or combinations thereof (Pettitt & Burnett, 2005).

Driving Simulators: Provide a safe, controlled and cost-effective virtual environment in which research and training issues related to driving can be considered. Simulators vary considerably in their fidelity (i.e., the extent to which they replicate aspects of real driving).

In-Car Computing Systems: Provide information to support the driving task or control some aspect/s of the driving task. In-car computing systems may also provide information and/or services that are unrelated to driving.

Keystroke Level Model: Is an established HCI method used to predict expert's task times with a user-interface. It can be used with in-car user-interfaces to predict static task time, that is, the time taken to achieve tasks in a stationary vehicle. Recently, the KLM has been extended to predict visual demand measures related to the occlusion protocol.

Overload: (Due to in-car computing systems) occurs when a driver's information processing resources are overwhelmed and performance on primary driving tasks inevitably suffers.

Underload: (Due to in-car computing systems) occurs when automation of core driving tasks (such as steering, braking, etc.) has led to a situation in which driving performance has deteriorated. This may have arisen because the driver has reduced awareness of other road users, has changed their behavior in negative ways or has inferior skills/ knowledge in driving.

The Occlusion Protocol: Is a user trial method used in the design and evaluation of in-car user-interfaces. Participants typically wear LCD glasses which restrict the visual experience by only enabling short (e.g., 1.5 seconds) chunks of visual attention with an in-car user interface. Measures related to the visual demand of an interface can be established.

Chapter XV
Speech-Based UI Design for the Automobile

Bent Schmidt-Nielsen
Mitsubishi Electric Research Labs, USA

Bret Harsham
Mitsubishi Electric Research Labs, USA

Bhiksha Raj
Mitsubishi Electric Research Labs, USA

Clifton Forlines
Mitsubishi Electric Research Labs, USA

ABSTRACT

In this chapter we discuss a variety of topics relating to speech-based user interfaces for use in an automotive environment. We begin by presenting a number of design principles for the design of such interfaces, derived from several decades of combined experience in the development and evaluation of spoken user interfaces (UI) for automobiles, along with three case studies of current automotive navigation interfaces. Finally, we present a new model for speech-based user interfaces in automotive environments that recasts the goal of the UI from supporting the navigation among and selection from multiple states to that of selecting the desired command from a short list. We also present experimental evidence that UIs based on this approach can impose significantly lower cognitive load on a driver than conventional UIs.

INTRODUCTION AND BACKGROUND

The US census bureau reported in 2005 that the average American spends over 100 hours driving to and from work every year and spends several hundred more driving on errands, vacations, to social engagements, and so on. A significant fraction of this driving is spent while engaged in concurrent activities, such as listening to the radio, listening to music on a personal music player, operating an in-car navigation system, and talking on or accessing information with a hands-free or hand-held cell phone. These secondary activities involve interactions between the driver and a device that can distract the driver from the primary task—that of driving safely to the destination. While it is understood that the safest option is for a driver not to engage in such activities and instead concentrate completely on driving, drivers seem intent on engaging in these distractions; thus, minimizing the impact on safety is a worthy area of research.

It has been estimated that at least 25% of police reported accidents in 1995 involved some form of driver inattention (Wang, Knipling, & Goodman, 1996). A study by Stutts et al. (2001) estimated that, of the drivers whose state was known at the time of the crash, at least 13% were distracted, with adjusting the audio system of the car accounting for 11% of these distractions. Since the advent of cellular phone technology, there has been a great deal of research on the effects of cellular phone use on driving performance (e.g., Ranney et al., 2004); however, only recently have studies begun to address the effects of use of other in-car systems on driving performance. In an analysis of the 100-Car Naturalistic Driving Study, Klauer et al. (2006) found that "Drivers who are engaging in moderate secondary tasks are between 1.6 and 2.7 times as likely to be involved in a crash or near-crash, and drivers engaging in complex secondary tasks are between 1.7 and 5.5 times as likely" (p. 28).

Since these studies, a number of electronics manufacturers have introduced products that incorporate personal digital music collections into automobile audio systems. Some automobile manufacturers have gone as far as bundling a personal digital music player with the purchase of a new car. Recent high-end car models also offer GPS-linked navigation systems. These systems offer functions such as address entry and point-of-interest search, both of which are usually implemented as multi-step tasks requiring significant attention from the user. Navigation and entertainment systems are among the first examples of highly complex automotive interfaces that are available for use while driving. We expect the amount of information available in the car to continue increasing drastically as more and more car systems become networked, and as car makers try to differentiate their products by offering new functionality.

Given this situation, it becomes necessary to design effective user interfaces that will enable drivers to operate devices such as radios, music players, and cellphones in a manner that distracts them minimally from driving, while still allowing them to obtain the desired response from their devices.

A compelling choice for UI design in the automotive environment is the speech-based user interface. By "speech-based" we mean an interface which uses utterances spoken by the user as a primary input mode. A speech based interface may also have other input modes, such as dedicated or softkey input, and may also have voice feedback and/or visual feedback. By being largely hands free, a speech-based interface can minimize the need for the driver to disengage their hands from the steering wheel. By presenting information aurally, it can allow a driver to keep their eyes on the road.

These qualities are by themselves not sufficient: automobile UIs must not only allow drivers to keep their hands on the wheel and their eyes on the road, but also must allow them to keep their mind on the task at hand—that of driving safely. Spoken input is typically used as substitute for tactile input. It is frequently unclear how tactile actions such as turning a knob, pressing a button, or selecting an item on a touch screen may best be replaced by simple spoken commands that can be recalled easily by the user. The inability to recall the correct command can lead to poor system response and driver distraction. Further, Automatic Speech Recognition (ASR) engines are error-prone—they will often fail to recognize spoken input correctly, or worse still mistakenly recognize an incorrect command. These problems are magnified in noisy environments such as the inside of a fast-moving automobile.

Common approaches to minimize the adverse effects of ASR errors include detailed dialog mechanisms, help menus, confirmatory prompts, and error correcting dialogs. Unfortunately, these mechanisms are problematic in an automotive environment in which it is important for interactions between a driver and a system to be short in order to create minimal distraction to the driver.

Interface designers must strive to minimize unnecessary cognitive load such as those arising from extended interactions with a system, frustration from poor task completion, and other distractions of the mind. Recognition errors can often be minimized by constraining the choices that the ASR engine must consider at any one time. When using this approach, the UI must be designed to restrict the number of spoken commands available at each state while at the same time ensuring that the currently available commands are evident to a user. Thus, there is a strong bi-directional coupling between ASR performance and usability: ASR problems can manifest themselves to the end-user as usability problems, and interface design or implementation problems can easily lead to reduced recognition accuracy. Our experience has been that speech-based user interfaces are most effective when designed around the constraints of both ASR and UI. It is probably unrealistic to expect application and interface designers to be well versed in the technical details of ASR; however we believe that using a set of reasonable design guidelines could help automotive designers to design more effective speech-based interfaces.

There are a number of published documents that give design guidance for telematics interfaces, (see UMTRI, 2006); however, very few of these give any specific guidelines for the design of speech interfaces for telematics systems. Nor do they specify which general guidelines should apply to speech-based UIs, and which should not. In fact, some sets of guidelines (for instance, Society of Automotive Engineers, 2004) specifically exclude speech interfaces. Of course, there are also many sets of design guidelines for voice user interfaces in general, but few of these were written for the specific issues of an automotive user interface.

In this chapter, we will begin by presenting a number of design principles for the design of speech-based UIs, that have been derived from several decades of our combined experience in the development of spoken user interfaces for automobiles. Our discussion addresses issues such as the cognitive load imposed on the driver, interaction time, task completion rate, feedback, and the appeal of the system to the user. Based on these principles, we then present a brief review of the spoken UIs in a few current car models and highlight the positive and negative facets of these interfaces.

We conclude the chapter with a description of a speech interface paradigm which we call SILO (Divi et al., 2004) that conforms to most of our design principles, as an alternative to highly modal tree structured menus for the selection of a item out of a large number of alternatives. We describe an experiment that indicates that the SILO paradigm has significantly lower driving interference than the menu-based interface for a music selection task (Forlines et al., 2005).

DESIGN PRINCIPLES FOR SPEECH-BASED AUTOMOTIVE UIS

Here, we introduce a short set of design goals and specific recommendations for consumer automotive interfaces. These are based on the collective experience of the authors in implementing speech interfaces for use in automobiles over the last ten years. Most of the work that forms the basis for these recommendations is unpublished; however we feel that it is useful to present these recommendations here in collected form. Many of these guidelines can be found individually in other sources—this combined set is based on the particular constraints of consumer automotive applications.

Note that these guidelines are written with consumer automobile driving in mind. Commercial and military applications have different constraints and thus should be treated differently (for instance, the military user base receives training prior to in-field use, and the risk vs. task success rate balance may be different).

General Design Goals for Automotive Speech Interfaces

- **Reduce driver cognitive load.** In our view, the highest priority in design of automotive interfaces is to reduce the risk associated with performing any secondary task while driving. Since interfaces in use by a passenger can be distracting to the driver, the driver's cognitive load should be considered when designing interfaces that will be used by other occupants.
- **Reduce interaction time.** Reduced interaction time should lead to reduced risk.
- **Increase task completion rate.** Increased task completion rate should lead to reduced risk (fewer repeated interactions) and better user experience. Of course, task completion rate is related to the underlying speech recognition accuracy, but is also strongly influenced by the affordances offered by the user interface.
- The feedback (visual and audio) should reinforce correct use by the user.
- The user should be able to mentally model the system behavior.
- System should be effective for an experienced user (e.g., a long time owner).
- System should be appealing for a new user (especially during a pre-sales test drive).

These goals are given in order of decreasing importance. Not all of these design goals are achievable at the same time, and in fact, some may be in opposition to each other for various tasks. For instance, it can be difficult to make a system that is effective for both a new user and a long-time user. Below are some design recommendations that we believe follow from these design goals

Design Recommendations

1. **Interactions should be user paced:** Speech interfaces for use in automobiles should be *entirely* user paced. The primary task of the driver is the safe operation of the automobile. The operation of other equipment in the vehicle is a secondary task. The driver must be available to respond to changing traffic conditions. Driver responses to the system must be timed when the driver can spare attention for the secondary task. Thus every voice response by the driver should require its own push-to-talk event. Systems which ask for further voice input without waiting for a signal from the driver can cause the driver to feel pressured to respond even under difficult traffic conditions. Many state of the art systems include dialogs that violate this recommendation.
2. **Use a Push and Release button with a listening tone:** After the user activates the Push-to-talk, the system should promptly produce a short pleasant listening tone to indicate that it is listening. Studies have indicated that in the absence of a listening tone, some users will start speaking prior to activating the push-to-talk and other users will delay speaking for a variable amount of time. Either of these changes in timing can confuse the speech recognizer, resulting in an overall reduction of recognition accuracy.
   In the presence of a listening tone, most users quickly learn to wait for the listening tone and to start speaking promptly after they hear the tone. However, the time between activation of the push-to-talk and production of the listening tone must be short and consistent; otherwise the cognitive load on the user is higher.
   Push and Release interfaces, where the user releases the PTT button immediately, usually impose a lower cognitive load on the user

than Push and Hold style interfaces, where the user is required to hold the button for the entire utterance. This is because the user's physical actions are sequential (press, then speak) in a Push and Release interface rather than simultaneous. It can be useful to indicate with a different tone that the system is no longer listening. This affordance helps the user to adapt to system constraints such as listening timeouts.

3. **Use physical input instead of voice for simple things:** There are numerous actions for which there are existing effective physical interfaces. In almost all instances keeping those physical interfaces is superior to the substitution of voice commands. For example, using up and down buttons to navigate through a list of options is much easier than saying commands such as "scroll down". In addition, the users are already familiar with such physical interfaces. There is also less new learning involved to be able to use the interface.
4. **Provide always-active commands including voice help:** A "Help" or "What can I say" command should always be available for users who are unsure as to what functions are currently available. Systems that have modal behavior should have mode-specific voice help, as well as commands to cancel the current mode and to backup.
5. **Use consistent grammars with minimal modality:** Consistency and predictability of the grammar is very important so that the user will have less to remember. Also the grammar should have minimal modality so that users will have access to the functions of the system with fewer interactions and will not need to remember the state of the system. Modal behavior should be associated with audio and visual cues so that the user can easily understand what mode the system is in.
6. **Visual cues should be consistent with the active grammar:** The use of visual cues that do not model the grammar usually causes an increased number of out-of-grammar utterances.
7. **Feedback should indicate the recognition result:** The visual and audio feedback given to the user should indicate what was heard by the speech recognizer. This helps to reduce confusion when the system does not behave as expected, either due to misrecognition, or to user confusion about the effect of the spoken command. Users have a strong tendency to mimic the sentences that they hear; thus, some systems echo the recognized utterance back to the user as a confirmation of what the system heard or echo the preferred form of the command to help the user learn the grammar.
8. **Reasonable behavior for out-of-grammar utterances:** Systems should attempt to detect out-of-grammar utterances and indicate that the last utterance was not understood. It is far better for the system to respond that the command was not understood than it is to perform an unexpected action. Unexpected actions cause user confusion (e.g., was the utterance in the grammar and misrecognized, or not in the grammar?). Lack of rejection raises the cost of an out-of-grammar utterance as the user must take whatever action is necessary to undo the undesired action, resulting in significantly longer task completion times.
9. **Provide reasonable backoff strategies:** In some cases a speech interface will consistently fail to recognize certain voice commands from the user. For example, a system may allow the entry of a street name as part of an address, a very difficult voice recognition problem for a locality with many streets. Thus, if the voice system fails to get the correct street after a couple of tries, it should offer an alternative method of entry, such a spelling. As another example, many systems allow the entry of long telephone numbers in one long utterance by the user. For some users the error rate for strings of ten or more digits can be too high. For such users the system should allow the entry and correction of the digits forming a long telephone number in smaller chunks.

EVALUATION OF RECENT AUTOMOTIVE SPOKEN USER INTERFACES

In April 2006, we conducted a review of three automotive speech-based interfaces from model year 2006. Rather than give an exhaustive review, here we highlight the system attributes that had the most impact on interface usability. In order to focus on the attributes of the interfaces rather than the identities of the manufacturers, we will call these interfaces "Model A," "Model B," and "Model C."

The state of the art for speech-based navigation systems at the time of the review was:

- Push and Release interface with listening tone
- Grammar based systems with varying degrees of modality
- Navigation Entry of part or all of an address by voice

There was significant variation in these features between the tested systems. There was also significant variation in the accuracy of the underlying automatic speech recognition engines, but we found the usability to be more affected by the UI design than the underlying ASR.

Model A

Model A had good ASR engine accuracy. However, there was no rejection of out-of-grammar utterances, so ASR errors (misrecognitions) frequently led to unexpected actions.

User Pacing

This system was entirely user-paced, with a push-to-talk button which was required for every interaction, and a prompt, pleasant listening tone.

Grammar Consistency and Modality

Model A had a fairly simple modal grammar structure, with some common always-active commands. The grammar format was consistent, and the audio feedback modeled the grammar. Each mode was associated with a visual state. Overall it was easy to understand what mode the system was in, and to guess what voice commands were available.

However, there were some modes with inconsistent design. Two interesting examples:

- **Voice help modes:** Each voice mode (v_n) had a corresponding help mode (vh_n) which was invoked through the always-active help command. Each help mode visually displayed the available commands for its parent mode in a numbered list.
  However, the help modes had their own very limited grammar. Thus in any help mode (vh_n) the commands for v_n were displayed but not active. The result of this was to induce users to utter unavailable commands in help mode, resulting in frequent and confusing misrecognition in help mode. In order to utter one of the displayed commands, users had to first dismiss the help screen, at which point the commands were no longer visible. This design was almost consistent with design recommendation 6 (*Visual cues should be consistent with the active grammar*), but the seeming minor detail, that the visual cues for these modes matched an inactive grammar, led to a major usability issue.
- **Setup modes:** This system had several preference modes for manipulating system settings. Typically, the system displayed a set of buttons and sliders when in one of these modes. There were voice commands for manipulating the values, but the commands were inconsistent and there was no visual indication of which objects could be manipulated by voice or what command language to use. There were physical/softkey methods

to change the settings, and it appeared that the speech interface had been grafted onto these modes as an afterthought. The interface could have been more effective without the additional voice commands, consistent with design recommendation 3 (*Use physical input instead of voice for simple things*).

Other Usability Issues

The interface included a 'Back' button, which was probably designed to move the interface to the previous state after a recognition event, but had inconsistent behavior. This was an interesting attempt to provide a reasonable backoff strategy for misrecognition. Prior speech user interface (SUI) implementations have shown that a 'Back' button or an "Undo that" command that undoes the previous action can be a very effective affordance for an ASR system, mitigating the inevitable ASR errors. In this particular case, the behavior was unreliable.

Model B

Model B had poor ASR engine accuracy, especially in high noise conditions typical of highway driving.

There was no rejection of out-of-grammar utterances, however, some modes accepted unavailable commands and then reported that that command was not available in the current mode.

User Pacing

This interface had a push-to-talk button which was required for every interaction. As noted in design recommendation 2 (*Use a Push and Release button with a listening tone*), this interface pattern usually improves usability, but not in this case.

Model B's implementation of push-to-talk degraded not only the usability of the interface, but also the ASR performance, all while increasing cognitive load: Activation of the push-to-talk was followed by a visual cue which indicated that the user could begin speaking. This visual cue was followed by a highly variable delay and then a listening tone. Sometimes the listening tone was produced promptly, other times there was a delay of 1-2 seconds after the visual cue. Our testing indicated that the variable listening tone delay frequently caused recognition errors (when the users spoke too quickly) and higher cognitive load (the user had to actively think about waiting for the tone). Furthermore, the listening tone was hard to hear in high noise conditions.

Grammar Consistency and Modality

Model B had a fairly simple modal structure. However, the voice modes were not well associated with visual cues. This made it very difficult to determine which commands were available at any given time. For instance, there were several modes where a map was visible on the screen, but only one of these was "map mode." Therefore the system sometimes responded "That command is only available in map mode" even when there was a map showing on the screen.

Model B also had a confusing lack of consistency between visual cues, voice commands, and voice feedback. Recognition feedback was aligned with, but did not match the grammar. When the user spoke a command of the form "action object," the voice response was "<action>-ing object." For instance:

- **User:** *"raise temperature"*
- **System:** *"raising temperature"*

This was a clear indication of what action the system was taking, but an indirect indication of what command had been recognized. The use of different language sometimes induced users to copy the response language (e.g., to say "raising temperature" by accident).

In those cases where users mimicked the response language, the lack of rejection of out-of-grammar utterances usually caused an unexpected action to occur, forcing the user to attempt to undo the unexpected action before re-trying the original task.

Model C

Model C had reasonably good ASR performance. It also had rejection of out-of-grammar utterances. Some out-of-grammar utterances were misrecognized, but many were rejected.

User Pacing

Model C was not user paced. An interaction with the system could consist of a single (conversational) turn by the user or an entire sequence of turns by user and system. Push-to-talk was used to initiate a task, but then the system controlled the turn-taking and pace until the end of the task. To mark the user's subsequent turns, the system generated a voice prompt followed by another listening tone and immediately expected more voice input from the user. This design was confusing to new users because they did not know when a conversational sequence would be finished.

When used during driving conditions, the interface created a high cognitive load on the user, who had to pay enough attention to the voice prompts to be able to respond to demands for more input. If the user ignored a prompt, the prompt and listening tone were repeated, thus during a task the system constantly demanded attention from the user regardless of the traffic conditions. Our experience indicates that this can be hazardous—see design recommendation 1 (*Interactions should be user paced*).

Grammar Consistency and Modality

In contrast to the other interfaces tested, Model C consisted of a tree of menus 3-5 levels deep. At any given time, the active grammar commands were based on the active node of the menu tree. In addition, the voice commands for the top two levels of the tree (including voice commands for navigation through the menu tree) were always active. Visually, up to three levels of the menu tree were displayed on a single screen.

The command language for Model C was inconsistent and hard to remember. The voice feedback for a command sometimes echoed the recognized command and sometimes told the user what action the system had taken. However, there was a visual display of the command recognized from the last voice utterance, which was useful. There was a good correspondence between the text of on-screen objects/icons and the command language used to select them, but there was no clear visual indication of which objects could be manipulated by voice.

The combination of menu complexity and grammar inconsistency made it difficult for the user to determine the active menu node and to infer what voice commands were available in a particular state. This system had a very high cost for recognition errors because a misrecognition would usually cause a transition to a state very far away in the menu tree, with no way to recover.

Summary of Evaluation of Current Interfaces

Model A was superior to the other interfaces that we tested. The consistency of the UI design and adherence to good design principles was the primary reason this was the best of the three tested systems, although the fact that this system had good ASR performance also contributed. In those areas of the system where the design had inconsistent grammars and/or visual cues, the resulting misrecognitions combined with the lack of rejection of out-of-grammar utterances to significantly degrade the usability. The 'back' functionality provided by this interface could have been a significant feature if it had worked consistently.

Model B was the worst of the interfaces tested. In spite of the poor ASR performance, the main weakness of this system was the implementation of the Push-and-Release with listening tone. This system provides an example of how an interface implementation problem can manifest as an ASR problem. Here, recognition performance was seriously degraded by the variable delay of the listening tone. In addition, the lack of rejection, and the relatively poor underlying ASR performance (especially in noise), all combined to make this interface unacceptably poor.

Model C had the most complicated modal structure. It also had system-paced turn-taking, which usually results in higher cognitive load on the user. Its speech recognition was reasonably good, with some notable exceptions.

In general, current state of the art systems show very distinct design philosophies. While we are in favor of innovative design, we hope that the design of future systems will more thoroughly consider the driver's cognitive load. It is interesting to note that the underlying speech recognition performance, while significant, was far from the most important factor in the usability of these interfaces—this shows the importance of good interface design in this application space.

A SPEECH-IN LIST-OUT APPROACH TO IN-CAR SPOKEN USER INTERFACES

Many in-car applications for which spoken UIs may be used deal with the selection of one of an enumerable set of possible responses, e.g., selecting one of a number of radio stations, retrieving a song from a music collection, selecting a point of interest (from the set of all points of interest), etc. The most common UI for these applications is through a hierarchy of menus. Even when the UI is speech-based, speech is used primarily as an input or output mechanism for the underlying menu-driven interface.

At Mitsubishi Electric Research Labs we have developed an alternative speech UI for selecting an element from a set, which we refer to as the "Speech-In List-Out" interface or SILO. In this section, we briefly describe the SILO interface. We also describe an experiment that indicates that the SILO interface can result in lower driving interference than the menu-driven interface.

Selection from a Set

The most common paradigm for retrieving a specific response from a large set is through menus; however, as the size of the selection set increases (e.g. for a UI to a digital music player with an ever increasing repertoire of songs), the tree of menus increases in depth and width, and can become problematic, particularly for users who are also simultaneously involved in other attention-critical tasks such as driving.

While the problem may be alleviated to some degree through voice output and spoken input (Leatherby & Pausch, 1992; Cohen et al., 2000), this by itself is not a solution. Spoken enumeration of menu choices cannot fully replace visual display; a deeply nested menu-tree, presented aurally, is very demanding in terms of cognitive load. Knowing that a quick glance at a screen can recover forgotten information relieves the user from having to keep close track of the system's state in their mind. Spoken-input-based interfaces must address the problem of misrecognition errors (e.g., in noisy environments) and, more importantly, the "what can I say" problem—users must be able to intuit what to say to the speech recognition system (which typically works from rigid grammars for such tasks). This latter issue can be particularly difficult when selecting from long lists.

The SILO interface (Divi et al., 2004) recasts the UI as a search problem. The set of all possible responses are viewed as documents in an index. The user prompts the system with a single spoken input, which is treated as a query into this index. The system returns a short *list* of possible matches to the query. The user must make the final selection from this list. While search-based speech UIs have previously been proposed (e.g., Cohen, 1991), SILO differs from them in that it places no restrictions on the user's language. The user is not required to learn a grammar of query terms. The simplicity of the resulting interactions between the user and the system is expected to result in a lower cognitive load on the user, an important consideration in the automotive environment.

The enabling technology for SILO is the SpokenQuery (SQ) speech-based search engine (Wolf & Raj, 2002). SQ is similar to text-based information retrieval engines except that users speak the query instead of typing it. Users may say whatever words they think best describe the desired items. There is no rigid grammar or vocabulary. The output is an ordered list of items judged to be pertinent to the query.

A major problem for speech UIs is misrecognition error, which can derail an interaction. The reason is that they attempt to convert the user's spoken input to an unambiguous text string, prior to processing. In contrast, SQ converts the spoken input to a set of words with associated probabilities, which is then used to retrieve documents from an index. The set of words and their associated probabilities are derived from the *recognition lattice* that represents candidate words considered by the recognizer. The lattice often includes the actual words spoken by the user even when they are not included in the disambiguated text output. As a result, SQ is able to perform well even in highly noisy conditions (such as automobiles) in which speech UIs that depend on accurate recognition fail. Consequently, SILO interfaces are able to perform robustly in noisy environments (Divi et al., 2004). Table 1 lists some example phrases and their (often poor) interpretation by the speech recognizer along with the performance of the SQ search.

Though the disambiguated phrase output by the speech recognition system is often wildly inaccurate, SQ manages to return the desired song near or at the top of the list. The right-most column shows the rank of the desired result in SQ's output.

The design of the SILO interface follows several of the design principles enumerated above. Specifically:

- **Interactions should be user paced:** The system always waits for the user to initiate the next interaction.
- **Appropriate use of speech:** Speech input is used only for choosing from very large sets where the use of buttons (for scrolling and selection) is inefficient or impossible. All choices from small sets are performed by direct manipulation.
- **Non-modal:** SILO is not modal, therefore the user does not need to remember the system state.

Experimental Evaluation of the SILO Interface

We designed an experiment to compare the SILO interface to a menu-based UI for in-car music selection. An effective in-car UI must not only allow users to find desired information quickly, but also affect their driving performance minimally. To evaluate both factors, we compared quantitative measurements of simulated steering and

Table 1. Example of SQ search to retrieve songs from a collection

User says...	System hears...	SILO search result
"Play Walking in my shoes by Depesh Mode"	layla [NOISE] issues [NOISE] [NOISE] load	1
"Depesh Mode, Walking in my shoes"	E [NOISE] looking [NOISE] night shoes	1
"Walking in my shoes"	law(2) pinion mae issues	1
"Walking in my shoes by Billy Joel" (partially incorrect information)	walking inn might shoes night billie joel	1
"um, uh, get me Credence Clearwater Revival... um... Who'll stop the Rain" (extra words)	fall(2) [UH] dead beat creedence clearwater revival [UM] long will stop it rains	1
"Credence Clearwater Revival, Who'll stop the Rain" (very noisy environment)	[NOISE] [COUGH] clearwater revival [COUGH] down [COUGH] [BREATH]	6

braking while searching for music with the two interfaces.

Our hypotheses were:

- **H1:** Subjects will more accurately track a moving target with a steering wheel while searching for songs using the Mediafinder SILO interface than while using the menu-driven interface.
- **H2:** Subjects will react faster to a braking signal while searching for songs using the SILO interface than while using the menu-driven interface.
- **H3:** Subjects will be able to find songs faster while using the SILO interface than while using the menu-driven interface while driving.

Experiments were conducted on a simple driving simulator, such as those in Beusmans et al. (1995) and Driving Simulators (2006), that mimicked two important facets of driving—steering and braking. The simulator had both a "windshield" and "in-dash" display. Subjects steered, braked, and controlled the interfaces with a steering wheel and gas and brake pedals. A microphone was placed on top of the main monitor. Steering was measured with a pursuit tracking task in which the subject used the wheel to closely frame a moving target (Strayer et al., 2001). The simulator recorded the distance in pixels between the moving target and the user-controlled frame 30 times a second. Braking was measured by recording subjects' reaction time to circles that appeared on screen at random intervals. Subjects were asked to only react to moving circles and to ignore stationary ones. Moving and stationary circles were equally probable.

We built two interfaces for this study. The first was a menu-based interface based on a sampling of currently available MP3 jukeboxes; the second was the SILO interface. Both interfaces ran on the same "in-dash" display and were controlled using buttons on the steering wheel. Both interfaces searched the same music database of 2124 songs by 118 artists, and both were displayed at the same resolution in the same position relative to the subject. Additionally, both interfaces displayed the same number of lines of text in identical fonts. Neither interface dealt with many of the controls needed for a fully functional in-car audio system, such as volume, power, and radio controls.

Fourteen subjects, eight male and six female, of ages ranging from 18 to 37, participated in this experiment. All but one were regular automobile drivers. Subjects were first instructed on how to correctly perform the steering and braking tasks and were given as much time as they wanted to practice "driving." Next, they were instructed to search for and playback specific songs while performing the driving task. Subjects completed 8 trials each with both the SILO and the menu-driven interfaces. Before each set of trials, subjects were instructed on how to use the current interface and allowed to practice searches while not driving. During each trial, the testing application displayed the steering and braking signals along with instructions asking the user to search for a specific song (e.g., "Please listen to the song Only the Good Die Young by Billy Joel from the album The Stranger"). Subjects were allowed to take a break between trials for as long as they wished. The order that the interfaces were used was balanced among participants, and the order of the requested songs was randomized. The application logged the distance between the moving target and the subject-controlled black box, as well as the reaction time to any brake stimulus presented during each trial. The task time was also logged, measured from the moment that the instructions appeared on the screen to the moment that the correct song started playing. To reduce learning effects, only the last 4 of each set of 8 trials contributed to the results.

Results

Our data supports hypotheses H1 and H3 and rejects H2.

- **H1:** Subjects were able to steer more accurately while searching for music using the SILO interface than with the menu-driven interface (on average, 9.2 vs. 11.6 pixels of

Figure 1. The SILO interface had both a significantly lower mean steering error (below left) and a significantly lower mean largest steering error (below right) than the menu-driven interface (Source: Forlines et al., 2005)

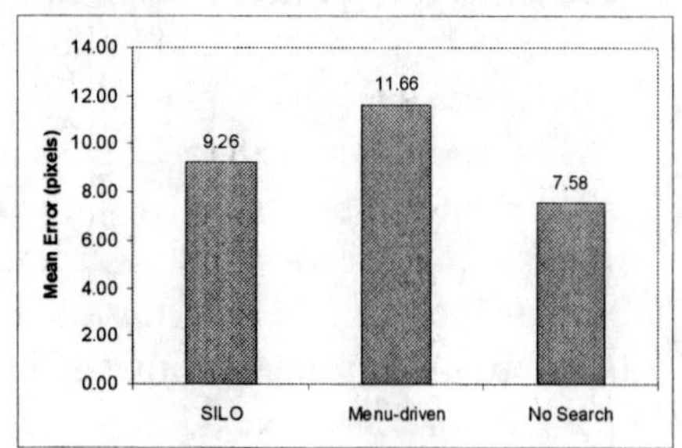

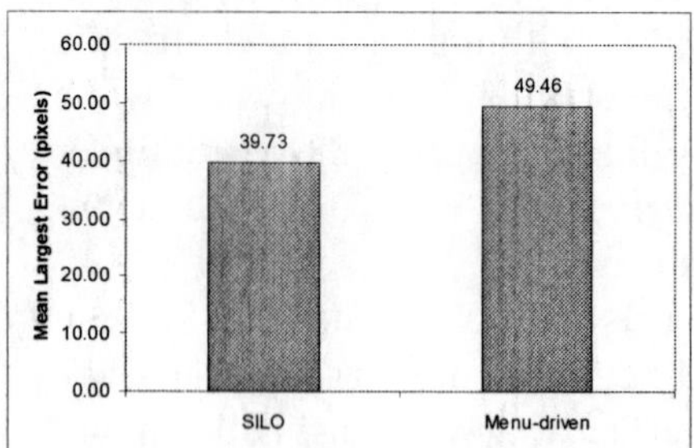

error respectively, t(13) = 3.15, p=0.003). However subjects steered most accurately while driving without searching (on average, 7.4 vs. 9.2 pixels for SILO, t(13)=2.5, p=0.013). The average error for each condition is shown in Figure 1 (*below left*). The SILO interface had a significantly lower *maximum* steering error as well (39.7 pixels vs. 49.4 pixels, t(13)=2.27, p=0.02). This measurement of error roughly corresponds to the point when the subject was most distracted from the steering task. If actually driving, this point would be the point of greatest lane exceedence. The average maximum error for the two interfaces is shown in Figure 1 (*below right*).

- **H2:** The mean breaking reaction times were indistinguishable between the SILO and menu-driven conditions (on average, 1196 ms vs. 1057 ms, t(13)=1.66, p=0.12); however, subjects were significantly faster at braking while not searching for music than while searching using the SILO (p=0.008) or the menu-driven (p=0.03) interface. The mean reaction time to the brake stimulus for each condition is shown in Figure 2.
- **H3:** Subjects were significantly faster at finding and playing a specific song while using the SILO interface than while using the menu-driven interface (on average, 18.0 vs. 25.2 sec., t(13)=2.69, p=0.009). The mean search time for each interface is shown in Figure 3. It is important to note that it was not unusual for the SILO interface to have a computational interruption of 3-6 seconds, which was included in the SILO search time. A faster CPU or better microphone could decrease this time.

Figure 2. There was no significant difference in mean break reaction times between the search conditions (Source: Forlines et al., 2005)

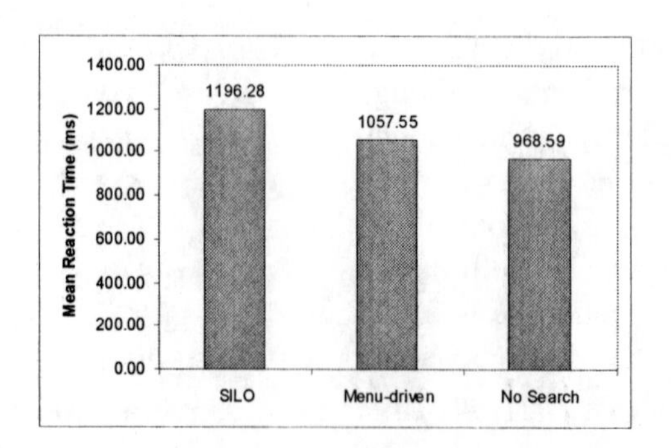

Figure 3. Subjects were significantly faster at finding songs with the SILO interface (Source: Forlines et al., 2005)

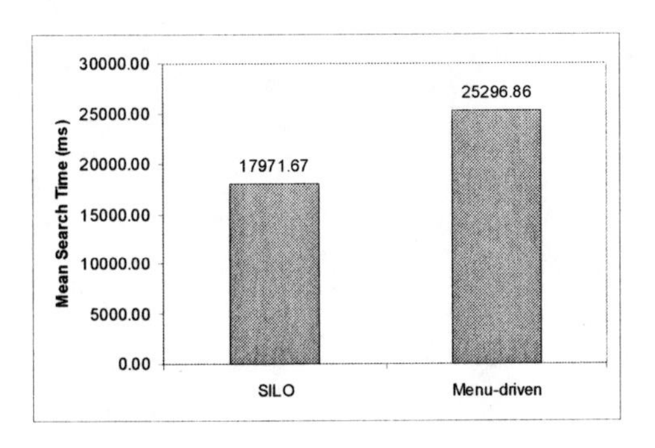

Since SILO returns a list of songs from which the user must select, an important factor is the position of the correct song in the returned list. In 35 out of the 56 trials SILO returned the correct song at the top of the list on the first try. The average position for the correct song for all SILO trials was 5.1.

Experimental Discussion

The evidence indicates that our SILO interface for music finding has measurable advantages over the standard menu-based approach for users operating a simulated automobile. The SILO interface scores better than the menu-based system on several of the design goals mentioned outlined above. It poses lower cognitive load, as evidenced by the improved driving accuracy, has lower interaction time, and is more effective for the experienced user. Although we observed no statistical difference in mean break reaction times between the SILO and menu-driven interfaces, closer inspection revealed that subjects were less likely to encounter a brake stimulus while using SILO due to the faster task completion. SILO effectively results in fewer opportunities for braking error.

The SILO interface has several additional advantages that were not explicitly evaluated in the experiment. We expect that these will be subjects of future study:

- **Flexibility:** The SILO interface was able to retrieve songs from only partial information, such as the song title. On the other hand, for the menu-based interface, it would not have been sufficient to ask subjects to find the song "Never Die" without telling them it is by the artist "Creed" on the album "Human Clay."
- **Scalability:** The song library used for the study contained only 2124 songs. The latest handheld music players can hold over 10,000 songs. As the number of available artists, albums, and songs grows, we expect the time needed to search through a menu-driven interface to grow as well. An informal evaluation of the SILO interface searching a database of 250,000 songs shows no noticeable differences in search time.
- **Robustness:** The metadata in the music files in our library was not always consistent. For example, music by the group "The B-52s" was erroneously split into many artists: "The B-52s," "B-52s," "The B52s," etc. While these inconsistencies were problematic for the menu-driven interface, they do not affect the SILO interface.

The experiments reported here only evaluated SILO in a limited music selection task. Many music jukeboxes can present their content in alternative fashions such as user defined playlists, favorites, etc. Mediafinder is easily modifiable to handle playlists and personalization; however a rigorous evaluation of its capabilities in this direction remains to be performed.

Other limitations of this study include the fact that an actual in-car environment that included environmental noise was not used. Other tests using automotive speech data have shown that the SpokenQuery information retrieval engine is very robust to high levels of environmental noise (e.g., Divi et al., 2004). We are therefore optimistic about the performance of the SILO interface in real in-car environments, but must confirm this expectation with future experiments. Finally, we look forward to a comparison between SILO and other speech based music selection systems.

FUTURE TRENDS

The current trend toward increasing the number and complexity of secondary tasks for automobile drivers is both worrisome and accelerating. The risks associated with these complex tasks raise two opportunities. First, standards bodies, government agencies, and the public must demand less distracting, safer interfaces for drivers. Second, manufacturers must provide these interfaces, and perhaps market not only their additional functionality, but also their safety advantages. Researchers and manufacturers must conduct studies measuring the degree of interference between the driving task and existing and future automotive user interfaces. Safety and liability dictate that these experiments be conducted in a driving simulator environment.

CONCLUSION

The advent of complex user interfaces in automobiles raises many issues relating to safety and usability, some of which can be mitigated by the appropriate use of speech in the UI. We have presented a set of design principles that can help mitigate some of the problems cited, and have applied these principles in a review of several existing automotive speech interfaces. Finally, we presented an in-car interface for the selection of items from a large collection and have shown that this method interferes less with driving than the current status quo. Much work remains to maximize the safety and usability of complex devices in automobiles, and we hope that this writing will aid the UI designer in this endeavor.

REFERENCES

Beusmans, J., & Rensink, R. (Eds.) (1995). *Cambridge Basic Research 1995 Annual Report* (Tech. Rep. No. CBR-TR-95-7). Cambridge, MA: Nissan Cambridge Basic Research.

Cohen, P. (1991). Integrated interfaces for decision-support with simulation. In *Proceedings of the 23rd Winter Conference on Simulation* (pp. 1066-1072).

Cohen, P., McGee, D., & Clow, J. (2000). The efficiency of multimodal interaction for a map-based task. In *Proceedings of the Applied Natural Language Processing Conference (ANLP'00)* (pp. 331-338). Seattle, WA: Morgan Kaufmann.

Divi, V., Forlines, C., Van Gemert, J.V., Raj, B., Schmidt-Nielsen, B., Wittenburg, K., Woelfel, J., Wolf, P., & Zhang, F. (2004, May 2-7). A speech-in-list-out approach to spoken user interfaces. In *Proceedings of the Human Language Technology Conference (HLT 2004)* (pp. 112-116). Boston, MA.

Driving Simulators. (2006). Retrieved October 24, 2006 from http://www.inrets.fr/ur/sara/Pg_simus_e.html

Forlines, C., Schimdt-Nielsen, B., Raj, B., Wittenburg, K., & Wolf, P. (2005). A comparison between spoken queries and menu-based interfaces for in-car digital music selection. *IFIP TC13 International Conference on Human-Computer Interaction (INTERACT 2005)*

Klauer, S.G., Dingus, T. A., Neale, V. L., Sudweeks, J.D., & Ramsey, D.J. (2006). *The Impact of Driver Inattention on Near-Crash/Crash Risk: An Analysis Using the 100-Car Naturalistic Driving Study Data.* (DOT HS 810 594) Washington, DC: United States Government National Highway Traffic Safety Administration, Office of Human-Vehicle Performance Research.

Leatherby, J.H., & Pausch, R. (1992) Voice input as a replacement for keyboard accelerators in a mousebased graphical editor: An empirical study. *Journal of the American Voice Input/Output Society, 11*, 2.

Ranney T., Watson, G. S., Mazzae, E. N., Papelis, Y. E., Ahmad O., & Wightman, J. R. (2004). *Examination of the distraction effects of wireless phone interfaces using the national advanced driving simulator—Preliminary report on freeway pilot study* (DOT 809 737). East Liberty, OH: United States Government National Highway Traffic Safety Administration, Vehicle Research and Test Center

Society of Automotive Engineers. (2004). *SAE recommended practice navigation and route guidance function accessibility while driving* (SAE 2364). Warrendale, PA: Society of Automotive Engineers.

Strayer, D.L., Drews, F. A., Albert, R. W., & Johnston, W. A. (2001). Cell phone induced perceptual impairments during simulated driving. In D. V. McGehee, J. D. Lee, & M. Rizzo (Eds.), *Driving assessment 2001: International symposium on human factors in driver assessment, training, and vehicle design.*

Stutts, J.C., Reinfurt, D.W., Staplin, L.W., & Rodgman, E.A. (2001). *The role of driver distraction in traffic crashes.* Washington, DC: AAA Foundation for Traffic Safety. Retrieved October 24, 2006, from http://www.aaafts.org/pdf/distraction.pdf.

US Census Bureau News. (2005). Americans spend more than 100 hours commuting to work each year. *Census Bureau Reports*, Retrieved October 24, 2006, from http://www.census.gov/Press-Release/www/releases/archives/ american_community_survey_acs/004489.html.

UMTRI Driver Interface Group. (updated 2006). *All the major telematics guidelines we've seen.* University of Michigan Transportation Research Institute, Driver Interface Group. Retrieved October 26, 2006, from http://www.umich.edu/~driving/guidelines/guidelines.html

Wang, J., Knipling, R. R., & Goodman, M. J. (1996). The role of driver inattention in crashes; new statistics from the 1995 crashworthiness data system (CDS). In *proceedings of the 40th Annual Proceedings: Association for the Advancement of Automotive Medicine* (pp. 377-392).

Wolf, P., & Raj, B. (2002). The MERL Spoken-Query Information Retrieval System: A System for Retrieving Pertinent Documents from a Spoken Query. In *Proceedings of ICME, Vol. 2* (pp. 317-320).

KEY TERMS

Cognitive Load: A measure of the mental effort required to carry out a given task.

Driver Distraction: A measure of the degree to which attention is taken away from the driving task.

Listening Tone: A sound generated by a speech-based user interface when it is ready to accept spoken input

Lombard Effect: The specific changes in style of speech caused by the presence of noise. In particular the speech gets louder and higher frequencies are emphasized

Misrecognition: A speech recognition result which does not accurately represent what was spoken by the user. In spoken command recognition, recognizing the exact words spoken is not necessary to avoid a misrecognition as long as the correct command is recognized.

Push and Hold: A type of speech interaction where the user must hold down a button while speaking to the system. This kind of system is familiar to most users as it is reminiscent of a walkie-talkie.

Push and Release: A type of speech interaction where the user must depress a button prior to the start of speech. This type of interaction is unfamiliar to some users, but provides an easy learning curve with the proper affordances.

Recognition lattice: A directed graph of candidate words considered by a speech recognizer. This graph will often contain alternate words with similar phonetics. It will also contain confidence weights.

SILO: A speech-based user interface which returns a shortlist of possible responses, from which the user must make a final selection. We refer to such interfaces as Speech-In List-Out, or SILO.

Speech-Based [User] Interface (SUI): A user interface which uses utterances spoken by the user as a primary input mode. A speech based interface may also have other input modes, such as dedicated or softkey input, and may also have voice feedback and/or visual feedback.

Telematics: Broadly, telematics refers to the combination of telecommunication and computation. More specifically telematics has come to refer to mobile systems which combine wireless data communications with local computation resources. Voice communication and/or location information provided by GPS are often assumed.

Chapter XVI
Design for Mobile Learning in Museums

Nikolaos Tselios
University of Patras, Greece

Ioanna Papadimitriou
University of Patras, Greece

Dimitrios Raptis
University of Patras, Greece

Nikoletta Yiannoutsou
University of Patras, Greece

Vassilis Komis
University of Patras, Greece

Nikolaos Avouris
University of Patras, Greece

ABSTRACT

This chapter discusses the design challenges of mobile museum learning applications. Museums are undoubtedly rich in learning opportunities to be further enhanced with effective use of mobile technology. A visit supported and mediated by mobile devices can trigger the visitors' motivation by stimulating their imagination and engagement, giving opportunities to reorganize and conceptualise historical, cultural and technological facts in a constructive and meaningful way. In particular, context of use, social and constructivist aspects of learning and novel pedagogical approaches are important factors to be taken in consideration during the design process. A thorough study of existing systems is presented in the chapter in order to offer a background for extracting useful design approaches and guidelines. The chapter closes with a discussion on our experience in designing a collaborative learning activity for a cultural history museum.

INTRODUCTION

Use of mobile devices spreads in everyday human activities. These devices offer portability, wireless communication and connectivity to information resources and are primarily used as mobile digital assistants and communication mediators. Thus, it is no surprise that various attempts to use mobile appliances for learning purposes have been reported either inside or outside school (Roschelle, 2003). The term *mobile learning* or *m-learning* has been coined and concerns the use of wireless technologies, portable appliances and applications in the learning process without location or time restrictions. Practitioners' reports (Perry, 2003; Vahey & Crawford, 2002) and scientific findings (Norris & Soloway, 2004; Roschelle, 2003; Zurita & Nussbaum, 2004) communicate promising results in using these applications in various educational activities. The related bibliography proposes various uses of mobile appliances for learning. These Activities might concern access and management of information and communication and collaboration between users, under the frame of various learning situations.

A particular domain related to collaborative learning is defined as the support provided towards the educational goals through a coordinated and shared activity (Dillenbourg, 1999). In such cases, peer interactions involved as a result of the effort to build and support collaborative problem solving, are thought to be conducive to learning. On the other hand, traditional groupware environments are known to have various technological constraints which inflict on the learning process (Myers et al., 1998). Therefore, mobile collaborative learning systems (mCSCL) are recognized as a potential solution, as they support a more natural cooperative environment due to their wireless connectivity and portability (Danesh, Inkpen, Lau et al., 2001). While the mobility in physical space is of primary importance for establishing social interaction, this ability is reduced when interacting through a desktop system. It is evident that, by retaining the ability to move around it is easier to establish a social dialogue and two discrete communication channels may be simultaneously established through devices: one physical and one digital. Additionally, a mobile device can be treated as an information collector in a lab or in an information rich space (Rieger & Gay, 1997), as a book, as an organizing medium during transportation or even as a mediation of rich and stimulating interaction with the environment (i.e., in a museum). Effective usage of mobile appliances has been reported in language learning, mathematics, natural and social sciences (Luchini et al., 2002).

Furthermore, various technological constraints need to be taken in consideration during the design of activities which involve mobile devices. Such an example is the small screen, which cannot present all the information of interest while the lack of a full keyboard creates constraints in relation to data entry (Hayhoe, 2001). There is a need to provide the user with the possibility to 'go large' by getting information from both the virtual and physical world, while simultaneously 'going small,' by retrieving the useful and complementary information and getting involved into meaningful and easy to accomplish tasks (Luchini et al., 2002). In addition, despite the fact that technological solutions are proliferating and maturing, we still have a partial understanding of how users take effectively advantage of mobile devices. Specifically, in relation to communication and interaction, we need to investigate how mobile technology can be used for development of social networks and how it can provide richer ways for people to communicate and engage with others. In public spaces, like museums, a crucial question is if the serendipitous exchanges and interactions that often occur should be supported through mobile technology, how and where the interaction between people takes place and how is affected by this novel technology. Clearly, a better understanding of social activities and social interactions in public spaces should emerge to answer these questions.

A number of the aforementioned issues are discussed next in the context of a museum visit. First, we analyse how the context can affect any activity and application design. Then, we outline the most promising mobile learning applications and finally, we present our experiences of introducing collaborative learning activities using a novel approach based on the best practices surveyed previously, in a large scale project for a cultural history museum.

INTERACTION DESIGN FOR MOBILE APPLICATIONS

Interaction design is one of the main challenges of mobile applications design. Direct transfer of knowledge and practices from the user-desktop interaction metaphor, without taking in consideration the challenges of the new interaction paradigm is not effective. A new conceptualization of interaction is needed for ubiquitous computing. The traditional definition (Norman, 1986) of the user interface as a "means by which people and computers communicate with each other," becomes in ubiquitous computing, the means by which the people and the environment communicate with each other *facilitated* by mobile devices. As a result, interaction design is fundamentally different. In the traditional case, the user interacts with the computer with the intention to carry out a task. The reaction of the computer to user actions modifies its state and results in a dialogue between the human and the machine.

On the other hand, the user interaction with mobile devices is triadic, as the interaction is equally affected not only by the action of the user and the system's response, but by the context of use itself. The level of transparency of the environment, taking into account the presence of the mobile device and the degree of support to 'environmental' tasks meaningful to the user, are new issues to be considered. Consequently, new interaction design and evaluation criteria are required, since the design should not only focus on the user experience but pay also attention to the presence of other devices or objects of interest, including the level of awareness of the environment. By building the virtual information space into the real, the real is enhanced, but conversely, by drawing upon the physical, there is the opportunity to make the virtual space more tangible and intuitive and lower the overall cognitive load associated with each task.

To summarize, a number of design principles are proposed for mobile applications design:

a. Effective and efficient *context awareness* methods and models, with respect to the concept of context as defined by Dey (2001): 'Context is any information that can be used to characterize the situation of an entity. An entity is a person, place, or object that is considered relevant to the interaction between a user and an application, including the user and applications themselves.'
b. Presentation of useful information to the user *complementary* to the information communicated by the environment.
c. *Accurate and timely update* of environmental data that affect the quality of interaction.
d. Contextualized and personalized information according to *personal needs.*
e. Information should be *presented to the user* rather than having the user searching for it.

Failure to look into these design issues can lead to erroneous interaction. For example, delays of the network, lack of synchronization between two artifacts of the environment or slight repositioning of the device can lead to misconceptions and illegal interaction states. In addition, information flow models should be aligned according to the information push requirement and relevant user modeling and adaptation techniques to support this flow of information should be defined. Finally, new usability evaluation techniques, concerning mobile applications should emerge to shape a novel interaction paradigm.

Dix et al. (2000) present a framework to systematically address the discussed design issues and successive context awareness elements are inserted in the design process: (a) the *infrastructure* level (i.e., available network bandwidth, displays' resolution), (b) the *system* level (type and pace of feedback and feed through), (c) the *domain* level (the degree of adaptability that a system must provide to different users) and (d) the *physical* level (physical attributes of the device, location method and the environment). All these elements should be tackled independently and as a whole in order to study the effect of every design decision to each other.

We formulate the interaction design aspects discussed through the problem of designing a mobile learning application for a museum. During the visit a user has only a partial understanding of the available exhibits. This situation can be supported by complementary information included in the physical environment, for example alternative representations, concerning the historical role of the people or the artifacts presented the artistic value of a painting (Evans & Sterry, 1999), and so forth. This cognitive process of immersion into the cultural context, represented by the museum exhibits, could be supported by drawing upon the stimuli produced during the visit using context aware mobile devices. Therefore, these devices should be viewed as tools to enhance the involvement of a user in the cultural discovery process, tools that challenge the user to imagine the social, historical and cultural context, aligning her to a meaningful and worthy experience.

It is not argued here that the infusion of mobile technologies in museums will necessarily result to meaningful learning processes. Our analysis involves the potential use of the technology when integrated in educational activities (Hall & Bannon, 2006), which will offer a structured learning activity according to the characteristics of museums' content and the functionalities of the technologies used. To better illustrate this point (a) we briefly present a set of selected exemplary cases which demonstrate different ways of integrating mobile educational applications in museums and (b) we provide a more detailed account of such an application that we designed for a museum in Greece.

In the next section a number of approaches supporting such a visit are reviewed and examined using the design aspects as guiding paradigm and point of reference. Since the goal of the visitor is to see and learn more and not to explicitly use technology, a deep understanding of visitors' needs is important during the design phase, to avoid disturbances that can destruct her from her objective. Therefore, decisions made for the technology used and the styles of interaction, with the involved devices, have to deal with user's patterns of visit. Having the above requirements in hand, we use the framework proposed by Dix et al. (2000) to organize a coherent characteristics inspection of some representative examples of mobile museum guides.

MOBILE DEVICES AS MUSEUM GUIDES

In this section, some representative design approaches for mobile museum applications are discussed. An extended survey is included in Raptis, Tselios and Avouris (2005). The first system named "Electronic Guidebook," deployed in the Exploratorium science museum (Fleck et al., 2002), tries to involve the visitors to directly manipulate the exhibits and provides instructions as well as additional science explanations about the natural phenomena people are watching. The system of the Marble Museum of Carrara (Ciavarella & Paterno, 2004) stores the information locally in the PDA's memory, uses a map to guide the visitors around the museum and presents content of different abstraction levels (i.e., room, section and exhibit). The "ImogI" system uses Bluetooth to establish communication between the PDAs and exhibits and presents the closest exhibits to the user, (Luyten & Coninx, 2004). The "Sotto Voce" system gives details about everyday things located in an old house (Grinter et al., 2002) by having pictures of the walls on the PDA's screen and asking from the user to select the exhibit she is interested in, by pressing it. The "Points of Departure" system (www.sfmoma.org) gives details in video and audio form by having 'thumbnails' of exhibits on the PDA's screen. It also uses 'Smart Tables' in order to enrich the interaction. A system, in the Lasar Segall Museum, Sao Paolo, Brazil (Dyan, 2004), automatically delivers information to the PDA, about more than 3,000 paintings. In the Tokyo University Digital Museum a system uses three different approaches to deliver content. The

PDMA, in which the user holds the device above the exhibit she is interested in, the Point-it, in which the visitor uses laser-pointer to select specific exhibits and finally the Museum AR in which visitors wear glasses in order to get details about the exhibits (Koshizuka & Sakamura, 2000).

The system developed in the C-Map project, (Mase, Sumi, & Kadobayashi, 2000), uses active badges to simulate the location of the visitor, allowing tour planning and a VR system, controlled by the gestures of the visitor. In a Tour guide (Chou, Lee, Lee et al., 2004), the information about the exhibits is automatically presented and there is no variation in the form of the visit, but subjective tour guides are used. A different approach is the one adopted in the Museum of Fine Arts in Antwerp (Van Gool, Tuytelaars, & Pollefeys, 1999), in which the user is equipped with a camera and selects exhibits, or details of an exhibit by taking pictures. A tour guide in the PEACH project, (Rocchi, Stock, Zancanaro et al., 2004), which migrates the interaction from the PDA to screens and uses a TV-like metaphor, using 'newscasters' to deliver content. Finally, a nomadic information system, the Hippie, developed in the framework of the HIPS project, (Oppermann & Specht, 1999), allows the user to access a personal virtual space during or after the visit. In the latter system, an electronic compass is used to identify the direction of a visitor.

The *infrastructure* context concerns the connections between the devices that comprise the system and influence the validity of the information that is provided through them to the users and needs not only to be addressed in problematic situations. It is also related with the validity and timely updates of available information. This can be clearly seen in collaboration activities where

Table 1. Design decisions affecting system context

	Location technology	Storage of information	Flow of information	Additional functions
"Antwerp project"	IrDA	In Server	Active	Cameras
C-Map	IrDA	In Server	Active, exhibit recommendations	Active Badges, Screens
Hippie	IrDA	In Server	Active, info presented based on the history of visit	
Imogl	Bluetooth	Info stored in Bluetooth transmitters	Active, proximity manager	
Lasar Segal Museum	IrDA	In Server	Passive	
Marble Museum	IrDA	Locally stored info, abstraction levels	Active, history of the visit	
PDMA, Point it, Museum AR	IrDA	In Server	Active	laser pointer, glasses
PEACH project	IrDA	In Server	Passive, task migration	Screens
Points of departure		Locally stored info	Active	Screens
Rememberer	RFID	In Server	Passive	Cameras
Sotto Voce		Locally stored info	Active	
Tour Guide System (Taiwan)	IrDA	In Server	Passive, subjective tour guides	

the user constantly needs to know the location of other users, the virtual space, the shared objects, and so forth. In the specific museum domain the results may not be so critical but can lead the user to various misunderstandings.

The mentioned systems use an indirect way of informing the user that her requests have been carried out: the user sees and hears the reflection of her requests on the PDA. There is no clear notification that the user's demands are executed successfully or not. Some of the systems use external factors, as signs of success, such as a led light ("Rememberer") and audio signals ("Marble Museum"). But in general terms, the user is on his or her own when problems occur and the systems leave it up to her to find it out, by observing that, there is no progress. We have to point that it could be very distracting and even annoying to have feedback messages in every state of interaction, but it is important for designers, to include a non-intrusive approach to inform that there is a problem and provide constructive feedback to overcome it.

Regarding the *system* context we can distinguish four different approaches as a means of awareness technology. In the first approach (Table 1), the PDA is the whole system. There are no other devices or awareness mechanisms involved and the information presented to the user is stored locally in the PDA. The second approach uses RFID tags to establish communication between the PDAs and the exhibits and the third which uses Bluetooth to establish communication with the exhibits and deliver content. The forth and most common approach uses IrDA technology to estimate the position of the visitor in space. Usually, IrDA tags are placed near every exhibit or in the entrance of each exhibition room and Wi-Fi derives the information to the PDA from a server Also, many different additional devices are built and integrated into these systems like screens (as a standalone devices or as interacting devices with the PDAs, where the user has the opportunity to transmit sequentially her interaction with the system from the PDA to a Screen). Regarding the *location*, all the studied systems use a topological approach to identify the position of a PDA, which informs approximately the system about the user's location. However, in the case of a museum with densely place exhibits, a more precise Cartesian approach can yield accurate user localization.

Domain context concerns aspects related to the situated interaction that takes place in the specific domain. Often in museum applications there is a lack of information about user profiles and characteristics. It is however important to consider that each visitor in a museum has different expectations, and is interested in different aspects regarding the exhibits. In the studied systems only in those that allow interaction of the users with servers there is a possibility for personalized interaction. Most of the systems require from the user to login, answer some specific questions, in order to build a model of the user and present the information in her PDA according to her language, her expertise level and her physical needs (i.e., bigger fonts for those with sight impediments). When domain context is absent from the design process the system operates as a tool suited for the needs of a single hypothetical 'ideal' user. In such an environment this 'ideal' user will likely represent the needs and expectations of a small fraction of real visitors.

The system may push information to the user or it may wait until the user decides to pull it from the system. In the first case, special consideration should be taken to the user's specific activity and objective. Questions related to situated domain context are the following: Does the system propose any relevant information based on the history of users interaction? Does it adapt to actions repeatedly made by the user? Does it present content in different ways? For example, the "ImogI" system rearranges the order of the icons putting in front the mostly used ones. Also, in PEACH and in 'Points of Departure' the user can change the interaction medium from PDAs to Screens, in order to see more detailed information.

The *physical* context lays in the relation of the system with the physical environment and in problems concerning the physical nature of the devices. However, in the studied systems there is not a single mechanism of identifying the physical conditions. For example, in a room full with people, where a lot of noise exists, it would be appropriate if the system could automatically switch from an audio to a text presentation.

From the survey of the mobile guides applications presented here it seems that efficient design approaches could be achieved by augmenting physical space with information exchanges, by allowing collaboration and communication, by enhancing interactivity with the museum exhibits and by seamlessly integrating instantly available information delivered in various forms. However, the synergy between technology and pedagogy is not straightforward especially if we take into account the need to tackle issues such as efficient context integration, transparent usage of the PDA, and novel pedagogical approaches to exploit the capabilities of mobile devices. As a result, after discussing in detail usages of a mobile device as a mean of museum guidance, in the following, we attempt to discuss explicit educational activities mediated by mobile devices and a specific example of a new Mobile Learning environment.

DESIGNING MUSEUM MOBILE EDUCATIONAL ACTIVITIES

The level of exploitation of mobile devices in a museum setting is increasing and part of this use may have educational value. In this section we will focus on the added value of integrating educational mobile applications in museums. We will start our analysis by posing two questions that we consider central to this issue: (a) what is changing in the learning process taking place in a museum when mediated by mobile technology and (b) why these changes might be of educational or pedagogical interest? We will attempt to address these questions by focusing on three aspects related not only to the characteristics of mobile technology but also to the results of its integration in a museum. Specifically we will discuss: (a) the types of interaction between the visitor and the learning environment (e.g., the museum), (b) the learning activities that these interactions can support and (c) the role of context and motion in learning.

One facet of the learning process when mediated by mobile technology in museum visits involves the tangibility of museum artifacts: distant museum exhibits that were out there for the visitors allowing them just to observe now can be virtually touched, opened, turned and decomposed. In this case, technology provides to the user the key to open up the exhibit, explore it and construct an experience out of it. The traditional reading of information and observation of the exhibit is considered as one-dimensional "information flow" from the exhibit to the user. Mobile technology facilitates the transformation of the one dimensional relationship to a dialectic relationship between the user and the exhibit. Furthermore, this relationship can now include another important component (apart of the exhibits) of the museum environment: the other visitors. By providing a record of user–exhibit interaction for other visitors to see, reflect upon and transform technology can support social activities of communication, co-construction, and so forth between the visitors. To sum up, mobile technology mediates three types of interaction between the learner and the learning environment of a museum: (a) "exhibit–user" interaction (b) "user –exhibit" interaction and (c) "between the users" interaction about "a" and "b."

The enrichment of interaction between the learner and the museum might result in more or different learning opportunities (Cobb, 2002) the characteristics of which are outlined here. Specifically, the dialectic relationship between the user and the museum artifacts, mediated by mobile devices, might offer chances for analysis of the exhibit, experimentation with it, hypothesis formulation and testing, construction of interpretation, information processing and organization, reflection and many more, according to the educational activity designed. Collaboration and communication about the exhibits and information processing about them makes possible socio-constructive learning activities. By comparing these elements of the learning process to the reading or hearing of information about the exhibits (which is a the starting point for a non technology mediated museum visit) we realize that mobile technology has the potential to offer an active role to the learner: she can choose the information she wants to see, open up and de-construct an exhibit if she is interested in it, see how other visitors have interacted with a certain exhibit, discuss about it with them, exchange information, store information for further processing and use and so on.

Up until now, we described the role of mobile technology in learning with respect to two characteristics of the museum as learning environment: the exhibits and the other visitors. Another characteristic of the museum, which differentiates it from other learning environments (e.g., classroom) is that learning in a museum takes place while the learner moves. Learning while moving, quite often takes place very effectively without the support of technology. However there are cases that further processing with appropriate equipment is needed or some structuring of this "mobile learning experience" is proved to be useful. Mobile technologies can find in museums an important area of implementation not only because museum visits are structured around motion but because we have to support visitors *during* and not just after or before the visit (Patten, Arnedillo Sanchez et al., 2006). But why is it important to support learning during the visit? The answer here comes from the theory of situated learning (Lave & Wenger, 1991) which underlines the role of context in learning. Specifically, context facilitates knowledge construction by offering the practices, the tools, and the relevant background along with the objectives towards which learning is directed and has a specific meaning or a special function (knowledge is used for something). Finally, the use of mobile devices provides a new and very attractive way of interacting with the museum content especially for young children (Hall & Bannon, 2006).

As mentioned previously a large number of mobile applications have been developed during previous years for use in the museums (Raptis et al., 2005). All these mobile applications can add educational value to a museum visit in various ways. A survey of mobile educational applications for use inside the museum, led us to a categorization according to the educational approach followed in every occasion. The first category includes applications that mainly deliver information to the visitor and concerns the vast majority of applications created for museums. Mobile devices take the place of the museums' docents and offer predetermined guided tours based upon certain thematic criteria. The aforementioned applications offer the museum visitor an enhanced experience which can support the learning process through a behaviorist approach. Enhancement is succeeded by supplying multimedia and context-related content.

The second category of applications, suitable for educational use in museums, consists of applications which provide tools that can support the learning process in a more profound way. Compared to the first category, they provide information about the exhibits of a museum but furthermore they include a series of functions that increase the interactivity with the user. Such an example is the Sotto Voce System (Grinter et al., 2002), which includes an electronic guide with audio content and the ability of synchronized sharing of this content between visitors. Thus, the users can either use individually the guide or "eavesdrop" to the information that another visitor listens.

Another example is the applications developed for the Exploratorium, a science museum in San Francisco (Fleck et al., 2002). In this museum, the visitor has the possibility to manipulate and experiment with the exhibits. Also, an electronic guidance was designed to provide information about the exhibits and the phenomena related with them, posing relative questions to provide deeper visitors' engagement. These applications are closer to social-cultural learning theories as they provide the user with tools to organize and control the provided information.

The third category of educational applications presents a specific educational scenario. Usually, game-based activities where the users, mostly children aged 5-15, are challenged to act a role and complete carefully designed pedagogical tasks. Such an example is the MUSEX application (Yatani, Sugimot & Kusunoki, 2004), deployed in the National Museum of Emerging Science and

Innovation in Japan. MUSEX is a typical drill and practice educational system in which children work in pairs and are challenged to answer a number of questions. Children select an exhibit with their RFID reader equipped PDA and a question is presented in the screen with four possible answers. The activity is completed when each pair collects twelve correct answers. Children may collaborate and communicate either physically or via transceivers and monitor each group progress through a shared screen. After the completion of the activity the participators have the possibility to visit a Website and track their path inside the museum. The users can deeply interact with the exhibits, review the progress of her partner or ask for help (Yatani et al., 2004).

DinoHunter project includes several applications for the transmission of knowledge through game-based and mixed reality activities in the Senckenberg museum, Frankfurt, Germany. Three of these applications, namely DinoExplorer, DinoPick and DinoQuiz, are being supported by mobile technologies (Feix, Gobel, & Zumack, 2004). DinoExplorer delivers information to the users as an electronic guide, DinoPick allows the users to pick one part of the body of a dinosaur and get more multimedia information about this specific part and DinoQuiz provides a set of questions for further exploration of the exhibits of the museum.

Mystery at the Museum is another mobile, game-based, educational activity created for the Boston Museum of Science. It engages visitors in exploring and thinking in depth about the exhibits, thus making connections across them and encourages collaboration (Klopfer, Perry, Squire et al., 2005). High School students and their parents are called to solve a crime mystery where a band of thieves has stolen one of the exhibits. The users try to locate the criminals by using a PDA and a walkie-talkie. The participants must select upon the role of a technologist, a biologist or a detective. Depending on the chosen role they can interview virtual characters, pick up and examine virtual objects by using virtual equipment (e.g., microscope), collect virtual samples via infrared tags or exchange objects and interviews through the walkie-talkies. A study confirmed deep engagement of the participants and extensive collaboration due to the roles set.

Another similar approach is presented through the Scavenger Hunt Game activity used in the Chicago Historical Society Museum (Kwak, 2004). In this case, the children are challenged to answer a series of questions related to the exhibits and the local history. They undertake the role of a historical researcher and they are called to answer 10 multiple choice questions while examining the exhibits. Each user is individually engaged into the activity and her progress is evaluated in a way similar to electronic games. The Cicero Project implemented in the Marble Museum of Carrara introduces a variety of games to the visitors (Laurillau & Paternò, 2004). The games vary from finding the missing parts of a puzzle to answering questions about the exhibits. Its main characteristic is the support it provides to the visitors to socially interact and collaboratively participate in activities concerning the exhibits of the museum, through peer-awareness mechanisms.

A series of mobile educational activities was also carried out in the frame of the Handscape Project in the Johnson Museum (Thom-Santelli, Toma, Boehner et al., 2005). The "Museum Detective" engage students in role-playing activities. Children working in pairs are called to locate an object described by one clue and learn as much as possible for it. A series of multiple-choice questions is presented for further exploration of the exhibit. Four types of interactive element are also provided for the exhibits: a painting, a drawing activity and a building activity and a multimedia narrative. The multiple-choice questions and the building activity were drill and practice activities and the rest were activities allowing children to make their own creations.

The systems of the latter category present coherent learning experiences comprised of planned and organized pedagogic activities, where an intervention has been purposefully designed to result a positive impact on children's cognitive and affective development. With respect to the contextual and interaction issues presented in the previous sections, we attempt to present in the next section an integrated application that involves children as role-playing characters by exploring the museum using a PDA.

AN EXAMPLE OF MOBILE ACTIVITY DESIGN FOR INDOOR MUSEUM VISIT

The "Inheritance" activity discussed here, is designed to support learning in the context of a cultural/historical museum visit. The application involves role-playing, information retrieval, data collecting and collaboration educational activities, suitable for children aged 10 or above working in teams of two or three members each. The activity scenario describes an imaginary story where the students are asked to help the Museum in finding the will of a deceased historian, worked for years in it. This will is hidden behind the historians' favorite exhibit. Clues to locate the document are scattered among the descriptions of some exhibits. If the children manage to find the will, all of the property of the historian will be inherited by the museum and not by his "greedy relatives." The scenario urges the students to read the description of the exhibits, find the clues and collaboratively locate the specific one.

During the design process of the activity we had to study the *museum context*, the *mobile technology* used and the *learning approach* to be followed in order to achieve the desired pedagogical outcome. The survey discussed in the previous section led us to adopt the following interaction design decisions. A PDA with wireless network capability is used and an RFID reader is attached to it to 'scan' the RFID tags used to identify the exhibits. Wi-Fi infrastructure is being used to deliver data and establish communication between the visitors. When an exhibit is scanned, the PDA sends a request for information to the server which delivers the appropriate content presented in the form requested by the user. Data exchange between two users is accomplished through alignment of their devices while pointing one to the other, which mimics the exhibit scanning procedure. We also opted for small chunks of text since reading at low resolution screens reduces reading comprehension significantly.

The educational design of the activity was inspired by the social and cultural perspective of constructivism. It was structured around a set of learning objectives relevant to the thematic focus of the museum, to the exhibits' information, to the age and previous knowledge of the students, and to the fact that involves a school visit (as opposed to individual museum visits). The basic elements which shaped the activity were:

a. **Engagement of interest:** Engagement and interest hold an important role in the learning process. Student interest in a museum should not be taken for granted, especially because a visit arranged by the school is not usually based on the fact that some students might be interested to the theme of the museum. In the inheritance activity we considered to trigger student interest by engaging them in a game. The setting, the rules and the goal of the game were presented in the context of a story.
b. **Building on previous knowledge:** The focus of the activity was selected with respect to the history courses that students were taught in school. They had a general idea about the specific period of the Greek history and the activity offered complementary information about certain issues of this period. Building on previous knowledge was expected to support students in problem solving and hypothesis formulation and testing.
c. **Selecting–processing–combining pieces of information:** The scenario is structured around the idea that students read the offered information, select what is relevant to their inquiry and combine it with other pieces of information that have selected and stored earlier. Thus the students are expected to visit and re-visit the relevant exhibits, go through the information that involves them as many times as they think necessary and not just retrieve that information but combine it and use it in order to find the favorite exhibit which is the end point of the game

d. **Hypothesis formulation and testing:** When students have selected enough information from the exhibits around one room of the museum they can attempt to use some of the clues they have selected in order to find the favorite exhibit. If they fail they can go around the room to collect more information and try again.
e. **Communication and collaboration:** The activity is designed to facilitate inter and intra-group collaboration. Specifically, two groups of students are expected to collaborate to determine which exhibit they will interact with, to exchange clues using their PDA and to discuss their ideas about the favorite exhibit.

During the activity, the participating teams are free to explore any exhibit. Each team is provided with a PDA to extract information related to the exhibits by reading the tags attached to each of the exhibits (Figure 1). Only some of the exhibits contain 'clues,' which give information about the favourite exhibit to be found. Children must locate them, store them in the PDAs notepad and after collecting all or most of the clues the teams are able to beam their clues to each other. After collecting all six clues the students are challenged to locate the favorite exhibit. When both teams agree that one exhibit is the favorite one, they can check the correctness of their choice by reading with both PDAs the RFID tag of the chosen exhibit.

After the development of a prototype application, a case study was conducted inside the museum in order to validate the design choices. Seventeen students, aged 11, participated in the study (Figure 2). Data concerning all involving elements were collected to study the activity in depth. The activity was videotaped, PDA screen recording has been used and voice recorders were used to record dialogues among the participants.

The goal of the data analysis was twofold. First, to identify problems children encountered during the process in relation to each of the activity's elements. Then, to identify the nature of the interactions occurred during the procedure. Our analysis is based on the Activity Theory, concerning mainly human practices from the perspective of consciousness and personal development. It takes into account both individual and collaborative activities, the asymmetrical relation between people and things, and the role of artifacts in everyday life. The activity is seen as a system of human processes where a subject works on an object in order to obtain a desired outcome. In order to accomplish a goal, the subject employs tools, either conceptual or embodiments. Activity is consisted by different components which are (Figure 3): (a) *subject*, (the persons engaged in the activity), (b) *object* (scope of the activity), (c) its *outcome* (c) *tools* used by the subjects (d) *rules-roles* that define the activity process, (e)*community* (context of the activity) and (f)*division of labour* (tasks division among the participants, Kuuti, 1995; Zurita & Nussbaum, 2004).

Figure 1. Screenshots of the "Inheritance" application: (a) Dialogue for RFID tags reading (b) information for a selected exhibit (c) clue selection (d) the notepad screen.

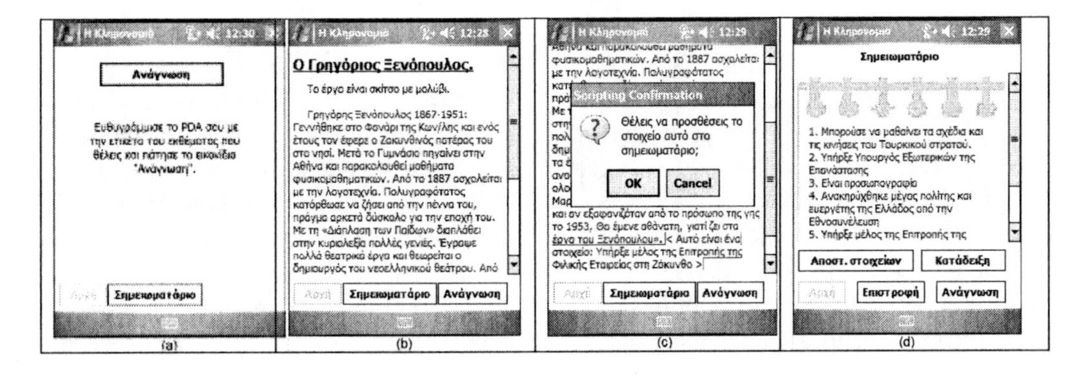

Figure 2. Children engaged in the activity

Activity Theory is of fundamental importance to deeper understand learning with mobile devices while visiting a museum, since in this case knowledge construction is mediated by cultural tools in a social context. The data collected were analyzed with the use of the Collaboration Analysis Tool (ColAT) environment which supports a multilevel description and interpretation of collaborative activities through fusion of multiple data (Avouris, Komis, Fiotakis et al., 2004).

In our analysis, an activity is a procedure during which objects become knowledge through three different levels-steps. Operation is the lower level where routine processes facilitate the completion of goal-oriented actions which in turn constitute the activity. Dialogues, user operations in the application and observations derived from the videos were transcribed in this first level of analysis. The actors, the operations and the mediating tools were noted in this level. Actors were the two participating teams and the researcher. The mediating tools were the dialogue among children and the researcher, texts of information (symbolic) and the application (technological). Some examples of operations in our case are text scrolling, RFID tag reading and transition from one screen to another. Analysis of these user operations led us to the identification of a problem in the use of the application. For example, due to data transfer delay from the server to the PDA, users in some occasions were frustrated and selected repeatedly an action due to lack of timely feedback.

In the second level, the different actions presented among the structural components of the activity are being studied. In order to identify and categorize the actions, a series of typologies were introduced. Typologies were set according to the goal and the mediating tool of each action. For example when children used the PDA to read the text information (mediating tools) their goal was not always the same. Three different typologies were used to describe the situation when the children read carefully the information provided ("Reading of information"), when they were reading the information and searched for clues also ("Reading and searching for clues"), and finally when they were "Searching for clues only." A "Reading and searching for clues" action example is presented in Table 2. Children scroll down the text and one of them states in the end of this action that they were unable to find a clue. When children search only for clues without paying attention to the information we observe rapid scrolling. A clear indication that they have already found all the clues is when children read only the information.

Other actions defined in our study were related to the dialogue between the children and the researcher aimed to overcome difficulties in using the application or understanding the rules-roles of the activity. Typologies where also introduced to describe the interaction between children related to the next step in the procedure (..."*Should we go there? ...ok")* and the exchange of thoughts about the solution of the activity (..."*Well, tell me, the first clue is? ...He could spy the Turkish army"...).* In the third level of analysis, patterns identified concerning the evolution of the procedure.

Figure 3. Description according to the activity theory model

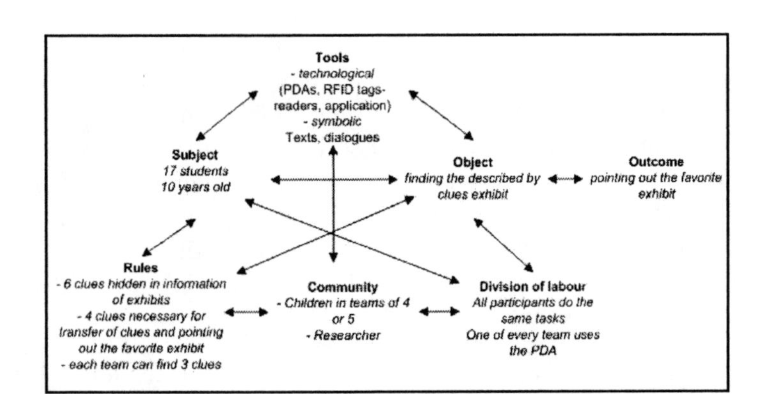

Table 2. An extract of the data analysis presenting action of the 'reading and searching for clues' class

Time	Actor	Tool	Events
00 : 08 : 50	Group1	PDA	Selection of "read"
00 : 08 : 52	Group1	texts	Information D. Stefanou
00 : 09 : 21	Group1	PDA	scrolling
00 : 09 : 33	Group1	PDA	scrolling
00 : 09 : 38	Group1	PDA	scrolling
00 : 09 : 42	Group1	PDA	scrolling
00 : 09 : 45	Group1	PDA	scrolling
00 : 09 : 49	Group1	PDA	scrolling
00 : 09 : 57	Group1	PDA	scrolling
00 : 10 : 02	Group1	PDA	scrolling
00 : 10 : 03	Group1	Dialogue	It doesn't have any (clues) here

Clearly, the basic goals of the activity as described previously in this section have been fulfilled. Data analysis indicated that children were highly motivated by the activity and collaborated in order to achieve their goal. As derived from the analysis, the teams adopted different strategies to accomplish the task. Collaboration was observed mainly while making the choice of the next exhibit to be examined. After completing the task of finding the clues, the two teams collaborated more closely. They divided the work needed to find the exhibit described by the clues and looked in different parts of the room while collaborating and sharing their thoughts and suppositions. They used the clues as information filters thus eliminating the ones that did not match. Additionally, the learning result of this activity, as derived from subsequent students' essays describing the visit experience, was a deeper understanding of the historical role of the persons represented in the exhibits and their interrelations.

CONCLUSION AND FUTURE WORK

This chapter attempted to present current design approaches for mobile learning applications in the context of a museum visit. In addition, thorough study of similar approaches took place, which lead to useful design patterns and guidelines. As discussed, design of mobile learning systems, is not a straightforward process. In addition to the challenge of integrating the concept of context into the design process and independently from context conceptualization, a comprehension of pedagogical goals, desired learning transfers, user typical needs and objectives should take place. We argue that proper design decisions should take into account a solid theoretical cognitive framework, as well as the special characteristics of the mobile devices used and the challenges of such an informal learning setting. A suitable activity should be properly supported by adequate interaction models, deeper understanding of the tasks involved to carry out the activity as a whole and their expectations while carrying out specific actions. For this reason, further validation of our proposed activity, took place in the actual museum. The activity was enjoyed by the students and enhanced their motivation to learn more about the cultural and historical context represented by the exhibits. The latter challenge has been better illustrated while discussing our experience of designing a collaborative learning activity in a cultural history museum and a case study validating its usefulness.

Clearly, the future of learning technology in museums lies in the blending, not the separation, of the virtual and the real world. That is because learning in a museum context could be conceived as the integration, over time, of personal, socio-cultural, and physical contexts. The physical setting of the museum in which learning takes place mediates the personal and socio-cultural setting. The so called 'interface transparency' should be treated as an effort to seamlessly integrate the computational device to our natural environment. This goal could be achieved by augmenting physical space with information exchanges, allowing collaboration and communication, enhancing interactivity with the museum exhibits and by seamless integrating instantly available information delivered in various forms. However, the synergy between technology and pedagogy is not straightforward, especially if we take into account the need to tackle issues such as efficient context integration, transparent usage of PDA, and novel pedagogical approaches to exploit the capabilities of the mobile devices. Therefore, further research effort should take place to experience established methods and practices.

REFERENCES

Avouris, N., Komis, V., Fiotakis, G., Dimitracopoulou, A., & Margaritis, M. (2004). Method and Tools for analysis of collaborative problem-solving activities. In *Proceedings of ATIT2004, First International Workshop on Activity Theory Based Practical Methods for IT Design* (pp. 5-16). Retrieved on February 28, 2007 from http://www.daimi.au.dk/publications/PB/574/PB-574.pdf

Chou, L., Lee, C., Lee, M., & Chang, C. (2004). A tour guide system for mobile learning in museums. In J. Roschelle, T.W. Chan, Kinshuk & S. J. H. Yang (Eds.), *Proceedings of 2nd IEEE International Workshop on Wireless and Mobile Technologies in Education—WMTE'04* (pp. 195-196). Washington, DC: IEEE Computer Society.

Ciavarella, C., & Paterno, F. (2004). The design of a handheld, location-aware guide for indoor environments. *Personal Ubiquitous Computing, 8,* 82–91.

Cobb, P. (2002). Reasoning with tools and inscriptions. *The Journal of the Learning Sciences, 11*(2-3), 187-215.

Danesh, A., Inkpen, K.M., Lau, F., Shu, K., & Booth, K.S. (2001). Geney: Designing a collaborative activity for the palm handheld computer. In *Proceedings of the SIGCHI conference on Human Factors in Computing Systems—CHI 2001* (pp. 388-395). New York: ACM Press.

Dey, A. (2001). Understanding and using context. *Personal and Ubiquitous Computing Journal, 5*(1), 4-7.

Dillenbourg, P. (Ed.) (1999). *Collaborative learning: Cognitive and computational approaches.* Oxford, UK: Elsevier Science.

Dix, A., Rodden, T., Davies, N., Trevor, J., Friday, A., & Palfreyman, K. (2000). Exploiting space and location as a design framework for interactive mobile systems. *ACM Transactions on Computer-Human Interaction, 7*(3), 285–321.

Dyan, M. (2004). *An Introduction to Art, the Wireless Way.* Retrieved on March 25, 2005 from http://www.cooltown.com/cooltown/mpulse/1002-lasarsegall.asp

Evans, J., & Sterry, P. (1999). Portable computers and interactive multimedia: A new paradigm for interpreting museum collections. *Journal Archives and Museum Informatics, 13,* 113-126.

Feix, A., Göbel, S., & Zumack, R. (2004). Dino-Hunter: Platform for mobile edutainment applications in museums. In S. Göbel, U. Spierling, A. Hoffmann, I. Iurgel, O. Schneider, J. Dechau & A. Feix (Eds.), *Proceedings of the Second International Conference on Technologies for Interactive Digital Storytelling and Entertainment: Conference Proceedings—TIDSE 2004* (pp. 264-269). Berlin: Springer.

Fleck, M., Frid, M., Kindberg, T., Rajani, R., O'Brien-Strain, E., & Spasojevic, M. (2002). From informing to remembering: Deploying a ubiquitous system in an interactive science museum. *Pervasive Computing, 1*(2), 13-21.

Grinter, R. E., Aoki, P. M., Szymanski, M. H., Thornton, J. D., Woodruff, A., & Hurst, A. (2002). Revisiting the visit: understanding how technology can shape the museum visit. In *Proceedings of the 2002 ACM Conference on Computer Supported Cooperative Work—CSCW 2002,* (pp. 146-155). New York: ACM Press.

Hall, T., & Bannon, L. (2006). Designing ubiquitous computing to enhance children's learning in museums. *Journal of Computer Assisted Learning, 22,* 231–243.

Hayhoe, G. F. (2001). From desktop to palmtop: creating usable online documents for wireless and handheld devices. In *Proceedings of the IEEE International Conference on Professional Communication Conference–IPCC 2001* (pp. 1-11).

Klopfer, E., Perry, J., Squire, K., Jan, M., & Steinkuehler, C. (2005). Mystery at the museum: a collaborative game for museum education. In T. Koschmann, T. W. Chan & D. Suthers (Eds.), *Proceedings of the 2005 conference on Computer support for collaborative learning: the next 10 years!* (pp. 316-320). Mahwah, NJ: Lawrence Erlbaum.

Koshizuka, N., & Sakamura, K. (2000). The Tokyo University Museum. In *Kyoto International Conference on Digital Libraries: Research and Practice* (pp. 85-92).

Kuuti, K. (1995). Activity theory as a potential framework for human-computer interaction research. In B. Nardi (Ed.), *Context and consciousness: Activity theory and human computer interaction* (pp. 17-14). Cambridge: MIT Press.

Kwak, S.Y. (2004). *Designing a handheld interactive scavenger hunt game to enhance museum experience.* Unpublished diploma thesis, Michigan State University, Department of Telecommunication, Information Studies and Media.

Laurillau, Y., & Paternò, F. (2004). Supporting museum co-visits using mobile devices. In S. Brewster & M. Dunlop (Eds), *Proceedings of the 6th International Symposium on Mobile Human-Computer Interaction—Mobile HCI 2004* (pp 451-455). Berlin: Springer.

Lave, J., & Wenger, E. (1991). *Situated learning: Legitimate peripheral participation.* New York: Cambridge University Press.

Luchini, K., Quintana, C., Krajcik, J., Farah, C., Nandihalli, N., Reese, et al. (2002). Scaffolding in the small: Designing educational supports for concept mapping on handheld computers. In *CHI 2002 Extended Abstracts on Human Factors in Computing Systems* (pp. 792-793). New York: ACM Press.

Luyten, K., & Coninx, K. (2004). ImogI: Take control over a context aware electronic mobile guide for museums. *In proceedings of the 3rd Workshop on HCI in Mobile Guides.* Retrieved on February 24, 2007 from http://research.edm.luc.ac.be/~imogi/

Myers, B. A., Stiel, H., & Gargiulo, R. (1998). Collaboration using multiple PDAs connected to a PC. In *Proceedings of the ACM 1998 Conference on*

Computer Supported Cooperative Work—CSCW '98 (pp. 285-294). New York: ACM.

Mase, K., Sumi, Y., & Kadobayashi, R. (2000). The weaved reality: What context-aware interface agents bring about. In *Proceedings of the Fourth Asian Conference on Computer Vision - ACCV2000* (pp. 1120-1124).

Norman, D. A. (1986). Cognitive engineering. In D. A. Norman & S. W. Draper (Eds.), *User centered systems design* (pp. 31-61). Mahwah, NJ: Lawrence Erlbaum.

Norris, C., & Soloway, E. (2004). Envisioning the handheld-centric classroom. *Journal of Educational Computing Research, 30*(4), 281-294.

Oppermann, R., Specht, M., & Jaceniak, I. (1999). Hippie: A nomadic information system. In H. W. Gellersen (Ed.), *Proceedings of the First International Symposium Handheld and Ubiquitous Computing - HUC'99* (pp. 330-333). Berlin: Springer.

Patten, B., Arnedillo Sanchez, I., & Tangney, B. (2006). Designing collaborative, constructionist and contextual applications for handheld devices. *Computers and Education, 46,* 294-308.

Perry, D. (2003). *Handheld Computers (PDAs) in Schools.* BECTA ICT Research Report. Retrieved on February 26, 2007 from http://www.becta.org.uk/ page_documents/research/handhelds.pdf

Raptis, D., Tselios, N., & Avouris, N. (2005). Context-based design of mobile applications for museums: a survey of existing practices. In M. Tscheligi, R Bernhaupt & K. Mihalic (Eds.), *Proceedings of the 7th international Conference on Human Computer interaction with Mobile Devices & Services- Mobile HCI 2005* (pp. 153-160). New York: ACM Press.

Rieger, R., & Gay, G. (1997). Using mobile computing to enhance field study. In R.P. Hall, N. Miyake & N. Enyedy (Eds.), *Proceedings of Computer Support for Collaborative Learning –CSCL 1997* (pp. 215–223). Mahwah, NJ: Lawrence Erlbaum.

Rocchi, C., Stock, O., Zancanaro, M., Kruppa, M., & Krüger, A. (2004). The museum visit: Generating seamless personalized presentations on multiple devices. In J. Vanderdonckt, N. J. Nunes & C. Rich (Eds.), *Proceedings of the Intelligent User Interfaces - IUI 2004* (pp. 316-318). New York: ACM.

Roschelle, J. (2003). Unlocking the learning value of wireless mobile devices. *Journal of Computer Assisted Learning, 19*(3), 260-272.

Thom-Santelli, J., Toma, C., Boehner, K., & Gay, G. (2005). Beyond just the facts: Museum detective guides. In *Proceedings from the International Workshop on Re-Thinking Technology in Museums: Towards a New Understanding of People's Experience in Museums* (pp. 99-107). Retrieved on February 25, 2007 from http://www.idc.ul.ie/museumworkshop/programme.html

Vahey, P., & Crawford, V. (2002). *Palm education pioneers program final evaluation report.* Menlo Park, CA: SRI International.

Van Gool, L., Tuytelaars, T., & Pollefeys, M. (1999). Adventurous tourism for couch potatoes. (Invited). In F. Solina & A. Leonardis (Eds.), *Proceedings of the 8th International Conference on Computer Analysis of Images and Patterns – CAIP 1999* (pp. 98-107). Berlin: Springer.

Yatani, K., Sugimoto, M., & Kusunoki, F. (2004). Musex: A System for Supporting Children's Collaborative Learning in a Museum with PDAs. In J. Roschelle, T.W. Chan, Kinshuk & S. J. H. Yang (Eds.), *Proceedings of 2nd IEEE International Workshop on Wireless and Mobile Technologies in Education – WMTE'04* (pp 109-113). Washington, DC: IEEE Computer Society.

Zurita, G., & Nussbaum, M. (2004). Computer supported collaborative learning using wirelessly interconnected handheld computers. *Computers and Education, 42,* 289-314.

KEY TERMS

Activity Theory: Is a psychological framework, with its roots in Vygotsky's cultural-historical psychology. Its goal is to explain the mental capabilities of a single human being. However, it rejects the isolated human being as an adequate unit of analysis, focusing instead on cultural and technical mediation of human activity.

Context: Context is any information that can be used to characterize the situation of an entity. An entity is a person, place, or object that is considered relevant to the interaction between a user and an application, including the user and applications themselves (Dey, 2001).

Context-Aware: The ability to sense context.

Interaction Design: Interaction design is a sub-discipline of the design notion which aims to examine the role of embedded behaviors and intelligence in physical and virtual spaces as well as the convergence of physical and digital products. In particular, interaction design is concerned with a user experience flow through time and is typically informed by user research design with an emphasis on behavior as well as form. Interaction design is evaluated in terms of functionality, usability and emotional factors.

Mobile Device: A device which is typically characterized by mobility, small form factor and communication functionality and focuses on handling a particular type of information and related tasks. Typical devices could be a Smartphone or a PDA. Mobile devices may overlap in definition or are sometimes referred to as information appliances, wireless devices, handhelds or handheld devices.

Mobile Learning: Is the delivery of learning to students who are not keeping a fixed location or through the use of mobile or portable technology.

Museum Learning: A kind of informal learning which is not teacher mediated. It refers to how well a visit inspires and stimulates people into wanting to know more, as well as changing how they see themselves and their world both as an individual and as part of a community. It is a wide concept that can include not only the design and implementation of special events and teaching sessions, but also the planning and production of exhibitions and any other activity of the museum which can play an educational role.

Chapter XVII
Collaborative Learning in a Mobile Technology Supported Classroom

Siu Cheung Kong
The Hong Kong Institute of Education, Hong Kong

ABSTRACT

This chapter introduces the migration of a Web-based cognitive tool (CT) for the generation of procedural knowledge about mathematical fractions from a desktop version to a mobile version. It aims to provide insight into the potential of human-computer interaction in mobile learning environments to encourage reciprocal tutoring and foster collaborative learning. A collaborative mobile learning environment is designed using a design-based research approach. A Web-based CT for learning the concept of fraction equivalence is improved and modified to suit the environment as applied to a mobile technology supported classroom. This chapter first delineates the theoretical design approach and empirical design methodology that underlie the migration exercise, and then discusses the architectural design of artifacts and the pedagogical design of learning activities to shed light on the development and application of mobile technology in a classroom learning environment.

INTRODUCTION

Procedural knowledge is the knowledge that guides the performance of a task in the absence of access to the knowledge that underlies the procedure (Anderson, 1976). To acquire procedural knowledge about the operation of mathematical fractions, it is necessary to first have knowledge of fraction equivalence, which comprises the concept of fraction equivalence and knowledge of the computation of equivalent fractions, both of which are of equal importance (Kong & Kwok, 2005). Procedural knowledge of adding fractions with unlike denominators is more likely to be generated if a conceptual understanding of fraction equivalence is initially developed (Kong & Kwok, 2005).

Early research shows that learners seldom easily understand the procedural knowledge that is associated with fraction operations, such as addition and subtraction (Huinker, 1998; Niemi, 1996; Pitkethly & Hunting, 1996). Traditional classroom instruction in this topic generally adopts the algorithmic approach, which suffers from the shortcoming of separating knowledge from meaning. To rectify this problem, a Web-based cognitive tool (CT) was developed to assist learners to generate procedural knowledge of adding fractions with unlike denominators (Kong, 2001) with the rationale that CTs are both mental and computational devices that can support, guide, and mediate the cognitive processes of learners (Derry & Lajoie, 1993; Kommers, Jonassen & Mayes, 1992).

Previous evaluation studies (Kong & Kwok, 2002, 2005) show that the adoption of reciprocal tutoring in a collaborative learning environment has the potential to increase learning effectiveness in this domain. As the portable nature of mobile devices offers the opportunity to promote reciprocal tutoring in a mobile technology supported classroom, the desktop version of the Web-based CT for comprehending procedural knowledge of mathematical fractions is adapted to create a mobile version for collaborative learning.

DESIGN FRAMEWORK

The goal of cognitive technology is to develop CTs that meet the needs of human users (Janney, 1999; Pea, 1985). The capability of the aforementioned Web-based CT to assist learners to generate procedural knowledge of adding fractions with unlike denominators has been validated, and experimental studies have revealed that it serves as a mediator that triggers discussion among learners. Slavin (1996) states that a collaborative learning context, such as discussion, is an important way of stimulating reflection among learners. From the perspective of cognitive science, peer discussion is a way of stimulating cognitive elaboration (Wittrock, 1979). In light of these views, the aim of the study presented in this chapter is to further improve the CT to meet the needs of learners who are learning naturally in a classroom setting by applying the tool to a collaborative learning environment in a mobile technology supported classroom. This section outlines the design framework of the new CT, which, when used in a mobile learning environment, promotes collaborative engagement and encourages the resolution of cognitive conflict by cognitive elaboration and reciprocal tutoring in the classroom.

Cognitive Elaboration

Cognitive elaboration is the process of forming associations between new information and prior knowledge. It is regarded as an essential process for facilitating comprehension and knowledge acquisition (Wittrock, 1986). In cognitive models, learners play an active, responsible, and accountable role in their generative learning. Newly learned materials are better retained and more easily recalled if learners undergo spontaneous cognitive elaboration to trace the relations between the new information and known information, because cognitive elaboration helps the transfer of new information from the short-term to the long-term memory (Doherty, Hilberg, Pinal et al., 2003; Wittrock, 1979).

An effective means of fostering the capability of learners to cognitively elaborate is to offer them opportunities to practice cognitive elaboration. Reciprocal tutoring is a good strategy for encouraging learners to practice cognitive elaboration in which learners take turns to tutor each other in a group learning context (Chan & Chou, 1997; Wong, Chan et al., 2003). The strategy enables students to learn from one another through the verbal elaboration of the new knowledge in a group learning context, thus allowing students who have gained insight into the new concept to reinforce their knowledge by providing explanations to others who need more opportunity to comprehend the knowledge. In this way, all group members benefit from engaging in the elaboration process.

Cognitive Conflict

Understanding mathematical ideas often involves the restructuring of the mathematical schema of learners. This restructuring process is intricately linked with the occurrence of cognitive conflict (Tall, 1977). Cognitive conflict is a tenet of the psychological theory of cognitive change, and is an inferred state of incompatibility between two inferred component states within the cognitive process (Cantor, 1983). In general, cognitive conflict is a perceptual state in which one notices the discrepancy between an anomalous situation and a preconception (Lee, Kwon, Park, et al., 2003). Since the 1980s, the exploitation of cognitive conflict has been regarded as a feasible teaching strategy because the cognitive change that such conflict creates induces introspection among learners about the newly learned conception and its incongruity with their preconceptions or misconceptions.

According to the cognitive conflict process model, there are three stages in the engagement of cognitive conflict in learning (Lee et al., 2003). The first stage is the preliminary stage, where an anomalous situation that differs from the preconceptions of learners is introduced. The second stage is the conflict stage, in which learners recognize and reappraise the anomalous situation and express anxiety or interest in resolving the conflict. The third stage is the resolution stage, in which learners try to resolve cognitive conflict in any way that they can.

Cognitive conflict has constructive, destructive, and meaningless potential. When learners clearly recognize an anomaly and reappraise a situation of cognitive conflict deeply by expressing strong interest or anxiety, the cognitive conflict has constructive potential. When learners do not recognize the anomaly or simply ignore it and express feelings of frustration or rejection, the cognitive conflict is regarded as destructive. When learners recognize the anomaly, but accept it passively without interest or cognitive reappraisal, the cognitive conflict is regarded as meaningless. Early studies show that the inducement of constructive cognitive conflict promotes positive outcomes in classroom learning (Limón, 2001). However, the creation of constructive cognitive conflict largely depends on the interdependence of learners, and is thus closely related to collaborative learning processes that provide learners with ample opportunity to learn from peer discussion.

From the perspective of cognitive science, cognitive elaboration has a positive effect on cognitive conflict. In cognitive models, individual differences between learners are important, especially when they involve cognitive processes (Wittrock, 1979), as they can be a source of cognitive conflict. In a discussion during which cognitive conflict occurs, learners learn from one another through peer interaction, and are provoked to reflect through argument. Understanding then emerges through mutual elaboration of the new concept. The process of cognitive elaboration therefore induces cognitive conflict with a positive potential.

As the construction of knowledge of fraction equivalence is often accompanied by the occurrence of cognitive conflict (Kong & Kwok, 2002), this study aims to design a collaborative learning environment to help learners to generate procedural knowledge of adding fractions with unlike denominators in this learning context.

Collaborative Learning

Collaborative learning is a process that encourages learners to participate in coordinated and synchronous learning activities with a number of other learners (Roschelle & Teasley, 1995). It emphasizes the concept that "every learner learns from everyone else" (Fischer, Bruhn, Gräsel et al., 2002, p. 215), and promotes self-directed and active learning through group learning activities that require interdependence among group members. The basic assumption behind collaborative learning is that learners are ready to interact with one another to offer help or share ideas. According to this rationale, learners value and encourage group members in the learning process, and thus depend on one another to achieve effective learning through collaborative interaction with their peers (Johnson & Johnson, 1999; Slavin, 1996).

There are several characteristics of collaborative learning, such as sharing knowledge among peers, mediation by teachers, and the arrangement of learners into heterogeneous groups (Dillenbourg, 1999). In collaborative learning, learners take the role of knowledge provider by sharing their own knowledge and learning strategies with other group members, and teachers play the role of facilitator by providing mediation for group learning, such as adjusting the information flow or level of interaction among groups and group members. The heterogeneous grouping of learners is important in collaborative learning, because it allows reciprocal tutoring and knowledge exchange, which helps learners to develop knowledge and interpersonal communication skills. Although collaborative learning has been found to be effective for learners at all learning achievement levels (Slavin, 1996), the two obstacles of problems with class control during active participation and the unsatisfactory participation of particularly quiet learners can decrease the effectiveness of the learning process (Roschelle, 2003).

Mobile Learning

Mobile learning refers to the use of mobile technology for learning and teaching. It is an emergent learning approach that has the potential to address the aforementioned two obstacles to collaborative learning. Portability and versatility make mobile devices a powerful medium for learning (Sharples, Taylor & Vavoula, 2005), and a number of design-based studies have shown the prospect of mobile learning in education (Roschelle, 2003). Mobile technology has two attributes that facilitate the design of collaborative learning activities in a classroom environment.

First, the portability of mobile devices offers learners a sense of ownership of individual mobile devices. Mobile learning enables learners to hold the mobile devices during the entire lesson in the classroom. This sense of ownership helps to provide incentives to learners to actively participate in collaborative learning activities. In addition, with the use of non-verbal communication features, such as graphical support in CT, of the mobile devices at various stages of the learning process, the use of mobile devices may trigger all learners, including the quiet ones, to participate in the learning process. This increases the opportunities for the inducement of cognitive conflict and cognitive elaboration.

Second, the versatility of the currently-available mobile devices facilitates collaborative learning activities in a coordinated manner. This can be achieved by creating continuous learning process to cater for learning diversity, and providing flexibility for teachers to rearrange the groupings of learners. By programming the mobile technology supported collaborative learning environment, learners may continually receive questions generated with appropriate difficulty levels and constantly receive judgment on the correctness of learners' responses, teachers are thus prevented from the chaotic class order caused by learning diversity of learners in the active learning process. With appropriate programming effort, such as data mining of collaborative learning, teachers may obtain information from the system to regroup learners into heterogeneous groups. These supports allow teachers to have more time to manage the learning progress of learners.

DESIGN METHODOLOGY

This study adopts a design-based research approach to designing the collaborative mobile learning environment. Design-based research is a fundamental mode of scholarly inquiry that is useful in many academic fields that in the past decade has become an increasingly accepted approach to theoretical and empirical study in the field of education (Bell, 2004). Design-based research is an attempt to combine empirical educational research with the theory-driven design of learning environments (Bell, 2004; Hoadley, 2004; Design-Based Research Collective, 2003) to design and explore a whole range of innovations. The key components of design-based research include architectural design, such as the artifacts that are involved in learning activities, and pedagogical design, such as the structure and scaffolding of learning activities (Design-Based Research Collective, 2003).

The study presented in this chapter is motivated by the results of empirical educational research on the effectiveness of a CT for learning the concepts of mathematical fractions. A qualitative pilot evaluation study and a quasi-experimental evaluation study that investigated how learners develop a concept of mathematical fractions and acquire procedural knowledge of fraction operations by using a CT named the "Graphical Partitioning Model" (GPM) form the design basis of the mobile CT.

The CT for supporting the generation of procedural knowledge about fraction addition that is introduced in this chapter originates from a cognitive task analysis of the domain (Kong, 2001) that initially led to the design of the GPM CT, which is a rectangular bar with partitioning capability.

The pedagogical benefit of the GPM is that it reveals the procedural structure for evaluating fraction expressions by linking the concrete manipulation of further partitioning the fraction bars to find a common fractional unit with the meaning of finding a common denominator in the process of adding fractions with unlike denominators. To create descriptive meanings and features that enable the learning effectiveness of the CT to be evaluated, a qualitative pilot evaluation study was conducted (Kong & Kwok, 2002). The evaluation study aimed to investigate how learners use the GPM to develop a concept of fractions and to understand and acquire procedural knowledge of fraction operations.

In the pilot evaluation study, 12 subjects with various levels of mathematical ability were selected to use the CT to learn about the subject domain for five successive two-hour sessions. Learners used the CT to acquire knowledge of fraction equivalence in the first three sessions of the course, and to develop knowledge of adding fractions with unlike denominators in the last two sessions. Learners used the partitioning capability of the CT to explore the concept of fraction equivalence and the procedural knowledge of adding fractions.

The results of the study indicate that the GPM only benefited learners who were already able. These learners gained knowledge of fraction equivalence by working with the CT, and acquired procedural knowledge about adding fractions with unlike denominators. The investigation of the knowledge profile of the learners in developing the concept of fraction equivalence (including cognitive artifacts such as diagrams that were drawn by the learners in the attainment tests, worksheets, and post-test performance results) and observation of the initiative of learners in using equivalent fractions to add fractions with unlike denominators yielded three key findings. First, half of the learners did not understand the inverse relationship between the number of parts and the size of a part of a unit. Second, almost 60 percent of the learners showed no intention of representing fractions to compare their equivalence. Third, around 70 percent of the learners showed no ability to use equivalent fractions to add fractions with unlike denominators, although some understood the concept of fraction equivalence but were unable to relate the concept to ways of finding equivalent fractions. These findings reveal that to improve the effectiveness of the CT in helping learners to learn the procedural knowledge of adding fractions with unlike denominators, its capability to teach fraction equivalence must be enhanced.

In response to the three key findings of the pilot evaluation study, the Web-based CT was improved to make it into a model of affordances to support the learning of mathematical fractions. This enhanced version of the Web-based CT forms the basis of the mobile version of the CT that is introduced in this chapter. From the constructivist perspective, the means of instruction should not be predetermined, because each learner constructs knowledge in a unique way, and thus "the pedagogical role of the system is to provide profitable spaces for interaction to the learner based on some model of the affordances of potential situations" (Akhras & Self, 2000, p. 24). Gibson (1979) introduces the notion of affordances and suggests that the perceptual task of human beings is to detect environmental aids that could be used in their attempts to interact with the environment to meet their needs. The function of a model of affordances is to make available profitable spaces, or provide the necessary scaffolding (Clark, 1997). Therefore, to enhance the effectiveness of the CT to support the learning of mathematical fractions, a model of affordances for the teaching of fraction equivalence to help develop procedural knowledge about adding fractions with unlike denominators was designed to give learners the means to interact in a way that meets their needs.

The enhanced CT underwent three modifications to improve its ability to stimulate knowledge of fraction equivalence in each learner according to that learner's particular needs (Kong & Kwok, 2003). The first modification was the addition of a space for partitioning that allows a choice to be made between an intentional slowed down animation that shows the partitioning or regrouping process and an instantaneous change that shows the results of the partitioning or regrouping process. The simulation of the partitioning strategy by the slowed animation addresses the lack of intention of representing fractions to compare their equivalence and the failure to recognize the inverse relationship between number of parts and the size of a part of a unit that were identified in the evaluation study. This modification allows learners to interact with the CT according to their needs, with capable learners being able to generalize the knowledge by rapidly calling up the results of partitioning and less capable learners being able to pick up the idea by activating the slowed down animation of the process of partitioning.

The second modification was the addition of a space for the comparison of the equivalence of fractions in response to the difficulties that learners who have no intention of representing fractions with the same unit to compare their equivalence encounter. An animation that shows the direct comparison of the equivalence of two fraction bars, which is triggered by dragging a fraction bar and dropping it onto another bar, was designed to allow an extra comparison of fraction equivalence in addition to the visual inspection of two separate fraction bars. This modification gives learners multiple opportunities to compare fraction equivalence in an interactive way.

The third modification was the addition of a space that consists of a hypothesis-testing interface $\frac{a}{b} = \frac{a \times c}{b \times d}$ to address the problem of learners who lack the ability to find equivalent fractions systematically. The hypothesis-testing interface asks learners to test possible fraction equivalent states by adjusting parameters c and d, and allows them to compare fraction equivalence using the aforementioned comparison animation by dragging the fraction bar of the hypothesized fraction and dropping it onto the fraction bar of the original fraction.

The enhanced CT was evaluated by a quasi-experimental evaluation study with a pre-test/post-test control group design (Kong & Kwok, 2005). The results of this evaluation study indicate that the model of affordances allows learners of varying learning abilities to develop a concept of fraction equivalence. It was observed that with the mediation of the enhanced CT, learners were able to generate procedural knowledge about adding fractions with unlike denominators.

Using the empirical findings on the effectiveness of the model of affordances for teaching fraction equivalence and the theoretical background that is discussed in the previous sections, in the following sections we discuss the architectural and pedagogical design of a collaborative learning environment in a mobile technology supported classroom.

DESIGN OF A COLLABORATIVE LEARNING ENVIRONMENT FOR COGNITIVE ENGAGEMENT

A Mobile Technology Supported Classroom

Collaboration is a coordinated, synchronous activity in which continued attempts are made to construct and maintain a shared conception of a problem (Roschelle & Teasley, 1995). In the study introduced in this chapter, a series of synchronous interactions in a mobile technology supported classroom is designed to encourage learners to engage in learning tasks, and a mobile platform is established that allows immediate interaction between learners working in pairs. This pair-wise grouping aims to induce in-depth discussion between group members. Figure 1 depicts the mobile technology supported classroom.

The learning activities take place in a wireless-networked classroom, and the mobile device that is used is a pocket PC, which is chosen for its

Figure 1. Mobile technology supported classroom for learning fraction equivalence in pairs

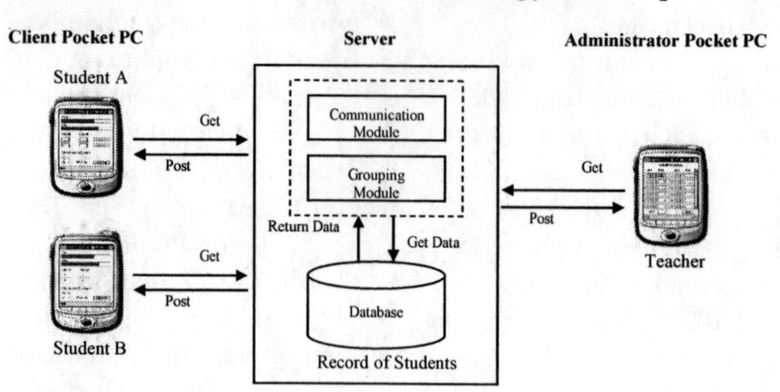

portability and relatively large screen. The teacher and learners are provided with pocket PCs. The teacher's pocket PC is installed with an interface for managing the pairs and for organizing the learning activities. The pocket PCs that are used by the learners are installed with a graphical tool for learning fraction equivalence. The learners interact in pairs through a server that is connected to an SQL database. The server coordinates the grouping instructions of the teacher and the synchronous interactions between paired learners.

A Model of Affordances for Teaching Fraction Equivalence

Knowledge of fraction equivalence is a prerequisite for the development of procedural knowledge of the addition of fractions with unlike denominators. Therefore, collaborative learning activities with graphical support are designed to develop the understanding of learners in this domain. In the learning activities, a graphical model of a rectangular bar is used to represent fractions (Kong & Kwok, 2003), with each fraction being represented by displaying shaded fractional parts of an equally partitioned rectangular bar according to the value of the fraction. The three spaces of the model of affordances that will be discussed are adapted from the desktop version of the Web-based CT.

First Space: Learner-Controlled Animation of Partitioning

The cognitive artifact of the first space is an animation that is adapted from the graphical partitioning capability of the Web-based CT and is designed to address the lack of understanding among learners of the part-whole concept and the inverse relationship between the number of parts in a unit and the size of a part. Figures 2a to 2d demonstrate this feature. The learner-controlled animation allows learners to partition fractions by clicking the graphical representation of the fractions. When learners click on the graphical representation (see Figure 2a), the fraction bar becomes blank to show the initial state of the fraction as a unit (see Figure 2b). When learners click on the blank rectangular bar, the bar is partitioned into fractional units (see Figure 2c). This allows learners to develop a concept of fractions as parts of a whole. When learners click on the calibrated fraction bar, the parts of a unit are shaded based on the value of the numerator (see Figure 2d). This feature helps learners build up the concept that the parts of the whole are equal, and to understand the inverse relationship between the number of parts and the size of a part of a unit, that is, that the larger the denominator, the smaller the size of a part in the unit.

Figure 2.

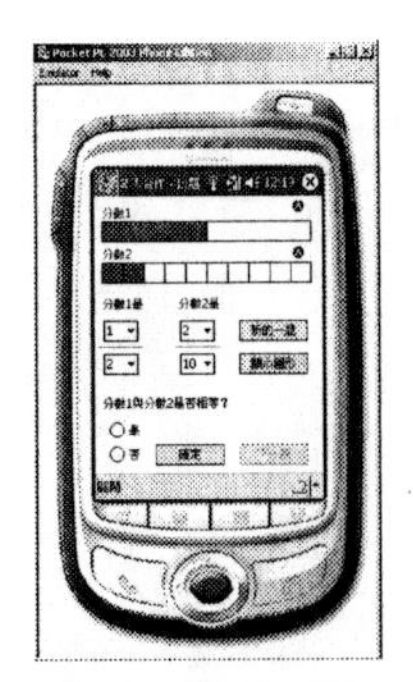

Figure 2a: Graphical representations of the display of two fractions on the interface.

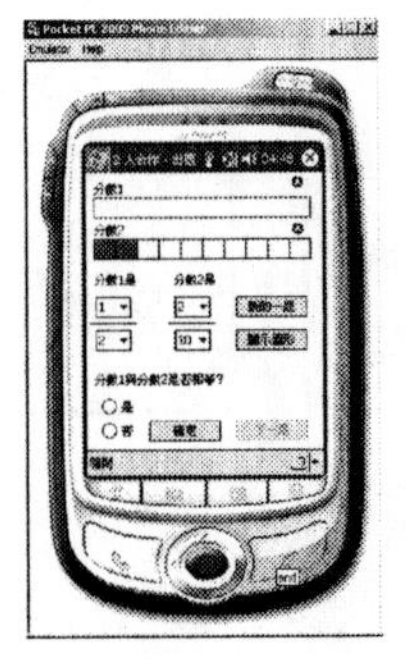

Figure 2b: A blank rectangular bar is shown when a learner clicks on a fraction bar.

Figure 2c: A calibrated fraction bar is shown when a learner clicks on the blank fraction bar.

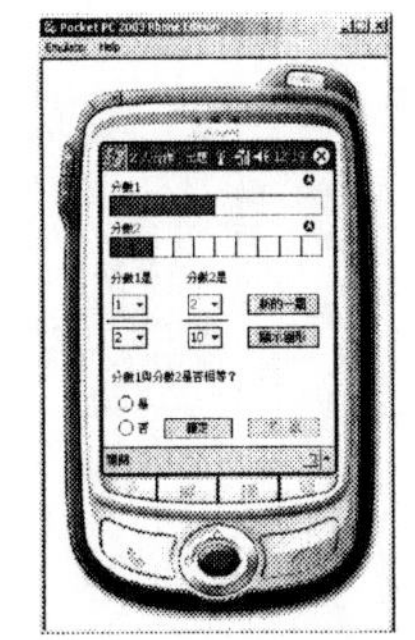

Figure 2d: The original fraction bar with shaded fractional part is shown when a learner clicks on the calibrated fraction bar.

This cognitive artifact is incorporated in all of the fraction bars that appear in the learning activities that involve graphical support. When learners work with only one fraction bar, the stepwise design helps them to develop the part-whole concept, and when they work with both fraction bars, the stepwise design helps them to understand the inverse relationship between the number of parts in a unit and the size of a part. This design is better than that of the desktop version in that it returns control of the learning process to the learners.

Second Space: Comparison of the Equivalence of Fractions

The mobile tool has two cognitive artifacts that are designed to address the difficulties that learners have in comparing the equivalence of fractions. The first is an animation that compares the equivalence of fractions that are represented in visual form, and is illustrated in Figure 3.

Button "A" in the top-right corner starts the comparison animation. Learners can press this button in order to obtain a graphical representation of a fraction, the bar of the selected fraction rolls over the other fraction bar to allow learners to compare the equivalence of the two fractions visually. This feature differs from the desktop version only in the positioning of the two fraction bars for comparison, in that the fraction bars in the desktop version are arranged in a row, whereas those in the mobile version are arranged in a column because of the relatively narrower screen of the mobile device.

The second artifact is the random display of fraction bars with non-comparable graphical representations that is given when learners are asked to compare fractions in the learning activities. An additional feature of an adjustable fraction bar is included to strengthen the concept of representing fractions using common units to compare their equivalence. Figures 4a to 4c demonstrate the adjustable fraction bar for representing fractions.

To compare the equivalence of two fractions, learners must represent fractions using the same unit. An adjustable bar that is 50 percent to 70 percent of the length of the bar of the other fraction is displayed. Button "B" in the top-right corner of the adjustable fraction bar adjusts the length of the bar, and elongates it to the length of the other fraction bar (see Figure 4b) when clicked by a learner. If the learner clicks button "B" again, then the adjustable fraction bar is shortened back to its original length (see Figure 4c). This feature acts as a random alert to assist learners to develop the awareness of the prerequisite of comparing fraction equivalence. There is a similar feature in the desktop version of the CT.

Figure 3. Animation for comparing the equivalence of two fractions

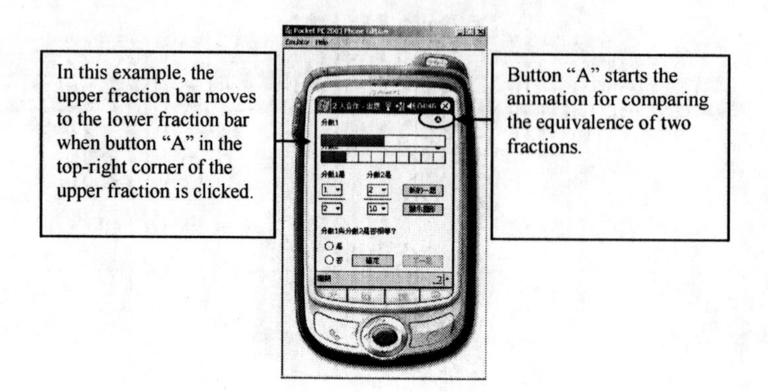
In this example, the upper fraction bar moves to the lower fraction bar when button "A" in the top-right corner of the upper fraction is clicked.
Button "A" starts the animation for comparing the equivalence of two fractions.

Figure 4.

Figure 4a: The lower fraction bar in this example is the adjustable fraction bar. Its length is 50% to 70% that of the upper fraction bar.

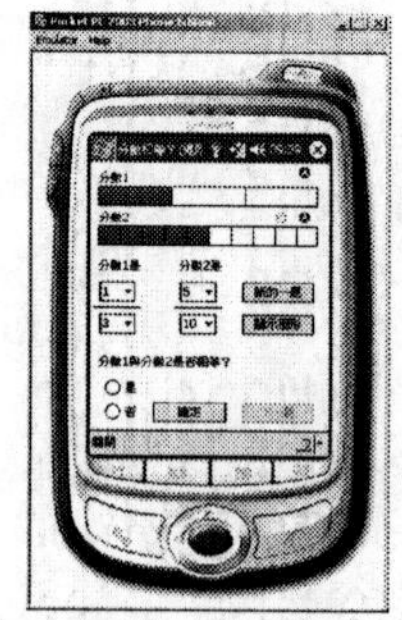

Figure 4b: The adjustable fraction bar elongates to the length of the upper fraction bar when button "B" in the top-right corner of the adjustable fraction bar is clicked.

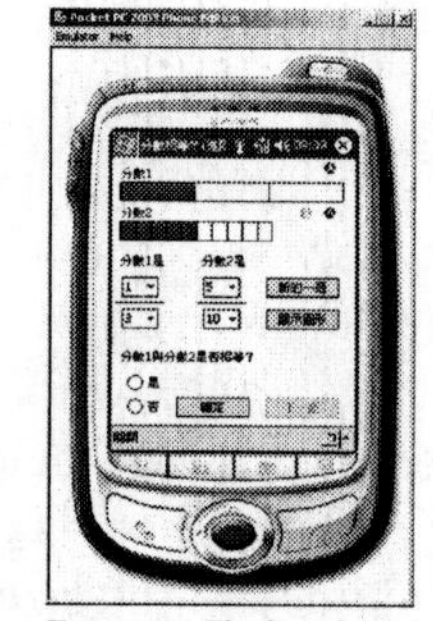

Figure 4c: The lengthened adjustable fraction bar reduces to its original length when button "B" in the top-right corner of the adjustable fraction bar is clicked again.

Third Space: Time-Keeping Hypothesis-Testing Competition

The third space of the model of affordances comprises a hypothesis-testing interface for finding equivalent fractions. This cognitive artifact is designed to assist learners to develop the capability to compute equivalent fractions. Knowledge of fractions, as aforementioned, consists of two parts: the concept of fraction equivalence and knowledge of the computation of equivalent fractions. To develop knowledge of the computation of equivalent fractions, a time-keeping learning activity is designed in which each pair of learners is engaged in a competition to find an equivalent fraction of a fraction that is assigned by the computer system.

This learning activity involves two steps. Step 1 is the process of finding the equivalent fraction. In this step, learners require to find an equivalent fraction with the use of a hypothesis-testing interface $\frac{a}{b} = \frac{a \times c}{b \times d}$. Learners are asked to adjust parameters c and d to test a possible fraction equivalent state. Graphical representations are generated to help learners to compare the equivalence of the two fractions by the instant change that takes place following the adjustment of parameters c and d. Once learners have decided on their answer, then they can click the "Confirm" button to send the answer to the server.

Step 2 is the process of judgment. In this step, the computer system measures the response time and judges the correctness of the answers that are provided by learners. For quick responses that are correct the words "Correct" and "Yeah!" are displayed on the screen (see Figure 5a); for slow responses that are correct, the words "Correct" and "Cheer up!" are displayed (see Figure 5b); and for incorrect answers the words "Incorrect" and "Cheer up!" (see Figure 5c) are displayed regardless of the response time. This learning activity offers learners the opportunity to engage in cognitive elaboration to find out the algorithm for the computation of equivalent fractions, and the use of peer competition helps learners to relate the concept of fraction equivalence to the algorithm for the computation of equivalent fractions. There is a similar interface in the desktop version of the CT, but the interface in the mobile version is improved by the use of competition between learners to encourage cognitive elaboration.

Pedagogical Design: Two Situations for Reflection and Cognitive Elaboration

In this section, we discuss how the pedagogical design of the collaborative learning environment stimulates cognitive conflict and cognitive elaboration. In a learning activity that involves understanding the concept of fraction equivalence, learners are placed in a situation in which they have to decide the equivalence of two fractions. The learners are arranged in pairs: one learner is the question-setter and the other the respondent. The learners alternate between the two roles.

The learning activity comprises three steps. Step 1 is the process of question-setting, in which the learners who are playing the role of question-setter set and send out questions about the equivalence of two fraction expressions. The question-setters have to state whether the two fraction expressions that they have chosen are equivalent by graphically representing the two fraction expressions at the top of the interface. Once the learners are satisfied with the question that they have set, they can click the "Confirm" button to send the question to their partners through the server. Step 2 is the response process. In this step, the learners who are playing the role of respondent receive the question in the form of two fraction expressions from their partner, and then have to decide whether the two expressions are equivalent with the help of visual representation. After indicating their decision, learners click the "Confirm" button to send their answer to the server. Step 3 is the process of judgment. In this step, the computer system assesses the correctness of the questions that are set by the question-setters and the answers that are provided by the respondents. The computer system then sends the messages in the form of the words "Correct" and "Incorrect" for right and wrong questions or answers, respectively.

Figure 5.

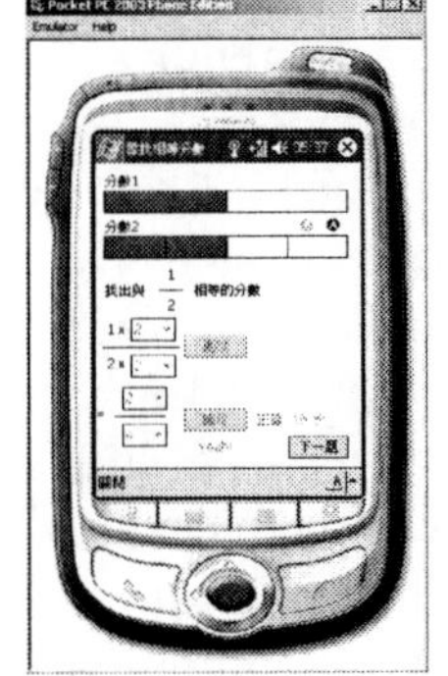

Figure 5a: The words "Correct" and "Yeah!" are displayed for correct and quick responses.

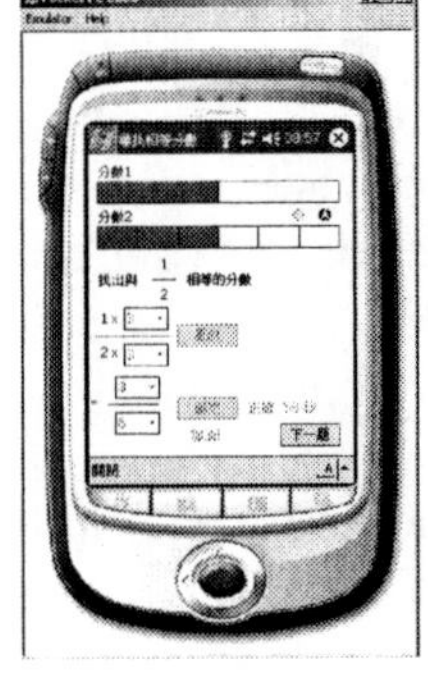

Figure 5b: The words "Correct" and "Cheer up!" are displayed for correct but slow responses.

Figure 5c: The words "Incorrect" and "Cheer up!" are displayed for incorrect response regardless of the response time.

Figure 6.

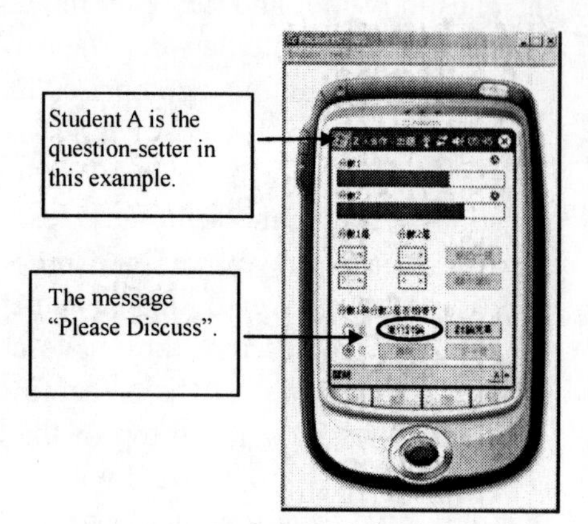
Student A is the question-setter in this example.
The message "Please Discuss".

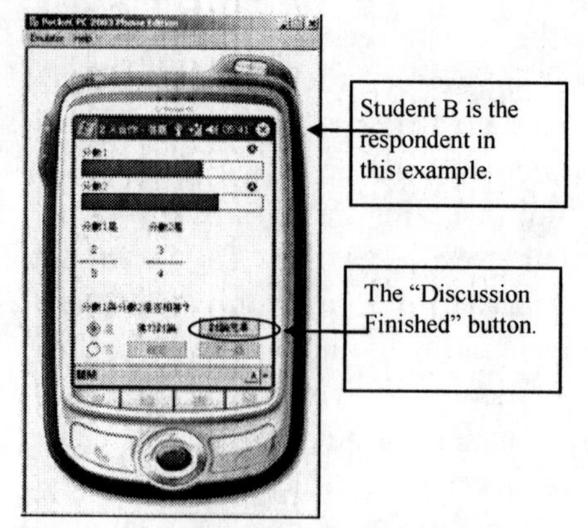
Student B is the respondent in this example.
The "Discussion Finished" button.

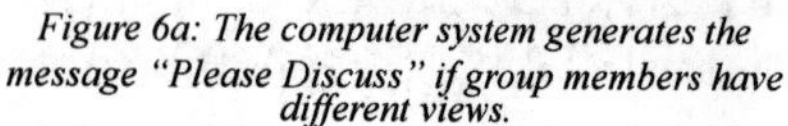

Figure 6a: The computer system generates the message "Please Discuss" if group members have different views.

Figure 6b: Each learner must click the "Discussion Finished" button after the discussion has ended.

Two types of cognitive conflict may be engendered by the learning activity: one is triggered by anomalies between learning peers and the other is triggered by anomalies between learners and the computer system. When one member of a pair of learners provides the correct question or answer and the other gives the wrong question or answer, then the computer system displays the message "Please Discuss" (see Figure 6a). This generates the first type of cognitive conflict and invites learners to share their understanding and to engage in self-reflection and negotiation through collaborative interaction.

When both members agree that they have finished their discussion, they click the "Discussion Finished" button (see Figure 6b) to inform the computer system. The computer system then tells the question-setter and respondent whether they are "Correct" or "Incorrect" (see Figures 7a and 7b). When they differ from the judgments of the learners, these authority judgments create the second type of cognitive conflict, which offers learners a second opportunity to engage in self-reflection and to share their understanding through a post-task discussion.

The aim of the learning activity is to equip learners with a basic knowledge of fraction equivalence through collaborative learning in a mobile technology supported classroom. The activity emphasizes the sharing of knowledge among learners using graphical support to aid the learning process. The different rates of progress of individual learners determine the learning progress of each group. Figure 8 depicts the different statuses of groups of learners in the process of learning the concept of fraction equivalence.

Case 1 is expected to occur commonly at the beginning of the learning process. In this case, both members of the group have a preconception or misconception about the equivalence of fractions, and always set and reply to questions incorrectly. In this situation, both learners encounter the first and second type of cognitive conflict, and such groups can be categorized as being at learning status 1. Cases 2 and 3 occur when one of the group members begins to grasp the concept of fraction equivalence better than his or her partner. The learner who has understood the concept of fraction equivalence begins to set correct questions and give correct answers, whereas his or her counterpart cannot always achieve this status. Both learners in this situation experience the first type of cognitive conflict, and in addition the learner who has yet to understand the concept of fraction equivalence experiences the second type of cognitive conflict, in which an anomalous situation exists between the learner and the judgment of the computer system. Such

Figure 7.

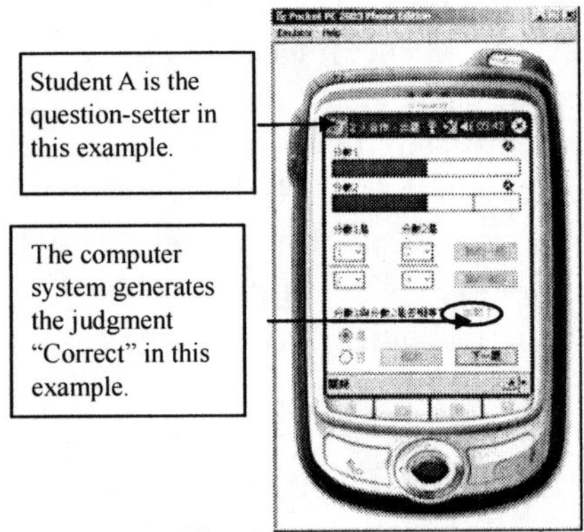

Figure 7a: The computer system generates the message "Correct" if a right question is set.

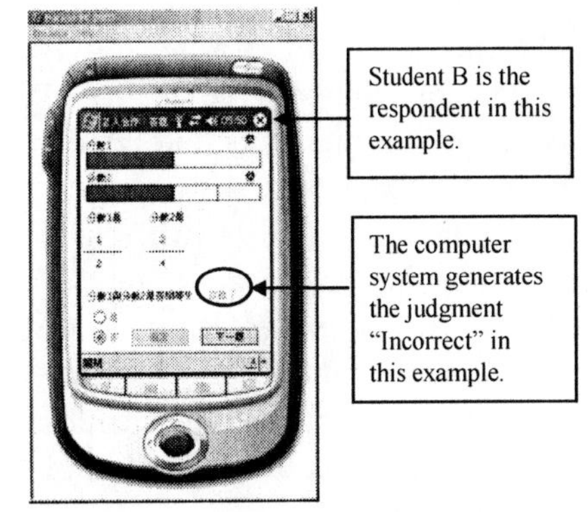

Figure 7b: The computer system generates the message "Incorrect" for a wrong answer.

groups can be categorized as being at learning status 2. Case 4 occurs when both learners in the group have a good understanding of the concept of fraction equivalence and always set and reply to questions correctly. In this case, cognitive conflict rarely occurs, and groups in this situation achieve learning status 3, which is the learning goal of all of the groups. Some groups may go through learning status 1 and 2 to reach learning status 3, and some may go directly from learning status 1 to status 3. The groups in learning status 2 are heterogeneous groupings in this study.

Pedagogical Design: Encouraging Reciprocal Tutoring

The goal of the learning activities that are designed to teach fraction equivalence is to help all of the groups of learners to attain learning status 3 through a collaborative learning environment. The groups in learning status 1 and 2 require the attention and mediation of teachers to promote productive knowledge sharing. In this collaborative learning environment, teachers play the role of mediator, rather than the authority that judges

Figure 8. Status of groups of learners in the process of learning the concept of fraction equivalence

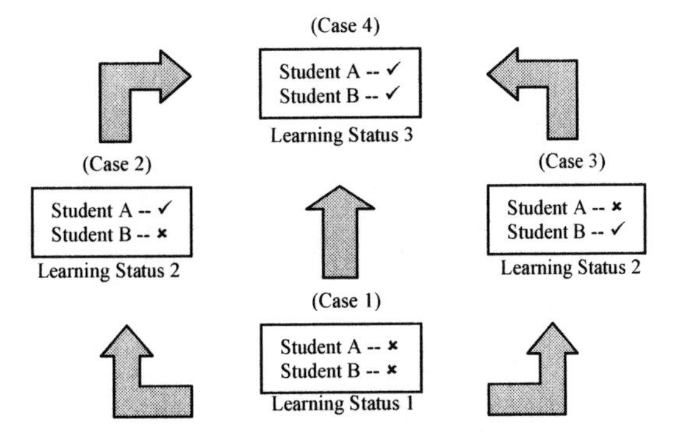

the correctness of answers, and serve to encourage reciprocal tutoring, which can be achieved more productively in a heterogeneous group context and by the promotion of tutoring activities. Reciprocal tutoring enhances the quality of arguments between group members and results in the inducement of more constructive cognitive conflict. Two pedagogical tools are therefore designed to encourage the reciprocal tutoring of learners. The first is the re-grouping of learners and the second is the alteration of the question-setting mode. Figure 9 shows the teacher's interface for the re-grouping of learners and the changing of the question-setting mode.

If teachers observe groups that are stuck at learning status 1 (in which both group members are struggling with a concept), then they can use the first pedagogical tool to swap one of the group members with a member from a group at learning status 3. This helps to achieve more heterogeneous groups, which in turn helps to encourage prolific reciprocal tutoring.

For groups at learning status 2 (in which one of the group members consistently designs incorrect questions or provides incorrect answers), teachers can use the second pedagogical tool to designate another learner as the sole question-setter by changing the mode of question-setting from "Turn-Taking" to "Designation." This creates an environment that allows learners with a better understanding to tutor learners who are still developing the required concept.

Once the teacher believes that all of the groups have reached learning status 3 (in which learners have a good understanding of the concept of fraction equivalence) then the mode of question-setting can be changed to "Random" for the entire class, which means that the role of question-setter is assigned randomly by the computer system. The "Random" mode provides learners with the opportunity to explore the concept further in a relaxed mode of inquiry, which helps to consolidate the learning outcomes.

Generating Procedural Knowledge of Adding Fractions with Unlike Denominators

The ultimate goal of the collaborative mobile learning environment is to support learners to generate procedural knowledge of adding fractions with unlike denominators. With a solid knowledge of fraction equivalence, learners will enter a proximal zone in which they can develop the procedural knowledge of adding fractions with unlike denominators.

The aforementioned three spaces of the model of affordances equip learners with a comprehensive understanding of fraction equivalence. The mobile version of the CT serves as a platform to stimulate the cognitive elaboration of learners to help them to derive the algorithm of adding fractions with unlike denominators. By putting learners through

Figure 9. Teacher's interface for the re-grouping of learners and the changing of the question-setting mode

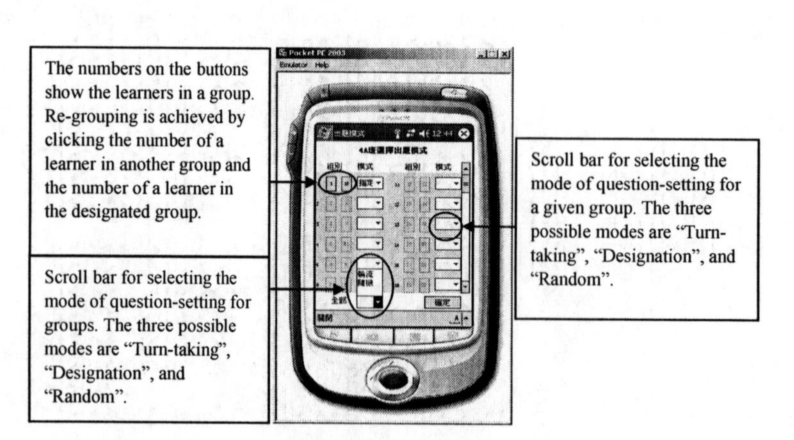

the steps of expanding the designated fractions to reach a common denominator and then adding the two expanded fractions with the support of graphical representation, the CT fosters learners to associate learning how to add fractions with the known concept of fraction equivalence.

CONCLUSION

The desktop version of a Web-based cognitive tool (CT) that supports the acquisition of procedural knowledge of mathematical fractions is migrated to a mobile version to increase its learning effectiveness by taking advantage of collaborative learning and mobile learning. Using the design-based research approach, a theory-driven design of a collaborative mobile learning environment is established based on two empirical evaluation research studies.

The mobile version CT is designed as a model of affordances for learning about fraction equivalence with three spaces that offer learners profitable learning opportunities. The first space comprises a learner-controlled animation of partitioning to help learners to develop the part-whole concept and understand the inverse relationship between the number of parts and the size of a part of a unit. The second space comprises an animation that compares the equivalence of fractions and features a random display of fraction bars in non-comparable representations to help learners to represent fractions in comparable forms to determine their equivalence. The third space comprises the hypothesis-testing interface $\frac{a}{b} = \frac{a \times c}{b \times d}$ for finding equivalent fractions, which helps learners to develop the capability to compute equivalent fractions by adjusting parameters *c* and *d*.

The designed learning environment offers profitable opportunities for learners to share knowledge with their peers, provides pedagogical designs for teachers to mediate cognitive elaboration, and allows teachers to organize learners into heterogeneous groups. Two pedagogical designs are suggested to realize these pedagogical benefits. The first aims to promote situations in which reflection and cognitive elaboration in the collaborative mobile learning environment take place, and involves grouping learners into pairs to engage in in-depth discussion about the learning activities. Learners are encouraged to learn together both as peers through tasks that are designed for collaborative engagement, and as partners through friendly competition to trigger in-depth cognitive elaboration.

Reciprocal tutoring is a key strategy for helping learners resolve cognitive conflict, and thus the second pedagogical design aims to promote an environment for reciprocal tutoring in a mobile technology supported classroom. Teachers in the designed learning environment act as facilitators to mediate and promote the sharing of knowledge among the class, and can reorganize learning pairs using the mobile device to promote reciprocal tutoring.

The architectural design of the artifacts and the pedagogical design of the learning activities enable learners to develop a concept of fraction equivalence using a collaborative interactive approach. With a solid knowledge of fraction equivalence, learners will enter a proximal zone in which they can develop and generate procedural knowledge of adding fractions with unlike denominators.

The use of mobile technology to encourage collaborative learning is a promising research direction that deserves further study, especially in terms of its effect on classroom learning environments. We have begun a case study of learning the concept of fraction equivalence in this collaborative learning environment, and further large-scale studies to investigate whether learners recognize and reappraise anomalies and the way in which they attempt to resolve cognitive conflict in the learning process will be attempted after the completion of the pilot case study.

REFERENCES

Akhras, F. N., & Self, J. A. (2000). Modeling the process, not the product, of learning. In S. P. Lajoie (Ed), *Computers as cognitive tools, volume two: no more walls* (pp. 3-28). Mahwah, NJ: Lawrence Erlbaum Associates.

Anderson, J. R. (1976). *Language, memory, and thought.* Hillsdale, NJ: Erlbaum.

Bell, P. (2004). On the theoretical breadth of design-based research in education. *Educational Psychologist, 39*(4), 243-253.

Cantor, G. N. (1983). Conflict, learning, and Piaget: comments on Zimmerman and Bloom's "Toward an empirical test of the role of cognitive conflict in learning." *Developmental Review, 3*, 39-53.

Chan, T. W., & Chou, C. Y. (1997). Exploring the design of computer supports for reciprocal tutoring. *International Journal of Artificial Intelligence in Education, 8*, 1-29.

Clark, A. (1997). *Being there: Putting brain and world together again.* Cambridge, MA: MIT press.

Derry, S. J., & Lajoie, S. P. (1993). A middle camp for (un)intelligent instructional computing: an introduction. In S. P. Lajoie & S. J. Derry (Eds), *Computers as cognitive tools* (pp. 1-11). Hillsdale, NJ: Lawrence Erlbaum Associates.

Design-Based Research Collective. (2003). Design-based research: An emerging paradigm for educational inquiry. *Educational Researcher, 32*(1), 5-8.

Dillenbourg, P. (1999). Introduction: What do you mean by "collaborative learning?" In P. Dillenbourg (Ed), *Collaborative learning: cognitive and computational approaches* (pp. 1-19). Amsterdam: Pergamon.

Doherty, R. W., Hilberg, R. S., Pinal, A., & Tharp, R. G. (2003). Five standards and student achievement. *NABE Journal of Research and Practice, 1*(1), 1-24.

Fischer, F., Bruhn, J., Gräsel, C., & Mandl, H. (2002). Fostering collaborative knowledge construction with visualization tools. *Learning and Instruction, 12*, 213-232.

Gibson, J. J. (1979). *The ecological approach to visual perception.* Boston: Houghton Mifflin.

Hoadley, C. M. (2004). Methodological alignment in design-based research. *Educational Psychologist, 39*(4), 203-212.

Huinker, D. (1998). Letting fraction algorithms emerge through problem solving. In L. J. Morrow & M. J. Kenney (Eds), *The teaching and learning of algorithms in school mathematics* (pp. 170-182). Reston, VA: NCTM.

Janney, R. W. (1999). Computers and psychosis. In J. P. Marsh, B. Gorayska & J. L. Mey (Eds.), *Humane interfaces: questions of method and practice in cognitive technology* (pp. 71-79). Amsterdam: Elsevier Science.

Johnson, D. W., & Johnson, R. T. (1999). *Learning together and alone: cooperative, competitive and individualistic learning (5th Ed.).* Boston: Allyn & Bacon.

Kommers, P. A. M., Jonassen, D. H., & Mayes, J. T. (1992). *Cognitive tools for learning.* Heidelberg: Springer-Verlag.

Kong, S. C. (2001). Interactive Web-based learning tools for knowledge construction on subject-specific contents. In K. S. Volk, W. M. So & P. T. Gregory (Eds.), *Science & Technology Education Conference 2000 Proceedings* (pp. 374-386). Hong Kong: Science & Technology Education Conference.

Kong, S. C., & Kwok, L. F. (2002). Modeling a cognitive tool for teaching the addition/subtraction of common fractions. *International Journal of Cognition and Technology, 1*(2), 327-352.

Kong, S. C., & Kwok, L. F. (2003). A graphical partitioning model for learning common fractions: designing affordances on a web-supported learning environment. *Computers and Education, 40*(2), 137-155.

Kong, S. C., & Kwok, L. F. (2005). A cognitive tool for teaching the addition/subtraction of common fractions: a model of affordances. *Computers and Education, 45*(2), 245-265.

Lee, G. H., Kwon, J. S., Park, S. S., Kim, J. W., Kwon, H. G., & Park, H. K. (2003). Development of an instrument for measuring cognitive conflict in secondary-level science classes. *Journal of Research in Science Teaching, 40*(6), 585-603.

Limón, M. (2001). On the cognitive conflict as an instructional strategy for conceptual change: a critical appraisal. *Learning and Instruction, 11*, 357-380.

Niemi, D. (1996). Assessing conceptual understanding in mathematics: Representations, problem solutions, justifications, and explanations. *Journal of Educational Research, 89*(6), 351-363.

Pea, R. D. (1985). Beyond amplification: using the computer to reorganize mental functioning. *Educational Psychologist, 20*(4), 167-182.

Pitkethly, A., & Hunting, R. (1996). A review of recent research in the area of initial fraction concepts. *Educational Studies in Mathematics, 30*, 5-38.

Roschelle, J. (2003). Keynote paper: Unlocking the learning value of wireless mobile devices. *Journal of Computer Assisted Learning, 19*, 260-272.

Roschelle, J., & Teasley, S. D. (1995). The construction of shared knowledge in collaborative problem solving. In C. O'Malley (Ed.), *Computer-supported collaborative learning* (pp. 145-168). Berlin: Springer-Verlag.

Sharples, M., Taylor, J., & Vavoula, G. (2005). Towards a theory of mobile learning. *Proceedings of the mLearn 2005 Conference* (pp. 9). Cape Town: mLearn 2005.

Slavin, R.E. (1996). Research on cooperative learning and achievement: What we know, what we need to know. *Contemporary Educational Psychology, 21*(1), 43-69.

Tall, D. (1977). Conflicts and catastrophes in the learning of mathematics. *Mathematical Education for Teaching, 2*(4), 2-18.

Wittrock, M. C. (1979). The cognitive movement in instruction. *Educational Researcher, 8*(2), 5-11.

Wittrock, M. C. (1986). Students' thought processes. In M. C. Wittrock (Ed.), *Handbook of research on teaching* (pp. 297-314). NY: Macmillan Publishing Co.

Wong, W. K., Chan, T. W., Chou, C. Y., Heh, J. S., & Tung, S. H. (2003). Reciprocal tutoring using cognitive tools. *Journal of Computer Assisted Learning, 19*, 416-428.

KEY TERMS

Affordances: Making available profitable spaces in which learners can interact in ways that meet their needs.

Cognitive Conflict: A perceptual state in which one notices the discrepancy between an anomalous situation and a preconception.

Cognitive Elaboration: The process of forming associations between new information and prior knowledge.

Collaborative Learning: A process that encourages learners to participate in coordinated and synchronous learning activities with a number of other learners.

Cognitive Tools: Mental and computational devices that can support, guide, and mediate the cognitive processes of learners.

Mobile Learning: The use of mobile technology for learning and teaching.

Procedural Knowledge: Knowledge that guides the performance of certain tasks in the absence of the knowledge that underlies the performance of the procedure.

Chapter XVIII
Design of an Adaptive Mobile Learning Management System

Hyungsung Park
Korea National University of Education, Korea

Young Kyun Baek
Korea National University of Education, Korea

David Gibson
The University of Vermont, USA

ABSTRACT

This chapter introduces the application of an artificial intelligence technique to a mobile educational device in order to provide a learning management system platform that is adaptive to students' learning styles. The key concepts of the adaptive mobile learning management system (AM-LMS) platform are outlined and explained. The AM-LMS provides an adaptive environment that continually sets a mobile device's use of remote learning resources to the needs and requirements of individual learners. The platform identifies a user's learning style based on an analysis tool provided by Felder & Soloman (2005) and updates the profile as the learner engages with e-learning content. A novel computational mechanism continuously provides interfaces specific to the user's learning style and supports unique user interactions. The platform's interfaces include strategies for learning activities, contents, menus, and supporting functions for learning through a mobile device.

INTRODUCTION

The rapid advancement of the global information infrastructure, mobile informational technologies, and intelligent applications is leading to a change of educational paradigm. The new paradigm is evolving similar to the way that "distance learning" evolved into "e-learning." Now e-learning is changing into m-learning (mobile learning) and is providing new possibilities for education.

Among those possibilities are increased ability to promote student motivation through personalization and a change from teacher-centered teaching to learner-centered learning. Adaptive m-learning can support these possibilities by considering and using a learner's diverse variables, such as abilities, attitudes, and learning styles, to promote effective learning and place the learner at the center of a more personalized experience.

Every teacher has witnessed how some students prefer visual information while others are surprised and perplexed when complex diagrams are given. Although one student may be weak in a speed test, he or she might understand a discourse more deeply than another student and be able to submit substantial and excellent reports. Learners also vary in their backgrounds and experience, and possess a diversity of abilities that cause them to learn in different ways. They are unique in their personalities and values, for example. In addition, they develop individual preferences for learning environments that support their favored modalities of learning. In general, students exhibit a wide variety of unique blends of strengths and weaknesses resulting in classrooms with a wide diversity of talents that need to be developed.

The benefits of personalizing learning are well documented in the literature on differentiation of learning (Brimijoin 2003; Stevens 1999; Tomlinson & National Association for Gifted Children 2004) and are also easy to illustrate with an example. If the classroom has as few as two different kinds of learners and only one kind of instruction used, there will be a "best and worst fit" among the students. If the same instruction is used repeatedly, then one of the students will be systematically denied access to the most effective instruction.

Understanding the different ways that children learn, interact with, and process information can help teachers modify instruction so that all students have an equal opportunity to succeed (Theroux, 2004). In order for teaching to be an intentional and planned activity that supports each student's academic success, it is necessary to accept and utilize each learner's features to foster the most effective learning. It follows that teachers, learning devices, and instructional programs that provide a variety of learning approaches have a greater chance of offering appropriate challenges to every student in the learning environment. However, with highly portable m-learning, the teacher's role needs to shift to the device.

When the learning environment involves mobile devices, the variety of learners' background, abilities, and learning styles are expected to be more diverse than in a traditional classroom environment. This is true because the mobile device can be picked up and used by anyone at anytime, with or without a teacher present. The handheld learning environment thus needs a great deal of adaptability. It must be able to support independent learning without expecting a teacher's support and guidance. As we envision it, the mobile device itself can play an adaptive role that shapes the learning environment on the basis of a learner's preferred style.

To capture this idea, Park developed a prototype adaptive mobile learning management system (AM-LMS) which assesses a user's learning style, creates a learner profile, and then provides content based on decisions the learner makes while interacting with the content, continuously updating the learner profile. The chapter presents background, rationale and a functional overview of the AM-LMS.

MOBILE LEARNING

Mobile learning is based on wireless Internet connections and uses devices such as notebook computers, cellular phones, personal communication system (PCS) phones, and personal digital assistants (PDAs). The important features of mobile devices are their portability, immediacy, individuality, and accessibility; features which are bringing about a change of paradigm in approaches to teaching (Shostsberger & Vetter, 2000).

Dye et al. (2003) define m-learning as "learning that can take place anytime, anywhere with the help of a mobile computer device. The device must be capable of presenting learning content and providing wireless two-way communication between teacher(s) and student(s)" (p. i). Figueiredo and Chabra (2002) emphasize device flexibility; not only does m-learning offer the ability to receive learning anytime, anywhere, but as important, on any device. Harris (2001) adds to the definition by defining m-learning as that point at which mobile computing and e-Learning intersect. The term "mobile learning" in this chapter will be defined broadly as a form of learning delivered through mobile devices such as mobile phones, PDAs, smart phones, tablet PCs and similar devices combined with e-learning content.

Mobile Learning Environment

Mobile devices are a familiar part of the lives of most teachers and students. They offer the opportunity to embed learning in a natural environment (Schwabe & Goth, 2004) and enable learning that is independent of time and location constraints and with increasingly customized contents (Abfalter et al., 2004). Currently there are increasing efforts to apply mobile technology to learning (see, for example, Gay, Reiger, & Bennington, 2001; Hoppe, Joiner, Milrad et al., 2003; Kristiansen, 2001; Lundin, Nulden, & Persson, 2001; Schwabe & Goth, 2005; Sharples, Corlett, & Westmancott, 2002). Sharples et al. (2002) point out that there may be a particular opportunity for mobile learning outside the traditional formal learning settings.

The last decade has seen far-reaching changes in living, learning, working, and collaboration, fundamentally influenced by information and communication technologies, specifically the World Wide Web. Projecting 10 years into the future we may ask what the new impact of wireless and mobile technologies will be. We should take up the challenge that the future is not "out there" to be discovered, but has to be invented and designed to meet new needs and possibilities that emerge as wireless and mobile technologies become widely available (Fisher & Konomi, 2005).

A mobile learning environment requires a wireless Internet service that provides content upon a request by a client who holds a mobile device (Figure 1). To supply learning content to the wireless Internet, Web services are provided through a WAP Gateway, which enables regular communication with mobile devices. The WAP Gateway allows both WAP and TCP/IP protocol. Through the network-enabled WAP Gateway, a mobile learning environment can provide students and teachers with the opportunity to obtain any and all class-related material on a handheld computer, such as a Palm Treo, through a simple process of point-and-connect using infrared.

Landers (2005), the forum administrator for 'From e-learning to m-learning,' presented options created by mobile learning methods in Europe as

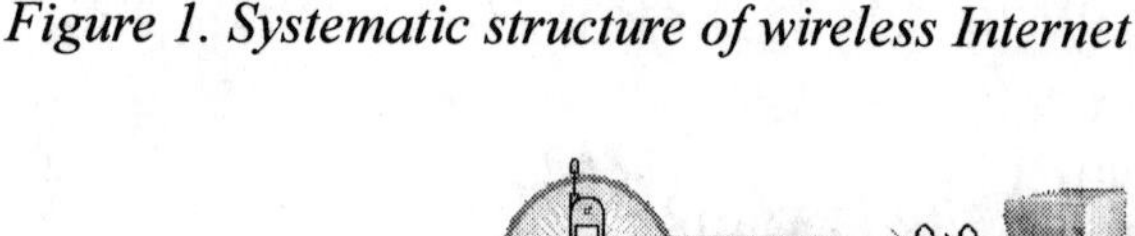

Figure 1. Systematic structure of wireless Internet

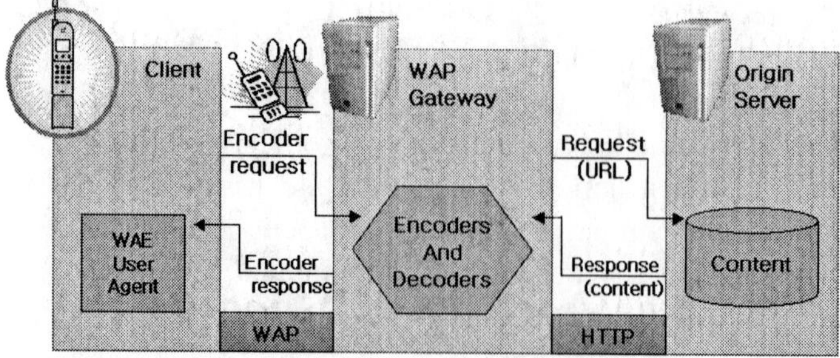

Figure 2. Expansion of mobile learning environment (Abfalter et al., 2004)

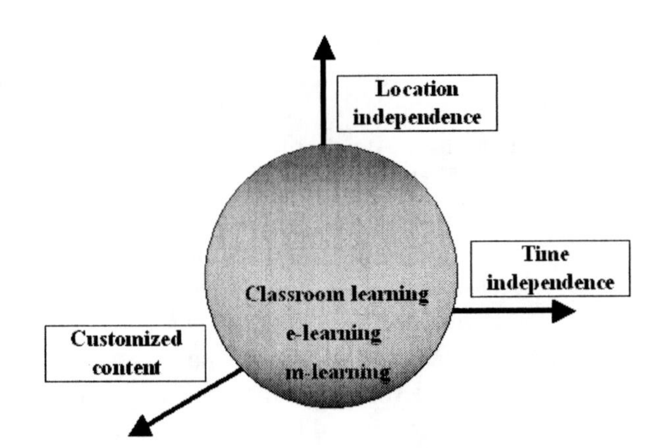

follows: first, learning content can be separated from mobile media (Ketamo, 2005). Second, content can be used to help students study a regular curriculum in the classroom (Vahey & Crowford, 2005). Third, they can be integrated into web-based learning and web based instruction (Heoyoung, 2002). Fourth, learning communities can share material and undertake self-directed study. Fifth, mobile technology can be used by students when taking trips or participating in field activities (Gay et al., 2002). Sixth, the mobile tools can be the main materials while others support the mobile materials.

Mobile learning with an adaptive learning management system supports learning that is independent of both time and location through contents provided on demand to the wireless device. The intersection of mobile computing and e-learning also provides strong search capabilities; rich interactions with users; powerful support for effective learning; and performance-based assessment. Wireless communication expands the learning space beyond the traditional classroom to unlimited cyber space. Abfalter et al. (2004) for example, explains that in a mobile learning environment, "Teaching and learning are no longer confined to time and place. With mobile communication technologies, the time and physical boundaries of the traditional classroom are being expanded (p. 2)." Figure 2 shows the independent characteristics of mobile learning.

Research Trends in Mobile Learning

Three primary trends can be found in the research on mobile learning: research on the interface contained in learning contents, research on the learning management system, and research on mobile learning patterns based on ubiquitous computing.

Research on the interface considers the special qualities of mobile devices: small screen size, slow text input facilities, small storage capacity, limited battery life, low bandwidth network capabilities and slow CPU speed. In particular, the limitation of small screen directly affects the user's learning process and behavior. Recent studies (Buyukkoten et al., 2000; Kawachiya & Ishikawa, 1999) on the effect of screen size on completing browser related tasks for example, show that mobile users tend to follow links less frequently than traditional Internet users (Antonellis et al., 2005).

Another theme of the research concerns learning management and supporting systems. Studies have found that teachers use mobile devices for attendance reporting, reviewing student marks, general access of central school data, and managing their schedules more effectively (Laura, 2006). Also, handheld computers have been found to bring important benefits to schools by assisting administration, supporting classroom management, and enabling personal and group learning (Perry, 2003). Research studies by the

Becta PDA (personal digital assistant) project (2003) were focused on two aspects: managing teacher's workload and supporting teaching and learning on PDA use. They found that PDAs have considerable potential for improving a teacher's management and presentation of information. At the same time, PDAs were found to bring important benefits to schools by enabling individualized learning and group learning.

A third type of m-learning research focuses on ubiquitous computing, in which a huge number of tiny computers are embedded into an invisible part of the fabric of everyday life—in watches, microwave ovens, cars, and clothes. Recently m-learning researchers have been proposing adapted learning contents for a variety of learning styles, joined with the idea of ubiquitous computing (Sakamura & Koshizukea, 2005). For example, Shindo et al. (2003) introduces the idea of a digital ubiquitous museum that embeds ubiquitous learning materials and tools into daily living environments. Also, Deng et al. (2005) proposed using wireless and mobile devices to support academic conferences with an "ask the author" application. Schwabe & Goth (2005) reports on the design of a mobile orientation game in a university setting. Gay et al. (2001) studied the use of mobile computing to enhance field study. MOBIlearn, a major European research project, is focusing on the context-aware delivery of content and services to learners with mobile devices (Lonsdale, Nudin & Persson, 2004). With these kinds of research efforts, it seems timely to suggest an adaptive "learning style" architecture for mobile learning device interfaces to a learning management system.

LEARNING STYLES

Felder (1996) and many others have made the point that students have different learning styles—characteristic strengths and preferences in the ways they take in and process information. Some students tend to focus on facts, data, and algorithms; others are more comfortable with theories and mathematical models. Some respond strongly to visual forms of information, like pictures, diagrams, and schematics; others get more from verbal forms-written and spoken explanations. Some prefer to learn actively and interactively; others function more introspectively and individually. Keefe (1982) explained that learning styles are the cognitive, affective, and psychological traits that serve as relatively stable indicators of how learners perceive, interact with, and respond to the learning environment. Dunn (2000) explained that learning style is the way a person processes, internalizes, and studies new and challenging material. The cornerstone of the learning style theory is that most people can learn, and each individual has his or her own unique ways of mastering new and difficult subject matter.

Kolb's (1984) point of view is that teaching with learning styles is a way shaping and intensifying learning by making the environmental demands of learning tasks coincide with individual preferences. Felder and Silverman (1996) defined learning styles as the process of acquiring and controlling information with the traits and preferred ways in which students study. Della-Dora and Blanchard (1979) also believe that "learning style can be defined as a personally preferred way of dealing with information and experience that crosses content areas." Many researchers have grasped the importance of the preference factor to improve learning (Kruzich, Friesen & Van Soest, 1986).

In brief, learning style can be described as the general characteristics of an individual's intrinsic procedures of information processing that lead to unique behavioral patterns, which tend to be durable and stabile in a variety of learning environments.

Index of Learning Styles

The Index of Learning Styles is an online instrument used to assess learning preferences on four dimensions (active/reflective, sensing/intuitive, visual/verbal, and sequential/global) of a learning style model formulated by Richard M. Felder and Linda K. Silverman. Richard M. Felder and Barbara A. Soloman of North Carolina State University developed the instrument.

The Felder and Soloman (2005) Index of Learning Style (ILS) provides critical guidance for determining the type of content when individual students are learning with the adaptive mobile learning management system (AM-LMS). The ILS analysis contributes to the AM-LMS system in two aspects. First, the analysis suggests one of sixteen learning styles identified by the Felder and Soloman model and second, those styles are sufficient to provide relevant learning contents according to the various learners' needs in the process of learning. On this theoretical foundation, the AM-LMS platform is structured to provide an interface for adaptive contents.

Four Dimensions of Learning Style

In a report by Felder and Silverman (1988) a model of learning style was originally described and is now classified by four dimensions (2005). The first dimension is related to the question, what type of information does the student preferentially perceive: sensory-sights, sounds, sensations, or intuitive-memories, ideas, insight? The second dimension concerns effective perception and is the answer to the question, through which modality is sensory information most effectively perceived: *visual*—pictures, diagrams, graphs, demonstrations, or *verbal*—written and spoken words and formulas? The third dimension is the pattern of processing information and the answer to the question, how does the student prefer to process information: *actively*—through engagement in a physical activity or discussion, or *reflectively*—through introspection? The fourth dimension is the type of the progress to understanding: How does the student progress toward understanding: *sequentially*—in a logical progression of small incremental steps, or *globally*—in large jumps, holistically? (Felder, 1995).

It is convenient to set up pairs of opposites on the ends of continua representing the complex mental processes by which perceived information is converted into knowledge. In what follows, four such pairs of opposites further refine the four learning style dimensions.

Active and Reflective Learners

An "active" learner is a person with a natural tendency toward active experimentation more than toward reflective observation. This type of learner tends to solve problems through discussion and group work and is extroverted. A "reflective" learner is conversely, a person with a natural tendency toward reflective observation more than toward active experimentation. This type of learner tends to solve problems through self-evaluation and reflection and is introverted.

Sensing and Intuitive Learners

"Sensing" learners prefer to accept information in a well-structured order. They tend to be concrete and methodical and they prefer facts, data, and experimentation. They are patient with detail but dislike complications and they rely more on memorization as a learning strategy. They are more comfortable learning and following rules and standard procedures than using their intuition. "Intuitive" learners, on the other hand, prefer to organize information in their memory according to their own rules. They tend to be abstract as well as imaginative and deal better with principles, concepts, and theories. They are apt to be bored by details and they welcome complications.

Visual and Verbal Learners

"Visual" learners prefer that information be presented in pictures, diagrams, flow charts, time lines, films, and demonstrations—rather than in spoken or written words. Meanwhile, "verbal" learners prefer spoken or written explanations to visual presentations.

Sequential and Global Learners

"Sequential" learners absorb information and acquire understanding of material in small, connected chunks, whereas "global" learners take in information in seemingly unconnected fragments and achieve understanding in large holistic leaps.

Index of Learning Style Questionnaire

The Index of Learning Style (ILS) is a 44-item, forced-choice instrument first developed in 1991 by Richard Felder and Barbara Soloman to assess learning preferences on the four scales of the Felder-Silverman model. Currently, the Felder and Silverman theory (2005) categorizes an individual's preferred learning style by a sliding scale of four dimensions: sensing-intuitive, visual-verbal, active-reflective and sequential-global (Table 1).

The ILS questionnaire consists of 44 questions and each with two possible answers: a or b. All questions are classified corresponding to the four pairs in the Felder and Silverman learning style theory. Each dimension has 11 questions and the 16 learning styles are classified based on the scores earned on each dimension. An example of a questionnaire based on the Felder and Silverman index of learning style is presented in a screen shot of the AM-LMS in Figure 3.

The scores earned on the dimensional scales of the questionnaire are explained as follows:

- If the score on any scale is 1-3, it means that the student is fairly well balanced on the two dimensions of that scale.
- If the score on a scale is 5-7, it means a moderate preference for one dimension of the scale and the students will learn more easily in a teaching environment that favors that dimension.
- If the score on a scale is 9-11, it means a very strong preference for one dimension of the scale. The student may have difficulties in learning environment that does not support that preference.

Table 1. Felder's learning dimensions (Carver et al., 1999)

Definitions	Dimension		Definitions
Do it	Active	Reflective	Think about it
Learn facts	Sensing	Intuitive	Learning concepts
Require Pictures	Visual	Verbal	Require reading or lecture
Step by Step	Sequential	Global	Big picture

Figure 3. Learning style results of ILS

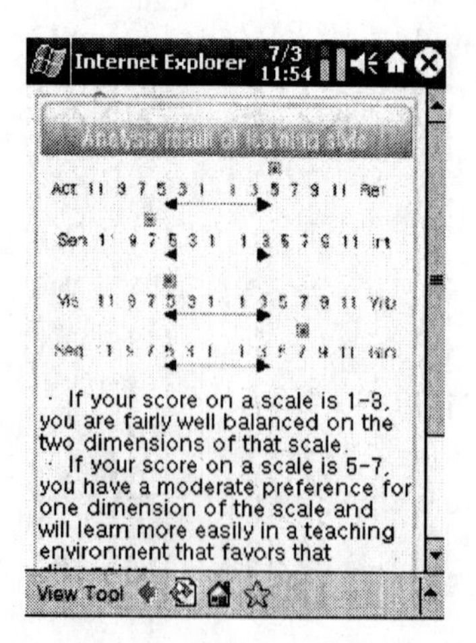

The Classification of Learning Style for Adaptation

The AM-LMS system classifies the learning style of the user by using the ILS assessment to build a profile of the learner. It then constructs adaptive contents for the learning process. It initially discriminates the user's learning style by a combination of the items selected with weak preference and the items with strong preference. Those learning styles are then used as criteria to provide adaptive contents for the learning process, the application selects from among the 16 possible adaptive content types that results from 4 pairs of opposites (8 dimensions of the learning style typology), as shown in Figure 4.

Here are the rules used by the AM-LMS for constructing adaptive contents based on learning styles:

- **Active:** An active learner asks questions frequently and is given answers accordingly. In addition, the system provides these learners with a discussion room, bulletin board, and chat room.
- **Reflective:** A reflective learner needs to contemplate the topic of a subject in advance before starting a lesson. During the lessons, the learner summarizes his or her learning and reviews the whole learning process. The AM-LMS system provides this style of learner with relevant tools, and confirms prerequisite knowledge before the lesson, then provides reflective learners with links for related learning, materials for downloading and performing evaluations.
- **Sensing:** A sensing learner needs an example of the facts, demonstrations and practical materials. The AM-LMS provides easily located learning contents and summarizes and structures the learning contents for sensing learners.
- **Intuitive:** An intuitive learner needs to be provided with concepts, abstract ideas, demonstrations, and theories before concrete examples. The AM-LMS provides a sequential order of exposition followed by examples to best fit the intuitive learner.
- **Visual:** A visual learner needs, and the AM-LMS provides, pictures, graphs, diagrams, flow charts, schematics, demonstrations, concept maps, color notes, and slides with multimedia.
- **Verbal:** For verbal learners, the AM-LMS presents content primarily as text and audio.
- **Sequential:** For sequential learners, the AM-LMS structures material in a logical, step by step, orderly outline.
- **Global:** A global learner needs to see the big picture before the details in order to view more of the context of a subject. The AM-LMS presents the big picture of the course and all the links are made available

Figure 4. Combinations of eight learning style dimensions to create 16 learning styles

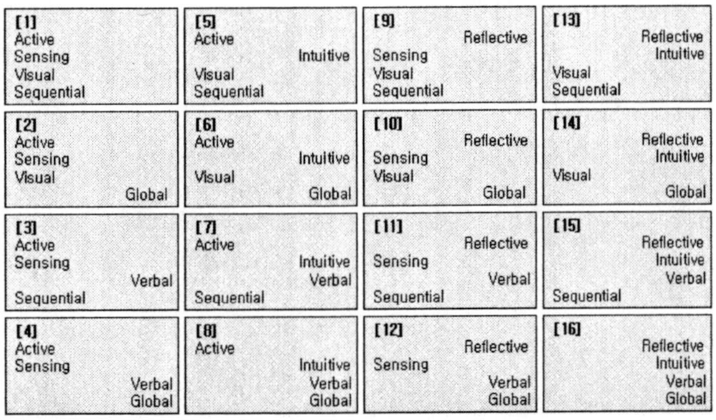

[1] Active Sensing Visual Sequential	[5] Active Intuitive Visual Sequential	[9] Reflective Sensing Visual Sequential	[13] Reflective Intuitive Visual Sequential
[2] Active Sensing Visual Global	[6] Active Intuitive Visual Global	[10] Reflective Sensing Visual Global	[14] Reflective Intuitive Visual Global
[3] Active Sensing Verbal Sequential	[7] Active Intuitive Verbal Sequential	[11] Reflective Sensing Verbal Sequential	[15] Reflective Intuitive Verbal Sequential
[4] Active Sensing Verbal Global	[8] Active Intuitive Verbal Global	[12] Reflective Sensing Verbal Global	[16] Reflective Intuitive Verbal Global

as flexible options to ensure free movement by the global learner.

Figure 5 shows a screen of content presented adaptively according to a corresponding learning style. The display in the left of Figure 5 is constructed for the ASVG (Active, Sensing, Verbal, Global) style among 16 styles presented in Figure 4. The display in the right of Figure 5 is constructed for the ASVS (Active, Sensing, Visual, Sequential) style.

User Interface for Mobile Learning

The user interface in mobile learning is different from other computing interfaces in its size of display. Display sizes range from 100 x 80 pixels of mobile phones to 240 x 320 pixels for personal digital assistants. That small display area, its color, and the amount of displayable information are major points which need to be considered when adapting contents.

The AM-LMS uses several strategies to deal with the smaller display. First, the amount of information has been reduced so that the mobile device can accept it even when the learner is traveling. The wireless Internet also limits the amount of information that can be sent and received. Thus, the scroll has been minimized in AM-LMS content. Second, the user interface avoids asking the learners for a lot of input because of the limitation of input devices on mobile devices. Thus simple response patterns such as pointing and clicking without moving around text areas has been adopted. Third, vertical navigation within a same topic has been confined to one or two hyperlink depths. Fourth, sentences are simple as opposed to compound and complex. Fifth, the location and the topic in current learning are displayed so that the learner can understand what he or she is doing under any topic.

To deliver the interface adapted to each learner's learning style, the AM-LMS analyzes past performance, including the initial survey information and the learner's subsequent choices and selections, and then generates new content and presentations based on that analysis. For example, reflective learners need to review their learning process. When given reflective opportunities, they can manage their own learning based on their judgment of what is working for them. Reflective users navigate their learning through continuous adjustments in their learning behaviors based on questions such as "What are the key characteristics of this material that help me understand?" "What is the criteria for mastery of what I am doing?" and "How am I conceptualizing this problem?" Schön (1983). Thus, in the interface for the reflective learners, elements prompting reflective thinking are introduced in display forms, inducing and stimulating questions which help them understand what they are doing and the relationships among previous and current topics.

Figure 5. Contents adaptive to learning styles (ASVG, ASVS)

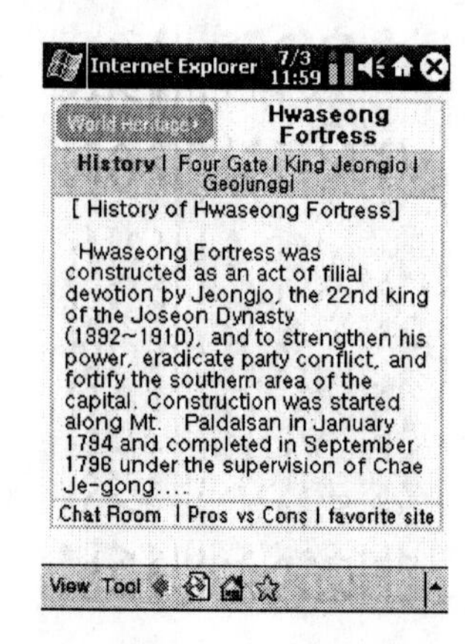

The AM-LMS establishes similar strategies of user interface for each of the eight elements of learning style (Table 2).

An outstanding question for future research concerns the peculiarities of mobile device interaction that may affect the applicability of the four dimensional model. For example, a highly visual learner may be more impacted by the screen limitations of the mobile device than a textual learner. Research designs are needed that explore the learning effectiveness of handhelds with and without adaptive LMS as well as the degree of change in a student's choices of learning styles over time. Comparisons of mobile adaptive versus traditional learning management systems and traditional systems with learning styles adaptability are needed.

STRUCTURE OF THE ADAPTIVE MOBILE LEARNING MANAGEMENT SYSTEM

As the previous sections have hopefully made clear, the adaptive mobile learning management system (AM-LMS) is a platform for providing learning contents adapted to a learner's learning style in a mobile learning environment. The AM-LMS manages the whole learning process, monitors the progress of a learner and presents learning contents adaptive to the user's learning style, including the potential change in learning preferences that may appear over time. The system plays its role following each learning stage by analyzing learning progress and giving feedback. Figure 6 shows the functional structure of the AM-LMS.

Table 2. Design strategies according to learning style (based on Felder & Silverman, 1988)

Learning Style	Characteristic	AM-LMS Interface Strategy
Active	Has outward character and prefers problem solving through discussion and cooperative/group work	Provides bulletin board and chatting room
Reflective	Has inward character and prefers problem solving through self assessment and reflections	Provides links to other references, downloadable materials, and performance evaluation materials
Sensing	Prefers the ways for understanding in order in organized pattern	Provides well organized and summarized information in balanced location in the display
Intuitive	Prefers the ways for organizing information with his or her freedom for easy memorizing	Describes text information in narration format
Visual	Prefers visual information such as pictures, graphs, drawings	Provides pictures and graphs with explanation
Verbal	Prefers audible information	Presents text with audible information
Sequential	Prefers sequential and structural ways of learning	Organizes content in order and makes navigation exploratory
Global	Prefers the ways of understanding content as a whole and overall perspective with freedom	Makes navigation free so that content can be selected with learner's will

Figure 6. Structure of the adaptive mobile-learning management system

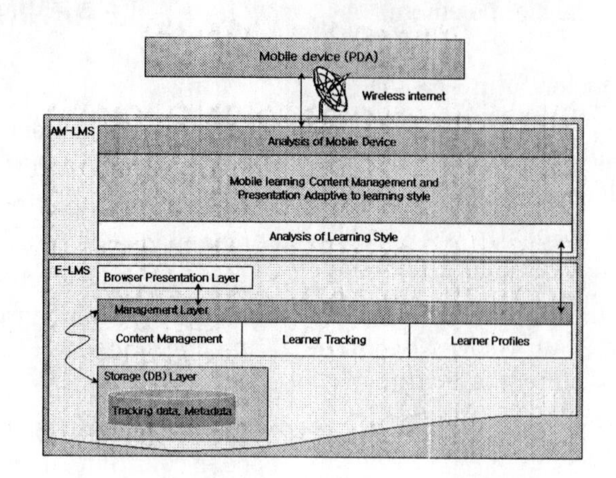

The AM-LMS consists of (1) Access device analysis, (2) Learning style analysis, (3) Mobile content management and presentation adaptation (Figure 6). The AM-LMS works on top of an e-learning platform that is responsible for the general management of learners and learning contents. The extra mobile learning management system analyzes mobile devices, converts e-learning contents into the m-learning contents suitable for each learner and presents the contents adaptively on screens and other devices. A learner interacts through question items and other products, which are provided to the WAP Gateway through a mobile device that transmits them to the AM-LMS.

The AM-LMS system aims to provide learners with specific contents suitable to their learning styles so that they can learn in ways that are based on their own needs. Learners at a distance cannot expect to get a teacher's support and guidance as they can in classroom settings, thus the AM-LMS system takes many possible characteristics into consideration including learners' backgrounds, capabilities, and learning styles. The AM-LMS has a stand-alone capability for building a rich and responsive learning environment without any teacher's support.

Analyzing Learning Style

The learning style analysis module evaluates the learner and stores the evaluation into a personal profile. When the learner comes back later, the profile provides basic data so that the AM-LMS can supply adaptive contents on the basis of the variously designed learning styles materials in a database. Figure 7 shows a screen to analyze the learning style of the user through a PDA.

Analysis Module

When the mobile device accesses the system, the analysis module automatically determines the type of device in use and transfers that information to the mobile learning management and presentation adaptation modules. This analysis module determines the device type, running

Figure 7. Screens analysis of learning style

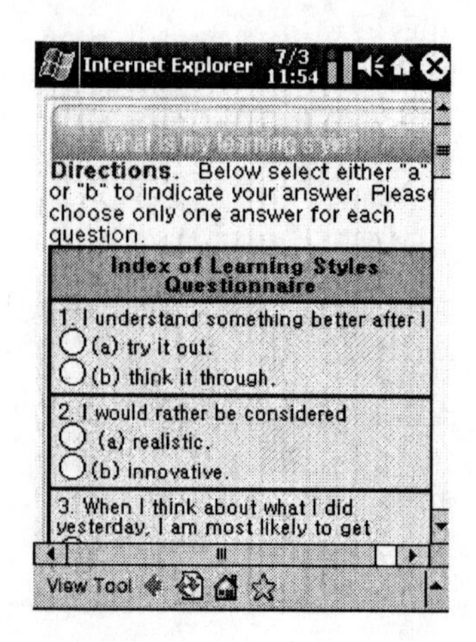

Table 3. Sample code to identify a browser

```
If (InStr(Request.ServerVariables("HTTP_USER_AGENT"), "Windows CE")) Then
" REDIRECT TO CODE FOR HANDHELD
Else
" YOUR NORMAL WEB SITE
End If
```

Figure 8. Choice of learning process through the analysis results of learning styles

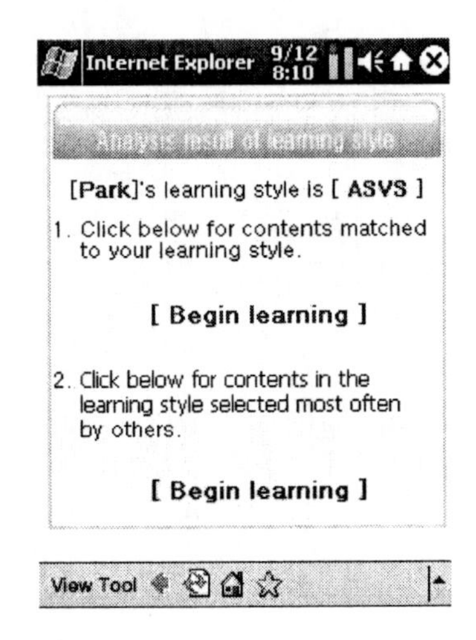

environment, features, and operating system of a mobile device that is accessing learning content by using the information in the header. In order to identify a handheld device, a Windows CE browser also identifies itself via the HTTP request header (Table 3).

Managing and Adapting Mobile Contents

A PDA's screen size is limited to 240 by 320 pixels and a cellular phone and smart phone have even smaller screen sizes. In addition to smaller amounts and types of information, the AM-LMS module presents two elements in response to the learning progress selected by a learner:

1. Learning path matched to the learner's learning style. This link functions as an opening for packaging and delivering contents suitable for each learner's learning style. Figure 8, the first screen for the learner, displays the learner's learning style on the first row and has options to get into either the learning path specific to his/her learning style or a universal learning path which most learners has taken before this session.
2. The learning path taken by most learners. This link functions as an opening for packaging and delivering contents preferred or taken by most learners. This link also is a path to frequently presented contents.

Flow of Adaptive Mobile Learning

In the AM-LMS, a learner advances his or her own learning by making frequent decisions. Then the learner is provided with learning content according to the analyzed learning style. Learning in the AM-LMS flows as presented Figure 9.

Figure 9. Flow of learning in AM-LMS

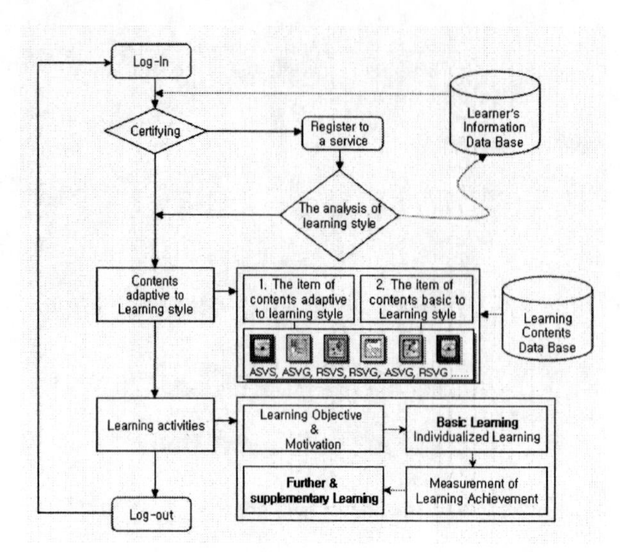

The learner enters the learning environment through the process of certification on the login screen (Figure 9). If the learner has his or her own learning style analyzed previously, he or she enters the learning process that matches his or her learning style. Otherwise, he or she advances into the learning process after the learning styles analysis. The AM-LMS system periodically keeps track of the progress of learners' activities and continuously stores the learners' profile and tracking information.

CONCLUSION

The AM-LMS design introduces the theoretical background for the development of an adaptive mobile learning management system that presents content in response to the learning style of a user. It enlarges the concept of adaptive learning system approaches in traditional LMS systems to include mobile devices. The Felder and Silverman (2005) learning style theory provides the platform with a framework to analyze a user's learning style and present contents adapted to that style. The system can thus promote individualized learning and learner-centered education while taking advantage of the unique features of mobile devices and network-based learning environments.

The AM-LMS is an ongoing research project. Learner variables such as cognitive style, task commitment and others will be added to the system in the future so that it can be more adaptive. At this point, interface research on content size, navigation, and display techniques are urgently needed in the area of mobile devices. In addition, issues such as cognitive overloading for the learners on mobile devices have not yet been adequately studied, although there are several related research findings for the Web and other information technologies. The AM-LMS conceptual model is thus expected to adopt more variables resulting from future research studies and will hopefully contribute to strong and effective ways of achieving individualized learning on mobile devices.

REFERENCES

Abfalter D., Mirski P. J., & Hitz, M. (2004). Mobile learning—Knowledge enhancement and vocational training on the move. In *The 5th European Conference on Organizational Knowledge, Learning and Capabilities.* Innsbruck: OKLC.

Antonellis, I., Bouras, C., & Poulopoulos, V. (2005, July 27-30). Game based learning for mobile users. In *Proceedings of the 6th International Conference on Computer Games: AI and Mobile Systems (CGAIMS 2005)*. Lousville, Kentucky.

Becta ICT Research (2003). *Handheld computers (PDAs) in schools*. Department for education and skills.

Brimijoin, K. (2003). Using data to differentiate instruction. *Educational Leadership, 60*(5), 70-3.

Buyukkokten, O., Garcia-Molina, H., & Paepcke, A., & Winograd, T. (2000). Power browser: Efficient web browsing for PDAs. In *proceedings of the Conference on Human Factors in Computing Systems, CHIOO* (pp. 43-437).

Carver, C.A., Howard, R.A., & Lane, W.D. (1999). Enhancing student learning through hypermedia courseware and incorporation of student learning style. *IEEE Transactions on Education, 42*(1), 33-38.

Chabra, T., & Figueiredo, J. (2002). How to design and deploy handheld learning. Retrieved on December 8, 2006, from http://www.empoweringtechnologies.net/eLearning/eLearning_expov5_files/frame.htm

Deng, Y. C., Chang, H. Z., & Chang, B., & Liao, H. C., & Chiang, M. C., & Chan, T. W.(2005). "Ask the author": An academic conference supported system using wireless and mobile devices. In *Proceedings of the third IEEE WMTE2005* (pp. 29-33).

Della-Dora, D., & Blanchard, L. (1979). *Moving toward self-directed learning: Highlights of relevant research and promising practice*. Alexandria, VA: Association for Supervision and Curricular Development.

Dunn, R. (2000). Learning styles: Theory, research, and practice. *National Forum of Applied Educational Research Journal, 13*(1), 3-22.

Dye, A., Solstad, B.E., & Joe, A. K.(2003). Mobile education—A glance at the future. Retrieved on July 15, 2006, from http://www.nettskolen.com/forskning/mobile_education.pdf

Felder, R.M. (1995). Learning and teaching styles in foreign and second language education. *Foreign Language Annals, 28*(1), 21-31.

Felder, R.M. (1996). Matters of style. *ASEE Prism, 6*(4), 18-23.

Felder, R., Silverman, L. (1988). Learning and teaching styles in engineering education. *Engr. Education, 78*(7), 674-681.

Felder, R.M., & Soloman, B.A. (2005). *Index of Learning Styles Questionnaire*. Retrieved on July 14, 2006, from http://www.ncsu.edu/felder-public/ILSdir/ilsweb.html

Fischer, G., & Konomi, S.(2005). Innovative Media in Support of Distributed Intelligence and Lifelong Learning. In *Proceedings of the 3rd IEEE WMTE2005* (pp. 3-10).

Gay, G., & Reiger, R. & Bennington, T. (2001). Using mobile computing to enhance field study. In Miyake, N., Hall R, & Koschmann, T. (Eds.), *Carrying the conversation forward*. Mahwah, NJ: Erlbaum 507528.

Gay, G., Stefanone, M., & Grace-Martin, M., & Hembrooke, H. (2002). The effect of wireless computing in collaborative learning environments. *International Journal of Human-Computer Interaction, 13*(2), 257-276.

Harris, P. (2001). Goin Mobile. Retrieved on July 3, 2006, from http://www.learningcircuits.org/2001/jul2001/harris.html

Heoyoung, K. (2002). *Development of wire and wireless learning system based on XML*. Dissertation, Keimyung University.

Hoppe, H.U., Joiner, R., & Milrad, M., & Sharples, M. (2003) Guest editorial: Wireless and mobile technology in rducation. *Journal of Computer Assisted Learning, 19*(3), 255-259.

Kawachiya, K., & Ishikawa, H. (1999). Improving Web interaction on Small Displays. In *Proceedings of 8th international WWW Conference* (pp. 51-59).

Keefe, J. W. (1982). Assessing student learning style: An overview. In J. W. Keefe (Ed.), *Student learning style and brain behavior* (pp. 43-53). Reston. VA: National association of Secondary School Principals.

Ketamo, H. (2002). mLearning for kindergarten's mathematics teaching. *IEEE International Workshop on Wireless and Mobile Technologies in Education.* Retrieved on March 14, 2005, from http://csdl.computer.org/comp/proceedings/wmte/2002/1706/00 1706toc.htm

Kolb, D. (1984). *Experiential learning: Experience as the source of learning and development.* Englewood Cliffs, NJ: Prentice-Hall.

Kristiansen, T. (2001). *M-learning. Experiences from the use of WAP as a supplement in learning.* Oslo: Fornebu Knowation.

Kruzich, J.M., Friesen, B.J., & Van Soest, D.(1986). Assessment of student and faculty learning styles: Research and application. *Journal of Social Work Education, 22*(3), 22-30.

Landers, P. (2005). From e-learning to m-learning. *e-Learning and m-Learning.* Retrieved on March 14, 2006, from http://learning.ericsson.net/mlearning2/project_one/index.html

Laura, N., Lonsdale, P., & Vavoula, G., & Sharples, M. (2006). *Literature Review in Mobile Technologies and Learning.* FUTERLAB SERIES. Retrieved on July 14, 2006, from http://www.futurelab.org.uk/research/lit_reviews.htm

Lonsdale, P., Baber, C., Sharples, M., Byrne, W., Arvanitis, T., Brundell, P., & Beale, H. (2004). Context-awareness for MOBIlearn: Creating an engaging learning experience in an art museum. Proceedings of MLEARN2004, Bracciano, Rome: LSDA.

Lundin, J., Nulden, U., & Persson, L. M. (2001). *MobilLearn: Competence development for Nomads.* CHI 2001, 7-8.

Perry, D. (2003). Handheld computers (PDAs) in schools. Becta ICT Research Report, March 2003.

Sakamura, K., & Koshizuka, N. (2005). Ubiquitous computing technologies for ubiquitous learning. In *Proceedings of the third IEEE WMTE2005* (pp. 11-18).

Shindo, K., Koshizuka, N., & Sakamura, K. (2003). Ubiquitous information system for digital museum sing smart cards. In *Proceedings SSGRR.*

Schwabe, G., & Goth, C. (2005). Mobile learning with a mobile game: Design and motivational effects. *Journal of Computer Assisted Learning, 21*(3), 204-216.

Schön, D. A. (1983). *The reflective practitioner.* NY: Basic Books.

Sharples, M., Corlett, D., & Westmancott, O. (2002). The design and implementation of a mobile learning resource. *Personal and Ubiquitous Computing, 6*, 220-234.

Shotsberger, P. G., & Vetter, R. (2000). How mobile wireless technologies will changes web-based instruction and training. *Educational Technology, 40*(5), 49-52.

Stevens, A. (1999). *Tracking, curriculum differentiation, and the student experience in english (high schools, language arts).* State University Of New York at Albany: 1053.

Theroux, P.(2004). *Enhance learning with technology.* Retrieved on July 15, 2006, from http://www.enhancelearning.ca/

Tomlinson, C. A. & National Association for Gifted Children (U.S.) (2004). *Differentiation for gifted and talented students.* Thousand Oaks, CA: Corwin Press.

Vahey, P., & Crawford, V. (2005). *Palm Education Pioneers Program Final Evaluation Report.* Retrieved on March 14, 2006, from http://www.palmgrants.sri.com

KEY TERMS

Adaptive Mobile Learning Management System: The system managing the whole learning process and progress of a learner and containing the function to make learning adaptive to learning styles.

Learning Management System: A software system designed to facilitate teachers in the management of online educational courses for their students. These services generally include access control, provision of e-learning content, communication tools, and administration of user groups.

Learning Style: Described as general characteristics showing individual differences in the intrinsic procedures of information processing.

Mobile Device: Any portable device used to access the Internet. For example, PDA, cellular phone, Tablet PC and so on.

Mobile Learning: A form of learning where mobile computing is combined with e-learning and as a form of teaching and learning delivered through mobile devices such as mobile phones, PDAs, smart phones, tablet PCs, and so on.

Mobile Learning Environment: Provide students and teachers with the opportunity to obtain any and all class-related material on a Palm handheld computer through the network-enabled WAP Gateway.

Wireless Internet: Wireless Internet grants access to the World Wide Web or Internet e-mail via wireless networks.

Chapter XIX
Adaptive Interfaces in Mobile Environments: An Approach Based on Mobile Agents

Nikola Mitrovic
University of Zaragoza, Spain

Eduardo Mena
University of Zaragoza, Spain

Jose Alberto Royo
University of Zaragoza, Spain

ABSTRACT

Mobility for graphical user interfaces (GUIs) is a challenging problem, as different GUIs need to be constructed for different device capabilities and changing context, preferences and users' locations. GUI developers frequently create multiple user interface versions for different devices. The solution lies in using a single, abstract, user interface description that is used later to automatically generate user interfaces for different devices. Various techniques are proposed to adapt GUIs from an abstract specification to a concrete interface. Design-time techniques have the possibility of creating better performing GUIs but, in contrast to run-time techniques, lack flexibility and mobility. Run-time techniques' mobility and autonomy can be significantly improved by using mobile agent technology and an indirect GUI generation paradigm. Using indirect generation enables analysis of computer-human interaction and application of artificial intelligence techniques to be made at run-time, increasing GUIs' performance and usability.

INTRODUCTION

Mobile computing is an increasingly important topic in today's computational environment, because the demand for ubiquitous access to information is constantly increasing. Furthermore, users want to increase their efficiency and process information when using mobile equipment. To support this demand, software applications face a number of challenges. One of the important challenges in mobile computation is user interaction.

The importance of user interfaces (UI) comes from the fact that UIs represent the first line of interaction between a user and a computer. A user's ability to execute a required task and his efficiency are directly impacted by the user interface.

In the past, user interfaces have been developed mostly for a specific device, for example a specific PDA variant or a work station. Such an interface was usually designed for a single platform. This was done in conjunction with specialized user interface libraries that were defined for a specific platform or programming language. For example, if we assume an application that is developed for Windows and UNIX platforms, the user interface for the windows platform would be designed and developed separately from the UNIX user interface. The cost and effort required for such a development are obviously high; such an approach frequently leads to other problems, for example GUI implementations on one (or more) platform(s) being at different levels of development due to the lack of resources required to maintain the same GUI version on multiple platforms.

In the mobile environment GUIs face additional challenges: a user could be using an application on a mobile phone and could require the same application on his PDA or WebTV. In addition, the user could be moving and requiring an application to move with him. For example: while in the car, a user could read his e-mail using car's on-board computer; when he steps out of the car he could prefer to continue working on his PDA until he gets to the office, where a desktop PC could be his preferred equipment to continue working.

Mobile devices have different capabilities and requirements: different processing power, screen size, supported colors, sound functionalities, keyboard, and so forth. In addition, mobile devices use an ever-increasing number of different hardware and OS solutions, and frequently rely on batteries for operation. Mobile applications use wireless networks; wireless networks are not stable, have limited capacity and performance, and are expensive (e.g., 3G networks).

In addition to this, application interface and functionality may change depending on a user's context. For example, a music player application should mute if the user is indoors and should turn on when outdoors. Furthermore, a user could prefer the speaker to be on a louder setting when in the car. These requirements could be either a user's preferences or rules associated with a particular location where the user is.

To meet such challenges, researchers in the user interface area have adopted a common approach—user interface abstraction. To be presented on a concrete platform, abstracted user interfaces are transformed and rendered to meet a concrete platform's requirements. This approach provides a single user interface definition that is later transformed to the target device's user interface. The abstraction level in such an abstract user interface definition varies. Some abstract user interface notations offer very abstracted descriptions of user interfaces, while others are more linked to specific user interface concepts, for example window-based user interfaces. An abstract user interface definition is usually delivered in XML (W3C, 2000) notation, which enables efficient processing and data exchange between multiple platforms. Some notations describe the user interface at a high level, for example, a button is required; others allow sophisticated definitions of constraints and additional parameters, such as requiring the button for some (specific) device(s) only.

This chapter presents different approaches to adapting user interfaces to devices, with specific interest focused on enabling architectures that adapt to users' preferences and contexts. We discuss difficulties with mobile user interface generation for wireless devices. Finally, we present an approach for user interface adaptation based on mobile agents and examine sample usage of such an approach. With this example we show a flexible and mobile generation of user interfaces in a wireless environment that allows monitoring of user interaction and application of knowledge.

BACKGROUND

User Interface Abstraction

Abstraction of user interfaces adds flexibility when generating user interfaces. It provides a single and comprehensive description of the user interface. It is a rich layer of information that describes a user's interaction with the computer. Such abstraction is formalized by using an abstract user interface definition language (Stottner, 2001) or task models (Limbourg & Vanderdonckt, 2003). This information can then be used to generate a user interface that meets a concrete platform's limitations and requirements. Such an approach develops a single set of information to support all variants of the user interface that should be created for different devices.

The user interface is typically abstracted through the use of design models (e.g., task models) or by using an XML-based abstract user definition language (Luyten & Coninx, 2001; Mitrovic & Mena, 2002; Molina, Belenguer & Pastor, 2003; Stottner, 2001). Task models (Limbourg & Vanderdonckt, 2003) provide information that is focused on tasks. Some of the task models can describe multi-modal tasks for different types of devices (Paterno & Santoro, 2002). However, task models do not necessarily specify the exact presentation of a user interface. On the other side, XML-based abstract user interface descriptions describe the user interface's presentation and constraints. Many different versions of such abstract user interface definition languages exist.

Such approaches' abstraction levels vary: some approaches include information specific to a device type (e.g., a mobile phone, or a specific mobile phone model), some are more generic and do not consider the specifics of any device type. In addition, user interface abstraction can differ conceptually—some models can define any kind of interaction (e.g., via voice or specialized interfaces), while some are more linked to specific concepts (e.g., window-based user interfaces). Examples of such abstraction languages include XUL, UIML, XIML and XForms (Stottner, 2001).

The ability to effectively adapt such user interface definitions to a concrete platform is a key factor in achieving mobile and efficient user interfaces. The resulting, concrete user interface must meet the specification and be functional on the target device. This requires that device capabilities and limitations be successfully addressed. In addition, user preferences and context frequently impact this adaptation.

Abstract User Interface Adaptation

User interface adaptation is a complex task, and includes not only adaptation to the specific device's capabilities, but also to the user. Mobile devices have different capabilities such as screen size, keyboard and support for particular user interface widgets, hardware platform or operating system. Adaptation to a user includes adaptation to the user's preferences and changing contexts, but sometimes includes factors such as previous knowledge or location.

Platforms may have exceptionally different user interface capabilities and requirements (see Figure 1). In many cases adapting a user interface simply as a per user interface specification is not sufficient. For example, a combo-box widget as specified in the user interface description may be available on a particular platform, or not; there could be a similar widget or this widget should be transformed into a set of different widgets. To address this and to maintain the user interfaces' plasticity (Thevenin & Coutaz, 1999) additional adaptation effort is required. User interface plasticity is a user interface's capacity to preserve usability regardless of variations in the system hardware specification or operating environment.

Adapting to users includes a wide range of considerations: users' preferences, context, location, ambient environment, and so forth. This is a more complex transformation than adaptation

Figure 1. Some of the target platforms for mobile applications (two versions of smart phone and a web TV)

to a device as it requires knowledge about the user. The key to this adaptation is the ability to understand the user's actions, location and other parameters of interest, and to apply different artificial intelligence techniques to such information. To achieve this goal it is very important to gather information surrounding the user, but also to gather information coming from the user himself. This information coming from the user could be in the form of predefined preferences (e.g., color scheme preference) but, more importantly, the data about interactions between the user and his computer are of great value for our purposes. By observing and analyzing the user's interaction, his corresponding interface could be modified according to the type of interaction (e.g., using stylus or keyboard), tasks previously executed by the user, or tasks executed by other users.

Adapting to device capabilities requires a program capable of processing or interpreting the abstract user interface definition in such a way that it can be rendered to a particular device. Two major approaches are used to achieve this:

- Using a design-time tool to create concrete user interfaces for the required platforms
- Using a run-time tool or process to render a concrete user interface

Design-time tools are closer than run-time techniques to traditional user interface development and frequently encompass initial modeling, analysis and later development of multiple user interfaces. Such approaches include design patterns (Seffah & Forbrig, 2002), task models (Limbourg & Vanderdonckt, 2003; Molina et al., 2003;Paterno & Santoro, 2002) or off-line analysis of user interaction (Pitkow & Pirolli, 1999). This approach provides good facilities for designing static user interfaces, but is challenged by mobility and unanticipated situations.

The design-time tools are usually specialized stand-alone tools used by the user interface or software developer. Run-time tools are usually more generic mechanisms built into the application or programming framework.

Run-time tools offer higher flexibility than design-time tools and reduce effort for user interface developers. However, they are less capable of producing fine-tuned user interfaces, which can be produced by design-time methods on a case-by-case basis. The run-time approach is better suited for applications that are accessed remotely or applications that require execution on mobile devices with varied GUI and processing capabilities.

In the following sections we will present the benefits and drawbacks of both design-time and run-time approaches, with a particular emphasis on techniques used to adapt user interfaces to various devices.

Adaptation to Devices

Adapting user interfaces to devices requires a program capable of processing an abstract user interface definition to a device's capabilities. Such a program must adapt the abstract specification to the target (concrete) user interface specification. The concrete user interface specification must represent user interface design well and remain functional on the target device.

Two major approaches are used for such adaptations:

- **Design-time:** This encompasses using a design-time tool to transform the abstract definition into a user interface adapted to a concrete user interface's specific capabilities. This approach offers flexibility for the user interface developer to fine-tune the user interface's generation; but requires additional effort and is challenged by mobility, as it is not portable to devices with different capabilities, which is a common situation in mobile computing.
- **Run-time:** This method uses a run-time tool to render an abstract user interface specification for the device on which such a GUI is needed. This method does not allow user interface developers to fine-tune user interfaces, but gives higher flexibility and mobility to developed user interfaces at a much lower effort than design-time approaches.

Design-Time Adaptation

The design-time adaptation approach is based on generating user interfaces at design-time, as opposed to the run-time approach. Typically, a user interface definition is created for an application. This definition is then transformed using a tool into a concrete user interface for a specific device make or model. Such transformations usually generate different program source code for different interfaces (e.g., Visual Basic, Java or others). Source code is then compiled for the required platform and then executed on the platform. The most common approach is to define a multi-platform task model (Paterno & Santoro, 2002) and then to generate different user interfaces from this model. User interfaces generated in this way can be executed only on a specific device and platform; in contrast, the run-time approach tends to be much more versatile with respect to device capabilities.

Molina et al investigated an approach to define user interface using models and then automatically generating different programs (program source code) for different platforms (Molina et al., 2003). Code generators in this work are created for a limited number of different programming languages/platforms; however, additional code generators can be added to accommodate additional languages or platforms.

A similar approach, using an abstract user interface definition, was developed by Microsoft Corporation (2005) for its Longhorn/Avalon platform—it defines a user interface using an abstract notation, and then programmers develop code for different devices. This centralizes the user interface design but still requires multiple code implementations for different devices. In addition, Microsoft's approach is supported only on Windows platforms.

In general, a design-time approach offers some flexibility for the user interface developer because the resulting user interface can be manually fine-tuned before it is executed on the platform. However, the effort and expertise required for inspecting and fine-tuning multiple GUI versions before compilation and execution can be time-consuming and costly. Some of the design-time approaches still require programmers to develop multiple code implementations to handle user interface rendering and interaction on different platforms.

Run-Time Adaptation

Run-time adaptation is performed by using a program to adapt user interface definition at the time of program execution. Several approaches are used for run-time adaptation:

- **Standalone adaptation:** A specialized program adapts an abstract user interface definition to a specific platform.
- **Client-server adaptation:** A client program communicates with the server program in order to generate and present the user interface.
- **Mobile agent adaptation:** Mobile agents within a mobile agent platform compose a mobile application that generates a mobile user interface.

In this section we will examine standalone and client-server adaptation; mobile agents and approaches based on them are detailed in the next section.

A standalone adaptation is delivered through a specialized program that adapts an abstract user interface definition to a specific platform. An example of such an adaptation is a Windows program that adapts an XML user interface definition to the Windows platform. Multiple implementations of adaptation programs can be developed for different platforms—for Java platform, Palm PDAs, and so forth. Luyten and Coninx (2001) developed a platform that utilizes an abstract user interface definition that is later rendered by multiple middleware software to various platforms. This approach can provide application functionality by a specialized proxy server, rather than by the mobile device itself. In this case data and program functionality are then handled using Web services. This could present a limitation in situations where a wireless connection is not available or is not performing well.

A similar approach was adopted by Microsoft Corporation (2002), with "mobile forms" which automatically render to platforms supported by Microsoft's underlying engine, but which offers only a limited set of UI widgets (X Org, 1984). In addition, Microsoft's approach is available only on platforms supporting Windows and Microsoft Mobile Forms. This also restricts the number of available (supported) hardware platforms.

Such an adaptation approach lacks the mobility and flexibility required for mobile applications. The user interface is derived from a single specification, but it is pre-loaded onto devices and each software update triggers a new software set-up. The applications and user interfaces are frequently locked for a single device make and model. Development effort is high as multiple programs should be developed and verified for multiple platforms. In addition, mobile devices may not be capable of executing application functionality and, in such cases, specialized designs have to be developed. This introduces additional complexity and makes software more difficult to develop, maintain, and support.

The MobiLife project (IST MobiLife, 2006) investigates the creation of user interfaces for mobile devices. This project studies the use of multimodal user interfaces defined in an XML-based notation with the use of Web services (Baillie et al., 2005). Among other areas, MobiLife focuses on context awareness, sharing and personalization (Kernchen et al., 2006; Salden et al., 2005). The MobiLife approach is based on a client-server architecture (Baillie et al., 2005) and is limited in terms of network mobility and autonomy: platform-specialized program implementations (clients) are used—network mobility is limited only to compatible platforms; autonomy of clients is restricted—autonomy depends on the availability of corresponding server components.

The client-server adaptation is based on the client-server computation model. A server platform works in conjunction with clients and usually provides computation for (less capable) clients. For example, a Web application that is capable of transforming abstract UI definitions to both HTML and WML follows a client-server adaptation model. Such an approach requires a specialized client-side program capable of interpreting server-side information. Thus, multiple client programs must be developed for different devices that introduce difficulties in maintaining client programs and their versions for a variety of devices. In addition, a level of anticipation is required in this approach in order to create constantly functional client programs. For example, if the user interface definition for a particular device is correctly specified, but requires more processing power (e.g., the minimal memory required for processing all UI elements exceeds the available memory for the specific mobile device), then it is anticipated that the client-program would fail unless this has been anticipated by the client-program developer at the time of development.

IBM (2002) developed the WebSphere Transcoding Publisher server, which could transform a user interface specification into HTML, WML and other formats. The program could also handle JavaScript and transformations to WmlScript. However, this tool does not provide true mobility and is designed for Web applications. A similar approach, aimed mostly at Web development, is present in the Java community under the name Java Server Faces (JSF) (Sun Microsystems, 2005). JSF is focused mainly on Web development and is server oriented. This framework provides a component-based framework for Web user interface development. In addition, this framework allows the specification and development of alternative renders—a rendering engine can potentially render a custom user interface definition to different platforms. For example, a render could transform a JSP or XUL UI definition to WML.

In the server-client model, a client-side program must be developed for all supported platforms, which introduces additional development, support, and management complexity, and could pose limitations when adapting to different platforms. This approach, however, provides limited mobility for the applications, as one application can be used from a number of different platforms. The client-server model relies on a good network connection, which is a limiting factor in costly, unstable or low-performing mobile networks.

Mobile Agent Adaptation

This adaptation approach is based on mobile agent technology (FIPA, 2002; Milojicic et al., 1998). The mobile agent technology eases automatic system adaptation to its execution environment. A mobile agent is a program that executes autonomously on a set of network hosts on behalf of an individual or organization. Software agents can easily adapt their behavior to different contexts. Mobile agents can bring computation wherever needed and minimize network traffic, especially in wireless networks (expensive, slow and unstable). In addition, mobile agents do not decrease the system's performance, even when communications are based on a wired network, as shown in (Mitrovic, Royo & Mena, 2005). Mobile agents can arrive at the users' device and show their GUIs to the users in order to interact with them. Mobile agents can be hosted by platforms that support different models of user interfaces or that have different processing capabilities. Because mobile agents are autonomous they can handle communication errors (unreachable hosts, etc.) by themselves. Also, they can move to the target device instead of accessing target devices remotely. For instance, agents can be sent to a home computer supporting Java and Swing, or they can play the role of a proxy server for a wireless device, such as mobile telephone or a Web terminal; in that case they should produce an adequate GUI—WML or HTML (Mitrovic & Mena, 2002). Such an adaptation is not limited to Web pages or mobile phones; other devices such as PDAs benefit from such architectures (Mitrovic, Royo & Mena, 2005).

Mobile agents' platforms (Bursell & Ugai, 1997; Grasshopper, 2000; Ilarri, Trillo & Mena, 2006) are usually based on Java due to its portability, but are not limited to any particular platform (Wang, Sørensen & Indal, 2003). Mobile agent platforms are being developed for both old and new platforms such as Java Micro Edition (J2ME) (Wang et al., 2003).

Mobile agents can incorporate various learning techniques and learn from past experiences (Mitrovic & Mena, 2003). This is particularly important in the mobile world, where the abilities to learn and adapt to contexts and users are some of the most important requirements.

The mobile agent approach brings higher mobility and flexibility of user interaction while reducing development complexity—only one version of a program is defined.

To summarize, user interface adaptation to different devices is difficult; approaches are challenged by true mobility, require multiple development efforts, and have various degrees of flexibility and transparency. Mobile agents eliminate the majority of these issues and provide a flexible platform for the development of adaptive and portable GUIs.

Adaptation to Users

Different approaches are used to adapt user interfaces to meet users' preferences and contexts. Similar to adaptation to devices, adaptation to users can be done at design-time or at run-time. The design-time techniques include design patterns (Seffah & Forbrig, 2002), analysis of UI usage (Pitkow & Pirolli, 1999) or usability tests followed by UI redesign. Run-time techniques are present mostly in the Web arena, where Web-page content is rearranged by users' preferences or the Website's context (Amazon Online bookshop, 2006). Run-time techniques are based mostly on statistical and data mining techniques (Zukerman & Albrecht, 2000), which apply various artificial intelligence techniques to interaction data.

An additional challenge for mobile applications and run-time adaptation is users' attention (Vertegaal, 2003). Users' attention is a limited resource that is required by many external stimuli, including software applications. It is important to prioritize and adapt requests coming from software applications toward user in order to most effectively utilize this resource. Therefore, user interaction timing and volume must be considered when designing attentive user interfaces. Monitoring users' attention through physical indicators such as eye movement, geographic location, or statistical data mining is crucial for prioritizing, adapting and designing interaction tasks (Vertegaal, 2003).

To be able to effectively adapt UIs to users at the run-time, it is crucial that systems provide facility for collecting information on computer-user interaction. Such facility enables later analysis and exchange of collected information, which could lead to changes in the application logic or user interface. Therefore, another requirement for successful UI adaptation to users is a facility for changing the user interface at run-time. Using this facility, or architecture, the user interface could be easily amended according to the collected information.

The majority of applications, providing such facilities, are not designed with interoperability in mind and focus only on the current application's ability to gather and analyze interaction information. Applications frequently cannot exchange interaction information, and applications programmed by different developers cannot use a common set of facilities. This leads to a multiplication of developments designed for just a single application or application vendor. However, some systems based on mobile agents can be used by multiple applications and different (independent) learning modules.

ADUS: AN APPROACH BASED ON MOBILE AGENTS

The following introduces ADUS, our proposal for indirect generation of adaptive and portable GUIs. ADUS—ADaptive User Interface System (Mitrovic et al., 2004) is a system based on mobile agent technology that helps with user interface generation and allows monitoring of user-computer interaction. ADUS is part of the ANTARCTICA system (Goñi, Illarramendi, Mena et al., 2001)—a multi-agent system that provides users with different wireless data services to enhance their mobile devices' capabilities.

ADUS has three main functions: (1) transparently adapting an abstract user interface definition to a concrete platform, (2) monitoring user interaction and communicating this information to other agents, and (3) communicating and collaborating with other agents in the ANTARCTICA platform. ADUS also performs a number of agent-based functions such as optimized network operation and collaboration with other agents. The ADUS system uses XUL (XUL Tutorial, 2002), an XML-based abstract user interface definition language.

Adaptive User Interface Generation in ADUS

In this section we present and discuss several approaches to generate adaptive user interface allowing the monitoring of the user behavior.

Option 1: The Visitor Agent Generates the GUI

The first approach is when the visitor agent arrives at the user's device it requests from the user agent the available resources, the user's preferences and the device's capabilities. Then the visitor agent creates the GUI by itself and interacts with the user directly.

This approach solves the generation of customized GUIs. However, it still has several problems:

- The user agent cannot monitor the user's behavior because the data provided to the GUI flow directly to the visitor agent.
- The user agent must trust the visitor agent to render a GUI according to the user's preferences and the device's capabilities. Visitor agents could ignore the user agent descriptions and try to show their own GUI (in that case, the type of GUI created by the visitor agent could not be executed on that device).
- All the visitor agents have to know how to process and apply the knowledge provided by the user agent (which implies that they all must know how to generate any kind of GUI).

Option 2: The User Agent Generates the GUI and Delegates Event Handling to the Visitor Agent

In this approach the visitor agent, after arriving at the user's device, provides the user agent with a specification of the needed GUI. Then the user agent generates a GUI according to the user's preferences, the device's capabilities and the visitor agent's requirements, and it delegates the GUI event handling to the visitor agent.

The user interface specification can be made in the Extensible User-interface Language (XUL) (XUL Tutorial, 2002). This interface definition can be later adapted by the user agent using XSL transformations to the required GUI representation language (HTML, WML, etc.). The XUL interpretation on Java-enabled platforms is interpreted by a Java XUL platform that renders XUL using standard AWT and Swing widgets.

The advantages of this approach are:

- The user agent guarantees that the visitor agents' GUIs will be generated correctly (according to the user's preferences and the device's capabilities) if they are specified in XUL.
- Visitor agents do not need to know how to generate GUIs in different devices.
- The user agent can deny permission to generate GUIs to all visitor agents (Mitrovic & Arronategui, 2002) in order to avoid direct GUI generation.

However, following this approach, the user agent cannot monitor the user's behavior because GUI events are handled directly by visitor agents. Therefore the user agent must trust the visitor agent to get the information about its interaction with the user.

Option 3: An Intermediate Agent Generates the GUI and Handles the Events

In this approach, first the visitor agent sends its XUL specification of the GUI to the user agent, then the user agent generates the GUI and handles all the events (it receives data from the user), and finally, it sends the user data back to the visitor agent.

This approach has all the advantages of the approaches presented above. Furthermore, it allows the user agent to monitor the user's behavior easily and efficiently as it handles the GUI events. Although this approach is interesting, its implementation faces a problem: the user agent must attend the different services executed on the user's device, and some tasks, such as GUI generation, could overload it. Therefore, a better approach is for the user agent to delegate the generation of adaptive GUIs to a specialized agent (the ADUS agent). Thus, the distribution of the service execution across the three agents (the ADUS agent, the user agent and the visitor agent) allows us to balance the system's load.

Indirect Generation of GUIs

This section describes in more detail the architecture needed to efficiently generate adaptive GUIs. To illustrate this process we use an example application. As shown in Figure 2, the system contains the following agents:

- **The visitor agent:** This is a mobile agent that brings a service requested by the user to the user's device. This agent can generate an XUL specification of the GUIs that it needs to interact with the user. Such XUL specifications are sent to the user agent on the user's device.
- **The user agent:** This is a highly specialized personalization agent that is in charge of storing as much information as possible about the user and his computer. For example, it knows the user's look and feel preferences and the GUI preferred by the user or imposed by the user's device or the operating system. This agent's main goals are: (1) to proxy the generation of user interfaces, (2) to help the user to use the visitor agent's services, (3) to modify the visitor agent's GUI specification according to the user's preferences, (4) to create an ADUS agent initialized with the static GUI features such as the user's device capabilities, and (5) to monitor user interactions by receiving such information from the ADUS agent.
- **The ADUS agent:** This agent's main activities are: (1) to adapt the user interface to the user's preferences and device capabilities, following the user agent's suggestions; (2) to generate GUIs for different devices according to the XUL specification; and (3) to handle GUI events and communicate them to the visitor and user agents (allowing the user agent to monitor the user's interaction). There is one ADUS agent per visitor agent.

The following describes the synchronization of the previously mentioned agents. As an example, we use a simple currency converter application that converts between currencies per a user's requests, and displays the results of the conversion. This application is executed by mobile agents that travel to the user's device when requested by the user. The main steps are (see Figure 2):

Figure 2. Indirect generation of GUIs

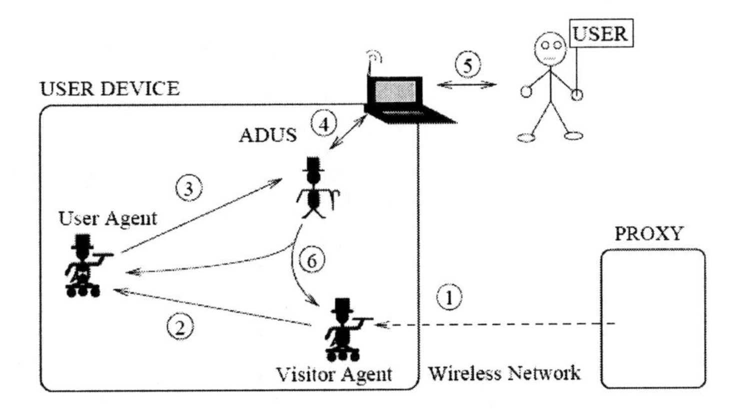

1. The visitor agent travels to the user's device. This step is only for approaches that are based on mobile agents. For example, it is equivalent to the call of a local application (in client-server architecture).
   The visitor agent requests the generation of its GUI. In this step the visitor agent sends the XUL description of its GUI to the user agent. In Figure 3 we show the XUL definition of the GUI for the currency converter service.
2. The user agent processes the GUI specification, transforms the GUI description to adapt it to the user's preferences, and creates the corresponding ADUS agent initialized with: (1) the transformed XUL description of the GUI to generate, and (2) the static information for the GUI such as the device's capabilities, screen resolution, representation language of the user's device (WML, HTML, Java Swing, etc.), among other information.
3. The ADUS agent generates the GUI according to the information provided by the user agent (static GUI information and specific information for this service). The ADUS agent can map any XUL description into GUIs for devices with different features; for example, a WAP device or a laptop with a Java GUI.

In the example, if the converter application were executed on a device with Java Swing capabilities (e.g., a home PC or laptop) the ADUS agent would generate a Swing GUI (see Figure 4a). When the converter application is executed on a Web terminal without applet support, the generated GUI would be based on HTML, as shown in Figure 4b. Finally, if it is executed on a WAP mobile phone, then the GUI is based on WML, as shown in Figure 4c. We point out that this functionality of the ADUS agent works for any XUL specification. The ADUS agent could be extended with mappings to other kinds of GUI languages, such as Macromedia Flash or J2ME.

4. The user interacts with the GUI by looking at the information presented on the screen device and by using the device's peripherals (keyboard, mouse, buttons, etc.) to enter data or select among different options.
5. The ADUS agent handles and propagates the GUI events. User actions trigger GUI events that are captured by the ADUS agent. This information is sent to: (1) the visitor agent, which reacts to user actions according to the service that it executes, perhaps by generating a new GUI (step 2), and (2) the user agent, which can store and analyze the information provided by the user in order to reuse it in future service executions. One of the advantages of the presented architecture is that both messages can be sent concurrently, so the load is balanced.

Figure 3. XUL definition for the currency converter service

```
 ...
  <vbox>  <hbox>
    <label control="lblQty" value="Quantity:"/>
    <textbox value="0.00" id="Qty" size="20"/>
  </hbox>  </vbox>
  ...
  <-- label that will be used for the Output -->
  <box>
   <label control="lblOutput" value="Result:  "/>
  </box>
  <box>
   <label id="Output" control="Output" value=""/>
  </box>
<!-- adding button -->
 <box>
   <button id="Convert" label="Convert"  oncommand="convert()"/>
  </box>
 ...
```

Figure 4. Currency converter service (swing, HTML and WAP transformation)

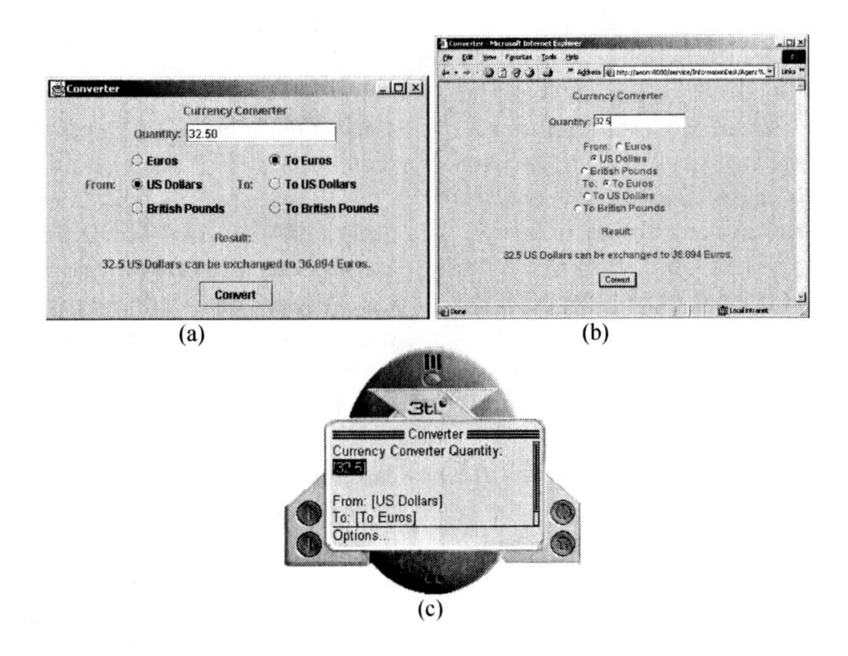

Finally, we would like to stress the relevance to the user agent of monitoring the user's interaction with visitor agents. By knowing the user's reactions and data entered on those services, the user agent can store such data locally and apply different artificial intelligence techniques to extract knowledge about the user's behavior. In the previous example, the next time that the currency converter service executes on the user's device, the user agent could select (in the visitor agent's XUL specification) U.S. dollars and Euros as the initial and target currencies, respectively, because that was the user's selection during the previous execution of that service. Even if the user now selects another configuration, the user agent could learn and improve its behavior for the next time. In addition, the user interface could be rearranged to meet the user's preferences, for example the "convert" button could be in a different part of the screen. Thus, the customization of GUIs can become really useful for the user, as the user agent can monitor, store and analyze his interactions with all the GUIs/applications. In this case, mobile agents would learn from interaction data of other applications and from interaction data collected by other users (Mitrovic & Mena, 2003).

FUTURE TRENDS

In our increasingly mobile world, human-computer interaction has become one of the most important topics. Mobile user interface systems are more and more required to be context and location-aware, and we should improve interfaces by applying knowledge-based techniques. The mobility of systems providing a mobile user interface is an important factor, and mobile agents are one of the promising technologies in this field. Significant improvement in mobile agent platforms' scalability and performance is underway (Ilarri et al., 2006).

From the adaptation perspective, trends are likely to be the further standardization of UI definition languages. Currently many different variants of XML-based user interface languages exist, which makes collective, community efforts less efficient than they would be if only one such language existed. Industry bodies such as the WWW Consortium have already begun proposing standard approaches, such as XForms (WWW Consortium, XForms, 2003). Improvements in mobile user interfaces' plasticity are a challenge yet to be fully addressed by researchers.

Consolidating mobile devices' capabilities and constantly increasing the performance of devices and infrastructure are two of the forces driving mobile systems and mobile user interface systems. Because processing power and different capabilities are two of the key issues for mobile user interfaces, this will positively affect both research efforts and developments. Although mobile devices are constantly being improved, certain device properties, such as screen size, will require methods that allow adaptation to users' devices.

In the agent-based user interfaces area, future trends are in increased interoperability across different multi-agent platforms. This will improve agent-based user interface systems' cross-platform mobility.

CONCLUSION

Generation of adaptive user interfaces in mobile environments is a complex task that faces many challenges. Mobile devices have different capabilities, work in different contexts, and have different kinds of users. Users require adaptation to their context and preferences. Mobile networks remain an expensive and unreliable medium, which presents additional difficulties. Many different approaches are used to solve user interfaces' problems in a mobile environment, from design-time considerations to various run-time solutions.

Design-time solutions are inclined more toward traditional approaches and are based more on multi-modal user interface modeling and code developments for different mobile devices than are run-time solutions. These approaches do not offer mobility and could suffer from mobile devices' low processing capability. Design-time approaches usually allow fine tuning of the user interface code (before deployment), which requires an additional development effort and expertise in the specific mobile system.

Run-time solutions offer a degree of mobility and higher flexibility than design-time solutions. The quality of the automatically generated interfaces is lower than in design-time approaches because of the lack of fine-tuning the user interface code before deployment. Run-time systems based on a client-server model give a degree of mobility but are limited by the requirement for mobile networks' constant availability. In addition, some kind of client software is required, which represents an additional effort.

However, mobile agents as a run-time approach provide good mobility for user interfaces and increased flexibility over other approaches. Mobile user interfaces based on mobile agents can transparently (to developers and users) adapt user interfaces to devices "on the fly." In addition, agents can distribute the processing load appropriately so that devices with low processing capabilities can still execute complex applications. Systems based on mobile agents can monitor and analyze computer-user interaction, and share such information between different program instances, users or systems. Such information can then be used to improve application logic and to create additional user interface adaptations. As mobile agents are fully mobile software entities, they enable mobile-agent based software and user interfaces to follow users wherever required.

Topics yet to be fully explored by researchers in the agent-based user interfaces area include inter-platform interoperability. Providing higher plasticity than is currently seen in mobile user interfaces remains a challenging task for all mobile user interfaces.

REFERENCES

Amazon online bookshop. (2006). Retrieved from Amazon: www.amazon.com

Baillie, L., Schatz, R., Simon, R., Anegg, H., Wegscheider, F., Niklfeld, G., & Gassner, A. (2005). Designing Mona: User interactions with multimodal mobile applications. *HCI International 2005, 11th International Conference on Human-Computer Interaction* (pp. 22-27).

Bursell, M., & Ugai, T. (1997). *Comparison of autonomous mobile agent technologies.* Retrieved from APM: www.ansa.co.uk/ANSATech/97/Primary/198901.pdf

FIPA, Foundation for Intelligent Physical Agents. (2002). Retrieved from www.fipa.org

Goñi, A., Illarramendi, A., Mena, E., Villate, Y., & Rodriguez, J. (2001). ANTARCTICA: A Multiagent System for Internet Data Services in a Wireless Computing Framework. *NSF Workshop on an Infrastructure for Mobile and Wireless Systems* Vol. 2538 (pp. 119-135). Germany: Springer Verlag Lecture Notes in Computer Science LNCS.

Grasshopper. (2002). Retrieved from http://www.cordis.europa.eu/infowin/acts/analysys/products/thematic/agents/ch4/ch4.htm

IBM. (2002). Retrieved from IBM WebSphere Transcoding Publisher: http://www-3.ibm.com/software/webservers/transcoding/

Ilarri, S., Trillo, R., & Mena, E. (2006, June). SPRINGS: A scalable platform for highly mobile agents in distributed computing environments. In *proceedings of the 4th International WoWMoM 2006 workshop on Mobile Distributed Computing (MDC'06).* (pp. 633-637). New York: IEEE Computer Society.

IST MobiLife. (2006). Retrieved from http://www.ist-mobilife.org/: IST MobiLife

Kernchen, R., Bonnefoy, D., Battestini, A., Mrohs, B., Wagner, M., & Klemettinen, M. (2006). Context-Awareness in MobiLife. *15th IST Mobile Summit, Mykonos, Greece.*

Limbourg, Q., & Vanderdonckt, J. (2003). Comparing task models for user interface design. *The Handbook of Task Analysis for Human-Computer Interaction.* Lawrence Erlbaum Associates.

Luyten, K., & Coninx, K. (2001). An XML runtime user interface description language for mobile computing devices. In Johnson, C. (Ed.), *Design, Specification and Verification of Interactive Systems - DSV-IS 2001* (pp 1-15), Germany: Springer Verlag Lecture Notes in Computer Science.

Microsoft Corporation. (2002). *Creating mobile web applications with mobile web forms in visual studio.NET.* Creating Mobile Web Applications with Mobile Web Forms in Visual Studio.NET. Retrieved from http://msdn.microsoft.com/vstudio/technical/articles/mobilewebforms.asp

Microsoft Corporation. (2005). *Longhorn Developer Center Home: Avalon.* Avalon. Retrieved from http://msdn.microsoft.com/Longhorn/understanding/pillars/avalon/default.aspx

Milojicic, D., Breugst, M., Busse, I., Campbell, J., Covaci, S., Friedman, B., Kosaka, K., Lange, D., Ono, K., Oshima, M., Tham, C., Virdghagriswaran, S., & White, J. (1998). MASIF, The OMG Mobile Agent System Interoperability Facility. *Mobile Agents,* Vol. 1477 (pp 50-68). Germany: Springer Verlag Lecture Notes in Artificial Intelligence - LNAI.

Mitrovic, N., & Arronategui, U. (2002). Mobile agent security using proxy-agents and trusted domains. *Second International Workshop on Security of Mobile Multiagent Systems (SEMAS 2002)* (pp. 81-83). Germany: German AI Research Center DFKI.

Mitrovic, N., & Mena, E. (2002, June). Adaptive user interface for mobile devices. In P. Forbrig, Q. Limbourg, B. Urban & J. Vanderdonckt (Eds.), *Interactive systems. Design, specification, and verification. 9th International Workshop DSV-IS 2002,* vol. 2545 (pp. 47-61). Germany: Springer Verlag Lecture Notes in Computer Science LNCS.

Mitrovic, N., & Mena, E. (2003). Improving user interface usability using mobile agents. In *Interactive Systems. Design, Specification, and Verification. 10th International Workshop DSV-IS 2003,* vol. 2844 (pp. 273-288). Madeira, Portugal: Springer Verlag Lecture Notes in Computer Science LNCS.

Mitrovic, N., Royo, J. A., & Mena, E. (2004). ADUS: Indirect Generation of User interfaces on Wireless Devices. *Fifteenth International Workshop on Database and Expert Systems Applications (DEXA)* (pp. 662-666). IEEE Computer Society.

Mitrovic, N., Royo, J.A., & Mena, E. (2005, September). Adaptive user interfaces based on mobile agents: Monitoring the behavior of users in a wireless environment. *Symposium on Ubiquitous Computing and Ambient Intelligence (UCAmI 2005)* (pp. 371-378). Spain: Thomson-Parninfo.

Molina, P. J., Belenguer, J., & Pastor, O. (2003, July). Describing just-UI concepts using a task notation. In J. Joaquim, N. Jardim Nunes & J. Falcao e Cunha (Eds.), *Interactive systems. Design, specification, and verification. 10th International Workshop DSV-IS 200,* vol. 2844 (pp. 218-230). Germany: Springer Verlag Lecture Notes in Computer Science LNCS.

Paterno, F., & Santoro, C. (2002). One model, many interfaces. In C. Kolski & J. Vanderdonckt (Eds.), *Fourth International Conference on Computer-Aided Design of User Interfaces* (pp. 143-154). Dordrecht: Kluwer Academics.

Pitkow, J., & Pirolli, P. (1999). Mining longest repeatable subsequences to predict world wide web surfing. In *proceedings of the 2nd Usenix Symposium on Internet Technologies and Systems (USITS)*, (pp. 139-150). USITS.

Salden, A., Poortinga, R., Bouzid, M., Picault, J., Droegehorn, O., Sutterer, M., Kernchen, R., Rack, C., Radziszewski, M., & Nurmi, P. (2005). Contextual personalization of a mobile multimodal application. *International Conference on Internet Computing (ICOMP)* (pp. 650-665). Las Vegas, USA: International Conference on Internet Computing (ICOMP).

Seffah, A., & Forbrig, P. (2002). Multiple user interfaces: Towards a task-driven and patterns-oriented design model. In *proceedings of the 9th International Workshop on Design, Specification and Verification of Interactive Systems (DSV-IS)* Vol. 2545 (pp. 118-133). Germany: Springer Verlag Lecture Notes in Computer Science LNCS.

Stottner, H. (2001). *A platform-independent user interface description language* (16). Linz, Austria: Institute for Practical Computer Science, Johannes Kepler University.

Sun Microsystems. (2005). *Java Server Faces.* Retrieved from JSF: http://java.sun.com/j2ee/javaserverfaces/

Thevenin, D., & Coutaz, J. (1999). Plasticity of user interfaces: Framework and research agenda. In *Proceedings of IFIP TC 13 International Conference on Human-Computer Interaction INTERACT'99, Vol. 1* (pp. 110-117). IOS Press.

Vertegaal, R. (2003). Attentive user interfaces. *Communications of the ACM, 46*(3), 30-33

W3C. (2000). Retrieved from Extensible Markup Notation: http://www.w3.org/TR/REC-xml

Wang, A.I., Sørensen, C.F., & Indal, E. (2003). A mobile agent architecture for heterogeneous devices. *3rd IASTED International Conference on Wireless and Optical Communications (WOC 2003, ISBN 0-88986-374-1* (pp. 383-386). Norway: ACTA Press.

WWW Consortium, XForms. (2003). Retrieved from X Forms: www.xforms.org

X Org. (1984). *X Windows.* Retrieved from X Org: www.x.org

XUL Tutorial. (2002). Retrieved from XUL: http://www.xulplanet.com/tutorials/xultu/

Zukerman, I., & Albrecht, D. (2000). Predictive statistical models for user modeling. In A. Kobsa (Ed.), *User Modeling and User Adapted Interaction (UMUAI)*, Vol. 11, (pp. 5-18). *The Journal of Personalization Research. Tenth Aniversary Special Issue.* The Netherlands: Kluwer Academic Publishers.

NOTE

This work is supported by the CYCYT project TIN2004-07999-C02-02.

KEY TERMS

Abstract User Interface Definition: Platform-independent and technology-neutral description of the user interface.

Design-Time User Interface Adaptation: Manual adaptation of the user interface by a designer, analyst or software developer.

Indirect User Interface Generation: A method in which the mobile agent requiring user interaction does not create a user interface directly, but passes the user interface definition to another agent (specialized for the user and his mobile device) that creates the above mentioned user interface and acts as an intermediary between the user and the agent that requires interaction with the user.

Mobile Agent: A program that executes autonomously on a set of network hosts on behalf of an individual or organization. One of the key features of such agents is mobility.

Multi-Agent System: A system that allows concurrent operation and communication of multiple (mobile) agents.

Run-Time User Interface Adaptation: Automatic adaptation of the user interface by a program during its execution.

Transparency: Automatic adaptation to specific conditions or circumstances without implicit or explicit intervention from the user, user interface designer or software developer.

User Interface Plasticity: A user interface's capacity to preserve usability regardless of variations in systems' hardware specification or operating environment.

Chapter XX
Intelligent User Interfaces for Mobile Computing

Michael J. O'Grady
University College Dublin, Ireland

Gregory M.P. O'Hare
University College Dublin, Ireland

ABSTRACT

In this chapter, the practical issue of realizing a necessary intelligence quotient for conceiving intelligent user interfaces (IUIs) on mobile devices is considered. Mobile computing scenarios differ radically from the normal fixed workstation environment that most people are familiar with. It is in this dynamicity and complexity that the key motivations for realizing IUIs on mobile devices may be found. Thus, the chapter initially motivates the need for the deployment of IUIs in mobile contexts by reflecting on the archetypical elements that comprise the average mobile user's situation or context. A number of broad issues pertaining to the deployment of AI techniques on mobile devices are considered before a practical realisation of this objective through the intelligent agent paradigm is presented. It is the authors hope that a mature understanding of the mobile computing usage scenario, augmented with key insights into the practical deployment of AI in mobile scenarios, will aid software engineers and HCI professionals alike in the successful utilisation of intelligent techniques for a new generation of mobile services.

INTRODUCTION

Mobile computing is one of the dominant computing usage paradigms at present and encapsulates a number of contrasting visions of how best the paradigm should be realized. Ubiquitous computing (Weiser, 1991) envisages a world populated with artefacts augmented with embedded computational technologies, all linked by transparent high-speed networks, and accessible in a seamless anytime, anywhere basis. Wearable computing (Rhodes, Minar, & Weaver, 1999) advocates a world where people carry the necessary computational artefacts about their actual person.

Somewhere in between these two extremes lies the average mobile user, equipped with a PDA or mobile phone, and seeking to access both popular and highly specialized services as they go about their daily routine.

Though the growth of mobile computing usage has been phenomenal, and significant markets exists for providers of innovative services, there still exist a formidable number of obstacles that must be surpassed before software development processes for mobile services becomes as mature as current software development practices. It is often forgotten in the rush to exploit the potential of mobile computing that it is radically different from the classic desktop situation; and that this has serious implications for the design and engineering process. The dynamic nature of the mobile user, together with the variety and complexity of the environments in which they operate, provides unprecedented challenges for software engineers as the principles and methodologies that have been refined over years do not necessarily apply, at least in their totality, in mobile computing scenarios.

How to improve the mobile user's experience remains an open question. One approach concerns the notion of an application autonomously adapting to the prevailing situation or context in which end-users find themselves. A second approach concerns the incorporation of intelligent techniques into the application. In principle, such techniques could be used for diverse purposes, however, intelligent user interfaces (IUIs) represent one practical example where such techniques could be usefully deployed. Thus the objective of this chapter is to consider how the necessary intelligence can be effectively realized such that software designers can realistically consider the deployment of IUIs in mobile applications and services.

BACKGROUND

Research in IUIs has been ongoing for quite some time, and was originally motivated by problems that were arising in standard software application usage. Examples of these problems include information overflow, real-time cognitive overload, and difficulties in aiding end-users to interact with complex systems (Höök, 2000). These problems were perceived as being a by-product of direct-manipulation style interfaces. Thus, the concept of the application or user interface adapting to circumstances as they arose was conceived and the terms "adaptive" or "intelligent" user interfaces are frequently encountered in the literature. How to effectively realize interfaces endowed with such attributes is a crucial question and a number of proposals have been put forward. For example, the use of machine learning techniques has been proposed (Langley, 1997) as has the deployment of mobile agents (Mitrovic, Royo, & Mena, 2005).

In general, incorporating adaptability and intelligence enables applications to make considerable changes for personalization and customization preferences as defined by the user and the content being adapted (O'Connor & Wade, 2006). Though significant benefits can accrue from such an approach, there is a subtle issue that needs to be considered. If an application is functioning according to explicit user defined preferences it is functioning in a manner that is as the user expects and understands. However, should the system autonomously or intelligently adapt its services based on some pertinent aspect of the observed behavior of the user, or indeed, based on some other cue, responsibility for the system behavior moves, albeit partially, from the user to the system. Thus, the potential for a confused user or unsatisfactory user experience increases.

A natural question that must now be addressed concerns the identification of criteria that an application might use as a basis for adapting its behavior. Context-aware computing (Schmidt, Beigl & Gellersen, 1999) provides one intuitive answer to this question. The notion of context first arose in the early 1990s as a result of pioneering experiments in mobile computing systems. Though an agreed definition of context has still not materialized, it concerns the idea that an application should factor in various aspects of the prevailing situation when offering a service. What these aspects might be is highly dependent on the application domain in question. However, commonly held aspects of context include knowledge of the end-user, for example through a user model; knowledge of the surrounding environment, for example through a geographic information system (GIS) model; and knowledge of the mobile device, for example through a suitably populated database. Other useful aspects of an end-user's context include an understanding of the nature of the task or activity currently being engaged in, knowledge of their spatial context, that is, location and orientation, and knowledge of the prevailing social situation. Such models can provide a sound basis for intelligently adapting system behavior. However, capturing the necessary aspects of the end-user's context and interpreting it is frequently a computationally intensive process, and one that may prove intractable in a mobile computing context. Indeed, articulating the various aspects of context and the interrelationships between them may prove impossible, even during system design (Greenberg, 2001). Thus, a design decision may need to be made as to whether it is worth working with partial or incomplete models of a user's context. And the benefit of using intelligent techniques to remedy deficiencies in context models needs to be considered in terms of computational resources required, necessary response time and the ultimate benefit to the end-user and service provider.

SOME REFLECTIONS ON CONTEXT

Mobile computing spans many application domains and within these, it is characterized by a heterogeneous landscape of application domains, individual users, mobile devices, environments and tasks (Figure 1). Thus, developing applications and services that incorporate a contextual component is frequently an inherently complex and potentially time-consuming endeavor, and the benefits that accrue from such an approach should be capable of being measured in some tangible way. Mobile computing applications tend to be quite domain specific and are hence targeted at specific end-users with specialized tasks or objectives in mind. This is in contrast to the one-size-fits-all attitude to general purpose software development that one would encounter in the broad consumer PC arena. For the purposes of this discussion, it is useful to reflect further on the following aspects of the average mobile user's context: end-user profile, devices characteristics, prevailing environment and social situation.

User Profile

Personalization and customization techniques assume the availability of sophisticated user models, and currently form an indispensable component of

Figure 1. An individual's current activity is a notoriously difficult aspect of an individual's context to ascertain with certainty

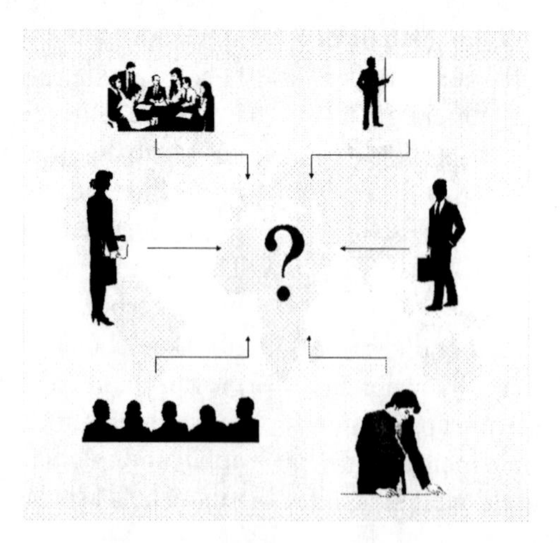

a number of well-known e-commerce related Web sites. Personalizing services for mobile computing users is an attractive proposition in many domains as it offers a promising mechanism for increasing the possibility that the end-users receive content that is of interest to them. Though this objective is likewise shared with owners of e-commerce sites, there are two issues that are of particular importance when considering the mobile user. Firstly, mobile interactions are almost invariably short and to the point. This obligates service providers to strive to filter, prioritize, and deliver content that is pertinent to the user's immediate requirements. The second issue concerns the question of costs. Mobile users have to pay for services, which may be charged on a KB basis, thus giving mobile users a strong incentive to curtail their use of the service in question if dissatisfied.

A wide number of features and characteristics can be incorporated into user models. As a basic requirement, some information concerning the user's personal profile, for example, age, sex, nationality and so on, is required. This basic model may then be augmented with additional sub-models that become increasingly domain-specific. In the case of standard e-commerce services, a record of the previous purchasing history may be maintained and used as a basis for recommending further products. Electronic tourist guides would require the availability of a cultural interest model, which as well as indicating cultural topics of interest to the user, would also provide some metric that facilitated the prioritization of their cultural interests.

Device Characteristics

Announcements of new devices are occurring with increasing frequency. Each generation successively increases the number of features offered, some of which would not be associated with traditional mobile computing devices, embedded cameras and MP3 players being cases in point. Though offering similar features and services, there are subtle differences between different generations, and indeed interim releases within the same generation, that make the life of a service provider and software professional exceedingly difficult and frequently irritating. From an interface perspective, screen size and support for various interaction modalities are two notable ways in which devices differ, and these have particular implications for the end-user experience. This problem is well documented in the literature and a number of proposals have been put forward to address this, the plasticity concept being a notable example (Thevenin & Coutaz, 1999). Other aspects in which mobile devices differ include processor, memory and operating system; all of which place practical limitations on what is computationally feasible on the device.

Prevailing Environment

The notion of environment is fundamental to mobile computing and it is the dynamic nature of prevailing environment in which the mobile user operates that most distinguishes mobile computing from the classic desktop usage paradigm. As an illustration, the case of the physical environment is now considered, though this in no way diminishes the importance of the prevailing electronic infrastructure. Scenarios in which mobile computing usage can occur are multiple and diverse. The same goes for physical environments. Such environments may be hostile in the sense that they do not lend themselves to easily accessing electronic infrastructure such as telecommunications networks. Other environments may experience extreme climatic conditions thus causing equipment to fail.

Developing a service that takes account of or adapts to the local physical environment is an attractive one. Two prerequisites are unavoidable, however. A model of the environment particular to the service domain in question must be available, and the location of the end-user must be attainable. In the former case, the service provider must construct this environmental model, possibly an expensive endeavor in terms of time and finance. In the latter case, an additional technological solution must be engaged—either one based on satellites, for example GPS, or one that harnesses the topology of the local wireless telecommunications networks. Each solution has its respective advantages and disadvantages, and a practical understanding of each is essential. However, by fulfilling these prerequisites, the service provider is in a position to offer services that take the end-users' physical position into account. Indeed, this vision, often termed location-aware computing (Patterson, Muntz & Pancake, 2003), has grasped the imagination of service providers and end-users alike. In essence, it is a practical example of just one single element of an end-user's context being interpreted and used as a basis for customizing services.

Social Situation

Developing a service that adapts to the end-user's prevailing social context is fraught with difficulty, yet is one that many people would find useful. What exactly defines social context is somewhat open to interpretation but in this case, it is considered to refer to the situation in which end-users find themselves relevant to other people. This is an inherently dynamic construct and capturing the prevailing social situation introduces an additional level of complexity not encountered in the contextual elements described previously.

In limited situations, it is possible to infer the prevailing social situation. Assuming that the end-user maintains an electronic calendar, the detection of certain keywords may hint at the prevailing social situation. Examples of such keywords might include lecture, meeting, theatre and so on. Thus, an application might reasonably deduce that the end-user would not welcome interruptions, and, for example, proceed to route incoming calls to voicemail and not alert the end-user to the availability of new email. Outside of this, one has to envisage the deployment of a suite of technologies to infer social context. For example, it may be that a device, equipped with a voice recognition system, may be trained to recognize the end-user's voice, and on recognizing it, infer that a social situation is prevailing. Even then, there may be a significant margin of error; and given the power limitations of the average mobile device, running a computationally intensive voice recognition system continuously may rapidly deplete battery resources.

ARTIFICAL INTELLIGENCE IN MOBILE COMPUTING

Artificial intelligence (AI) has been the subject of much research, and even more speculation, for almost half a century by now. Though failing to radically alter the world in the way that was envisaged, nevertheless, AI techniques have been successfully harnessed in a quite a number of select domains and their incorporation into everyday applications and services continues unobtrusively yet unrelentingly. Not surprising, there is significant interest amongst the academic community in the potential of AI for addressing the myriad of complexity that is encountered in the mobile computing area. From the previous discussion, some sources of this complexity can be easily identified. Resource management, ambiguity resolution, for example, in determining contextual state and resolving user intention in multimodal interfaces, and adaptation, are just some examples. Historically, research in AI has focuses on various issues related to these very topics. Thus, a significant body of research already exists in some of the very areas that can be harnessed to maximum benefit in mobile computing scenarios. A detailed description of these issues may be found elsewhere (Krüger & Malaka, 2004).

One pioneering effort at harnessing the use of intelligent techniques on devices of limited computational capacity is the Ambient intelligence (AmI) (Vasilakos & Pedrycz, 2006) initiative. AmI builds on the broad mobile computing vision as propounded by the ubiquitous computing vision. It is of particular relevance to this discussion as it is essentially concerned with usability and HCI issues. It was conceived in response to the realization that as mobile and embedded artefacts proliferate, demands for user attention would likewise increase, resulting in environments becoming inhabitable, or more likely, people just disabling the technologies in question. In the AmI concept, IUIs are envisaged as playing a key role in mediating between the embedded artefacts and surrounding users. However, AmI does not formally ratify the use of any particular AI technique. Choice of technique is at the discretion of the software designer whose selection will be influenced by a number of factors including the broad nature of the domain in question, the requirements of the user, the capability of the available technology and the implications for system performance and usability.

Having motivated the need for AI technologies in mobile contexts, practical issues pertaining to their deployment can now be examined.

STRATEGIES FOR HARNESSING AI TECHNIQUES IN MOBILE APPLICATIONS

It must be reiterated that AI techniques are computationally intensive. Thus, the practical issue of actually incorporating such techniques into mobile applications needs to be considered carefully. In particular, the implications for performance must be determined as this could easily have an adverse effect on usability. There are three broad approaches that can be adopted when incorporating AI into a mobile application and each is now considered.

Network-Based Approach

Practically all mobile devices are equipped with wireless modems allowing access to data services. In such circumstances, designers can adopt a kind of client/server architecture where the interface logic is hosted on the mobile devices and the core application logic deployed on a fixed server node. The advantage of such an approach is that the designer can adopt the most appropriate AI technologies for the application in question. However, the effect of network latency must be considered. If network latency is significant, the usability of the application will be adversely affected. Likewise, data rates supported by the network in question must be considered. Indeed, this situation is aggravated when it is considered that a number of networks implement a channel sharing system where the effective data rate at a given time is directly proportional to the number of subscribers currently sharing the channel. It is therefore impossible to guarantee an adequate quality of service (QoS) making the prediction of system performance difficult. Often, the worst case scenario must be assumed. This has particular implications where the AI application on the fixed server node needs either a significant amount of raw data or a stream of data to process.

One key disadvantage of placing the AI component on a fixed server node concerns the issue of cost. There is a surcharge for each KB of data transferred across the wireless network, and though additional revenue is always welcome, the very fact that the subscriber is paying will affect their perception of application in question and make them more demanding in their expectations.

A network–based AI approach is by far the most common and has been used in quite a number of applications. For example, neural networks have been used for profiling mobile users in conversational interfaces (Toney, Feinberg & Richmond, 2004). InCa (Kadous & Sammut, 2004) is a conversational agent that runs on a PDA but uses a fixed network infrastructure for speech recognition.

Distributed Approach

In this approach, the AI component of the service may be split between the mobile device and the fixed network node. The more computationally expensive elements of the service are hosted on the fixed network node while the less expensive elements may be deployed on the device. Performance is a key limitation of this approach as the computational capacity of the devices in question as well as the data-rates supported by the wireless network can all contribute to unsatisfactory performance. From a software engineering perspective, this approach is quite attractive as distributed AI (DAI) is a mature research discipline in its own right; and a practical implementation of DAI is the multi-agent system (MAS) paradigm.

One example of an application that uses a distributed approach is Gulliver's Genie (O'Grady & O'Hare, 2004). This is a tourist information guide for mobile tourists, realized as a suite of intelligent agents encompassing PDAs, wireless networks and fixed network servers. Agents on the mobile device are responsible for manipulating the user interface while a suite of agents on the fixed server collaborate to identify and recommend multimedia content that is appropriate to the tourist's context.

Embedded Approach

As devices grow in processing power, the possibility of embedding an AI based application on the actual physical device becomes ever more feasible. The key limitation is performance, which is a direct result of the available hardware. This effectively compromises the type of AI approach that can be usefully adopted. Overtime, it can be assumed that the capability and variety of AI techniques that can be deployed will increase as developments in mobile hardware continue and the demand for ever-more sophisticated applications increases. From an end-user viewpoint, a key advantage of the embedded approach concerns cost as the number of connections required is minimized.

One example of an application that uses the embedded approach is iDorm (Hagras et al., 2004), a prototype AmI environment. This environment actually demonstrates a variety of embedded agents including fixed motes, mobile robots and PDAs. These agents collaborate to learn and predict user behavior using fuzzy logic principles and, based on these models, the environment is adapted to the inhabitant's needs.

Deployment Considerations

Technically, all three approaches are viable, but the circumstances in which they may be adopted vary. For specialized applications, the networked AI approach is preferable as it offers greater flexibility and maximum performance, albeit at a cost. For general applications, the embedded approach is preferable, primarily due to cost limitations, but the techniques that can be adopted are limited. The distributed approach is essentially a compromise, incorporating the respective advantages and disadvantages of both the networked and embedded approach to various degrees. Ultimately, the nature of the application domain and the target user base will be the major determinants in what approach is adopted. However, in the longer term, it is the embedded approach that has the most potential as it eliminates the negative cumulative effect of network vagrancies, as well as hidden costs. Thus, for the remainder of this chapter, we focus on the embedded approach and consider how this might be achieved.

So what AI techniques can be adopted, given the inherent limitations of mobile devices? Various techniques have been demonstrated in laboratory conditions but one paradigm has been demonstrated to be computationally tractable on mobile devices: intelligent agents. As well as forming the basis of mobile intelligent information's systems, a number of toolkits have been made available under open source licensing conditions thus allowing software engineers access to mature platforms at minimum cost. Before briefly considering some of these options, it is useful to reflect on the intelligent agent paradigm.

THE INTELLIGENT AGENT PARADIGM

Research in intelligent agents has been ongoing since the 1970s. Unfortunately, the term agent has been interpreted in a number of ways thereby leading to some confusion over what the term actually means. More precisely, the characteristics that an arbitrary piece of software should possess before applying the term agent to it are debatable. In essence, an agent may be regarded as a computational entity that can act on behalf of an end-user, another agent or some other software artefact. Agents possess a number of attributes that distinguish them from other software entities. These include amongst others:

- **Autonomy:** The ability to act independently and without direct intervention from another entity, either human or software-related
- **Proactivity:** The ability to opportunistically initiate activities that further the objectives of the agent

Figure 2. Architecture of a BDI agent

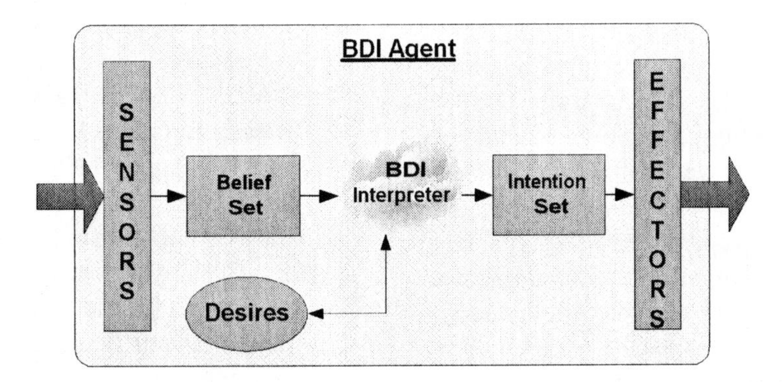

- **Reactivity:** The ability to respond to events perceived in the agent's environment;
- **Mobility:** The ability to migrate to different nodes of a network as the need to fulfill its objectives dictates; and
- **Social ability:** The ability to communicate with other agents using a shared language and ontology leading to shared or collaborative efforts to achieve individual and shared objectives.

To what extent an agent possesses or utilizes each of those attributes is at the discretion of the designer. For clarity purposes, it is useful to consider agents as existing on a scale. At the lower end are so-called reactive agents. Such agents act in a stimulus-response manner, and a typical usage scenario might involve the agent monitoring for user interaction and reacting to it. Such agents are generally classified as weak agents (Wooldridge & Jennings, 1995). At the other end of the scale are so-called strong agents. Such agents maintain a sophisticated model of their environment, a list of goals or objectives, and plans detailing how to achieve these objectives. Such agents support rational reasoning in a collaborative context and are usually realized as multi-agent systems (MAS). This strong notion of agenthood is synonymous with the view maintained by the AI community.

One popular interpretation of the strong notion of agency is that of the belief-desire-intention (BDI) paradigm (Rao & Geogeff, 1995). This is an intuitive and computationally tractable interpretation of the strong agency stance. To summarize: beliefs represent what the agent knows about its environment. Note that the term environment can have diverse meanings here and may not just relate to the physical environment. Desires represent the objectives of the agent, and implicitly the raison d'être for the application. However, at any moment in time, an agent may be only capable of fulfilling some of its desires, if even that. These desires are then formulated as intentions and the agent proceeds to fulfill these intentions. The cycle of updating its model of the environment, identifying desires that can be fulfilled, and realizing these intentions is then repeated for the duration of the agent's lifecycle (Figure 2).

When should agents be considered for realizing a software solution? Opinion on this is varied. If the solution can be modeled as a series of dynamic interacting components, then agents may well offer a viable solution. However, many see agents as being particular useful in situations that are inherently complex and dynamic as their native capabilities equip them for handling the myriad of situations that may arise. Naturally, there are many situations that fulfill the criteria but, for the purposes of this discussion, it can be easily seen that the mobile computing domain offers significant opportunities for harvesting the characteristics of intelligent agents.

Intelligent Agents for Mobile Computing

As the capability of mobile devices grew, researchers in the intelligent agent community became aware of the feasibility of deploying agents on such devices, and perceived mobile computing as a potentially fertile area for the intelligent agent paradigm. A common approach was to extend the functionality of existing and well-documented MAS environments such that they could operate on mobile devices. It was not necessary to port the entire environment on to the device; it was just necessary to develop an optimized runtime engine for interpreting the agent logic. In this way, the MAS ethos is persevered and such an approach subscribes to the distributed AI approach alluded to previously. A further benefit was that existing agent-oriented software engineering (AOSE) methodologies could be used. In the case of testing, various toolkits have been released by the telecommunications manufacturers that facilitate the testing of mobile applications. A prudent approach is of course to test the application at various stages during its development on actual physical devices, as this will give a more accurate indication of performance, the look and feel (L&F) of the application and so on. For a perspective on deploying agents on mobile devices, the interested reader should consult Carabelea and Boissier (2003).

While a number of environments may be found in the literature for running agents on mobile devices, the following toolkits form a useful basis for initial consideration:

1. **LEAP (Lightweight Extensible Agent Platform)** (Bergenti, Poggi, Burg, et al., 2001) is an extension of the well-documented JADE platform (Bellifemine, Caire, Poggi et al., 2003). It is FIPA (http://www.fipa.org/) compliant and capable of operating on both mobile and fixed devices.
2. **MicroFIPA-OS** (Laukkanen, Tarkoma & Leinonen, 2001) is a minimized footprint of the FIPA-OS agent toolkit (Tarkoma & Laukkanen, 2002). The original FIPA-OS was designed for PCs and incorporated a number of features that did not scale down to mobile devices. Hence, MicroFIPA-OS minimizes object creation, reduces computational overhead and optimizes the use of threads and other resource pools.
3. **AFME (Agent Factory Micro Edition)** (Muldoon, O'Hare, Collier & O'Grady, 2006) is derived from Agent Factory (Collier, O'Hare, Lowen, & Rooney, 2003), a framework for the fabrication and deployment of agents that broadly conform to the BDI agent model. It has been specifically designed for operation on cellular phones and such categories of devices.
4. **JACK** is, in contrast to the three previous frameworks, a commercial product from the Agent Oriented Software Group (http://www.agent-software.com). It comes with a sophisticated development environment, and like AFME, conforms to the BDI agent model.

A detailed description of the each of these systems is beyond the scope of this discussion. However, the interested reader is referred to (O'Hare, O'Grady, Muldoon & Bradley, 2006) for a more advanced treatment of the toolkits and other associated issues.

FUTURE TRENDS

As mobile devices proliferate, and each generation surpasses its predecessor in terms of raw computational capacity and supported features, the potential for incorporating additional AI techniques will increase. In a similar vein, new niche and specialized markets for mobile services will appear. If a more holistic approach is taken towards mobile computing, it can be seen that developments in sensor technologies, fundamental to the ubiquitous and pervasive vision, will follow a similar trajectory. Indeed, the possibility of deploying intelligent agents on sensors is being actively investigated in widespread expectation that the next generation of sensors will incorporate processors of a similar capability to the current range of PDAs. Such a development is essential if the AmI vision to reach fruition.

As the possibility of incorporation of ever more sophisticated AI techniques increases, the potential for extending and refining the adaptability and IUI constructs for the support of mobile users increase. Indeed, adaptability may reach its fulfillment through the incorporation of autonomic computing precepts (Kephart & Chess, 2003). Self-configuring, self-healing, self-optimizing and self-protecting are the key attributes of an autonomic system, and it can be seen that incorporation of AI techniques may make the realization of these characteristics more attainable.

Finally, the practical issues of engineering mobile AI solutions must be considered. Mobile computing poses significant challenges to the traditional software engineering process, and the broad issue of how best to design for mobile services still needs to be resolved. The situation is exacerbated when AI technologies are included. However, it may be envisaged that as experience and knowledge of the mobile computing domain deepens and matures, new methodologies and best practice principles will emerge.

CONCLUSION

Mobile computing scenarios are diverse and numerous, and give rise to numerous challenges that must be overcome if the end-user experience is to be a satisfactory one. IUIs offers one viable approach that software designers can adopt in their efforts to make their systems more usable in what is frequently a hostile environment. However, the pragmatic issue of realizing mobile applications that incorporate intelligent techniques is of critical importance and gives rise to significant technical and design obstacles.

In this chapter, the broad issue of realizing an intelligent solution was examined in some detail. At present, the intelligent agent paradigm offers an increasingly viable proposition for those designers who wish to include intelligent techniques in their designs. To illustrate the issues involved, the intelligent agent paradigm was discussed in some detail.

As mobile developments continue unabated, the demand for increasingly sophisticated applications and services will likewise increase. Meeting this demand will pose new challenges for software and HCI professionals. A prudent and selective adoption of intelligent techniques may well offer a practical approach to the effective realization of a new generation of mobile services.

ACKNOWLEDGMENT

The authors gratefully acknowledge the support of the Science Foundation Ireland (SFI) under Grant No. 03/IN.3/1361.

REFERENCES

Bellifemine, F., Caire, G., Poggi, A., & Rimassa, G. (2003). *JADE—A white paper*. Retrieved January 2007 from http://jade.tilab.com/papers/2003/WhitePaperJADEEXP.pdf

Bergenti, F., Poggi, A., Burg, B., & Caire, G. (2001). Deploying FIPA-compliant systems on handheld devices. *IEEE Internet Computing, 5*(4), 20-25.

Collier, R.W., O'Hare, G.M.P., Lowen, T., & Rooney, C.F.B., (2003). Beyond prototyping in the factory of agents. In J. G. Carbonell & J. Siekmann (Eds.), *Lecture Notes In Computer Science*, 2691, 383-393, Berlin: Springer.

Carabelea, C., & Boissier O. (2003). Multi-agent platforms on smart devices: Dream or reality. Retrieved January, 2007, from http://www.emse.fr/~carabele/papers/carabelea.soc03.pdf.

Greenberg, S. (2001). Context as a dynamic construct. *Human-Computer Interaction*, 16, 257-268.

Hagras, H., Callaghan, V., Colley, M., Clarke, G., Pounds-Cornish, A., & Duman, H. (2004).

Creating an ambient-intelligence environment using embedded agents. *IEEE Intelligent Systems, 19*(6) 12-20.

Höök, K. (2000). Steps to take before intelligent user interfaces become real. *Interacting with Computers, 12*, 409-426.

Kadous, M.W., & Sammut, C. (2004). InCA: A mobile conversational agent. *Lecture Notes in Computer Science*, 3153, 644-653. Berlin: Springer.

Kephart, J.O., & Chess, D.M. (2003). The vision of autonomic computing. *IEEE Computer, 36*(1), 41-50.

Kruger, A., & Malaka, R. (2004). Artificial intelligence goes mobile. *Applied Artificial Intelligence, 18*, 469-476.

Langley, P. (1997). Machine learning for adaptive user interfaces. *Lecture Notes in Computer Science*, 1303, 53-62. Berlin:Springer.

Laukkanen, M., Tarkoma, S., & Leinonen, J. (2001). FIPA-OS agent platform for small-footprint devices. *Lecture Notes in Computer Science*, 2333, 447-460. Berlin:Springer.

Mitrovic, N., Royo, J. A., & Mena, E. (2005). Adaptive user interfaces based on mobile agents: Monitoring the behavior of users in a wireless environment. *Proceedings of the Symposium on Ubiquitous Computation and Ambient Intelligence* (pp. 371-378), Madrid: Thomson-Paraninfo.

Muldoon, C., O'Hare, G.M.P., Collier, R.W., & O'Grady, M.J. (2006). Agent factory micro edition: A framework for ambient applications, *Lecture Notes in Computer Science*, 3993, 727-734. Berlin:Springer.

O'Connor, A., & Wade, V., (2006). Informing context to support adaptive services, *Lecture Notes in Computer Science*, 4018, 366-369. Berlin: Springer.

O'Grady, M.J., & O'Hare, G.M.P. (2004). Just-in-time multimedia distribution in a mobile computing environment. *IEEE Multimedia, 11*(4), 62-74.

O'Hare, G.M.P., O'Grady, M.J., Muldoon, C., & Bradley, J.F. (2006). Embedded agents: A paradigm for mobile services. *International Journal of Web and Grid Services, 2*(4), 355-378.

Patterson, C.A., Muntz, R.R., & Pancake, C.M. (2003). Challenges in location-aware computing. *IEEE Pervasive Computing, 2*(2), 80-89.

Rao, A.S., & Georgeff, M.P. (1995). BDI agents: from theory to practice. In V. Lesser and L. Gasser (Eds.), *Proceedings of the First International Conference on Multiagent Systems* (pp. 312-319). California: MIT Press.

Rhodes, B.J., Minar, N. & Weaver, J. (1999). Wearable computing meets ubiquitous computing: Reaping the best of both worlds. *Proceedings of the Third International Symposium on Wearable Computers* (pp. 141-149). California: IEEE Computer Society

Schmidt, A., Beigl, M., & Gellersen, H-W, (1999). There is more to context than location. *Computers and Graphics, 23*(6), 893-901.

Tarkoma, S., & Laukkanen, M. (2002). Supporting software agents on small devices. In *Proceedings of the First International Joint Conference on Autonomous Agents and Multi-Agent Systems (AAMAS)* (pp. 565-566). NewYork: ACM Press.

Thevenin, D., & Coutaz, J. (1999). Plasticity of user interfaces: Framework and research agenda. In (M. A. Sasse & C. Johnson (Eds.), *Proceedings of IFIP TC 13 International Conference on Human-Computer Interaction (INTERACT'99)*, (pp. 110-117). Amsterdam: IOS Press.

Toney, D., Feinberg D., & Richmond, K. (2004). Acoustic features for profiling mobile users of conversational interfaces, *Lecture Notes in Computer Science*, 3160 (pp. 394-398). Berlin: Springer.

Vasilakos, A., & Pedrycz, W. (2006). *Ambient intelligence, wireless networking, ubiquitous Computing*. Norwood:Artec House.

Weiser, M. (1991). The computer for the twenty-first century. *Scientific American, 265*(3), 94-100.

Wooldridge, M., & Jennings, N.R. (1995). Intelligent agents: Theory and practice. *Knowledge Engineering Review, 10*(2), 115-152.

KEY TERMS

Ambient Intelligence: (AmI) was conceived by the Information Society Technologies Advisory Group (ISTAG) as a means of facilitating intuitive interaction between people and ubiquitous computing environments. A key enabler of the AmI concept is the intelligent user interface.

BDI Architecture: The Belief-Desire-Intention (BDI) architecture is an example of a sophisticated reasoning model based on mental constructs that can be used by intelligent agents. It is allows the modeling of agents behaviors in an intuitive manner that complements the human intellect.

Context: Context-aware computing considers various pertinent aspects of the end-user's situation when delivering a service. These aspects, or contextual elements, are determined during invocation of the service and may include user profile, for example language, age, and so on. Spatial contextual elements, namely location and orientation, may also be considered.

Intelligent Agent: Agents are software entities that encapsulate a number of attributes including autonomy, mobility, sociability, reactivity and proactivity amongst others. Agents may be reactive, deliberative or hybrid. Implicit in the agent construct is the requirement for a sophisticated reasoning ability, a classic example being agents modeled on the BDI architecture.

Intelligent User Interface: Harnesses various techniques from artificial intelligence to adapt and configure the interface to an application such that the end-user's experience is more satisfactory.

Mobile Computing: A computer usage paradigm where end-users access applications and services in diverse scenarios, while mobile. Mobile telephony is a popular realization of this paradigm, but wearable computing and telematic applications could also be considered as realistic interpretations of mobile computing.

Multi-Agent System: A suite of intelligent agents, seeking to solve some problem beyond their individual capabilities, come together to form a multi-agent system (MAS). These agents collaborate to fulfill individual and shared objectives.

Ubiquitous Computing: Conceived in the early 1990s, ubiquitous computing envisages a world of embedded devices, where computing artefacts are embedded in the physical environment and accessed in a transparent manner.

Chapter XXI
Tools for Rapidly Prototyping Mobile Interactions

Yang Li
University of Washington, USA

Scott Klemmer
Stanford University, USA

James A. Landay
University of Washington & Intel Research Seattle, USA

ABSTRACT

We introduce informal prototyping tools as an important way to speed up the early-stage design of mobile interactions, by lowering the barrier to entry for designers and by reducing the cost of testing. We use two tools, SUEDE and Topiary, as proofs of concept for informal prototyping tools of mobile interactions. These tools address the early stage design of two important forms of mobile interactions: speech-based and location-enhanced interactions. In particular, we highlight storyboarding and Wizard of Oz (WOz) testing, two commonly used techniques, and discuss how they can be applied to address different domains. We also illustrate using a case study: the iterative design of a location-enhanced application called Place Finder using Topiary. In this chapter we hope to give the reader a sense of what should be considered as well as possible solutions for informal prototyping tools for mobile interactions.

INTRODUCTION

The iterative process of prototyping and testing has become an efficient way for successful user interface design. It is especially crucial to explore a design space in the early design stages before implementing an application (Gould et al., 1985). Informal prototyping tools can speed up an early-stage, iterative design process (Bailey et al., 2001; Klemmer et al., 2000; Landay et al., 2001; Li et al., 2004; Lin et al., 2000). These tools are aimed at lowering the barrier to entry for interaction designers who do not have technical backgrounds, and automatically generating early-stage prototypes that can be tested with end users. The informal look and feel of these tools and their fluid input techniques, for example using pen sketching (Landay et al., 2001), encourage both designers and end users to focus on high level interaction ideas rather than on design or implementation details (e.g., visual layouts or colors). These details are often better addressed at a later stage. In this chapter, we focus on informal tool support for the early stage design of interactive mobile technologies. In particular, we describe informal prototyping tools that we developed for two increasingly important forms of mobile interaction: speech-based interactions (Klemmer et al., 2000) and location-enhanced interactions (Li et al., 2004).

The first of these two types of interactions, speech-based, works well on mobile phones, the major platform of mobile computing. These devices often have tiny screens and buttons to increase mobility, which makes speech interaction an important alternative. Although the accuracy of speech recognition is an important concern for a successful speech-based UI, the real bottleneck in speech interface design is the lack of basic knowledge about user "performance during computer-based spoken interaction" (Cohen et al., 1995). Many interaction designers who could contribute to this body of knowledge are excluded from speech design by the complexities of the core technologies, the formal representations used for specifying these technologies, and the lack of appropriate design tools to support iterative design (Klemmer et al., 2000). SUEDE (Klemmer et al., 2000) demonstrates how tool support can be used in the early stage design of speech-based user interfaces.

The second of these two types of interactions, location-enhanced, is important because of its implicit nature. While the explicit input channels (e.g., keyboarding or mouse pointing) available on mobile technology are more limited than on the desktop, the bandwidth of implicit input (using contextual information) is greatly expanded on mobile platforms. Mobile technology is more available in our context-rich, everyday lives than traditional desktop computing. One especially promising form of context-aware computing that has begun to see commercialization is location-enhanced computing, applications that leverage one's current location as well as the location of other people, places, and things (Li et al., 2004). For example, mobile phone services allow users to locate friends and family (LOC-AID), provide real-time navigation (InfoGation) and monitor and motivate users toward their fitness goals by using phone-based GPS to measure the user's speed, distance and elevation (BonesInMotion). E911 transmits a mobile phone user's current location when making emergency calls. However, location-enhanced applications are hard to prototype and evaluate. They employ sophisticated technologies such as location tracking and their target environment is mobile and in the field. Topiary (Li et al., 2004) demonstrates how high-level tool support can be provided for lowering the threshold and cost for designers to design and test location-enhanced applications.

Using SUEDE and Topiary as proofs of concept, we highlight two techniques commonly used in informal prototyping tools: storyboarding and Wizard of Oz (WOz) testing. To overcome the technical barrier for design, both SUEDE and Topiary employ a storyboarding-based approach for specifying interaction logic. To allow easy testing of prototypes, both tools employ WOz approaches where a human wizard simulates a sophisticated, nonexistent part of the prototype such as location tracking or speech recognition. To demonstrate how these types of tool can actually help prototype and test mobile technology, we introduce a case study using Topiary to design the Place Finder application.

BACKGROUND

User interface tools have been a central topic in HCI research. An extensive review of user interface tools can be found in (Myers et al., 2001). A large number of research prototypes and commercial products have been developed for rapid prototyping of user interfaces (Apple, 1987; Bailey et al.,

2001; Hartmann et al., 2006; Klemmer et al., 2000; Landay et al., 2001; Li et al., 2004; Lin et al., 2000; Macromedia; MacIntyre et al., 2004).

In particular, informal prototyping tools are aimed at the early stages of a design process, and are used to create early-stage prototypes for testing key design ideas rather than building full-fledged final systems (Landay et al., 2001). They often result in example-based interface mockups that are able to demonstrate exploratory interactive behaviors but ignore other non-exploratory aspects of a desired system. Informal tools have shown great potential to facilitate the early stages of a design process and have been developed for various domains. For example, SILK is a tool for designing graphical user interfaces (Landay et al., 2001) that allows designers to create GUI prototypes by sketching and storyboarding. DENIM (Lin et al., 2000), a tool for the early stage design of Web sites, has become one of the most popular informal prototyping tools (downloaded over 100,000 times since 2000). Informal prototyping tools are often grounded in current practices of designers, e.g., paper prototyping (Rettig, 1994; Snyder, 2003), and lower the barrier to entry by maintaining the affordance of an existing practice. At the same time, informal tools provide extra value by allowing the easy editing and maintenance of a design, and by generating testable prototypes.

MAIN FOCUS OF THE CHAPTER

In our research, two features have emerged as being particularly valuable for rapidly prototyping mobile interactions. The first is storyboarding, which is inspired by traditional paper prototyping where designers draw key interaction flows visually on paper. Storyboarding is enhanced by electronic tool support to create the states and transitions. Many systems have been influenced by Harel's Statecharts model (Harel, 1987). Storyboarding is employed by both SUEDE and Topiary to lower the technical barrier for creating early-stage prototypes.

The second valuable feature is Wizard of Oz (WOz) testing, where a designer simulates part or all of the application logic by manipulating the interface in response to user input. This significantly reduces the time and labor required to create a testable prototype. As both speech-based interfaces and location-enhanced computing involve a necessary but sophisticated component, that is speech recognition and location tracking, respectively, both SUEDE and Topiary employed a WOz approach to avoid the complexity of introducing these components. To give an example of how this type of tool can help design and evaluate mobile technology in practice, we describe a case study for the iterative design of a PDA-based mobile Place Finder application using the Topiary.

Prototyping with Storyboards

In the early stages of design, it is important that tools allow designers to focus on the high-level concerns of interaction design, rather than forcing designers to also specify how these interactions are implemented. Storyboarding is an efficient way for designers to describe how a user interface should behave by enumerating concrete interaction sequences including both user input and interface output. These sequences should cover the key interaction paths of a proposed system in a particular design space. The concerns of early-stage prototyping are distinct from those of constructing an actual system, which focus more on completeness than exploration of the design space.

SUEDE allows two kinds of storyboarding: linear (conversation examples) and non-linear (design graphs of an actual interface) storyboarding. Designers start a design by creating simple conversation examples (see the Script Area at the top of Figure 1). These examples then evolve into the more complex, graph structure representing the actual interface design (see the design graph at the bottom of Figure 1) (Klemmer et al., 2000). The process of creating linear examples first and then forming more general design graphs is based on the existing practices of speech UI designers: we have found that often, designers begin the design process by writing linear dialog examples and then use those as a basis for creating a flowchart representation of the dialog flow on paper.

Designers lay out linear conversation examples horizontally as cards in the script area. *Prompts*, colored orange, represent the system's speech prompts. They are recorded by the designer for the phrases that the computer speaks. *Responses*, colored green, represent example responses of the end user. They are the phrases that participants make in response to prompts. System prompts alternate with user responses for accomplishing a task. A designer can record her own voice for the speech on both types of cards, as well as type in a corresponding label for each of the cards. By playing the recordings from left to right, the designer can both see and hear the example interaction. For example, in Figure 1, a designer has recorded a conversation example with the following alternating prompts and responses: "message from James," "erase it," "Are you sure," "Yes." After constructing example scripts, a designer can construct an actual design of a speech-based interface using the design graph (see Figure 1). A design graph represents a dialog flow based on the user's responses to the system's prompts. To create a design graph, designers can drag prompt or response cards from a script onto the design area, or create new cards on the design area, and link them into the dialog flow. SUEDE's storyboard mechanism embodies both the input and output of a speech interface in cards that can be directly manipulated (e.g., via drag & drop), and hides the complexity of using speech recognition and synthesis. This abstraction allows designers to focus on high-level design issues.

Topiary's storyboards also embed the specification of input and output interactions into a storyboard. Before introducing Topiary's storyboards, we first discuss Topiary's *Activity Map* workspace, a component designed for creating scenarios describing location contexts of people, places and things by demonstration (see Figures 2 and 3). The created scenarios can be used as input by Topiary storyboards when prototyping location-enhanced interactions (see Figure 4). Modeling implicit input, location context in this case, is a new challenge posed by mobile computing.

Topiary's Activity Map workspace employs an intuitive map metaphor for designers to demonstrate location contexts describing the spatial relationship of people, places and things. Designers can create graphical objects on the map to represent people, places and things (see Figure 2). Designers can move people and things on the map to demonstrate various spatial relationships. For example, in Figure 2, Bob is out of the library, the astronomy building and the café. However, Bob is close to the library because Bob's proximity region, indicated by the red circle around Bob, intersects with the library. The proximity region

Figure 1. SUEDE allows designers to create example scripts of speech-based interactions (top) and speech UI designs (bottom) by storyboarding.

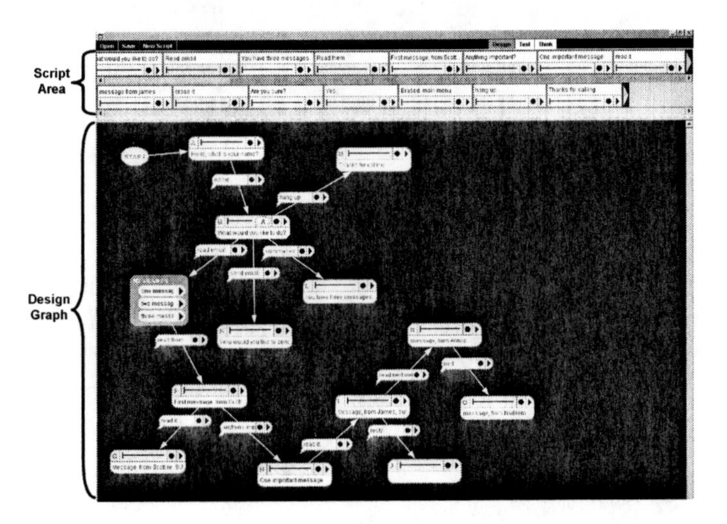

Figure 2. The active map workspace of Topiary is used to model location contexts of people, places and things and to demonstrate scenarios describing location contexts.

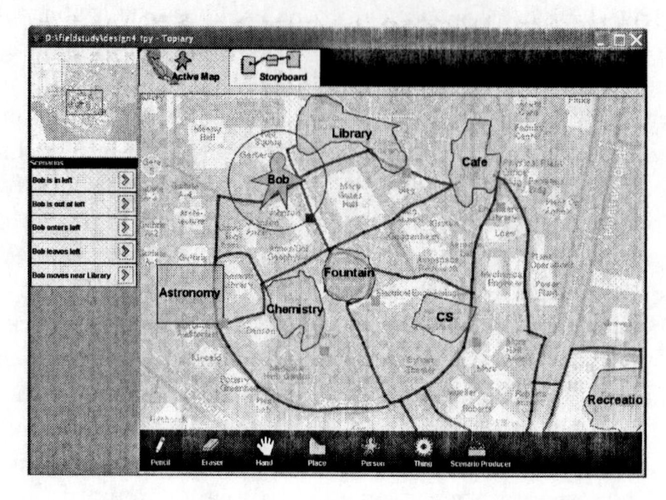

Figure 3. The designer drags Bob into the Library, with the context changing from "Bob is out of Library" to "Bob enters Library." As the entity CS (building) is unchecked, all related contexts to this place are filtered out.

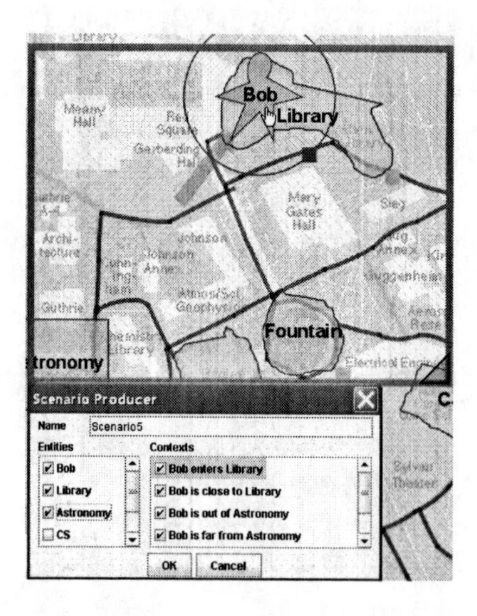

can be resized by dragging the rectangular handle. These spatial relationships can be captured via Topiary's *Scenario Producer*. Like a screen capture tool, a designer can position a *Scenario Producer* window over entities of interest to capture spatial relationships (see Figure 3). A dialog box is then brought up that allows designers to select contexts of interest. Designers can demonstrate dynamic contextual transitions such as "entering" or "leaving" by moving entities within the recording window. For example, dragging Bob into the Library changes the event "Bob is out of Library" into "Bob enters Library" (see Figure 3).

Based on the location scenarios captured in the Active Map workspace, designers can create application prototypes in the Storyboard workspace (see Figure 4). In Topiary, a storyboard page represents a screen of visual output and a link represents transitions between pages. The key innovation in Topiary's storyboards is that scenarios created in the *Active Map* workspace can be used as conditions or triggers on links (Li et al., 2004). Designers create pages and links by sketching. Topiary has two kinds of links (see Figure 4). *Explicit links*, denoted in blue, start on ink within a page and they represent GUI elements that users have to click on, for example buttons or hyperlinks. *Implicit links*, denoted in green, start on an empty area in a page. They represent transitions that automatically take place when scenarios associated with that link occur. Explicit links model explicit interactions taken by end-users though they can be conditioned by sensed information, whereas implicit links model purely sensed data such as locations. One or more scenarios can be added to a link and multiple scenarios represent the logical AND of the scenarios. Multiple links starting from the same source represent the logical OR of transitions.

The Activity Map abstraction allows designers to focus on location contexts of interest rather than how these contexts can be sensed. Topiary's graphical storyboarding allows designers to specify rich interactions by drag & drop or sketching instead of specifying complex rules or Boolean logic expressions. From both SUEDE and Topiary, we conclude that the key to a successful informal tool is to devise an appropriate abstraction that matches designers' conceptual model for design and hides the less important aspects of exploring target interactions. Storyboards, as a meta-design metaphor, should be adapted and developed to fit within a specific domain when being applied.

Testing Using WOz Approaches

Speech-based or location-enhanced interactions resist rapid evaluation because the underlying technologies require high levels of technical expertise to understand and use, and a significant amount of effort to tune and integrate. For example, location-tracking infrastructures are not always available (e.g., GPS does not work well indoors) and they require a great deal of effort to deploy and configure. Incorporating these technologies too early in a design process may distract designers from fully exploring the design space. Consequently, we employed WOz approaches in these tools for testing early-stage prototypes. That is, a wizard (played by a designer or experimenter) simulates what these technologies would do in the final application.

Figure 4. Topiary's Storyboard workspace allows application prototypes to be created. The lower three links (in blue) are explicit links, representing the behavior of the OK button depending where "Bob" is. The top link (in green) is an implicit link, representing an automatic transition from the Map page to the Nearest Friends page when "Anyone moves near Bob"

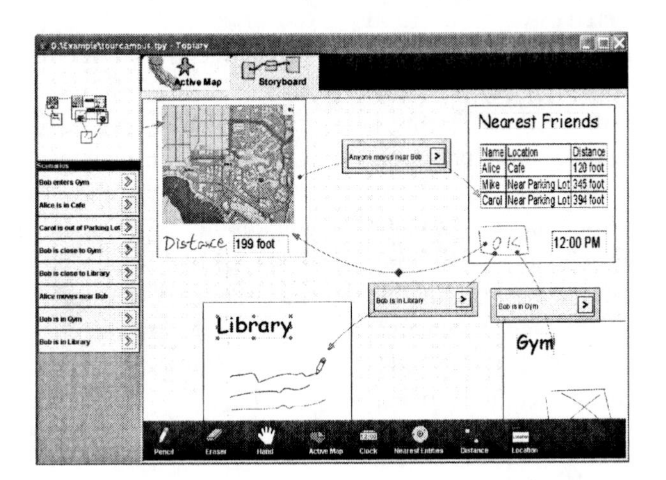

Wizard of Oz (WOz) testing has been widely employed in user interface design. In a WOz test, a wizard (often played by a designer) fakes an incomplete (or nonexistent) system component to conduct early user evaluations (Dahlbäck et al., 1993). In its most basic form, a WOz test works by the wizard simulating the machine behavior entirely. There is no computation in the loop at all. Examples of this form include testing paper prototypes by having the wizard physically move around the paper-based windows and menus (Rettig, 1994) and testing potential speech interface interaction flows by having a human operator on the other side of the telephone, following a pre-specified interaction graph. When an interactive prototype has been created (at least partially), the wizard can simply use the implemented interface. As a variant of this approach, a programmer can implement a functionally complete but suboptimal interface, and have the wizard control this interface during testing as a means of eliciting users' conceptual models of the task for example (Akers, 2006).

Significant gains beyond these basic approaches can be achieved through tools designed explicitly to support a Wizard of Oz approach. The fundamental insight behind a WOz-enabled tool is that the wizard is provided with a distinct user interface from that of the end user, and that the primary goal for this interface is to enable the wizard to rapidly specify to the system what the user's input was. In SUEDE, the interaction flow and audio prompts are specified by the designer ahead of time, and the user's responses to the speech prompts are interpreted by the wizard and specified to the system using a graphical interface that is runtime-generated based on the user's current state within the interaction flow. During a test, a wizard works in front of a computer screen. The participant performs the test away from the wizard, in a space with speakers to hear the system prompts and a microphone hooked up to the computer to record his responses. During the course of the test session, a transcript sequence is generated containing the original system audio output and a recording of the participant's spoken audio input.

When the wizard starts a test session, SUEDE automatically plays the pre-recorded audio from the current prompt. The wizard interface in SUEDE displays hyperlinks that represent the set of possible options for the current state (see Figure 5); the wizard waits for the test participant to respond, and then clicks on the appropriate hyperlink based on the response. Here, the wizard is acting as the speech recognition engine. Additionally, effective wizard interfaces should provide a display of the interaction history (as well as capture this for subsequent analysis); global controls for options generally available in an interface genre but independent of a particular interface or interface state (these globals can be defined by the tool or specified by the designer); and support for simulated recognition errors. This set of functionality enables the wizard to customize the test as she sees fit, handle user input beyond what was originally designed, and test whether the application is designed in such a way that users can understand and recover from "recognition errors."

Location-enhanced interfaces introduce the additional challenge that, almost by definition, a test must be conducted while moving to be ecologically valid. To address this, Topiary's WOz interface was specifically designed for a wizard to interact with the interface while walking. Topiary automatically generates user interfaces for testing, including the Wizard UI and the End-User UI, based on the Active Map and the Storyboard workspace. The Wizard UI (see Figure 6) is where a wizard simulates location contexts, as well as observes and analyzes a test. The End-User UI is what an end user interacts with during a test and it is also shown in the End User Screen window of the Wizard UI (see Figure 6) so that designers can monitor user interactions. The designer can also interact with the End-User Screen window for debugging purposes. The Wizard UI and End-User UI can be run on the same device (to let a designer try out a design) or on separate devices (one for the Wizard, the other for the user).

During a test, the wizard follows a user; each carries a mobile device, and these devices are connected over a wireless network. The wizard

Figure 5. SUEDE's Test mode is presented in a web browser and allows the wizard to focus on the current state of the UI (top) and the available responses for that state (bottom).

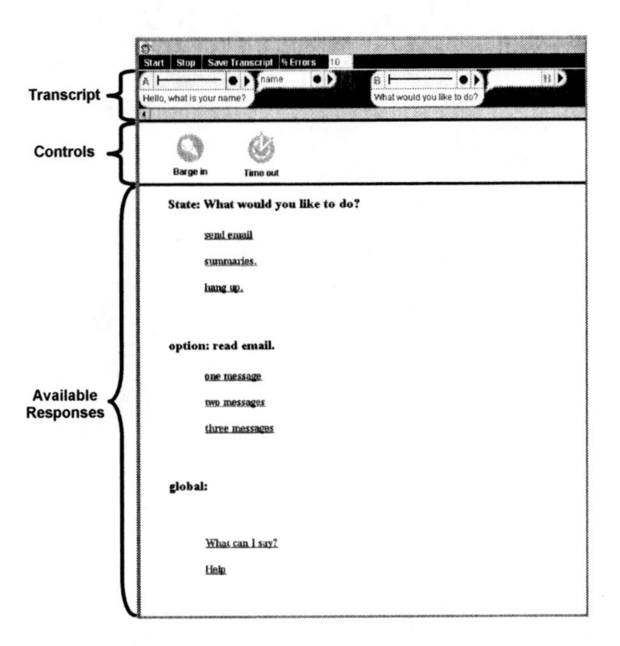

simulates location contexts by moving people and things around on the Active Map to dynamically update their location. The location changes of people and things on a map may trigger implicit transitions in the storyboard that will update the End-User UI. Topiary can also employ real location data if it is available, for more realistic testing at larger scales. A designer can choose to turn on a built-in location-tracking engine, based on Place Lab (LaMarca et al., 2005), which allows a WiFi-enabled or GSM-enabled device to passively listen for nearby access points to determine its location in a privacy-sensitive manner. In addition, a designer can analyze a design by recording a test and replaying it later. Topiary capture users' actions, like mouse movements and clicks, as well as physical paths traveled. The Storyboard Analysis window (see the bottom of Figure 6) highlights the current page and the last transition during a test or a replay session, which can help designers to figure out interaction flows.

Through our experience building SUEDE and Topiary, we have learned that effective tool support for Wizard of Oz testing comprises several key elements: the current state of the user interface (e.g., what is the current page in both tools), the current state of the user (e.g., the user's current location in Topiary) and the set of available actions (e.g., available responses in SUEDE). These elements should be provided to the wizard in an effective manner that allows the wizard to easily grasp and rapidly react. An effective Wizard interface should minimize the wizard's cognitive load by proactively maintaining a visible representation of state and having the displayed (and hence selectable) options for future action tailored to the state at hand.

A Case Study

To demonstrate how an informal prototyping tool can help at an early stage of the design process and how informal prototyping can inform the later design or development process, we report on our experience with the iterative design of a location-enhanced Place Finder using Topiary. A location-enhanced Place Finder embodies many features of location-enhanced, mobile applications. It allows users to find a place of interest more efficiently by leveraging the user's location (e.g., showing a

Figure 6. The Wizard UI has four major parts: The Active Map (a clone of the Active Map workspace) for simulating location contexts, the End User Screen for monitoring a user's interaction or debugging a design, the Storyboard Analysis Window for analyzing interaction logic and the Radar View for easy navigation of the Activity Map.

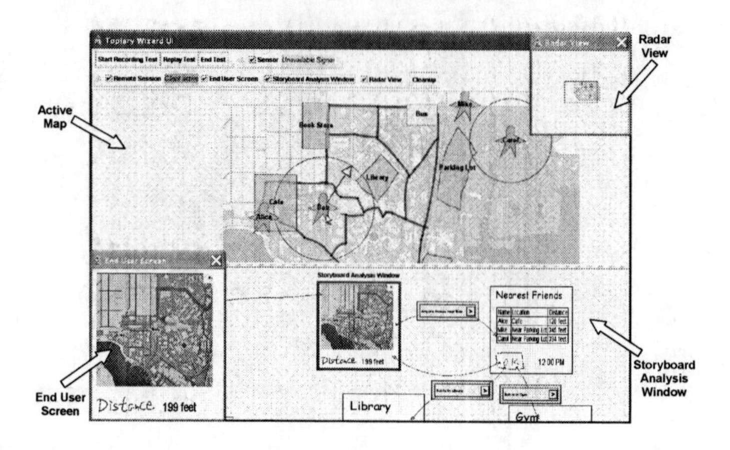

path to the destination). With the help of Topiary, we were able to efficiently explore the usability issues of map-based navigation techniques on a PDA held by a user walking in the field. Map-based navigation is a key component of a Place Finder application. Based on two design iterations that involved creating five different designs and testing them with four participants in the field as well as an analysis of implementation issues, we built a high fidelity prototype of the Place Finder.

The first iteration included four different user interface designs that shared the same the underlying map of places and paths in the Active Map workspace (see Figure 2). At each iteration, a user test was conducted in the field on a college campus, using a Toshiba Tablet PC and an HP iPAQ™ Pocket PC. During each test, the wireless communication between the two devices was based on a peer-to-peer connection so that the connection was not affected by the availability of access points in the field.

Iteration #1

It took us only three hours in total to create four prototypes, each using a different navigation technique. The first design shows a map of the entire campus (see Figure 7a). The second design shows an area centered on the user and lets the user manually zoom in and out (see Figure 7b). The third design uses the user's current location to show different regions of the campus (see Figure 7c). The last design is similar to the second, except it automatically zooms in or out based on the user's current speed (see Figure 7d). This last design was based on the idea that people are reluctant to interact with a device while walking. All four designs showed the user's current location and shortest path (see the thick pink lines in Figure 7) to the target, both of which are updated dynamically by Topiary.

Four navigation segments were included in the test of Iteration #1, one segment for each of the four designs. These four segments were selected based on two principles. First, to smoothly connect the four experimental segments, the target of a segment should be the starting point of the following segment. Second, each segment should cover an area that requires a moderate walk, not too long or too short (e.g., an eight minute walk), and can produce a path with enough complexity to avoid simple paths (e.g., the entire path is a straight line.)

We had three participants try all four designs on a PDA in the field, with a wizard updating their location on a Tablet PC. Each experimental ses-

Figure 7. Storyboard fragments of the four designs in Iteration #1. A page, which holds maps and sketches, represents a screen of visual output of the user interface. Arrows (links) between pages represent transitions. The blue links represent GUI elements such as buttons for which scenarios can be used as conditions (not shown here). The green links represent transitions that can automatically take place when the associated scenarios occur.

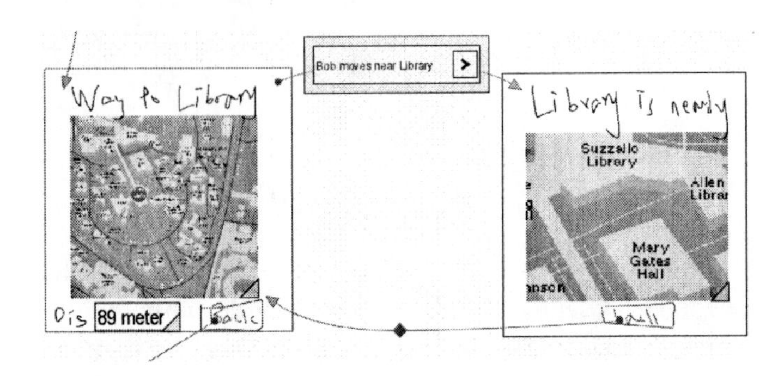

(a) Design #1 shows the entire campus and a detailed map is automatically shown when a user gets close to a target. Here the scenario "Bob moves near library" triggers showing a detailed map around the library.

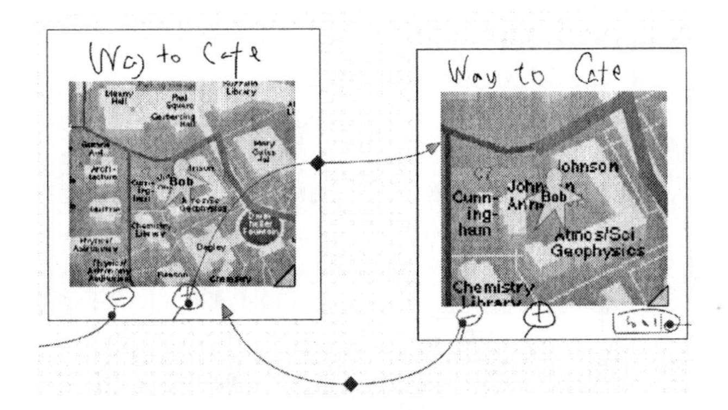

(b) Design #2 shows an area automatically centered on the user and lets the user manually zoom in or out by clicking on the sketched "+" or "–" buttons.

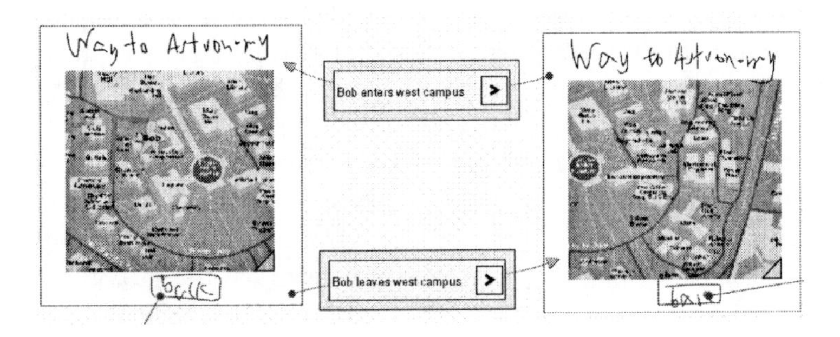

(c) Design #3 uses the user's current location to show different regions of the campus. Here the scenario "Bob enters (or leaves) west campus" triggers showing the west (or east) region of the campus.

continued on following page

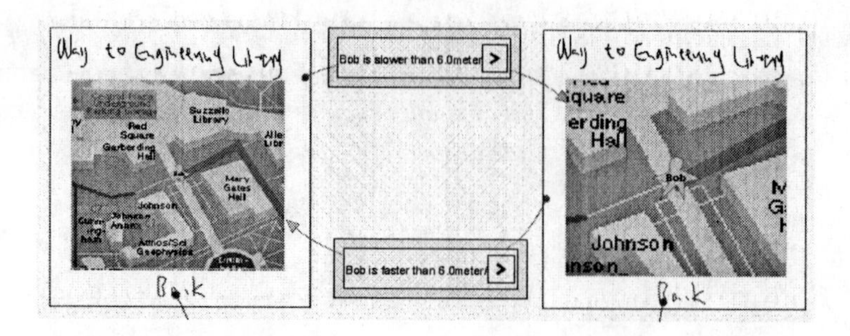

(d) Design #4 is similar to Design #2 except it automatically zooms in or out based on the user's current speed. Here the scenario "Bob is slower (or faster) than 0.6 meter/s" triggers showing maps at different zoom levels.

sion lasted about one hour and each segment took about fifteen minutes to complete. During the test, we were able to make some minor changes to the design instantly in response to the participant's suggestions.

All three participants preferred the map centered on the user's current location (#2 and #4). The problem with the first design is that it shows the entire campus on a small PDA screen, which turned out to be hard to read. The third design does show more detail but it does not give a global view of the campus and the participants complained that they could not see the target until they were physically in that region, although they were still able to see the path.

Two participants preferred manual zooming to automatic zooming as they thought manual zooming gave them more control over the zoom levels. However, the other participant thought both kinds of zooming were good to use. All our participants thought the distance label from Design #1 was useful and they also suggested that we should flash the target when users get close to it.

One common problem with the four designs was that there was not enough orientation information provided. We originally thought users could figure out their orientation by referring to nearby buildings and the continuous change of their location on the map.

Iteration #2

Based on participant feedback and our observations during Iteration #1, we spent *one hour* creating a new design combining the best features of the four designs (see Figure 8). We added a page for users to choose automatic or manual zooming (see Figure 8a). We explored different ways of showing orientation on a map, including rotating the map, showing an orientation arrow, and showing trajectory arrows (see Figure 8b). These orientation representations are provided by Topiary. In addition, in response to the participants' request, we added the feature of flashing a target when it is nearby. We tested this new design again with three people[1]. Each test session lasted about half an hour in total. In the middle of the test, we turned on the sensor input that is built into Topiary to see how sensor accuracy affected our participants.

Our participants gave us many useful comments. For example, two of them suggested showing a movement trail to help to indicate orientation. Also, the inaccurate update of the user's location, either by the Wizard or by the sensor input (while it was turned on), did confuse the participants. As a result, one person suggested showing a region for the possible location instead of just a point. They also gave us some other suggestions, such as placing the distance label at the top of the screen rather than at the bottom.

Interestingly, some of our participants did not realize their location was being updated by a wizard rather than by real sensors. It was also observed that the prototype showed an optimal path to a participant who had spent three years on the campus but did not know the existence of this path. We did not know this path either and we simply drew a road network in Topiary by

Figure 8. Two screens (pages) of the new design

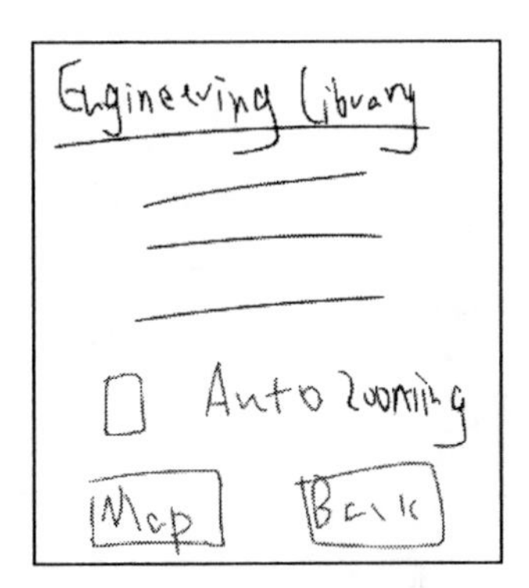

(a) A user can select or deselect the checkbox to choose automatic or manual zooming.

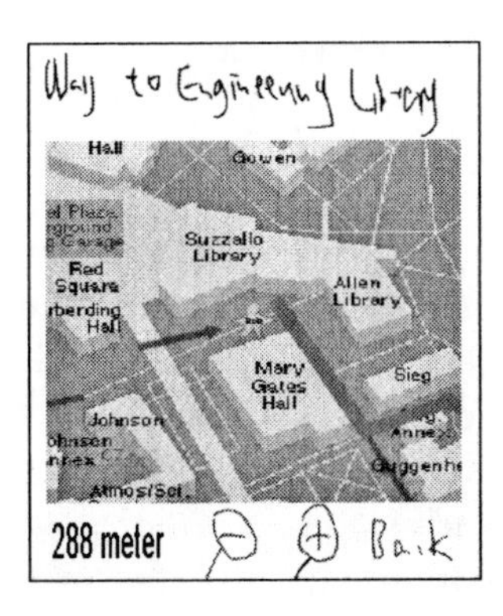

(b) A map screen with zooming buttons and a trajectory arrow

which this path was automatically constructed by the tool.

Building a High Fidelity Prototype

Through these two iterations of informal prototyping and testing, we got a rough view of what the Place Finder should be like. Then it was the time to consider implementation issues and to create a high fidelity prototype. Because we did not want to add an extra device, like GPS, for the Place Finder PDA, we chose to use Place Lab for location sensing, since it requires only WiFi. However, Place Lab, like GPS, cannot provide precise orientation. As a result, we decided to show a movement trail (feedback from the earlier study) instead of showing potentially inaccurate directional arrows or employing map rotation. In addition, because the movement speed cannot be accurately measured, we cut the automatic zooming feature, although one participant showed interest in it. This also helped improve application performance on the PDA.

Based on the earlier tests and an analysis of the implementation issues, we built a high fidelity prototype in Java, using the IBM J9 SWT Java toolkit, in about *two weeks* (see Figure 9). We have used this prototype in the field for hours and it has helped us to find places that we had never been to before. We also got positive feedback from people to whom we demoed this prototype. However, performance on the PDA is still a major issue with this prototype and more profiling is necessary before widely testing it.

Figure 9. The high fidelity prototype was built based on the informal prototypes and an analysis of implementation issues

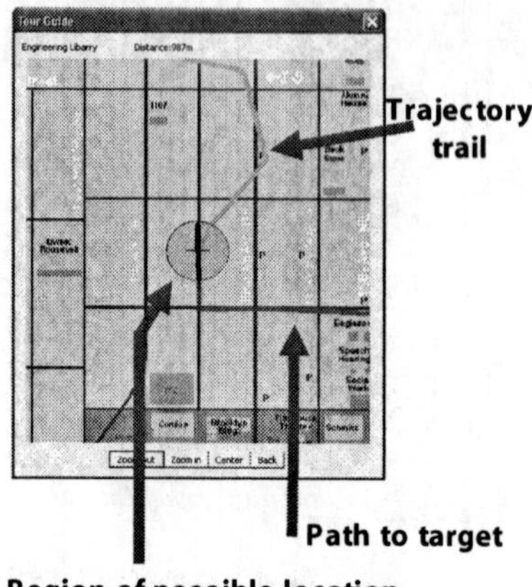

Lessons Learned

This study offers lessons in two areas. First, it identified usability issues as well as solutions for building map-based navigation techniques in a location-enhanced Place Finder application. Second, the study gives an example of how early stage design iteration can be conducted using informal tools. The study showed obvious advantages over traditional paper prototyping since we were able to test our ideas in the field and leverage those results for later stage development.

Informal prototyping and testing in *hours* was much less expensive than directly building a high fidelity prototype over a period of *weeks* and then testing it with users. The tools allowed us to focus on interaction rather than implementation details. It turned out that little feedback from our participants was related to the informal look of the interface. *Focusing on key interactions* rather than specifying the behaviors of the entire application is important to efficiently conducting early stage design because prototyping tools often employ example-based approaches. In our study, only five places were modeled for testing the five low-fi designs. Once the early usability issues were solved, the design was scaled up to 35 places in the high fidelity prototype.

Carefully testing in the field is important for a successful early stage design because the field is where a mobile application design will be used. Testing in the field requires extra consideration when compared to controlled experiments in a lab setting. The *Wizard of Oz* technique was extremely useful in testing an early stage design since it can reasonably approximate realistic situations. On the other hand, using real sensor input, if not expensive, might help find more usability problems due to the inaccuracy of sensors in a test.

FUTURE TRENDS

Sensors such as accelerometers are becoming available on an increasing number of mobile devices to detect a user's context (e.g., movement, lighting or ambient noise) as well as other peripheral input (e.g., digital compass for the orientation of the device). With these sensors, richer interactions can be constructed. It is important for informal prototyping tools to support interaction design based on the available sensors of the platform. Multimodal interaction that combines multiple interaction modalities has shown promise. Speech-based interaction enhanced by location context is an extremely promising research direction. By leveraging location context, a system can optimize speech recognition by focusing on phrases that have meaning in a particular context. This brings new research opportunities to the rapid prototyping of mobile technology. The two tools that we discussed in this chapter address speech-based interaction and location-enhanced computing separately. It would be interesting to combine the strengths of these types of tools for prototyping location-enhanced speech user interfaces.

CONCLUSION

Informal prototyping tools play an important role in the early stage design of interactive mobile technology. They lower the threshold for entry and reduce the cost for prototyping and testing. As a proof of concept of informal prototyping tools for mobile interaction, we discussed how SUEDE and Topiary address the design of speech-based interaction and location-enhanced interaction, respectively, the two representative types of interaction for mobile technology. We highlight two common features of these tools: graphical storyboarding and Wizard of Oz testing. To show how these tools can help an iterative design process, we reported on our experience in iteratively prototyping a location-enhanced Place Finder application, and testing its prototypes with real users in the field. The study indicated that this type of tool allowed a designer to effectively explore a design space in the early stages of design. As mobile computing becomes more powerful and prevalent, there will be more opportunities for research on informal prototyping tools for the design and evaluation of interactive mobile technology.

REFERENCES

Akers, D. (2006). CINCH: A cooperatively designed marking interface for 3D pathway selection. In *UIST'06* (pp. 33-42).

Apple (1987). *HyperCard User's Guide*. Apple Computer, Inc.

Bailey, B. P., Konstan, J. A., & Carlis, J. V. (2001). DEMAIS: designing multimedia applications with interactive storyboards. In *ACM Multimedia* (pp. 241-250).

BonesInMotion. BiM Active. Retrieved from http://bonesinmotion.com/corp/

Cohen, P. R., & Oviatt, S. L. (1995). The role of voice input for human-machine communication. In *Proceedings of the National Academy of Sciences, 92*(22), 9921-9927.

Dahlbäck, N., Jönsson, A. & Ahrenberg, L. (1993). Wizard of Oz studies—Why and how. In *Intelligent User Interfaces '93* (pp. 193-200).

Gould, J. D., & Lewis, C. (1985). Designing for usability: Key principles and what designers think. *Communications of the ACM, 28*(3), 300-311.

Harel, D. (1987). Statecharts: A visual formalism for complex systems. *Science of Computer Programming, 8*(3), 231-274.

Hartmann, B., Klemmer, S. R., Bernstein, M., Abdulla, L., Burr, B., Robinson-Mosher, A., & Gee, J. (2006). Reflective physical prototyping through integrated design, test, and analysis. In *UIST'06* (pp. 299-308).

InfoGation. Odyssey Mobile. from http://www.infogation.com/

Klemmer, S. R., Sinha, A. K., Chen, J., Landay, J. A., Aboobaker, N., & Wang, A. (2000). SUEDE: A wizard of oz prototyping tool for speech user interfaces. In *CHI Letters: 2*(2), *UIST'00* (pp. 1-10).

LaMarca, A., Chawathe, Y., Consolvo, S., Hightower, J., Smith, I. E., Scott, J., Sohn, T., Howard, J., Hughes, J. Potter, F., Tabert, J., Powledge, P, Borriello, G., & Schilit, B. N. (2005). Place lab: Device positioning using radio beacons in the wild. In *Proceedings of Pervasive'05* (pp. 116-133).

Landay, J. A., & Myers, B. A. (2001). Sketching interfaces: Toward more human interface design. *IEEE Computer, 34*(3), 56-64.

Li, Y., Hong, J. I., & Landay, J. A. (2004). Topiary: A tool for prototyping location-enhanced applications. In *CHI Letters: 6*(2), *UIST'04* (pp. 217-226).

Lin, J., Newman, M. W., Hong, J.I., & Landay, J. A. (2000). DENIM: Finding a tighter fit between tools and practice for Web site design. In *CHI Letters: 2*(1), *CHI'00* (pp. 510-517).

LOC-AID. LOC-AID People Service. from http://www.loc-aid.net/people_en.htm

MacIntyre, B., Gandy, M., Dow, S., & Bolter, J. D. (2004). DART: A Toolkit for Rapid Design Exploration of Augmented Reality Experiences. In *CHI Letters: 6*(2), *UIST'04* (pp. 197-206).

Macromedia. Director. from http://www.macromedia.com/software/director/

Myers, B., Hudson, S. E. & Pausch, R. (2001). Past, present and future of user interface software tools. In J. M. Carroll (Ed.), *The new millennium*, (pp. 213-233) New York: ACM Press, Addison-Wesley.

Rettig, M. (1994). Prototyping for tiny fingers. *Communications of the ACM, 37*(4), 21-27.

Snyder, C. (2003). *Paper prototyping: The fast and easy way to design and refine user interfaces.* Morgan Kaufmann.

KEY TERMS

Graphical Storyboarding: A technique that informal prototyping tools often employ for designers to describe how an interface should behave. Like a state transition diagram (STD), it has the concepts of states and transitions. However, in graphical storyboarding these states and transitions represent high level UI components or events rather than the computational elements found in a traditional STD.

Informal Prototyping: A type of user interface prototyping used in the early stages of design in which designers explore a design space by focusing on key interaction ideas rather than visual (e.g., color or alignment) or implementation details. These details are often better considered when creating **hi-fidelity** prototypes at a later stage. Paper prototyping is a representative form of informal prototyping in which designers draw interfaces as well as interaction flows on paper.

Informal UI Prototyping Tools: A type of UI prototyping tool that fluidly supports an informal UI prototyping practice. These tools maintain an "informal" look and feel, use fluid input techniques (e.g., sketching) and can automatically generate testable, interactive prototypes.

Location-Enhanced Applications: Computer applications that leverage the location of people, places and things to provide useful services to users. For example, based on the user's current location, show the nearby restaurants or friends. By using the location context, this type of application reduces explicit input required from a user (such as mouse clicks or typing).

Sketch-Based User Interfaces: A type of user interface in which users interact with a computer system by drawing with a pen. The drawings can be recognized and interpreted as commands, parameters or raw digital ink. This type of interface has shown promise in supporting domains such as UI design, mechanical design, architectural design and note-taking.

Speech-Based Interfaces: A type of user interface in which the user input is submitted mainly via speech. A computer system responds based on either recognized words or vocal variation of the speech. The interface output is typically auditory (e.g., when it is on a phone) or visual.

User Interface Prototyping: A practice of creating user interface mockups to test some aspects of a target interactive system.

UI Prototyping Tools: Electronic tools supporting a user interface prototyping process.

Wizard of Oz Testing: A technique for testing an incomplete interface mockup, named after the movie the *Wizard of Oz*. In this technique, a wizard (often played by a designer) fakes an incomplete (or nonexistent) system component to conduct early user evaluations, (e.g., a wizard can simulate speech recognition when testing a speech-based interface or location tracking when testing a location-enhanced application).

Chapter XXII
Modelling and Simulation of Mobile Mixed Systems

Emmanuel Dubois
University of Toulouse III, France

Wafaa Abou Moussa
University of Toulouse III, France

Cédric Bach
University of Toulouse III, France

Nelly de Bonnefoy
University of Toulouse III, France

ABSTRACT

Interactive systems are no longer expected to be used in confined and predefined places. By increasingly taking advantage of the physical environment, interactive systems are becoming mixed, that is, merging physical and digital worlds. Moreover, they support user's mobility and thus can be referred to as "mobile mixed systems." To overcome technology-driven development processes and to take into account their physical nature and mobile dimensions, specific design approaches are required. From this perspective, we present the interweaving of an existing design model (ASUR) for mixed systems, and a 3-D environment (SIMBA) for simulating modelled mobile mixed system. The aims are to support the investigation of mobile mixed system design through the dedicated modelling approach, and to better understand the limit of the modelled solutions through their simulation. This constitutes a first step toward an iterative method of design for mobile mixed systems, based on "midfidelity" prototyping.

INTRODUCTION

Interacting with a computer system through keyboard, mouse, screen, and speaker in a fixed and predefined working environment is no longer the only available solution. Although such static interactive situations are useful for individual and acontextual applications, mobile interactive systems are tightly interwoven with the existing user's activity, physical artefacts, and application domain resources (Rodden, 1998). They provide a good support to information dissemination, opportunistic share or collection of information, intuitive manipulation, and so forth. Simple examples include the Wii console interaction device, museum guide overlaying exhibits with digital information, and so forth.

Recent advances in the technological, software, and communication infrastructure domains facilitate the implementation of different forms of mobile interactive systems (Renevier, 2004):

- Nomadic systems are carried by a mobile user. Wireless, light, and small devices are required.
- Ubiquitous systems offer services in any places, thus supporting the user's mobility, but remains invisible. It relies on the understanding of the use and manipulation of physical artefacts.
- Context-sensitive systems are influenced by physical properties of the user. User's location and orientation are the two main characteristics of importance.

In this chapter we particularly focus on the two last categories, and use the term "mobile mixed systems" to refer to them. The MARA prototype constitutes a good example of such systems (Kähäri & Murphy, 2006). In this denomination, the "mobile" dimension covers the third form listed (context-sensitive) and "mixed systems" depicts interactive systems that combine the use of physical and digital entities (Dubois, Nigay, Troccaz, Chavanon, & Carrat, 1999) and covers the second case (ubiquitous). Mixed systems include interaction paradigms such as tangible interfaces, augmented and mixed reality, and so forth.

The design of mobile mixed systems introduces many new aspects such as physical artefacts and properties description, links between physical and digital entities, identification of the multiple interaction facets, and so forth. However, traditional HCI approaches leave out most of these specific aspects. Therefore, most contributions in this domain consist of the development of empirical and ad hoc systems. Other contributions have recently introduced models and approaches supporting early design phases: physical objects and interaction description, context models, spatial organisation, and so forth.

The work presented in this chapter aims to take advantage of both kinds of approaches, empirical developments and modelling approaches, to support the design of mobile mixed systems. By coupling these approaches, the goal is twofold:

- To add some rationale in the currently ad hoc solutions in order to justify and document design choices, and to better support the reusability of technical implementation parts;
- To settle direct links between early design and development phases, in order to anchor the reasoning about interactive technique design and its software implementation.

To do so, we have chosen to adopt an iterative HCI design process: participatory design. As presented in Mackay et al. (Mackay, Ratzer, & Janecek, 2000), this iterative process is based on four steps, each of them instrumented with different methods:

- **Analysing:** It refers to the requirements analysis and problems identification. This step relies on user's observations *in situ* with probes, or *in vitro* within labs.
- **Designing:** It consists of generating ideas to solve aspects of the interaction techniques being designed. This second step can be based on a combination of informal techniques (brainstorming, focus group, interviews, etc.) and formal techniques (models, notation, diagrams, etc.), respectively used to generate ideas and then describe the solutions.

- **Prototyping:** It leads to the development of a part of the interactive systems, covering one or more requirements or problems identified in the first step. Tools for prototyping include paper, video, or mock-ups, but also high-fidelity prototypes based on programming languages.
- **Evaluating:** It aims at studying usability aspects of the prototype, and potentially identifying new requirements or problems to trigger a new loop through the participatory design process. This last step involves user's test or speak aloud for example.

The first step is similar to traditional HCI observation and analysis. The second one must address specificities of a mobile mixed system and in particular, its physical nature. The third step must be adapted to this specific area: on one hand, usual low-fidelity prototyping tools (pen and papers, storyboard, etc.) do not reflect all the specificities of these systems (physical properties of the manipulated objects, the interaction context, user's position in the interactive environment…); on the other hand, high-fidelity prototyping tools are very costly (expensive equipment, complexity of software combination and communication, etc.). Finally, the fourth step is still in a very prospective state.

We therefore concentrate on the coupling of the two intermediary steps (design and prototyping), and propose to interweave 1) an extension of an existing modelling approach specific to mixed systems (ASUR) and, 2) a 3-D environment (SIMBA) used to simulate a mobile mixed system that has been previously modelled with ASUR.

This combination constitutes a support to the design and cheap, yet relevant, prototyping of a mobile mixed system. As a result it allows a designer of such interactive systems to:

- Investigate possible interaction techniques in a mobile mixed interaction context thanks to a systematic design model
- Better understand the limits of the modelled solution thanks to a cheap "midfidelity" prototype.

After a brief overview of existing mixed systems design approaches and simulation supports, we present the two facets of our contribution. We then sum up the main aspects of our mixed systems modelling approach, ASUR, and present its extension to suit the mobile mixed system specificities. In the following section, we introduce our ASUR-based simulation environment, called SIMBA, and illustrate the use of SIMBA on a concrete aeronautical application. Finally, we identify a set of perspectives and future trends for this work.

BACKGROUND

Designing Mixed Systems

Definition and Classification

In order to face the profusion of terms, such as augmented or mixed reality, tangible user interfaces, user's augmentation, and so forth, we introduced the generic term mixed systems. It includes all kinds of interactive systems involving physical and digital entities (Dubois et al., 1999). Different aspects have been used to compare mixed interaction techniques: type of data provided (Azuma, 1997; Feiner, MacIntyre, & Seligmann, 1993; Noma, Miyasato, & Kishino, 1996) (3-D graphics, sound, haptic), entity being enriched (Mackay et al., 1996) (user, objects, environment), information representation (Milgram & Kishino, 1994).

As opposed to these implementation-driven approaches, we refined mixed systems domain into two classes (Dubois et al., 1999) according to the type of interaction:

- **Augmented reality systems** (AR) enhance the interaction between the user and her/his physical environment by providing additional digital capabilities and/or information
- **Augmented virtuality systems** (AV) enhance the interaction between a user and a computer by enabling and interpreting the use of physical objects.

So far, no clear consensus has been raised to adopt a common and unique definition and similarly, a large number of prototypes have been developed to illustrate the technical feasibility to use and combine new technologies. To overcome this exploratory process, different design approaches have been investigated. They can be organised into two separate types: implementation support that aims at assisting the development of mixed systems, and modelling approaches that aim at improving the understanding and the exploration of mixed system design solution.

Implementation Support

Many ready-to-use libraries have been developed to integrate specific features of mixed systems such as video-based marker tracking (Kato & Billinghurst, 1999), gestures recognition (Hong & Landay, 2000) physical data sensing, and so forth.

More than a support to the integration of various technologies, development environments have been worked out. For instance, AMIRE (Haller, Zauner, Hartmann, & Luckeneder, 2003) and DWARF (Bauer, Bruegge, Klinker, MacWilliams, Reicher, Riß, Sandor, & Wagner, 2001) offer a set of predefined components, patterns, and connection facilities. Extension mechanisms are not clearly stated, but such approaches provide the developers with a structured view of the application.

Finally, additional works intend to connect these advances with existing standardised tools (Director), or format (SVG).

Modelling Approaches

TAC paradigm (Shaer, Leland, Calvillo-Gamez, & Jacob, 2004) and MCPrd (Ishii& Ullmer, 2000) architecture describe the elements required in tangible user interfaces: one focuses on the description of physical elements while the second focuses on the software structure of TUI.

Other models support the exploration of mixed systems design space: they are based on the identification of artefacts, entities, characteristics, and tools relevant to a mixed system. They are based on a set of HCI and context of use models (Trevisan, Vanderdonckt, & Macq, 2005), on the description of the different interaction facets of the user's interaction with a mixed system (Dubois, Gray, & Nigay, 2003), on the definition of the software elements required by the mixed modalities (Coutrix & Nigay, 2006).

More recent works in mixed systems try to link design and implementation steps by projecting scenarios on software architecture models (Delotte, David, & Chalon, 2004; Renevier et al., 2004) or combining Petri Nets and DWARF components (Hilliges, Sandor, Klinker, 2005).

High level of abstraction, component-based approach, tools interoperability, and implementation support constitutes the main challenges of today's mixed system design approaches. However, models remain static and do not easily address evolutions in the interactive situation, especially when the user is mobile. 3-D simulation is one cheap alternative to the final system development and gives a simple and realistic representation of the mobile dimension of mobile mixed systems.

3-D Simulation Environment

Among numerous fast 3-D generation platforms, we quote the popular VR Juggler (Cruz-Neira, Bierbaum, Hartling, Just, & Meinert, 2002), Avengo (Tramberend, 1999), VIPER (Caubet & Torguet, 1995), NPSNET, DIVE, Bamboo, and Blue-c (Rodriguez, Jessel, & Torguet, 2001). They can render distributed environments and enable interaction through various traditional input devices. However, generated 3-D worlds can rarely be considered as reusable.

In contrast, Balcisoy et al.'s (Balcisoy, Kallmann, Fua, & Thalmann, 2000) platform's main drawbacks are that multiple prototypes must be developed to test the usability of different interaction techniques, and most of the questions related to the interaction mode (voice commands, motion capture, etc.) remain untackled. With SPI (Bernard, Chevalier, & Baudoin, 2004) human behaviour is modelled, but other user's interaction dimensions are lacking. Finally, all these software tools cannot easily fit into a development process, thus making design and prototyping appear as distinct and separate tasks.

More recently, a modelling of an interactive situation can be used to simulate an interactive system displaying information on different surfaces (screen, wall, etc.) (Molina Masso, Vanderdonckt, Gonzalez Lopez, Caballero, & Lozano Perez, 2006). The user can freely modify the use and organisation of the surfaces. However, there is no way to study other interaction aspects than these surfaces position and no clear possibilities for populating the simulation environment with other components.

Using 3-D environments as a support for simulating systems constitutes one of the current challenges of interactive system designers. So far, no simulators are sufficient to support a complete simulation of a mobile mixed interaction system and offer extension mechanisms.

Before presenting and illustrating our simulation environment of mobile mixed systems, we present the UI design model on which the simulation platform is based.

ASUR: A MIXED SYSTEM MODELLING APPROACH AND AN EXTENSION

The ASUR model (Dubois et al., 2003) adopts a user's interaction point of view on the design of mixed systems. The following sections detail and illustrate this model, highlight some limits, and present an extension based on ergonomic criteria.

ASUR Basic Principles

To introduce the ASUR model, we consider an augmented museum scenario. Information is provided according to the exhibit in front of which the visitor is standing. To display this information, the visitor carries a PDA. Information is stored in a database to which the user is automatically connected when entering the museum. Finally, the museum is equipped with a localisation system.

The first step of the ASUR modelling consists in identifying entities involved in this task.

ASUR Components

ASUR distinguishes different component types:

- The **S component** depicts the computer system, including computational and storage capabilities (the museum database), and data acquisition and delivery.
- The **U component** refers to the user of the system: the visitor.
- **R_{object} and R_{tool} components** denote physical entities involved when performing the task. **R_{object}** designate real focus of the task such as the exhibit, and, **R_{tool}** play the role of intermediary entities required to perform the task, for example a wand manipulated by the user to point an exhibit of interest.
- **A_{in} and A_{out} components** represent adaptors conveying data from the physical to the digital world (**A_{in}**, e.g. a localiser) or conversely (**A_{out}**, e.g. a PDA).

These components are not autonomous and need to communicate during the task realisation. Such communication is modelled with ASUR relationships.

ASUR Relationships

We identified three different types of ASUR relationships. A **data exchange (A→B)** means that component B may perceive information rendered by component A. In our example, the visitor observes the exhibit (R_{Object}→U) and data displayed on the PDA (A_{Out}→U). Two distinct devices localise the exhibit and the user (R_{Object}→A_{In1}, U→A_{In2}), and transmit the positions to the computer system (A_{In1}→S, A_{In2}→S). After processing the data, updates are sent to the PDA (S→A_{Out})

Physical proximity (A==B) denotes the physical link that exists between two entities. For example, it represents the fact that the user is holding the PDA in his hand.

Figure 1. ASUR modelling of the augmented museum

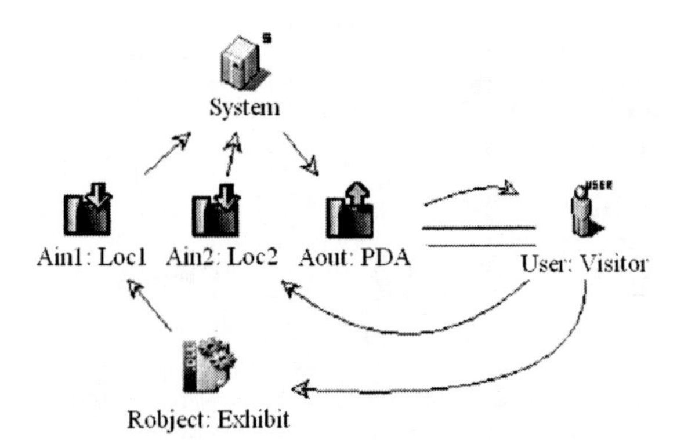

Triggers (A⇨B) is always linked to a data exchange (C→D): the data transfer from C to D will only occur when a specific spatial condition is reached between A and B. No such link is used in the museum example.

ASUR Characteristics

Figure 1 is a diagrammatic representation of this first part of the ASUR modelling of the augmented museum. Additional characteristics are used to refine this modelling (Dubois et al., 2003): **location** and **perception sense** indicate where the user has to focus to get the information and through which human sense it is perceivable, **dimension** (1-D, 2-D, 3-D) and **point of view** refine the description of information transfer.

ASUR Outcomes

ASUR constitutes a good support for comparing and reasoning about the specificities of mixed systems in early design phases. However, the S component includes every digital entity. Typically, it is not possible to accurately model ubiquitous interfaces: links between the physical nature of the system and its digital complement are hidden in relationships to a unique digital entity (S component), and every digital interaction facet takes its origin in the same global digital entity.

An ASUR refinement is thus required to develop a more precise model of ubiquitous systems that support user's mobile activity. We present this extension in the next section.

ASUR Extension

The main goal of this extension is to refine the S component that aggregates every digital aspect involved in a user's task with a mixed system. Refining this component requires the identification of the relevant objects involved in this component. According to our knowledge, no domain model describing the digital part of a mixed interactive system has been studied so far. Therefore, we chose to rely on a domain model of virtual environments. We present this domain model, the principles of its combination with ASUR, and the resulting ASUR extension.

Combining ASUR and a Virtual Environment Domain Model

Our definition of a virtual environment domain model has been triggered by the elaboration of a method for the definition and execution of user's experiments in the field of human virtual environment interaction (HVEI). To overcome the current lack of methodological support when performing experiments in HVEI and analysing the results, we adapted the existing ergonomic criteria (EC) (Bastien & Scapin, 1993) to the specificities of virtual environment (VE) (Bach & Scapin, 2003). The adaptation process is similar to the one already used and validated in the context of Web applications (Leulier, Bastien, & Scapin, 1998). The first phase of this adaptation consisted in a thorough analysis of the literature that led to the compila-

Figure 2. The eight main groups of the list of elements constituting a virtual environment (V.E.)

1 – **User's profile**: user's preferences and abilities management
2 – **Represented objects**: general description and role of the objects present in the EV
3 – **Actions**: availability of services on objects
4 – **Spatial organisation** of the V.E.
5 – **3D Decor** of the V.E.
6 – **3D Boundaries** of the V.E.
7 – **Autonomous elements**
8 – **V.E. behaviour**: general description and actions of the V.E.

tion of 170 ergonomic recommendations, ordered according to a set of 20 ergonomic criteria. The second phase consisted in identifying elements to which these recommendations applied. Finally, we organised the elements identified through this process to produce a list of elements constituting a virtual environment: the list is structured into eight main groups listed in Figure 2.

Since at least one ergonomic recommendation is attached to each of these 73 elements, elements of the list are relevant when performing an ergonomic analysis of an interactive situation with a virtual environment: a model of interactive systems involving a virtual environment must then include these elements. Mixed systems being partly physical, partly digital (or virtual), integrating this list of elements into our ASUR model will improve its ability to model the digital part, that is, the S component.

To proceed with this extension, we first identified among the 73 elements those which corresponded to existing component or characteristics of the ASUR model: "Actions" (3rd point) refers to the ASUR relationships, "Represented objects" (2nd point) includes the interaction mode (gesture, haptic, audio, etc.) and corresponds to the characteristic "perception sense" of the ASUR adapters, and so forth.

We then focused on the remaining elements of the list to refine our model by integrating these complementary elements.

ASUR Model Refinement

The analysis of the list of elements constituting a V.E. resulted into two main changes in the initial ASUR model: the first one is a split in the S component that leads to the identification of three distinct components and their own characteristics; the second one consists in introducing a new kind of relationship.

Split in the S Component

Initially this component included every digital aspect involved when interacting with the system. The first split consists in separating the fundamental digital resources such as the operating system, drivers, storage capacity, rendering and communication capabilities, and so forth, from the digital entities relevant for the interaction with the system.

Former aspects are now grouped into the "***S meta-componen,***" which is implicitly included in every ASUR model. Indeed these aspects do not have an impact on the user's interaction with the system or cannot be directly modified by the designer. Describing them in more detail is thus postponed to the software design step.

To better describe the latter aspects, we first rely on a traditional approach in HCI that consists in separating the interaction into an execution and a perception phase (Norman, 1986). As a result, we introduce two different kinds of S components: S_{tool} and $S_{presentation}$ components.

S_{tool} component. S_{tool} depicts a digital entity used as a tool to perform an action. Its activation has an effect on another digital entity. It corresponds to an articulatory subtask required to apply the desired action onto a digital object. For example, a menu offering several commands to apply to a 3-D object is considered as an S_{tool}: selecting the command is not the finality of the interaction, it is rather intended to apply the command on the 3-D object.

The list of elements constituting a V.E. highlighted two relevant characteristics for the S_{tool} components. They were both derived from the second items of the list ("Represented objects"), and its precise nature (simple, group, etc.) and role (input interaction or navigation support).

$S_{presentation}$ component. $S_{presentation}$ symbolise digital entities carrying information relevant to the user's interaction. The set of $S_{presentation}$ components present in an ASUR model forms the digital-object domain: it represents the concepts manipulated by the application kernel. The analysis of the list of elements constituting a V.E. highlighted subtype of $S_{presentation}$ components:

- **S_{object} component:** Symmetrical to the R_{object} component, it designates a digital object of the task, the object with which it is associated, and the purpose of the task. For example, in a word-processing situation, the file containing the document constitutes the digital object of the task.
- **S_{info} components:** They represent domain objects relevant to the task being performed, without being the focus of the task. Such components contain information related to the internal state of some part of the system: when perceivable, they help the user to build a correct mental representation of the state of the system. In the word-processing example, the amount of pages of the document is an S_{info} component.

The list of elements constituting a V.E. highlighted two relevant characteristics for the S_{info} and S_{object} components. Subitem of "represented objects" reveals the need for the S_{object} and S_{info} to precise if they accept commands or information from other digital component or not. The second important characteristic is derived from the second, fourth, fifth, sixth, and eighth items, and is again the role played by a S_{info} in the interactive task: it can be one of interaction feedback, help, decors/boundaries, or data.

An Additional Relationship

ASUR relationships are used to represent physical relations (proximity or trigger) or data exchanges between two components. Relationships between physical and digital entities are mediated by an adaptor and correspond to input and output of the interactive system. For example in the case of the mediaBlocks (Ullmer & Ishii, 1999), a user (U) manipulates physical cubes (R_{tool}) to modify the sequence order of video sequences (S_{object}): a localiser (Ain) is used to detect the position of the cubes and apply the changes to the video sequences ($R_{tool} \rightarrow A_{in}, A_{in} \rightarrow S_{object}$). This part of the modelling describes concisely data flows between entities involved in the interaction, but does not highlight that the digital video sequences are physically represented by the physical cubes. This aspect is a design-significant aspect of a user's interaction with a mixed system, and has to be added to the initial model.

The "**Representation link**" is thus a new kind of ASUR relationship that connects a physical and a digital entity. It does not represent any information exchange, but depicts a semantic link between two entities. Characteristics of this relationship include the dynamic or static aspect of its existence and the analogy, in term of representation and behaviour of the entity, and the one representing it. These characteristics have been identified on the basis of the list of elements constituting a V.E., the "reproduction fidelity" property introduced by Milgram and Kishino (1994) and the link described by Renevier (2004).

This extended version of the ASUR model, presented in a more concise way with the ASUR Metamodel (Dupuy-Chessa & Dubois, 2005), now supports the detailed modelling of the user's interaction when performing a task with an interactive system merging physical and digital artefacts, such as mobile and ubiquitous systems. Based on this model and its relationship to ergonomic criteria, ergonomic properties studies may also be performed (Dubois et al., 2003). However, testing some physical aspects of the interaction, such as the environment constraints, the availability of the entities, and so forth, is still not easy to conduce on a static model. We thus develop SIMBA, an environment that supports the simulation in a 3-D space of a mobile mixed system modelled with ASUR.

SIMBA: AN ASUR-BASED 3-D SIMULATION ENVIRONMENT

Our primary goal is to develop a prototyping platform that enables closer observation of some aspects of the user's interaction with a mobile mixed system through simulation. To build upon past experiences in terms of simulation and overcome their limitations, we identified a set of requirements:

- Taking advantage of the outcomes of earlier design phases to seamlessly integrate this prototyping approach into the development cycle.
- Avoiding the need for an extensive knowledge in 3-D programming in order to let the designer concentrate on usability rather than technical issue solving.
- Including extension mechanisms for inserting new simulation elements.

In the following sections we present SIMBA functionalities, the software model on which SIMBA's components are based and an illustration of its use on a concrete application.

SIMBA Overall Process

Loading the Required Information

SIMBA stands for "**SIM**ulation **B**ased on **A**sur." Its use is based on:

- An ASUR model and a representation of the physical organisation of the interactive environment
- Predefined SIMBA elements that simulate physical objects, devices, or digital objects during the simulation

GUIDE-Me, the graphical environment for the manipulation of ASUR models (Guide-Me, 2004), generates XML descriptions of ASUR models. Such a file is loaded in SIMBA and interacting entities, as well as relationships between them, are extracted. One unique ID is attached to each entity.

Secondly, a 2-D map is loaded: it represents the physical environment in which the interaction takes place. The position of any pixel encodes the position of one entity in this environment. The RGB-coded colour of the pixel is used to express additional information: the Red value holds the ID provided by the XML description of the ASUR model, the Green and Blue values identify the existing SIMBA element to associate

Figure 3. Extract of an XML file corresponding to an ASUR model (left) and the associated 2-D map (right)


```
- <component>
    <name>Localisateur</name>
    <id>3</id>
    <description />
  - <diagramPosition>
      <xPosition>109.0</xPosition>
      <yPosition>132.0</yPosition>
    </diagramPosition>
  - <charactComponent>
      <shareLevel>1</shareLevel>
      <location>Salle</location>
    </charactComponent>
  - <typeOfComponent>
    - <adapter>
      - <charactAdapter>
        - <sensAction>
            <physicalAction />
            <language>Physical action</language>
          </sensAction>
          <sensPerception>visuel</sensPerception>
        </charactAdapter>
      - <adapterType>
        - <adapterIn>
            <charactAdapterIn />
          </adapterIn>
        </adapterType>
      </adapter>
    </typeOfComponent>
  </component>
```


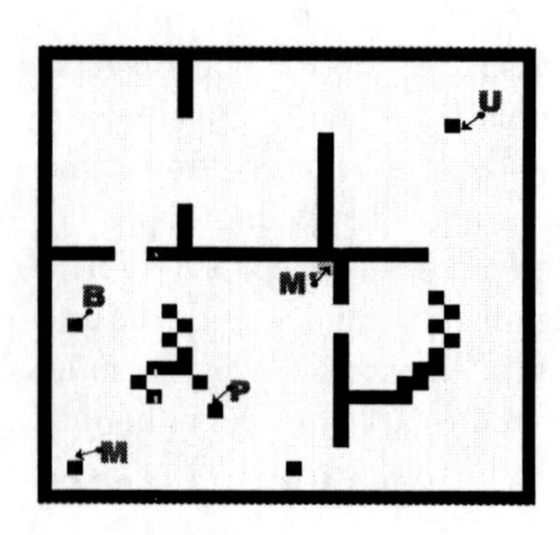

to this ASUR entity during the simulation. Finally, SIMBA interface dialogs allow the user to provide additional information.

Starting and Using the Simulation

Once the simulation is started, SIMBA generates a Doom-like 3-D environment. The designer using the SIMBA simulation environment manipulates an avatar, inside this virtual world, through the standard input devices. SIMBA also handles collision detection. Simulated visual output devices, such as monitors, projectors, PDAs, and so forth, are represented in a separate 2-D window, the "output emulation window," to enable WIMP-style interaction with menus and dialogs. Interaction between the 3-D environment and the output emulation window is handled by SIMBA.

Figure 4 illustrates the simulation of a visitor in the augmented museum modelled in ASUR in Figure 1.

Running a simulation thus entirely relies on the loading of an ASUR model of the system to simulate and a 2-D map representing the physical settings, including the positioning of the different objects involved. In order to facilitate the development of additional SIMBA elements, a SIMBA element model has been adopted and is presented in the next section.

SIMBA Element Model

The implementation of SIMBA, our 3-D simulation environment, is based on Java and Java3D as a rendering engine. This provides the portability and the high level of programming essential to simplify the development of additional SIMBA entities usable in the simulation.

Following most conceptual decompositions in the domain (Sanchez-Segura, 2005), SIMBA elements include two components:

- **"Presentation Components":** Usually programmed by 3-D experts, they handle the loading and rendering of the objects.
- **"Functionality Components":** Usually programmed by interaction experts and application domain specialists, they are in charge of the data processing and communication between the different entities.

Next sections detail the different facets of the functionality component.

Interaction Between SIMBA Entities

To handle the communication between entities, SIMBA uses the relationships provided by the ASUR model. Let us consider a camera (A_{in}) that localises a user ($U \rightarrow A_{in}$) and transmits this information to a digital component ($A_{in} \rightarrow S_{object}$) in charge of storing the user's path.

Figure 4. SIMBA simulation output: visitor's view on museum (right) and "output emulation window" (left)

The functionality component associated with this camera is developed to respond to entity movements inside its visual range, and to identify the moving entity. If the detected entity corresponds to the user, the functionality component associated with the camera posts an event that is caught by SIMBA and forwarded to each entity connected to the camera via an ASUR relationship (the S_{object} in our case). The callback mechanism is triggered inside the receiving entity's functionality component, and the HandleAsurPost() method is automatically invoked upon receipt of the event. This method only receives a reference to the event emitter.

As a result, developing a new functionality component only requires writing the HandleAsurPost() method, that is, the code to execute when this component receives information from another ASUR component. The core of SIMBA environment is in charge of managing the broadcasting of events with respect to the relationships expressed in the ASUR model.

Additional Functionalities

Other methods are included in the functionality component model to handle other kinds of events. Three methods may be automatically invoked by SIMBA, based on the triggering event:

- HandleProximity()is required to simulate physical world constraints that are not clearly expressed by the ASUR model. For example, when the avatar comes close enough as to satisfy certain physical constraints (i.e., to press a button, it must be within the user's arm reach), this method will have to activate or deactivate the device.
- HandleEvent() is called to respond to JAVA events independent of the user's interaction inside the simulated environment.
- AlternativeHandle() is the default method invoked whenever a condition different from those described above is met (e.g., Java advanced functionality).

Most functionality components will only implement one of the mentioned methods. Using distinct function calls for different types of events leads to a simpler event handling

Any entity of a mobile mixed system implemented with SIMBA adheres to this model. The SIMBA element library is thus made of a hierarchical set of SIMBA elements. Of course presentation and functionality component of each element are subclassable, thus promoting code-reuse.

Figure 5. Different aspects of the SIMBA platform

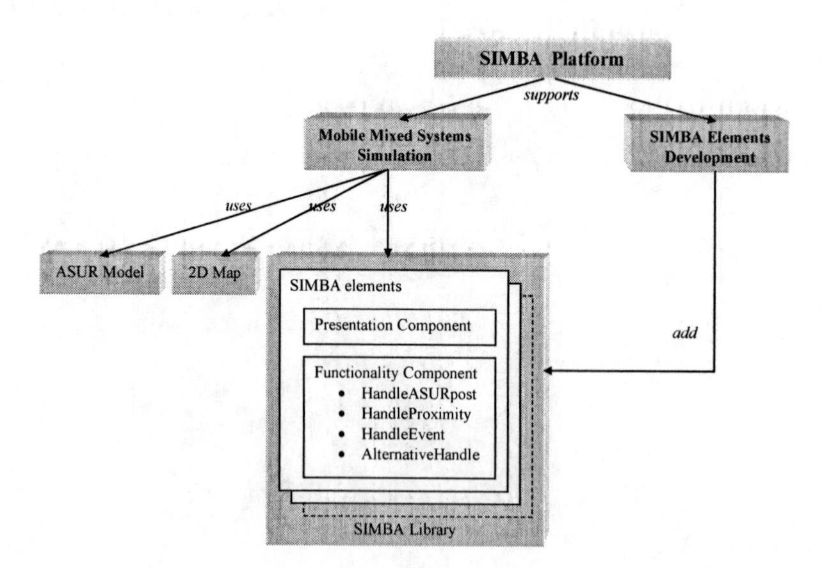

A Concrete Illustration

Aeronautical Maintenance Context

The case study takes place in the aeronautical maintenance process. Maintenance activities are expanding into many aspects of the aeronautical domain. The maintenance process is thus getting richer and more complex; as a result, the cognitive workload of the operators is extremely high. To reduce this workload and provide a better support to the maintenance process, ubiquitous and mixed information systems are promising because they are able to provide mobile operators with appropriate information, at the right place and time and, in the best suitable way.

Such systems are based on physical artefacts (element of planes, tools, manuals, etc.) and the sensing of contextual parameters. Among these dynamic aspects, the localisation of the operator is required to provide information relevant to the maintenance process, such as availability of services (wifi, localisation, etc.) and collaborators. However, before implementing such a system, design decisions must be taken with respect to:

- The technology (type and physical organisation)
- The rendered information (content and representation)
- Data flows among entities

To validate such design decisions, full-scale experiments are highly suitable but remain very expensive: many human and material resources in the operating maintenance areas are required, technologies are not easy to implement, planes immobilisations are too expensive, and so forth.

To overcome the high cost of such experiments, we used SIMBA to study the first two design aspects mentioned: impact of the use of cameras rather than RF-ID to localise the operator, and the nature of the representation used to visualise available services. Next section illustrates the whole simulation process.

ASUR Modelling of the Aeronautical Maintenance Situation

This mobile mixed system involved different ASUR components listed and represented in Figure 6:

- The operator in charge of the maintenance (U)
- The plane on which the maintenance is performed (R_{object})

Figure 6. ASUR model of an aeronautical maintenance situation involving a mobile operator

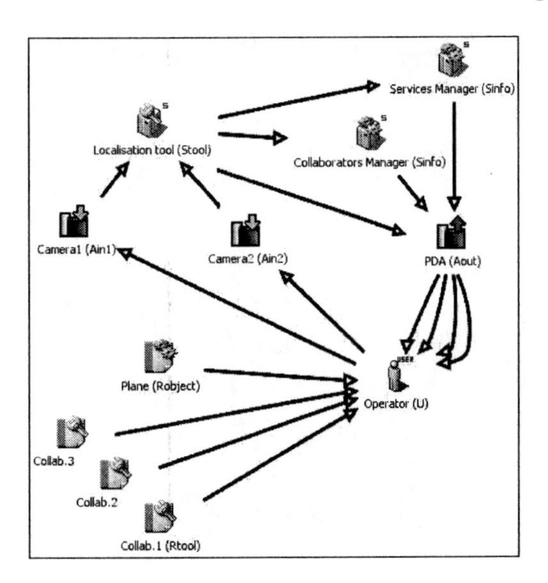

- Collaborators, that is, persons helping the operator during the maintenance process (set of R_{tool}), but not localised in this simulation
- Two cameras detecting the operator's position and transmitting it to the system ($A_{in}1$, $A_{in}2$)
- A PDA displaying to the operator, appropriate computer-generated information (A_{out})
- A localisation tool that converts localisation data coming from the cameras into coordinates useful for the computer system (S_{tool})
- A services manager in charge of identifying the available services ($S_{info}1$)
- A collaborators manager in charge of identifying the available collaborators ($S_{info}2$)

Simulation with SIMBA

To prototype this situation into our 3-D simulation environment, three SIMBA elements were specifically developed: the localisation tool and the services and collaborators managers (S_{tool}, $S_{info}1$, $S_{info}2$). The SIMBA element corresponding to the PDA derives from an existing class of PDA to which a video display capability has been added. To simulate the cameras, two instances of an existing SIMBA_Camera element are used. Finally, 3DS max models are attached to the presentation components associated to the planes and the collaborators.

Once the simulation is started, the designer can move an avatar around the plane and collaborators, and get an avatar-centred view on the 3-D environment (Figure 7, left). Cones are used to simplistically represent the cameras (red cones) and the avatar (pink cone).

Finally, according to the avatar's position in the environment, a PDA emulation displays the avatar's coordinates, the available collaborators' names, even those who are not in eye contact, the activated services (WiFi, camera detection), and the camera's video feedback that used to indicate that the avatar is visible (pink cone visible) or where an area of localisation exists. This simulation provides the designer with a mean to evaluate the adequacy of:

- The content of the information provided to a future user with the task considered
- The data representation with the future user's expertise: label and/or video-feedback can be used to represent the availability of the video-based localisation service
- The position of the camera required for any localisation-based service: if the designer notices that the avatar is always in places where services are unavailable, the designer will be encouraged to modify the initial design of the considered mobile mixed system. For example, solutions may be the definition of a

Figure 7. SIMBA windows when simulating the mobile mixed aeronautical maintenance situation

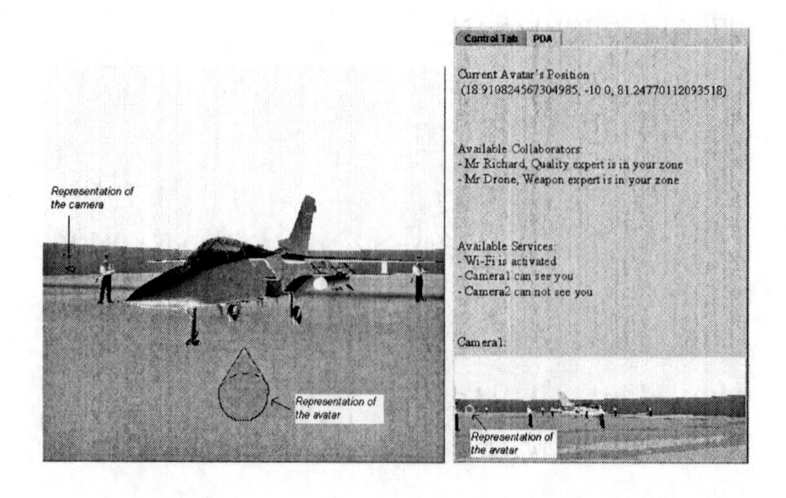

different disposition of the physical space or, the use of different or additional resources (e.g., adding more camera to support the video-based localisation or replacing it with an RF-ID based localisation).

Modifying elements of the ASUR model of the mobile mixed systems allows the testing of different localisation techniques, content, and representations of the dialog. Following such modification, SIMBA elements clearly associated with the modified ASUR elements must then be adapted while the other SIMBA elements remain unchanged. Based on this simulation environment, advantages and limitations of video-based localisation techniques vs. an RF-ID based system for example can be assessed in flexible manner.

Through this example, we have illustrated that our three requirements are fulfilled:

- The simulation is relying on an ASUR design model of the system
- SIMBA elements of the simulation have been directly reused from previous use of the platform: wall, humans, plane, camera-based localisation
- SIMBA elements have been specifically developed for entities used in this interaction situation: PDA emulation, application specific functions

Practice and Experience Feedback of the Use of SIMBA

Five different persons have been using SIMBA to build an ASUR-based simulation. Informal interviews were conducted to retrieve their comments.

SIMBA 3-D scene management and easy insertion of elements in the digital environment through the 2-D map are the two major advantages of SIMBA. The mechanism in charge of the automatic event forwarding to relevant entities, according to the ASUR model, is also really helpful: ASUR adds an abstract level for reasoning and identifying data exchanges to the simulation environment. Moreover, SIMBA elements model enforces to decouple presentation and functional aspects of the application. Finally, the predefined methods of the functionality component are well suited to the needs of the simulation.

However, when starting the simulation, the creation and loading order of the different SIMBA elements has a very big impact on the simulated application. This clearly appears to be restrictive in some cases. For example, it is not straightforward to create a generic PDA because its emulation depends on the data that will be displayed. Other identified drawbacks are the heavy link with Java3D and the lack of a clear documentation.

Regarding the outcomes in terms of evaluation of the simulated systems, moving the avatar and understanding the relationship between output devices in the 3-D scene and their emulation displaying the information appears to be less disturbing than we thought: the emulation window is considered as a tool-palette, which is a usual way for presenting options in interactive systems. Users of the simulation were thus not disturbed by this aspect. In addition, playing with the avatar gives a better representation of the scenario that has to be performed. Alternatives would have required paper prototypes or 2-D interfaces, which would have cut them from the physical environment.

The main drawback reported is that the user does not feel completely immersed in the simulated mixed interaction space. In fact, we do not consider this as a negative aspect on our approach. We are not chasing a completely realistic environment for the simulated system, but rather an environment that supports the analysis of aspects specific to mobile systems: localisation techniques, presentation mode, impact on the context or other persons, and so forth. Finally, interacting through the keyboard is tiresome: a wider range of input devices should be provided according to the 3-D situations and the skills of the users.

As a result, SIMBA is based on outcomes of early design phases, and constitutes an important first step towards the simulation of mobile mixed systems in order to prototype and evaluate such systems.

FUTURE TRENDS

To build upon this coupling of design results and implementation process, improvements to the simulating environment and the model on which it is based are already envisioned.

In Terms of the ASUR Model

We have illustrated that ASUR now covers the design of every kind of mixed systems, from augmented reality to augmented virtuality. But ASUR only provides a part of the description of the user's interaction with a mixed system: it highlights domain concepts, partially represents presentation and dialog aspects, but does not really cover task sequences, system and platform models, device models, and so forth. Linking this model with complementary tools or notations would lead to a more complete modelling of the user's interaction with a mobile mixed system. These additional aspects could then be taken into account into the simulation step, thus enhancing the insight of the simulation. For example, coupling ASUR with a task-based approach is already ongoing, and would help a designer in analysing the impact on the user of changing from one mixed interaction technique to another one.

To support the articulation of these different models, we believe that adopting a model driven engineering (MDE) approach will greatly enhance and support a more systematic design method. MDE approach is also useful to transform an abstract model into a more concrete model such as software component description, for example (Dubois, Gauffre, Bach, Salembier, 2006).

In Terms of SIMBA

We ran different simulations of our application scenario with SIMBA, using various sensing technologies. The required SIMBA elements have thus been added to the SIMBA library. However, more elements must be developed to a sufficient basis for the rapid simulation of mobile mixed systems.

Technically, some improvements are required to better address the creation process of the different elements when starting a simulation, and to enable the use and development of more generic elements. To do so, technologies such as runtime loading of classes and CorbaCCM are envisioned.

We also reported, in this chapter, results of a limited evaluation. A more detailed experiment, based on the same scenario, is planned for this year. Real operators and engineers will be involved so we can learn, from end-users, the benefit of this simulation.

In terms of use of the platform to prepare the simulation, some effort is required to support, in a more usable way, the definition of characteristics not listed in the ASUR models, properties of the 3-D environment, and so forth.

Finally, our ultimate goal is to be able to substitute part of the simulated interaction by their tangible counterparts: the simulation of the mixed system would then gradually change into a mixed system instead of a virtual application. Indeed, one can imagine that the PDA does not have to be simulated and can be directly handled by the user of the SIMBA simulation; manipulating a physical wand to point at a given place on the screen can also be extracted from the simulation, and so forth. This would support an incremental prototyping approach of mobile mixed systems.

CONCLUSION

This chapter introduced an extension of ASUR, a mixed system model, and a platform to simulate mobile mixed systems previously modelled with ASUR. The extended version of ASUR provides designers with a tool to reason about entities, different interaction facets, and data flows involved in mixed systems. The extension is based on established results in ergonomics: predictive analysis of ergonomic properties based on the model is thus supported.

Based on this modelling approach, SIMBA constitutes a support for running a simulation of an ASUR-modelled mobile mixed system:

preexisting SIMBA elements ready to use in a simulation can be very simply associated to elements of the ASUR model, so that the simulation of the mobile mixed system becomes independent of the programmer's skills to manage the available virtual reality platforms, 3-D toolkits, and authoring systems. For more experienced users, SIMBA also offers a predefined format for creating additional SIMBA elements.

This form of prototyping clearly complements the predictive analysis performed on the model by tackling technological, topographical, and dynamic aspects of the system that are not easily visible on a modelling approach or traditional low-fidelity prototyping approaches.

Combining these two approaches thus constitutes a first step toward the tight coupling of designing and implementing steps of a mobile mixed system development. As highlighted in this chapter, interlacing a model-based approach and the prototyping approach only covers two steps of the traditional participatory design approach. Further work will lead to articulate tools and methods in order to equip the whole iterative cycle: improving the SIMBA architecture model will allow its connection to other component platforms; usability property expression in terms of ASUR and SIMBA will support the evaluation of mobile mixed systems specific aspects. This work is thus a first step towards a well-founded set of models, well integrated with one another, and that fit into a development process of mobile mixed systems.

ACKNOWLEDGMENT

Part of this work has been financially supported by the French National Research Center (Grant CNRS–GDR-I3–AS153) and the Région Midi-Pyrénées (Grant for the project Protopraxis). Authors would like to thank Isabelle Fonquernie for help improving SIMBA and Amélie Rault-Azeem for her thorough proofreading.

REFERENCES

Azuma, R. T. (1997). A survey of augmented reality. *Presence: Teleoperators and Virtual Environments, 6*(4), 355-385.

Bach, C., & Scapin, D. L. (2003). Adaptation of ergonomic criteria to human-virtual environments interactions. In *Proceedings of Interact'03* (pp.880-883). Amsterdam: IOS Press.

Balcisoy, S., Kallmann, M., Fua, P., & Thalmann, D. (2000). A framework for rapid evaluation of prototypes with augmented reality. In *Proceedings of VRST'00* (pp. 61-66). NY: ACM Press.

Bastien, J. M. C., Scapin, D. L. (1993). *Ergonomic criteria for the evaluation of human-computer interfaces* (Tech. Rep. No. 156). Rocquencourt, France: INRIA.

Bauer, M., Bruegge, B., Klinker, G., MacWilliams, A., Reicher, T., Riß, S., Sandor, C., & Wagner, M. (2001). Design of a component-based augmented reality framework. In IEEE (Ed.), *Proceedings of ISAR01* (pp.45-54).

Benard, R., Chevalier, P., & Baudoin C. (2004). Simulation participative et immersive.Retrieved January 12, 2006, from http://www.cerv.fr/fr/activites/SPI.php

Caubet, R., & Torguet, P. (1995). Viper: A virtual reality applications design platform. In *Proceedings of Eurographics Workshop on Virtual Environments*. Monaco.

Coutrix, C., & Nigay, L. (2006). Mixed reality: A model of mixed interaction. In *Proceedings of AVI06* (pp.45-53). NY: ACM Press.

Cruz-Neira, C, Bierbaum, A., Hartling, P., Just, C., & Meinert, K. (2002). *VR Juggler - An open source platform for virtual reality applications*. Paper presented at the AIAA Aerospace Sciences Meeting and Exhibit, Reno, NV.

Delotte, O., David, B., & Chalon, R. (2004). Task modelling for capillary collaborative systems based on scenarios. In *Proceedings of TAMODIA'04* (pp.25-31). NY: ACM Press.

Dubois, E., Gauffre, G., Bach, C., & Salembier, P. (2006). Participatory design meets mixed reality design models - Implementation based on a formal instrumentation of an informal design approach. In *Proceedings of CADUI'06* (pp.71-84). Berlin: Springer-Verlag.

Dubois, E., Gray, P. D., & Nigay, L. (2003). ASUR++: A design notation for mobile mixed systems. *Interacting with Computers, 15*(3), 497-520.

Dubois, E., Nigay, L., Troccaz, J., Chavanon, O., & Carrat, L. (1999). Classification space for augmented surgery, an augmented reality case study. In *Proceedings of Interact'99* (pp. 353--359). Amsterdam: IOS Press.

Dupuy-Chessa, S., & Dubois, E. (2005). Requirements and impacts of model driven engineering on mixed systems design. In *Proceedings of IDM'05* (pp.43-65), Paris, France.

Feiner, S., MacIntyre, B., & Seligmann, D. (1993). Knowledge-based augmented reality. *Communication of the ACM, 36*(7), 53-61.

GUIDE-Me. Retrieved from http://liihs.irit.fr/guideme

Haller, M., Zauner, J., Hartmann, W., & Luckeneder, T. (2003). *A generic framework for a training application based on Mixed Reality* (Tech. Rep). Upper Austria University of Applied Sciences, Vienna.

Hilliges, O., Sandor, C., & Klinker, G. (2005). *Interaction management for ubiquitous augmented reality user interfaces*. Diploma Thesis, Technische Universität München.

Hong, I. J., & Landay, J. A. (2000). SATIN: A toolkit for informal ink-based applications. In the *Proceedings of UIST'00* (pp.63-72). NY: ACM Press.

Ishii, H., & Ullmer, B. (2000). Emerging frameworks for tangible user interfaces. *IBM Systems Journal, 39*(3/4), 915-931.

Kähäri, M., & Murphy, D. J. (2006). MARA – Sensor based augmented reality system for mobile imaging. In *Proceedings of ISMAR'06*. NY: ACM Press.

Kato, H., & Billinghurst, M. (1999). *Marker tracking and HMD calibration for a video-based augmented reality conferencing system*. Paper presented at IWAR'99, San Francisco, CA.

Leulier, C., Bastien, J. M. C., & Scapin, D. L. (1998). *Compilation of ergonomic guidelines for the design and evaluation of Web sites* (Commerce & Interaction Report). Rocquencourt, France: INRIA.

Mackay, W. E. (1996). Réalité Augmentée : le Meilleur des Deux Mondes. *La Recherche (285)*, 80-84.

Mackay, W. E., Ratzer, A., & Janecek, P. (2000). Video artifacts for design: Bridging the gap between abstraction and detail. In *Proceedings of DIS2000*. NY: ACM Press.

Milgram, P., & Kishino, F. (1994). A taxonomy of mixed reality visual displays. *IEICE Transactions on Information Systems, E77-D* (12), 1321-1329.

Molina Masso, J. P., Vanderdonckt, J., Gonzalez Lopez, P., Caballero, A. F., & Lozano Perez, M. D. (2006). Rapid prototyping of distributed user interfaces. In *Proceedings of CADUI'06* (pp.153-168). Berlin: Springer-Verlag.

Noma, H., Miyasato, T., & Kishino, F. (1996). A palmtop display for dextrous manipulation with haptic sensation. In *Proceedings of Conference on Human Factors in computing systems* (pp.126-133). NY: ACM Press.

Norman, D. (1986). *Cognitive engineering in user centered system design: New perspectives on human-computer interaction*. Lawrence Erlbaum Associates.

Renevier, P. (2004). *Systèmes Mixtes Collaboratifs sur Supports Mobiles: Conception et Réalisation*. Unpublished doctoral dissertation, University Joseph Fourier, France.

Renevier, P., Nigay, L., Bouchet, J., & Pasqualetti, L. (2004). Generic interaction techniques for mobile collaborative mixed systems. In *Proceedings of CADUI'04* (pp.307-320). NY: ACM.

Rodden, T., Cheverst, K., Davies, N., & Dix, A. (1998). Exploiting context in HCI design for mobile systems. In C. Johnson (Ed.) *Proceedings of the First Workshop on Human Computer Interaction with Mobile Devices.* University of Glasgow.

Rodriguez, N., Jessel, J. P., & Torguet, P. (2001). Asset: A testbed for teleoperation systems. In *Proceedings of SIM01.* WSES-IEEE.

Sanchez-Segura, M. (2005). *Developing future interactive systems.* Hershey, PA: Idea Group Publishing.

Shaer, O., Leland, N., Calvillo-Gamez, E. H., & Jacob R. J. K. (2004). The TAC paradigm: Specifying tangible user interfaces. *Personal and Ubiquitous Computing, 8*(5), 359-369.

Tramberend, H. (1999). Avocado: A distributed virtual reality framework. In *Proceedings of VR'99* (pp.75-81). Washington DC: IEEE Computer Society.

Trevisan, D. G., Vanderdonckt, J., & Macq B. (2005). Conceptualising mixed spaces of interaction for designing continuous interaction. *Virtual Reality, 8*(2), 83-95.

Ullmer, B., & Ishii, H. (1999). MediaBlocks: Tangible interfaces for online media. In *Proceedings of CHI'99*, vol. 2 (pp.31-32). NY: ACM Press.

KEY TERMS

Aeronautical Maintenance: Activity that consists in preserving planes from failure, decline, or accident. Due to the amount of documentation, space to observe, and operator's constraints, mobile mixed systems constitute promising solutions to support the aeronautical maintenance.

ASUR Model: Design model describing the entities involved in the user's interaction with a mixed system.

Component Model: Structure adopted to develop any software element of a given application or library.

Mixed Systems: Interactive systems involving physical and digital entities. It covers interactive systems in which digital objects enrich the user's interaction with physical objects and interactive systems in which physical objects are manipulated to support the user's interaction with a digital application. Other terms are also used to cover part of mixed systems: augmented reality, tangible user interfaces, and mixed reality systems.

Mobile Mixed Systems: Interactive systems supporting the user's mobility and making extensive use of the physical nature of the user's environment.

Model-Based Design: Early step of a development process that consists in describing the object of the design (software, interaction techniques, etc.) with a formalised notation. The notation can be textual, graphical, or a combination of both. A notation is formalised if it conforms to a metamodel.

Model-Driven - Simulation-Based Prototyping (MD-SBP): Simulation-based prototyping of an interactive application tightly coupled with the description of the application using a given model.

SIMBA: Platform for model-driven – simulation-based prototyping of mobile mixed systems. This platform, based on the ASUR model, provides extension facilities to add new simulated elements.

Simulation-Based Prototyping (SBP): Design step that aims at producing a functional form of an interactive application that can be manipulated through a 3-D virtual environment. Cheaper and easier to produce than a high-fidelity prototype, a simulation-based prototype includes the dynamic aspects of the interactive application and the physical aspects and constraints of the interactive situation.

Chapter XXIII
Engineering Emergent Ecologies of Interacting Artefacts

Ioannis D. Zaharakis
Computer Technology Institute, Greece

Achilles D. Kameas
Computer Technology Institute, Greece
Hellenic Open University, Greece

ABSTRACT

Nowadays, our living environments already provide ubiquitous network connectivity and are populated by an increasing number of artefacts (objects enhanced with sensing, computation, and networking abilities). In addition, people are increasingly using mobile devices as intermediaries between themselves and the artefacts. In order to create, manage, communicate with, and reason about ubiquitous computing environments that involve hundreds of interacting artefacts and cooperating mobile devices, we propose to embed, in these entities, social memory, enhanced context memory, and shared experiences. In this context, we describe an engineering approach and a framework to deal with emergent ecologies of locally interacting artefacts that provide services not existing initially in the individuals, and exhibiting them in a consistent and fault-tolerant way. Because they are emergent, their structure or availability are not predefined or known before hand; we draw from swarm intelligence methods to describe such ecologies.

INTRODUCTION

Already, an increasing number of sensors are becoming embedded in the everyday objects or in the environment at a low cost. As a result of this continuing trend, an elementary ambient intelligence (AmI) infrastructure has become installed (though still fragmented), information appliances are commercially available, and ubiquitous computing (UbiComp) applications (currently in the form of games and informative services) are being deployed. As Norman (1998) anticipated, with the proliferation of networks, information appliances, and artefacts, large amounts of data start being diffused in our living environment, and knowledge about patterns and context of human activities are generated. In addition, new generations of mobile devices (such as mobile phones, tablet PCs, PDAs) are being developed having increased capabilities and resources. These devices can now be considered as powerful information processing, storage, and access tools that can be used as facilitators between people and a smart environment, as they can be aware of the artefacts in their vicinity (Lopez de Ipina, Vaszquez, Garcia, Fernandez, & Garcia, 2005).

These developments have the potential to greatly enhance human activities (i.e., by automating dull or ordinary tasks; by speeding up time to exchange data, records, and files; by providing ubiquitous access to services and infrastructures, etc.). We need, however, to overcome the current limitations of distributed and context sensitive computing basically due to the classical client-server approach that is embodied in most, if not all, object-oriented environments and move towards reusable components and peer-to-peer (P2P) networks. By encouraging the adoption of agents and services as a building block for future advanced applications, full delegation of tasks in societies of interconnected computational and human resources can be achieved.

As a consequence of the availability of new technologies, the nature of the human activities eventually assisted by artefacts is rapidly changing. People have to (consent to) build new task models or adapt the ones they have already been using, a task not trivial at all. Execution of new tasks may become difficult due to the inherent systemic complexity of UbiComp applications that, among others, result from device incompatibility, and a huge number of interactions among visible and nonvisible actors. As ambient intelligence becomes widespread, people with low levels of IT literacy will be increasingly asked to interact with smart objects. Humans, with their "analogue" way of thinking and acting, have difficulties in using digital systems, because the latter demand precision, cannot tolerate misuse, and are unable to adapt to changes in operating environments (Norman, 1998). In addition, people will have to adjust to task execution involving high degrees of interruption and task switching. This situation might lead to the social exclusion of those not able to cope with this complexity, and to possible failure of realizing the AmI vision (ISTAG, 2001). Mobile devices are expected to play an important role in the adoption of ambient intelligence because we have already become familiar with using them, albeit in a simpler context.

This work builds upon the envisaged structure of the AmI environment as one populated by thousands of communicating tangible objects and virtual entities (Kameas, Bellis, Mavrommati, Delaney, Colley, & Pounds-Cornish, 2003). Following an agent-oriented approach and adopting principles of Nouvelle AI (an alternative to the symbolic representation of internal models of the world, promoting that intelligence, as expressed by complex behaviour, "emerges" from the interaction of a few simple behaviours), the next sections describe a conceptual framework, which has been inspired by biological structures and is capable of dealing with phenomena emerging in such an environment. Finally, an engineering approach related to new research issues and requirements is introduced.

On the road to realising UbiComp applications and AmI spaces, several technical issues need to be resolved in order to make these systems adoptable and usable. Some of the major requirements a UbiComp system has to confront are: mask the *heterogeneity* of networks, hardware, operating systems, and so forth; tackle *mobility* and *unavailability* of nodes; support component *composition* into applications; *context awareness*; preserve object *autonomy* even for *resource constraint* devices; be *robust*, *fault tolerant*, and *scalable*; *adapt* to environmental changes; and *be usable by novice users* via understandable designed models (Drossos, Mavrommati, & Kameas, 2007). The approach followed by current efforts assumes a network infrastructure that allows direct communication of application components, that is, UbiComp systems are largely treated as (a) distributed systems with resource constrained nodes,

(b) software components that interact using message exchange or parameter passing, or (c) mobile ad-hoc networks whereby some nodes may fail or be unavailable. Within such systems, objects have computerized interfaces (i.e., containing screens, keyboards, command buttons etc.), which make them much different from the objects that populate our everyday spaces.

For example, in Humble *et al.* (Humble, Crabtree, Hemmings, Akesson, Koleva, Rodden, & Hansson, 2003), the "jigsaw puzzle" metaphor is adopted in the interface. Objects are represented as puzzle-piece-like icons that the user "snaps" together to build an application. While this metaphor is comprehensible, the interactions are simplified to sequential execution of actions and reactions depending on local properties (e.g., sensor events), which limits the potential to express many of the user's ideas. In Truong et al. (Truong, Huang, & Abowd, 2004), a pseudonatural language interface based on the fridge magnet metaphor is proposed, while in the browser approach of Speakeasy (Edwards, Newman, Sedivy, Smith, & Izadi, 2002), components are connected using a visual editor based on file-system browsers. ZUMA (Baker, Markovsky, van Greunen, Rabaey, Wawrzynek, & Wolisz, 2006) is a platform, based on a set of clean abstractions for users, content, and devices, that supports the configuration and organization of content and networked heterogeneous devices in a smart-home environment. ZUMA employs the notion of multiuser optimal experience and attempts to achieve optimization by migrating applications to different environments. However, the system operates using fixed rules, without taking into consideration user's feedback.

In this chapter, we intend to present a different conceptualization that is based on the way natural, living systems use local interactions to self-organize and behave coherently. In our framework, the environment is used as the communication medium; objects need not be aware of each other, and self-organization results from a multitude of local interactions. People are part of this ecology, using objects to perform tasks, only being aware of object affordances. Thus, there need not exist a computer-like screen-based interface for the ecology; instead, the ecology exchanges information with people using the most suitable object (suitability depends on context) in their environment.

BACKGROUND

An AmI space is a hybrid (i.e., physical and digital) environment populated by a large number of communicating tangible objects and virtual entities. An AmI space provides infrastructure to support the composition, deployment, and usage of distributed applications and services. As computational power is diffused in our living/working environment, and the number of everyday devices that are capable of sensing, processing, and communicating continuously grows rapidly, new requirements are posed by the heterogeneity of the involved devices, the complexity of interactions, and the need for simple usage models.

Although the constituent components may have restricted resources, the huge number of interactions gives rise to emergent phenomena. Thus, new research issues arise concerning i) the system complexity that results from the thousands of local interactions and their effect on system stability; ii) the need for flexible and dynamic system architecture capable to evolve and adapt to new situations and configurations; iii) the context dependence of system operation; and iv) the human involvement, which calls for new, more natural, human-machine interaction schemes.

The study of emergence as an inherent property of AmI spaces, as the complexity of AmI systems increases, is of paramount importance if one wants to be able to regulate their behaviour. The proliferation of embedded computers (i.e., in mobile phones, CCTVs, PDAs, smart appliances, etc.) and the huge increase in the available network bandwidth will lead to the development of large computation platforms (i.e., "global computers") capable of supporting complex applications. Artificial intelligence (AI) may prove the decisive success factor by contributing techniques to help us deal with research issues at all levels, including bio-inspired computation models, services aimed at end users, as well as machine to machine services, dynamic composition and adaptability, context awareness, autonomy, and semantic interoperability.

Bio-Inspired Approaches, Complexity, and Emergent Behaviours

When studying natural systems consisting of many living organisms, properties such as stability, coherence, flexibility, and adaptability can be observed. These natural organisms exhibit these characteristics because they are integrated and optimized with respect to their computation and control strategies, morphology, materials, and their environment (Knoll & de Kamps, 2004). Moreover, the capabilities of the ecology are distributed over the whole system, and the physical phenomena are created by the interaction of the participants with their environment. Examples of such natural systems include insect colonies, flocks of birds, schools of fishes, herds of mammals, and so forth. Interestingly, these systems demonstrate coherent collective behaviours, although there is no evidence of some kind of centralised control or leadership.

The ICT systems that exhibit global complex behaviour emergent by the local interactions of their simple components are referred to as complex systems (Bullock & Cliff, 2004). Depending on the analysis level, these systems can be described either as simple components that interact with each other in relatively simple ways, influencing only their neighbours, or as a system exhibiting a complex overall behaviour. Thus, when engineering an artificial complex system to operate "intelligently," the increased task complexity can experience significant reductions if each component in the control loop solves a simplified problem while relying on the other components to create the conditions that make these simplifying assumptions valid (Brooks, 2002).

In current AI research, there is an increasing interest in how to engineer autonomous entities with limited capabilities in both peripheral (sensors, actuators) and computational resources (processors, memory, communication, etc.), yet simultaneously, exhibiting robustness and behavioural agility. The swarm intelligence field contributes to this effort by focusing on the emergent collective intelligence of (unsophisticated) agents that interact locally with their environment (Bonabeau, Dorigo, & Theraulaz, 1999; Kennedy & Eberhart, 2001). These unsophisticated agents are referred to, in the literature, as simple reflex agents (Russell & Norvig, 2003) or purely reactive agents (Wooldridge, 1999), thought the latter are more hardware oriented.

Dealing with Symbiotic AmI Spaces

Based on the above-mentioned analysis and focusing on the creation, management, communication with, and reasoning about UbiComp environments that involve hundreds of interacting and cooperating devices, we propose a bio-inspired engineering approach, and a framework to deal with emergent ecologies of locally interacting artefacts with computing and effecting capabilities that provide services not existing initially in the individuals, and exhibiting them in a consistent and fault-tolerant way. Furthermore, in order to reduce the difficulty of carrying out everyday activities in an AmI environment, we consider the delegation of certain tasks to a digital alter-ego (roughly related to ISTAG, 2001) that continuously observes user's interactions with digital objects in different contexts, can learn the user's interests and habits, and can evolve to take initiative in well-known and harmless activities. This system may have a wide family of forms and means to serve the user; it should pervade the living space, and manifest itself unobtrusively by using the essential available artefacts. As delegation becomes the key issue, it is evident that an agent-oriented or service-oriented view, as recently described, among others, by Foster *et al.* (Foster, Jennings, & Kesselman, 2004), are the most suitable.

Interaction

Within such an AmI space, people still have to realize their tasks, ranging from mundane everyday tasks (i.e., studying, cooking, etc.) to leisure or work-related tasks, or even tasks that relate to emergency situations (i.e., home care, accident, unexpected guests, etc.). To do so, they have at their disposal the objects that surround them. These, in fact, are new or improved versions of existing objects that, by using information and communication technology (ICT) components (i.e., sensors, actuators, processor, memory, wireless communication modules), can receive, store, process, and transmit information, thus allowing people to carry out new tasks or old tasks in new and better ways. In the following, we shall use the term "artefacts" for this type of augmented objects. Turning an object into an artefact is a process that aims at enhancing its characteristics, properties, and abilities so that the new affordances will emerge. In practical terms, it is about embedding in the object the necessary hardware and software modules.

From the interaction point of view, we are mostly concerned with the interface of the artefacts and the collective interface of UbiComp applications. The former shall directly affect or depend upon the physical form and shape of the artefacts. The latter can exist in the "digital space" of a computer, that is, a PDA that runs specific software representations of the artefact services.

Thus, people interact with an AmI environment in order to (Kameas, Mavrommati, & Markopoulos, 2004):

- **Engineer a UbiComp application within the environment**, as a composition of artefacts that collectively serve a specific purpose or satisfy a declared set of needs.
- **Use an application to satisfy their needs:** Such an application may be composed by people themselves, or could be bought and installed.

Interaction takes place in two levels:

- **Artefact-to-artefact:** The objects themselves may form an "underlying" layer of interactions, mainly in order to exchange data and to serve their purpose better. Such interactions may use wired channels or any of the available wireless protocols (in a peer-to-peer or broadcast manner), or even the Internet.
- **User-to-application:** The user interacts a) with any single artefact b) with a collection of cooperating devices. Moreover, one has to consider the case where many users interact with the same application.

The degree of visibility and control that people may have on these interactions may vary depending on people's ability to perceive the system state: any of these two types of interaction may happen either explicitly or implicitly.

An explicit interaction that happens under the control of people always provides feedback about its state to them. Although this may seem desirable, it may also become very annoying if one takes into account that there will be hundreds of artefacts in our environment. People interact explicitly with objects or services (or collections of these). In the case of individual objects (or, preferably, services), interaction can be supported by the object affordances. When people interact with a UbiComp application composed of a collection of objects, its "affordances" have to emerge and be made explicit.

Implicit interactions are usually under the control of actors other than people; these could be processes, artefacts, intelligent agent mechanisms, or even artefact owners. Implicit interactions can only be acceptable if they can be trusted, and do not violate privacy or ethics. People need not be directly aware of the communication among objects. Moreover, even certain interactions with UbiComp applications should happen unobtrusively, that is, people should be made aware of the state changes of the application components without being disrupted from their current tasks.

Finally, one should also consider the context of interaction, which ranges from public to private (with respect to disclosure), from individual to shared (with respect to stakeholders) and from closed to open (with respect to space).

A symbiotic AmI space, as described in the next section, should enable people to simply act upon its objects to use their services; no special collective interface should be required.

CONCEPTUAL FRAMEWORK

When considering AmI spaces, such as those described (e.g., an everyday living or working space), populated by many actors with digital selves (artefacts and human), one has to deal with undoubtedly complex emergent phenomena (Lindwer, Marculescu, Basten, Zimmermann, Marculescu, Jung, & Cantatore, 2003). By viewing the user as part of the system, one has to ensure an unobtrusive symbiosis between human and artificial entities, and to establish the user acceptance and confidence towards the UbiComp applications. Our approach is based on a human-centric autonomic system (Mitchell Waldrop, 2003) constituting of self-managing ecologies that are diffused in the everyday living space. In such ecologies, artificial entities coexist unobtrusively with humans, and perform collaborative tasks through a continuous evolvable process concerning both their physical and social cognitive growth (we call them "*ambient spheres*").

It is apparent that the social dimension is a significant factor that characterises such systems. Technically, the reproduction of social behaviours and the handling of complex tasks with an equal agility, as the one exhibited by natural intelligent systems, can be achieved by i) considering that all the necessary information lies out in the environment and surrounds the participants (according to Brooks, 1991, 2002), and ii) using bio-inspired approaches in designing intelligent systems, in which autonomy, emergence, and distributed functioning are promoted (Bonabeau *et al.*, 1999; Kennedy & Eberhart, 2001). By distributing the individual physical/computational/cognitive capabilities over the entire ecology, and by immersing the ecology into a UbiComp environment, we can generate theory and technology for the understanding of the own self and its relation with the surrounding world. In order to deal with the collective behaviour of large societies in situated domains, the system has to perform analysis and synthesis of primitive behaviours that result from individual interactions. This is originated by the engineering methodology proposed by Brooks (1991) (and followed by Mataric, 1995 in the multiagent domain), who decomposes the system into parts, builds the parts, and then interfaces them into a complete system. The decomposition is by activity, and the advantage of this approach is that it gives an incremental path from very simple systems to complex autonomous intelligent systems. At each step of the way, it is only necessary to build one small piece and interface it to an existing, working, complete intelligence. We extend the above consideration by additionally i) attributing AmI objects with physical expression (dimensions, shape, texture, colour, plugs, sockets, connectors, etc.), and ii) dealing with the provided services as basic behavioural building blocks of the overall system behaviour.

According to our approach, a living/working AmI space is populated with many heterogeneous objects with different capabilities and provided services. All these objects and services are regarded as basic building blocks, having an internal part that encapsulates the internal structure and functionality, and an external part that manifests the capabilities and influences the surroundings. Additionally, every basic building block has several predefined functions (we call them *basic behaviours*). Some of the basic behaviours are just reactions to external events, and some are continuously pursued to be fulfilled. The former type of basic behaviours imitates the reflex actions of the living organisms, while the latter the preservation instincts. The interrelationships between the basic building blocks and the associated environment form an *ecology*. Definitely, until now, our approach does not differ a lot from a natural ecosystem, or from (purely reactive) collective robotics, except that we consider services as part of the swarm, too. However, we use another ingredient, called *ambient system* (AmS), with the following characteristics:

- It acts as the "glue" between the tangible and nontangible basic building blocks of the ecology by providing an interface definition language (IDL) and thus, integrating the basic building blocks into a common interoperability framework. A similar approach is adopted by the common component architecture (Larson, Norris, Ong, Bernholdt, Drake, El Wasif, *et al.*, 2004) (CCA) that uses the Babel (Dahlgren, Epperly, & Kumfert, 2003) language to allow components written in various languages to interoperate. One step to this end is the design and implementation of a hierarchy of multidimensional ontologies that will include both nonfunctional descriptions, and rules and constraints of application, as well as aspects of dynamic behaviour and interactions. A core ontology will be open and universally available and accessible. During the ecology lifetime, the core ontology will be supplemented with higher-level goal, application, and context specific ontologies. These ontologies, describing specific application domains, can be proprietary. Emerging behaviour, in this context, will be considered as a result of interactions among heterogeneous, seemingly incompatible or non-predefined entities. Moreover, all higher-level constructs will be inherently able to use all the knowledge they will be able to access.

- It is responsible for the observation and collection of the interactions between AmI objects, provided services, and users. The collected data are used to create best-practice ecology configurations that will help the gradual accumulation of social memory. Aspects of social intelligence are embodied with the ecology configurations as basic social behaviours aiming to regulate the group interactions. These social behaviours are provided as ontological constructions that are also subject to evolution.
- It uses the collected information as input to appropriate reinforcement learning algorithms (e.g., Q learning algorithm, (Watkins & Dayan, 1992)) in order to learn the suitable configurations in association to the task to be accomplished. On the other hand, it utilises a genetic algorithm responsible for the ecology evolution in terms of resource availability. To this end, the modelling of "user perceived quality" (or user comfort (Mozer, 2005)) in the AmI space is necessary, and this could be a parameter of the fitness (evaluation) functions that could be used in the evolution and learning processes.
- It provides feedback to the members of the ecology (social memory in association to "user perceived quality") favouring the best configurations and implicitly assigning cognition to the whole ecology. It is mentioned that the feedback information is provided to the ecology as another stimulus or stimuli and thus, does not require extra (complicated) sensors mounted to the artefacts. This would require a different modelling and engineering approach comparatively to the one described so far. Instead, and along the lines of swarm intelligence, where the environment is a stimulus for the swarm, we treat the AmS as another (special) basic building block; the environment building block.

The AmS is realised as a distributed platform that supports the instantiation of ambient spheres, each of which are formed to support human activity (Figure 1). Examples of candidate AmS technologies are the currently available distributed component frameworks and service-oriented architectures. An ambient sphere is an integrated autonomous system realised as a set of configurations between the AmI objects and the provided services into the AmI environment. In our model of discourse, the end-users are placed inside the ambient sphere, as this allows us to model them as another basic building block that generates events and changes the environment. Then the sphere can evolve on its own (in a sense, it develops a "self," as it becomes aware of its capabilities and the context of operation, while it maintains a set of goals to achieve), by (self-) configuring and (self-) adapting to better satisfy certain user needs.

Summarising, the ambient system, which includes networking, middleware, learning, and evolution algorithms and mechanisms, ontologies describing structural and (both individual and

Figure 1. Individual and Social levels correspond to the basic building blocks and ecologies, respectively. Abstract level encloses the social memory of the ecologies; such knowledge must be transferred to the ecologies implicitly, for example, as stimuli of the environment, since individuals and, consequently, the emergent ecologies do not contain any knowledge representation scheme nor reasoning mechanism. Infrastructure level provides system designers with the appropriate tools to develop a system. Definition level is the user interface with the final user.

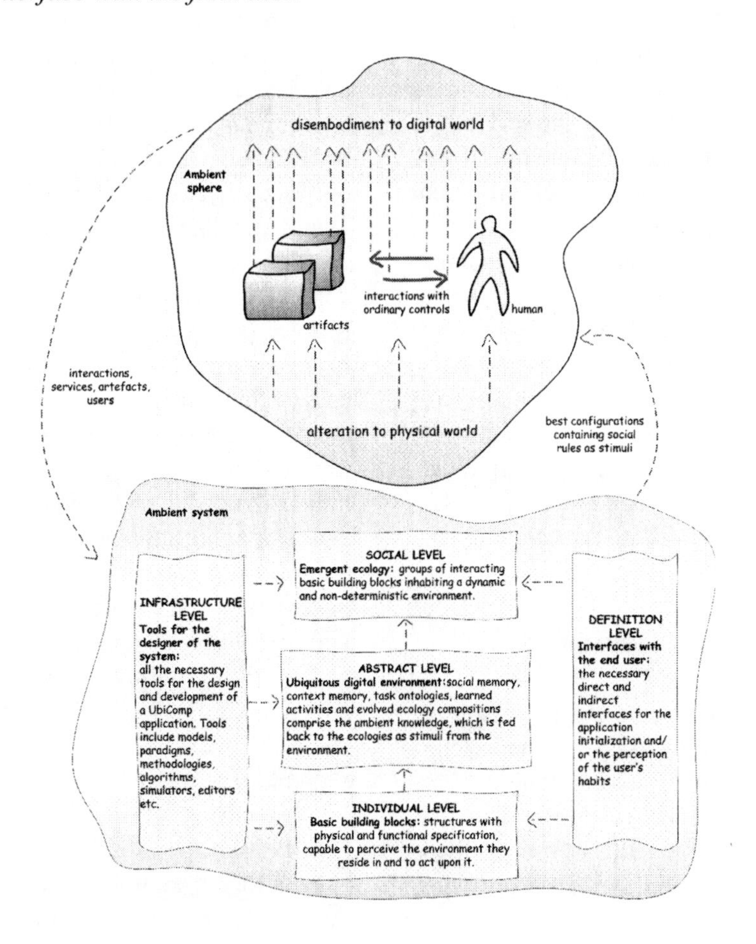

social) behavioural aspect, and so forth, provides an integration framework to collaborating objects and services in order to support user activities by forming ambient spheres. The driving force behind the sphere formation is the selfishness of objects and services; they try to "survive" by operating within a given set of resources. The antagonism of the members is reflected on the spheres, which are formed in an ad-hoc (emergent) way, and they persistently try to meet the "success criterion": user satisfaction.

There are many potential benefits of such an approach including greater flexibility and adaptability of the system to the environment, robustness to failures, and so forth. The swarm comprises different typologies of ecologies; thus it is heterogeneous, also, from the point of view of the provided services. Such differences contribute to the overall capabilities of the system. As a general principle, the services are as simple as possible. The swarm system aggregates the capabilities of extremely simple members by increasing the number of agents, supporting the sharing of resources and maximizing the effectiveness of communication. Ideally, the composition emerges based on previous interactions, and on the context (time and place) they took place. The proposed framework adopts an innovative approach in that:

- It applies to software and services domain an approach previously applied only to robotic applications (which perform local decision making, are inherently distributed, and consist of interacting parts)
- It constitutes an attempt to build integrated UbiComp applications from the services of an AmI environment (without modifying the services, but by using the environment as the integration medium)
- It deals with different and heterogeneous actors (objects, services, social rules, spheres, and people) by applying a hierarchy of semantically-rich representations

Together with the advantages and the benefits of the proposed framework goes a series of technological issues that entail engineering and evaluation. The most important of them are considered in the following section.

An Example: Virtual Residence

As we move from physical to digitized spaces, some of the existing real-world components, concepts, or metaphors will have to be adapted. One important component of human life is residence. It does not only represent the space where one lives, but also a sphere that encompasses one's activities, a place where one can seek refuge, a repository of one's objects and experiences, a private space where more intimate interactions can take place, and many more. It is for these reasons that people feel uncomfortable when they are "away from home," something that happens very often in modern society, which is based on collaboration and "facilitates" (almost demands) mobility.

The concept of virtual residence (Beslay & Punie, 2002) describes the evolution of the physical home into the smart, digital home that will exist within an AmI space. It consists of the smart and connected home infrastructure and the objects therein; the online lives of people, families, households; and the services that support interaction, mobility, and interoperability between different AmI environments. Note that the concept uses most of the notions that apply to "physical" residence, such as borders, markers, activities, and so forth, and expands them to the digital space. Thus, in a virtual residence, one has to define digital borders to delimit the use of digital information, markers that describe allowable interactions with digital media, bridges (i.e., sensors and actuators) between the physical and digital spaces, and policies that ensure a balance between privacy and security, and enhance users' identities (Daskala & Maghiros, 2006).

Mobility can be supported by the virtual residence infrastructure. The aim is to make people feel "at home," even when they are physically located in a different place, by providing them access to content and services in such a way that, in the new location, they can either port or continue the tasks they usually perform at home with minimum disruption, or develop the feeling of being aware of or with the other home residents. These requirements extend beyond mere remote control of home services, which are currently supported via Internet-based applications.

Types of mobility that this concept can support include:

- **Migration of tasks:** The user can continue executing a task, even when he/she moves between AmI spheres. A simple example is the handover service offered by mobile phone operators. Within a rich AmI environment, this would require the restoration of the processes used by the task using the services available in every sphere.
- **Environment porting:** The user reconstructs his/her working, living, and so forth, environment within a new sphere using the objects in the sphere.
- **Mobility of people:** The user, being either on the move or away from the virtual residence, still has access to the services it offers (i.e., conceptually, the virtual residence borders "expand" to still contain the user).

ENGINEERING APPROACH

In order to build UbiComp systems from autonomous, resource-constrained, possibly heterogeneous components making optimum use of distributed intelligence embedded in the periphery, one must develop theories, technologies, and scientific communities that are undoubtedly interdisciplinary. These systems are modelled as self-aware and self-reconfigurable symbiotic ecologies where artificial beings and humans coexist. The applications include autonomous software running on autonomous devices. Social interactions arise among the different elements and adaptation to unforeseen (at design time) situations encountered in dynamic environments is needed.

Basic Building Blocks and Emergent Behaviour

When trying to define the basic building block, one is confronted with questions on i) which should be the basic building block, ii) what structural and functional properties it should encompass, iii) how it could interact with the others, and iv) how it could be realised. From a technology point of view, one would consider as basic building block every self-sustained digital (h/w or s/w) artefact with certain functionality that (a) can operate without the contribution of others, and (b) can interact with others. This definition includes robots with predefined specialised capabilities, but also sensors, motors, computational sources, and so forth. In all cases, as a result of interactions among basic building blocks, it is possible for emergent ecologies to be formed, which exhibit capabilities not found in the individuals.

One approach is to consider basic building blocks as hyperobjects (Mavrommati & Kameas, 2003) and treat everyday objects as communicating tangible components. This approach has been implemented in the gadgetware architectural style (GAS) (Kameas et al., 2003), which provides a platform (GAS-OS) that supports the composition of applications from interacting autonomous artefacts (called eGadgets) with the use of the plug-synapse metaphor (to be described later). This approach scales both "upwards" towards the assembly of more complex objects, and "downwards" towards the decomposition of eGadgets into smaller parts.

In dynamic environments, an individual must be reactive, that is, it must be responsive to events that occur in its environment, where these events affect either the individual's goals or the assumptions, which underpin the procedures that the individual is executing in order to achieve its goals. As a result, it is hard to build a system that achieves an effective balance between goal-directed and reactive behaviour (Wooldridge, 1999). Furthermore, when the construction of the individuals is based on the composition of primitive behaviours, the issues of how to select potentially the correct behaviours in different circumstances, and how to resolve conflicts between them, are raised.

The primitive behaviours approach considers that all the (individual) behaviours run in parallel and, depending on the stimuli of the environment, some of them manifest themselves by enabling a suppression mechanism and taking control of the actuators. However, this technique requires a predefined and exhaustively tested set of implicit rules (usually encoded into finite state automata) of firing priorities. Thus, this technique does not scale well even in a moderate number of primitive behaviours, and it lacks learning even in very often tasks.

In order to apply the well-established primitive behaviours approach in swarm societies that can learn and evolve, component-oriented principles and practices could be employed. Synthetic behaviour control mechanisms could be developed based on bio-inspired approaches like spiking neural networks. These behaviour control mechanisms responsible for the arbitration and/or the composition of the primitive behaviours could also be the subject of learning and evolution. The individuals may exhibit varying behaviour: perceiving/exploring their environment, selectively focusing attention, initiating and completing several tasks. The learning and evolution could be studied and investigated at both the individual and social levels. In this case, the focal point must be the components of behaviour control mechanisms. The outcome could contribute to a novel dynamic and adaptive architecture of swarm systems that exploits the global effects through local rules/behaviour.

AmI Spheres and Collective Behaviour

The realisation of an AmI sphere and the supporting computational environment raises several issues fundamentally pertaining to sphere architectural design and real-time perception-deliberation-action loop. The key features of an AmI sphere are its omnipresence via emergence, polymorphism, and adaptation.

In building a swarm system, communication plays a pivotal role. A flexible and lightweight approach is the indirect (stigmergic) communication. The essence of stigmergy is that the individual modifies a local property of the environment that, subject to environmental physics, should persist long enough to affect the individual's behaviour later in time. It is the temporal aspect of this phenomenon that is crucial for emergent collective behaviour (collaborative exploration, building and maintenance of complex insect nest architectures, etc.) in societies of ants, agents, and robotics. Thus, the individuals could be provided with the proper periphery (actuators/sensors), enabling them to emit/perceive electromagnetic signals and emulate the biological "quorum sense" signals. Such a quorum sense communication may be based on an application-specific vocabulary that will be encoded in the signal. The specifics of the temporal modulation aspect of this "quorum sense signal" will come from theoretical biology and existing results. Additionally, the frequency of the signal will be determined after studying the combined influence of the physical medium properties, the range and interference constraints, power requirements, and the size of available hardware components.

Awareness and Presence

Stigmergic signals are used to denote presence. In Sheridan, (1992) there are proposed three categories of determinants of presence: (1) the extent of sensory information presented to the participant, (2) the level of control the participant has over the various sensor mechanisms, and (3) the participant's ability to modify the environment. In Lombard and Ditton, (1997) presence is defined as the "perceptual illusion of nonmediation", that is, the extent to which a person fails to perceive or acknowledge the existence of a medium during a technologically mediated experience. One can roughly identify two types of presence: physical and social. *Physical presence* refers to the sense of being physically located in mediated space, whereas *social presence* refers to the feeling of being together, of social interaction with a virtual or remotely located communication partner.

Presence is used in our approach as a universal concept that applies to all actors (i.e., people, agents, objects) within a sphere, although each actor will recognize different types of signals, use different mediums, and employ different mechanisms to perceive presence. For example, digital entities will exchange stigmergic signals using digital traces, while people will prefer visual or auditory cues. When people to object mediated interaction must be supported, the "valid" stigmergic signals will be constrained by the sensors that the object embeds. To achieve object to people stigmergic communication, we adopt principles from awareness systems and social intelligence. Awareness systems are a class of computer-mediated communication (CMC) systems that support individuals to maintain, with low effort, a peripheral awareness of each other's activities. The Casablanca project (Hindus et al., 2001) and the ASTRA project (Markopoulos, Romero, van Baren, Ijsselsteijn, de Ruyter, & Farschian, 2004) are examples of early awareness systems for the home.

INTERACTING WITH AMI SPHERES

The AmI spheres constitute a dynamic distributed system composed of artefacts with finite sets of capabilities (services) offered usually through proprietary user interfaces. People interact with an AmI sphere in two levels:

- The task level, whereby they will have to use each individual artefact in order to make use of the collective AmI sphere capabilities
- The metatask level, whereby they will have to compose, decompose, or otherwise edit AmI spheres

When interacting with an AmI sphere, people are in fact using the artefacts that compose it (i.e., they are simply acting, not interacting). This is as close as we can get to the notion of calm technology promoted by M. Weiser, who stated that the most profound technologies are those that disappear in the background (Weiser, 1993). This view is directly inspired from Heidegger's theory of "dasein," which states that people are thrown in the world and are always engaged with acting within it to accomplish their tasks. In this view, technological tools disappear in the background in favour of tasks-at-hand; tools only appear when the task accomplishment procedure breaks down, that is, when something goes wrong.

Artefacts have to demonstrate their affordances, both in the physical environment (for people to be able to use them) and to the digital space (so that other artefacts, agents, and so forth, will be able to interact with them). Then, the state of each artefact must be made visible/available for the same reasons (although the procedure used to compute the state should be internal to the artefact). AmI spheres introduce two factors that cause the breakdown of the existing task models: people will have to make sure that they can still carry on with ordinary tasks, and they will have to become familiar with the new affordances of the artefacts. In addition to adapting their skills for using artefacts, people will have to develop skills for using the computing properties of their new environments as well (Mavrommati & Kameas, 2004).

The GAS Approach

To deal with this, GAS adopts a layered architecture that transparently supports composing and using AmI spheres (called eGadgetWorlds) from autonomous artefacts (called eGadgets), which can be objects, services, or both (Kameas et al., 2003). To enable composition of AmI spheres, GAS proposes the plug-synapse model: a "plug" is the manifestation of a property, capability, or service in a semantically-rich way, and a "synapse" is a communication established between compatible plugs. For example, a TV set may offer the display service and a digital camera may establish a synapse in order to output images; a chair may offer the capability to recognize whether a person is seated on it and a table lamp may use it to switch itself on, and so forth. Clearly, this concept scales well, as more complex "plugs" can be defined as compositions of simpler "plugs," either at an artefact or sphere level. For example, a "reading plug" for an office AmI sphere (Figure 2) may be defined by combining specific plugs from a chair, a table, a lamp, and a book (in fact, we refer to their artefact counterparts) in such a way that when someone is seated on the chair and the chair is located close enough to the table and a book is opened on the table, then the lamp is switched on. With the help of the room (considered as an artefact), the system could also recognize who is seated on the chair and switch on automatically his/her reading profile; then, the "reading plug" could be used by the room to redirect phone calls so as not to disturb the user unless necessary.

An interesting case appears when the AmI sphere breaks down. Consider, for example, the case where the desk lamp is broken. Then the system can either inform the user and wait for his/her action, or search for a similar service in the environment (for example, the sphere, with the help of the AmS system, can locate the room lamp and switch it on). In the latter case, it is necessary that all artefacts hold an internal description of their services and goals; and that these descriptions are compatible. GAS includes a multilayered ontology that describes artefact "plugs" and rules of usage (i.e., constraints) using a commonly available core ontology of basic terms. The use of ontology makes possible the communication between heterogeneous eGadgets, and helps in achieving a shared understanding (as described in Habermas, 1984). Emergent behaviour of this type is a direct result of the ability of eGadgets to communicate

Figure 2. An example eGadgetWorld that implements the "reading office" sphere

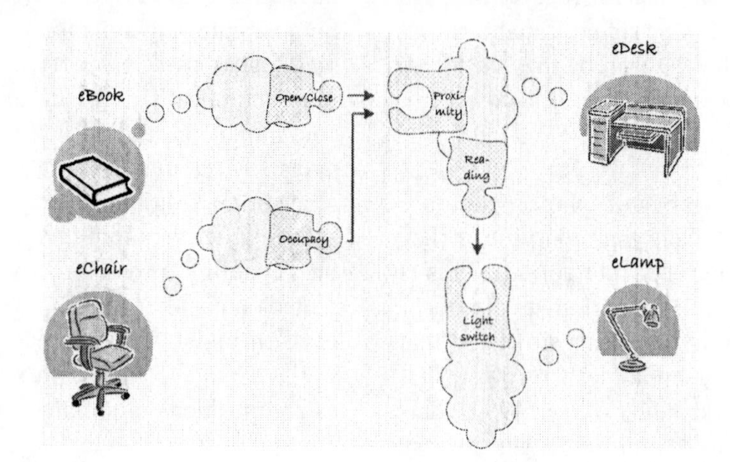

in socially meaningful ways, as described in their hierarchy of basic behaviours.

This definition supports emerging functionality because (a) artefacts are self-sufficient and their plugs are described in a functionally independent way, (b) not all synapses need to be known from the start, (c) new synapses may be added or existing synapses may be deleted, for example, as an artefact may move outside the sphere (that is, outside the range of the wireless network), and (d) experience may be recorded in local artefact ontologies and appear in the form of higher level plugs (the use of a common core ontology available in the sphere's environment ensures the compatibility of plug definitions).

Plugs and synapses are managed by GAS-OS, a distributed middleware platform that takes care of resource management and communication (Figure 3). Thus, an AmI sphere is defined as a GAS-OS application; all eGadgets in it run GAS-OS; compatible plugs of these eGadgets are engaged in synapses to provide collective sphere functionality.

GAS regards each artefact or mobile device as an autonomous component of an AmI sphere. Although GAS-OS has to run on each artefact or mobile device to ensure compatibility, each artefact can locally and transparently manage its resources. Other approaches support the downloading of software representatives of artefacts, either close (Siegemund & Krauer, 2004) or remote (Lopez de Ipina *et al.*, 2005), into a mobile device, thus making the device to assume the role of a superartefact capable of running a component framework. GAS aims to maintain functional autonomy of artefacts; moreover, mobile devices are also considered as artefacts from an interaction perspective.

Supporting Tasks and Metatasks

Within an AmI sphere composed as an eGadgetWorld, a user may perform his/her tasks simply by using the artefacts or mobile devices therein. We do not propose to embed screen-based interfaces on every artefact, or to use a computer as a sphere master, as these would greatly alter the affordances of the artefacts and consequently, have a negative effect on people's capacity to form new task models. Another unwanted consequence is that these artefacts would no longer be functionally autonomous.

The issues that pertain to the individual artefact user interfaces and the ways their affordances can be manifested will not concern us here; a treatise of these can be found in Mavrommati and Kameas (2003). Instead, we are concerned with the collective sphere interface. To this end, we adopt the basic notions and goals of UI plasticity (Calvary, Coutaz, & Thevenin, 2002), although we are not

Figure 3. GAS-OS layered design

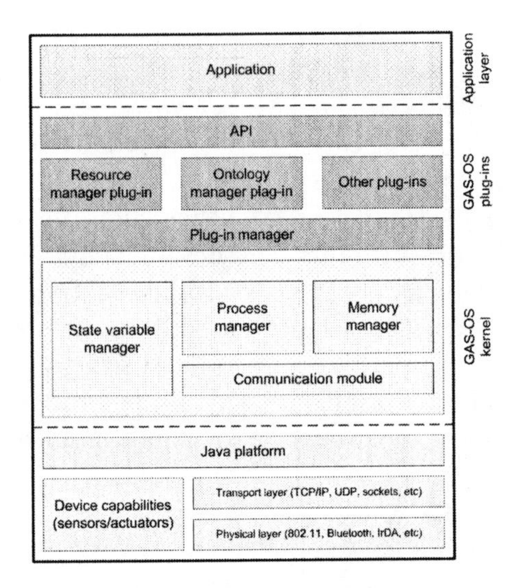

Figure 4. AmI sphere configuration tools

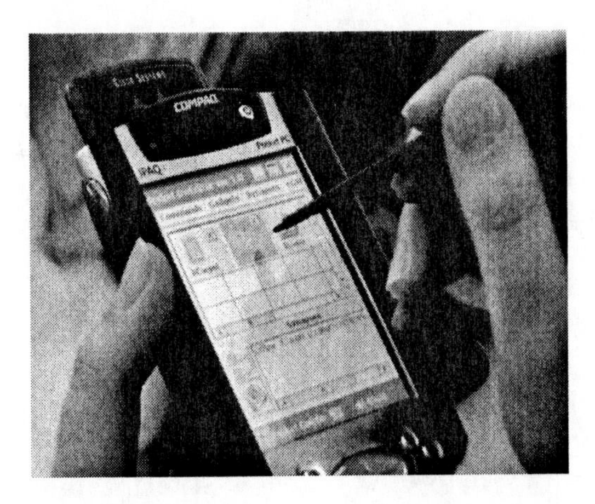

concerned about UI migration and distribution, as these apply to centralized UIs. In the case of AmI spheres, the issues are to conduct a coherent dialogue composed of user actions within the sphere and eGadgets' responses, and to preserve a distributed but meaningful dialogue state. To achieve this, each eGadget must be aware of the state of other eGadgets in the sphere. This can be achieved by exchanging information through the synapses that compose the sphere. By processing the combined perception of the states of itself, peer artefacts (connected via synapses), and the environment (through stigmergic/awareness signals), and applying its architecture of subsumed behaviours, each artefact is able to locally maintain a dialogue state that is compatible with the AmI sphere dialogue state.

GAS offers a set of tools that support the metatasks of creating and editing spheres (Figure 4). These editors run on mobile devices (i.e., PDAs and laptops), and have been positively evaluated by nonexpert users (Mavrommati, Kameas, & Markopoulos, 2004).

The Subsumption Architecture

The proposed conceptual framework extends the GAS approach by allowing the eGadgetWorld management tasks to be dynamically performed by the AmS (though the direct user involvement is not prohibited) based on the observed user habits, the available artefacts and services, and the social rules in context. The issue that arises with this approach is the uniform support for individual functionality and "social interactions" between the sphere members (artefacts, agents, and people).

Our modelling approach achieves uniformity and coherence because it uses the notion of basic behaviours to represent both functional and social capacities of eGadgets. Thus, the decision-making module of the middleware of each eGadget is composed of two kinds of basic behaviour modules: those that implement its core functionality, and those that realize the context of social intelligence. The former use the data gathered by the eGadget sensors to calculate the object's state and to decide a set of (re)actions. The latter use sensor data, as well as synapse input, to determine context of operation and to select the most appropriate action in the list. All basic behaviours are placed in the same hierarchy, with social intelligence behaviours having greater precedence over functional behaviours; thus, allowing the eGadget to realise the most socially intelligent response. Using this two-level selection mechanism, we ensure independence in the determination of local state and response, while we achieve a socially driven eGadget behaviour.

Examples of functional basic behaviours are "turn light on," "produce specific sound," "move towards a specific direction," and so forth; these depend on the actuators of the eGadget and determine the affordances the eGadget offers. This set also contains the basic behaviours "form synapse" and "learn", which ensure that affordances such as composeability and changeability are supported. To deal with possible heterogeneity in signal definition, each eGadget uses a local ontology to translate incoming signals through synapses.

Regarding social behaviour, we consider basic social behaviours drawn from the social intelligence studies,for example, benevolence, nonintrusion, altruism, responsibility, antagonism, empathy, emergency, and so forth. The subsumption scheme contains, in a hierarchical structure, several degrees of sociality, from nonsociality at all in the lower level to high sociality in the top level, as well as the suppression between the social behaviours. The mechanism that implements the subsumption scheme and is responsible for the arbitration of the social behaviour enablement, according to the context it has to deal with, resides in the abstract level of the AmS. Thus, the sociality of the basic building blocks participating in an emergent society is defined without requiring representation or reasoning capabilities from the artefacts.

The GAS Symbiotic Interaction Metaphor

In order to describe our proposed metaphor for interacting with UbiComp applications composed from communicating artefacts, let us first make explicit some basic assumptions:

- User inhabits an AmI space that contains artefacts, having a physical presence and offering digital services
- User forms a plan to achieve a goal he/she has in mind; in this stage, the plan probably consists of steps and subgoals, some of which may not be conscious at all
- User tries to realise his/her plan by combining services offered by the artefacts in his/her environment; in this endeavour, user can only be aware of the affordances of the artefacts and tries to use them accordingly

First of all, GAS supports the following metatasks, using the GAS editor:

- User can query the services and capabilities of each eGadget
- User composes an eGadgetWorld by combining eGadgets capabilities using the plug/synapse model

- User can have an overview of the existing eGadgetWorlds and even edit or delete any of them

Then, we propose the following interaction metaphor:

- The state of each eGadget is communicated using its actuators
- User simply uses each eGadget based on its affordances, directly affecting its local state
- As a consequence, the eGadget communicates the new state to the eGadgets it is connected to via its synapses
- Peer eGadgets calculate new local states (thus user indirectly affects them) and communicates them using their actuators; each eGadget decides the form of communication using its local subsumption architecture, its local state, and the context it perceives via its local sensors and peer eGadgets states (via its synapses)
- A new global eGadgetWorld state emerges as a consequence of local state changes of all the eGadgets in the eGadgetWorld
- The new global state is communicated to the user by all eGadgets in the eGadgetWorld

In this approach, we must make a few remarks. Firstly, because this is a symbiotic ecology, there is no centralized decision-making component. All eGadgets are considered as peers, and each one is responsible for local decision making and acting by taking into account local and global information (here "global" is restricted to those eGadgets that have synapses with it). However, the eGadgets that compose an eGadgetWorld, if they have this basic behaviour, may choose to elect representatives, that is, eGadgets that will act as eGadgetWorld "leaders." This process is supported by contemporary communication protocols (i.e., Wi-Fi) and ensures that the eGadgetWorld will remain functional, even if some secondary or weak eGadgets are not occasionally operating.

Secondly, because the local eGadget state is communicated to other peer eGadgets via the synapses, and triggers changes in their local states, which are also communicated to peer eGadgets, there exists the risk of the eGadgetWorld falling in an infinite loop of recursive global state changes. This falls within the scope of our modelling approach and can be avoided using two measures. Firstly, synapses are directed: if a synapse exists from eGadget A to eGadget B, this means that only changes in the state of eGadget A are communicated to eGadget B, and not the other way round. In addition, when composing or editing an eGadgetWorld, the GAS tools offer the user the ability to send a "ping" message that propagates to all eGadgets in the eGadgetWorld, thus making clear if any loops exist.

AN EXAMPLE SCENARIO

Sonia is a 36-year old single mother who lives in Athens with here two children. She is a hard-working expert employee who is overcommitted with her children. These days she is working on a report, and today is the deadline for its submission due to a company meeting at 20:00 local time in Brussels. However, she could not travel far from Athens and thus, she could not participate in the meeting. Instead, her colleague, Steve, will present her report. Undoubtedly, she is anxious about the results. As the meeting will take long, she cannot stay awake, but she needs to hear the news from Steve as soon as she wakes up in the morning.

It is 16:30 in Athens (+1:00 CET) and Sonia works in her office. She makes the last minute changes in her report, but she realises that she forgot some important handwritten notes in her home. Additionally, she must take her children to their ballet class. She must urgently leave her office, but also she must finalise and submit the report. Fortunately, she has a little more time until 21:00 local time.

Additionally, Sonia has "an ace in her sleeve"; she likes technology, and has created several eGadgetWorld as AmI spheres to make her overcommitted life easier (Figure 5):

- Her "work-in-home/office" sphere includes the eBook, eNotebook, eChair, eDesk, eLamp, and ePC eGadgets. The functionality of this sphere is similar to the one described in "The GAS Approach" section with the extension that the ePC eGadget provides a synchronisation service concerning a shared repository between different and remote PCs (e.g., home PC and office PC).
- The "wake up" sphere consists of the following eGadgets: an alarm clock, a bed mattress, the window blinds, and a room light. The eGadgets in this sphere have plugs through which they offer access to their properties and services. For example, the clock is equipped with a light sensor whose reading is made available through the plug "luminosity"; the bed mattress has weight and pressure sensors and can decide whether there is someone "lying upon" or not; the window blinds offer the plug "open," which can be used to lift them to a specified height; finally, the room light offers the plug "on-off." This eGadgetWorld, clearly a ubiquitous computing application, is set up to gradually increase the amount of light in the room (until Sonia gets up from bed) when it is time for her to wake up, first by opening the window blinds to let natural light come in, and, if it is still dark (i.e., consider a winter's day when the lighting remains lower than a certain threshold), switch on the room light. All Sonia had to do is to synapse the "luminosity" plug of the alarm clock with "open" plug of the blinds and the "on-off" plug of the room light, synapse the "alarm on-off" plug with the "person lying" plug of the mattress (so that when she stands up from the bed, the alarm is switched off and the blinds stop opening), and finally set the alarm clock.
- In a third, simpler eGadgetWorld, she has already synapsed the mattress with her slippers, and the latter with the coffee maker, thus starting coffee brewing when she gets up and steps into the slippers. Now, she needs to extend it, with a few more synapses between her bathroom mirror and Steve's avatar. She names this eGadgetWorld "coffee 'n' news." It is worth mentioning that Sonia uses this mirror in another eGadgetWorld as a display to read the morning news and the weather as she gets ready for work.

It is 18:00 local time and Sonia prepares dinner. She has already put the children in their room to play. She exploits the time until the dinner is ready and moves to her desk in the penthouse. The "work in home/office" sphere is active and when she sits on her eChair, the ePC pops up a sign to inform her of an urgent task to be completed. In fact, the awareness service between the home and office spheres triggered the synchronisation synapse, and the home PC informed about the task by the office PC and also downloaded the unfinished report into the shared repository. Sonia takes a look at her handwritten notes, completes the report, and sends it to Steve; the home PC closes the task, and the awareness service will inform the office PC about it when Sonia goes to her office and starts using her PC. Now, it is time for dinner. A couple of hours later, the children are in their beds and Sonia feels tired and sleepy.

The "wake up" sphere is active and the next morning, at 06:45 local time, as the sun starts rising, the light sensors are triggered and the luminosity plugs arrange the bedroom illumination. As Sonia gets up and steps into the slippers, the coffee starts brewing, as the "coffee 'n' news" sphere is active. Now Sonia is in her bathroom in front of the mirror and brushes her teeth. The "news" synapse with the eMirror, which she made yesterday, performs a connection with Steve's avatar; he is already awake as he must catch the morning flight to Athens and also cannot wait to inform Sonia about the yesterday's meeting. Clearly, he did not forget to synapse his avatar with Sonia's eHome sphere. Although the eMirror is able to provide live video services, it does not do so this time; it is not socially proper to "intrude" on Sonia's private spaces. The social behaviour service, endowed with the AmS, informs the avatar about the pri-

Figure 5. AmI spheres and interactions

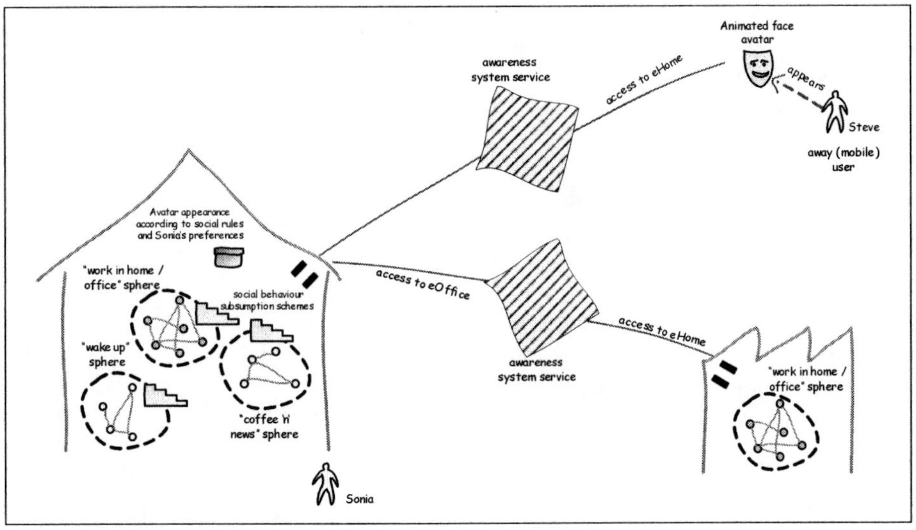

vacy restrictions and allows it to appear only as a "talking box." Now, Sonia gets the good news about the Brussels meeting. She can enjoy her morning coffee. A good day just started.

CONCLUSION

As everyday objects are being enhanced with sensing, processing, and communication abilities, the near future of our everyday living/working is indicated by a high degree of complexity. The emergent complexity concerns the machine-machine and human-machine interactions as well as the provided services aimed at end users and at other machines. Into this rapidly changing AmI environment, new requirements and research issues arise, and the need for a conceptual analysis framework is apparent. This work attempts to introduce a bio-inspired world model that draws features from natural systems, and applies them into symbiotic ecologies inhabited by both humans and artefacts. Furthermore, it introduces a high-level framework of AmI spaces that encloses the fundamental elements of bio-inspired self-aware emergent symbiotic ecologies.

REFERENCES

Baker, C., R., Markovsky, Y., van Greunen, J., Rabaey, J., Wawrzynek, J., & Wolisz, A. (2006 July). ZUMA: A platform for smart-home environments. In *Proceedings of the 2nd IET Conference on Intelligent Environments.* Athens, Greece,

Beslay, L., & Punie, Y. (2002). The virtual residence: Identity, privacy and security. *The IPTS Report, Special Issue on Identity and Privacy, 67*, 17-23.

Bonabeau, E., Dorigo, M., & Theraulaz, G. (1999). *Swarm intelligence: From natural to artificial systems.* Oxford University Press.

Brooks, R. A. (1991). Intelligence without representation. *Artificial Intelligence, 47*, 139–159.

Brooks, R. A. (2002). *Robot: The future of flesh and machines.* London: Penguin Books.

Bullock, S., & Cliff, D. (2004). *Complexity and emergent behaviour in ICT systems* (Tech. Rep. No HP-2004-187). Semantic & Adaptive Systems, Hewlett-Packard Labs.

Calvary, G., Coutaz, J., & Thevenin, D., (2002). A unifying reference framework for the development of plastic user interfaces. In *Proceedings of EHCI01* (pp. 173-192). Toronto, LNCS 2254.

Dahlgren, T., Epperly, T., & Kumfert, G. (2003). *Babel users' guide*, version 0.8.4. CASC Technical Publication, Lawrence Livermore National Laboratory.

Daskala, B., & Maghiros, I. (2006 July). Digital territories. In *Proceedings of the 2nd IET Conference on Intelligent Environments*. Athens, Greece.

Drossos, N., Mavrommati, I., & Kameas, A. (2007). Towards ubiquitous computing applications composed from functionally autonomous hybrid artefacts. In N. Streitz, A. Kameas, & I. Mavrommati (Eds.), *The disappearing computer book*. Springer-Verlag, (to appear in 2007).

Edwards, W., K., Newman, M., W., Sedivy, J., Smith, T., & Izadi, S. (2002 September). Challenge: recombinant computing and the speakeasy approach. In *Proceedings of the Eighth Annual International Conference on Mobile Computing and Networking* (MobiCom 2002) (pp. 279-286). ACM Press, New York.

Foster, I., Jennings, N., & Kesselman, C. (2004 July). Brain meets brawn: Why grid and agents need each other. In *Proceedings of the 3rd International Joint Conference on Autonomous Agents and Multi-Agent Systems, (AAMAS 04)*. New York.

Habermas, J. (1984). *The theory of communicative action (Vol. 1-2)* (translator T. McCarthy). Boston: Beacon Press.

Hindus, D., Mainwarin, S. D., Leduc, N., Hagstro A. E., & Bayley, O. (2001, March). Casablanca: Designing social communication devices for the home. In *Proceedings of the SIG-CHI on Human factors in computing systems (CHI 2001)* (pp 325–332). Seattle, WA.

Humble, J., Crabtree, A., Hemmings, T., Akesson, K. P., Koleva, B,. Rodden, T., & Hansson, P. (2003 October). Playing with the bits: User-configuration of ubiquitous domestic environments. In *Proceedings of UBICOMP 2003* (pp. 256-263). Springer-Verlag.

IST Advisor Group. (2001). *Scenarios for ambient intelligence in 2010*. European Commission Community Research. Retrieved on September 12, 2006, from http://cordis.europa.eu/ist/istag-reports.htm

Kameas, A., Bellis, S., Mavrommati, I., Delanay, K., Colley, M., & Pounds-Cornish, A. (2003 March). An architecture that treats everyday objects as communicating tangible components. In *Proceedings of the First IEEE International Conference on Pervasive Computing and Communications*, (PERCOM2003). Texas-Fort Worth.

Kameas, A., Mavrommati I., & Markopoulos, P. (2004). Computing in tangible: Using artifacts as components of ambient intelligent environments. In G. Riva, F. Vatalaro, F. Davide, & M. Alcaniz (Eds.), *Ambient intelligence: The evolution of technology, communication and cognition* (pp. 121-142). IOS Press.

Kennedy, J., & Eberhart, R. (2001). *Swarm intelligence*. Morgan Kaufmann Publishers.

Knoll, A., & de Kamps, M., (Eds). (2004). *Roadmap of neuro-IT development.* EU Neuro-IT Network of Excellence. Retrieved on September 12, 2006, from http://www.neuro-it.net/NeuroIT/Roadmap

Larson, W., Norris, B., Ong, E. T., Bernholdt, D. E., Drake, J. B., El Wasif, W. R., Ham, M. W., Rasmussen, C. E., Kumfert, G., Katz, D. S., Zhou, S., DeLuca, C., & Collins, N. S. (2004). Components, the common component architecture, and the climate/ocean/weather community. In *Proceedings of the 20th International Conference on Interactive Information and Processing Systems (IIPS) for Meteorology, Oceanography, and Hydrology*, 84th American Meteorological Society Annual Meeting.

Lindwer, M., Marculescu, D., Basten, T., Zimmermann, R., Marculescu, R., Jung, S., & Cantatore, E. (2003 March). Ambient intelligence visions and achievements: Linking abstract ideas to real-world concepts. In *Proceedings Design Automation & Test in Europe (DATE)*. Munich, Germany.

Lombard, M., & Ditton, T. (1997). At the heart of it all: The concept of presence. *Journal of Computer Mediated-Communication [On-line], 3.* Retrieved on January 17, 2007, from http://www.ascusc.org/jcmc/vol3/issue2/lombard.html

Lopez de Ipina, D., Vaszquez, I., Garcia, D., Fernandez, J., & Garcia, I. (2005 October). A reflective middleware for controlling smart objects from mobile devices. In *Proceedings of the 3rd European joint Symposium on Systems on Chip and Ambient Intelligence (sOc-EUSAI)* (pp. 213-218). Grenoble, France.

Markopoulos, P., Romero, N., van Baren, J., Ijsselsteijn, W., de Ruyter, B., & Farschian, B. (2004 April). Keeping in touch with the family: Home and away with the ASTRA awareness system. In *Proceedings of the International Conference for Human-Computer Interaction, (CHI2004)* (pp. 1351-1354), Vienna, Austria.

Mataric, M., J. (1995). Designing and understanding adaptive group behaviour. *Adaptive Behaviour, 4*(1):51-80.

Mavrommati, I. & Kameas, A. (2003). The evolution of objects into hyper-objects. *Personal and Ubiquitous Computing 779, ACM, 7*(3-4), 176-181.

Mavrommati, I., & Kameas, A. (2004 April). The concepts of an end user enabling architecture for ubiquitous computing. In *Proceedings of the 2nd International Conference on Pervasive Computing (Pervasive 2004).* Vienna.

Mavrommati I., Kameas A., & Markopoulos P. (2004 May). An editing tool that manages device associations in an in-home environment. In *Proceedings of the 2nd International Conference on Appliance Design (2AD)* (pp. 104-111), HP Labs, Bristol, UK, 11-13 May 2004.

Mitchell Waldrop, M. (2003). *Autonomic computing—The technology of self-management.* Retrieved on September 12, 2006, from http://www.thefutureofcomputing.org

Mozer, M. C. (2005). Lessons from an adaptive house. In D. Cook R. & Das, R., (Eds.), *Smart environments: Technologies, protocols, and applications* (pp. 273-294). Hoboken, NJ: J. Wiley & Sons.

Norman, D. A. (1998). *The invisible compute.* MIT Press.

Russell, S., & Norvig, P. (2003). *Artificial intelligence: A modern approach* (2nd ed.). Prentice Hall.

Sheridan, T. B. (1992). Musings on telepresence and virtual presence. *Presence: Teleoperators and Virtual Environments, 1,* 120-125.

Siegemund, F., & Krauer, T. (2004 November). Integrating handhelds into environments of cooperating smart everyday objects. In *Proceedings of the 2nd European Symposium on Ambient Intelligence (EUSAI 2004)* (pp.160-171). Eindhoven, The Netherlands: Springer-Verlag.

Thorndike, E. L. (1920). Intelligence and its uses. *Harper's Magazine, 140,* 227-235.

Truong, K. N., Huang, E. M., & Abowd, G. D. (2004 September). CAMP: A magnetic poetry interface for end-user programming of capture applications for the home. In *Proceedings of the Sixth International Conference on Ubiquitous Computing, (Ubicomp 2004)* (pp. 143-160). Nottingham, England.

Vernon, P. E. (1933). Some characteristics of the good judge of personality, *Journal of Social Psychology, 4,* 42–57.

Watkins, C. J. C. H., & Dayan, P. (1992). Q learning. *Machine Learning, 8,*279–292.

Weiser, M. (1993). Some computer science issues in ubiquitous computing. *Communications of the ACM, 36*(7),75–84.

Wooldridge, M. (1999). Intelligent agents. In G. Weiss (Ed.), *Multiagent systems: modern approach to distributed artificial intelligence* (pp. 27-77). MIT Press.

KEY TERMS

Ambient Intelligence: A set of [emergent] properties of an environment that we are in the process of creating; it is more an imagined concept than a set of specified requirements (IST Advisor Group, "Ambient intelligence: From vision to reality." Retrieved on September 16, 2006, from http://cordis.europa.eu/ist/istag-reports.htm). In particular, AmI puts the emphasis on user friendliness, efficient and distributed services support, user empowerment, and support for human interactions. This vision assumes a shift away from PCs to a variety of devices that are unobtrusively embedded in our environment, and that are accessed via intelligent interfaces (Retrieved on September 16, 2006, from http://en.wikipedia.org/wiki/Ambient_intelligence).

Ambient Sphere: Ecology of artificial entities coexisting unobtrusively with humans and performing collaborative tasks through a continuous evolvable process concerning both their physical and social cognitive growth.

Ambient System: A distributed platform that supports the instantiation of ambient spheres.

Basic Behaviours: Predefined functions of the artefacts that are either enabled as reactions to external events or are continuously pursued to be fulfilled.

Basic Building Block: Modelling abstraction representing self-sustained entities that are members of an AmI sphere.

Hyper Objects (eGadgets): Ordinary objects that are commonly used for everyday, even mundane tasks (objects such as tables, chairs, cups, shelves, lights, carpets, etc..), and which in the future, can be enhanced with communication, processing, and sometimes sensing abilities (Mavrommati & Kameas, 2003).

Plug-Synapse Model: A conceptual abstraction that enables uniform access to eGadget services/capabilities/properties and allows users to compose applications that realize a collective behaviour in a high-level programming manner (Mavrommati & Kameas, 2004)

Social Intelligence: "The ability to understand and manage men and women, boys and girls—to act wisely in human relations." (Thorndike, 1920). According to a broader definition, social intelligence is "… a person's ability to get along with people in general, social technique or ease in society, knowledge of social matters, susceptibility to stimuli from other members of a group, as well as insight into the temporary moods of underlying personality traits of strangers" (Vernon, 1933).

Swarm Intelligence: "… an alternative way of designing intelligent systems, in which autonomy, emergence and distributed functioning replace control, preprogramming, and centralisation" (Bonabeau et al., 1999).

Ubiquitous Computing: "…the method of enhancing computer use by making many computers available throughout the physical environment, but making them effectively invisible to the user" (Weiser, 1993).

Section II
Novel Interaction Techniques for Mobile Technologies

This section focuses on the innovative possibilities for interaction with mobile technologies. Starting with a potential classification scheme for mobile interaction techniques, this section looks at a number of novel interaction techniques such as text entry, speech-based input, and audio and haptic interaction for mobile devices. Chapters are included that introduce the concept of unobtrusive interaction and the use of EMG signals to achieve subtle interaction. This section concludes with a look at visual means of interaction, from camera-based input, through 3-D visualisation and the presentation of large data sets using starfield displays, to projected displays for collaborative interaction.

Chapter XXIV
The Design Space of Ubiquitous Mobile Input

Rafael Ballagas
RWTH Aachen University, Germany

Michael Rohs
Deutsche Telekom Laboratories, Germany

Jennifer G. Sheridan
BigDog Interactive Ltd., UK

Jan Borchers
RWTH Aachen University, Germany

ABSTRACT

The mobile phone is the first truly pervasive computer. In addition to its core communications functionality, it is increasingly used for interaction with the physical world. This chapter examines the design space of input techniques using established desktop taxonomies and design spaces to provide an in-depth discussion of existing interaction techniques. A new five-part spatial classification is proposed for ubiquitous mobile phone interaction tasks discussed in our survey. It includes supported subtasks (position, orient, and selection), dimensionality, relative vs. absolute movement, interaction style (direct vs. indirect), and feedback from the environment (continuous vs. discrete). Key design considerations are identified for deploying these interaction techniques in real-world applications. Our analysis aims to inspire and inform the design of future smart phone interaction techniques.

INTRODUCTION

Today, mobile phones are used not just to keep in touch with others, but also to manage everyday tasks, to share files, and to create personal content. Consequently, our mobile phones are always at hand. Just as Mark Weiser suggested in his vision of ubiquitous computing, the ubiquitous nature of mobile phones certainly does make them "*blend into the fabric of our everyday lives*" (Weiser, 1991).

Technology trends show an increasing number of features packed into this small, convenient form factor. Smart phones already have eyes (camera), ears (microphone), and sensors to perceive their environment. However, their real power, as Weiser pointed out, comes not just from one device, but from the interaction of all of them. Our interest is in showing how modern mobile phones, which resemble Weiser's "*tabs,*" can be used as interaction devices for our environment. Within this environment, emphasis will be placed on interactions with public and situated displays (O'Hara, Perry, & Churchill, 2004) – what Weiser called "*boards.*"

The range of input and output (I/O) capabilities for modern mobile phones is broad. Keypad, joystick, microphone, display, touch-screen, loudspeaker, short-range wireless connectivity over Bluetooth, WiFi, or infrared, and long-range wireless connectivity via GSM/GPRS and UMTS all provide multiple ways of interacting with our phones. These multiple I/O capabilities have increased our ability to use mobile phones to control resources available in our environment, such as public displays, vending machines, and home appliances.

Could this ubiquity mean that mobile phones have become the default input device for ubiquitous computing applications? If so, then mobile phones are positioned to create new interaction paradigms, similar to the way the mouse and keyboard on desktop systems enabled the WIMP (windows, icons, menus, pointers) paradigm of the graphical user interface to emerge and dominate the world of personal desktop computing. However, before this potential is realized, first we must consider which input techniques are intuitive, efficient, and enjoyable for users and applications in the ubiquitous computing domain.

EXAMINING THE DESIGN SPACE OF INPUT DEVICES

Recent research demonstrates a broad array of mobile phone input techniques for ubiquitous computing application scenarios. To make sense of the cumulative knowledge, we systematically organize the input techniques to give insights into the design space. The design space is an important tool for helping designers of ubiquitous computing applications to identify the relationships between input techniques, and to select the most appropriate input technique for their interaction scenarios. Design spaces can also be used to identify gaps in the current body of knowledge and suggest new designs (Zwicky, 1967).

Looking to Foley, Wallace, and Chan's classic paper (Foley et al., 1984), we find a taxonomy of desktop input devices that are structured around the graphics subtasks that they are capable of performing (Position, Orient, Select, Path, Quantify, and Text Entry). These subtasks are the elementary operators that are combined to perform higher-level interface tasks, and will be elaborated upon in later sections. In this chapter, we structure our analysis of smart phones as ubiquitous input devices using this taxonomy. This analysis builds on classic design spaces (Buxton, 1983; Card, Mackinlay, & Robertson, 1991) and extends our own previous work (Ballagas, Ringel, Stone, & Borchers, 2003, Ballagas, Rohs, M., Sheridan, J., and Borchers, 2006) on the design space of input techniques. In our analysis, we blur the line between smart phones and personal digital assistants (PDAs) because their feature sets continue to converge.

Although Foley et al.'s analysis was completed with the desktop computing paradigm in mind, the subtasks in their analysis are still applicable to ubiquitous computing today. They naturally apply to situated display interactions; however, their applicability is not limited to graphical interactions.

In the following sections, each of Foley et al.'s subtasks will be examined in the context of mobile phone interactions. Foley et al.'s taxonomy uses the following input characteristics to further classify input techniques:

- **Feedback:** Continuous interactions describe a closed-loop feedback, where the user continuously gets informed of the interaction progress as the subtask is being performed. For example, when using a mouse, the current cursor position is continually fed back to the user. Discrete interactions describe an open-loop feedback, where the user is only informed of the interaction progress after the subtask is complete. For example, when selecting an object on a touch panel, the progress of the selection is not displayed until after the finger meets the surface to complete the selection of the desired item.
- **Interaction Style:** In direct interactions, input actions are physically coupled with the user-perceivable entity being manipulated (such as an image on a display). Physical coupling can be achieved when the feedback spatially coincides with the input action, or can be achieved at a distance if the user is manipulating a 3-D ray (such as with a laser pointer) that intersects directly with the entity being manipulated. To the user, this appears as if there is no mediation, translation, or adaptation between input and output.

In indirect interactions, user activity and feedback occur in disjoint spaces (e.g., using a mouse to control an on-screen cursor). Scaling and abstraction between input actions and feedback are often necessary in indirect interactions.

Position

During a POSITION task, the user specifies a position in application coordinates, often as part of a command to place an entity at a particular position. Positioning techniques can either be *continuous,* where the object position is continually fed back to the user, or *discrete,* where the position is changed at the end of the positioning task. Positioning tasks can further be differentiated using the directness of the interaction. In *direct* interactions, input actions are physically coupled with the object being positioned; in *indirect* interactions, user activity and feedback occur in disjoint spaces. We note that position could refer to screen position, or physical position in the real world. For example, the height of motorized window blinds can be adjusted using the position subtask.

The mobile phone has been used for positioning tasks in a variety of ways:

Continuous Indirect Interactions

1. **Trackpad:** A trackpad is a touch-sensitive surface that is used as a relative pointing device, standard in modern laptops. Remote Commander (Myers, Stiel, & Gargiulo, 1998) enables individuals to use the touch screen on a PDA as a trackpad to control the relative position of a cursor on a remotely situated display. In this interaction, the user's attention is concentrated on the situated display and no application-level feedback is provided on the PDA; thus, the functionality of the PDA is essentially reduced to an input device.
2. **Velocity-controlled joystick:** A return-to-zero joystick controls the velocity of an object (such as a cursor) that is continuously repositioned on the display. Zero displacement of the joystick corresponds to no motion (zero velocity). Positioning with a velocity-controlled joystick (a temporally and spatially constrained task) has been shown to be inferior to positioning with a mouse (a spatially constrained task) for desktop pointing scenarios (Card, English, & Burr, 1978). Silfverberg et al. (Silfverberg, MacKenzie, & Kauppinen, 2001) have done an in-depth study of isometric joysticks on handheld devices to control the cursor on a situated public display. Many of today's mobile phones are shipping with simple joysticks with a push button for menu navigation.

3. **Accelerometers:** Accelerometers are beginning to emerge in handheld devices. For example, Samsung's SCH-S310 mobile phone comes with an integrated 3-D accelerometer. Several researchers (Bartlett, 2000; Harrison et al., 1998; Hinckley and Horvitz, 2001) have proposed interactions that allow users to scroll (e.g., through an electronic photo album) by tilting the handheld device. The scrolling is typically activated through a clutch mechanism, such as squeezing the sides of the device (Harrison, Fishkin, Gujar, Mochon, & Want, 1998). The degree of tilting controls the speed of scrolling, making this a temporally constrained positioning task similar to the velocity-controlled joystick. Although these techniques were used to interact with an application directly on the device, they could clearly be extended to positioning tasks in ubiquitous computing environments.
4. **Camera tracking:** C-Blink (Miyaoku, Higashino, & Tonomura, 2004) rapidly changes the hue of a color phone screen to allow an external camera system to track the phone's absolute motion for cursor control on a large public display. The hue sequence encodes an ID to allow multiple users to interact simultaneously and control independent cursors.

   The Smart Laser Scanner uses a laser combined with a wide-angle photo detector (see Figure 1) to detect relative finger motion in 3-dimensional space (Cassinelli, Perrin, & Ishikawa, 2005). The laser beam is steered with a two-axis micro-mirror. The tracking principle is based on the backscatter of a laser beam. When the backscatter is disrupted, the motion is deduced from the angle of the backscatter, and the laser is repositioned for the next measurement. Like other tracking techniques, it is possible for the device to lose track if the finger moves too fast, but input can easily be resumed by repositioning the finger to the laser. The research prototype of the tracker is fast enough to track the motion of a bouncing ping-pong ball.
5. **Motion detection:** With the Sweep (Ballagas, Rohs, Sheridan, & Borchers, 2005) interaction technique, the phone is waved in the air to control relative cursor motion on a remote screen (see Figure 2). This is accomplished using motion detection, an image processing technique involving rapidly sampling successive images from the phone's camera and sequentially comparing them to determine relative motion in the (x, y, θ) dimensions. No visual tags are required. The screen on the phone can be ignored, and the camera does not even need to be pointed at the display. A

Figure 1. The Smart Laser Scanner: A 3-D input technique for mobile devices using laser tracking (Cassinelli et al., 2005). Reprinted with permission from the authors.

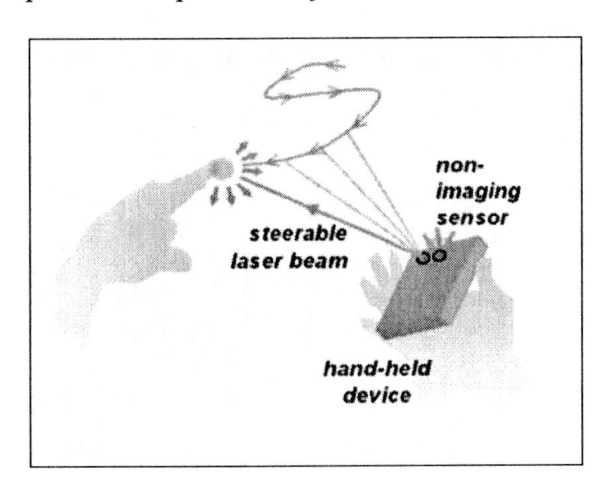

clutch mechanism, such as a button press, is used to activate the Sweep interaction. The clutch can be used to reposition the arm, similar to the way a mouse can be lifted to be repositioned without additional cursor motion.

6. **Location detection:** Location of the phone can also be used as input, where the user moves through physical space. Mogi (Licoppe & Inada, 2006), for instance, is a phone-based persistent item collection and trading game where the absolute geo-position of a subscriber correlates to the position in the game world. Mogi combines GPS (global positioning system) technology built into the phone with information from different mobile infrastructure towers from the network service provider to determine the player's position.

Continuous Direct Interactions

7. **Camera tracking:** Madhavapeddy, Scott, Sharp, and Upton (2004) present camera-based interactions involving tagging interactive GUI elements such as sliders and dials (see Figure 3). In manipulating the position and orientation of the phone camera, the user can position a graphical slider, or orient a graphical dial. Similarly, Direct Pointer (Jiang, Ofek, Moraveji, & Shi, 2006) uses a handheld camera to track the standard cursor on the display. An analogy can be drawn to the classic light pen with a tracking cross. As the light pen moves to a new position, the cross follows the motions of the pen. Tracking may be lost if the pen is moved too fast, but can be easily resumed by repositioning the pen back to the tracking cross. Madhavapeddy et al.'s interactions rely on the tagged GUI widget instead of a cross for tracking; in Direct Pointer, the mouse cursor is the modern equivalent of the tracking cross.

   In these tracking examples, the handheld device is responsible for tracking. An alternative is to use a tracker in the environment to track the output from a handheld device. For example, smart phones have been augmented with laser pointers, as in Patel and Abowd (2003), making them suitable for positioning tasks, described by Dan, Olsen, and Nielsen (2001), that use a camera in the environment to track the laser.

   The mobile phone can also be passively tracked using a camera in the environment, such as in VisionWand (Cao & Balakrishnan, 2003). The user holds a passive handheld device that is augmented with distinctive markings (such as colored balls) at each end. Using two fixed cameras to perform stereo tracking, a 3-D ray can be deduced from the orientation of the markings in the stereo view, assuming the distance of the markings on the device is known *a priori*. This allows using a projection of the ray as a pointing device for a fixed remote screen.

Figure 2. The Sweep technique uses camera input and optical flow image processing to control a cursor (Ballagas et al., *2005). © 2006 IEEE. Adapted with permission.*

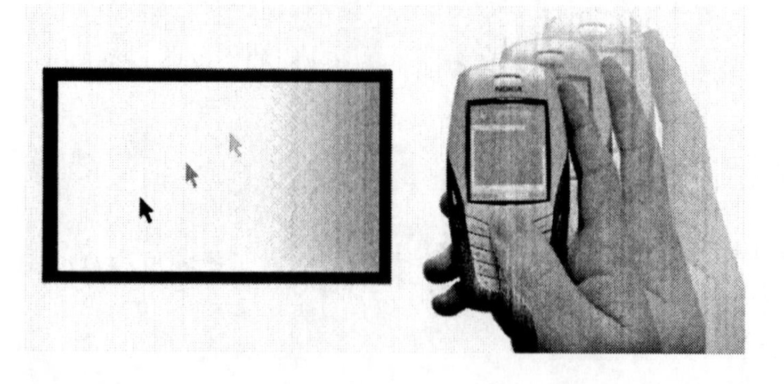

The result is an interaction that is very similar to pointing using a laser pointer, except the ray is not a visible beam of light. This technique has an advantage over the standard laser pointer in that it provides an extra dimension of information: the distance to the display. The disadvantage of this interaction is that it is vulnerable to occlusion (e.g., by the users' own body) bringing into question the robustness of tracking in practical scenarios, although different camera configurations (such as from overhead facing downward) may solve these issues for certain interaction scenarios.

Discrete Indirect Interactions

8. **Directional step keys:** The location of an object is controlled using up, down, left, and right step keys for 2-D applications, plus in and out for 3-D. In the Blinkenlights project (Chaos Computer Club, 2002), users played the arcade classic "Pong" using the side of a building as a large public display. Each window equaled one pixel on the 18x8 pixel display. Players connected to the display by making a standard voice call to a phone number. Pressing the number 5 on the phone keypad moved the paddle up, and the number 8 moved it down. The server controlling the "Pong" application would decode the tones generated from the key activity during the phone call and use them as application input. One of the notable things about this interaction is that it used the lowest common denominator of phone technologies. The communications channel was the standard voice channel, and the input was the numeric keypad, requiring no additional hardware or software besides what standard phones provide.

Discrete Direct Interactions

9. **Camera image:** Using the Point and Shoot (Ballagas et al., 2005) interaction technique, the user can specify an absolute position on a public display using a cross-hair drawn over a live camera image on the mobile phone. To make a selection, the user presses a button while aiming at the desired target.[1] The button press triggers a brief overlay of a grid of 2-D tags over the large display contents, as can be seen in the middle of Figure 4. The grid allows the phone to derive a perspective-independent coordinate system on the large display that is enabled by the special properties of the Visual Code tags (Rohs, 2005a). Only one visual tag is required to establish a coordinate system, but a grid is used to increase the probability of having one tag entirely in the camera view. The drawback of the current implementation is that the tag grid is disruptive in multiuser scenarios, but future implementations could, for example, display the tags in infrared so

Figure 3. Using the phone to manipulate tagged widgets such as buttons, dials, and sliders (Madhavapeddy et al., 2004). Reprinted with permission from the authors.

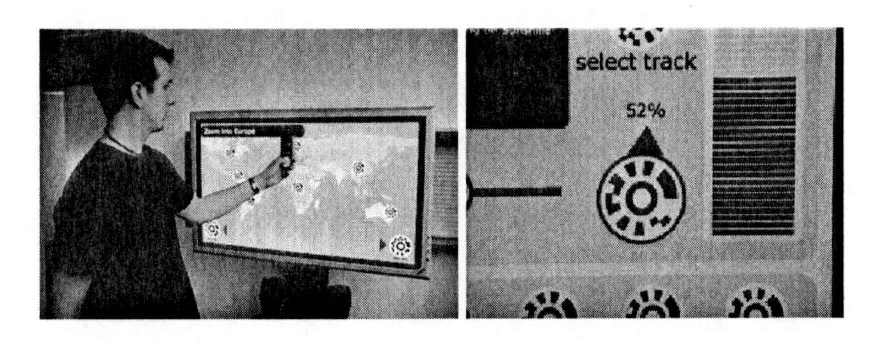

that they are visible to the camera but not to other users.

Point and Shoot is related to the classic light pen, where position is discretely determined by displaying a raster scan when the user clicks a button on the light pen. When the raster scan is temporally sensed by the pen, the position of the pen is known because of a tight coupling between the pen clock and display clock. In Point and Shoot, a visual tag grid replaces the functionality of the raster scan except its mechanics are spatial rather than temporal. The lack of temporal dependencies makes Point and Shoot robust to different display technologies and the loose coupling between camera and display.

The breadth of positioning techniques is relatively large, making it difficult to choose which technique is most appropriate for a particular application scenario. To help with this selection, it is important to examine different figures of merit for each device.

Evaluating Positioning Techniques

There have been only a handful of thorough evaluations of the different ubiquitous mobile input techniques (Ballagas et al., 2005; Myers, Bhatnagar, Nichols, Peck, Kong, Miller, & Long, 2002; Silfverberg et al., 2001; Wang, Zhai, & Canny, 2006), as the field is still relatively new. These studies are difficult to directly compare, since they each used different experimental parameters, and some evaluations were not done in the context of ubiquitous computing interaction scenarios. Therefore, rough estimates for a variety of ergonomic measures are used to create a high-level comparison table for the positioning task presented in Figure 6. These rough estimates are derived using our knowledge of the interaction techniques for mobile phones and the collective knowledge of their desktop computing counterparts. The ergonomic parameters are mostly borrowed from Foley et al.'s survey of interaction techniques.

The evaluation measures are grounded in psychological and physiological foundations. Card et al. (Card, Newell, & Moran, 1983) provide an integrated survey of the various fundamental theories in a way that makes them more accessible and easier to use during analysis. Central to this work is the human processor model, which brings knowledge of the perceptual, cognitive, and motor processes of a human together under a single model. Ideally, a user interface minimizes the work required for each of these basic psychological processes.

The comparison table also incorporates various ergonomic measures designed to capture the efficiency of users executing the subtask, the accuracy they can achieve, and the pleasure the user derives from the process. The individual measures used in our comparison table are as follows:

- **Perceptual load** refers to the difficulty for the user to recognize, with their own senses, the physical stimuli and feedback of the interaction. For example, in the Point and Shoot interaction, users need to shift their perceptual attention between a large display and the phone screen to isolate a target in the phone camera view, leading to a comparatively high perceptual load.
- **Cognitive load** refers to the difficulty for the users to organize and retrieve information related to the interaction technique.
- **Motor load** refers to the number of motor steps required to execute the action after the appropriate action has been determined in the cognitive process. For example, Mogi is classified as a high-motor-load technique because the user needs to physically move at the city scale to specify the necessary position.
- **Motor acquisition time** characterizes the amount of time for the processes involved in the interaction technique (i.e., reaching for an object, moving to a certain target area, rotating to a certain orientation, etc.)
- **Visual acquisition time** characterizes the amount of time it takes to perceive the physical stimuli of the interaction technique.

Figure 4. Point & Shoot interaction: (Left) The phone screen is used to aim at a puzzle piece on a large situated display. (Middle) Pressing the joystick indicates a selection and a Visual Code grid is briefly superimposed to compute the target coordinates in the captured photo. (Right) The grid has disappeared and the target puzzle piece is highlighted on the large display, indicating successful selection (Ballagas et al., 2005). © 2006 IEEE.

Figure 5. Summary of positioning techniques using a smart phone as an input device. © 1984, 2006 IEEE. Adapted by permission.

	Foley	Mobile Phone Interactions	
	In Environment	On Phone	In Environment
Direct Pick	**Search for Light Pen** (Raster Scan)	**Camera + On-Screen Cursor +**	**Visual Tags** [Point & Shoot]
		Laser Pointer +	**Camera** [Olsen]
	Light Pen Tracking	**Camera Tracking +**	**Visual Tags** [Madhavapeddy]
		Camera Tracking +	**On-Screen Cursor** [Direct Pointer]
		Vision Markers +	**Camera** [VisionWand]
		Camera + Touch-Screen	
Direct with Locator Device	**Touch Panel**		
Indirect with Locator Device	**Mouse**	**Camera** [Sweep]	
		Display +	**Stationary Camera** [C-blink]
	Joystick (Velocity)	**Joystick** (Velocity) [Silfverberg]	
	Tablet	**Trackpad** [Remote Commander]	
	Trackball		
	Cursor Control Keys with Auto-Repeat	**Cursor Control Keys with Auto-Repeat**	
	Joystick (absolute)	**Accelerometer** [Harrison + Rock 'n' Scroll]	
		Steerable Laser + Wide Angle Photodetector [Cassinelli]	
Indirect with Directional Commands/ Button Push	**Up-Down-Left-Right Arrow Keys**	**Directional Step Keys** [Blinkenlights]	
	(See Selection)	(See Selection)	
Location Detection		**GPS +**	**Cell Network Towers** [Mogi]
Numerical Value/ Numerical Coordinates/ Character String Name	(See Text Input)	(See Text Input)	

Figure 6. Rough estimates of ergonomic measures to compare mobile-phone-based positioning techniques (small circle = low, medium circle = medium, large circle = high). © 2006 IEEE. Adapted by permission.

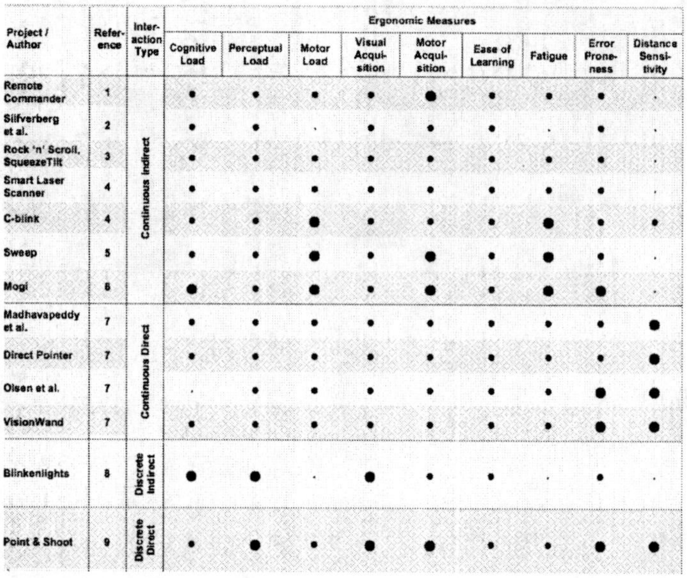

Project / Author	Reference	Interaction Type	Ergonomic Measures								
			Cognitive Load	Perceptual Load	Motor Load	Visual Acquisition	Motor Acquisition	Ease of Learning	Fatigue	Error Proneness	Distance Sensitivity
Remote Commander	1	Continuous Indirect	•	•	•	•	●	•	•	•	·
Silfverberg et al.	2		•	•	·	•	•	•	·	•	·
Rock 'n' Scroll, SqueezeTilt	3		•	•	•	•	•	•	•	•	·
Smart Laser Scanner	4		•	•	•	•	•	•	•	•	·
C-blink	4		•	•	●	•	•	•	●	•	•
Sweep	5		•	•	●	•	●	•	●	•	·
Mogi	6		●	•	●	•	●	•	●	●	·
Madhavapeddy et al.	7	Continuous Direct	•	•	•	•	•	•	•	•	●
Direct Pointer	7		•	•	•	•	•	•	•	•	●
Olsen et al.	7		·	•	•	•	•	•	•	●	●
VisionWand	7		•	•	•	•	•	•	•	●	●
Blinkenlights	8	Discrete Indirect	●	●	·	●	•	•	·	•	·
Point & Shoot	9	Discrete Direct	•	●	•	●	●	•	•	●	●

- **Ease of learning:** Characterizes the level of skill that is required to use the device.
- *Fatigue* characterizes how tiring the interaction technique is to perform.
- **Error proneness:** Characterizes the susceptibility for errors of the input technique, the degree to which the interaction technique, by its design, allows/avoids errors, for example, if possible movement trajectories match the degrees of freedom of the required input then certain errors can be avoided.
- **Sensitivity to distance:** Users in ubiquitous computing scenarios typically have freedom of motion, making the amount of separation between the user and the target in the environment (such as a large display or other device) dynamic and unpredictable. Thus, the range of distances the interaction will support is an important design consideration. Interactions that are based on aiming, such as laser pointers, become more difficult to perform when further away, where targets are perspectively smaller. Other techniques, such as the Sweep technique, are not significantly affected by distance of interaction.

Orient

The ORIENT subtask involves specifying a heading or direction instead of a position. Like POSITION, ORIENTATION is also not limited to graphics subtasks as it can relate to physical orientation in the real world, such as a security camera, a spotlight, or a steerable projector. Some of Foley et al.'s original graphics interactions carry over directly to ubiquitous computing, including *indirect continuous orientation with velocity-controlled joystick* and *discrete orientation with angle type-in*. The remaining techniques observed in our survey include

Continuous Indirect Interactions

1. **Locator device:** The user can specify the angle of orientation by using a continuous quantifier or one axis of a positioning device. The Sweep technique supports detection of rotation around the Z-axis (perpendicular to the display), allowing interactions like rotating a puzzle piece in a jigsaw puzzle

application, where the phone is used like a ratchet to adjust orientation. The image processing used by Sweep also detects rotation around the X and Y-axis. However, for better performance as a positioning device, rotation around the Y-axis is mapped to translation along the X-axis, and rotation around the X-axis is mapped to translation along the Y-axis.

2. **Camera tracking:** VisionWand (Cao & Balakrishnan, 2003) uses a set of cameras in the environment to track the absolute orientation of a marked handheld device. The technique requires that at least two markers are visible in at least two camera viewpoints to determine the orientation in 3-dimensional space.

Continuous Direct Interactions

3. **Camera tracking:** Madhavapeddy's tagged GUI dials (Madhavapeddy et al., 2004) can be oriented using the phone camera to track rotation movement. Similar to the Sweep technique, the phone is used like a rachet to adjust orientation.
4. **Compass:** Electronic compasses, such as the Honeywell HMC1052 magnetometer, can be used to detect the physical orientation of the phone with a +/-3° error, enabling a continuous and direct ORIENT task. This or similar sensors could be easily incorporated into future mobile phone applications.

Discrete Direct Interactions

5. **Camera image:** The Point & Shoot technique supports discrete orientation along the Z-axis. As the user aims at a target, they rotate the phone to specify the desired Z-orientation using the aiming cross-hair as an axis of rotation.

Select

In many interaction scenarios, the user must choose from a set of alternatives, such as a menu of icons. The SELECTION subtask addresses this style of interaction. The SELECTION subtask is commonly accomplished by arranging the items spatially in a graphical user interface, allowing the user to complete the selection using a cursor controlled through the positioning subtask. Instead of icons, the set of alternatives might be a list of commands. However, selection is not limited to graphical interactions, as a user may select a physical object

Figure 7. Summary of ORIENT techniques using a smart phone as an input device. © 1984, 2006 IEEE. Adapted by permission.

	Foley	Mobile Phone Interactions	
	In Environment	On Phone	In Environment
Direct Pick		**Camera** [Point & Shoot]	
		Camera Tag Tracking +	**Visual Tags** [Madhavapeddy]
		Vision Markers +	**Camera** [VisionWand]
Indirect with Cursor Match/ Locator Device		**Camera** [Sweep]	
	Joystick (Velocity)	**Joystick** (Velocity)	
	Joystick (absolute)		
Numerical Value/ Numerical Coordinates/ with Character String Name	(See Text Input)	(See Text Input)	

to operate upon, such as selecting a lamp to adjust its setting. Many selection techniques carry over directly from Foley et al.'s earlier analysis, such as *character string name type-in* common for command prompts, or *button push–soft keys,* where buttons are located on the edge of the display area with their labels displayed on screen. The remaining selection techniques are as follows:

Continuous Indirect Interactions

1. **Gesture recognition:** The user makes a sequence of movements with a continuous positioning device such as the joystick, camera, trackpad, or accelerometers. For example, Patel, Pierce, & Abowd (2004) used gesture recognition of accelerometer data from the handheld device to authenticate users that wanted to access data on their mobile phone through an untrusted public terminal. Using this technology, users could securely bring up data on the public terminal from their phone without removing it from their purse.

Continuous Direct Interactions

2. **Tagged objects:** RFIG Lamps (Raskar, Beardsley, van Baar, Wang, Dietz, Lee, Leigh, & Willwacher, 2004) allows a handheld projector to be used to select objects with photosensitive RFID tags in the physical world. The handheld projector emits a graycode pattern that allows the tags to determine their relative position in the projected view. Waving the handheld projector around, you can navigate a cursor in the center of the projected view to select individual physical objects.

Discrete Indirect Interactions

3. **Voice recognition:** The user speaks the name of the selected command, and a speech recognizer determines which command was spoken. The Personal Universal Controller (Nichols & Myers, 2006) supports automatic generation of speech interfaces (as well as graphical interfaces) to issue commands to objects in the real world.
   VisionWand (Cao & Balakrishnan, 2003) also demonstrates a rich gesture vocabulary using stereovision to track a passive wand. For example, a tapping gesture is used to allow selection of the current cursor position specified by the orientation of the wand. It should be noted that information from any continuous positioning technique can be used for gesture recognition, as long as there is a mechanism to specify when a gesture begins and ends.

Discrete Direct Interactions

4. **Tagged objects:** Tagged objects can be used to present information on a wireless mobile computer equipped with an electronic tag reader, as demonstrated by the early E-tag project (Want, Fishkin, Gujar, & Harrison, 1999). For example, selecting a book by scanning its embedded RFID tag would activate a virtual representation of the object on the screen, such as a Web reference to the book allowing it to be purchased. Similar interactions have also been proposed for visual tags in the environment (Rohs, 2005a) and tagged GUI elements (Madhavapeddy et al., 2004; Rohs, 2005b), where a camera is used to acquire an image to decode the selected tag. Patel and Abowd (2003) present a physical world selection method for mobile phones in which a modulated laser pointer signal triggers a photosensitive tag placed in the environment, allowing users to bring up a menu to control the object on their handheld device.
5. **Laser pointer:** Myers et al. (2002) proposed a multilayer selection technique, called "semantic snarfing," that combines multiple devices in consecutive actions. First, a laser pointer integrated with a handheld computer is used to make a coarse-grained selection of a screen region on a display in the environment. A camera, also in the environment, detects

Figure 8. Summary of SELECTION *techniques using a smart phone as an input device (Continued on next page). © 1984, 2006 IEEE. Adapted by permission.*

	Foley	Mobile Phone Interaction	
	In Environment	On Phone	In Environment
Direct Pick		**Camera + On-Screen Cursor** [Point & Shoot]	
		Laser Pointer +	**Light Sensor** (e.g. camera) [Olsen, Semantic Snarfing, Patel]
	Light Pen Tracking	**Camera +**	**Visual Tags** [Madhavapeddy]
		Camera Tracking +	**On-Screen Cursor** [Direct Pointer]
		Camera +	**Pen Input**
		Vision Markers +	**Camera + Tapping Gesture Recognition** [VisionWand]
		Handheld Projector + RFID Reader	**Light Sensitive RFID Tags** [RFIG]
		RFID Reader +	**RFID Tags** [Want]
	Touch Panel		
Indirect with Cursor Match/ Locator Device	**Mouse**	**Camera** [Sweep]	
		Display +	**Stationary Camera** [C-blink]
	Joystick (Velocity)	**Joystick** (Velocity) [Silfverberg]	
	Tablet	**Trackpad** [Remote Commander]	
	Trackball		
	Cursor Control Keys	**Cursor Control Keys**	
	Joystick (absolute)	**Accelerometer** [Harrison, Rock 'n' Scroll]	
		Steerable Laser + Wide Angle Photodetector + Button Push [Cassinelli]	

Figure 9. Summary of SELECTION *techniques using a smart phone as an input device (Continued from previous page). © 1984, 2006 IEEE. Adapted by permission.*

	Foley	Mobile Phone Interaction	
	In Environment	On Phone	In Environment
Indirect with Directional Commands/ Button Push/ Time Scan	**Programmed Function Keyboard**	**Programmed Function Keyboard**	
	Soft Keys	**Soft Keys**	
	Alphanumeric Keyboard		
Gesture Recognition/ Sketch Recognition		**Camera + Button Push Clutch** [TinyMotion]	
		Display +	**Camera**
		Accelerometer [Patel]	
	Light Pen	**Pen Input**	
		Steerable Laser + Wide Angle Photodetector + Button Push for Clutch [Cassinelli]	
		Laser Pointer + [Olsen, Semantic	**Light Sensor** (e.g. camera) Snarfing, Patel]
		Vision Markers + [Vision	**Camera** Wand]
	Tablet + Stylus		
Location Detection		**GPS +**	**Cell Networks Towers** [Mogi]
Voice Input		**Microphone +**	**Voice Recognizer** [PUC]
Numerical Value/ Numerical Coordinates/ with Character String Name	(See Text Input)	(See Text Input)	

laser activity on the display. The system then transmits the details of the selected screen region to the handheld device, which composes a GUI on the handheld screen to make the fine-grained selection with a stylus.

Path

The PATH subtask involves specifying a series of positions and orientations over time. The PATH subtask has different requirements than POSITION and ORIENT because the movement is governed by the speed-accuracy tradeoff (Schmidt, Hawkins, Frank, & Quinn, 1979). Despite this, PATH adheres to the same taxonomy as the corresponding POSITION and ORIENT techniques, because a PATH task can be specified using the more primitive subtasks.

Quantify

The QUANTIFY task involves specifying a value or number within a range of numbers. This technique is used to specify numeric parameters such as time or speaker volume. In ubiquitous applications, QUANTIFY tasks using phone input were typically accomplished through the GUI using 1-D POSITION or ORIENT subtasks.

Text

TEXT ENTRY for mobile phones is a well-studied area (MacKenzie & Soukoreff, 2002) as it is central to text-based mobile communications like SMS (short messaging service) and personal information management functionality. Text entry also has many applications for ubiquitous applications, for example, the Digital Graffiti (Carter, Churchill, Denoue, Helfman, & Nelson, 2004) project seeks to annotate public content on large public displays. This section is not intended to be a comprehensive survey of mobile text entry techniques, but we have selected a few examples to illustrate the design space. All of the techniques listed were originally designed for text input directly on the mobile phone, but could clearly be used for text entry for a ubiquitous computing application.

Keyboard

Although some mobile phones and handheld devices feature a full QWERTY keyboard (albeit much smaller than their desktop counterparts), miniaturization trends make this type of keyboard impractical in a majority of mobile phone form factors. The most well known text entry techniques for mobile phones use a standard numeric keypad. For text entry from a 26-character alphabet using this keyboard, a mapping with more than one character per button is required. Following the classification by Wigdor and Balakrishnan (2004), there are two fundamental types of disambiguation: *consecutive*, where the user first selects a letter grouping and then an individual letter, or *concurrent*, where the user simultaneously selects the letter grouping and the individual letter.

Consecutive approaches are the most common today. One approach to disambiguate text entry is MultiTap, which requires users to make multiple presses to select a single letter from the characters associated with a certain key. Another solution is to use a two-key disambiguation, where the first key selects the letter group, and the second key specifies the letter in the group. Dictionary-based techniques, such as T9[2], deduce the word being typed, based on the different possibilities for combining the groups of characters. When multiple words match the key sequence, the user selects the intended word from a list (typically ordered by probability or frequency of use).

Concurrent approaches, however, demonstrate a lot of promise. For example, TiltText (Wigdor & Balakrishnan, 2003) combines the standard 12-key keypad with an accelerometer. To disambiguate which character is intended when a key is pressed, TiltText uses the tilt orientation of the handset. A keypress with the phone tilted to the left enters the first character on the key, forward tilt enters the second character, right tilt enters the third character, tilting towards the user enters the fourth character (if one exists for the key), and no tilt enters the numeric character.

ChordTap (Wigdor & Balakrishnan, 2004) combines the standard numeric keyboard with additional "chording" buttons on the back of the phone. A user selects an individual letter by selecting the key group on the numeric keyboard while pressing the appropriate "chord" key on the back of the phone.

If miniaturization trends continue, TiltType (Partridge, Chatterjee, Sazawal, Borriello, & Want, 2002) represents an interesting point in the design space that combines chord button presses to specify a letter grouping and tilting to allow the user to specify a particular character within that grouping. Using only four buttons and a two-axis accelerometer, the technique supports an alphabet of 55 characters in a watch-sized form factor. Expert users can memorize the character positions, allowing the letter grouping and individual character within the grouping to be specified concurrently.

Speech Recognition

Text entry by speech recognition is not yet technically viable on mobile platforms, but we list it here for completeness. Technology is making rapid advances in the realm of speech processing. For example, system on a chip designs for speech processing (Ravindran, Smith, Graham, Duangudom, Anderson, & Hasler, 2005) have the potential to bring speech input to interactive text entry on mobile phones. Karpov et al. (Karpov, Kiss, Leppanen, Olsen, Oria, Sivadas, & Tian, 2006) have developed a short message (SMS) dictation system for Symbian phones with a vocabulary of 23,000 words. The language model is adapted to words typically used in SMS messages.

Speech recognition could also be achieved in a compound architecture where the speech is recognized through an external computer (i.e., connected through a voice call) and sent back to the mobile phone.

Stroked Character Recognition

Pen-based techniques, such as Graffiti, are very common in the PDA form factor, and are also available on a small portion of the handsets on today's market. However, any of the continuous positioning tasks discussed earlier are capable of generating stroke information necessary for stroked-character recognition. For example, TinyMotion (Wang et al., 2006) demonstrates both English and Chinese stroked character recognition using camera-based motion estimation (similar to the Sweep technique).

Menu Selection

On-screen keyboards are common for touch sensitive displays, where the letters of the alphabet are displayed as a menu of buttons, commonly in a spatial layout similar to the QWERTY keyboard. If the screen size of the mobile phone is not large enough to depict a keyboard layout, items in the environment could be used to display the menu, where users select the characters using the SELECTION subtask previously discussed.

Figure 10. Summary of TEXT ENTRY *techniques using a smart phone as an input device. © 1984 IEEE. Adapted by permission.*

	Foley	Mobile Phone Interactions	
	In Environment	On Phone	In Environment
Keyboard	**Alphanumeric**	**Alphanumeric** [Multitap, T9]	
		Alphanumeric + Accelerometer [TiltText]	
	Chord	**Alphanumeric + Chord** [ChordTap]	
		Chord + Accelerometer [TiltType]	
Stroked Character Recognition	**Tablet with Stylus**	(See Continuous Positioning)	
Voice/Speech Recognition	**Voice Recognizer**	**Speech Recognizer**	
		Microphone +	**Speech Recognizer**
Direct Pick from Menu with Locator Device	**Light Pen**	(See Selection)	
	Touch Panel	(See Selection)	
Indirect Pick from Menu with Locator Device	(See Positioning)	(See Positioning)	

SPATIAL LAYOUT OF THE DESIGN SPACE

Our interaction taxonomy is summarized in Foley-style graphs in Figures 5, 7, 8, 9, and 10. Card et al. (1991) point out that this format is somewhat ad hoc and lacks a notion of completeness. Card then builds on the work of Buxton (1983) to create a systematic spatial layout of the design space of input devices that captures the physical properties of manual devices very well. However, it does not capture many aspects that are relevant to ubicomp interactions such as modality or feedback (Ballagas et al., 2003).

Using Foley et al.'s taxonomy, we propose a five-part spatial layout, shown in Figure 11, for mobile phone interaction tasks discussed in our survey including supported subtasks (POSITION, ORIENT, and SELECTION), dimensionality, relative vs. absolute movement, interaction style (direct vs. indirect), and feedback from the environment (continuous vs. discrete). Feedback and interaction style have been previously defined in the introduction to Foley et al.'s taxonomy. We describe the remaining dimensions in more detail in the remainder of this section.

Supported Subtasks

When choosing the most appropriate input device for a particular interaction scenario, the subtasks an interaction supports are the primary consideration. By including the subtask directly in the design space, it becomes more useful as a design tool.

Dimensionality

Dimensionality refers to the number of dimensions the interaction supports. Dimensionality can indicate spatial dimensions (*X,Y,Z*) or rotational dimensions (*rX,rY,rZ*). This distinction is visible in our design space by observing the subtask of the dimension. Following Card et al. (1991), if a particular interaction uses a combination of dimensions across different points in the design space, the relationship is indicated using a merge composition operator (a solid line). In contrast to Card's notation, our merge composition operators are connecting subtasks, not spatial sensor dimensions.

Relative vs. Absolute

Relative input is specified with respect to interaction history: the input technique provides information about the amount of change from the previous state. Relative input can be specified regardless of the current physical properties, such as position and orientation. For example, standard desktop mouse input is specified through motion across the desktop regardless of the physical position of the mouse on the desktop.

Absolute input is specified with respect to current physical properties, and can be specified independently of any interaction history. For example, stylus input can be used to provide absolute positional information on a screen space.

Other Relevant Attributes of Interaction Devices

It should be noted that this set of dimensions is not comprehensive, and other dimensions, such as resolution, direction (input vs. output), and modality, may provide further insights into the design space. However, the design space depicted in Figure 11 does provide an interesting overview of the interaction techniques covered in this chapter. Using this graphical layout, we are able to pinpoint gaps in the breadth of the interaction techniques surveyed, and can anticipate opportunities for future work. For example, our space shows no interaction that supports 3-dimensional relative direct orientation. An alternative layout might include direction and modality, which would demonstrate the sparse usage of auditory and haptic feedback in these techniques.

Designing for Serendipity

One key design consideration is the ease and speed of setting up a data connection between the phone and the environment or the device it is controlling. In some of the interactions surveyed,

Figure 11. Classification of different mobile phone interactions that have been implemented in the projects surveyed. Inspection of the diagram reveals opportunities for future work, for instance, developing interaction techniques that support 3-D relative direct orientation. In the listing of techniques, (P) indicates capabilities of the phone, and (E) indicates capabilities of the environment. © 2006 IEEE. Adapted by permission.

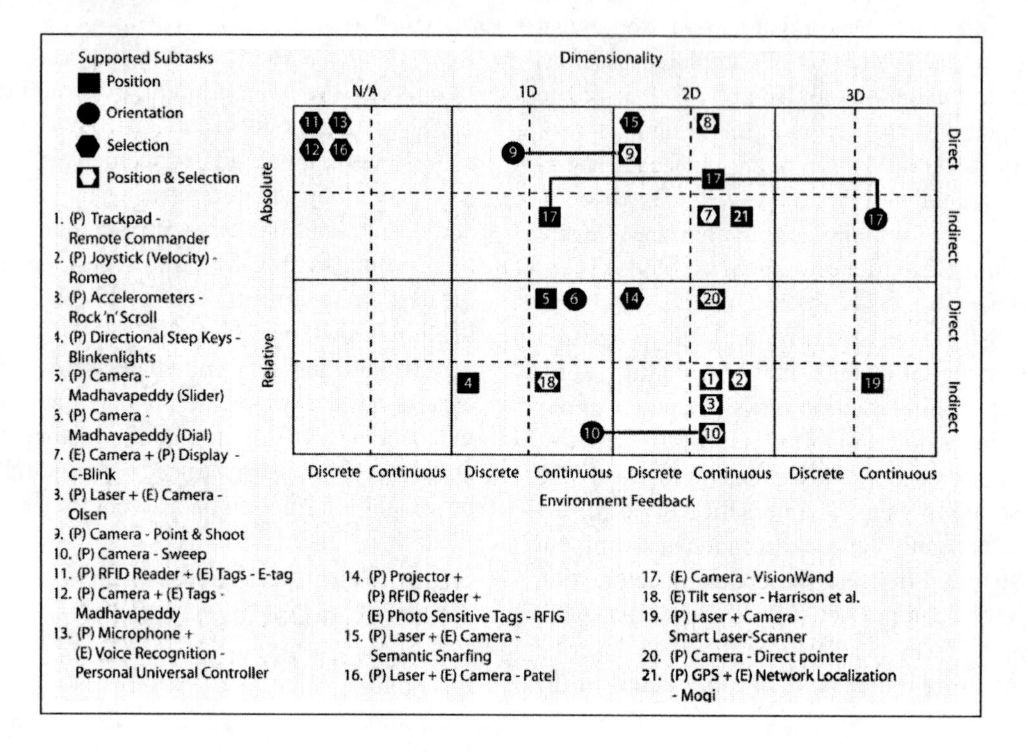

the data connection is inherent in the physical properties of the device. The VisionWand, for example, is a completely passive system, so it requires no additional action on the user's part to start the interaction.

The C-Blink interaction is classified as highly serendipitous, as the users merely launch an application on their mobile phone to interact with a display; no network connection or handshaking is required. The RFIG Lamps project also falls into this category because RFID tags are so simple in terms of communications protocol that no connection need to be established before data can be transferred.

For projects that use short-range wireless communications models, such as Bluetooth, visual or RFID tags can be used to encode the connection information for the environment, creating a very low threshold of use.

Social Acceptance

Smart phones today are social devices. While smart phone ubiquity seems inevitable, social acceptance will influence the success of these new interactions. Remind yourself, for example, of the first time you came across a person using a wireless headset to communicate via their mobile phone. For many people, this communication technique is still awkward and strange, particularly in public places. Smart phone interaction will require users to perform particular actions and behaviors that might feel unintuitive and awkward to them. Furthermore, they will perform these actions in the presence of passive or active others, both familiars and strangers. On one hand, outside observers might find these interactions disturbing or embarrassing, but on the other hand, these kinds of interaction have the potential to raise your social status, similar to the way phones themselves are status symbols for part of our society.

DESIGN SPACES IN THE DESIGN PROCESS

Design spaces are a particularly useful design tool as a part of a human-centered iterative design process (Nielsen, 1993). One of the pitfalls of iterative human-centered design is that if you pick a poor starting point, you may reach a peak in the usability of a particular design without reaching your desired usability goals. In this case, it may be necessary to throw the design away and start over. False starts are relatively painless early in the design process, but can be extremely expensive if determined late in the design process. In order to minimize the risk of false starts, a parallel design strategy (Nielsen & Faber, 1996) can be used, where multiple designs can be explored independently early in the design process. As the designs mature, the best design becomes clear, or the strengths of the top designs can be merged to a unified design. Using the design space, designers can more easily reason about alternative input techniques in a parallel design process.

As a concrete example, REXplorer (Ballagas, Walz, Kratz, Fuhr, Yu, Tann, Borchers, & Hovestadt, 2007) is a pervasive spell-casting game that allows tourists to explore the history of the medieval buildings in Regensburg, Germany. The game premise is that historical spirits are trapped inside of medieval buildings. Players need to interact with the spirits to learn their stories and perform quests on their behalf to earn points in the game. The game design called for spell-casting as the primary interaction metaphor; in order to awaken a spirit, one of four spells must be cast.

Choosing one spell out of four can be characterized as a SELECT subtask. The design space was used to identify a set of design alternatives that we initially considered:

1. Four dedicated spell buttons
2. Selecting one of four spells from on-screen menu
3. Recognition of spell gestures. We noted that gestures are actually specified using the path subtask. Then we came up with gesture input alternatives including:
    a. Pen trace across a touch screen
    b. Path using camera-based motion detection to allow the phone to be used like a magic wand.

After preliminary analysis with our target group (students aged 15-25), we decided to go with the

Figure 12. REXplorer uses the Sweep technique to allow players to cast spells using the PATH *subtask (Ballagas et al., 2007). Reprinted with permission from the authors.*

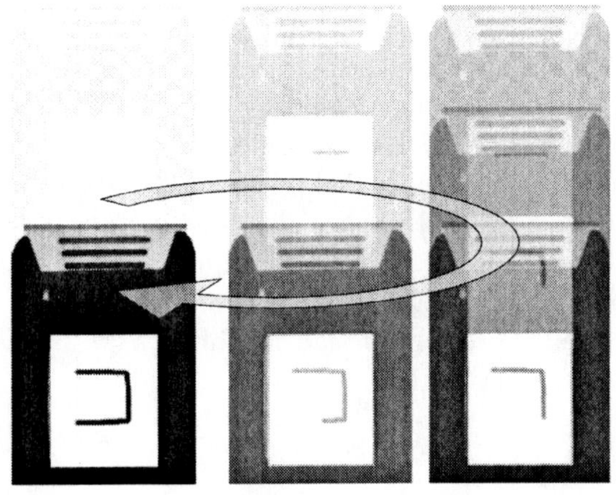

camera-based motion detection solution (see Figure 12). Waving the phone through the air is not the most efficient technique, but is the most similar to the spell-casting metaphor. Also, this physical style of gesture was more likely to create an engaging experience (Hummels, 2000).

Later in the design process, after a working-gesture recognition system was created, we did a full playability test. Most of the test players found the gestures to be an important element of gameplay. They found it heightened the sense of magic and mysteriousness. However, we also discovered during the playability tests that a few of our players (especially our older participants) found the gestures awkward. As a compromise, we created a unified design, where an alternative gesture selection mechanism through an on-screen menu can be used anytime an invalid gesture is performed, effectively allowing people to avoid gestures altogether if desired. This final design encouraged the use of gestures for spell selection to promote engagement, but allowed an alternative selection mechanism to those who preferred to avoid gestures.

CONCLUSION

Our structured tour illustrates the state of the art in using smart phones to interact with and control our environments. The taxonomy organizes the range of techniques into families that help make functional relations between the mobile phone techniques and their desktop counterparts. The design space addresses the lack of a sense of completeness in the taxonomy, and structures the range of interactions in a way that helps visually identify gaps and predict future interaction techniques. The design space can be used as a part of a human-centered iterative design process to help generate parallel or alternative designs. These methods of thought are intended to inspire new applications that use the mobile phone for interaction with the environment, as well as inform the design of future smart phone interaction techniques.

REFERENCES

Ballagas, R., Ringel, M., Stone, M., & Borchers, J. (2003). iStuff: A physical user interface toolkit for ubiquitous computing environments. In *CHI '03 Proceedings of the SIGCHI Conference on Human Factors in Computing Systems* (pp. 537–544). New York, NY: ACM Press.

Ballagas, R., Rohs, M., Sheridan, J. G., & Borchers, J. (2005). Sweep and point & shoot: Phonecam-based interactions for large public displays. In *CHI '05 Extended abstracts of the SIGCHI Conference on Human Factors in Computing Systems* (pp. 1200–1203). New York, NY: ACM Press.

Ballagas, R., Rohs, M., Sheridan, J., & Borchers, J. (2006). The smart phone: A ubiquitous input device. *IEEE Pervasive Computing*, *5*(1), 70–77.

Ballagas, R., Walz, S. P., Kratz, S., Fuhr, C., Yu, E., Tann, M., Borchers, J., & Hovestadt, L. (2007). REXplorer: A mobile, pervasive spell-casting game for tourists. In *To appear in CHI '07 Extended Abstracts on Human Factors in Computing Systems*. San Jose, CA: ACM Press.

Bartlett, J. F. (2000). Rock 'n' scroll is here to stay. *IEEE Comput. Graph. Appl.*, *20*(3), 40–45.

Buxton, W. (1983). Lexical and pragmatic considerations of input structures. *SIGGRAPH Comput. Graph.*, *17*(1), 31–37.

Cao, X. & Balakrishnan, R. (2003). VisionWand: Interaction techniques for large displays using a passive wand tracked in 3-D. In *Proceedings of the 16th annual ACM Symposium on User Interface Software and Technology* (pp. 173–182). ACM Press.

Card, S. K., English, W. K., & Burr, B. J. (1978). Evaluation of mouse, rate controlled isometric joystick, step keys and text keys for text selection on a CRT. *Ergonomics*, *21*, 601–613.

Card, S. K., Mackinlay, J. D., & Robertson, G. G. (1991). A morphological analysis of the design space of input devices. *ACM Transactions Information Systems*, *9*(2), 99–122.

Card, S. K., Newell, A., & Moran, T. P. (1983). *The psychology of human-computer interaction.* Lawrence Erlbaum Associates, Inc.

Carter, S., Churchill, E., Denoue, L., Helfman, J., & Nelson, L. (2004). Digital graffiti: Public annotation of multimedia content. In *CHI '04: Extended abstracts of the SIGCHI Conference on Human Factors and Computing Systems* (pp. 1207–1210). ACM Press.

Cassinelli, A., Perrin, S., & Ishikawa, M. (2005). Smart laser-scanner for 3-D human-machine interface. *CHI '05: Extended abstracts of the SIGCHI Conference on Human Factors and Computing Systems* (pp. 1138–1139).

Chaos Computer Club. (2002). *Blinkenlights.* Retrieved from http://www.blinkenlights.de

Dan, R. Olsen, J., & Nielsen, T. (2001). Laser pointer interaction. In *CHI '01 Proceedings of the SIGCHI Conference on Human Factors in Computing Systems* (pp. 17–22). New York, NY: ACM Press.

Foley, J. D., Wallace, V. L., & Chan, P. (1984). The human factors of computer graphics interaction techniques. *IEEE Computer Graphics and Applications, 4*(11), 13–48.

Harrison, B. L., Fishkin, K. P., Gujar, A., Mochon, C., & Want, R. (1998). Squeeze me, hold me, tilt me! An exploration of manipulative user interfaces. In *CHI '98 Proceedings of the SIGCHI Conference on Human Factors in Computing Systems* (pp. 17–24). New York, NY: ACM Press/Addison-Wesley Publishing Co.

Hinckley, K., & Horvitz, E. (2001). Toward more sensitive mobile phones. In *UIST '01 Proceedings of the 14th annual ACM Symposium on User Interface Software and Technology* (pp. 191–192). New York, NY: ACM Press.

Hummels, C. C. M. (2000). *An exploratory expedition to create engaging experiences through gestural jam sessions.* Ph.D. thesis, Delft University of Technology.

Jiang, H., Ofek, E., Moraveji, N., & Shi, Y. (2006). Direct pointer: direct manipulation for large-display interaction using handheld cameras. In *CHI '06 Proceedings of the SIGCHI Conference on Human Factors in Computing Systems* (pp. 1107–1110). New York, NY: ACM Press.

Karpov, E., Kiss, I., Leppanen, J., Olsen, J., Oria, D., Sivadas, S., & Tian, J. (2006). Short message dictation on Symbian Series 60 mobile phones. In *ICMI '06 Proceedings of the 8th international conference on Multimodal interfaces* (pp. 126–127). New York, NY: ACM Press.

Licoppe, C., & Inada, Y. (2006). Emergent uses of a multiplayer location-aware mobile game: The interactional consequences of mediated encounters. *Mobilities, 1*(1), 39–61.

MacKenzie, I., & Soukoreff, R. (2002). Text entry for mobile computing: Models and methods, theory and practice. *Human-Computer Interaction, 17*(2), 147–198.

Madhavapeddy, A., Scott, D., Sharp, R., & Upton, E. (2004). Using camera-phones to enhance human-computer interaction. In *Sixth International Conference on Ubiquitous Computing (Adjunct Proceedings: Demos).* Springer-Verlag.

Miyaoku, K., Higashino, S., & Tonomura, Y. (2004). C-blink: A hue-difference-based light signal marker for large screen interaction via any mobile terminal. In *UIST '04 Proceedings of the 17th annual ACM Symposium on User Interface Software and Technology* (pp. 147–156). ACM Press.

Myers, B. A., Bhatnagar, R., Nichols, J., Peck, C. H., Kong, D., Miller, R., & Long, A. C. (2002). Interacting at a distance: measuring the performance of laser pointers and other devices. In *CHI '02 Proceedings of the SIGCHI Conference on Human Factors in Computing Systems* (pp. 33–40). New York, NY: ACM Press.

Myers, B. A., Stiel, H., & Gargiulo, R. (1998). Collaboration using multiple PDAs connected to a PC. In *Proceedings of the 1998 ACM Conference on Computer Supported Cooperative Work* (pp. 285–294). ACM Press.

Nichols, J., & Myers, B. A. (2006). Controlling home and office appliances with smart phones. *IEEE Pervasive Computing, 5*(3), 60–67.

Nielsen, J. (1993). Iterative user-interface design. *Computer, 26*(11), 32–41.

Nielsen, J., & Faber, J. (1996). Improving system usability through parallel design. *Computer, 29*(2), 29–35.

O'Hara, K., Perry, M., & Churchill, E. (2004). *Public and situated displays: Social and interactional aspects of shared display technologies (Cooperative Work, 2).* Norwell, MA: Kluwer Academic Publishers.

Partridge, K., Chatterjee, S., Sazawal, V., Borriello, G., & Want, R. (2002). TiltType: Accelerometer-supported text entry for very small devices. In *UIST '02: Proceedings of the 15th annual ACM Symposium on User Interface Software and Technology* (pp. 201–204). New York, NY: ACM Press.

Patel, S. N., & Abowd, G. D. (2003). A 2-way laser-assisted selection scheme for handhelds in a physical environment. In *Ubicomp '03 Proceedings of the 5th International Conference on Ubiquitous Computing* (pp. 200–207). LNCS 2864, Springer.

Patel, S. N., Pierce, J. S., & Abowd, G. D. (2004). A gesture-based authentication scheme for untrusted public terminals. In *Proceedings of the 17th annual ACM Symposium on User Interface Software and Technology* (pp. 157–160). ACM Press.

Raskar, R., Beardsley, P., van Baar, J., Wang, Y., Dietz, P., Lee, J., Leigh, D., & Willwacher, T. (2004). RFIG lamps: interacting with a self-describing world via photosensing wireless tags and projectors. *ACM Transactions on Graphics, 23*(3), 406–415.

Ravindran, S., Smith, P., Graham, D., Duangudom, V., Anderson, D., & Hasler, P. (2005). Towards low-power on-chip auditory processing. *EURASIP Journal on Applied Signal Processing, 7*, 1082–1092.

Rohs, M. (2005a). Real-world interaction with camera phones. In H. Murakami, H. Nakashima, H. Tokuda, & M. Yasumura (Eds.), *Second International Symposium on Ubiquitous Computing Systems (UCS 2004), Revised Selected Papers* (pp. 74–89), Tokyo, Japan. LNCS 3598, Springer.

Rohs, M. (2005b). Visual code widgets for marker-based interaction. In *IWSAWC '05 Proceedings of the 25th IEEE International Conference on Distributed Computing Systems – Workshops (ICDCS 2005 Workshops* (pp. 506–513). Columbus, OH.

Schmidt, R. A., Zelaznik, H. N., Hawkins, B., Frank, J. S., & Quinn, J. T. (1979). Motor-output variability: A theory for the accuracy of rapid motor acts. *Psychological Review, 86*, 415–451.

Silfverberg, M., MacKenzie, I. S., & Kauppinen, T. (2001). An isometric joystick as a pointing device for handheld information terminals. In B. Watson & J. W. Buchanan (Eds.), *Proceedings of Graphics Interface 2001* (pp. 119–126). Canadian Information Processing Society.

Wang, J., Zhai, S., & Canny, J. (2006). Camera phone based motion sensing: Interaction techniques, applications and performance study. In *UIST '06 Proceedings of the 19th annual ACM Symposium on User Interface Software and Technology* (pp. 101–110). New York, NY: ACM Press.

Want, R., Fishkin, K. P., Gujar, A., & Harrison, B. L. (1999). Bridging physical and virtual worlds with electronic tags. In *Proceedings of the SIGCHI Conference on Human Factors in Computing Systems* (pp. 370–377). ACM Press.

Weiser, M. (1991). The computer for the 21st century. *Scientific American, 265*, 94–104.

Wigdor, D., & Balakrishnan, R. (2003). TiltText: Using tilt for text input to mobile phones. In *UIST '03 Proceedings of the 16th annual ACM Symposium on User Interface Software and Technology* (pp. 81–90). New York, NY: ACM Press.

Wigdor, D., & Balakrishnan, R. (2004). A comparison of consecutive and concurrent input text entry techniques for mobile phones. In *CHI '04 Proceedings of the SIGCHI Conference on Human Factors in Computing Systems* (pp. 81–88). New York, NY: ACM Press.

Zwicky, F. (1967). The morphological approach to discovery, invention, research and construction. In *New Methods of Thought and Procedure* (pp. 273–297). Berlin: Springer.

KEY TERMS

Continuous Interaction: Interactions with a closed-loop feedback, where the user continuously gets informed of the interaction progress as the task is being performed.

Design Space: Design spaces provide a formal or semiformal way of describing and classifying entities along different dimensions, each listing relevant categories or criteria.

Direct Interaction: Input actions are physically coupled with the user-perceivable entity being manipulated (such as an image on a display). To the user, this appears as if there is no mediation, translation, or adaptation between input and output. Physical coupling can be achieved when the feedback spatially coincides with the input action, or at a distance if the user is manipulating a 3-D ray (such as with a laser pointer) that intersects directly with the entity being manipulated.

Discrete Interaction: Interactions with an open-loop feedback, where the user is only informed of the interaction progress after the task is complete.

Indirect Interaction: User activity and feedback occur in disjoint spaces, where scaling and abstraction between input actions and feedback are often necessary.

Input Technique: A specific way of providing data input to a computer through a combination of input devices and software for visual, auditory, or haptic feedback.

ENDNOTES

1. An alternative implementation of the Point & Shoot technique could use pen input instead of the cross-hair image so that the user repositions the cursor by selecting the desired position directly on the live camera image displayed on the phone screen.
2. www.tegic.com

Chapter XXV
Text Entry

Mark David Dunlop
University of Strathclyde, UK

Michelle Montgomery Masters
University of Strathclyde, UK

ABSTRACT

Text entry on mobile devices (e.g., phones and PDAs) has been a research challenge since devices shrank below laptop size: mobile devices are simply too small to have a traditional full-size keyboard. There has been a profusion of research into text-entry techniques for smaller keyboards and stylus input: some of which have become mainstream, while others have not lived up to early expectations. This chapter will review the range of input techniques, together with evaluations, that have taken place to assess their validity: from theoretical modelling through to formal usability experiments. Finally, the chapter will discuss criteria for acceptance of new techniques, and how market perceptions can overrule laboratory successes.

INTRODUCTION

Although phones have traditionally been used for voice calls, with no need for text entry, many services such as text messaging, instant messaging, e-mail and diary operations require users to be able to enter text on phones; text messaging has even overtaken voice calling as the dominant use of mobile phones for many users. Phones and palmtop computers (or electronic organisers/personal digital assistants, PDAs) are too small for a standard desktop or laptop keyboard, thus requiring miniaturisation of the input methods. Furthermore, handheld screen technologies are making it increasingly convenient to read complex messages or documents on handhelds, and cellular data network speeds are now often in excess of traditional wired modems and considerably higher in wi-fi hotspots. These technological developments are leading to increased pressure from users to be able to author complex messages and small documents on their handhelds. Researchers in academia and industry have been working since the emergence of handheld technologies for new text-entry methods that are small and fast but easy-to-use, particularly for novice users. This chapter will look at different approaches to keyboards, different approaches to stylus-based entry, and how these approaches have been evaluated to establish which techniques are actually faster or less error prone. The focus of the chapter is both to give a perspective on the breadth of research in text entry, and also to look at how researchers have evaluated their work. Finally, we will look at perceived future directions, attempting to learn from the successes and failures of text-entry research.

KEYBOARDS

The simplest and most common form of text entry on small devices, as with large devices, is a keyboard. Several small keyboard layouts have been researched that try to balance small size of overall device against usability. These approaches can be categorized as unambiguous, where one key-press unambiguously relates to one character, or ambiguous, where each key is related to many letters (e.g., the standard 12-key phone pad layout where, say, *2* is mapped to *ABC*). Ambiguous keyboards rely on a disambiguation method that can be manually driven by the user or semiautomatic with software support and user correction. This section looks first at unambiguous mobile keyboard designs, then at ambiguous designs and, finally, discusses approaches to disambiguation for ambiguous keyboards.

Unambiguous Keyboards

Small physical keyboards have been used in mobile devices from their very early days on devices such as the Psion Organiser in 1984 and the Sharp Wizard in 1989, and have seen a recent resurgence in devices targeting e-mail users, such as most of RIM's Blackberry range. While early devices tended to have an alphabetic layout, the standard desktop layout was soon adopted (e.g., QWERTY for English language countries, French AZERTY, German QWERTZ, and Italian QZERTY layouts – to reduce ambiguity we will, casually, refer to this family of keyboards as QWERTY keyboards). When well designed, small QWERTY keyboards can make text entry fast by giving the users good physical targets and feedback. However, there is a strong design trade-off between keys being large enough for fast, easy typing and overall device size, with large-fingered users often finding the keys simply too small to tap individually at speed. Physical keyboards also interact poorly with touch screens, where one hand often needs to hold a stylus, and they reduce the space available on the device for the screen.

The QWERTY keyboard layout was designed as a compromise between speed and physical characteristics of traditional manual typewriters: the layout separates commonly occurring pairs of letters to avoid head clashes on manual typewriters, and is imbalanced between left and right hands. Clearly, head clashes and manual carriage returns are not an issue with desktops nor handhelds, but their history in the design leads to a suboptimal layout, where users have to move their fingers more often than they would on an ideal layout. Faster keyboards have been designed, the most widely known being the Dvorak keyboard (or American Simplified Keyboard) for touch typists of English-language documents (Figure 1). While significantly faster than QWERTY keyboards, these have not been widely adopted, primarily because of the learning time and invested skill set in QWERTY keyboards. This investment has been shown to carry over into smaller devices, where the suboptimality issue is even stronger, as users tend to type with one or two thumbs, not the nine fingers envisaged of touch typists. There is strong evidence that alphabetic layouts on desktop computers give no benefits even for novice users, and hinder people with any exposure to the QWERTY layout (Norman, 2002; Norman & Fisher, 1982). It can be reasonably assumed this is also true of palmtops, although some research has shown that experience of using a desktop QWERTY keyboard gives no benefit when moved to a very new environment (McCaul & Sutherland, 2004). While optimal layouts could be designed around two-thumb entry, these are likely to be so different from users' experiences that initial use would be very slow and, as with the Dvorak, rejected by end users. Furthermore, these would

Figure 1. Dvorak keyboard

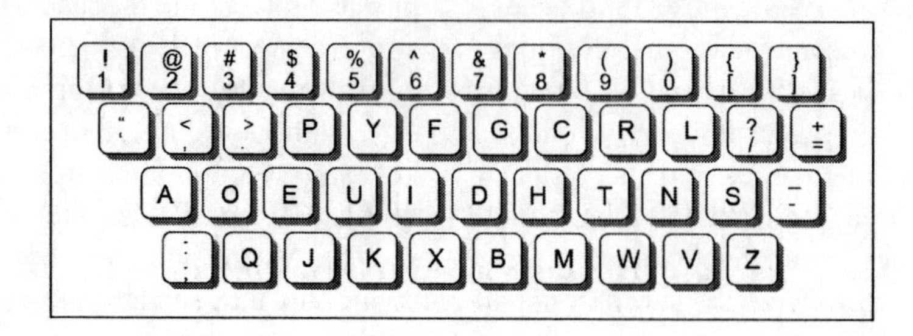

still be suboptimal for one-thumb use (or vice-versa if designed for one thumb).

One approach to small keyboard design that builds on QWERTY skills, and the imbalance between left and right hands on the QWERTY pad, is the half-QWERTY keyboard (Matias, MacKenzie, & Buxton, 1996). Here the keyboard is physically halved down the centre with only the left set of the letters given as physical keys, the user holds the space bar to *flip* the keyboard to give the right side letters (Figure 2-left), and permits fast one-handed entry. In experiments, users of the half-QWERTY keyboard were shown to quickly achieve consistent speeds of 30 words per minute or higher (when using a keyboard with desktop-sized keys). The FrogPad™ is a variant, using an optimised keyboard, so that use of the "right side" of the keyboard is minimised (Figure 2-right). Matias et al. (1996) predicted an optimised pad would lead to a speed increase of around 18% over the half-qwerty design, but at a cost of lost transferable skills. FrogPad™ Inc. now manufactures an optimised keyboard along these lines and claims 40+ words–per-minute typing speeds. Neither approach has yet to make it onto handheld devices, but the FrogPad™ is marketed as a separate keyboard for PDAs.

The FastTap™ keyboard, however, has been targeted at mobile devices from initial conception. This patented technology takes a different approach to miniaturisation by including an alphabetic keyboard as raised keys between the standard numeric keys of a phone pad, giving direct nonambiguous text entry on a very small platform while preserving the standard 12-key keypad currently used by over 90% of mobile users globally (see Figure 3. Experiments (Cockburn & Siresena, 2003) have shown that FastTap™ is considerably faster and easier to use for novice users than more standard predictive text approaches (discussed next), and the two approaches perform similarly for expert users (once practiced, FastTap users, in their trial, achieved 9.3 words per minute (wpm) with T9™ users achieving 10.8wpm, somewhat slower than in other trials, see next section for discussion of T9).

A drastically different unambiguous keyboard approach is to use chords, multiple simultaneous key presses mapping to a single character, either using one or both hands. One-handed chord

Figure 2. Simplified half-QWERTY and FrogPad™

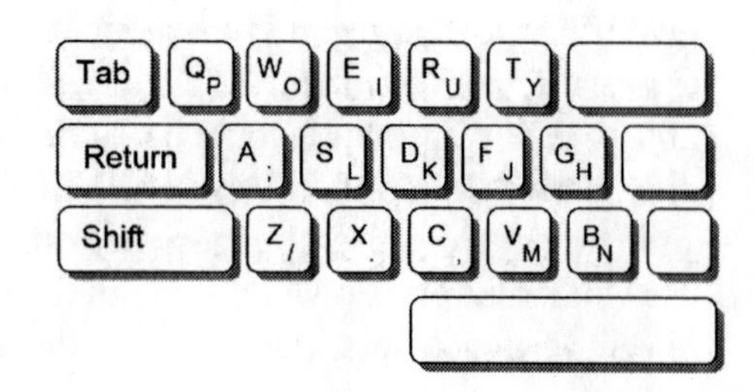

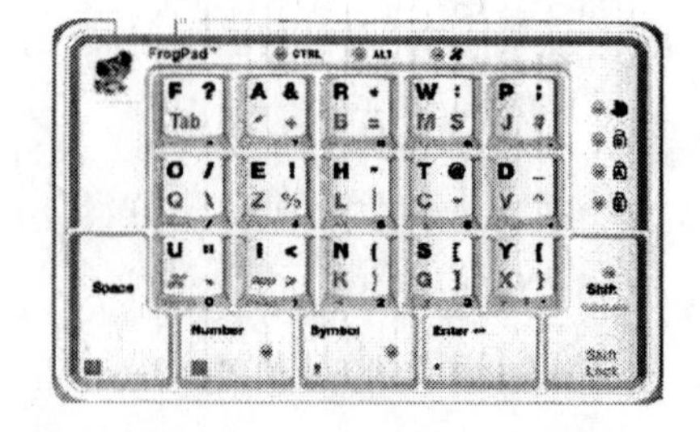

Figure 3. FastTap™ phone keyboard

keyboards were originally envisaged as the ideal partner to the mouse (Engelbart & English, 1968), allowing users to enter text and point at the same time. Chord keyboards can give extremely fast entry rates, with court stenographers reaching around 225 words per minute using a two-handed chord keyboard (compared to skilled QWERTY touch-typing speeds of around 135wpm and hand-writing of about 15wpm). Single-handed chord keyboards are, by definition, palm sized, as users have to be able to have one digit on each key, and have been used in mobile devices (Figure 4 right, shows the AgendA organiser including an alphabetic keyboard surrounded by a chord keyboard). However, the learning time is prohibitive with few users willing to learn the chords required to use these keyboards. Furthermore, the keyboards are not usable without training, users cannot guess how to use them when first picking up a device. Thus, despite size and speed advantages, chord keyboards are generally considered too alien for mainstream devices, and rarely appear on consumer products.

Ambiguous Keyboards

The most common ambiguous keyboard, and the dominant keyboard for mobile phones, is still the telephony ISO/IEC standard 12-key phone keypad (e.g., Figure 5 left). Originally envisaged for name-based dialling of telephone area codes, not for text entry, this keyboard groups three or four letters on each of the physical keys *2* through *9* (with the *1*, ***, *#*, and *0* keys typically acting as space, shift, and other control keys on mobile phones, depending on handset). The method of disambiguating the multiple letters per key is discussed later. However, to further complicate text entry, numbers are typically entered either in a special number-only mode or by pressing and holding the appropriate key. Recently some phones have been released with a slightly stretched mobile phone pad, typically with two extra columns, to give at most two letters per key plus extra space for control keys (Figure 5 right).

While the 12-key mobile phone pad is the smallest commonly found keyboard layout, there has been a history of research into very small

Figure 4. Sample chord keyboards (Douglas Engelbart and Microwriter AgendA)

Figure 5. 12-key phone pad (Nokia N73) and stretched phone pad (Blackberry™ 7100)

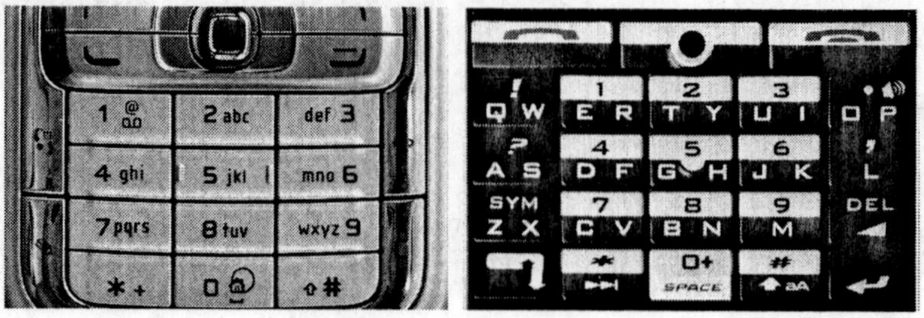

keyboards with as few as three keys. Text entry on these usually involves cursor movement through the alphabet. The most common approach is the 3-key date-stamp method widely used in video games (left and right scroll through an alphabetic strip of letters with fire entering the current letter) and 5-key variant using a joystick with a 2-D keyboard display. Experiments have been conducted to compare simple alphabetic layouts with layouts that are optimised for text entry in English, and ones that dynamically adjust to the most likely next letter (Bellman & MacKenzie, 1998; MacKenzie, 2002b). These studies showed that, once practiced, users of a 1-D alphabetic strip achieved around 9 words per minute while 2-D QWERTY users reached around 10-15 words per minute. Unfortunately, dynamic adjustment of the layout based on probabilities of next letter did not have the expected speed up in either 1-D or 2-D, due to attention load from the user slightly dominating the reduced time to select a letter. An alternative approach is to use short codes representing the letters, for example, short sequences of cursor keys. Evreinova, Evreino, and Raisamo (2004) showed that users could achieve good entry speeds with 3-key combinations of cursor keys, for example, left-up-left for *A*, and that, despite high initial error rates, users could learn the codes quickly.

Disambiguation

The traditional approach to disambiguating text entry on a mobile phone keypad is the manual multitap approach: users press keys repeatedly to achieve the letter they want, for example, on a standard phone keypad, *2* translates to *A* with *22* translating to *B*, and so forth. This approach has also been adopted in many other domestic devices such as audio/video remote controls. Multitap leads to more keystrokes, as users have to repeatedly click for most letters, and to a problem with disambiguating a sequence of letters on the same key, for example, *CAB* is *222222*. Users typically manually disambiguate this by either waiting for a timeout between subsequent letters on the same key or hitting a *time-out kill* button (say *); clearly an error-prone process and one that slows users down. Wigdor and Balakrishnan (2004) refer to multitap as an example of *consecutive* disambiguation; the user effectively enters a key then disambiguates it. An alternative manual disambiguation approach is *concurrent* disambiguation; here users use an alternative input method, for example, tilting the phone (Wigdor & Balakrishnan, 2003) or a small chord-keyboard on the rear of the phone (Wigdor & Balakrishnan, 2004), to disambiguate the letter as it is entered. While clearly potentially much faster than multitap and relatively easy to use, this approach has not yet been picked up by device manufacturers.

Aimed at overcoming the problems of multitap, predictive text-entry approaches use language modelling to map from ambiguous codes to words so that users need only press each key once, for example, mapping the key sequence *4663* directly to *good*. While there are clearly cases where there are more than one match to the numeric key sequence (e.g., *4663* also maps to *home* and *gone*, amongst others), these are surprisingly rare for common words. The problem of multiple matches can be alleviated to a large extent by giving the most likely word as the first suggestion, then allowing users to scroll through alternatives for less likely words. Based on a simple dictionary of words and their frequency of use in the language, users get the right word suggested first around 95% of the time (Gong & Tarasewich, 2005). AOL-Tegic's T9 (Grover, King, & Kushler, 1998; Kushler, 1998) industry-standard entry method is based around this approach and is now deployed on over two billion handsets. Controlled experiments have shown this form of text entry considerably outperforms multitap (Dunlop & Crossan, 2000; James & Reischel, 2001), with text entry rising from around 8 words per minute to around 20 for T9. While predictive text entry is very high quality, it is not perfect, and can lead to errors that are undetected, as users tend to type without monitoring the screen (e.g., a classic T9 user error is sending the message *call me when you are good* rather than *are home*). The main problem, however, with any word prediction system is handling out-of-vocabulary words: words that are not known to the dictionary cannot be entered using this form of text entry. The usual solution is to force users into an "add word" dialogue, where the new word is entered in a special window using multitap, clearly at a considerable loss of flow to their interaction and reduction in entry speed. As most people do not frequently enter new words or place/people names, this is not a major long-term problem. However, it does considerably impact on initial use and can put users off predictive text messaging, as they constantly have to teach new words to the dictionary in the early days of using a new device. This in turn impacts on consumer adoption, with many people not using predictive text, despite it clearly being faster for experienced users.

An alternative approach to dictionary and word-level disambiguation is to use letter-by-letter disambiguation, where letters are suggested based on their likelihood, given letters already entered in the given word or likely letters at the start of a word (e.g., in the clearest case in English, a *q* is most likely to be followed by a *u*). This gives the user freedom to enter words that are not in the dictionary, and considerably reduces the memory load of the text-entry system (no longer an issue with phones but still an issue on some devices). Experiments using this approach (MacKenzie, Kober, Smith, Jones, & Skepner, 2001) showed keystrokes halved and speed increased by around 36% compared with multitap. They also claim that this speed is inline with T9 entry, and that their approach outperforms T9 by around 30% when as few as 15% of the least common words are missing from the predictive dictionary. Predicting letters based on previous letters is actually a specific implementation of Shannon's approach to prediction based on n-grams of letters (Shannon, 1951): predicting the next letter based on the previous letters (or next word based on previous words). Some work has been carried out to extend this to the word level and shows good promise: for example bigram word prediction in Swedish with word completion reduced keystrokes by between 7% and 13% when compared with T9 (Hasselgren, Montnemery, Nugues, & Svensson, 2003).

Some work has been conducted on very small ambiguous keyboards following the same approach as predictive text on 12-key keypads. In work on watch-top text entry we (Dunlop, 2004) found that moving to a 5-key pad reduced accuracy from around 96% to around 81%, with approximately 40% reduction in text-entry speed. While this is a considerable drop in speed, this is a reasonable speed on such a small device, and it is considerably faster than picking from a virtual keyboard with a 5-way keypad (Bellman & MacKenzie, 1998). An interesting alternative input method for small devices is to use a touch-wheel interface, such as those on iPods™. Proschowsky, Schultz, and Jacobsen (2006) developed a method where users are presented with the alphabet in a circle, with a predictive algorithm hidden from the user. This algorithm increases the target area for letters based on the probability of them being selected next, so that users are more likely to hit the correct target when tapping on the touch wheel. User trials show around six to seven words per minute entry rates for novices, about 30% faster than the same users using a date-stamp approach on a touch wheel.

As with the QWERTY layout, the letters on an ambiguous phone keypad do not need to be laid out alphabetically. Here, however, the disambiguation method introduces an additional aspect to designing an optimal layout: the letters can be rearranged to minimise the level of ambiguity for a given language, in addition to looking at minimising finger movement. However, experiments predict that a fully optimised 8-key layout[1] would improve text-entry rates by only around 3% for English (Gong & Tarasewich, 2005). We found a larger but still small reduction of around 8% in keystrokes for a pseudo-optimised layout with the alphabet on four keys when compared against alphabetic ordering (Dunlop, 2004). This modelling does, however, show that stretching the standard phone pad from 8 to 12 keys for text entry is likely to result in a decrease of between 3% and 13% of keystrokes, depending on the style of language used (e.g., Figure 5 right) (Gong & Tarasewich, 2005).

STYLUS-BASED TEXT ENTRY

Compared to mobile phones, personal organisers (PDAs) have made more use of touch screens and stylus interaction as the basis of interaction, and this is now emerging on high-end phones such as Apple's iPhone. This frees up most of the device for the screen and leads to natural mouse-like interaction with applications. Lack of a physical keyboard has led to many different approaches for stylus-based text entry on touch-sensitive screens, which will be discussed in this section: on-screen (or soft) keyboards, handwriting recognition and more dynamic gesture-based approaches.

On-Screen Keyboards

A simple solution to text entry on touch screens is to present the user with an on-screen keyboard that the user can tap on with their stylus, or on particularly large touch screens with their finger. The most common implementation is to copy the QWERTY layout onto a small touch-sensitive area at the bottom of a touch screen. However, as with physical keyboards and keypads, there has been research into better arrangements of the keys. Mackenzie and his team have conducted a series of experiments on alternative on-screen keyboard layouts that are optimised for entry using a single stylus while one hand is holding the device (as opposed to desktop assumption of eight fingers and a thumb being used for entry). They investigated both unambiguous keyboards and an optimised 12-key ambiguous keypad, inspired by the success of T9™ and the fundamental rule of interaction that large targets are quicker to hit than small ones (Fitts, 1954). Their results estimate that an expert user could achieve 40+ wpm on a soft QWERTY keyboard with novice soft-keyboard users achieving around 20 wpm (MacKenzie, Zhang, & Soukoreff, 1999). The alternative layouts were predicted to give higher entry rates for expert use: the unambiguous Fitaly layout was predicted to reach up to 56wpm, and ambiguous JustType, 44wpm (Figure 6). However, novice users achieved only around 8wpm using these alternative keyboard layouts, highlighting the carry-over effect of desktop QWERTY layout.

While simple and fast, the on-screen keyboard approach can be tiring for users, as they are required to repeatedly hit very small areas of the screen. The patented technology underlying the

Figure 6. Fitaly and JustType keyboard layouts

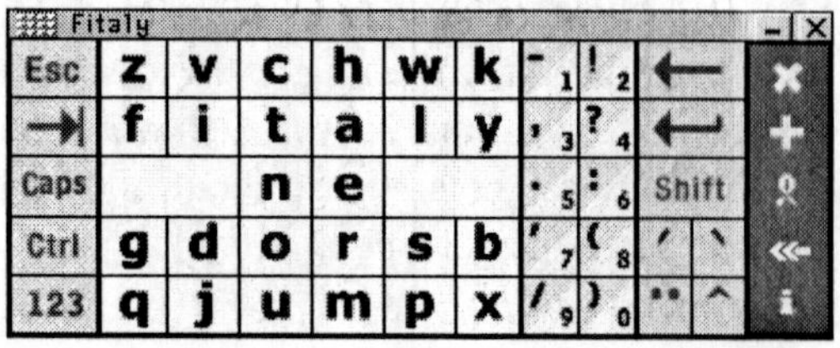

Fitaly

Esc	z	v	c	h	w	k	- 1	! 2	←	×
→|	f	i	t	a	l	y	, 3	? 4	↵	+
Caps			n	e			. 5	: 6	Shift	
Ctrl	g	d	o	r	s	b	' 7	(8	' `	«
123	q	j	u	m	p	x	/ 9	) 0	·· ^	i

Shift	R P Q	A D F	N B Z
Delete	O L X	E W V	I M G
Space	C Y K	T H J	S U

Figure 7. Sample XT9™ Mobile Interface

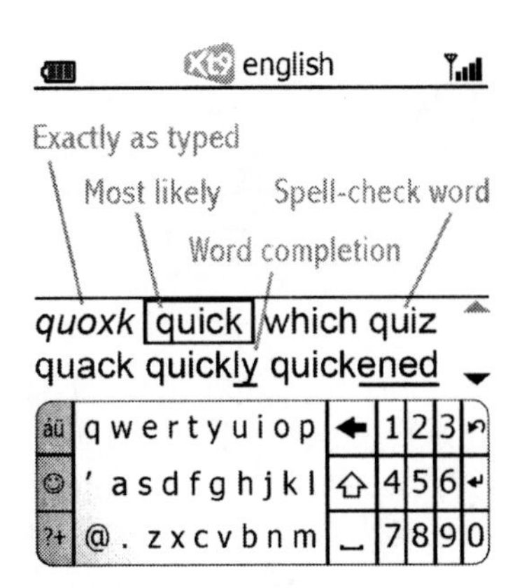

XT9™ Mobile Interface from Tegic Communications attempts to address this problem by including a level of disambiguation in an otherwise unambiguous keyboard (Robinson & Longe, 2000). For example, if the user taps a letter adjacent to the letter in the intended word, then the "intended" letter is used instead of the letter tapped. Their approach defaults to the most likely full word given the approximate letters entered, while offering alternative corrections and word completions as well as the letters actually typed (Figure 7). XT9 technologies have been developed by Tegic for multiple platforms, including handprinting and small physical keyboards.

Handwriting

To many, the obvious solution to text entry on handheld devices is handwriting recognition. However, there are many problems with basing text entry around handwriting,: most obvious being the slow rate at which people write (about 15 wpm (Card, Moran, & Newell, 1983)) and the wide variability of individual's handwriting styles. Modern handwriting recognition systems, for example on Windows XP Tablet edition, are extremely good at recognising in-dictionary words, but struggle on words that are not previously known, and are still inherently limited by writing speeds. Cursive handwriting recognition also requires a reasonably large physical space for the user to write on (governed by both touch-screen resolutions and human dexterity). Furthermore, cursive handwriting recognition still requires considerable processing power that is more in line with modern laptops/tablets than phones. Simplified alphabets to reduce the processing complexity and space needed for writing were originally introduced with the unistroke (Goldberg & Richardson, 1993) approach. Here each letter is represented as a single stroke, with letters typically drawn on top of each other in a one letter wide slot. This approach forces users to learn a new alphabet (Figure 8) but makes recognition computationally easier and more accurate, while also reducing the time it takes to draw each letter for skilled users. Palm popularised a more intuitive version, Graffiti™, on their palmtops: a mostly unistroke alphabet, Graffiti™, was composed mostly of strokes with high similarity to standard capital letters. CIC's Jot™ alphabet provides a mix of unistroke and multistroke letters and is deployed on a wide range of handhelds. Experiments comparing handprinting with other text-entry methods are rare, but a comparison between handprinting, QWERTY-tapping, and ABC-tapping on pen-based devices (MacKenzie, Nonnecke, McQueen, Riddersma, & Meltz, 1994) showed that a standard QWERTY layout can achieve around 23wpm while hand-

Figure 8. Unistroke, Graffiti™ and Jot™ sample letters

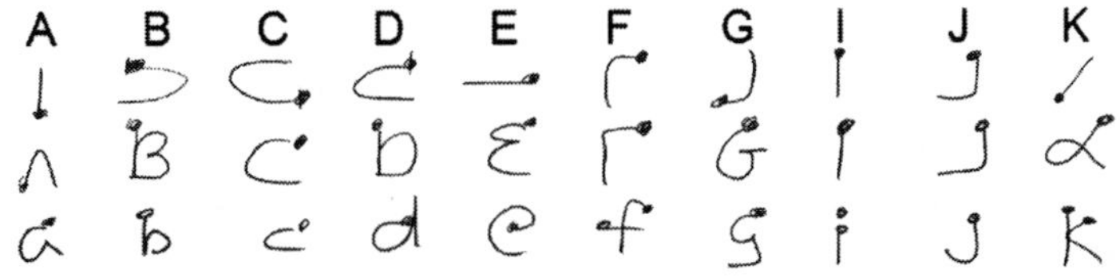

A B C D E F G I J K

Figure 9. WordComplete™ (left) and AdapTex™ (centre and right)

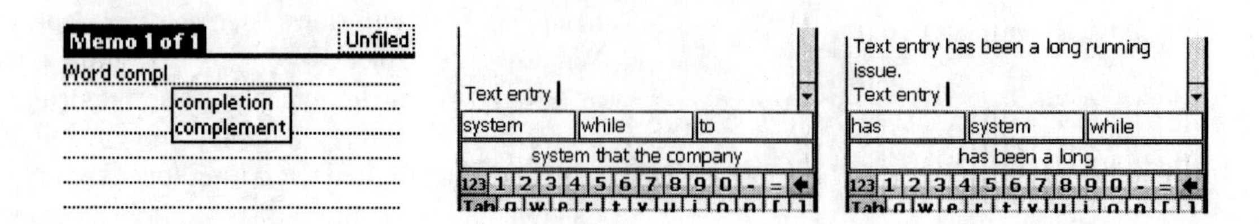

printing achieved only 17wpm and alphabetic soft-keyboard only 13wpm.

Word completion can also be used to help users with text entry, and sits most comfortably with touch screens where users can point quickly at word completion suggestions. For example, CIC's WordComplete™ (Figure 9 left) suggests short phrases or word completions based on what the user has written. Similar technologies are used on the eZiType™[2] and XT9[3] technologies deployed on some mobile phones. While tempting, word completion needs to perform very well in order to give users a benefit: users rapidly get into a flow of text entry, which word completion interrupts, so the saving in terms of letters entered has to be considerable before a real saving in time is achieved. We (Dunlop & Crossan, 2000) estimated that simple word completion would reduce keystrokes by 17%, but our model-based evaluation (see section 4.1) predicted an approximate halving of entry speed once user interruption time was taken into account. Some recent advances, however, have shown that when based on more complex language models, word completion can be beneficial with novice users increasing typing speed by around 35% when using a soft keyboard (Dunlop, Glen, Motaparti, & Patel, 2006) (Figure 9 centre right). AdapTex™ performance also increases over time by learning patterns of use in the user's language to tune suggestions to the individual user and his/her context of use.

Gesture-Based Input

Gesture-based interaction attempts to combine the best of visual keyboards with easy-to-remember stylus movements to gain faster and smoother, while still easy-to-learn, text entry. Building on our motor memory for paths, approaches such as Cirrin (Mankoff & Abowd, 1998), Quikwriting (Perlin, 1998) and Hex (Williamson & Murray-Smith, 2005) are based on the user following a path *through* the letters of the word being entered (Figure 10). For on-screen approaches, this achieves faster entry rates than single character printing with reduced stress and fatigue when writing. Furthermore, in the case of Hex, the approach can be used one-handed on devices with accelerometers/tilt sensors.

Gestures can be combined with more conventional soft keyboards so that users can choose to tap individual letters, improving pick-up-and-use usability, or to enter words in one gesture by following the path of the letters on the touch keyboard (experts can then enter the gestures anywhere on screen) (Zhai & Kristensson, 2003).

Dasher is a drastically different approach to text entry that attempts to exploit interactive displays more than traditional text entry approaches. In Dasher (Ward, Blackwell, & MacKay, 2000) (Figure 12), letters scroll towards the user and (s)he picks them by moving the stylus up and down as the letters pass the stylus. The speed of scrolling

Figure 10. Quikwriting (left) and the Hex entry for "was" (omitting letter display)(right)

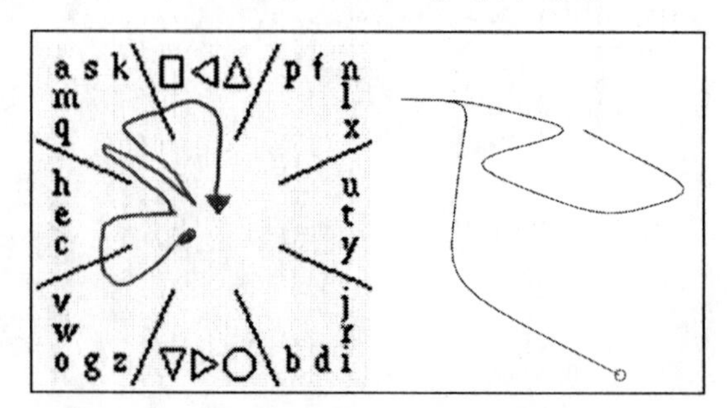

Figure 11. The ATOMIK keyboard with SHARK shortcuts

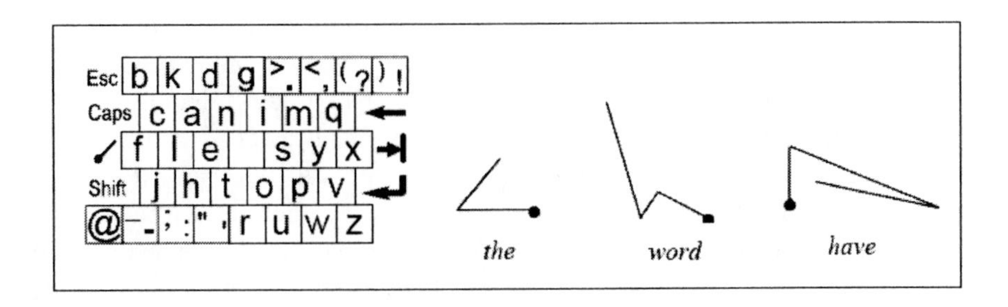

Figure 12. Dasher

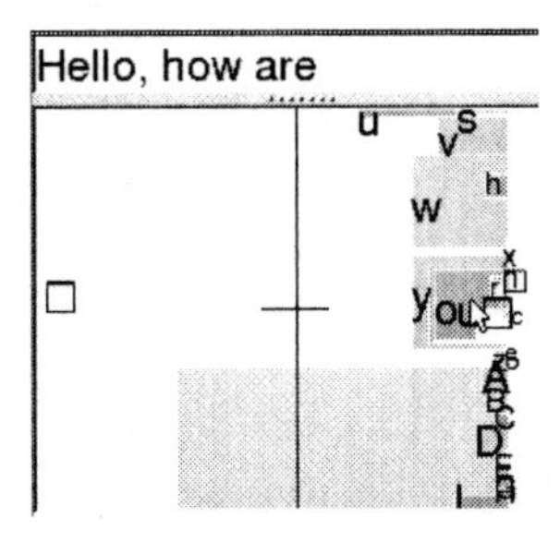

is controlled by the user moving the stylus left and right with predictive text-entry approaches dynamically changing the space allocated to each letter (so that likely next letters are given more space than less likely ones, but all letters are available at each stage). Experiments show that users can enter at over 30 words per minute.

EVALUATION

Unlike many areas of mobile technology, where market forces and commercial ingenuity dominate, the field of text entry has benefited from considerable scientific study to establish the benefits of one method over another. These studies have been conducted by academic and industrial research groups, often in collaboration, and are used both to compare techniques and to tune their usage to how users actually enter text. Much of the related evaluation work and results have already been discussed previously; in this section, we focus on the evaluation methods themselves.

Technical Evaluation

The literature commonly uses three methods for reporting the performance of text entry: *average ranked list position* (ARP), *disambiguation accuracy* (DA), and *keystrokes per character* (KSPC).

The average ranked-list position (e.g., Dunlop & Crossan, 2000) for evaluating ambiguous text-entry methods is calculated in two phases. First language models, for example, in the simplest case, word frequencies, are learned from a corpus appropriate to the target language. Once trained, the second phase involves processing the same corpus one word at a time. Each word taken from the corpus is encoded using the ambiguous key coding for the target keypad (e.g., *home* is encoded as *4663*) and a ranked list of suggested words produced for that encoding based on the learned language model. The position of the target word in this list is averaged over all words to give the average ranked-list position for that corpus and keypad. An ARP value of 1.0 indicates that the correct word was always in the first position in the ranked list of suggestions, a value of 2.0 that, on average, the correct word was second in the ranked list. We predicted an ARP value of around 1.03 for a large corpus of English language newspaper articles using a standard phone keypad layout. ARP naturally biases the averaging process so that words are taken into account proportionally to their occurrence in the text corpus.

Disambiguation accuracy (e.g., Gong & Tarasewich, 2005) reports the percentage of times the first word suggested by the disambiguation process is the word the user intended: a DA value of 100% implies the disambiguation process always gives the correct word first, while 50% indicates that it only manages to give the correct word first half of the time. Gong and Tarasewich reported DA of 97% for written English corpus and 92% for SMS messages (both on a phone pad). This is a more intuitive and direct measure than ARP, but does not take into account the performance of words that do not come first in the list.

KSPC (MacKenzie, 2002a) reports the average number of keystrokes required to enter a character, for example, *home* followed by a space on a standard T9™ mobile phone requires six keystrokes – 4663*# where * is the next suggestion key and # space, giving a KSPC for *hello* of 6/4=1.5. As with ARP and DA, the value is normally averaged over a large corpus of appropriate text for the target language. A KSPC value of 1.0 indicates perfect disambiguation, as the user never needs to type any additional letters, while a higher figure reflects the proportional need for the next key in disambiguation (and a lower level, successful word completion). Full-sized nonambiguous keyboards achieve KSPC=1.00, standard date stamp method for entering text on three keys achieves KSPC=6.45, date stamp like interaction on five keys achieves KSPC=3.13 and multitap on a standard 9-key mobile phone achieves a KSPC of around 2.03 (MacKenzie, 2002a). Hasselgren et al. (2003) reported KSPC of 1.01 and 1.08 for T9 using Swedish news and SMS corpora respectively, improving to 1.01 and 0.88 respectively for their bigram model with word completion. KSPC does take into account ranked list position for all words, and compares easily with nonpredictive text-entry approaches; however, it is a rather abstract measure being based on letters, especially for dictionary-based approaches that are inherently word-based methods.

To gain an insight into potential expert user behaviour with different keyboards, different approaches have been taken to modelling interaction in order to predict expert (trained, error-free) performance. There are two basic approaches: physical movement modelling and keystroke level modelling. We (Dunlop & Crossan, 2000) proposed a keystroke level model based on Card, Moran and Newall's work (Card, Moran, & Newell, 1980). Our model was based on predicting the time *T(P)* taken by an expert user to enter a given phrase. The model calculates this in an equation that combines a set of small time measurements for elements of the user interaction. In the case of text entry, the relevant factors are the homing time for the user to settle on the keyboard T_h (0.40 seconds); the time it takes a user to press a key T_k (0.28s); the time it takes the user to mentally respond to a system action T_m (1.35s); the length of an average word k_w (4.98); and the number of words in the phrase w (10). In addition, for predictive text entry, where disambiguation occurs by the user moving through the ranked list of suggestions, the ARP value is required, here given as l=1.03. The overall time equation for entering a phrase is given in Equation 1 (Dunlop and Crossan's keystroke model):

$$T(P) = T_h + w\ (k_w T_k + l(T_m + T_k))$$

This model, as corrected by Pavlovych and Stuerzlinger (Pavlovych & Stuerzlinger, 2004), predicts a text-entry time for a 10-word phrase at 31.2 seconds, equating to a speed of 19.3 words per minute. This prediction matches closely with focused user experiments on experienced users of T9 of 20.4 wpm (James & Reischel, 2001).

We modelled keystroke speed at 0.28s based on a fixed figure from Card et al.'s work that is equivalent to "an average nonsecretary typist" on a full QWERTY keypad. This works well, however, it cannot take into account fine-grained keyboard design elements that can have a considerable impact on typing speed in practice: for example, different keyboard layouts clearly affect the average time it takes a user to move his/her fingers to the correct keys. Mackenzie's group have conducted considerable work using Fitt's law (Fitts, 1954) to calculate the limit of performance given distance between keys (e.g., Silfverberg, MacKenzie, & Korhonen, 2000). The basic form of their distance-based modelling predicts 40.6 wpm for thumb-based predictive input, assuming no *next* key operations (essentially equivalent to no thinking or homing times in equation 1). Later work modifies the Fitt's distance models to take into account two inaccuracies that can noticeably affect predictions: repeated letters on the same key (Soukoreff & MacKenzie, 2002), and parallel finger movements, where users move one finger at the same time as pressing with another (a main design criteria for QWERTY desktop keyboards that impacts on expert text users who use two thumbs) (MacKenzie & Soukoreff, 2002).

These models are useful in predicting performance but focus on expert error-free performance. More complex modelling approaches have been researched to support novices to model more complete interaction, and to model error behaviour (e.g., How & Kan, 2005; Pavlovych & Stuerzlinger, 2004; Sandnes, 2005). Although users studies are the traditional acid test for any interactive system, these models are valuable either in the early stages of design or to understand methods where user experiments are difficult (or biased by users' prior experience of current technologies).

User Studies

Models that predict text-entry performance only give us part of the picture, proper user studies often give a truer indication of how text-entry methods perform in reality. While there are many parameters that can affect the design of user studies, the three prominent ones for text-entry experiments are the environment in which the experiments are conducted, the platform on which experiments are conducted, and the source of the phrases that users enter.

Most user studies into text entry have been conducted in laboratories. A laboratory is a controlled environment that leads to a more consistent user experience and, thus, considerably easier statistical analysis as there are fewer confounding variables from the environment to interfere with measurements taken. However, conducting experiments on people entering text on mobile phones in quiet office settings where they can focus exclusively on the text-entry tasks is arguably not representative of normal use. There is a growing debate in mobile HCI research on the validity of laboratory experiments with some researchers arguing that, while the focus of most common errors is different in the real world, laboratory experiments do not miss errors that are found in real-world experiments (Kaikkonen, Kekäläinen, Cankar, Kallio, & Kankainen, 2005) while others claim a wider range of errors were found in the real world than in laboratories (Duh, Tan, & Chen, 2006). Kjeldskov and Graham (2003) report that "71% [of studied evaluations were] done through laboratory experiments, 19% through field experiments and the remaining 10% through surveys." As a specific example, Brewster (2002) showed usability and text-entry rates were significantly reduced for users performing an outdoor walking circuit, while entering on a soft numeric key pad, than those conducting the same experiment in a traditional laboratory. Whereas Mizobuchi, Chignell, and Newton (2005) showed that, while walking was slowed down when using a device, it did not impinge upon the text-entry rate or accuracy.

Experiments are either conducted on a real device or on a simulator; either running on a desktop PC or on a touch screen PDA. Obviously, experimenting on real devices is preferable in terms of appropriateness of experimental setting, however, development on real devices was very difficult until recently, as programming these by anyone but the manufacturer was difficult. However, most modern phones and PDAs have powerful Java platforms available, as standard, that can be used for controlled experiments. Conducting experiments on desktops leads to problems of users typing too fast: desktop keys are typically bigger and give a more positive response than many mobile phone keyboards, thus, potentially significantly increasing key speeds, which may bias some text-entry methods over others. There is also a problem with computer numeric pads being vertically inverted compared to phone pads; a historic difference that most users are unaware of that can become very visible when forced to use a computer keypad for phone operations. Emulating on a touch-screen handheld is a tempting alternative as users can hold the device naturally, and the interface designer has full freedom of key layout, however, without physical feedback of key presses and physically distinguished hit areas, typing can be considerably slower, again biasing some input methods over others.

Finally, users are typically required to enter a set of phrases on devices to measure their text-entry speed. These phrases are usually the same for all users in a trial so that variations in phrases can be excluded in statistical analysis. While there is no widespread agreement on phrases that are used, MacKenzie and Soukoreff (2003) proposed a standard set of short phrases that has been used by other researchers and provides a valuable baseline for comparisons. One problem with mobile phone text entry is that it is often used for short casual messages and testing with formal phrases from a traditional text corpus is not appropriate (see differences discussed previously for those who have experimented with formal English and SMS). This is compounded by the original multitap text-entry approach and short length of text messages[4] leading to considerable use of, often obscure, abbreviations that are not normally found in a corpus. To address this, How and Kan (2005) developed a large set of phrases extracted from SMS users' real text conversations. Although somewhat skewed to local Singaporean phrases and abbreviations (much of *SMS speak* is heavily localised and even personalised within a group of friends), the corpus is a valuable insight into the language often used on mobile phones. It should also be noted that entry speeds of 33wpm for users when transcribing text on desktop keyboards have been found to drop to around 19wpm for composing new text (Karat, Halverson, Horn, & Karat, 1999), so most results from text-entry experiments can be assumed to be over-inflating speeds by around 40% as they are typically based on transcription.

CONCLUSION AND FUTURE TRENDS

This chapter has reviewed a large number of text-entry methods that range from standard methods, through slight variations, to completely different approaches. We have looked at different hardware keyboard designs, different on-screen keyboard layouts, handwriting-based approaches, and more novel approaches such as gestures. We have also looked at ambiguous and unambiguous designs and the related approaches to disambiguation. Much of the work reported has experimental backing to show the potential benefits of each approach. However, when comparing the wide diversity of approaches in the literature to widely available implementations on real devices, the overriding message we see is that guessability, the initial pick-up-and-use usability of hardware/software, is paramount to success.

It is extremely hard to predict future trends for mobile devices: while there is considerable research showing the benefits and strengths of different approaches, market forces and the views of customers and their operators have a major role in deciding which techniques become widely adopted. Predicted gains in expert text-entry performance are of no use if people do not understand how to use the text-entry approach out of the box. To this end, we see considerable scope for entry methods that provide a smooth transition from novice to expert performance: XT9™ is one successful example of novice-to-expert support, as users get faster they will learn to be sloppier and type faster, without necessarily being consciously aware of why. Context-aware word completion that learns about individuals is another area that shows good potential: good for slow novice typists as they start, but building context and personalising as they gain proficiency.

Finally, looking at current market directions and the increasing desire to enter more text on small devices, we see the 12-key keypad slowly disappearing from phones, to be replaced with less number-centric entry methods. Despite its suboptimality and problems on small devices, both market trends and some user tests point to the QWERTY keyboard taking on this role, either as a physical or an on-screen keyboard.

PROJECTS AND DISCUSSION TOPICS

To build your experience of the material covered in this chapter, you could try the following small experiments:

1. Ask friends/colleagues/family how they normally enter text messages on their mobile phone, then time them entering a common phrase, say a nursery rhyme. Record times and error rates. Swap your users to another input method and see how their time and error rates compare. You can now compare expert vs. (semi-) novice performance on a range of techniques.
2. A variation of (1) would be to have a shoot-out, where you get a group together, agree a phrase, then shout "start" to get the whole group entering the phrase at once then raising their device when finished. You might want to control error policy by saying that all errors should be corrected before hands are raised. You might also want to include a word that is not in the dictionary, say an unusual local place name.
3. Search for some word completion software for your mobile device platform (e.g., Windows Mobile, PalmOS, or Symbian). Install this and work consistently, using it for 1 week. Record your views on the software after 10 minutes, 1 hour, 1 day, and 1 week. You can now compare your views as you learned the software (and it possibly learned your vocabulary/language).

You might also consider the following reflective comments/discussion topics (they are intentionally aggressive!):

1. T9™ is only for softies, real texters use multitap
2. QWERTY keypads on phones are included mainly so that business users do not look like teenagers sending text messages
3. Technical usability evaluation is pointless, only real user studies are valid for assessing different text-entry techniques
4. If the Fitaly keyboard is so much faster for expert users and is usable out of the box, then phone manufacturers should be bold and drop the QWERTY keyboard.

ACKNOWLEDGMENT

Our thanks are extended to our reviewers, whose comments have strengthened this chapter. We are also grateful to those organisations and companies that have given permission to use images throughout the chapter.

All trademarks are used with appropriate reference to their holders. FrogPad is a trademark of FrogPad Inc.—Figure 2 (right) is used with permission of FrogPad. FastTap is a trademark of Digit Wireless Inc. Figure 3 used with permission of Digit Wireless. Douglas Engelbart image (Figure 4 left) is used courtesy of SRI International, Menlo Park, CA. MicroWriter AgendA image (Figure 4 right) used with permission of Bellaire Electronics. Blackberry is a trademark of Research In Motion Limited. Fitaly keybord image (Figure 6 left) used with permission of Textware Solutions. XT9 is a registered trademark of Tegic Communications, Inc. The Tegic Communications, Inc. name, icons, trademarks, and image in Figure 7 are used with permission. Microsoft Windows is a registered trademark of Microsoft Corp. Unistroke is patented technology of Xerox. Graffiti is a registered trademark of Palm Inc; Jot and WordComplete of CIC. Figure 9 (centre and right) images used courtesy of KeyPoint Technologies, of which AdapTex is a trademark. Figure 10, 11, and 12 images used courtesy of related paper authors. Figures 1, 2 (left), 6 (right), and 8 drawn by authors and may be altered from original layouts for clarity. Figures 9 (left), 10 (left), and 12 screen images taken by authors. Images in Figure 5, original photographs by authors.

REFERENCES

Bellman, T., & MacKenzie, I. S. (1998). *A probabilistic character layout strategy for mobile text entry.* Paper presented at the Graphics Interface '98, Toronto, Canada.

Brewster, S. A. (2002). Overcoming the lack of screen space on mobile computers. *Personal and Ubiquitous Computing, 6*(3), 188-205.

Card, S. K., Moran, T. P., & Newell, A. (1980). The keystroke-level model for user performance time with interactive systems. *Communications of the ACM, 23*(7), 396-410.

Card, S., Moran, T., & Newell, A. (1983). *The psychology of human-computer interaction.* Lawrence Erlbaum Associates.

Cockburn, A., & Siresena, A. (2003). *Evaluating mobile text entry with the Fastap keypad.* Paper presented at the People and Computers XVII (Volume 2): British Computer Society Conference on Human Computer Interaction, Bath, England.

Duh, H. B., Tan, G. C., & Chen, V. H. (2006). *Usability evaluation for mobile device: A comparison of laboratory and field tests.* Paper presented at the 8th Conference on Human-Computer interaction with Mobile Devices and Services: MobileHCI '06, Helsinki, Finland.

Dunlop, M. D. (2004). *Watch-top text-entry: Can phone-style predictive text-entry work with only 5 buttons?* Paper presented at the 6th International Conference on Human Computer Interaction with Mobile Devices and Services: MobileHCI 04, Glasgow.

Dunlop, M. D., & Crossan, A. (2000). Predictive text entry methods for mobile phones. *Personal Technologies, 4*(2).

Dunlop, M. D., Glen, A., Motaparti, S., & Patel, S. (2006). *AdapTex: Contextually adaptive text entry for mobiles.* Paper presented at the 8th Conference on Human-Computer interaction with Mobile Devices and Services: MobileHCI '06, Helsinki, Finland.

Engelbart, D. C., & English, W. K. (1968). *A research center for augmenting human intellect.* Paper presented at the AFIPS 1968 Fall Joint Computer Conference, San Francisco, USA.

Evreinova, T., Evreino, G., & Raisamo, R. (2004). *Four-key text entry for physically challenged people.* Paper presented at the 8th ERCIM workshop: User Interfaces for All (Adjunct workshop proceedings), Vienna, Austria.

Fitts, P. M. (1954). The information capacity of the human motor system in controlling the amplitude of movement. *Journal of Experimental Psychology, 47*(6), 381-391.

Goldberg, D., & Richardson, C. (1993). *Touch-typing with a stylus.* Paper presented at the INTERCHI93.

Gong, J., & Tarasewich, P. (2005). *Alphabetically constrained keypad designs for text entry on mobile devices.* Paper presented at the CHI '05: SIGCHI Conference on Human Factors in Computing Systems, Portland, USA.

Grover, D. L., King, M. T., & Kushler, C. A. (1998). *USA Patent No. 5818437. I. Tegic Communications.*

Hasselgren, J., Montnemery, E., Nugues, P., & Svensson, M. (2003). *HMS: A predictive text entry method using bigrams.* Paper presented at the Workshop on Language Modeling for Text Entry Methods, 10th Conference of the European Chapter of the Association of Computational Linguistics, Budapest, Hungary.

How, Y., & Kan, M.-Y. (2005). *Optimizing predictive text entry for short message service on mobile phones.* Paper presented at the Human Computer Interfaces International (HCII 05), Las Vegas, USA.

James, C. L., & Reischel, K. M. (2001). *Text input for mobile devices: Comparing model prediction to actual performance.* Paper presented at the SIGCHI conference on Human factors in computing systems, Seattle, Washington, USA.

Kaikkonen, A., Kekäläinen, A., Cankar, M., Kallio, T., & Kankainen, A. (2005). Usability testing of mobile applications: A comparison between laboratory and field testing. *Journal of Usability Studies, 1*(1), 4-17.

Karat, C., Halverson, C., Horn, D., & Karat, J. (1999). *Patterns of entry and correction in large vocabulary continuous speech recognition systems.* Paper presented at the ACM CHI99: Human Factors in Computing Systems.

Kjeldskov, J., & Graham, C. (2003). *A review of mobile HCI research methods.* Paper presented at the Fifth International Symposium on Human Computer Interaction with Mobile Devices and Services: MobileHCI 03, Udine, Italy.

Kushler, C. (1998). *AAC using a reduced keyboard.* Paper presented at the Technology and Persons with Disabilities Conference, USA.

MacKenzie, I. S. (2002a). *KSPC (keystrokes per character) as a characteristic of text entry techniques.* Paper presented at the MobileHCI 02: Fourth International Symposium on Human-Computer Interaction with Mobile Devices, Pisa, Italy.

MacKenzie, I. S. (2002b). *Mobile text entry using three keys.* Paper presented at the Proceedings of the second Nordic conference on Human-computer interaction, Aarhus, Denmark.

MacKenzie, I. S., Kober, H., Smith, D., Jones, T., & Skepner, E. (2001). *LetterWise: Prefix-based disambiguation for mobile text input.* Paper presented at the 14th annual ACM symposium on User interface software and technology Orlando, USA.

MacKenzie, I. S., Nonnecke, R. B., McQueen, C., Riddersma, S., & Meltz, M. (1994). *A comparison of three methods of character entry on pen-based computers.* Paper presented at the Factors and Ergonomics Society 38th Annual Meeting, Santa Monica, USA.

MacKenzie, I. S., & Soukoreff, R. W. (2002). *A model of two-thumb text entry.* Paper presented at the Graphics Interface 2002.

MacKenzie, I. S., & Soukoreff, R. W. (2003). *Phrase sets for evaluating text entry techniques.* Paper presented at the ACM Conference on Human Factors in Computing Systems–CHI 2003, New York.

MacKenzie, I. S., Zhang, S. X., & Soukoreff, R. W. (1999). Text entry using soft keyboards. *Behaviour & Information Technology, 18*, 235-244.

Mankoff, J., & Abowd, G. D. (1998). *Cirrin: A word-level unistroke keyboard for pen input.* Paper presented at the 11th Annual ACM Symposium on User interface Software and Technology: UIST '98, San Francisco.

Matias, E., MacKenzie, I. S., & Buxton, W. (1996). One-handed touch typing on a QWERTY keyboard. *Human-Computer Interaction, 11*(1), 1-27.

McCaul, B., & Sutherland, A. (2004). *Predictive text entry in immersive environments.* Paper presented at the IEEE Virtual Reality Conference 2004 (VR 2004), Chicago, USA.

Mizobuchi, S., Chignell, M., & Newton, D. (2005). *Mobile text entry: Relationship between walking speed and text input task difficulty.* Paper presented at the MobileHCI 05: 7th international Conference on Human Computer interaction with Mobile Devices and Services Salzburg, Austria.

Norman, D. A. (2002). *The design of everyday things.* Basic Books.

Norman, D. A., & Fisher, D. (1982). Why alphabetic keyboards are not easy to use: Keyboard layout doesn't much matter. *The Journal of the Human Factors Society, 25*(5), 509-519.

Pavlovych, A., & Stuerzlinger, W. (2004). *Model for non-expert text entry speed on 12-button phone keypads.* Paper presented at the CHI '04: SIGCHI Conference on Human Factors in Computing Systems, Vienna, Austria.

Perlin, K. (1998). *Quikwriting: Continuous stylus-based text entry.* Paper presented at the 11th Annual ACM Symposium on User interface Software and Technology: UIST '98, San Francisco.

Proschowsky, M., Schultz, N., & Jacobsen, N. E. (2006). *An intuitive text input method for touch wheels.* Paper presented at the SIGCHI Conference on Human Factors in Computing Systems: CHI '06, Montréal, Canada.

Robinson, B. A., & Longe, M. R. (2000). *USA Patent No. 6,801,190/7,088,345.*

Sandnes, F. E. (2005). *Evaluating mobile text entry strategies with finite state automata.* Paper presented at the MobileHCI 05: 7th international Conference on Human Computer interaction with Mobile Devices and Services, Salzburg, Austria.

Shannon, C. (1951). Prediction and entropy of printed english. *Bell System Technical Journal, 30*, 50-64.

Silfverberg, M., MacKenzie, I. S., & Korhonen, P. (2000). *Predicting text entry speed on mobile phones.* Paper presented at the SIGCHI Conference on Human Factors in Computing Systems (CHI'00), The Hague, The Netherlands.

Soukoreff, R. W., & MacKenzie, I. S. (2002). *Using Fitts' law to model key repeat time in text entry models.* Paper presented at the Graphics Interface 2002.

Ward, D. J., Blackwell, A. F., & MacKay, D. J. (2000). *Dasher—A data entry interface using continuous gestures and language models.* Paper presented at the 13th Annual ACM Symposium on User interface Software and Technology: UIST '00, San Diego, USA.

Wigdor, D., & Balakrishnan, R. (2003). *TiltText: Using tilt for text input to mobile phones.* Paper presented at the 16th annual ACM Symposium on User Interface Software and Technology (UIST '03), New York.

Wigdor, D., & Balakrishnan, R. A. (2004). *Comparison of consecutive and concurrent input text entry techniques for mobile phones.* Paper presented at the SIGCHI Conference on Human Factors in Computing Systems (CHI '04), New York.

Williamson, J., & Murray-Smith, R. (2005). *Dynamics and probabilistic text entry.* Paper presented at the Hamilton Summer School on Switching and Learning in Feedback systems.

Zhai, S., & Kristensson, P. (2003). *Shorthand writing on stylus keyboard.* Paper presented at the SIGCHI Conference on Human Factors in Computing Systems: CHI '03, Ft. Lauderdale, Florida, USA.

KEY TERMS

Ambiguous Keyboards: A keyboard layout where each key is related to many letters (e.g., the standard 12-key phone pad layout where, say, 2 is mapped to ABC)

Evaluation: Method for assessing and measuring the performance of a text-entry system in terms of either usability or technical performance.

Handwriting Recognition: Method for interpreting text that has been entered using handwriting via stylus

Predictive Text Entry: Text-entry method that attempts to predict the user's intended words from an ambiguous input (sometimes extended to predict word or phrase completions).

Text Entry: Method of inputting text to a mobile device

Unambiguous Keyboards: A keyboard layout where one key-press unambiguously relates to one character (e.g., 2 is mapped to A)

Usability Techniques: Series of methods and tools for designing and evaluating the usefulness and effectiveness of a text-entry system.

User Studies: Evaluations that are conducted to assess the performance of a system with real end users, generally conducted in usability laboratories under controlled settings.

ENDNOTES

[1] The standard 12-key phone key pad uses 8 keys for letters
[2] http://www.zicorp.com/eZiType.htm
[3] http://www.tegic.com/products/xt9.asp
[4] Original SMS (Short Message Service) messages were limited to 160 characters.

Chapter XXVI
Improving Stroke-Based Input of Chinese Characters

Min Lin
University of Maryland, Baltimore County (UMBC), USA

Andrew Sears
University of Maryland, Baltimore County (UMBC), USA

Steven Herbst
Motorola, USA

Yanfang Liu
Motorola China Electronics Ltd., PR China

ABSTRACT

This chapter presents a case study of the redesign of the mobile phone keypad graphics that support the Motorola iTap™ stroke-based Chinese input solution. Six studies were conducted to address problem identification, proof of concept evaluation, usability testing in both US and China, and design simplification to support business objectives. Study results confirmed that a new abstract-with-examples design helped users to develop more accurate knowledge regarding stroke-to-key mappings and lead to significant improvements in both text-entry speed and accuracy. The data also showed that, when using the new keypad graphics, the stroke-based input method could outperform the popular Pinyin technique after about 1 hour of casual usage, making the stroke method a competitive alternative for Chinese entry on mobile phones.

INTRODUCTION

The capabilities of modern mobile phones are far beyond the literal combination of "mobile" and "phone." Rapid improvements in both hardware and software have turned mobile phones into personal multichannel communication centers. Mobile phones now serve many roles including music player, camera, camcorder, voice recorder, game player, calendar, notepad for short notes, and text, e-mail, and IM messaging device. As a result, the global mobile phone market is growing rapidly. According to iSuppli, over 800 million mobile phones were shipped in 2005, a 14% increase compared to 2004. China is the largest mobile phone market in the world. Due to the growing popularity of short message service (SMS), increased support for personal information management (PIM), and Internet browsing capabilities, effective text-entry techniques are becoming more and more important. Researchers have investigated a variety of alternatives for entering English text using the limited number of keys available on mobile phones, confirming that existing solutions can be awkward and slow (James and Reischel, 2001; MacKenzie, Kober, Smith, Johns, & Skepner, 2001; Silfverberg, MacKenzie, & Korhonen, 2000). While entering English text can be challenging, entering Chinese characters using mobile phone keypad is much more difficult. Instead of a relatively modest 26 letters plus numbers and a few symbols, Chinese entry requires the user to learn how to enter thousands of characters, a task for which professionals may use a keyboard with as many as 4,000 keys (Archer, Chan, Huang, & Liu, 1988).

In this chapter, we present the redesign of the keypad graphics (the symbols printed on the keys as the legends for Chinese strokes) for the Motorola iTap™ stroke-based input solution as a case study. This chapter expands significantly on an abbreviated version published earlier (Lin & Sears, 2005b). This chapter includes more detail and covers the entire process, from problem identification to solution development, proof of concept evaluation, and a series of user studies. This collaborative effort involved UMBC faculty and students, as well as Motorola employees in both the US and China.

The underlying problem was defined collaboratively, with input from Motorola and UMBC personnel. Motorola provided financial support, allowing UMBC faculty and students to conduct the initial studies that lead to the design of the new keypad graphics. Once the efficacy of the graphics was confirmed through US-based studies, Motorola provided resources and personnel to replicate the UMBC studies in Beijing, China. Raw data from the Beijing studies was sent to UMBC for analysis. UMBC and Motorola personnel subsequently collaborated to develop presentations describing the results of these studies, which ultimately resulted in the new keypad graphics being formally adopted as the preferred solution for future Motorola mobile phones for the Chinese market. These graphics have already been used in several new phones.

This chapter starts by presenting a brief introduction to Chinese characters and various text techniques for entering Chinese text on mobile platforms. Next, a case study is presented that describes the redesign of the keypad graphics that support Motorola's iTap™ software. The case study involved (1) a comparison of the original iTap™ solution and the popular Pinyin method; (2) a test of an alternative design that was developed based on observations from the initial study; (3) a 6-day, longitudinal hands-on study of the proposed design; (4) a duplication of the longitudinal test that was conducted in China; (5) a study designed to simplify the proposed design to allow it to fit on smaller keypads; and (6) a final evaluation of the simplified design. For each study, we present the experimental design, data analysis, and a discussion of the results. We conclude by summarizing the experience presented in this case study.

CHINESE CHARACTERS

Chinese differs significantly from western languages such as English. Chinese is an ideographic language, with the shape of each character playing a critical role in presenting the meaning of the

Figure 1. Examples of the sound, shape, and meaning for Chinese characters

抱 饱

character. The shape often determines pronunciation as well. For example, the two characters on figure 1 have the same component on the right, differing only on the left. These two characters mean "hold in arms" and "well fed" (left and right characters, respectively), matching the meanings of the differing components on the left side of the characters ("hand" and "food", respectively). The shared right component is pronounced "bao," while both characters are pronounced as "bao" with different tones.

In Chinese, a stroke is the minimum structural unit and a character is the minimum functional unit. While there are thousands of Chinese characters, there are only 36 unique strokes. Most characters are composed of two or more strokes. For example, the two characters in figure 1 are both written using eight strokes. In addition, for any given character, there is a predefined "correct" order in which the strokes should be written. Therefore, if thinking of strokes as letters in English, then writing characters using strokes is conceptually the same as writing words using letters.

MOBILE INPUT SOLUTIONS

Handwriting Recognition

For mobile phones with a touch screen, such as smart phones, handwriting recognition is widely used for input, including tools like DragonPen™ and PenPower for Chinese input. With incremental recognition techniques, these tools may recognize a character without writing all the strokes (Matić, Platt, & Wang, 2002). However, like speech, the technology is unlikely to provide error-free input. More importantly, handwriting requires both hands: one to hold the device and the other to write. As a result, when one hand is occupied, handwriting can be difficult or impossible to use for effective input. In addition to usability, economic concerns may limit the use of handwriting recognition, since touch-sensitive screens can increase the cost of producing mobile phones.

Pinyin Method

Due to the nature of the Chinese language, there is no natural mapping between Chinese characters and the letters printed on the standard QWERTY keyboard or the standard telephone keypad. Therefore, the process of "Romanization" was used to define the pronunciation of Chinese characters using the letters of the Roman alphabet. Although some people considered such a process "peculiar" (Sacher, Tng, & Loudon, 2001), an official standard system, Pinyin, was created based on a northern dialect (i.e., Mandarin) and has been taught in primary schools in China for decades.

With Pinyin, every character can be entered using the letters of the standard mobile phone keypad. However, two major issues hinder performance when using Pinyin. Although the Mandarin pronunciation of each Chinese character is used to define Pinyin, it is not spoken by everyone in China. There are more than 50 dialects in China, and a large number of people either do not speak Mandarin or speak it in a nonstandard way. As a result, these people face significant difficulty in "translating" their pronunciations into Pinyin. For example, about 99% Hong Kong people did not speak Mandarin as their primary language (Census and Statistics Department, 2001). Even if this translation is not an issue, using Pinyin on a mobile phone is not as straightforward as entering English.

While the standard telephone keypad provides 12 keys, just 8 of these keys are used to represent the 26 letters of the Roman alphabet. As a result, each key represents several letters, resulting in ambiguity when a key is pressed. For English text entry, several input techniques have been developed to address this challenge and to speed up the process of entering English text. For example, Multitap allows a user to select one of several letters represented by a single key by pressing the key multiple times (Silfverberg et al., 2000). Less-Tap rearranges the letters on each key to allow for faster entry speeds (Pavlovych & Stuerzlinger, 2004). Predictive solutions, such as T9® by Tegic Communications, the iTap™ solution by Motorola, and eZiText® by Zi Corporation, automatically present a set of words that can be generated based on a sequence of key presses. Using these techniques, users press just one key per letter and an internal dictionary is used to determine which words the user might have been entering (James & Reischel, 2001). Like standard word completion applications, LetterWise guesses the next letter based on letters that have already been entered (MacKenzie et al., 2001). TiltText utilizes the orientation of the phone to resolve ambiguity. Users tilt the phone in one of the four directions to choose a specific letter following each key press (Wigdor & Balakrishman, 2003). Fastap™ from Digit Wireless uses more keys so each letter and number is represented by a unique key (Cockburn & Siresena, 2003). While results appear promising, adding more keys can become problematic as mobile phones become smaller.

By allowing users to enter English text more quickly, each of the techniques described could directly or indirectly speed up the process of entering Chinese text using Pinyin. Some techniques, such as TiltText and Fastap™, could provide direct benefits because they allow users to enter any random sequence of Roman characters more quickly (i.e., they are not based on a dictionary of English words). Other techniques, such as T9®, Less-Tap, and LetterWise are based on dictionaries and would need to be adapted to work with Pinyin scripts (the sequence of Roman letters used to represent the sound of a Chinese character) instead of English words. However, entering Pinyin is more complex than entering a simple sequence of Roman letters. First, the ambiguity introduced by multiple Roman letters sharing a key must be addressed. This is typically accomplished by choosing the desired Pinyin script from a set of possibilities associated with the sequence of keys that were pressed. Most Pinyin implementations facilitate this process using predictive capabilities similar to those provided by T9® and eZiText®. With these implementations, as users enter Roman letters, possible Pinyin scripts are presented, and the desired script can often be selected before the entire script has been entered. However, a single Pinyin script can still produce multiple Chinese characters. As a result, once the Pinyin script has been entered, the user must choose the correct character from a list of alternatives. The ambiguity introduced by multiple Roman letters sharing keys, combined with the fact that each Pinyin script may correspond to multiple Chinese characters, significantly increases the number of keystrokes required, and also forces users to shift their attention between the alternative lists and keys. Given the number of steps involved, mistakes can be hard to find and difficult to correct.

Structure-Based Methods

Unlike the many-to-many relationship between characters and pronunciations, the relationship between characters and structures is one-to-one. More importantly, the structure of every character is independent from its pronunciation, so structure-based solutions are not affected by the dialect that an individual speaks. Therefore, solutions based on the structure or shape of the character, instead of sound produced when it is spoken, can be used by people who do not speak Mandarin.

The primary structure-based method for normal QWERTY keyboard is Wubi (means "five keystrokes," literally). Wubi uses some arbitrary rules to decompose characters into a set of substructures that are mapped onto the 26 Roman letters. Although expert users of Wubi can type much faster than those using Pinyin method, Wubi is very difficult to learn, as users have to memorize numerous, arbitrary, decomposition rules. Wubi is not practical for mobile phone users since they are not likely to spend multiple hours to learn an input method.

MOTOROLA ITAP™ STROKE INPUT METHOD

The Motorola iTap™ stroke input method ("stroke method" thereafter) enables users to enter Chinese characters stroke by stroke. For historical reasons, many Chinese characters can be written in two ways, using either simplified Chinese (official version of Chinese in the People's Republic of China) or traditional Chinese (widely used elsewhere). While some characters remain the same in both versions, most differ significantly in appearance. The major difference is that simplified Chinese uses fewer strokes than traditional Chinese when writing the corresponding character (Figure 2), but both character sets are written using the same set of strokes. Therefore, any stroke-based input method can be used to write both simplified and traditional Chinese characters. While Pinyin can also be used to enter both sets of characters, people who use traditional Chinese typically use a different pronunciation annotation system, called Bopomofo, rather than Pinyin. Finally, the process of using stroke-based input may sound intuitive but a challenge exists, since the number of strokes is larger than the number of keys on a standard mobile phone keypad. As a result, multiple strokes must be assigned to each key.

In the stroke method, strokes are grouped and then assigned to the number keys, with a legend placed next to the number (Figure 3). Using predictive technology, a list of possible characters is presented and updated after each stroke input. Two keys (left and right arrow keys) are used to navigate through the list for character selection.

Figure 2. Three examples of corresponding characters in simplified Chinese and traditional Chinese

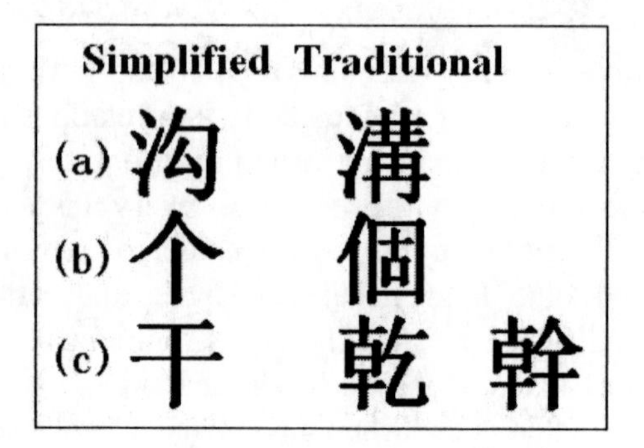

Figure 3. Motorola keypad design for iTap™ stroke input method circa 2000

PRELIMINARY COMPARISON

In theory, structure-based methods should provide advantages over Pinyin, but no studies have been reported comparing data entry rates for the two solutions when used on a mobile phone. Therefore, our first study compared the stroke-based solution to Pinyin to provide a foundation for this project. While both Pinyin and stroke-based solutions can support the entry of both traditional and simplified Chinese characters, our study focuses on the use of simplified Chinese.

Participants

Thirty native Chinese speakers living in the Maryland area volunteered to participate in the study. Participants were randomly assigned to two groups. The first group completed the task

Table 1. The gender and age distribution of the two groups of participants

Input Method	Gender		Age Range (mean)
	Male	Female	
Pinyin	9	6	24-35 (27)
Stroke	5	10	24-31 (27)

using Pinyin, while the second used the stroke method (Table 1).

Task

The experimental task involved entering an e-mail message, comprised of 59 Chinese characters and 8 punctuation marks, using a mobile phone that supported both Pinyin and stroke method. Before starting the task, participants were given a brief verbal explanation of the keys used for navigation and character selection. The explanation did not address how characters or strokes were entered. Participants were given 10 minutes to practice using the phone. The e-mail message was presented on paper and was available throughout the task.

Results

The data revealed that participants using the Pinyin method were able to complete the task significantly faster than those using the stroke method [$t(28) = -4.1$, $p<0.001$]. The character entry speeds for Pinyin and the stroke method were 5.5cpm (characters per minute) and 2.7cpm, respectively. In addition, the error rate for the Pinyin group was significantly lower than that of the stroke method group [$t(28) = -3.0$, $p<0.01$]. The error rate of the Pinyin group was 1.5% while the error rate for the stroke group was 7.6%. The slow and error prone results of the participants using the stroke method raised significant concerns and motivated an in-depth investigation, with the goal of determining the factors that lead to such poor performance.

Since Pinyin is taught in primary schools in China and has become the most popular Chinese input method on PC platform, the significant difference between Pinyin and the stroke-based method could arguably be attributed to the participants' familiarity with Pinyin. Although Pinyin on mobile phones differs somewhat from the standard PC implementation, the general rules are the same. On the other hand, the stroke-based method was completely new to our participants. We suspect that the 10 minutes of practice that was allowed was not sufficient to allow these individuals to develop an effective mental model of this new solution. While additional practice may allow for improved performance, we believe that design changes could produce more dramatic changes while reducing the time required to become proficient with the stroke-based solution.

PROBLEM IDENTIFICATION

Video recordings of participants interacting with the stroke method during the preliminary comparison were reviewed in detail. This analysis confirmed that the participants pressed the wrong key for more than one third of all strokes they entered. Clearly, the keypad graphics did not provide our participants with the information they needed to map individual strokes to keys, resulting in numerous errors as participants completed the experimental task.

Three types of mapping errors were identified:

- **Correct stroke/incorrect key:** These errors occurred when participants decomposed the character into valid strokes, but selected the wrong key when entering the stroke.
- **Stroke subdivision:** These errors occurred when participants divided a single stroke into more than one piece and tried to enter each piece separately.
- **Stroke combination:** These errors occurred when participants combined multiple strokes and attempted to enter the combination of strokes with a single key press.

These observations indicated that although the character decomposition process that forms the foundation of the stroke method was familiar to participants, the specific symbols printed on the keypad caused confusion and errors. With the current symbols, participants were not able to learn the stroke-to-key mappings. Participants had more difficulty entering strokes associated with the 1, 3, 7, and 9 keys due to the diversity of strokes associated with these keys.

REDESIGNING THE GRAPHICS

The goal for this redesign was to help users better understand the stroke-to-key mappings. All of the strokes entered using any single key share common characteristics, but these common characteristics were not effectively highlighted by the original graphics. Therefore, our first goal was to more effectively convey this information to the user. Abstract symbols were designed for the 1, 3, 7, and 9 keys that highlighted the characteristics that were used to group the strokes. These abstract symbols were designed such that they would not be confused with any specific, real strokes. In addition to displaying the abstract symbols, several example strokes were also presented to highlight the diversity of strokes that are represented by individual keys. Three new designs were generated: abstract, abstract-with-examples, and original-with-examples (Figure 4).

INITIAL EVALUATION OF NEW KEYPAD GRAPHICS

A study was designed to evaluate the effectiveness of the three new designs illustrated in Figure 4 as well as the original design. The goal of this study was to assess how well the various keypad graphics matched users' mental models when they first encountered the keypad. This was assessed by measuring how accurately users could map strokes to specific keys using the various graphics.

Participants

A power analysis based on pilot data suggested that 32 participants would be adequate for a between-group study. Thirty-two native Chinese speakers living in the Maryland area volunteered to participate in this study. Participants were familiar with simplified Chinese and had no difficulties listening to Mandarin. Participants were randomly assigned to use one of the four designs. The gender and age information for each group were listed in Table 2.

Tasks

The experimental task was to enter all of the strokes necessary to input 40 Chinese characters. The characters were carefully selected to ensure that participants would enter all possible strokes at least one time if they completed the task correctly. The 40 characters were presented using audio recordings. To help ensure that participants could determine exactly which character was to be entered, the target character was placed in the context of a two-character phrase. The target character was always the first of the two characters in the phrase. The character set was selected to ensure that all possible strokes would be entered, while keeping the number of strokes required for any given character reasonably low. On average, 4.2 strokes were required to enter each character.

Figure 4. Three new keypad symbol designs: (a) Abstract (b) Abstract-with-examples (c) Original-with-examples

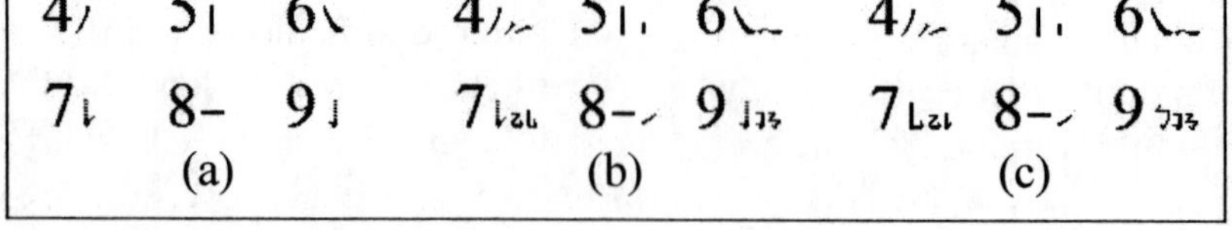

Table 2. The gender and age distribution of the four groups of participants

Design	Gender		Age Range (mean)
	Male	Female	
Original	4	4	23-35 (27)
Abstract	3	5	22-40 (28)
Original-with-examples	3	5	24-45 (30)
Abstract-with-examples	6	2	22-36 (28)

Figure 5. Java application for the blind test (abstract-with-examples design was shown)

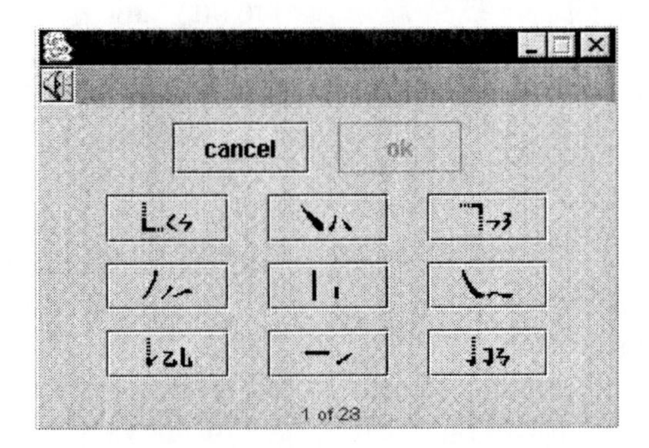

Since the primary goal was to assess the participants' initial mental models regarding the relationship between individual strokes and keys, we designed a Java application (Figure 5) that provides visual feedback regarding the key selected, but no information regarding whether the input would produce the desired character. This ensured that our participants did not learn from their interactions with the application, and that our results reflect the participants' initial mental models regarding the stroke-to-key mappings.

Results

Both character- and stroke-level accuracy were measured. Stroke-level accuracy assessed whether each key corresponded to the desired stroke. Character-level accuracy assessed whether or not the individual could have entered the character correctly using the character prediction technique provided by the iTap™ technique. Using these character prediction capabilities, users typically enter a subset of the strokes required to write the character. After each key is selected, a set of possible strokes is presented, based on the complete sequence of keys that had been entered. Character-level accuracy was assessed by assuming that the participant would scan the list of possible characters after each key is pressed, selecting the desired character as soon as it appeared.

Keypad design had a significant effect on the character-level accuracy [$F(3,28)=21.42, p<0.001$]. Post hoc tests revealed no statistical difference between the original design and the abstract design. The original-with-examples design resulted in significantly higher accuracy than the original design [$t(14)=3.35, p<0.01$]. Similarly, the abstract-with-examples design resulted in significantly higher accuracy than the abstract design [$t(14)=7.07, p<0.001$]. Finally, the abstract-with-examples design resulted in significantly higher accuracy than the original-with-examples design

Figure 6. The character-level error rate (using character prediction)

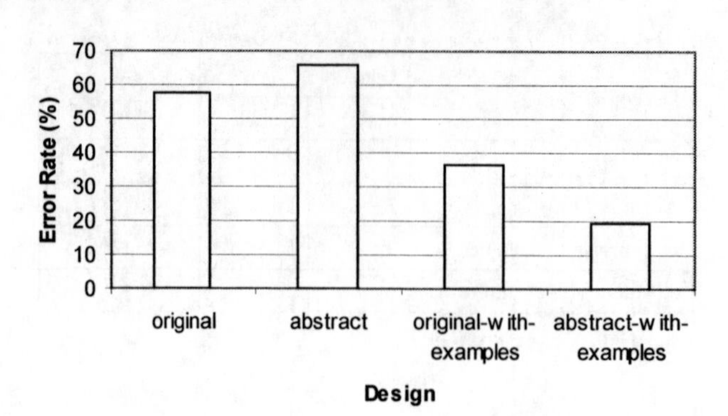

[t(14)=2.61, p<0.05]. The character-level accuracy results confirm that even without feedback regarding the stroke-to-key mappings, the abstract-with-examples solution allowed users to correctly enter almost twice as many characters as compared to the original keypad design (Figure 6).

The stroke-level accuracy was considered more important, since it directly reflects the stroke-to-key mappings that participants develop, given the graphics presented on the keypads. Figure 7 shows overall error rates for the four keypad designs, as well as the frequency of the three different types of errors that occurred. Overall, the pattern was the same as was observed for character-level accuracy: no significant difference was found between the original and abstract designs; adding examples significantly reduced error rates [original-with-examples vs. original: t(14)=-4.39, p<0.002; abstract-with-examples vs. abstract: t(14)=-6.24, p<0.001]; the abstract-with-examples design resulted in fewer errors than the original-with-examples design [t(14)=-2.70, p<0.02].

When looking at the three different types of mapping errors, we found that adding examples reduced the number of errors where a stroke was subdivided and entered using multiple key presses. We also found that both the abstract and the abstract-with-examples designs resulted in significantly fewer errors where participants tried to enter multiple strokes by combining them and using a single key press. Finally, the abstract-with-examples design resulted in fewer errors than any other design that involved situations where participants knew the stroke but selected the wrong key. These results indicated that both the abstract symbols and the example strokes were useful in

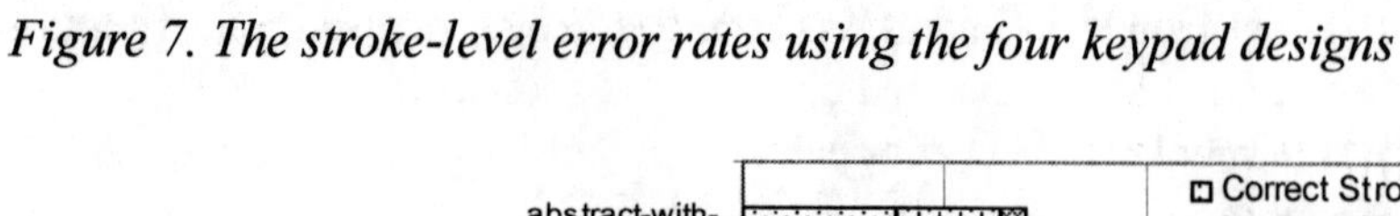

Figure 7. The stroke-level error rates using the four keypad designs

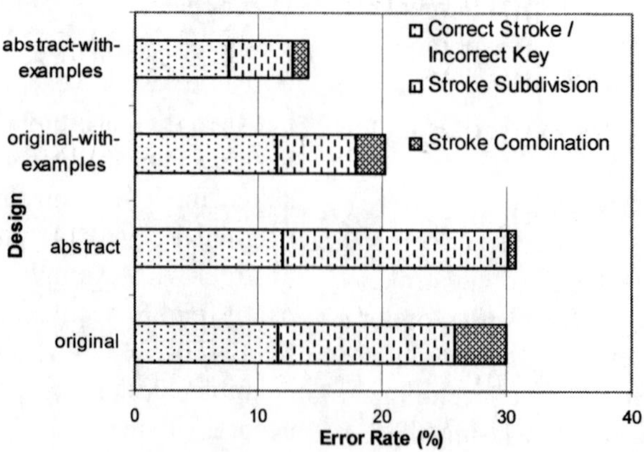

reducing mapping errors. Combining the abstract symbols with the example strokes provided additional benefits, reducing the frequency of all three types of errors. A more detailed treatment of this study can be found in Lin and Sears (2007).

HANDS-ON TEST

Our initial evaluation confirmed that combining abstract symbols with examples improved the participants' understanding of stroke-to-key mappings. However, the design of this initial study did not provide feedback regarding the efficacy of the proposed solution when individuals are using a mobile phone to enter text. Therefore, a hands-on longitudinal study was conducted to examine both performance with, and the learnability of, the proposed solution.

Participants

All participants were born and raised in China and had been living in the United States for no more than 4 years. A total of 26 people from the Baltimore/Washington area participated. The first two participants took part in a pilot study that allowed us to determine how many days the actual study would last. The other 24 participants took part in the actual study.

The participants were comfortable understanding Madarin and writing simplified Chinese, but none had experience using iTap™ solutions. Half of the participants had a minimum of a Bachelor's degree while the other half had earned a Master's degree. Participants were randomly assigned into two task groups to use either the original keypad design or the abstract-with-examples design (see table 3).

Tasks

Motorola produced the new abstract-with-examples keypad for this experiment with the same finish as the already-released keypad so that the mobile phone using the new keypad would have the same look and feel as a mature product rather than a prototype. Participants used one of two phones: the standard phone with the original keypad graphics or the new phone with the abstract and example graphics. Our pilot study indicated that 6 days would provide sufficient time for our participants to learn how to use the phones to enter text effectively and for performance to level off.

The primary task for each participant was to use the assigned mobile phone to enter five 17-character sentences. The complete set of 30 sentences was generated based on the headlines of a popular Chinese news Web site, with the topics spread over five categories: international affairs, economy, education, technology, and sports. As a result, each participant entered a total of 510 characters using 361 unique characters during the 6-day study. As in the earlier study, the sentences were presented using audio recordings. After hearing the audio, participants wrote the sentence on paper before entering it using the mobile phone. Participants were instructed to balance the input speed and accuracy as they normally would.

Each day, after entering the five sentences, participants completed a calculation task that consisted of adding 20 sets of three 2-digit numbers. This task was included with the goal of distracting the participants so that they were prevented from rehearsing the stroke-to-key mappings. Since rehearsal is required to retain information in short-term memory, this process ensured that responses during the subsequent

Table 3. The gender and age distribution of the two groups of participants

Keypad	Gender		Age Range (mean)
	Male	Female	
Original	6	6	22-35 (28)
Abstract-with-examples	6	6	23-29 (26)

character entry task were based entirely on the participants' long-term memories. Following the calculation task, participants entered 28 characters using the application from our first study. The characters were selected to include all the strokes at least twice, with the exception of 2 strokes that only exist in one character. The average number of strokes per character was 5.2. The order in which participants entered the characters was randomized each day. This task was included to assess how the participants' mental models of the stroke-to-key mappings evolved as they gained more experience using the phone. At the conclusion of the 6-day study, each participant completed a simple key-pressing task, providing a measure of how fast each individual normally pressed the keys on the mobile phone. This measure of keystroke speed was integrated into our subsequent analyses to address any individual differences with regard to how fast participants normally pressed the buttons.

Results

ANCOVA analysis, using keystroke speed as a covariate, revealed significant effects of keypad design [$F(1, 21) = 5.2$, $p<0.04$] and experiment trial [$F(5, 105) = 10.2$, $p<0.001$] (Figure 8). Both groups entered text more quickly as they gained experience, but participants using the abstract-with-examples design were consistently faster than those using the original design.

During our preliminary evaluation, individuals entered text using Pinyin at a rate of 5.5cpm. Since this is the most frequently used method for entering Chinese text when using a full keyboard (Yuan, 1997), and all participants in the preliminary evaluation were experienced using Pinyin on PCs, we use this as a baseline for comparisons. Our data indicate that with practice, the stroke method did allow for faster data entry than Pinyin. When using the original keypad design, data entry rates surpassed Pinyin on the fourth trial, with a speed of 5.7cpm. Data entry rates for the new abstract-with-examples design exceeded those achieved with Pinyin during the third trial, with an average speed of 6.0cpm. While this appears promising, it is more important to know how much time users must invest before they achieve this level of performance. Figure 9 illustrates the cumulative time spent interacting with the two mobile phones across the six trials. With the original design, individuals invested a total of 95 minutes interacting with the system before they were able to enter text more quickly than the baseline rate for Pinyin. In contrast, only 62 minutes were required to achieve the same goal with the new abstract-with-examples keypad.

An ANOVA analysis with repeated measures showed significant effects of both trial [F(5, 110)

Figure 8. Entry speeds of the two keypad designs

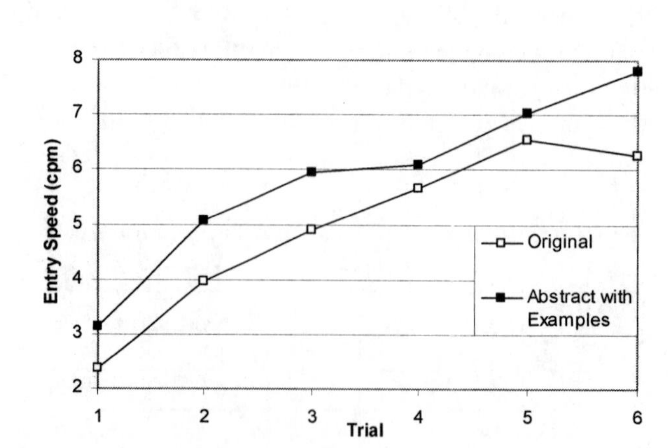

Figure 9. The accumulative time spent for text-entry tasks when using the two designs

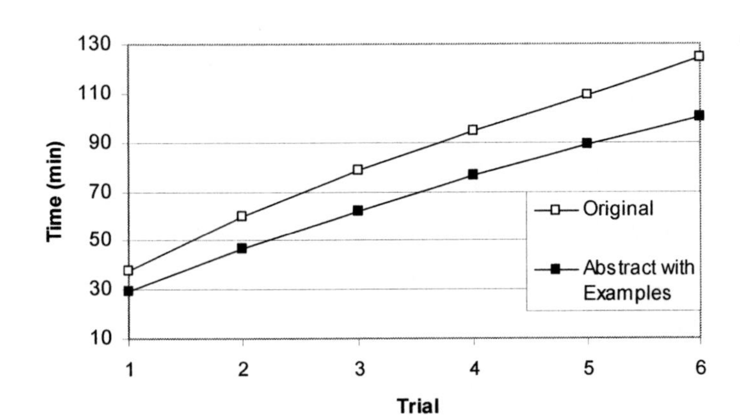

Figure 10. Character-level input error rates when using the two designs

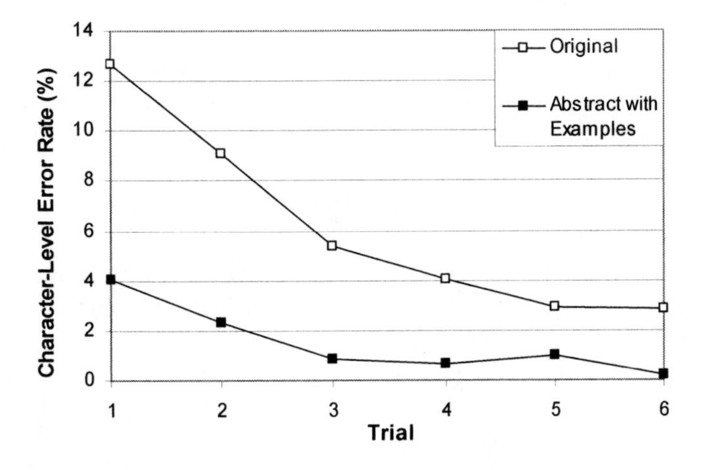

= 16.3, p<0.001] and keypad design [F(1, 22) = 6.7, p<0.02] on the character-level error rates (Figure 10). A significant interaction between trial and keypad design was also detected [F(5, 110) = 3.8, p<0.004], indicating that performance when using the abstract-with-examples design was more stable than performance with the original design. With practice, error rates decreased from almost 13% to approximately 3% for the original design. Interestingly, error rates for the abstract-with-examples keypad started off at just 4% and decreased to 0.2% with practice. Throughout the 6-day period, error rates for the abstract-with-examples design were always lower. In fact, even with 6 days of practice, the error rate for the original design never dropped below the day two error rate for the abstract-with-examples design.

During our preliminary evaluation, the error rate for the stroke-based solution was 7.6%, which was lower than the day-one error rate for the current study. We believe that several factors may have contributed to the increased error rate in the current study. First, unlike the current study, participants in the preliminary study were given 10 minutes to practice using the phone. More importantly, a larger character set was used in the current study as compared to the preliminary study. Even if an error rate of 7.6% were used for comparison, individuals using the new abstract-with-examples design performed much better during their first

Figure 11. Learning curves of the stroke-to-key mappings by the two groups

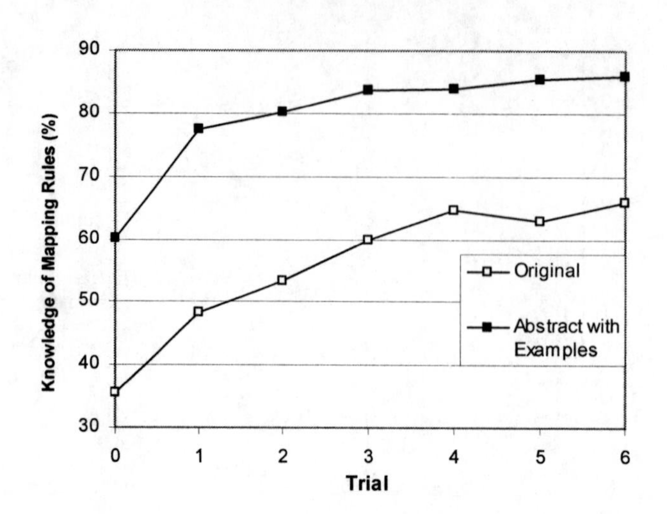

interactions with the phone (i.e., 4% error rate). When compared to Pinyin, where we observed a 1.5% error rate, users of the new abstract-with-examples design were able to reach a comparable error rate during day three, while error rates when using the original design were still 60% higher than the Pinyin baseline even during day six. These findings strongly supported the use of the new abstract-with-examples design.

While the data entry and error rates were very encouraging, we were also interested in understanding the users' mental models regarding stroke-to-key mappings and how these models evolved with practice. Using the data gathered at the end of each day, we are able to determine how many strokes each individual could map to the correct key. Figure 11 highlights these results, showing how the participants' mental models became more accurate as they interacted with the mobile phones. When using the original design, participants mapped approximately 35% of the strokes to the correct keys before using the phones, almost 50% after using the phones just once, and stabilized around 65% after day four. In contrast, when using the new design, participants mapped 60% of the strokes correctly before using the phone, nearly 80% after using the phones once, and stabilized around 85% after day three.

Significant effects of trial [$F(5, 110) = 18.6$, $p<0.001$] and keypad design [$F(1, 22) = 30.0$, $p<0.001$] on the understanding of stroke-to-key mappings were observed. The interaction between trial and keypad design was also significant [$F(5, 110) = 2.6$, $p<0.03$], consistent with the character-level error analysis (Figure 10). Regressions between the knowledge of mapping rules and character-level error rates confirmed strong linear correlations. The regression coefficients for the original design and the abstract-with-examples design were 0.97 ($p<0.001$) and 0.94 ($p<0.002$), respectively. Therefore, reductions in character-level error rates can be associated with improved knowledge of the stroke-to-key mappings. A more detailed description of this study can be found in Lin and Sears (2005a).

CONFIRMATION TEST

The longitudinal study described confirmed that the new abstract-with-examples design allowed for more rapid learning, faster data entry, and lower error rates. While the participants were all from China, they were currently living in the United States. Therefore, an additional study was conducted in China to ensure that the results were not inappropriately biased by the participant's current location or the culture in which they were currently living. The instructions and materials from the previous study were used to conduct this

Table 4. The gender and age distribution of the two groups of participants

Keypad	Gender		Age Range (mean)
	Male	Female	
Original	6	6	22-36 (27)
Abstract-with-examples	6	6	23-36 (27)

follow-up study to ensure that the results could be compared.

Participants

Twenty-four individuals in Beijing were recruited to participate in the study (table 4). As with the US-based study, half of the participants had a Bachelor's degree and the other half had earned a Master's degree. Participants were randomly assigned to two groups, one using the original keypad and the other using the new abstract-with-examples design.

Results

The patterns observed in the US-based study were also apparent in the results from the follow-up study in China. Figure 12 shows the data-entry speeds achieved over the 6-day study.

The character-level error rates also resulted in a similar pattern (Figure 13). The results for the abstract-with-examples design were almost identical to those from the US-based study. Error rates for the original design were somewhat lower than in the US-based study, but the pattern was similar. As in the US-based study, error rates for the original keypad were always higher than those for the new abstract-with-examples design.

SIMPLIFICATION TEST

The level of consistency between the US- and China-based studies was encouraging, confirming the robustness of our results. While these results reaffirmed the belief that the abstract-with-examples design should replace the original design, one practical challenge still had to be addressed: mobile phones in the Asian market tend to be small and continue to get smaller. As a result, the keys on these phones are quite small, and may make it difficult to fit both the abstract symbol and two examples on each key. To ensure that the end result addressed both usability and business goals, we studied the possibility of simplifying the abstract-with-examples design. More specifically, we focused on the possibility of using just a single example stroke.

Figure 12. Data entry speeds using the two keypad designs when repeated in China

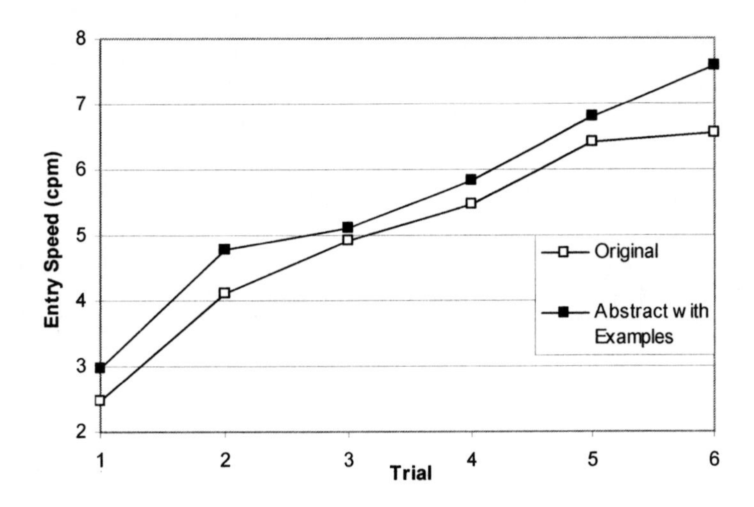

Figure 13. Character-level input error rates when repeated in China

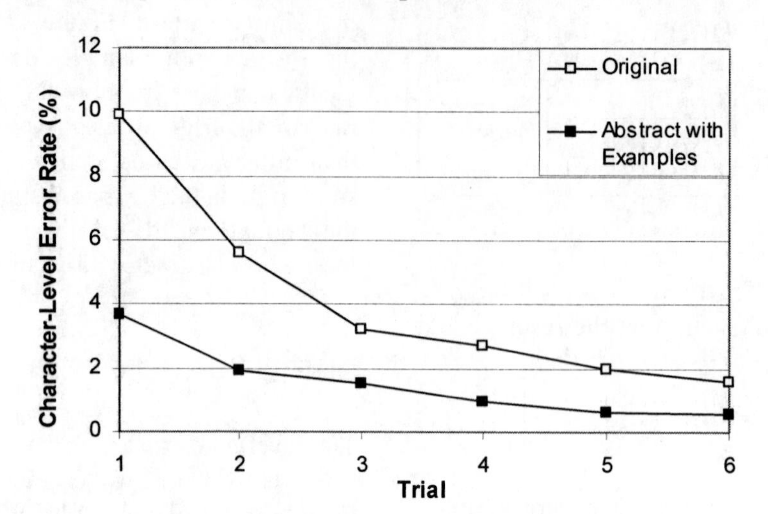

As described earlier, the primary goal for the abstract-with-examples keypad design was to enhance the users' understanding of the stroke-to-key mappings for four specific keys. Our simplification efforts also focused on these four keys (i.e., 1, 3, 7, and 9). Each key originally had two example strokes. In the current study, two alternative designs were considered for each of these keys. The alternatives paired the abstract symbol with one of the example strokes used in the initial abstract-with-examples design with the goal of determining which example stroke produced better results.

Table 5. The gender and age distribution of the eight groups of participants

Group	Gender		Age Range (mean)
	Male	Female	
1 (key 1 design #1)	6	14	22-35 (27)
2 (key 1 design #2)	15	5	24-37 (29)
3 (key 3 design #1)	11	9	24-34 (28)
4 (key 3 design #2)	9	11	18-33 (27)
5 (key 7 design #1)	8	12	21-34 (27)
6 (key 7 design #2)	7	13	22-34 (28)
7 (key 9 design #1)	7	13	22-37 (27)
8 (key 9 design #2)	10	10	25-41 (31)

Participants

One hundred and sixty participants were recruited in Beijing (Table 5). Participants were randomly assigned to one of the eight groups, with each group evaluating one design for a single key.

Tasks

A Web-based study was used, with each participant being assigned a unique user id and password to limit access to the experimental materials. Each user id and password worked one time, ensuring that participants did not complete the study more than once. Thirty six characters were presented to the participants one at a time. A single stroke was highlighted in each character. The set of 36 characters ensured that every possible stroke was presented one time. For each character, the participant was also presented a single key design that was under consideration. The specific key design presented was determined based on the group the participant had been assigned to. The participant had to judge whether the design should be used to enter the highlighted stroke.

Results

Since each key was presented in isolation, participants had to make decisions without the insights

Table 6. Error rates of the four pairs of one-example design alternatives

Design	Key 1		Key 3		Key 7		Key 9	
	#1	#2	#1	#2	#1	#2	#1	#2
Error rate (%)	68	76	74	68	82	78	82	76

Figure 14. Abstract-with-one-example design

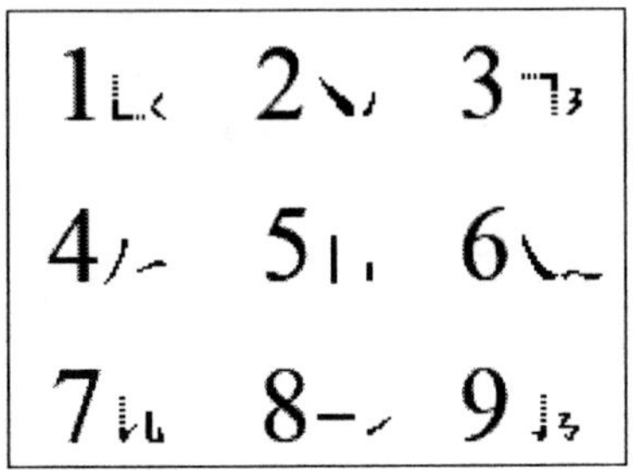

provided by the eight additional keys that would be available when interacting with a full keypad. The current task is further complicated by the fact that during character entry tasks, individuals judge which one of the nine keys is most appropriate, but in the current task they must determine if a specific stroke matches a specific key. Given the more complex nature of the current task, we expected stroke-level error rates to be substantially higher than would be experienced during the use of a real phone keypad. The mean stroke-level error rates are reported in table 6. Using paired t-tests, we did not identify any significant differences between the pairs of alternative designs for any of the four keys in question. While the differences were not statistically significant, we chose to use the design with the lower error rate for use in the abstract-with-one-example design (Figure 14).

FINAL EFFECTIVENESS TEST

As discussed earlier, new mobile phones developed by Motorola for the Chinese market continue to get smaller, severely limiting the amount of space available on individual keys. To balance usability goals and the reality of having very limited space available on each key for graphics, our last study focused on simplifying the abstract-with-examples solution to reduce the space required for the keypad graphics. The final evaluation focused on assessing the impact of the abstract-with-one-example design on the users' ability to determine correct stroke-to-key mappings. For this study, three designs were evaluated: the original keypad, the new abstract-with-examples design, and the abstract-with-one-example design.

Participants

Sixty participants were recruited in Beijing and randomly assigned to one of three groups (table 7). Each group interacted with a single keypad design.

Tasks

The 36 characters used in the previous test were used once again in this final evaluation that used the similar Web-based approach as the last study. When a character was presented the full keypad was displayed, and the participant had to indicate which key would be used to enter the highlighted stroke.

Table 7. Gender and age distribution of the three groups

Keypad	Gender		Age Range (mean)
	Male	Female	
Original	8	12	21-40 (30)
Abstract-with-examples	7	13	23-38 (29)
Abstract-with-one-example	8	12	26-34 (30)

Figure 15. The overall error rates for the three keypad designs

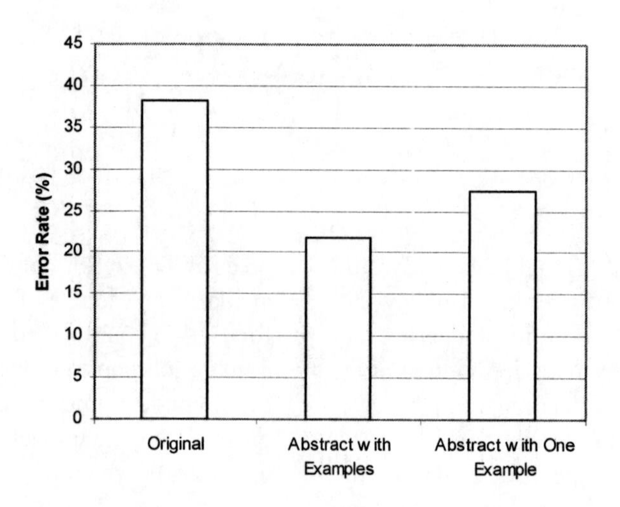

Figure 16. Error rates for the four critical keys

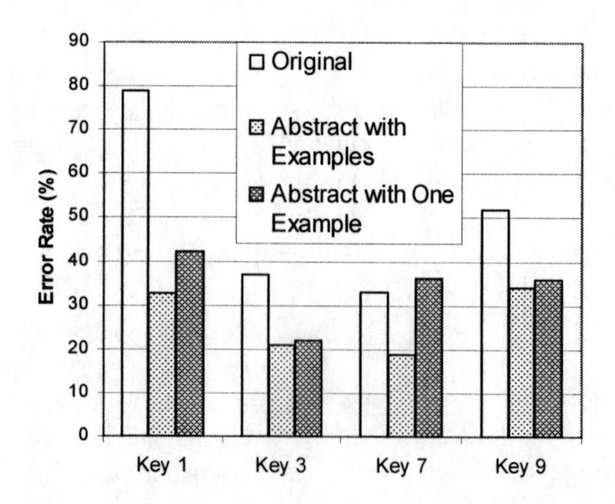

Figure 17. Keypads of the Motorola A732 and E2 phones (showing the abstract-with-one-example design). Both images © 2007, Motorola, Inc. Reproduced with Permission from Motorola, Inc.

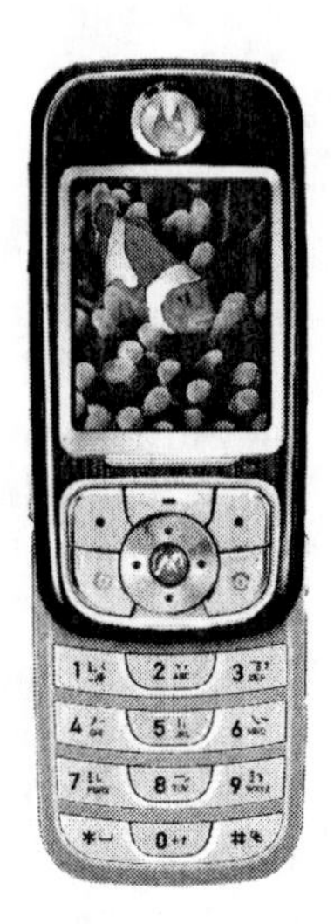

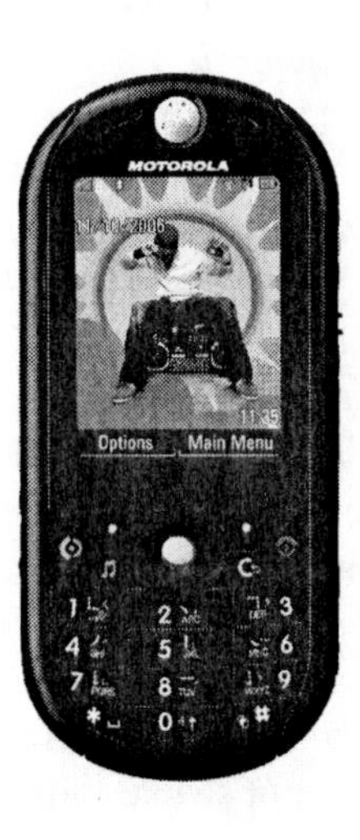

Figure 18. Keypads of the Motorola K1 and V3xx phones (showing the abstract-with-one-example design). Both images © 2007, Motorola, Inc. Reproduced with Permission from Motorola, Inc.

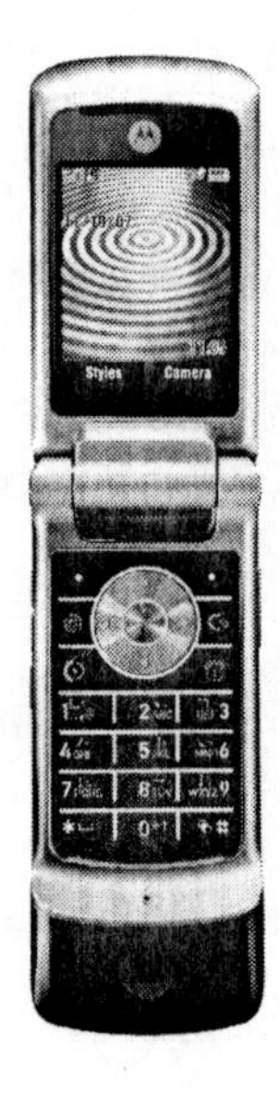

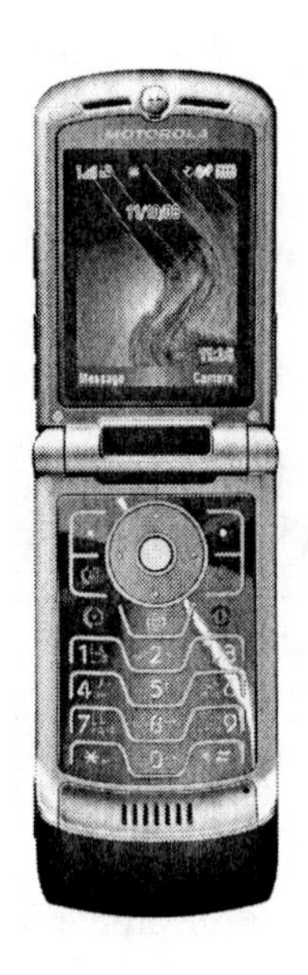

Results

One outlier was identified in each group using a QQ plot. Outliers were excluded from the subsequent analyses, which revealed a significant effect of group on stroke-to-key mapping error rates (Figure 15) [$F(2, 54) = 13.6$, $p<0.001$]. Group also had a significant effect on the error rates for the 1, 3, and 9 keys [$F(2, 54) = 32.6$, $p<0.001$; $F(2, 54) = 4.2$, $p<0.021$; $F(2, 54) = 3.6$, $p<0.034$].

Figure 16 shows the error rates of the four critical keys that were changed when the simplified design was developed. For 1, 3, and 9 keys, the error rate for the original design was significantly higher than for the abstract-with-examples design or the abstract-with-one-example design. At the same time, no significant differences were detected between the abstract-with-examples design and the abstract-with-one-example design for these three keys. These findings indicate that the abstract-with-one-example design provides benefits that are comparable to the abstract-with-examples design even after one example stroke has been removed. In contrast, removing one example from the abstract-with-examples solution for the 7 key resulted in the error rate increasing back to the level observed for the original design. We believe that this is due, in large part, to the fact that this key represents the largest number of strokes, and that the diversity of these strokes is difficult to capture with just a single example. Figures 17 and 18 provide several examples of Motorola phones that have adopted the new keypad graphics and are now available in various markets.

CONCLUSION

In this case study, we presented the step-by-step development of a new keypad design for use with Motorola's iTap™ stroke input technique. Most current research on Chinese character input focuses on key sequence optimization, including various methods of character prediction. The current study is unique in that it focused on the process of entering individual strokes more efficiently by helping users more effectively map strokes to specific inputs. The success of this project is due to the close collaboration between UMBC and Motorola personnel in both the US and China. Motorola helped motivate the problem and UMBC completed many of the initial studies with guidance and input from Motorola personnel. Motorola then lead the effort to replicate the UMBC studies in China, with UMBC providing guidance and developing new studies to address new issues such as simplifying the keypad designs. UMBC then worked with Motorola to develop the necessary materials that would allow for an informed decision regarding the adoption of this new design. Motorola personnel completed the final, internal steps to address certain questions that were raised during the process of formally adopting the new keypad graphics.

The resulting solutions are cost-effective, since the only change is to the graphics that are printed on the keypads, but they are also effective in that they allow for rapid learning, more efficient text entry, and reduced errors. More importantly, the new graphics make the stroke-based method a competitive alternative to the popular pronunciation-based input method (i.e., Pinyin), especially for individuals who do not speak Mandarin or speak Mandarin in a nonstandard way. Our studies confirmed that approximately 1 hour of casual practice with the new keypad is sufficient to allow for text-entry and error rates that are comparable or superior to those that have been observed with Pinyin. Mobile phones that use these new designs began shipping in the Chinese market in 2006.

ACKNOWLEDGMENT

We want to thank Mr. Jun Chen from Motorola China. Mr. Chen helped conduct the usability testing that examined the effectiveness of simplified design. This research was sponsored by Motorola.

REFERENCES

Archer, N. P., Chan, M. W. L., Huang, S. J., & Liu, R. T. (1988). A Chinese-English microcomputer system. *Communications of the ACM, 31*(8), 977-982.

Census and Statistics Department, Hong Kong Special Administration Region, PRC (2001). *Population of five-year-old and above, categorized by primary speaking language in year 1991, 1996, and 2001.* Retrieved April 15, 2004, from http://www.info.gov.hk/censtatd/chinese/hkstat/fas/01c/cd0062001c_text.htm

Cockburn, A., & Siresena, A. (2003). Evaluating mobile text entry with the Fastap keypad. In *Proceedings of British Computer Society Conference on Human Computer Interaction* (pp. 77-80). London: Springer-Verlag.

James, C. L., & Reischel, K. M. (2001). Text input for mobile devices: Comparing model prediction to actual performance. In *Proceedings of CHI 2001* (pp. 365-371). New York: ACM Press.

Lin, M., & Sears, A. (2005a). Chinese character entry on mobile phones: A longitudinal investigation. *Interacting with Computers*, *17*, 121-146.

Lin, M., & Sears, A. (2005b). Graphics matter: A case study of mobile phone keypad design for Chinese input. In *Proceedings of CHI 2005* (pp. 1593-1596). New York: ACM Press.

Lin, M., & Sears, A. (2007). Constructing Chinese characters: Keypad design for mobile phones. *Behavior and Information Technology.*

MacKenzie, I. S., Kober, H., Smith, D., Johns, T., & Skepner, E. (2001). LetterWise: Prefix-based disambiguation for mobile text input. In *Proceedings of UIST 2001* (pp. 111-120). New York: ACM Press.

Matić, N. P., Platt, J. C., & Wang, T. (2002). QuickStroke: An incremental on-line Chinese handwriting recognition system. In *Proceedings of 16th International Conference on Pattern Recognition*, vol. 3 (pp. 435-439). New York: IEEE Computer Society Press.

Pavlovych, A., & Stuerzlinger, W. (2004). Model for non-expert text entry speed on 12-button phone keypads. In *Proceedings of the SIGCHI Conference on Human Factors in Computing Systems* (pp. 351-358). New York: ACM Press.

Sacher, H., Tng, T.-H., & Loudon, G. (2001). Beyond translation: Approaches to interactive products for Chinese consumers. *International Journal of Human-Computer Interaction*, *13*(1), 41-51.

Silfverberg, M., MacKenzie, I. S., & Korhonen, P. (2000). Predicting text entry speed on mobile phones. In *Proceedings of CHI 2000* (pp. 9-16). New York: ACM Press.

Wigdor, D., & Balakrishnan, R. (2003). TiltText: Using tilt for text input to mobile phones. In *Proceedings of the 16th Annual ACM Symposium on User Interface Software and Technology* (pp. 81-90). New York: ACM Press.

Yuan, C. (1997). *Chinese language processing.* Shanghai, China: Shanghai Education Publishing Company.

KEY TERMS

Chinese Character: A Chinese character is the minimum functional unit of Chinese language.

iTap™ Software: iTap™ software was developed by Lexicus and Motorola, and enables predictive text entry in mobile phones. iTap™ software supports both Pinyin and stroke-based entry.

Mandarin: Mandarin is a northern Chinese dialect that is the basis for the official pronunciation of each Chinese character for Pinyin.

Mental Model: A mental model is the users' internal representation of how a system works.

Pinyin: Pinyin is the official Romanization system for Mandarin. It uses the 26 letters of the Roman alphabet to define the pronunciation of Chinese characters so that they can be entered using the standard western keyboard.

Stroke: Stroke is the minimum writing unit of Chinese language. Each Chinese character is constructed by writing one or more strokes in a specific order while following specific spatial relations.

Text Entry: Text entry refers to the process of creating messages composed of characters, numbers, and symbols using mobile devices. Text entry can be performed using small physical keys, virtual keyboards presented on touch sensitive screens, gesture or handwriting recognition, speech recognition, and various other technologies.

Chapter XXVII
Voice-Enabled User Interfaces for Mobile Devices

Louise E. Moser
University of California, Santa Barbara, USA

P. M. Melliar-Smith
University of California, Santa Barbara, USA

ABSTRACT

The use of a voice interface, along with textual, graphical, video, tactile, and audio interfaces, can improve the experience of the user of a mobile device. Many applications can benefit from voice input and output on a mobile device, including applications that provide travel directions, weather information, restaurant and hotel reservations, appointments and reminders, voice mail, and e-mail. We have developed a prototype system for a mobile device that supports client-side, voice-enabled applications. In fact, the prototype supports multimodal interactions but, here, we focus on voice interaction. The prototype includes six voice-enabled applications and a program manager that manages the applications. In this chapter we describe the prototype, including design issues that we faced, and evaluation methods that we employed in developing a voice-enabled user interface for a mobile device.

INTRODUCTION

Mobile devices, such as cell phones and personal digital assistants (PDAs), are inherently small, and lack an intuitive and natural user interface. The small keyboards and displays of mobile devices make it difficult for the user to use even the simplest of applications. Pen input is available on PDAs, but is difficult to use on handheld devices.

Voice input and output for mobile devices with small screens and keyboards, and for hands- and eyes-free operation, can make the user's interaction with a mobile device more user friendly. Voice input and output can also facilitate the use of Web Services (Booth, Hass, McCabe, Newcomer, Champion, Ferris, & Orchard, 2004) from a mobile device, making it possible to access the Web anytime and anywhere, whether at work, at home, or on the move. Global positioning system (GPS) technology (U.S. Census Bureau, 2006) can provide location information automatically for location-aware services.

Many everyday applications can benefit from voice-enabled user interfaces for a mobile device. Voice input and voice output for a mobile device are particularly useful for:

- Booking theater and sports tickets, making restaurant and hotel reservations, and carrying out banking and other financial transactions
- Accessing airline arrival and departure information, weather and traffic conditions, maps and directions for theaters, restaurants, gas stations, banks, and hotels, and the latest news and sports scores
- Maintaining personal calendars; contact lists with names, addresses, and telephone numbers; to-do lists; and shopping lists
- Communicating with other people via voice mail, e-mail, short message service (SMS), and multimedia message service (MMS).

It is important to provide several modes of interaction, so that the user can use the most appropriate mode, depending on the application and the situation. The prototype system that we have developed supports client-side, voice-enabled applications on a mobile device. Even though the applications support multimodal input, allowing keyboard and pen input, we focus, in this chapter, on voice input and on multimodal output in the form of voice, text, and graphics. The prototype includes a program manager that manages the application programs, and six voice-enabled applications, namely, contacts, location, weather, shopping, stocks, and appointments and reminders.

BACKGROUND

A multimodal interface for a mobile device integrates textual, graphical, video, tactile, speech, and/or other audio interfaces in the mobile device (Hjelm, 2000; Oviatt & Cohen, 2000). With multiple ways for a user to interact with the applications, interactions with the device become more natural and the user experience is improved. Voice is becoming an increasingly important mode of interaction, because it allows eyes- and hands-free operation. It is essential for simplifying and expanding the use of handheld mobile devices. Voice has the ability to enable mobile communication, mobile collaboration, and mobile commerce (Sarker & Wells, 2003), and is becoming an important means of managing mobile devices (Grasso, Ebert, & Finin, 1998; Kondratova, 2005).

The increasing popularity of, and technological advancements in, mobile phones and PDAs, primarily mobile phones, is leading to the development of applications to fulfill expanding user needs. The short message service (SMS) is available on most mobile phones today, and some mobile phones provide support for the multimedia messaging service (MMS) to exchange photos and videos (Le Bodic, 2002). The mobile phone manufacturers are no longer focused on making a mobile phone but, rather, on producing a mobile device that combines phone capabilities with the power of a handheld PC. They recognize that the numeric keypad and the small screen, common to mobile phones of the past, do not carry over well to handheld PCs (Holtzblatt, 2005).

With the emergence of Web Services technology (Booth et al., 2004), the Web now provides services, rather than only data as it did in the past. Of the various Web Services available to mobile users today, the map application seems to be the most popular, with online map services available from Google (2006) and Yahoo! (2006b). Much progress has been made in creating the multimodal Web, which allows not only keyboard and mouse navigation but also voice input and output (Frost, 2005).

GPS technology (U.S. Census Bureau, 2006) already exists on many mobile devices, and can be used to provide location-aware services (Rao & Minakakis, 2003), without requiring the user to input geographical coordinates, again contributing to user friendliness.

Speech recognition technology (Rabiner & Juang, 1993) has been developed over many years, and is now very good. Other researchers (Kondratova, 2004; Srinivasan & Brown, 2002) have discussed the usability and effectiveness of a combination of speech and mobility. Currently, handheld voice-enabled applications use short commands that are translated into functional or navigational operations. As observed in Deng and Huang (2004), speech recognition technology must be robust and accurate, and close to human ability, to make its widespread use a reality. Noisy environments present a particular challenge for the use of speech recognition technology on mobile devices and, therefore, multimodal interactions are essential. For example, the MiPad system (Deng, Wang, Acero, Hon, Droppo, Boulis, et al., 2002; Huang, Acero, Chelba, Deng, Droppo, Duchene, Goodman, et al., 2001) uses a strategy where the user first taps a "tap & talk" button on the device and then talks to the device.

Distributed speech recognition (Deng, et al., 2002), in which the speech recognition happens at a remote server exploits the power of the server to achieve fast and accurate speech recognition. However, studies (Zhang, He, Chow, Yang, & Su, 2000) have shown that low-bandwidth connections to the server result in significant degradation of speech recognition quality. In contrast, *local speech recognition* (Deligne, Dharanipragada, Gopinath, Maison, Olsen, & Printz, 2002; Varga, Aalburg, Andrassy, Astrov, Bauer, Beaugeant, Geissler, & Hoge, 2002) utilizes speech recognition technology on the mobile device, and eliminates the need for high-speed communication. Local speech recognition limits the kinds of client handsets that are powerful enough to perform complicated speech processing and, thus, that can be used; however, the computing power of mobile handsets is increasing.

THE PROTOTYPE

The prototype that we have developed allows mobile applications to interact with the user without the need for manual interaction on the part of the human. Speech recognition and speech synthesis software are located on the mobile device, and make the interaction with the human more user friendly. The prototype that we have developed processes natural language sentences and provides useful services while interacting with the user in an intuitive and natural manner. A user need not form a request in a particular rigid format in order for the applications to understand what the user means.

For our prototype, we have developed six application programs and a Program Manager. These applications are Contacts, Location, Weather, Shopping, Stocks, and Appointments and Reminders applications. The Program Manager evaluates sentence fragments from the user's request, determines which application should process the request, and forwards the request to the appropriate application.

The prototype is designed to interact with a human, using voice as the primary means of input (keyboard, stylus, and mouse are also available but are less convenient to use) and with voice, text, and graphics as the means of output. The speech recognizer handles the user's voice input, and both the speech synthesizer and the display are used for output. Characteristics of certain applications render a pure voice solution infeasible. For example, it is impossible to convey the detailed contents of a map through voice output. However, voice output is ideal when it is inconvenient or impossible for the user to maintain visual contact with the display of the mobile device, and it is possible to convey information to the user in that mode. Voice output is also appropriate when the device requests confirmation from the user.

Thus, an appropriate choice of speech recognition and speech synthesis technology is vital to the success of our prototype. Our choices were constrained by:

- The processing and memory capabilities of typical mobile devices
- The need for adaptability to different users and to noisy environments

The use of speech recognition and speech synthesis technology on a mobile device is different from its use in call centers, because a mobile device is associated with a single user and can learn to understand that particular user.

The Underlying Speech Technology

The prototype uses SRI's DynaSpeak speech recognition software (SRI, 2006) and AT&T's Natural Voices speech synthesis software (AT&T, 2006). It currently runs on a handheld computer, the OQO device (OQO, 2006). We chose this device, rather than a cell phone, because it provides a better software development environment than a cell phone.

Speech Recognition

The DynaSpeak speech recognition engine (SRI, 2006) is a small-footprint, high-accuracy, speaker-independent speech recognition engine. It is based on a statistical language model that is suitable for natural language dialog applications. It includes speaker adaptation to increase recognition accuracy for individuals with different accents or tone pitches. It can be configured so that it performs speech recognition specific to a particular individual. DynaSpeak is ideal for handheld mobile devices, because of its small footprint (less than 2 MB of memory) and its low computing requirements (66 MHz Intel x86 or 200 MHz Strong Arm processor).

DynaSpeak supports multiple languages, adapts to different accents, and does not require training prior to use. It incorporates a Hidden Markov Model (HMM) (Rabiner & Juang, 1993). In an HMM, a spoken expression is detected as a sequence of phonemes with a probability associated with each phoneme. A probability is also associated with each pair of phonemes, that is, the probability that the first phoneme of the pair is followed by the second phoneme in natural speech. As a sequence of phonemes is processed, the probability of each successive phoneme is combined with the transition probabilities provided by the HMM. If the probability of a path through the HMM is substantially greater than that of any other path, the speech recognizer recognizes the spoken expression with a high level of confidence. When the response is below an acceptable confidence threshold, the software seeks confirmation from the user or asks the user questions.

The HMM is augmented with grammars for the particular applications that are required for understanding natural language sentences (Knight, Gorrell, Rayner, Milward, Koeling, & Lewin, 2001). When the user says a new word, the word can be added to the vocabulary dynamically. The HMM is also extended by adapting the vocabulary of the speech recognizer to the current and recent past context of interactions of the user with the applications.

Accuracy of the speech recognition system can be increased by training it for the voice of the particular user. There are two kinds of training, explicit and implicit. *Explicit training* requires the user to read a lengthy script to the device, a process that is likely to be unpopular with users. *Implicit training* allows the device to learn to understand better its particular user during normal use. Implicit training can be provided in two modes, confirmation mode and standard mode.

In *confirmation mode*, the system responds to a user's sentence, and the user confirms or corrects the response. If the user corrects the sentence, the learning algorithm tries to match a rejected, lower probability, interpretation of the original sentence with the user's corrected intent. If a match is found, the learning algorithm adjusts the HMM transition probabilities to increase the probability of selecting the user's intent. Initially, a new user of the system will probably prefer confirmation mode.

In *standard mode*, the system does not confirm sentences for which there is one interpretation that has a much higher probability than any other interpretation. If no interpretation has a high probability, or if several interpretations have similar probabilities, the speech recognition system responds as in confirmation mode. More experienced users of the system are likely to use standard mode.

The success of implicit training strategies depends quite heavily on starting with a speech recognizer that is well matched to the individual speaker. It is possible, from relatively few sentences, to classify a speaker and then to download, to the mobile device, an appropriate initial recognizer for subsequent implicit training.

DynaSpeak can be used with either a *finite-state grammar* or a *free-form grammar*. We used the finite-state grammar because it offers greater control over parsed sentences. The tendency for DynaSpeak to accept or reject spoken sentences is heavily influenced by the complexity of the grammar. The *complexity of the grammar* is quantified by the number of paths by which an accepting state can be reached. The greater the complexity of the grammar, the higher is its tendency to accept an invalid spoken request. Conversely, the lower the complexity of the grammar, the higher is its tendency to reject a valid spoken request. To minimize the complexity of the grammar and to improve speech recognition accuracy, each application has its own relatively simple grammar. The program manager determines which applications are involved in a sentence and then reparses the sentence using the appropriate grammars.

Speech Synthesis

Natural Voices (AT&T. 2006) is a speech synthesis engine that provides a simple and efficient way of producing natural (rather than electronic) sounding device-to-human voice interactions. It can accurately and naturally pronounce words and speak in sentences that are clear and easy to understand, without the feeling that it is a computer that is speaking.

Natural Voices supports many languages, male and female voices, and the VoiceXML, SAPI, and JSAPI interface standards. Using Natural Voices, we created text-to-speech software for our prototype that runs in the background and accepts messages in VoiceXML format. Each message contains the name of the voice engine (i.e., "Mike" for a male voice and "Crystal" for a female voice) and the corresponding text to speak.

Managed Applications

For the prototype we developed six multimodal applications (contacts, location, weather, shopping, stocks, appointments, and reminders) that use speech as the main form of input. The stocks, maps, and weather applications exploit existing Web Services on the Internet. Communication with those Web Services uses a local WiFi 802.11 wireless network. The program manager controls the operation of the applications. The graphical user interface for the program manager with the six applications is shown in Figure 1. We now present an explanation of the functionality of each application and its role in the overall system.

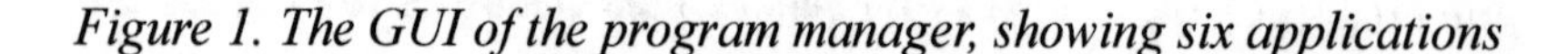

Figure 1. The GUI of the program manager, showing six applications

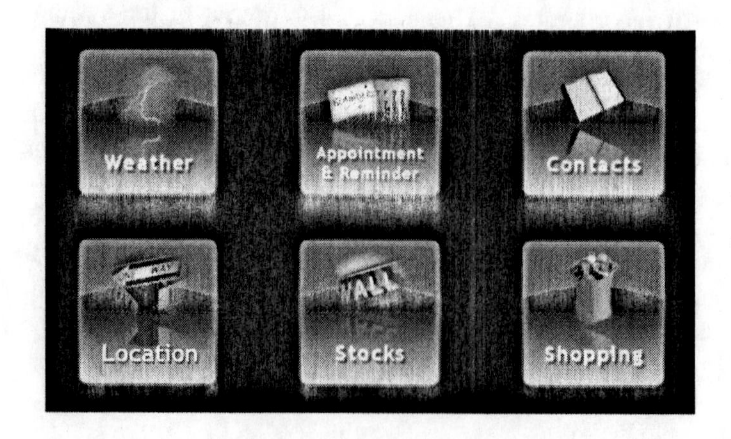

Contacts

The contacts application stores personal information regarding friends and acquaintances in a database, including their addresses and phone numbers. The contacts application is a mobile extension of a physical contact list or address book that is controlled by voice input. It retrieves data from Microsoft Office Outlook® to populate the database when in docking mode. After using the mobile device and possibly entering new contact information, the user can synchronize information on the mobile device with that on a desktop or server computer. The contacts application is configured to interact with other applications that require information about names, addresses, phone numbers, and so forth. The contacts grammar is the least complex of the application grammars that we developed. The contacts vocabulary grows linearly as contacts are added to the user's contact list.

Location

The Location application allows the user to search for restaurants, movie theaters, banks, and so forth, in a given area, using the Yahoo! LocalSearch Web Service (2006b). For example, if the user says to the mobile device "Search for a Mexican restaurant in 95131," the location application on the mobile device sends a Web Service request to Yahoo! LocalSearch, gets back the results, and presents up to 10 results to the user in list form. The user can then view additional information about a single location by indicating the location's number in the presented list. For example, the user can choose to view additional information about "Chacho's Mexican Restaurant" by speaking, "Get more information about number one." On processing this request, the location application presents the user with detailed information about the restaurant including its phone number, address, and a detailed street map showing its location. Figure 2 shows a screen shot of the graphical user interface for the location application.

The location application is loosely coupled with the contacts application to provide responses related to individuals listed in the user's contact list. For example, the request, "Search for a movie theater around Susan's house" uses the contacts grammar to determine the location of Susan's house and replaces the phrase "Susan's house" with the specific address so that the actual search request looks something like this: "Search for a movie theater around 232 Kings Way, Goleta, CA, 93117." The location application then searches for a movie theater in the vicinity of that address.

The location application is also loosely coupled with a GPS module that is contacted when the user has a question related to the user's current location. For example, if the user says "Look for a pizza place around here.", the word "here" is recognized by the application and replaced with the GPS coordinates of the user's current location.

Figure 2. An example graphical user interface for the location application

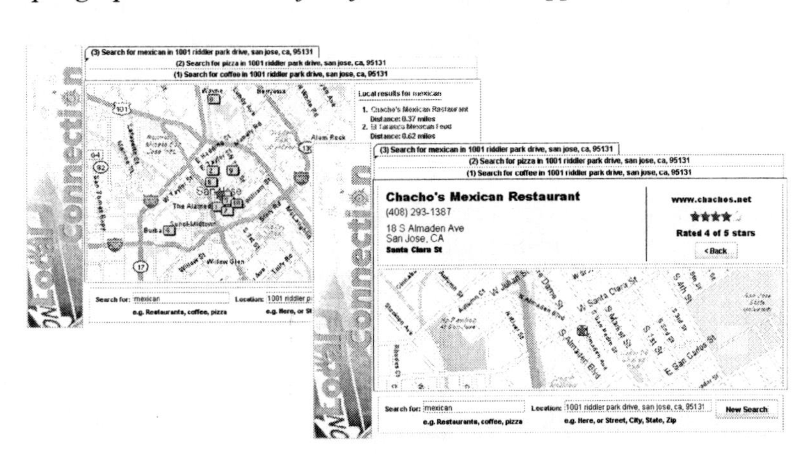

Figure 3. An example graphical user interface for the weather application

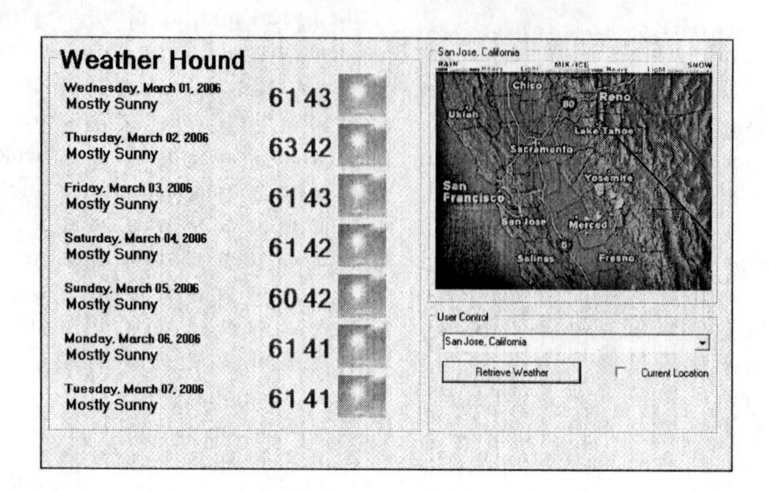

The location application then sends a Web Service request to Yahoo! LocalSearch, which returns a map of the user's current location, indicating where the user is, along with the 10 nearest pizza places. The Yahoo! LocalSearch Web Service is ideal to use with GPS because of its ability to locate positions on the map on the basis of longitude and latitude. With GPS, the user is no longer limited to requests involving a particular city or zip code. The user now has the ability to create requests that are truly location-aware.

Compared to the grammars of the other applications, the location grammar is one of the most complex. For information like maps and lists, it is desirable to use a graphical or textual display, as well as speech output, in a multimodal user interface. Thus, the most appropriate kind of output can be chosen, depending on the kind of information, the capabilities of the mobile device, and the context in which the user finds himself or herself.

Weather

The weather application supplies weather forecasts obtained from the Web Service provided by the National Weather Service (NOAA, 2006). It allows the user to query for weekly, daily, and 3-day weather information in major U.S. cities using voice input. It allows the user either to select a city or to use the user's current location, as the location for which the weather forecast is to be retrieved from the National Weather Service. The weather application knows the geographical coordinates of dozens of cities in the continental United States. It references those coordinates when the user requests a weather forecast from the National Weather Service for one of those cities.

A user can say "Tell me the weather forecast in San Jose," which then uses "today" as the starting time of the forecast, and produces the graphical user interface for the weather application shown in Figure 3.

Because the weather application operates on a mobile device, it is necessary to be able to determine the user's location dynamically. If the user asks "What's the weather like here two days from now?", the weather application consults the GPS module to obtain the geographical coordinates of the user, contacts the Web Service, and responds with the high and low predicted temperatures and an indication that there is a change to cloudy in Santa Barbara. Thus, the user does not need to provide his/her current location or to obtain the weather forecast for that location.

Our prototype takes into account the many ways in which a person can convey, semantically, equivalent requests in English. For example, a user can ask for the weather in many ways including "What is the weather in Boston like?" or "Tell me what the forecast is like in Boston." These two requests are semantically equivalent because they both contain the same essential parameter, namely the Boston location.

Shopping

The shopping application provides the user with a service capable of reducing the time that the user spends on grocery shopping and the associated stress. The shopping application maintains shopping lists, recipes, and floor plans of supermarkets. The multimodal interface includes speech, text, and graphics, which makes the shopping application easy to use. Figure 4 shows a screen shot of the graphical user interface for the shopping application.

The shopping application allows a user to update his/her shopping list and to forward it to another user. When a user issues a command, like "Remind John to go grocery shopping," the contacts application is used to find John's phone number or e-mail address in the user's contact list. A dialog box then appears asking the user if he/she wants to send, to John, not only a reminder to go shopping but also the shopping list. If so, the shopping list, consisting of the product ids and the quantities of the items needed, is formatted in XML, and appended to the message containing the reminder. The message is then sent to John's shopping application.

The shopping application also displays graphically the floor plan of the supermarket and the location of items in the store, as shown in Figure 4. This feature provides assistance to the user without the need for the user to contact an employee of the supermarket. The shopping application also allows the user to retrieve recipes while shopping, possibly on impulse, for an item that is on sale. A newly chosen recipe is cross-referenced with the current shopping list, so that needed items can be added automatically. The shopping application has the largest grammar of the applications that we developed, with a vocabulary that depends on the items that the user has purchased recently.

Stocks

The stocks application allows the user to manage his/her stock portfolio using voice input and output. The objective of the stocks application is to monitor stock fluctuations, rather than to trade stocks. The stocks application exploits the Yahoo! Finance Web service (2006a) to store and update stock information in a database. It stores the most recent stock information in the database so that it can reply to the user's requests when connectivity to the Yahoo! Finance Web Service is limited. Although such stored data can be somewhat stale, it allows the user to obtain information whenever the user requests it. The vocabulary of the stocks application grows to match the user's portfolio each time the user adds a new stock.

Figure 4. An example graphical user interface for the shopping application

Figure 5. An example graphical user interface for the stocks application

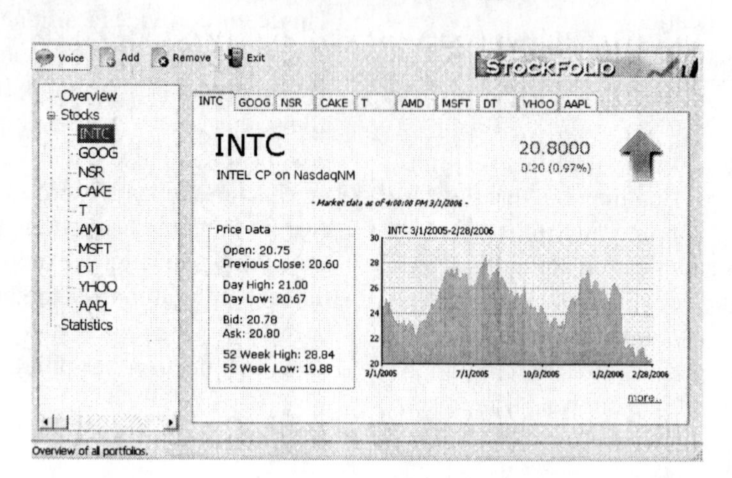

Appointments and Reminders

The appointments and reminders application manages the user's calendar and allows the user to send reminders to other people. It supports time-based requests of various forms, for example, "Remind me to go to the dentist on Monday," "Remind me to see the dentist on August 15th," and "Remind me to see the dentist a week from today." It displays an easily readable schedule, so that the user can recall what is planned for the day. The appointments and reminders application interacts with other applications, such as the shopping application. For example, the request "Remind John to go shopping on Monday" sends a reminder to John, along with the current shopping list, if the user wishes to forward that information. It also supports reminders to the user that are location-aware using GPS, for example, if the user is in the vicinity of a supermarket. The appointments and reminders application is an extension of a calendar service. It links to Microsoft Office Outlook®, and updates scheduled appointments and reminders when in the vicinity of the user's desktop.

Program Manager

The program manager evaluates sentence fragments from a user's request, identifies keywords that determine which application or applications should process the request, reparses the sentence using the grammars for those applications, and forwards the parsed request to the appropriate application. If more than one user is involved, the program manager on one user's mobile device sends messages to the program manager on another user's mobile device, which then handles the request.

The program manager leverages DynaSpeak and a weighted keyword recognition algorithm to break down recognized sentences into application-specific fragments. Those fragments are then processed by the appropriate applications, and are subsequently merged to form the final sentence meaning. This process allows the program manager to handle requests that involve more than one application, for example, "Search for a gas station around Paul Green's house." The parsing of this sentence, using the location grammar, requests a search centered on a location that the location grammar cannot itself provide. The program manager must recognize a keyword from the contacts grammar, parse the sentence using that grammar, and query the contacts application for the address of Paul Green's house. The response to the query is then sent to the location application to obtain the location of the gas station nearest his house.

Graphical User Interface

The graphical user interface (GUI) of thepProgram manager, shown in Figure 1, displays the current running application programs and allows the user to select an application by using voice or keyboard input. The GUI provides buttons that appear gray when an application has not been started and blue after startup. If the user makes a spoken request that requires an application to display a result, the display for that application is topmost and remains topmost until the user issues another request or a timeout occurs. Whenever the GUI is displayed, the user must provide a keyword in a spoken request to wake up the program manager, or click on one of the application-specific buttons on the display.

EVALUATION

Several experiments were performed to collect qualitative and quantitative data to evaluate the prototype system. Although it is difficult to determine a clear boundary between the user interface and the speech recognizer, it is important to evaluate the user interface and the speech recognizer separately, so that the qualitative and quantitative data gathered from the experiments are not mixed, leading to inconclusive results.

Thus, the experiments were designed as a classical "Don't mind the man behind the curtain" study. In this type of study, the user interacts with a system that is identical to the actual system except that the experiment is being controlled by someone other than the user. The man behind the curtain controls what is spoken as responses to the user's requests and changes the current screen to an appropriate graphical response. This method was used, so that the responses to the qualitative questions would not be biased by the accuracy of the speech recognizer.

To evaluate the system quantitatively, the program manager was instrumented with time segment metrics and data were collected for several performance metrics, including:

- Total time a participant took to complete all tasks
- Overhead of the DynaSpeak speech recognizer during live and batch recognition
- Runtime overhead of the program manager without DynaSpeak
- Spoken length of a request vs. processing time

The results are shown in Figure 6. The time segment metrics represent the runtime complexity of the code associated with the speech recognition and processing. The amount of time taken by each segment adds to the delay associated with the user's request. If any of the time segments has a large duration, the user might become irritated. By measuring each segment separately, the bottleneck in the system can be determined.

The speech processing time increases with the size of the grammar. However, by means of a multiphase procedure that uses keywords organized and weighted by application relevance, the grammar size and the speech processing time can be improved. After live recognition, the system provides a keyword-associated request, which it processes for application weights and then reprocesses using an application-specific grammar, possibly more than once with different grammars. This procedure increases both the speed and the accuracy of the speech recognition, by decreasing the size of the grammar size in the initial phase.

An alternative approach (Kondratova, 2004) is to force the user to make repeated requests, possibly from a menu, with responses by which the device asks for the next step or for more information, so that the device arrives at a better understanding of the user's request. Such an approach introduces navigational complexity for the user. Reducing the speech processing time by creating a complex navigational structure is not the best way to improve usability of the system.

The speech recognizer works better for some speakers than for other speakers. The accuracy of the results can be improved by tuning the speech recognition parameters and enabling learning capabilities. However, the developers of DynaSpeak advise against modification of the speech recognition parameters and use of learning until a relatively high success rate is achieved. For appropriately selected users, quite good speech recognition and understanding can be achieved

Figure 6. Processing overhead per task

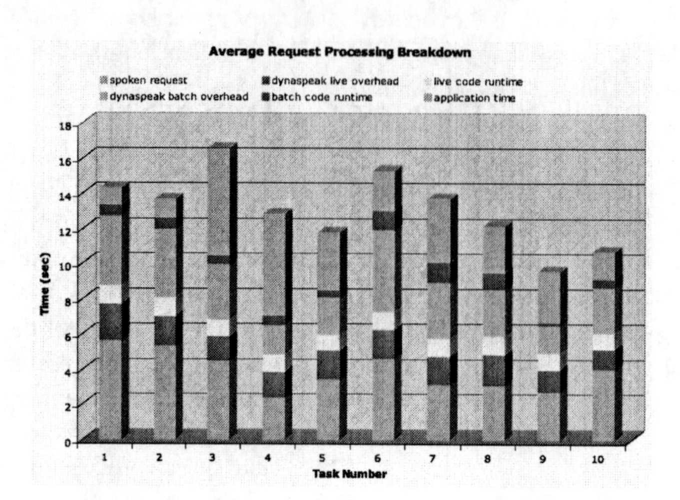

without using learning capabilities. However, speech recognition accuracy can only improve if voice profiling is combined with learning.

Ambient noise and microphone quality also affect speech recognition accuracy. The internal microphone in the OQO device is of rather poor quality. To ameliorate this problem, a Jabra© Bluetooth headset, was used to provide noise cancellation and reduce the distance between the microphone and the user's mouth. In addition, when the confidence score from DynaSpeak falls below an acceptable threshold, the program manager seeks confirmation from the user or asks for clarification. These mechanisms greatly improve the accuracy of the speech recognizer.

The accuracy of speech recognition is degraded when the grammar contains words that are phonetically similar. During preliminary experiments for the shopping application, we had problems recognizing differences between similar sounding requests like "Add lamb to my shopping list" and "Add ham to my shopping list." These problems arise particularly when users are non-native English speakers or when they have accents. Creating more specific requests can reduce the phonetic similarity, for example, by saying "Add a lamb shank to my shopping list" and "Add a ham hock to my shopping list." However, modifying requests in such a way is undesirable because the requests are then less intuitive and natural.

The location, weather, and stocks applications all use Web Services and require communication over the Internet and, thus, have longer application runtimes than the other Web Services. The location application is written in Java, which runs more slowly than C#. Both the weather application and the stocks application cache data associated with previous requests to take advantage of timing locality. Location requests are different because the caching of maps can involve a large usage of the memory, and users are not inclined to perform the same search twice. Memory is a precious commodity on a handheld device and needs to be conserved; thus, the location application is coded so that it does not cache maps resulting from previous queries.

To evaluate the qualitative aspects of the system, we performed a user study with participants from diverse backgrounds of education, ethnicity, and sex. The user study was completed with 10 individuals performing 10 tasks resulting in 100 request results. The participants were given a questionnaire that assessed their general impressions about the prototype, with the results shown in Table 1.

After analyzing the averaged responses of the participants, we found several trends. The participants' scores are not strongly correlated with speech recognition accuracy. Participant G gave the system a high score, but was one of the two

Table 1. Responses to the questionnaire

Questions	A	B	C	D	E	F	G	H	I	J	Mean
Was it comfortable talking to the device as if it were a human?	3	3	4	3	4	3	4	5	3	5	3.7
Was the GUI aesthetically pleasing?	5	4	4	5	5	5	5	5	5	4	4.7
Were the request responses appropriate and easy to understand?	3	3	5	5	5	4	4	5	4	4	4.2
Were the spoken responses relevant to your requests?	5	4	5	5	5	3	4	5	4	5	4.5
Was the system easy to use?	4	5	3	4	4	5	4	5	4	5	4.3
Do you think the services would be helpful in your daily life?	4	4	4	5	4	4	4	4	4	5	4.2
Would you recommend a system like this to your friends?	3	3	4	5	4	4	5	5	4	5	4.2
Would you buy the software if it were available for your phone?	3	2	4	5	3	3	5	5	3	5	3.8

participants who encountered the most speech recognition problems. Participant B gave the system a low score despite good speech recognition.

The participants agreed that speaking to a mobile handheld device as if it were a human is not comfortable. It is difficult to get used to interacting with a computer that can understand tasks that would be commonplace for humans. The participants were relatively pleased with the GUI interface design and felt the system is relatively easy to use. However, the ease-of-use metric needs to be taken lightly. Ease of use can be assessed more concretely by measuring the number of times a user must repeat a command.

The scores for response appropriateness and relevance are high, indicating that the spoken responses of the applications were well crafted. The scores related to recommending the service to friends and daily life helpfulness are relatively high, from which one might infer that the participants would purchase a device providing the speech-enabled applications. However, this conclusion is not necessarily justified. The participants were not enthusiastic about having to pay for such a device or for such services. However, most participants in the study were quite pleased with the prototype system and found the user interface helpful and easy to use.

FUTURE TRENDS

Integration of multiple applications, and multiple grammars, is not too difficult for a small number of applications that have been designed and programmed to work together, as in our prototype. However, future systems will need to support tens or hundreds of applications, many of which will be designed and programmed independently. Integration of those applications and their grammars will be a challenge.

Currently, speech-enabled applications typically use short commands from the human that are translated into navigational or functional operations. More appropriate is speech recognition technology that supports a more natural, conversational style similar to what humans use to communicate with each other (McTear, 2002).

A mobile device that listens to its owner continuously can provide additional services, such as populating the user's calendar. For example, when a user agrees to an appointment during a conversation with another person, the mobile device might recognize and automatically record the appointment, possibly confirming the appointment later with its user. Similarly, the mobile device might note that the user habitually goes to lunch with the gang at noon on Mondays, or that the user leaves work promptly at 5pm on Fridays. With existing calendar systems, the user often does not record appointments and other commitments, because it is too much bother using the human interfaces of those systems, greatly reducing the value of the calendar.

A useful capability of speech recognition systems for mobile devices is being able to recognize intonation and emotional overtones. "The bus leaves at 6" is, overtly, a simple declaration, but appropriate intonation might convert that declaration into a question or an expression of disapproval. Existing speech recognition systems do not yet recognize and exploit intonation. Similarly, the ability to recognize emotional overtones of impatience, uncertainty, surprise, pleasure, anger, and so forth, is a valuable capability that existing speech recognition systems do not yet provide.

Speech recognition requires a relatively powerful processor. Typical cell phones contain a powerful digital signal processor (DSP) chip and a much less powerful control processor. The control processor operates continuously to maintain communication with the cellular base stations. The DSP processor uses a lot of power and imposes a significant drain on the battery and, thus, analyzes and encodes speech only during calls. The DSP processor is capable of the processing required for speech recognition, although it might need more memory.

For mobile devices, battery life is a problem, particularly when speech recognition or application software requires a powerful processor. The limit of 2 hours of talk time for a cell phone is caused at least as much by the power drain of the DSP processor as by the power needed for wireless transmission. The DSP processor might be needed for speech processing for more than 2 hours per day. There are several possible solutions to this problem, namely, larger batteries, alcohol fuel cells, and DSP processors with higher speeds, reduced power consumption, and better power management.

Background noise remains a problem for speech recognition systems for mobile devices, particularly in noisy environments. The quality of the microphone, and the use of a headset to decrease the distance between the microphone and the speaker's mouth, can improve speech recognition accuracy.

CONCLUSION

The use of voice input and output, in addition to text and graphics and other kinds of audio, video, and tactile interfaces, provides substantial benefits for the users of mobile devices. Such multimodal interfaces allow individuals to access information, applications, and services from their mobile devices more easily. A user no longer has to put up with the annoyances of a 3-inch keyboard, nested menus, or handwriting recognition, nor does the user need to have a tethered desktop or server computer in order to access information, applications, and services. Providing multiple ways in which the users can interact with the applications on mobile devices brings a new level of convenience to the users of those devices.

REFERENCES

AT&T. (2006). *Natural voices*. Retrieved from http://www.natural voices.att.com/products/

Booth, D., Hass, H., McCabe, F., Newcomer, E., Champion, M., Ferris, C., & Orchard, D. (2004). *Web services architecture*. Retrieved from http://www.w.3.org/Tr/WS-arch

Deligne, S., Dharanipragada, S., Gopinath, R., Maison, B., Olsen, P., & Printz, H. (2002). A robust high accuracy speech recognition system for mobile applications. *IEEE Transactions on Speech and Audio Processing, 10*(8), 551-561.

Deng, L., & Huang, X. (2004). Challenges in adopting speech recognition. *Communications of the ACM, 47*(1), 69-75.

Deng, L., Wang, K., Acero, A., Hon, H., Droppo, J., Boulis, C., Wang, Y., Jacoby, D., Mahajan, M., Chelba, C., & Huang, X. D. (2002). Distributed speech processing in MiPad's multimodal user interface. *IEEE Transactions on Speech and Audio Processing, 10*(8), 605-619.

Frost, R. A. (2005). Call for a public-domain SpeechWeb. *Communications of the ACM, 48*(11), 45-49.

Google. (2006). *Google Maps API.* Retrieved from http://www.google.com/apis/maps

Grasso, M. A., Ebert, D. S., & Finin, T. W. (1998). The integrality of speech in multi-modal interfaces. *ACM Transactions on Computer-Human Interaction, 5*(4), 303-325.

Hjelm, J. (2000). *Research applications in the mobile environment. Wireless information service.* New York, NY: John Wiley & Sons.

Holtzblatt, K. (2005). Designing for the mobile device: Experiences, challenges, and methods. *Communications of the ACM, 48*(7), 33-35.

Huang, X., Acero, A., Chelba, C., Deng, L., Droppo, J., Duchene, D., Goodman, J., Hon, H., Jacoby, D., Jiang, L., Loynd, R., Mahajan, J., Mau, P., Meredith, S., Mughal, S., Neto, S., Plumpe, M., Stery, K., Venolia, G., Wang, K., & Wang, Y. (2001). MiPad: A multimodal interaction prototype. In *Proceedings of the International Conference on Acoustics, Speech and Signal Processing, 1*, 9-12.

Knight, S., Gorrell, G., Rayner, M., Milward, D., Koeling, R., & Lewin, I. (2001). Comparing grammar-based and robust approaches to speech understanding: A case study. In *Proceedings of Eurospeech 2001, Seventh European Conference of Speech Communication and Technology* (pp. 1779-1782). Aalborg, Denmark.

Kondratova, I. (2004, August). Speech-enabled mobile field applications. In *Proceedings of the IASTED International Conference on Internet and Multimedia Systems*, Hawaii.

Kondratova, I. (2005, July). Speech-enabled handheld computing for fieldwork. In *Proceedings of the International Conference on Computing in Civil Engineering*, Cancun, Mexico.

Le Bodic, G. (2002). *Mobile messaging, SMS, EMS and MMS.* John Wiley & Sons.

McTear, M. (2002). Spoken dialogue technology: Enabling the conversational user interface. *ACM Computing Surveys, 34*(1), 90-169.

National Oceanic and Atmospheric Administration (NOAA). (2006). *National Weather Service.* Retrieved from http://www.weather.gov/xml/

OQO. (2006). *The OQO personal computer.* Retrieved from http://www.oqo.com

Oviatt, S., & Cohen, P. (2000). Multi-modal interfaces that process what comes naturally. *Communications of the ACM, 43*(3), 45-53.

Rabiner, L., & Juang, B. H. (1993). *Fundamentals of speech recognition.* Upper Saddle River, NJ: Prentice Hall.

Rao, B., & Minakakis, L. (2003). Evolution of mobile location-based services. *Communications of the ACM, 46*(12), 61-65.

Sarker, S., & Wells, J. D. (2003). Understanding mobile handheld device use and adoption. *Communications of the ACM, 46*(12), 35-40.

SRI (2006). *DynaSpeak.* Retrieved from http://www.speechatsri.com/products/sdk.shtml

Srinivasan, S., & Brown, E. (2002). Is speech recognition becoming mainstream?. *Computer Magazine*, (April), 38-41.

U.S. Census Bureau. (2006). *Precision of GPS.* Retrieved from http://www.census.gov/procur/www/fdca/library/mcd/7-29%20 MCD_WG_hardware_subteam_report.pdf

Varga, I., Aalburg, S., Andrassy, B., Astrov, S., Bauer, J. G., Beaugeant, C., Geissler, C., & Hoge, H. (2002). ASR in mobile phones—An industrial approach. *IEEE Transactions on Speech and Audio Processing, 10*(8), 562-569.

Yahoo! LocalSearch. (2006a). Retrieved from http://www.local.yahooapis.com/LocalSearchService/V3/localSearch

Yahoo! Finance. (2006b). Retrieved from http://finance.yahoo.com/rssindex

Zhang, W., He, Y., Chow, R., Yang, R., & Su, Y. (2000, June). The study on distributed speech recognition system. In *Proceedings of the IEEE International Conference on Acoustical Speech and Signal Processing* (pp. 1431–1434), Istanbul, Turkey.

KEY TERMS

Global Positioning System (GPS): A system that is used to obtain geographical coordinates, which includes a GPS satellite and a GPS receiver.

Hidden Markov Model (HMM): A technique, based on a finite state machine that associates probabilities with phonemes, and pairs of phonemes, that is used in speech recognition systems, to determine the likelihood of an expression spoken by a user of that system.

Location Aware: An application that is based on a particular physical location, as given by geographical coordinates, physical address, zip code, and so forth, that determines the output of the application.

Mobile Device: For the purposes of this chapter, a handheld device, such as a cell phone or personal digital assistant (PDA), that has an embedded computer and that the user can carry around.

Multimodal Interface: The integration of textual, graphical, video, tactile, speech, and other audio interfaces through the use of mouse, stylus, fingers, keyboard, display, camera, microphone, and/or GPS.

Speech Recognition: The process of interpreting human speech for transcription or as a method of interacting with a computer or a mobile device, using a source of speech input, such as a microphone.

Speech Synthesis: The artificial production of human speech. Speech synthesis technology is also called text-to-speech technology in reference to its ability to convert text into speech.

Web Service: A software application identified by a Uniform Resource Indicator (URI) that is defined, described, and discovered using the eXtensible Markup Language (XML) and that supports direct interactions with other software applications using XML-based messages via an Internet protocol.

Chapter XXVIII
Speech-Centric Multimodal User Interface Design in Mobile Technology

Dong Yu
Microsoft Research, USA

Li Deng
Microsoft Research, USA

ABSTRACT

Multimodal user interface (MUI) allows users to interact with a computer system through multiple human-computer communication channels or modalities. Users have the freedom to choose one or more modalities at the same time. MUI is especially important in mobile devices due to the limited display and keyboard size. In this chapter, we provide a survey of the MUI design in mobile technology with a speech-centric view based on our research and experience in this area (e.g., MapPointS and MiPad). In the context of several carefully chosen case studies, we discuss the main issues related to the speech-centric MUI in mobile devices, current solutions, and future directions.

INTRODUCTION

In recent years, we have seen steady growth in the adoption of mobile devices in people's daily lives as these devices become smaller, cheaper, more powerful, and more energy-efficient. However, mobile devices inevitably have a small display area, a tiny keyboard, a stylus, a low speed (usually less than 400 million instructions per second) central processing unit (CPU), and a small amount (usually less than 64MB) of dynamic random-access memory. Added to these limitations is the fact that mobile devices are often used in many different environments, such as dark and/or noisy surroundings, private offices, and meeting rooms. On these devices, the traditional *graphical user interface* (GUI)-centric design becomes far less effective than desired. More efficient and easy-to-use user interfaces are in urgent need. The *multimodal user interface* (MUI), which allows users to interact with a computer system through multiple channels such as speech, pen, display, and keyboard, is a promising user interface in mobile devices.

Multimodal interaction is widely observed in human-human communications where senses such as sight, sound, touch, smell, and taste are used. The research on multimodal human-computer interaction, however, became active only after Bolt (1980) proposed his original concept of "Put That There." Since then, a great amount of research has been carried out in this area (Bregler, Manke, Hild, & Waibel 1993; Codella, Jalili, Koved, Lewis, Ling, Lipscomb, et al., 1992; Cohen, Dalrymple, Moran, Pereira, Sullivan, Gargan, et al., 1989; Cohen, Johnston, McGee, Oviatt, Pittman, Smith, et al., 1997; Deng & Yu, 2005; Fukumoto, Suenga, & Mase, 1994; Hsu, Mahajan, & Acero 2005; Huang, Acero, Chelba, Deng, Droppo, Duchene, et al., 2001; Neal & Shapiro, 1991; Pavlovic, Berry, & Huang, 1997; Pavlovic & Huang, 1998; Vo, Houghton, Yang, Bub, Meier, Waibel, et al., 1995; Vo & Wood, 1996; Wang, 1995). Importantly, the body of this research work pointed out that MUIs can support flexible, efficient, and powerful human-computer interaction.

With an MUI, users can communicate with a system through many different input devices such as keyboard, stylus, and microphone, and output devices such as graphical display and speakers. MUI is superior to any single modality where users can communicate with a system through only one channel. Note that using an MUI does not mean users need to communicate with the system always through multiple communication channels simultaneously. Instead, it means that users have freedom to choose one or several modalities when communicating with the system, and they can switch modalities at any time without interrupting the interaction. These characteristics make the MUI easier to learn and use, and is preferred by users in many applications that we will describe later in this chapter.

MUI is especially effective and important in mobile devices for several reasons. First, each modality has its strengths and weaknesses. For this reason, single modality does not permit the user to interact with the system effectively across all tasks and environments. For example, speech UI provides a hands-free, eyes-free, and efficient way for users to input descriptive information or to issue commands. This is very valuable when in motion or in natural field settings. Nevertheless, the performance of speech UI decreases dramatically under noisy conditions. In addition, speech UI is not suitable when privacy and social condition (e.g., in a meeting) is a concern. Pen input, on the other hand, allows users to interact with the system silently, and is acceptable in public settings and under extreme noise (Gong, 1995; Holzman, 1999). Pen input is also the preferred way for entering digits, gestures, abbreviations, symbols, signatures, and graphic content (Oviatt & Olsen, 1994; Suhm, 1998). However, it is impossible for the user to use pen input if he/she is handicapped or under "temporary disability" (e.g., when driving). MUI, on the other hand, allows users to shift between modalities as environmental conditions change (Holzman, 1999), and hence, can cover a wider range of changing environments than single-modal user interfaces.

Second, different modalities can compensate for each other's limitations and thus provide users with more desirable experience (Deng & Yu, 2005; Oviatt, Bernard, & Levow, 1999; Oviatt & vanGent, 1996; Suhm, 1998). For example, the accuracy of a resource-constrained, midsized vocabulary speech recognizer is low given the current speech technology. However, if the speech recognizer is used together with a predictive T9 (text on 9 keys) keyboard, users can greatly increase the text input throughput compared with using the speech modality or T9 keyboard alone (Hsu et al., 2005). The gain is obtained from the mutual disambiguation effect, where each error-prone modality provides partial information to aid in the interpretation of other modalities. Another reason for the improved user experience is users' active error avoidance, where users tend to select the input modality that they judge to be less error prone for a particular task and environment (Oviatt & vanGent, 1996), and tend to switch modalities to recover from system errors (Oviatt et al., 1999). Mutual compensation is very important for mobile devices because the ability of every single modality in the devices is extremely limited (e.g., a limited display and keyboard size, and limited speech recognition accuracy).

Despite the importance of MUI in mobile devices, designing effective MUIs is far from trivial. Many MUIs in mobile devices are *speech centric*, where speech is the central and main modality. In this chapter, we will focus on main issues on the design of effective speech centric MUIs in mobile devices based on our research and experience in developing MapPointS (Deng & Yu, 2005) and MiPad (Deng, Wang, Acero, Hon, Droppo, Boulis, et al., 2002; Huang, Acero, Chelba, Deng, Droppo, Duchene, et al., 2001). In Section 2, we describe a generic MUI architecture in mobile setting that consists of various recognizers for different input modalities, semantic parsers, a discourse manager, and a response manager. In Section 3, we discuss special considerations related to speech modality. In particular, we discuss the approaches to overcoming resource limitations on mobile devices, noise robust speech front-ends, noise robust modality switching interfaces, and context-aware language model. In section 4, we introduce the issues related to robust natural language understanding including construction of robust grammars. We discuss the problem of modality fusion, including modality-neutral semantic representation, unification approach, and modality integration, in Section 5. We discuss possible future directions and conclude this chapter in Section 6.

A GENERIC MUI ARCHITECTURE

The ultimate goal of an MUI is to fulfill the needs and requirements of the users. This principle is one of many emphasized in *user-centered design* (Gould & Lewis, 1985, Norman & Draper, 1986). According to the user-centered design principle, the acceptability of an MUI can be judged using three main attributes (Dybkjaer & Bernsen, 2001; Hone & Graham, 2001; Nielsen, 1993): effectiveness, efficiency, and learnability. The *effectiveness* assesses whether users can complete the tasks and achieve the goals with the predefined degree of perceived accuracy. It is usually measured on the targeted user population, over a specified range of tasks and environments. The *efficiency* judges how much effort (cognitive demand, fatigue, stress, frustration, discomfort, and so on) and resources (time) are needed for users to perform specific tasks. It is usually measured with the total time (including time for error corrections) taken to complete a task. The *learnability* measures whether users can easily discover the system's functionality and quickly learn to use the system.

Figure 1 depicts a typical speech-centric MUI architecture that is aimed to achieve a high level of effectiveness, efficiency, and learnability. As shown in the figure, users can communicate with the system through speech, keyboard, and other modalities such as pen and camera. Modality fusion usually is the center of an MUI system. There are two typical ways of fusing information from different input modalities, namely, early fusion and late fusion. With the *early fusion*, signals are integrated at the *feature* level and hence, the recognition process in one modality would affect that in another modality (Bregler et al., 1993, Pavlovic et al., 1997; Pavlovic & Huang, 1998; Vo et al., 1995,). Early fusion is suitable for highly coupled modalities such as speech and lip movements (Rubin, Vatikiotis-Bateson, & Benoit, 1998; Stork & Hennecke, 1995). However, early fusion can greatly increase the modeling complexity and computational intensity due to its nature of intermodality influence in the recognition phase. With the *late fusion*, information is integrated at the *semantic* level. The benefit of late fusion is its isolation of input modalities from the rest of the system. In other words, individual recognizers trained using unimodal data can be directly plugged into the system without affecting the rest of the system. This feature makes the late fusion easier to scale up to more modalities in the future than the early fusion. The architecture shown in Figure 1 utilizes the late fusion approach that has been widely adopted, for example, by a variety of systems including Put-That-There (Bolt, 1980), MapPointS (Deng & Yu, 2005), MiPad (Huang et al., 2001), ShopTalk (Cohen, et al., 1989), QuickSet (Cohen, Johnston, McGee, Oviatt, Pittman, Smith, et al., 1997), CUBRICON (Neal & Shapiro, 1991), Virtual World (Codella, Jalili, Koved, Lewis, Ling, Lipscomb, et al., 1992), Finger-Pointer (Fukumoto et al., 1994), VisualMan (Wang, 1995), and Jeanie (Vo & Wood, 1996).

In the late-fusion approach depicted in Figure 1, the input signals received by the system are first processed by *semantic parsers* associated with the corresponding modality into the *surface semantics* representation. Note that although each modality has its own semantic parser, the resulting surface semantics are represented in a common semantic representation and is thus independent of the modality. The surface semantics from all the input modalities are then fused by the *discourse manager* component into the *discourse semantics* representation (more discussions on this issue in Section 4). In order to generate discourse semantics, the discourse manager uses the semantic modal and interacts with the context manager to utilize and update such information as dialog context, domain knowledge, user's information, and user's usage history. The updated context information can be used to adapt the language model, which can improve speech recognition accuracy and enhance the quality of semantic parsers for the next user-computer interaction.

The discourse semantics, which is the output of the discourse manager, is then fed into the *response manager* to communicate back to the user. The response manager synthesizes the proper responses, based on the discourse semantics and the capabilities of the user interface, and plays the response back to the user. In this process, behavior model provides rules to carry out the required actions. The combination of discourse manager and response manager is usually referred to as the dialog manager.

Note that the components shown in Figure 1 may reside on the mobile devices, or distributed on other servers in real implementations. In addition, many MUI systems use an agent-based software solution in which a facility or hub is used to pass information to and from different components (or agents) (Kumar & Cohen, 2000; Schwartz, 1993).

Figure 1. A typical speech-centric MUI architecture and its components

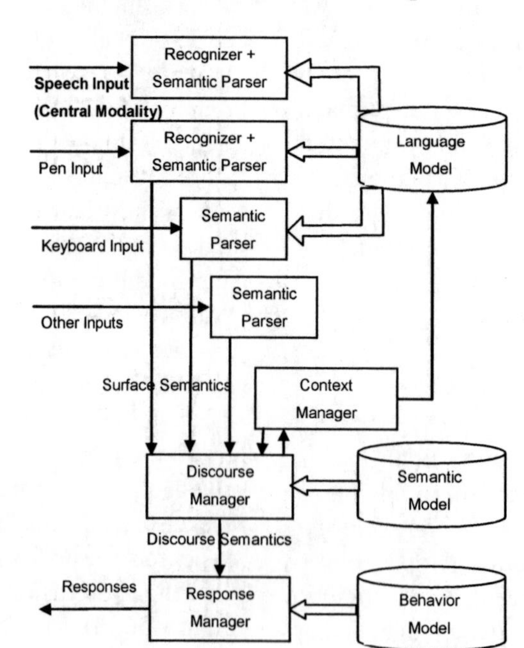

Many best practices and design principles have been developed for the speech-centric MUI design in the past decades (Becker, 2001; Dybkjaer & Bernsen, 2001; Ravden & Johnson, 1989; Reeves, Lai, J., Larson, J.A., Oviatt, S., Balaji, T.S., Buisine, et al. 2004), which we summarize next.

First, the system should explicitly inform the user about its state through appropriate feedback within a reasonable amount of time, so as to avoid state errors, that is, the user's perceived state is different from the system's perceived state. The feedback can be in different modalities, but must be clear and accurate. If speech feedback is used, recorded speech is usually preferred over the synthesized speech, due to its higher degree of naturalness. Note that the recorded speech usually takes a larger amount of resources than the synthesized speech. Since the memory and storage available in mobile devices is very limited, designers should strike a balance between the use of synthesized speech and of recorded speech. The system should follow real-world conventions, and use the words, phrases, and concepts that are familiar to the users. The system should also ensure that the output modalities be well synchronized temporally. For example, the spoken directions should be synchronized with the map display.

Second, the system should provide sufficient flexibility so that users can select the modalities that are best for the task under the specific environments. For example, the user should be able to switch to a nonspeech modality when inputting sensitive information such as personal identification numbers and passwords. A good MUI design should also allow users to exit from an unwanted state via commands that are global to the system, instead of having to go through an extended dialog. The system should provide enough information (e.g., through prompts) to guide novice users to use the system, yet at the same time allow barge-ins and accelerators for the expert users to reduce the overall task completion time.

Third, the system should be designed to allow easy correction of errors. For example, the system should provide context sensitive, concise, and effective help. Other approaches include integrating complementary modalities to improve overall robustness during multimodal fusion; allowing users to select a less error-prone modality for a given lexical content, permitting users to switch to a different modality when error happens; and incorporating modalities capable of conveying rich semantic information.

Fourth, the system's behavior should be consistent internally and with users' previous experiences. For example, a similar dialog flow should be followed and the same terms should be used to fulfill the same task. Users should not have to wonder whether the same words and actions have different meaning under different context.

Fifth, the system should not present more information than necessary. For example, dialogues should not contain irrelevant or rarely needed information, and the prompts should be concise.

While the best practices summarized are common to all speech-centric MUIs, some special attention needs to be paid to speech modality and multimodality fusion due to the great variations of mobile device usage environments. We address these special considerations next.

SPECIAL CONSIDERATIONS FOR SPEECH MODALITY

There are two main challenges for the use of speech modality on mobile devices. First, the resources on mobile devices, in particular, CPU speed, memory, and communication bandwidth, are very limited. Second, speech recognition accuracy degrades substantially in realistic noisy environments, where there are abrupt changes in noise, or variable phase-in phase-out sources of noise as the user moves. For example, the recognition accuracy may drop 30-50% inside a vehicle and cafeteria from that in a quiet environment (Das, Bakis, Nadas, Nahamoo, & Picheny, 1993; Lockwood & Boudy, 1992). Since the mobile devices will be used in these real-field settings without a close-talk microphone, robustness to acoustic environment, that is, immunity to noise and channel distortion, is one of the most important aspects to consider when designing speech-centric MUIs on mobile devices. Speech recognition accuracy and robustness can usually be improved with a noise-robust speech front-end, a noise-robust modality-switching interface, and a context aware language model.

Resource Constrained Speech Recognition

Speech recognition on mobile devices is typically carried out with two options: the *distributed recognition* (Deng et al., 2002) where the recognition happens at a remote server (Figure 2) and the *local recognition* (Deligne, Dharanipragada, Gopinath, Maison, Olsen, & Printz, 2002; Varga, Aalburg, Andrassy, Astrov, Bauer, Beaugeant, et al., 2002) where the recognition is carried out completely on the mobile device. The distributed recognition can take advantage of the power of the remote server to achieve a fast and accurate recognition, while the local recognition can eliminate the requirement of the device to have a fast data connection.

In the distributed architecture, the main consideration is the latency required to send data to and from the server. The latency is typically determined by the communication bandwidth and the amount of data sent. To reduce the latency, a typical approach is to use a standard codec on the device to transmit the speech to the server where the coded speech is subsequently decompressed and recognized (as depicted in Figure 3). However, since speech recognizers only need some features of the speech signal (e.g., Mel-cepstrum), an alternative approach is to put the speech front end on the mobile device and transmit only speech features to the server (Deng et al. 2002), as shown in Figure 4. Transmitting speech features can further save bandwidth because the size of the features is typically much less than that of the compressed audio signals.

Figure 2. Illustration of distributed speech recognition where the actual recognition happens at the server (e.g., PC)

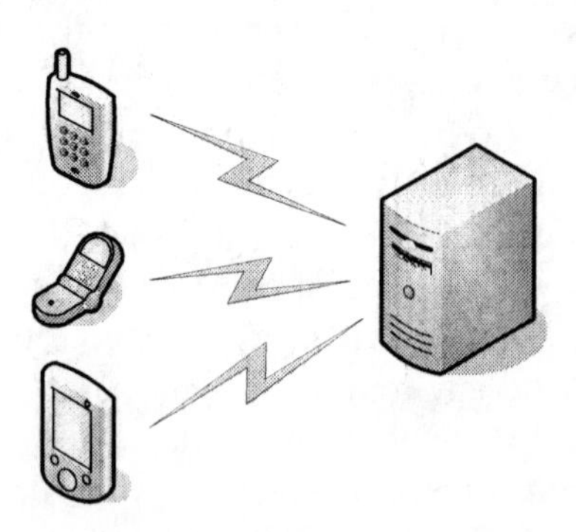

Besides the advantage of using the computing power at the server to improve speech recognition accuracy, there are other benefits of using server-side recognition. One such benefit is its better maintainability compared to the local recognition approach because updating software on the server is much easier and more cost effective than updating software on millions of mobile devices. It, however, does require the recognizer on the server to be front end or codec agnostic in order to materialize this benefit. In other words, the recognizer should make no assumptions on the structure and processing of the front end (Deng et al., 2002). Another benefit of using distributed recognition is the possibility for the server to personalize the acoustic model, language model, and understanding model all at the server, saving the precious CPU and memory on mobile devices. In the past, distributed recognition is unquestionably the dominant approach due to the low CPU speed and small amount of memory available on the mobile devices. Nowadays, although the CPU speed and memory size are increasing dramatically, distributed recognition is still the prevailing approach over local recognition due to the advantages discussed previously.

The major issue of the local recognition architecture is the low recognition speed and accuracy due to the slow CPU speed and low memory available on mobile devices. Speech recognizers running on mobile devices need to be specially designed (Deligne et al., 2002, Li, Malkin, & Bilmes, 2006; Varga, Aalburg, Andrassy, Astrov, Bauer, Beaugeant, 2002) to fit the requirement since speech recognizers designed for the desktop or telephony systems cannot be directly deployed to mobile devices. The greatest benefit of using the local recognition approach is its independency of the network connection and the server and

Figure 3. Distributed speech recognition architecture: speech input is encoded and sent to the server. Speech feature extraction happens at the server side

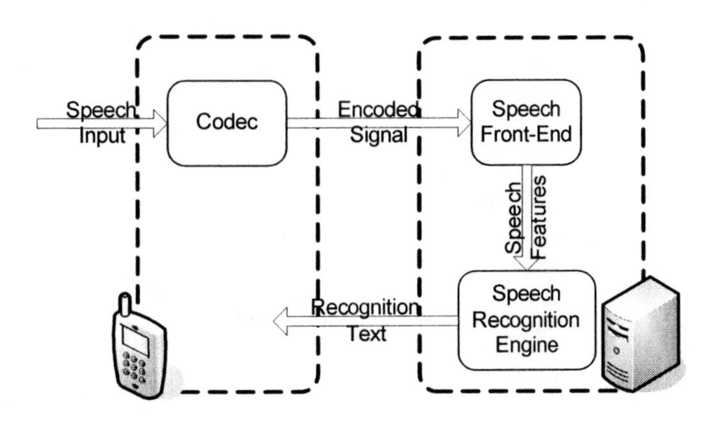

Figure 4. Distributed speech recognition architecture alternative: the speech feature extraction happens on the mobile devices. Only the features are sent to the server

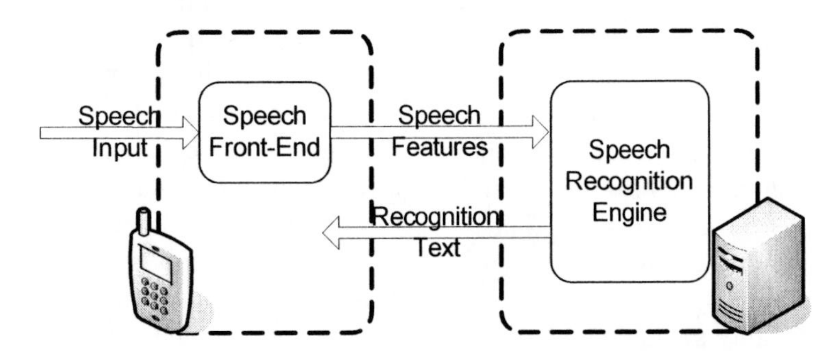

hence, can be used everywhere under any conditions. Given the consistent improvement of the CPU speed and memory on the mobile device hardware, in the future, the local recognition approach is expected to become more and more popular for simple tasks such as name dialing and media playing.

Noise Robust Speech Front End

Noise robustness is one of the most important requirements for speech-centric MUI on mobile devices. It has attracted substantial attention in the past several years. Many algorithms have been proposed to deal with nonstationary noises. A popular one is an advanced feature extraction algorithm (jointly developed by Motorola Labs, France Telecom and Alcatel) that was selected in February of 2002 as a standard in distributed speech recognition by the European telecommunications standards institute. The algorithm defines the extraction and compression of the features from speech that is performed on a local, terminal device, for example, a mobile phone. These features are then sent over a data link to a remote "back-end processor" that recognizes the words spoken. The major components of this algorithm are noise reduction, waveform processing, cepstrum calculation, blind equalization, and voice-activity detection. The noise reduction component makes use of two-stage Wiener filtering (Macho, Mauuary, Noé, Cheng, Ealey, Jouvet, et al., 2002).

The stereo-based piecewise linear compensation for environments (SPLICE), which has been used in the MiPad system (Deng et al., 2002), is another effective algorithm for noise robust speech feature extraction. SPLICE is a cepstrum enhancement algorithm dealing with additive noise, channel distortion, or a combination of the two. It is a dynamic, frame-based, bias-removal algorithm with no explicit assumptions made on the nature of the noise model. In SPLICE, the noise characteristics are embedded in the piecewise linear mapping between the "stereo" clean and distorted speech cepstral vectors. SPLICE has a potential to handle a wide range of distortions, including nonstationary distortion, joint additive and convolutional distortion, and nonlinear distortion (in time-domain), because SPLICE can accurately estimate the correction vectors without the need for an explicit noise model.

Modality Switching

One of the problems in speech recognition under noisy environment is modality switching. If the speech recognition engine is always on, noises and by-talks may be misrecognized as a legitimate user input and hence, can erroneously trigger commands.

A widely used modality switching approach is called "push to talk," where the user presses a button to turn on the speech recognizer, and releases the button to turn off the recognizer. Another approach is called "tap & talk" (Deng et al., 2002; Huang, Acero, A., Chelba, C., Deng, L., Duchene, D., Goodman, et al., 2000, Huang et al., 2001), where the user provides inputs by tapping the "tap & talk" field and then talking to it. Alternatively, the user can select the tap & talk field by using the roller to navigate and holding it down while speaking. Tap & talk can be considered as a combination of push-to-talk control and indication of where the recognized text should go. Both the push-to-talk and tap & talk avoid the speech detection problem that is critical to the noisy environment under which the mobile devices are typically deployed.

Figure 5 shows an example of the tap & talk interface used in the MiPad (Deng et al., 2002). If the user wants to provide the attendee information for a meeting scheduling task, he/she taps the "attendees" field in the calendar card. When that happens, the MUI will constrain both the language model and the semantic model based on the information on the potential attendees. This can significantly improve the accuracy and the throughput. Note that tap & talk functions as a user-initiative dialog-state specification. With tap & talk, there is no need for the mobile devices to include any special mechanism to handle spoken dialog focus and digression.

Figure 5. An example of the Tap & Talk interface (Deng et al., 2002, © 2002 IEEE)

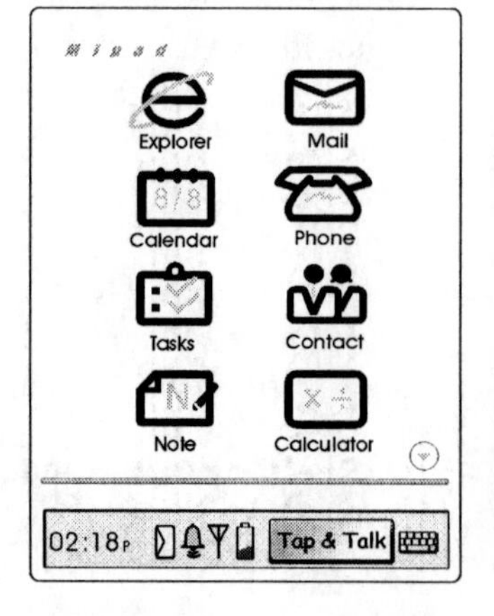

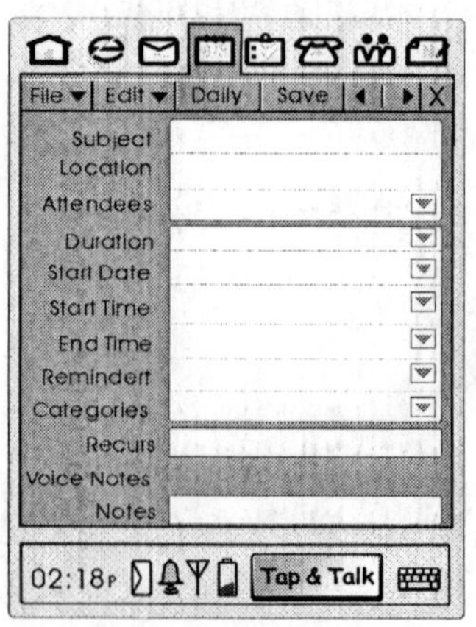

Context-Aware Language Model

Here, *context* refers to any information that can be used to characterize the situation related to human-computer interaction. It typically includes the surrounding environment (e.g., location and noise condition), the user (e.g., age and gender, preferences, past interaction experiences, and the interaction history in the current session), and the devices (e.g., remaining battery life, available memory, screen-size, screen-contrast, and speaker volume). Although context-awareness can be beneficial to all components in an MUI, it is especially important for improving speech recognition accuracy under noisy environments.

Context information can be utilized in many different ways in speech modality. One particular approach is to construct the language model based on the context. For example, the tap & talk approach (Deng et al., 2002) customizes the language model depending on the field the user is pointing to, as mentioned in section 3.3.

Language model can also be customized, based on the user information and the dialog state. For example, if the system is expecting the recipient information, the language model can include only the names in the global address book. If the user information is also used, the language model can also include user's contact list and people who have exchanged e-mails with the user in the past. An even more effective language model would weight different names differently, depending on the frequencies the user exchanged e-mail with the person, and the recentness of the interaction (Yu, Wang, Mahajan, Mau, & Acero, 2003). Another example of constructing the language model based on the context and user information is described in the speech enabled MapPoint (Deng & Yu, 2005). Without context information, the speech recognizer needs to load all location names and business names in the North America. This is definitely beyond the ability of most state-of-the-art speech recognizers. However, if the user's location information and/or the interaction history are known, the system can load only the location names and business names around the user's current location, and weight all the names based on the popularity of the names as well as the user's interaction history.

A more advanced context-aware language model construction technique is discussed by Wang (2004). This detection-based technique is used in the second generation of the MiPad (Wang, 2004). The basic idea of this approach is to detect the context cues from the user's partial utterances sequentially, and adjust the language model dynamically for the next part of the utterances. This approach has achieved excellent user experience.

LANGUAGE UNDERSTANDING

Good speech recognition accuracy does not always translate to good understanding of users' intents, as indicated by Wang, Acero, and Chelba (2003). A robust language-understanding model is needed to obtain good user experience for speech-centric MUI applications, especially since speech recognition errors will affect the understanding.

The first issue to address in language understanding is constructing the semantic grammar. Since the importance of each word to the understanding is different, the words need to be treated differently. A typical approach is to introduce a specific type of nonterminals called semantic classes to describe the schema of an application (Wang, 2001; Yu, Ju, Wang, & Acero, 2006). The semantic classes define the concepts embedded in the linguistic structures, which are usually modeled with probabilistic context-free grammars. The advantage of introducing the semantic classes is to make the linguistic realization of semantic concepts independent of the semantic concepts themselves. Once the semantic classes are defined, a robust linguistic grammar can be built using the approaches similar to the one described by Yu, et al. (2006).

The transformation from the recognized text to the semantic representation is usually done using a semantic parser. For example, in MiPad, this transformation is done using a robust chart parser (Wang, 2001). In this parser, "the robustness to ungrammaticality and noise can be attributed to its ability of skipping minimum unparsable segments in the input. The algorithm uses dotted rules, which are standard context free grammar rules in Backus Naur form plus a dot in front of a right-hand-side symbol. The dot separates the symbols that already have matched with the input words from the symbols that are yet to be matched." (Wang, 2001, pp. 1556) Since the language models used in MiPad are dynamically generated based on the current user information and the tap & talk field, the parser used in MiPad supports dynamic grammars. Given that some part of the user's utterances is in the free-style form (e.g., the topic of a meeting to be scheduled), they are modeled as dictation grammar rules. Since speech recognition is not perfect, the MiPad robust parser takes into account the N-best list, together with the associated confidence scores returned from the speech recognition engine, and combines the speech recognition score with the parsing score to obtain the best parsing result. More recent progress includes using maximum entropy models to classify the tasks and to disambiguate the meaning of the slots in the recognition result.

MODALITY FUSION

One strong advantage of using MUIs is the improved accuracy and throughput through modality integration. There are typically two fusion approaches: early fusion and late fusion. Given that late fusion has many superior properties over the early one, as discussed in Section 2, it will be the focus of our discussion in this section. There are two tasks in the late fusion: Process and convert the input signals into a common surface semantic representation using the semantic parsers (one specific to each modality), and fuse the surface semantics into discourse semantics using the discourse manager.

Semantic Representation and Unification

The semantic fusion operation requires a meaning representation framework that is common among modalities, and a well-defined operation for combining partial meanings.

Many semantic representation formats have been proposed in the past. For example, in Bolt's (1980) pioneering paper, only very limited modality fusion is required and hence, a simple semantic representation was used. In the past decade, researchers (Cheyer & Julia, 1995; Pavlovic & Huang, 1998; Shaikh, Juth, Medl, Marsic, Kulikowski, & Flanagan, 1997; Vo & Wood, 1996) have converged to using a data structure called typed *feature structures* (Kay, 1979) to represent meanings. Typed feature structure can be considered as an extended, recursive version of attribute-value-type data structures, where a value can, in turn, be a feature structure. It extends *frames* (Minsky, 1975) that represent objects and relations as nested sets of attribute/value pairs, by using shared variables to indicate common substructures. A typed feature structure indicates the kind of entity it represents with a type, and the values with an associated collection of feature-value or attribute-value pairs. In the typed feature structure, a value may be nil, a variable, an atom, or another typed-feature structure.

The primary operation on typed feature structure is *unification*. "*Typed-feature-structure unification* is an operation that determines the consistency of two representational structures and, if they are consistent, combines them into a single result." (Oviatt, Cohen, Wu, Vergo, Duncan, Suhm, et. al., 2000, online version pp. 21) Unification can combine complementary input from different modalities and rule out contradictory input (Johnston, 1998).

Note that users' multimodal inputs may involve *sequentially integrated* or *simultaneously delivered* signal fragments. In other words, temporal relationships between different input channels are very important. To fuse modalities, we need to first determine whether two input fragments are related. In most of the systems reported, this is achieved by considering all input contents that lie within a predefined time window. To do this, all input fragments need to be time stamped as soon as they are generated to remove the errors due to transit delays.

For example, the speech input "Show me the restaurants around here." might have a gesture-input accompanying it either "before," "during," or "after" the actual utterance, and all these three possibilities should provide the same result. Usually the term "before" represents a timeframe of up to several minutes, "during" represents a timeframe of 4 to 5 seconds, and "after" represents a timeframe of 500ms to 750ms. If these values are too small, many multimodal inputs will be considered as unimodal inputs and will not be integrated. If the values are too large the chances of an old or invalid user input are likely being accepted as part of a valid multimodal input.

To determine whether two input fragments should be treated as parts of a multimodal construction or separate unimodal commands, knowledge gained from a user study is very helpful. For example, it has been shown in Oviatt, DeAngeli, and Kuhn (1997) that users' written input precedes speech during a sequentially integrated multimodal command. They have also clarified the distribution of typical intermodal lags.

Semantic Fusion with Uncertain Inputs

The challenge of semantic fusion with uncertain inputs is to determine the unified meaning based on multimodal input fragments associated with probabilities. This is especially important for speech-centric MUI because the output of a speech recognizer is never certain. Note that the unification operation on the typed feature structure assumes that all input modalities are certain, and so they cannot be directly applied here. To fuse modalities with uncertainties, a *hybrid symbolic/statistical* architecture that combines statistical processing techniques with a symbolic unification-based approach is in need. This combined approach involves many factors when fusing the semantics. These factors include recognition accuracy of the individual modalities, the way of combining posterior probabilities, and the prior distribution of multimodal commands.

Note that a multimodal input gives rise to three different types of information overlay: nonoverlayed, overlayed and nonconflicting, and overlayed and conflicting. Nonoverlayed information indicates that the input (unimodal or multimodal) does not have any of the same information represented multiple times. This is the simplest condition. Overlayed and nonconflicting information refers to information segments that may have been represented multiple times without a conflict. The overlayed and conflicting information refers to the case that the information has been provided multiple times and conflicts. There are many approaches to resolving conflicting information in typed feature structure if no uncertainty is involved. The "unification" approach simply returns the value null when a conflict is detected. The "overlay" method returns the first argument when conflicting information is present. However, given that the semantic information from different modalities should not be equally trusted, a better conflicting information resolving approach can be found to handle input signals that may or may not be overlapped in their temporal delivery (Oviatt et al., 1997). Note that overlayed information may arise when inputs are from different modalities (e.g., speech and gesture), or when the same-type modality information occurs multiple times over an extended time frame. Both these two conditions need to be handled.

Conventionally, the probability of the merged feature structures is the cross product of the probabilities of individual feature structures based on the assumption that inputs are statistically independent with each other. In this section, we describe an alternative statistical approach that has been used in QuickSet (Wu, Oviatt, & Cohen, 1999). This approach uses the associative map to reduce the unification pairs and members-teams-committee (MTC) model to refine the multimodal integration process so that different weights are assigned to different modes and different constituents.

Associative map defines all semantically meaningful mapping relations that exist between different sets of constituents for each multimodal command. In its simplest form, it can be considered as a simple process of table lookup. For example, if an MUI consists of only the speech modality and the pen modality, we can build a two-dimensional table. If two inputs from different modalities can be fused, the value at the corresponding cell is 1; otherwise, the value is 0. The purpose of the associative map is to rule out considerations of those feature structures that cannot possibly be unified semantically.

Members-teams-committee weighs the contributions derived from different modality recognizers based on their empirically-derived relative reliabilities. MTC consists of multiple members, multiple teams, and a committee. "*members* are the individual recognizers that provide a diverse spectrum of recognition results (local posterior probabilities). Member recognizers can be on more than one team. Members report their results to their recognizer *team* leader, which then applies various weighting parameters to their reported scores. Furthermore, each team can apply a different weighting scheme, and can examine different subsets of data. Finally, the *committee* weights the results of the various teams, and reports the final recognition results. The parameters at each level of the hierarchy are trained from a labeled corpus." (Oviatt, et al., 2000, online version, p. 24).

CONCLUSION AND FUTURE DIRECTIONS

In this chapter, we discussed the importance of using the MUI in mobile devices, and described the state-of-the-art technologies in designing speech-centric MUI in mobile devices. Specifically, we discussed the noise robustness technologies, the reliable modality switching methods, the context-aware language model, and the robust language-understanding technologies that contribute to the usability of the speech modality. We also described the modality integration technologies that are important to improving the accuracy and throughput of the MUI. Although these technologies have greatly advanced the speech centric MUI design and development in the mobile devices, future research is needed in the following areas.

Microphone Array Processing

Noise robustness is still a challenging research area for speech-centric MUIs. Although many single-microphone noise robustness technologies (e.g., Deng, et al., 2002; Macho, et al. 2002) have been proposed to improve speech recognition accuracy under noisy environments, the progress so far is still limited. Given the continuous decrease in the hardware price, using microphone array on mobile devices is a trend to combat noisy acoustic conditions and to further decrease speech recognition errors. Microphone array algorithms, which take advantage of the received signal differences between microphones, can achieve noise suppression of 10-15 db effectively (Tashev & Malvar, 2005). Future research is needed for more efficient and effective algorithms using low-cost, low-quality microphone arrays that may be equipped in speech-centric mobile devices.

Error Handling Techniques

Fragile error handling continues to be a top interface problem for speech-centric MUI (Karat, Halverson, Horn, & Karat, 1999; Rhyne & Wolf, 1993; Roe & Wilpon, 1994). A great amount of research work needs to be done in developing graceful error-handling strategies in speech-centric MUI. First, new statistical methods need to be developed to reduce errors through mutual disambiguation between modalities. Second, new dialog strategies (e.g., mixed initiative) need to be developed to allow easy correction of the errors. Third, the system needs to be able to adapt to different environments and challenging contexts to reduce errors. Fourth, better robust speech recognition technologies need to be developed to increase the speech recognition accuracy under a wide range of environments.

Adaptive Multimodal Architectures

In most current MUI systems, their behaviors are predesigned by the developers. The system does not automatically learn to improve the performance as users use the system. Given that mobile devices are usually used by a single user, it is very important to develop adaptive MUI architectures.

For example, Oviatt (1999) showed that any given user's habitual integration pattern (simultaneous vs. sequential) is apparent at the beginning of their system interaction. When the user uses the system, the interaction pattern remains the same. An adaptive MUI system that can distinguish and utilize these patterns to improve the modality fusion could potentially achieve greater recognition accuracy and interactive speed. Another example is for the system to gradually change the behavior (e.g., automatically predict the user's next action) when the user changes from a novice to an experienced user.

Future research in this area would include what and when to adapt, as well as how (e.g., through reinforcement learning) to adapt MUI systems so that their robustness can be enhanced.

Mixed Initiative Multimodal Dialog

Most current speech-centric MUI systems are user initiative, where the user controls the dialog flow (for example, through push to talk). A user-initiative system can be modeled as a set of asynchronous event handlers. In a more advanced system, the system should also actively interact with the user to ask for missing information (which is called mixed initiative). For example, if the user wants to search for the phone number of a business using a mobile device and he/she forgets to mention the city and state information, the dialog system should automatically ask the user for that information through the multimodal output devices.

Future research should address the design and development of consistent and efficient conversational interaction strategies that can be used by different multimodal systems. Multimodal dialogue systems should be developed within a statistical framework (Horvitz, 1999) that permits probabilistic reasoning about the task, the context, and typical user intentions.

REFERENCES

Becker, N. (2001). *Multimodal interface for mobile clients*, Retrieved July 16, 2006, from http://citeseer.ifi.unizh.ch/563751.html

Bolt, R. A. (1980). Put-that-there: Voice and gesture at the graphics interface. *Computer Graphics, 14(3),* 262-270.

Bregler, C., Manke, S., Hild, H., & Waibel, A. (1993). Improving connected letter recognition by lipreading. *Proceedings of the International Conference on Acoustics, Speech and Signal Processing, 1,* 557-560.

Cheyer, A., & Julia, L. (1995). Multimodal maps: An agent-based approach. *International Conference on Cooperative Multimodal Communication* (pp. 103-113).

Codella, C., Jalili, R., Koved, L., Lewis, J., Ling, D., Lipscomb, J., Rabenhorst, D., Wang, C., Norton, A., Sweeney, P., & Turk, C. (1992). Interactive simulation in a multi-person virtual world. *Proceedings of the Conference on Human Factors in Computing Systems* (pp. 329-334).

Cohen, P. R., Dalrymple, M., Moran, D. B., Pereira, F. C. N., Sullivan, J. W., Gargan, R. A., Schlossberg, J. L., & Tyler, S. W. (1989). Synergistic use of direct manipulation and natural language. *Proceedings of the Conference on Human Factors in Computing Systems* (pp. 227-234).

Cohen, P. R., Johnston, M., McGee, D., Oviatt, S., Pittman, J., Smith, I., Chen, L., & Clow, J. (1997). Quickset: Multimodal interaction for distributed applications. *Proceedings of the Fifth ACM International Multimedia Conference* (pp. 31-40).

Das, S., Bakis, R., Nadas, A., Nahamoo, D. & Picheny, M. (1993). Influence of background noise and microphone on the performance of the IBM TANGORA speech recognition system. *Proceedings of the IEEE International Conference on Acoustic Speech Signal Processing* (pp. 71-74).

Deligne, S., Dharanipragada, S., Gopinath, R., Maison, B., Olsen, P., & Printz, H. (2002). A robust high accuracy speech recognition system for mobile applications. *IEEE Transactions on Speech and Audio Processing, 10*(8), 551-561.

Deng, L., Wang, K., Acero, A., Hon, H., Droppo, J., Boulis, C., Wang, Y., Jacoby, D., Mahajan, M., Chelba, C., & Huang, X.D. (2002). Distributed speech processing in MiPad's multimodal user interface. *IEEE Transactions on Speech and Audio Processing, 10*(8), 605-619.

Deng, L, & Yu, D. (2005). A speech-centric perspective for human-computer interface - A case study. *Journal of VLSI Signal Processing Systems (Special Issue on Multimedia Signal Processing), 41*(3), 255-269.

Dybkjaer, L., & Bernsen, N.O . (2001). Usability evaluation in spoken language dialogue system. *Proceedings of the Workshop on Evaluation for Language and Dialogue Systems, Association for Computational Linguistics 39th Annual Meeting and 10th Conference of the European Chapter*,,(pp. 9-18).

Fukumoto, M., Suenaga, Y., & Mase, K. (1994). Finger-pointer: Pointing interface by image processing. *Computer Graphics, 18(5),* 633-642.

Gong, Y. (1995). Speech recognition in noisy environments: A survey. *Speech Communication, 16*, 261-291.

Gould, J. & Lewis, C. (1985). Design for usability: Key principles and what designers think. *Communications of the ACM, 28*(3): 300-301.

Holzman, T. G. (1999). Computer-human interface solutions for emergency medical care. *Interactions, 6(3),* 13-24.

Hone, K. S., & Graham, R. (2001). Subjective assessment of speech system interface usability. *Proceedings of the Eurospeech Conference* (pp. 2083-2086).

Horvitz, E. (1999). Principles of mixed-initiative user interfaces. *Proceedings of the Conference on Human Factors in Computing Systems* (pp. 159-166).

Huang, X., Acero, A., Chelba, C., Deng, L., Droppo, J., Duchene, D., Goodman, J., Hon, H., Jacoby, D., Jiang, L., Loynd, R., Mahajan, M., Mau, P., Meredith, S., Mughal, S., Neto, S., Plumpe, M., Stery, K., Venolia, G., Wang, K., & Wang. Y. (2001). MIPAD: A multimodal interaction prototype. *Proceedings of the International Conference on Acoustics, Speech, and Signal Processing, 1*, 9-12.

Huang, X., Acero, A., Chelba, C., Deng, L., Duchene, D., Goodman, J., Hon, H., Jacoby, D., Jiang, L., Loynd, R., Mahajan, M., Mau, P., Meredith, S., Mughal, S., Neto, S., Plumpe, M., Wang, K., & Wang, Y. (2000). MIPAD: A next generation PDA prototype. *Proceedings of the International Conference on Spoken Language Processing, 3*, 33-36.

Hsu B.-J., Mahajan M., & Acero A. (2005). *Multimodal text entry on mobile devices.* The ninth bi-annual IEEE workshop on Automatic Speech Recognition and Understanding (Demo). Retrieved July 20, 2006, from

http://research.microsoft.com/~milindm/2005-milindm-ASRU-Demo.pdf

Johnston, M. (1998). Unification-based multimodal parsing. *Proceedings of the International Joint Conference of the Association for Computational Linguistics and the International Committee on Computational Linguistics* (pp. 624-630).

Karat, C.-M., Halverson, C., Horn, D., & Karat, J. (1999). Patterns of entry and correction in large vocabulary continuous speech recognition systems. *Proceedings of the International Conference for Computer-Human Interaction* (pp. 568-575).

Kay, M. (1979). Functional grammar. *Proceedings of the Fifth Annual Meeting of the Berkeley Linguistics Society* (pp. 142-158).

Kumar, S., & Cohen, P. R. (2000). Towards a fault-tolerant multi-agent system architecture. *Fourth International Conference on Autonomous Agents* (pp. 459-466).

Li, X., Malkin J., & Bilmes, J. (2006). A high-speed, low-resource ASR back-end based on custom arithmetic. *IEEE Transactions on Audio, Speech and Language Processing, 14*(5), 1683-1693.

Lockwood, P., & Boudy, J. (1992). Experiments with a nonlinear spectral subtractor (NSS), hidden Markov models and the projection for robust speech recognition in cars. *Speech Communication, 11*(2-3), 215-28.

Macho, D., Mauuary, L., Noé, B., Cheng, Y. M., Ealey, D., Jouvet, D., Kelleher, H., Pearce, D., & Saadoun, F. (2002). Evaluation of a noise-robust DSR front-end on Aurora databases. *Proceedings of the International Conference on Spoken Language Processing* (pp. 17-20).

Minsky, M. (1975). A framework for representing knowledge. In P. Winston (Ed.), *The psychology of computer vision* (pp 211-277). New York: McGraw-Hill.

Neal, J. G., & Shapiro, S. C. (1991). Intelligent multimedia interface technology. In J. Sullivan & S. Tyler (Eds.), *Intelligent user interfaces* (pp.11-43). New York: ACM Press.

Nielsen, J. (1993). *Usability engineering.* San Diego: Academic Press .

Norman, D. A. & Draper, S. W. (Eds.). (1986). *User-centered system design: New perspectives on human-computer interaction.* Hillsdale, NJ: Lawrence Erlbaum Associates.

Oviatt, S. L. (1999). Ten myths of multimodal interaction, *Communications of the ACM, 42* (11), 74-81.

Oviatt, S. L., Bernard, J., & Levow, G. (1999). Linguistic adaptation during error resolution with spoken and multimodal systems. *Language and Speech* (special issue on *Prosody and Conversation), 41*(3-4), 415-438.

Oviatt, S. L., Cohen, P. R., Wu, L., Vergo, J., Duncan, L., Suhm, B., Bers, J., Holzman, T., Winograd, T., Landay, J., Larson, J., & Ferro, D. (2000). Designing the user interface for multimodal speech and gesture applications: State-of-the-art systems and research directions. *Human Computer Interaction*, 263-322. (online version), Retrieved July 20, 2006, from http://www.cse.ogi.edu/CHCC/Publications/designing_user_interface_multimodal_speech_oviatt.pdf

Oviatt, S. L., DeAngeli, A., & Kuhn, K. (1997). Integration and synchronization of input modes during multimodal human-computer interaction. *Proceedings of Conference on Human Factors in Computing Systems (CHI'97)* (pp. 415-422).

Oviatt, S. L. & Olsen, E. (1994). Integration themes in multimodal human-computer interaction. In Shirai, Furui, & Kakehi (Eds.), *Proceedings of the International Conference on Spoken Language Processing, 2*, 551-554.

Oviatt, S. L., & vanGent, R. (1996). Error resolution during multimodal human-computer interaction. *Proceedings of the International Conference on Spoken Language Processing, 2*, 204-207.

Pavlovic, V., Berry, G., & Huang, T. S. (1997). Integration of audio/visual information for use in human-computer intelligent interaction. *Proceedings of IEEE International Conference on Image Processing* (pp. 121-124).

Pavlovic, V., & Huang, T. S., (1998). Multimodal prediction and classification on audio-visual features. *AAAI'98 Workshop on Representations for Multi-modal Human-Computer Interaction*, 55-59.

Ravden, S. J., & Johnson, G. I. (1989). *Evaluating usability of human-computer interfaces: A practical method.* Chichester: Ellis Horwood.

Reeves, L. M., Lai, J., Larson, J. A., Oviatt, S., Balaji, T. S., Buisine, S., Collings, P., Cohen, P., Kraal, B., Martin, J. C., McTear, M., Raman, T. V., Stanney, K. M., Su, H., & Wang, Q. Y. (2004). Guidelines for multimodal user interface design. *Communications of the ACM – Special Issue on Multimodal Interfaces, 47*(1), 57-59.

Rhyne, J. R., & Wolf, C. G. (1993). Recognition-based user interfaces. In H. R. Hartson & D. Hix (Eds.), *Advances in Human-Computer Interaction, 4,* 191-250.

Roe, D. B., & Wilpon, J. G. (Eds.). (1994). *Voice communication between humans and machines*. Washington, D.C: National Academy Press.

Rubin, P., Vatikiotis-Bateson, E., & Benoit, C. (1998). Special issue on audio-visual speech processing. *Speech Communication, 26* (1-2).

Schwartz, D. G. (1993). *Cooperating heterogeneous systems: A blackboard-based meta approach*. Unpublished Ph. D. thesis, Case Western Reserve University.

Shaikh, A., Juth, S., Medl, A., Marsic, I., Kulikowski, C., & Flanagan, J. (1997). An architecture for multimodal information fusion. *Proceedings of the Workshop on Perceptual User Interfaces*, 91-93.

Stork, D. G., & Hennecke, M. E. (Eds.) (1995). *Speechreading by humans and machines*. New York: Springer Verlag.

Suhm, B. (1998). *Multimodal interactive error recovery for non-conversational speech user interfaces*. Ph.D. thesis, Fredericiana University.

Tashev, I., & Malvar, H. S. (2005). A new beamformer design algorithm for microphone arrays. *Proceedings of International Conference of Acoustic, Speech and Signal Processing, 3*, 101-104.

Varga, I., Aalburg, S., Andrassy, B., Astrov, S., Bauer, J.G., Beaugeant, C., Geissler, C., & Hoge, H. (2002). ASR in mobile phones - an industrial approach, *IEEE Transactions on Speech and Audio Processing, 10*(8), 562- 569.

Vo, M. T., Houghton, R., Yang, J., Bub, U., Meier, U., Waibel, A., & Duchnowski, P. (1995). Multimodal learning interfaces. *Proceedings of the DARPA Spoken Language Technology Workshop*. Retrieved July 20, 2006 from

http://www.cs.cmu.edu/afs/cs.cmu.edu/user/tue/www/papers/slt95/paper.html

Vo, M. T., & Wood, C. (1996). Building an application framework for speech and pen input integration in multimodal learning interfaces. *Proceedings of IEEE International Conference of Acoustic, Speech and Signal Processing, 6*, 3545-3548.

Wang, J. (1995). Integration of eye-gaze, voice and manual response in multimodal user interfaces. *Proceedings of IEEE International Conference on Systems, Man and Cybernetics* (pp. 3938-3942).

Wang, K. (2004). A detection based approach to robust speech understanding. *Proceedings of the International Conference on Acoustics, Speech, and Signal Processing, 1*, 413-416.

Wang, Y. (2001). Robust language understanding in MiPAD. *Proceedings of the Eurospeech Conference* (pp. 1555-1558).

Wang, Y., Acero, A., & Chelba, C. (2003). Is word error rate a good indicator for spoken language understanding accuracy? *Proceedings of the Workshop on Automatic Speech Recognition Workshop and Understanding* (pp 577-582).

Wu, L., Oviatt, S., &. Cohen, P. (1999). Multimodal integration-A statistical view. *IEEE Transactions on Multimedia, 1*(4), 334-341.

Yu, D., Ju, Y. C., Wang, Y., & Acero, A. (2006). N-gram based filler model for robust grammar authoring. *Proceedings of the International Conference on Acoustics, Speech, and Signal Processing1*, 565-569.

Yu, D., Wang, K., Mahajan, M., Mau, P., & Acero, A. (2003). Improved name recognition with user modeling. *Proceedings of Eurospeech* (pp. 1229-1232).

KEY TERMS

Modality: A communication channel between human and computer, such as vision, speech, keyboard, pen, and touch.

Modality Fusion: A process of combining information from different input modalities in a principled way. Typical fusion approaches include early fusion, in which signals are integrated at the feature level, and late fusion, in which information is integrated at the semantic level.

Multimodal User Interface: A user interface with which users can choose to interact with a system through one of the supported modalities, or multiple modalities simultaneously, based on the usage environment or preference. Multimodal user interface can increase the usability because the strength of one modality often compensates for the weaknesses of another.

Push to Talk: A method of modality switching where a momentary button is used to activate and deactivate the speech recognition engine.

Speech-Centric Multimodal User Interface: A multimodal user interface where speech is the central and primary interaction modality.

Typed feature Structure: An extended, recursive version of attribute-value type data structures, where a value can, in turn, be a feature structure. It indicates the kind of entity it represents with a type, and the values with an associated collection of feature-value or attribute-value pairs. In the typed feature structure, a value may be nil, a variable, an atom, or another typed feature structure.

User-Centered Design: A design philosophy and process in which great attention is given to the needs, expectations, and limitations of the end user of a human-computer interface at each stage of the design process. In the user-centered design process, designers not only analyze and foresee how users are likely to use an interface, but also test their assumptions with actual users under real usage scenario.

Chapter XXIX
Model-Based Target Sonification in Small Screen Devices:
Perception and Action

Parisa Eslambolchilar
University of Wales, UK

Andrew Crossan
University of Glasgow, UK

Roderick Murray-Smith
University of Glasgow, UK
Hamilton Institute, NUI Maynooth, Ireland

Sara Dalzel-Job
University of Glasgow, UK

Frank Pollick
University of Glasgow, UK

ABSTRACT

In this work, we investigate the use of audio and haptic feedback to augment the display of a mobile device controlled by tilt input. The questions we answer in this work are: How do people begin searching in unfamiliar spaces? What patterns do users follow and which techniques are employed to accomplish the experimental task? What effect does a prediction of the future state in the audio space, based on a model of the human operator, have on subjects' behaviour? In the pilot study we studied subjects' navigation in a state space with seven randomly placed audio sources, displayed via audio and vibrotactile modalities. In the main study, we compared only the efficiency of different forms of audio feedback. We ran these experiments on a Pocket PC instrumented with an accelerometer and a headset. The accuracy of selecting, exploration density, and orientation of each target was measured. The results quantified the changes brought by predictive or "quickened" sonified displays in mobile, gestural interaction. Also, they highlighted subjects' search patterns and the effect of a combination of independent variables and each individual variable in the navigation patterns.

INTRODUCTION

One of the main goals of interaction design is to make the interfaces as intuitive as possible. In our everyday environments, humans receive a variety of stimuli playing upon all senses, including aural, tactile, and visual, and we respond to these stimuli. Even though hearing and vision are our two primary senses, most of today's interfaces are mainly visual.

Visual interfaces have crucial limitations in small-screen devices. These devices have a limited amount of screen space on which to display information. Designing interfaces for mobile computers/phones is problematic, as there is a very limited amount of screen resource on which to display information, and users' need to focus on the environment rather than the interface (so that they can look where they are going) so output is limited (Blattner, Papp, & Glinert, 1992; Brewster, 1997; Brewster & Murray, 1998; Johnson, Brewster, Leplatre, & Crease, 1998; Kramer, Walker, Bonebright, Cook, Flowers, Miner, 1999; Rinott, 2004; Smith & Walker, 2005; Walker & Lindsay, 2006); also, low graphics resolution and further constrain the freedom of interface designers. In new generations of mobile phones (e.g., iPhone) with high graphics resolution, power consumption for graphics rendering is high, which can adversely affect battery life; also, large screens can lead to physical robustness issues, as well as being very demanding of user attention in mobile scenarios.

One way around these problems would be sonically enhanced interfaces that require less or no visual attention and therefore, the size of the visual display and the portable device can be decreased; also, auditory interfaces potentially interfere less in the main activity in which the user is engaged. Consequently, the user may be able to perform more than one task at a time, such as driving a car while using a telephone or grabbing a cup of coffee while waiting for a mobile phone to finish downloading an image. Auditory feedback can often be a necessary complement, but also a useful alternative to visual feedback. When designing a mobile electronic device, it is difficult to predict all possible scenarios when it might be used. Obviously, visual feedback is preferred in many situations such as in noisy environments or when the user has to concentrate on a listening task. However, as there might be numerous occasions when a user cannot look at a display, versatile devices such as mobile phones or handheld computers benefit from having flexible interfaces.

Novel Interaction and Continuous Control

In the past 10 years many researchers have focused on tilt-based inputs, and audio and haptic outputs in mobile HCIs (Dong, Watters, & Duffy, 2005; Fallman, 2002a, 2002b; Harrison & Fishkin, 1998; Hinckley, Pierce, Horvitz, & Sinclair, 2005; Oakley, Ängeslevä, Hughes, & O'Modhrain, 2004; Partridge, Chatterjee, Sazawal, Borriello, & Want, 2002; Rekimoto, 1996; Sazawal, Want, & Borriello, 2002; Wigdor & Balakrishnan, 2003). The results of these researches have proved one-handed control of a small screen device needs less visual attention than two-handed control and multimodality in the interaction can compensate for the lack of screen space. So these novel interaction techniques, that is, gesture recognition, and audio and haptic devices, are characterised by the significance of the temporal aspect of interaction and in such an emerging environment, the interaction is no longer based on a series of discrete steps, but on a continuous input/output exchange of information that occurs over a period of time at a relatively high rate, somewhat akin to vision based or audio/haptic interfaces, which we may not model appropriately as a series of discrete events (Doherty & Massink, 1999; Faconti & Massink, 2001).

Novel interaction techniques with computers and handheld devices are examples of interactive dynamic systems, and development of these systems explores a range of possible solutions for overcoming some problems of development on computing devices, including the limited source of input/output devices, adaptability, predictability, disturbances, and individual differences. We explicitly include dynamics because we experience our environment in the way we want it by our actions or behaviour. Thus, we control what we perceive and while, in principle, interaction with handheld devices is rich in the variety of tasks supported, from computation and information storage to sensing and communication, we are dependent on the display of feedback (either visual, audio, or haptic) to help us pursue our sometime constantly changing goals feedback, which may influence a user's actions as more information becomes available (Doherty & Massink, 1999; Faconti & Massink, 2001). So developing interaction for such devices is closely related to the engineering of mobile interfaces based on dynamics.

Control Theory and Fitts' Law

A branch of control theory that is used to analyse human and system behaviour when operating in a tightly coupled loop is called *manual control theory* (Jagacinski & Flach, 2003; Poulton, 1974). The theory is applicable to a wide range of tasks involving vigilance, tracking and stability, and so forth. The general approach followed in manual control theory is to express the dynamics of combined human and controlled element behaviour as a set of linear differential equations in the time domain (Poulton, 1974). Several models include human-related aspects of information processing explicitly in the model, such as delays for visual process, motor-nerve latency, and neuromotor dynamics (Jagacinski & Flach, 2003). Control theory can be linked to Fitts' Law (Fitts, 1954; MacKenzie & Ware, 1993; Mackinlay, Robertson, & Card, 1991) by viewing the pointing movements towards the target as a feedback control loop based on visual input, and the limb as a control element (Bootsma, Fernandez, & Mottet, 2004; Crossman & Goodeve, 1983; Hoffmann, 1991; Jagacinski & Flach, 2003; Langolf, Chaffin, & Foulke, 1976).

This research outlines the use of model-based sonification to shape human action when users interact with small devices based on auditory feedback. In this work, we investigate the usability of nonspeech sounds and haptic feedback to augment the display of a mobile device controlled by a gesture input. Nonspeech sound has advantages over speech in that it is faster as well as language independent. We look at control strategies of users in browsing the audio/haptic state space. We also suggest one possible way of improving performance based on models of human control behaviour in a few example applications.

BACKGROUND

The single audio output channel has been little used to improve interaction in mobile devices. Speech sounds are, of course, used in mobile phones when calls are being made, but are not used by the telephone to aid the interaction with the device (Blattner et al., 1992; Brewster, 2002; Gaver, Smith, & O'Shea, 1991; Smith & Walker, 2005; Walker & Lindsay, 2006). Nonspeech sounds and vibrotactile devices are used for ringing tones or alarms but again, do not help the user interact with the system beyond this. Some signals provide feedback that some event has been successful, such as when buttons are pressed or devices are switched on. Selecting items with a stylus in PDAs without tactile feedback is often confusing for users because it is hard to know whether they have hit the target or not, especially if used in a mobile setting (Brewster, 2002). In this case, vibrators in mobile phones could be a good haptic feedback. It assures the user that s/he is in the target, and if the user wants to select a target, s/he can then press a key in the vibration area to select it.

If using continuous sounds as opposed to the more common brief signals, auditory interfaces do not need to be more transitory than visual interfaces. However, such sounds probably benefit from being quite discreet. While, most existing sound feedback today occurs in the foreground of the interface, subtle background sounds can be a useful complement in advanced auditory interfaces. Films and computer games generally make use of music and sound effects. Film sound theorist Michel Chion (Chion, 1994) has made the following statement concerning sound in film: *there is no soundtrack*. An extreme statement coming from a researcher of sound, Chion means that there is no way to separate the auditory and visual channels of a film. We experience them only through a unified sense, which he terms "audio-vision."

In a similar way, an interface that uses sound cleverly can enhance the user's immersion and improve interaction. Gaver (1997) found that during an experimental process control task, the participants' engagement increased when provided with relevant sound feedback. There is now evidence that sound can improve interaction and may be very powerful in small screen devices (Brewster, 2002). If the possibility of conveying information sonically were used to its full potential, it would be a powerful complement to visual interfaces (Brewster & Murray, 1998). A strong argument against the use of sound in interfaces is that it easily can become annoying, both for the user and other people around them, since it is more intrusive than visual impressions. It is not useful in noisy environments, for instance, train stations, undergrounds, and so forth. However, by skilfully designing auditory interfaces or using haptic feedback, this can be avoided, and interaction with machines can become easier and hopefully more pleasant.

The most advanced auditory/haptic feedback seems to exist in computer games and multimedia products. Gaver (1997) claims that memory limitations in the technical product are one reason why sound feedback has not been used on a larger scale. Until quite recently, it has been too expensive computationally to use sound of good quality in computers and handheld devices. Today, only light-weight electronic devices, such as mobile phones or handheld computers, have limited memory capacities, although this is rapidly changing with the development of memory cards and effective compression algorithms for sound. However, nowadays these devices give various choices of discrete audio/haptic ring tones and alarms and to their users. The potential to use sound and haptic in small electronics is growing fast.

MODEL-BASED SONIFICATION

As there are many ways in which sound can be employed in interfaces, it is important to define the purposes of every sound at an early stage in the design process. A sound that conveys crucial information should have different attributes to one that serves as a complement to visual information. It is important to distinguish between two very different approaches (Chion, 1994): the practical and the naturalistic approach. The "practical" approach to auditory interfaces deals with sound as the main feedback. This can be the case when designing interfaces for visually impaired people, who must rely on sound feedback to provide sufficient assistance in performing a task. Furthermore, sound is often the only means of communication when using a portable hands-free device with a mobile phone. Auditory interfaces based on a practical approach should be comprehensive and simple (Brewster, 2002; Brewster & Murray, 1998; Smith & Walker, 2005; Walker & Lindsay, 2006). The drawback of this approach is sound might be noisy and tiresome over time. The "naturalistic" view regards sound mainly as a complement to a visual interface. A naturalistic interface combines sound and vision in a way as similar as possible to corresponding phenomena in the natural world. Such auditory interfaces are supposed to enhance interaction between the user and a machine, especially in situations where the visual interface is ineffective on its own. Sounds that complement a visual interface can generally be subtle background events that do not disturb. In a way, such sounds correspond to the background music of films, since they convey information to the audience without interfering with the main events. Sound feedback based on the naturalistic strategy is thus very subtle, and might only be recognised subconsciously. The focus of this work is on the "practical" approach.

Sonification is a method suggested in "practical" domain, which is defined as the use of nonspeech audio to convey information. More specifically, sonification is the transformation of data relations into perceived relations in an acoustic signal for the purposes of facilitating communication or interpretation (Gaver, 1989). Many of the major current research areas in sonification are similar in that they focus on the identification of applications for which audition provides advantages over other modalities, especially for situations where temporal features are important or the visual modality is overtaxed. The main issues that will move sonification research forward include (1) mapping data onto appropriate sound features like volume, pitch, timbre, (2) understanding dynamic sound perception, (3) investigating auditory streaming, (4) defining and categorising salience in general auditory contexts, and understanding where highly salient sonic events or patterns can surpass visual representations in data mining, and (5) developing multimodal applications of sonification (Kramer et al., 1999); sonification is a way to help in the exploration of complex data. Various kinds of information can be presented using sonification, simply by using different acoustic elements. This information has been organised in Hermann, Hansen, and Ritter (2000).

Studies such as Cook et al. (2002) and Cook and Lakatos (2003) have investigated the human ability to perceive various physical attributes of sound sources, and have proved that feature-based synthesis is of use in studying the low-level acoustical properties that human listeners use to deduce the more complex physical attributes of a sound's source. The generated sounds from a set of features are correlated with the listener's perception of, for example, size, speed, or shape of the source. Two methods of sonification have been used in this chapter, the Doppler effect and derivative volume adaptation. Both of these methods create a continuous sound for each data point. Thus, the relative position to the targets is perceived by a change of volume when passing the data point and pitch shift for Doppler effect as well. From the data points obtained in this way, we may be able to discover consistent relationships between acoustical and human-generated features that can be used to predict how a sound manifesting certain acoustic feature values will be perceived.

Quickening

"Quickening" is a method for reducing the difficulty of controlling second-order or higher-order systems, by changing the display to include predictions of future states, that was proposed by Birmingham and Taylor (1954), and is reviewed in Jagacinski and Flach (2003). A quickened display for an acceleration control system like the system described in this chapter shows the user a weighted combination of position and velocity (see Figure 1). This weighted summation effectively anticipates the future position of the system. It can greatly improve human performance in controlling these

Figure 1. A block diagram for a second-order system with a quickened display. The output to the quickened display is the sum of position and velocity. Effectively, the quickened display projects the output into the future based on the current velocity.

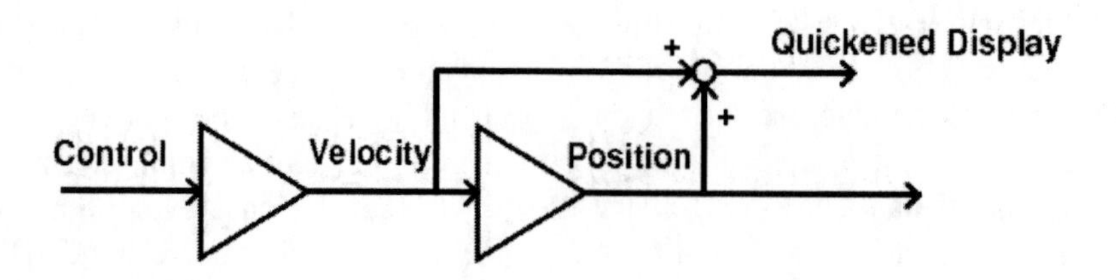

systems. Quickening in general is a prediction of the future state of the system based on the current state vector (for example position, velocity, acceleration) and a model of system behaviour and expected user action.

An example of this is based on the Doppler effect, which highlights the user's approach to a target, or a target's movement from the current state. Another example could be derivative of volume of sound source. When the user is further from the audio source, the sound is quieter than when the user is close to it. Another predictive method that has been investigated in Williamson, Strachan, and Murray-Smith (2006) is *Monte Carlo* simulation in a tilt-controlled navigation system.

Doppler Effect

The auditory system is responsible for constructing a map of the auditory scene around us, using information from audio input, that is, sound localisation (Bregman, 1990; Smith, 2004). There are various types of cues that humans can use to localise the position of a sound source. These cues can be divided into monaural and binaural cues. The two different types of monaural cue are loudness and Doppler shift. The loudness cue relies on the fact that when a sound source is far away, it is quieter than when it is close by. The Doppler shift corresponds to a frequency shift associated with a sound source moving through a homogeneous medium (Smith, 2004). Pressure wave crests emerge from the sound source at intervals corresponding to the acoustic wavelength. Each crest spreads spherically out from the point of origin at the speed of sound c (Figure 2). The successively generated spheres of wave crests are closer together ahead of the sound source but farther apart behind the source. For a stationary observer, the measured frequency corresponds to the number of crests per unit time, so the composite frequencies will be higher when the observer is in front of the moving sound source, and less when behind the moving sound source (Hermann et al., 2000; Hermann & Ritter, 1999). A familiar example is the shift in frequency of an ambulance siren as the vehicle approaches, passes, and then recedes. The well-known lawful dependence of the Doppler shifted frequency, here denoted Ψ_t, on velocity of the sound source relative to an observer is:

$$\Psi_t = f(1 + \frac{v}{c}\cos\Phi_t) \qquad (1)$$

where f is the intrinsic frequency of the sound source, v is the velocity magnitude (speed), and c is the speed of sound. The shifted frequency Ψ_t depends only on the velocity component directed toward the observer with angle Φ_t (see Figure 2). The shifted frequency has the maximum value when Φ_t is zero. As this angle reaches 90°, all motion is across the line of hearing and the Doppler shift is zero. This result holds true regardless of

Figure 2. The geometry for the Doppler shift of a moving sound source relative to an observer.

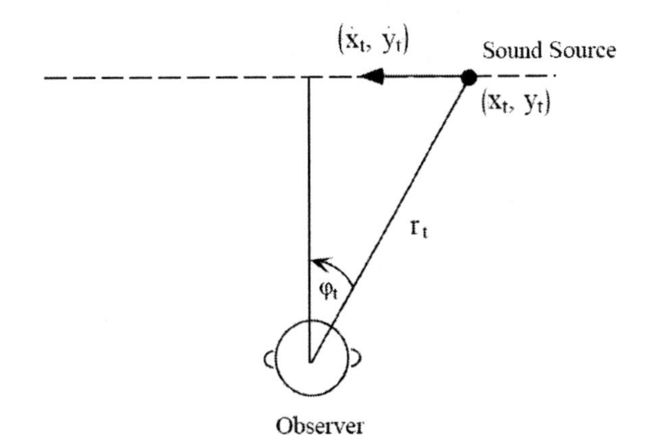

the time history of the trajectory (Jenison, 1997). These aural cues can be used to navigate through the virtual environment on a Pocket PC.

In the next sections we present pros and cons of different quickened methods and control strategies in browsing the state-space on a mobile device using tilt-input.

EXPERIMENT

Goals

There is a concept of accuracy explored in this work. The type of accuracy that is under primary consideration in this study is the capability of subjects to accurately identify audio sources in large audio data sets with a PDA and tilt-sensor using sound only. In navigating a computer display of data visually, accuracy is seldom a concern. Using a scrollbar or clicking a 10×10 pixel icon using one's vision is trivial from the perspective of the accuracy needed to accomplish this task (Holmes, 2005). Designers of auditory displays, on the other hand, are in need of research into the accuracy that is possible in this environment. Establishing the accuracy with which humans can navigate using sound alone is an early step in integrating sound into a multi-modal information system.

The other questions we answer in this work are: How do people begin searching in unfamiliar spaces? What patterns or techniques are employed to accomplish the experimental task? How will predicting the future state in the audio space change subjects' accuracy in targeting?

Apparatus

The experiment was conducted on a Pocket PC (hp5450), running windows CE, with a 240×320 resolution, colour display, an accelerometer Xsens P3C, 3 degree-of-freedom, attached to the serial port, which allows the users to navigate through the environment by tilting the device, and a stereo headset (Figure 3). The built-in vibrator unit in the Pocket PCs provides the haptic feedback in the experiment.

The experiment was written using the *FMOD* API (version 3.70CE)(FMOD, 2004), a visual programming environment with an object-oriented language (Embedded Visual C++) used primarily to manipulate and control sound production and *GapiDraw* (version 2.04) (GAPIDraw, 2004), a runtime add-in to *FMOD* used to generate real-time Pocket PC graphics. *FMOD,* and *GapiDraw* are available for free under the condition of the GNU public license (GPL).

Using *FMOD* and *GAPI,* an interface was developed with the following parameterisations: speed of sound, 340ms^{-1}, Doppler factor, 1.0, distance scale 100.0, minimum audible distance 80m, full volume (255)(minimum volume is 0

Figure 3. Left- Pocket PC, Accelerometer and experiment I running on the system (target sound sources displayed, for illustrative purposes). Right- A user interacting with the system.

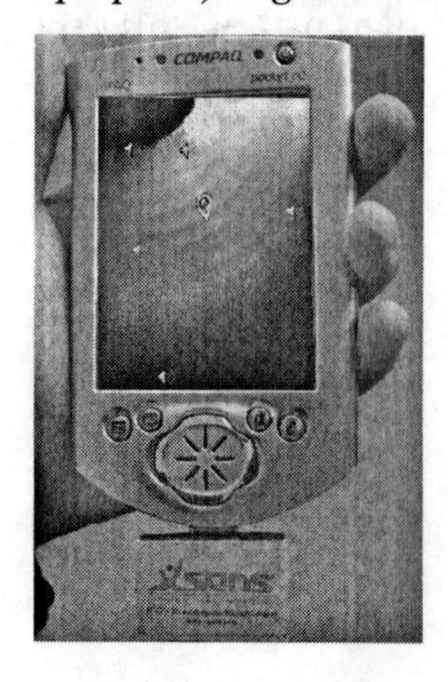

and max volume is 255 in *FMOD*), and maximum audible distance 8000m. Each pixel on the display represents 100 metres. An empty window (240 × 320 pixels) was centred on the screen. Audio sources represented by small (10 × 10) speaker icons are shown on the screen only for training (Figure 3). In the main experiment, sound sources are hidden and an empty window is shown on the screen. Only the cursor, represented by a small (10 × 10) ear icon, is visible in both training and main experiment.

Experiment I

We first conducted a pilot study with 12 subjects, 3 women and 9 men, all sighted, with a mean age of 29 years. Four participants were research fellows, and the rest were postgraduate students at the NUIM campus. All but one of the participants had neither experience of using Pocket PCs nor with accelerometer-based interfaces. Two of them were left-handed (Eslambolchilar, Crossan, & Murray-Smith, 2004).

Task and Stimuli

The task in this study was to select the centre of individual targets that appear (in audio but not visually) in different locations on the screen as accurate as possible. The individual targets are audible when the cursor is in their locality, and they have full volume only in the centre of the target (imagine a Gaussian distribution of the volume centred on the target). For each target, a vibration feedback has been assigned and whenever the user is in very close distance to the target, 10 pixels, s/he feels the vibration continuously. Our aim in using the vibration in this task is the vibration assures the user that s/he is very close to the centre of the target.

First, participants were asked to sit on a chair in a quiet office and were equipped with a headset and a Pocket PC in their palm. Then they were informed about the functioning of the accelerometer, Doppler effect, and the procedures of the experiment, in order to reduce the chance of any terminological misunderstanding. Subjects were asked to move the cursor to audio targets by tilting the PDA, and to select them by pressing a key on a small keyboard of the PDA. They were told to emphasise accuracy over speed.

Design

There were four experimental conditions: (1) No Doppler effect, no vibration feedback (2) No Doppler effect, vibration feedback, (3) Doppler effect, no vibration feedback, and (4) Doppler effect, vibration feedback. The participants performed the conditions in a counterbalanced order. This resulted in 12 different orders of experiments for participants. In each experiment, seven audio sources were used (a selection of different music) summarised in Table 1.

Visualisation

Matlab was used for visualising the logged experimental data. We use a number of techniques for investigating the users' behaviour in these experiments.

Audio and exploration Density Plots

These plots show the audio density (in pixels) at different points in the 2-D space (Figure 4 (Left)). The contour indicates the density of the sum of the amplitude of the mixture components associated with the different audio tracks. The exploration

Table 1. Audio sources in the first experiment in all conditions

Target Index	Music Type
1	Hip-Hop
2	Celtic
3	Arabic
4	Country
5	Jazz
6	Farsi
7	Opera

density plot for visualisation of cursor trajectories, used previously in Williamson and Murray-Smith (2004), has been used here, which plots a density around the trajectory, which is a function of the position and the length of time spent in that position. These plots give some indication of how users navigated when completing the task. An example is given in Figure 4(Right). This plot is created by placing a Gaussian distribution centred on the (x,y) position of the cursor for each point in the log file, with standard deviation proportional to that used in the audio sources. The Gaussians are summed for each pixel, and the resulting image gives an impression of the areas of the input space that were explored, and how long the user spent in them. The image can be summarised numerically by counting the percentage of pixels greater than a selected threshold e. In this experiment e=5.0. The image's resolution is 240 by 320 pixels (Eslambolchilar et al., 2004).

Distance to the Target

Whenever the user feels s/he is at the target, s/he presses a key indicating the selection of the target. For each selection made by the user, the distance to the nearest target is calculated as below, and recorded.

Figure 4. Left - The cursor trace of the 4th participant in the "no Doppler-no vibration" condition, is plotted over the density of the local audio amplitude of the different tracks. Right - the density contour plot and cursor trajectory density indicating the exploration of the space by the same participant in the same condition.

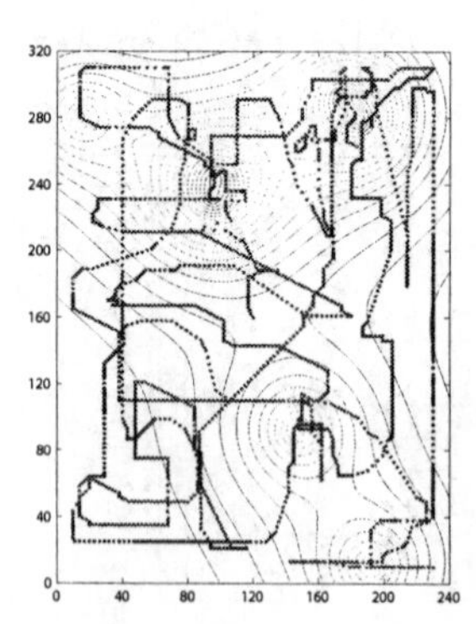

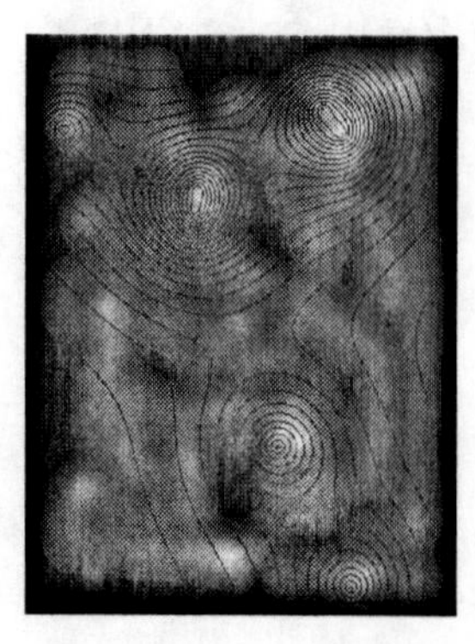

$$Dist = \sqrt{(x_{source} - x_{selected})^2 + (y_{source} - y_{selected})^2} \quad (2)$$

An example plot is shown in Figure 5. The distance to the location of the target (in pixels) gives some insight into the acuity with which the location can be perceived with the given display.

Results

Search Patterns Observed

In looking at the audio and exploration density plots, we are not attempting to establish a link between the search pattern used and the resulting measurement of accuracy. We simply make a subjective classification and qualitative assessment of the types of search patterns employed to accomplish the task. A subject may employ one of the search techniques and still not be very accurate, or they may be very accurate in spite of using no detectable systematic pattern. However, this factor gives an indication about the ease with which the audio environment could be clearly perceived by participants. In a clear and easy to navigate environment, with appropriate feedback, this should be similar to the density of targets, and linked to the smoothing used.

Figure 5. Hidden target positions (circles), and points selected by user 4 in the "no Doppler-no vibration" condition, as the best guess (crosses)

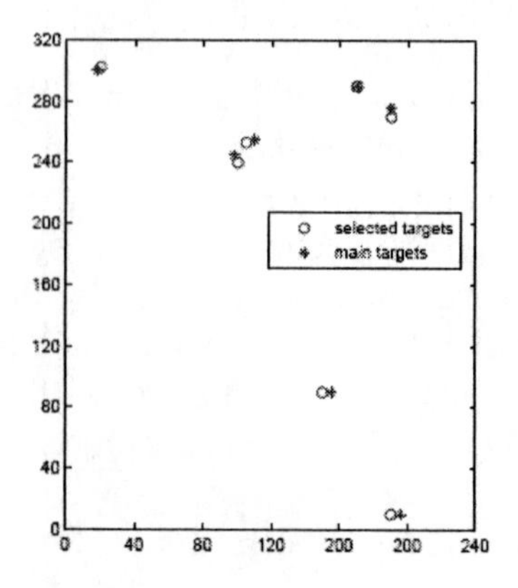

Some of the terms and their basic definitions used here are taken from search theory, a subfield within operations research (Patrol, 1999). The patterns developed by search theory are visual search patterns of physical space, but there is some crossover in the types of patterns used in the auditory interface used in the experiment to search in a virtual space.

1. **Parallel sweep:** The parallel sweep is used when uniform coverage of an area is desired and the area is unfamiliar. It is an efficient method of searching a large area in a minimum amount of time. Several subjects used the horizontal parallel sweep, "raster scan," similar to the one seen in Figures 6(a) or 7. This pattern can be related to the text-reading pattern we learn in the childhood.
2. **Quadrant search:** The quadrant search pattern is one in which the searcher mentally breaks down the screen into quadrants to divide the area into a more manageable size. Within the quadrants, the searcher may use another pattern to search each quadrant, such as a parallel sweep (Figure 8(a)).
3. **Sector search:** A sector search pattern begins once the approximate location of the target is located. In this pattern, the searcher explores out from the approximate location of the target and returns again, then conducts another exploration in another area, and returns again. This is repeated until they

Figure 6. (Left) The traces of the cursor for participant 12 in "no Doppler with vibration" experiment (a) and its exploration density plot (c), (Right) The traces of the cursor for participant 6 in "Doppler with vibration" experiment (b) and exploration density plot of this experiment (d)

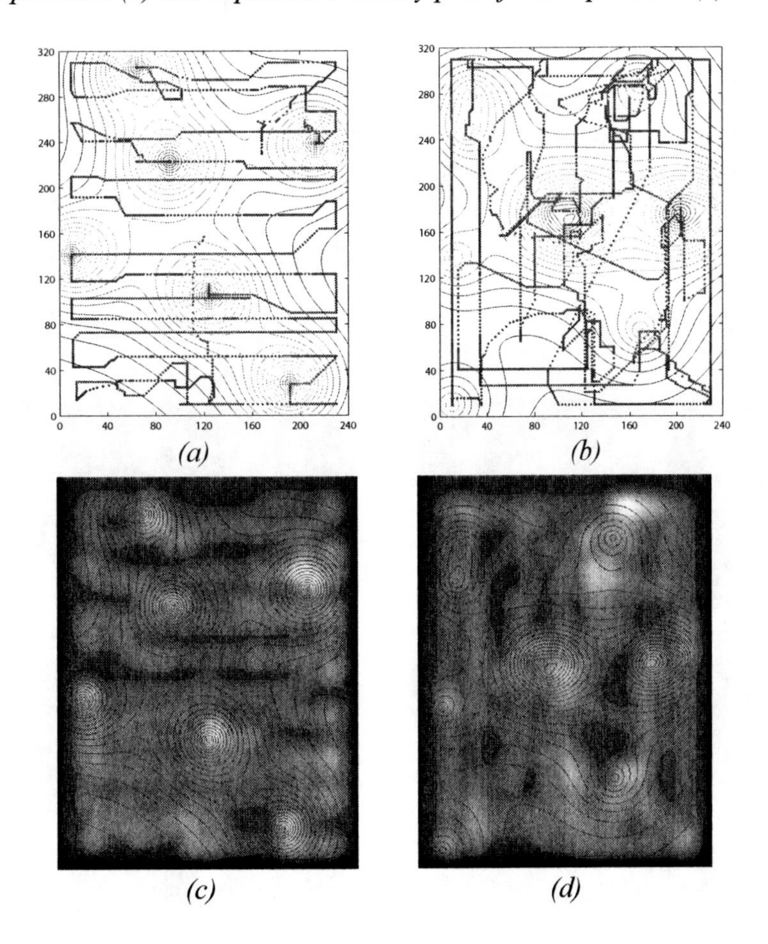

are confident that the space is adequately explored (Figure 8(b)).

4. **Perimeter search:** The perimeter search is one in which the boundaries of the space are explored, but little or none of the middle is traversed. The pattern of search can be a circle to circumscribe the border or a square-shaped pattern turning at a 90° angle. This type of search pattern would typically lead to inaccuracy given that none of the targets are located at the perimeter. This search pattern was not observed in this research.
5. **No formulaic search:** For some searchers, no discernable systematic technique was employed in exploring the space to accomplish the task. For these search patterns, there is no attempt to thoroughly explore the information space. Figure 8(c) illustrates the path used in the only trial to actually select the target exactly.

The search patterns of each subject were analysed to see if there were any tendencies based on demographic characteristics; 48 total patterns were analysed. The most common technique employed

Figure 7. (Left) The traces of the cursor for participant 5 in "Doppler with vibration" experiment, (Right) The traces of the cursor for participant 9 in "no Doppler with vibration" experiment.

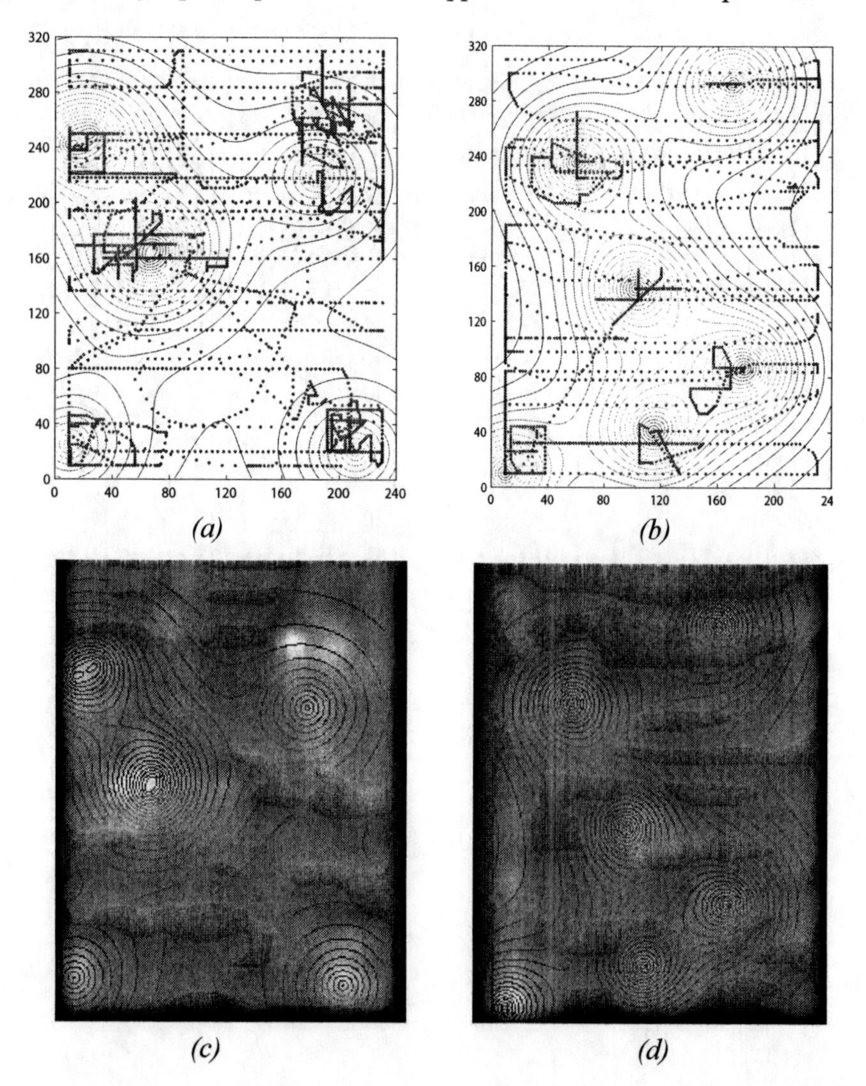

Figure 8. Examples of few search patterns in different conditions

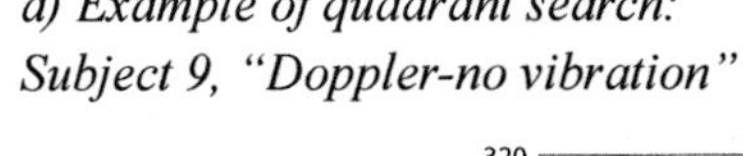

a) Example of quadrant search: Subject 9, "Doppler-no vibration"

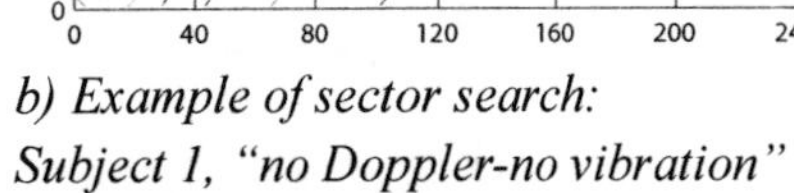

b) Example of sector search: Subject 1, "no Doppler-no vibration"

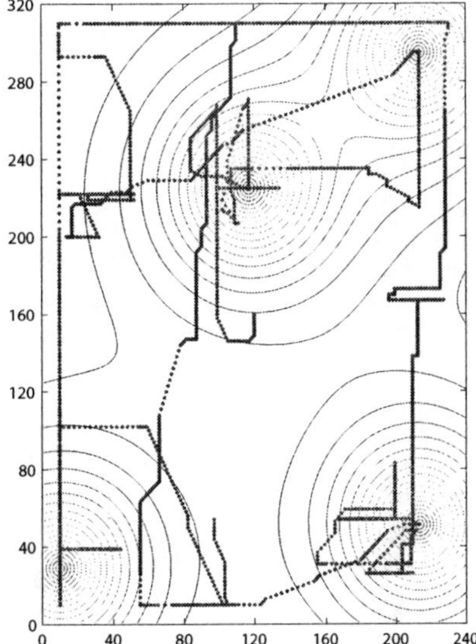

c) Example of no formulaic search: Subject 7, "Doppler-no vibration"

was the sweep search (76%). The next most common was the no distinguishable pattern (14%), followed by quadrant (6%), and sector (4%). Participants' audio and exploration density plots show "Doppler-no vibration" has the least covered space with 34.5%, and the rest have similar percentage of coverage, 37.6%.

Chosen Songs

The accuracy relative to the number of song chosen is another factor in improving audio interfaces. Because the type of songs may affect the perception of distortion due to the Doppler effect, and affect the users' ability to recognise and locate them. We measured the number of audio sources

Figure 9. Mean distance in pixels from target in different tasks

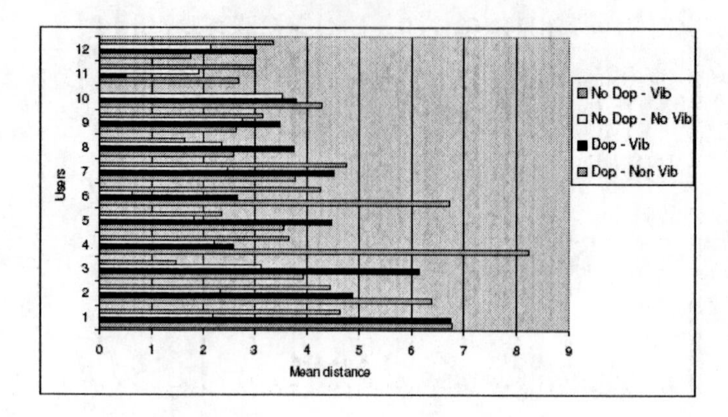

Figure 10. Count of most accurately chosen songs in different conditions for all users

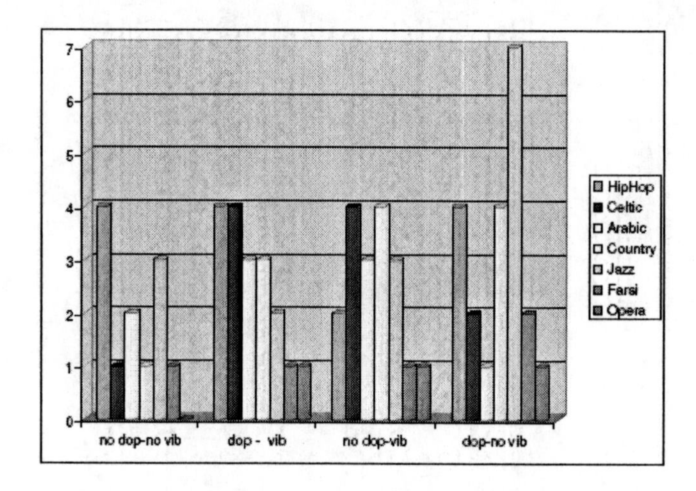

Figure 11. Mean distance (pixels) of selected songs in all conditions for all users.

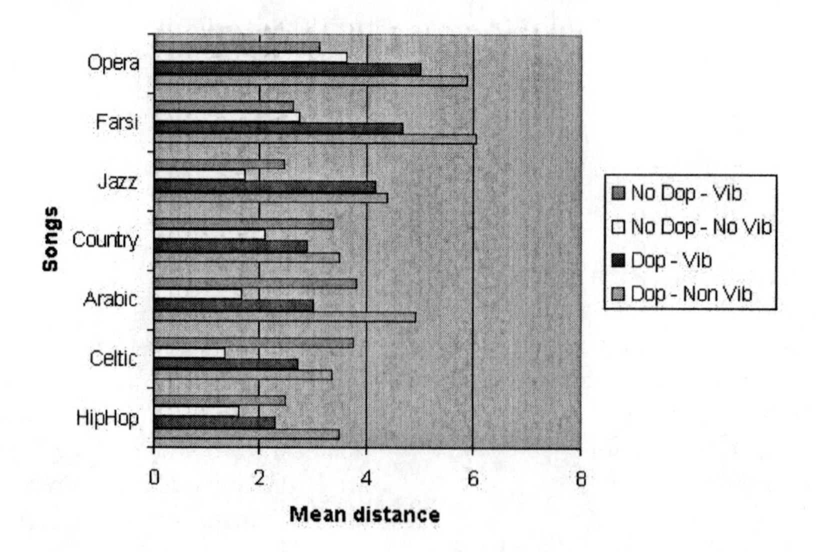

Table 2. Accuracy scores for audio sources in the first experiment

-	no dop- no vib	Dop-vib	no dop-vib	dop-no vib
Hip-Hop	4	4	2	4
Celtic	1	4	4	2
Arabic	2	3	3	1
Country	1	3	4	4
Jazz	3	2	2	7
Farsi	1	1	1	2
Opera	0	1	1	1

participants have selected. The mean accuracy for each of these sources has been summarised in Table 2. Figure 10 shows the mean accuracy count of songs in all conditions for all participants. This result is based on the number of times each source was selected with the smallest distance to the target in each condition. There is a large amount of variability in the results. Jazz music was selected more than others, on average, but Hip-hop music was chosen more accurately in the "no Doppler-no vibration" condition. Figure 11 shows mean error for songs in all conditions. In general "no Doppler-no vibration" has the lowest error among others and "Doppler-no vibration" has the highest error. Farsi and Arabic sources had high mean and maximum errors in the Doppler case.

Discussions

Post hoc examinations of the cursor's trace in this experiment showed that the subjects tended to use the same technique regardless of the sounds they heard and the audio condition. Six subjects (50%) used one search technique exclusively. Of these six subjects, five used the sweep technique, one used no distinguishable pattern exclusively. Another six subjects (50%) used the same technique in two out of the four conditions. This consistency in the application of a searching technique has several notable points. First, the same technique was employed regardless of the sound treatment. This would indicate that the subjects brought with them a technique that was not altered by the change in the treatments used in the auditory interface. The subjects were given no experimental feedback that might prompt them to change their search pattern to one that might be more effective. Left to their own means, the subjects tended to continue with the application of the search pattern with which they felt most comfortable. Second, the most common type of search pattern (sweep search) was also the least effective, given that the target in all four conditions was located towards the interior of the information space. In these cases, the subject was less likely to notice a change in the sounds they were hearing because of the low intensity of the sounds generated at the borders of the information space. Because they typically did not explore the interior, they would not hear the more intense sounds that might lead them to the target. In conditions with vibration feedback sweep search is combined with circular movements around the vibration source (Figures 6(a), 7(a) and 7(b)) and has led the users to the target. This suggests that the vibration was more important for the users in locating a target, and whenever they felt they were close to the song, they looked for the vibration source before clicking; so, feeling a vibration source meant they were at the centre of the audio source. This might also explain the fact that errors were not smaller, as the user may often have selected the location as soon as vibration was perceived, at the edge of the circle, rather than at the centre of the target itself.

The "no formulaic search" is the least thorough of the systematic techniques. Even though there was essentially no effort involved in exploring more thoroughly by applying a different search pattern, the subjects tended to use the no distinguishable search pattern. This could be accounted

Figure 12. The state space in experiment II and corresponding angles. Top pictures show the training application and bottom ones show the main application. Left pictures show the screen before pressing the button for Jazz music and right ones show the screen after pressing the button (colour changes and covered speakers are indicated in training application).

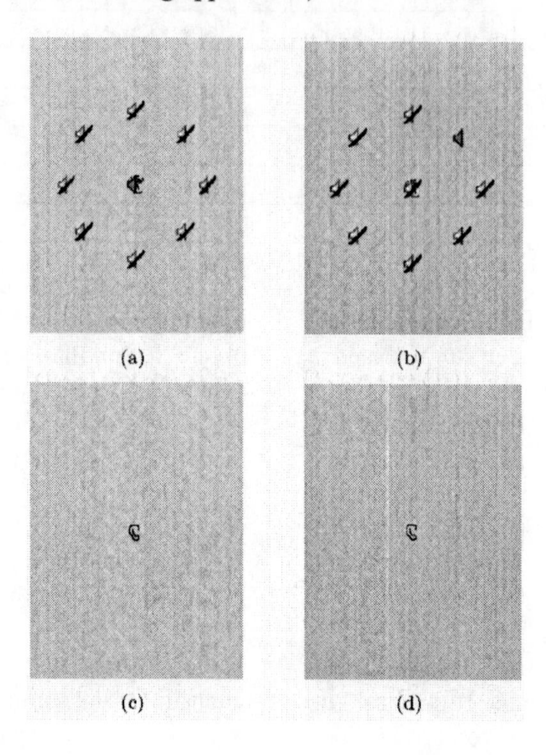

for by assumptions the subjects made about the nature of the information space. It would seem that some subjects took the experimental task seriously by systematically exploring the information space. Other subjects did not seem to be interested in exploration, but instead made a quick "stab" in the general direction of the high point. The case could be made that those subjects who explored liked the sonic interface, and those who did not explore did not like the interface. It may well be the case that auditory display is not for everyone. Some will like it and make use of it, others will not.

The results show that the mean distance from the selected position to the target in "no Doppler-no vibration" is less than other experiments (Figure 9). The extra clicks and navigation activities in the cursor trajectories for Doppler might be an effect of the extra sensitivity of the feedback to movement, which makes the users explore by varying their velocity vector. Variability in localisation accuracy is greater with the Doppler effect for the Farsi and Arabic sources, suggesting that for the mainly western European participants, their poorer familiarity with these sources made the distortions introduced by the Doppler effect more difficult to perceive. Opera also had larger errors, again suggesting that less familiarity with the target sources can affect the usefulness of this approach. The large number of falsely placed points for the Doppler method might be because of the amplification involved in moving towards something and potentially, frequency and speed of sound, which makes people feel they are getting a stronger response, and they over-interpret the quickened signal, believing they are already at the point - a common cited risk associated with quickened displays (Poulton, 1974).

Experiment II

Twenty-four paid participants (8 male, 16 female), all sighted with normal or corrected vision and normal hearing with mean age 24, were recruited through sign-up sheets in Glasgow University Psychology department and through e-mail. Two of the participants were left handed. Three participants had experience using a Pocket PC, and of these, one had frequent use. None had experience using an accelerometer as an input device.

Design

Given that the results of the pilot study showed we had possible confounding factors, that is, four different audio conditions, haptic feedback, few subjects, unfamiliar songs, and random located targets, the next experiment reduced the number of independent variables. This resulted in three different types of audio feedback without haptic feedback:

1. Doppler feedback
2. Derivative volume adaptation
3. No quickening

There were eight possible audio sources (targets) arranged in a circle (radius = 100 pixels) around the centre at 45° intervals (Figure 12). The audio feedback at the centre was jazz music, which played continuously for all conditions. When an outside target was to be located, the audio feedback was "Hotel California" played in a loop. The audio source to be located alternated between the centre (jazz music) and one of the outside targets ("Hotel California"), and always began with the centre target. The audio sources around the outside were presented in a random order, twice for each target for the training session (16 trials in all), and five times for each target for the experimental trial (40 trials altogether). Once one target had been located, a button was pressed. For each key pressing, there was a screen colour change and a short "beep" sound. Audio was noticeable within a radius of 90 pixels from sources (3s) in conditions 1 and 2, but audio was noticeable just within a radius of 15 pixels from the sources in condition 3, and there was no feedback at any other locations in this condition. Each participant was tested individually, and participants were told to commence the training session when they felt ready. After the training session, it was ensured that they understood the procedure fully and that they felt comfortable using the equipment. They were given a break between the training and experimental sessions if they wanted one.

Visualisation

In addition to exploration and audio density plots and distance to the target, we used another visualisation method that measures the orientation of each target with respect to the centre point, showing which angles in the state space have had the most accurate data in selecting targets. This measurement is important in this experiment to see whether the orientation of audio sources has any effect on the targeting task. Results in this experiment were analysed using a *GLM ANOVA* test.

Results

Proportion of Distance to the Target

Figure 13 shows the box plot of medians, means, and measures of spread of the distance between the audio sources and the position selected for each of the three audio feedbacks. The red triangle indicates the mean and the blue square indicates the median (50% of the observations lie below this line). The top of each box indicates the upper quartile (75% of the observations lie below this point) and the bottom indicates the lower quartile (25% of the observations lie below this point). The tops of the lines above the boxes indicate the highest observation, and the bottom of the lower line indicates the lowest observation. The blue stars indicate outliers: those observations that differ significantly from the mean. Figures 13 and 14 show a difference in the average distances between the actual audio source and the target selected (accuracy) for the three audio conditions. The most accurate target selection

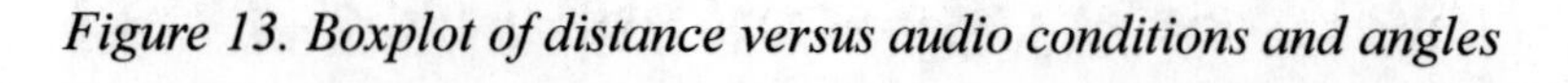

Figure 13. Boxplot of distance versus audio conditions and angles

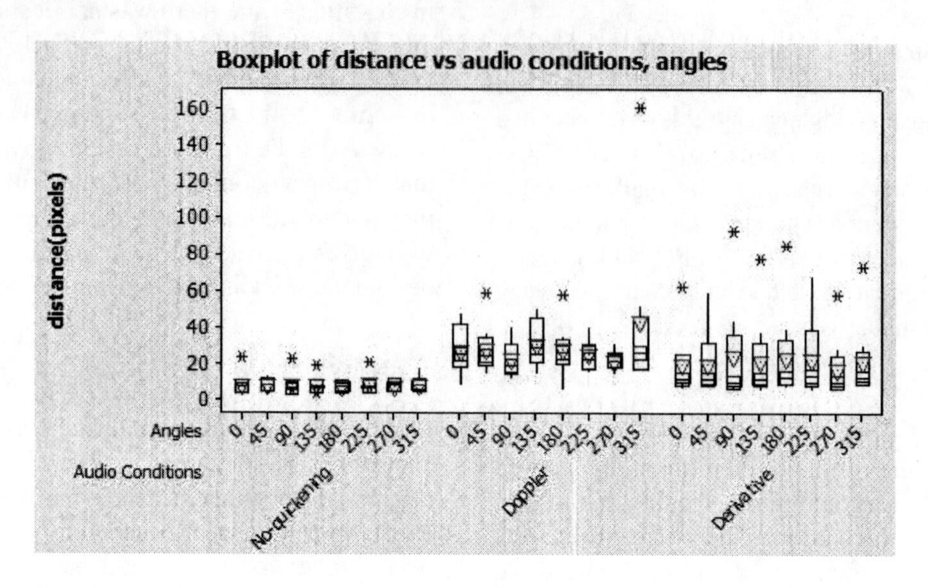

Figure 14. Average of position error in pixels for all participants in 3 audio conditions

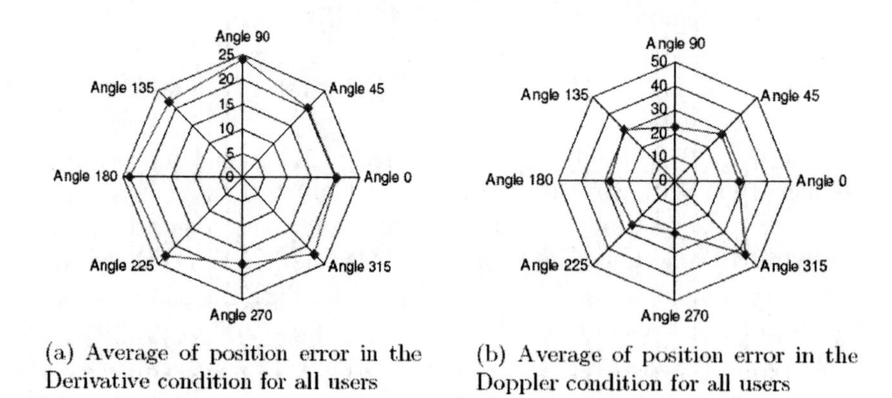

(a) Average of position error in the Derivative condition for all users

(b) Average of position error in the Doppler condition for all users

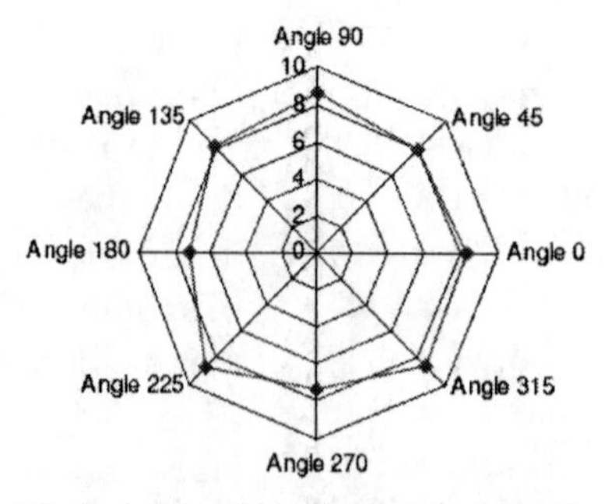

(c) Average of position error in the No-Quickening condition for all users

occurred in the no-quickening condition, and there was little difference between the levels of accuracy for each orientation. The spread for no quickening was very small, and the five outliers are not very far away from the median, suggesting that, overall, most people in this condition took approximately the same length of time to select a target. The derivative condition takes longer overall than the no-quickening condition. The spread is larger (largest of all three conditions) and the outliers further away from the median than in the no-quickening condition. The participants in the Doppler condition took longer on average to select the targets, compared with those in the other two conditions. The spread in the Doppler condition is smaller than in the derivative condition, but larger than in the no-quickening condition. This shows that angles 90° and 270° have higher accuracy. A *GLM ANOVA* analysis found that there was a significant effect of audio type on the distance ($F(2,21)=4.345$; $p<0.05$). There were, however, no significant differences between the eight audio orientations. Post-hoc analysis using a *Tukey* test showed that there was a significant difference between the estimated mean distances of the selected targets away from the audio sources in two of the three audio conditions. It was found that the mean distances were not significantly different between two quickening feedbacks and nonquickening condition ($p<0.087$). The *Tukey* test showed that there was no significant difference between the estimated mean distances of the selected targets away from the audio sources in the derivative condition. Running this test on Doppler results revealed that a significant difference ($p<0.025$) between the estimated mean distances of the two selected targets away from the eight audio orientations (90° and 270°). In the no-quickening condition, the *Tukey* test showed that there is no significant difference among the estimated mean distances of the selected targets away from the audio sources.

Search Patterns and Covered Space

In this experiment, audio targets had fixed positions in the state space so the observed search patterns were different than those we found in the pilot study. Twenty-four total patterns were analysed in this experiment. From the training sessions, subjects knew the approximate location of the audio sources. In 100% of search patterns in the no-quickening condition, the subjects moved to the edge of a circle in the size of the actual radius of the points and started circling to find the active target. In the Doppler condition, 87% of subjects could guess in which direction the target was located and after doing a few back and forth movements, they landed on the target. Subjects, therefore, followed a sector search pattern. Search patterns of subjects who worked with derivative volume adaptation were mixtures of the patterns of the no-quickening and Doppler conditions. Figure 15 shows some of the subjects' trajectories and density plots in different conditions. Figure 16 shows the percentage of the screen covered by participants' movement in three conditions. In the derivative condition, the top-left sections (90°-180°) were explored more than other parts. In the Doppler condition, the top-right (0°-90°) sections were popular to explore, and in "no quickening" there was no significant difference in the sections covered by participants' movement and all of the partitions were explored equally. These plots show the Doppler condition had the most covered space with 44.5% and the rest had a similar percentage of coverage, 39%.

Discussions

In this experiment, it was found that there was an effect of audio condition on the level of accuracy. When the feedback was no quickening, participants were more accurate than when the feedback was Doppler or derivative. This is due to the fact that with the no-quickening condition, the only time that audio feedback is heard is when the cursor is directly over the target, and the difference between hearing and not hearing audio feedback is larger than the difference between hearing different levels of audio feedback. There was also found to be no effect of angle. The level of accuracy was the same, irrespective of the orientation of the target.

Figure 15. Trajectories of different subjects in 3 audio conditions

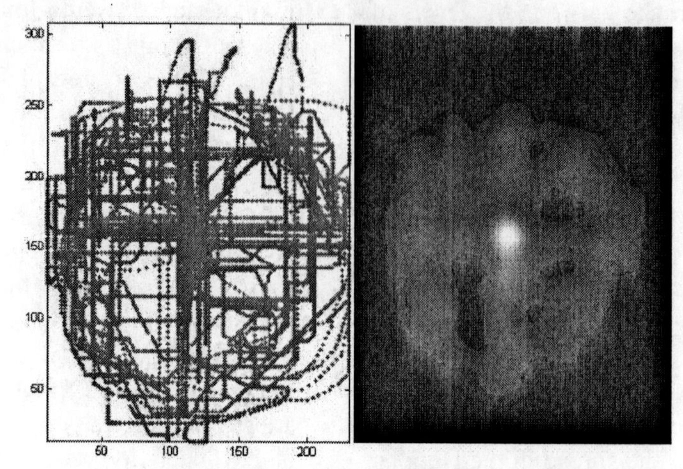

a) Trajectory and density plot of subject 4 in Derivative

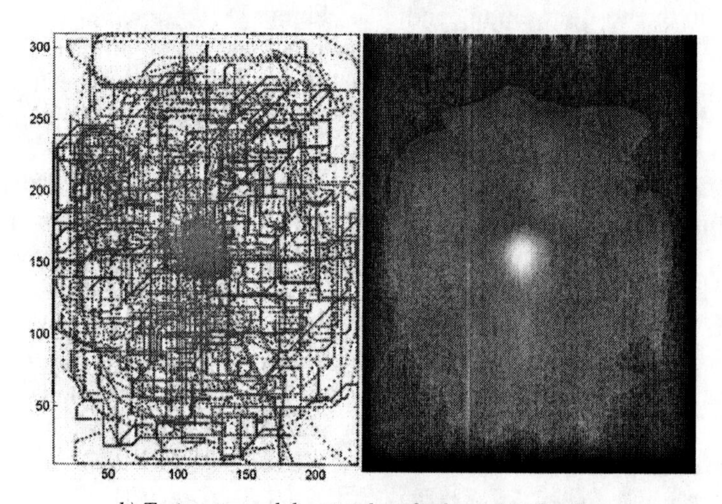

b) Trajectory and density plot of subject 4 in Dopler

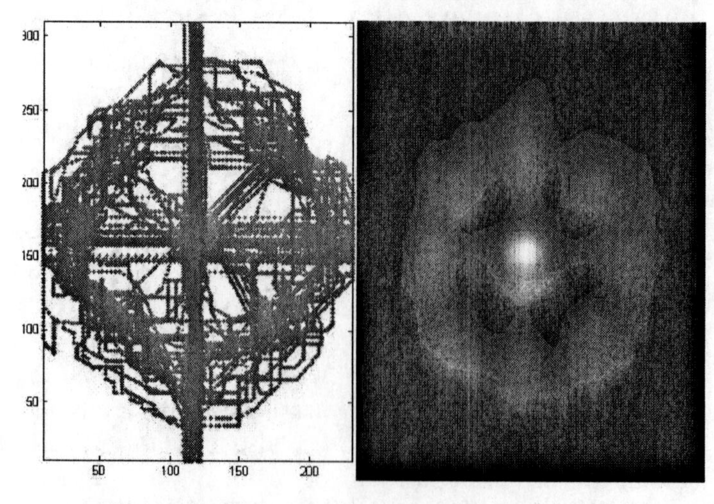

c) Trajectory and density plot of subject 4 in no quickening

Figure 16. Percentage of the screen covered by users' movement in different conditions. The variance of the coverage in the derivative condition is 0.0023, in the Doppler condition is 0.0017 and in the no-quickening condition is 8.5714e-005.

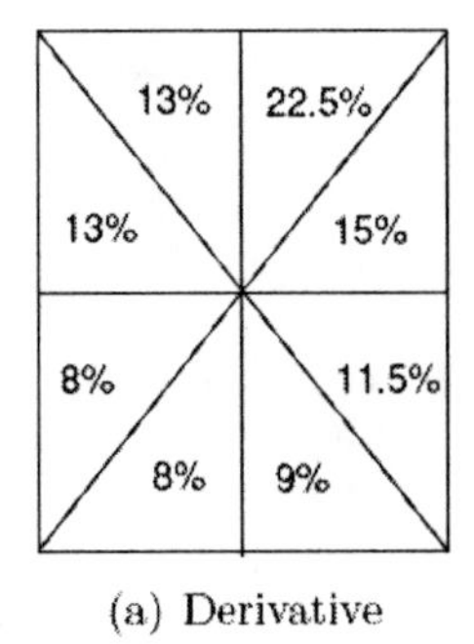

(a) Derivative

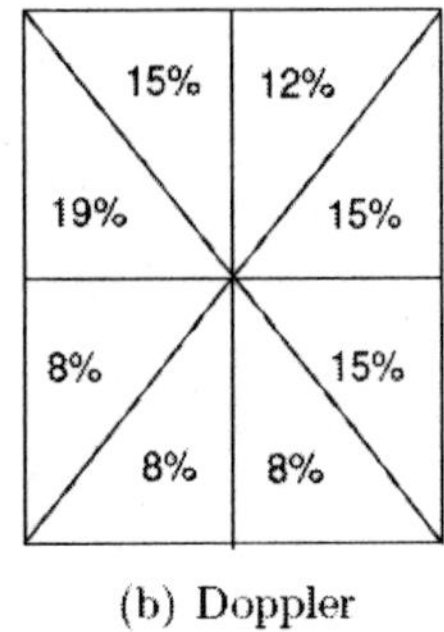

(b) Doppler

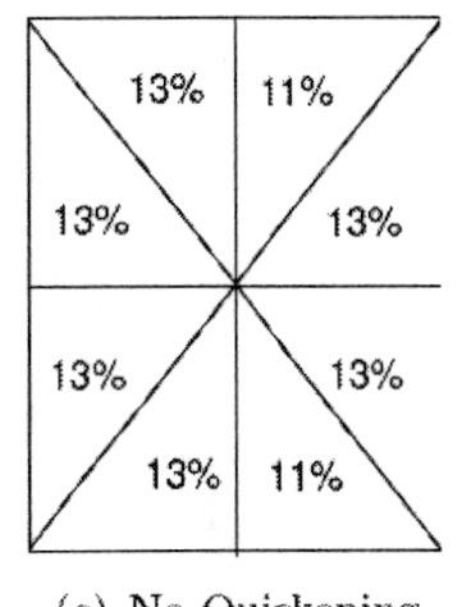

(c) No-Quickening

Many participants reported that sometimes they would just "land" on the audio source by chance, and at other times they would search for a long time and still not feel they had located the point accurately. This was an especially common complaint by participants in the Doppler condition. Disorganised search patterns observed in the Doppler condition, for instance in Figure 15(b), may correspond with the claim made by some participants that by the time they had established where the target was through the audio feedback, they had already passed the audio source, and had to go back to it. We plotted users' trajectories when they had moved from the centre of the screen to any active audio source around the centre. Figure 17(b) and other users' trajectories in the Doppler condition highlight that in the first moments after audio source activation, the subjects could guess the direction (left or right side of the space) and approximate position of the target and consequentially, moved towards the target correctly, but it was difficult to establish a correct target acquisition, and they made some back and forth or up-down movements to land on the target, which is compatible with the observed sector search pattern. This is shown more clearly in Figure 18(a-left) and the user's trajectories as time series in Figure 18 (b-left) in an individual target acquisition task when the user has moved from the centre to the target in angle 45°, which has been activated.

Figures 15(a) and 17(a) show a trajectory that is fairly typical for most participants in the derivative condition. It can be seen that the trajectory is far more ordered, with participants moving in the horizontal and vertical directions (in the directions of 0°, 90°, 180°, and 270°), more so than in the Doppler condition. It becomes obvious that participants moved in a circular motion that they learnt during the training session, far more so than those in the Doppler condition. From this, they have established that the audio sources in the experimental session were also arranged in a circle. This suggests that they were not necessarily using only audio feedback, but also prior knowledge about the probable locations of the audio sources. Since this circle is not as clearly defined in the Doppler trajectories, it suggests that participants in the Doppler condition were using predictive information, but were also less able to control their movements efficiently. There is a risk that any significant effects were masked by prior knowledge of the way the audio sources were arranged, and visual feedback has affected the users' behaviour in exploring the audio space.

Figures 17(a) and 19(a-left) provide a clearer picture of the users' browsing behaviour in this condition. As a result of the first impressions that the users have received from the volume of the audio source, they have chosen vertical or horizontal directions. Whenever they have not found the target in these directions, for instance in an individual target acquisition in angle 45° presented in Figure

Figure 17. Trajectories of different subjects in 3 audio conditions when they have moved from the centre to outlying active audio sources

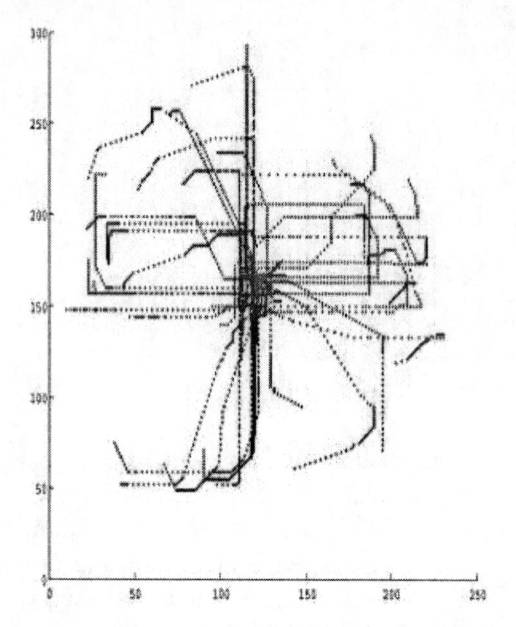

a) Trajectory and density plot of subject 1 in Derivative case

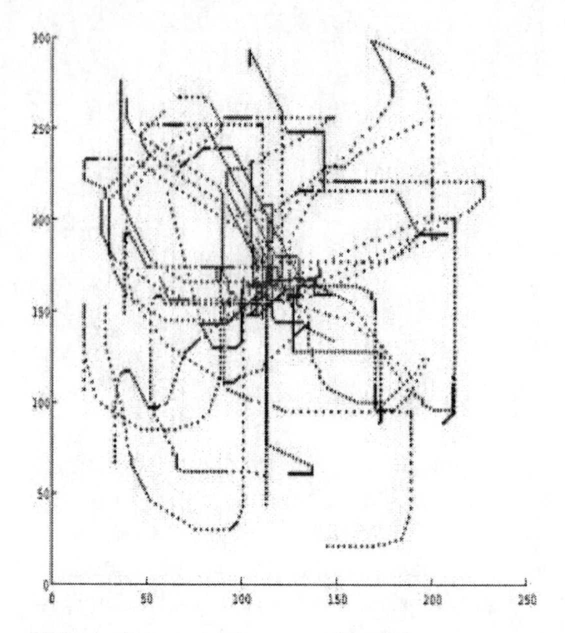

b) Trajectory of subject 4 in Doppler case

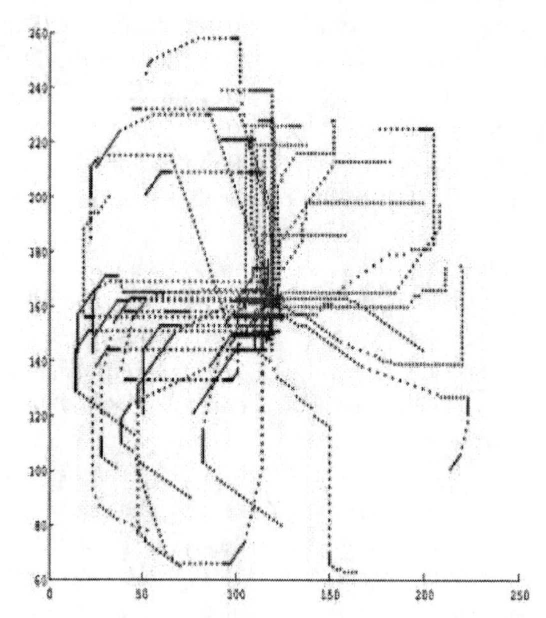

c) Trajectory of subject 4 in no quickening case

Figure 18. (a) One of the participants' trajectories in the Doppler condition with and without predictive feedback in an individual target acquisition task when the user has moved from the centre to an activated target in angle 45°. (b) The time series of the participant's X and Y position error in the same task.

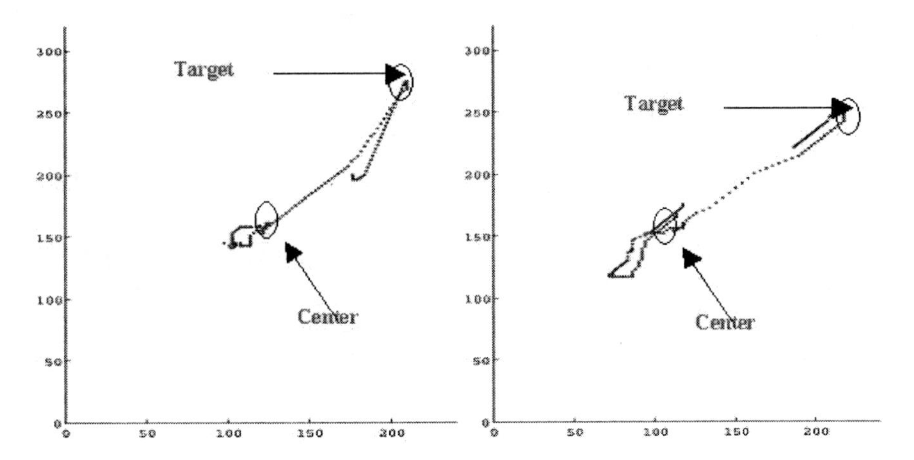

(a) An individual target acquisition task in the Doppler condition. (left) standard feedback, (right) predictive feedback

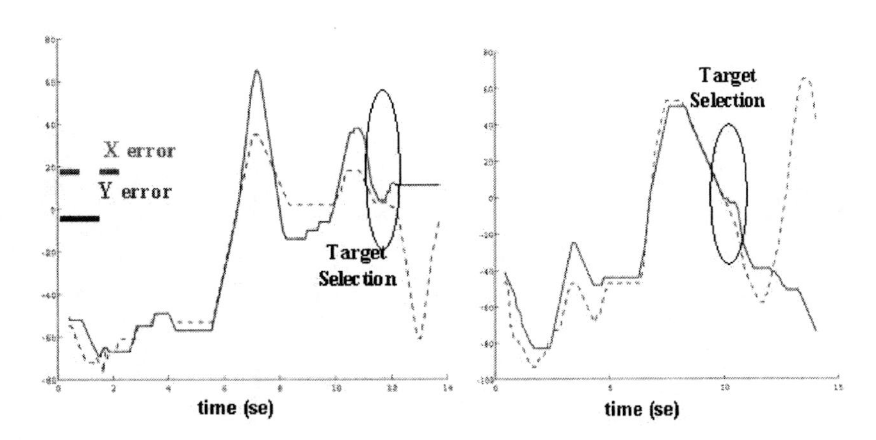

(b) The time series of one of the participants' position error in X and Y axis in the target acquisition task shown above, in the Doppler condition. (left) standard feedback, (right) predictive feedback

19, they have moved around the circle using prior knowledge of the landscape. Figure 15(c) shows a more pronounced circle and cross-shape for the no-quickening condition. Since the participants in the no-quickening condition were presented with no aural feedback except when directly over the target, it is, most likely that they were relying on the circular target distribution previously seen in the training. This led to a systematic search strategy, less "browsing around" because of the lack of predictive ability without quickening. All users who participated in this condition claimed that this was not an exciting method of exploring the auditory space.

Human Operator Modeling

In continuous control tasks, for instance browsing and finding audio targets in the audio space, the human operator can be modeled using the tools of

Figure 19. (a) One of the participants' trajectories in the derivative condition with and without predictive feedback, in an individual target acquisition task when the user has moved from the centre to an activated target in angle 45°. (b) The time series of the participant's X and Y position error in the same task.

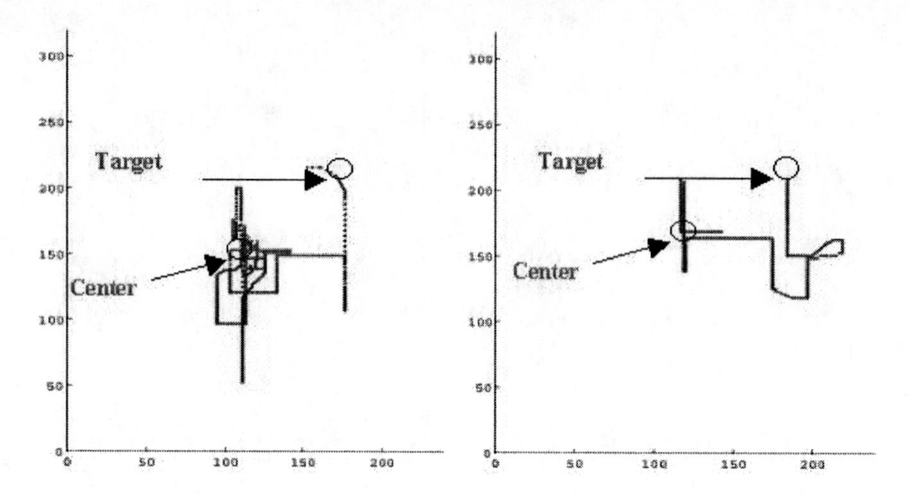

(a) An individual target acquisition task in the Derivative condition (left) standard feedback, (right) predictive feedback

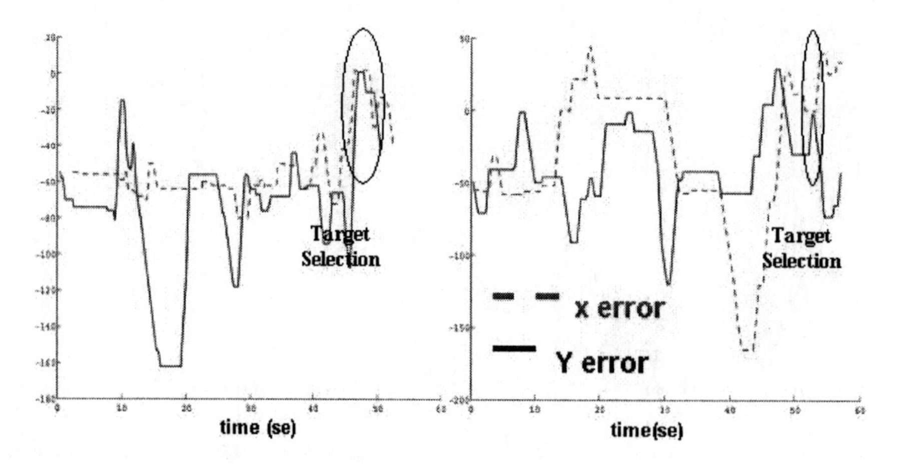

(b) The time series of one of the participants' position error in X and Y axis in the target acquisition shown above, in the Derivative condition (left) standard feedback, (right) predictive feedback

manual control theory. Quantitative models of the human operator may provide predictions and insights into basic properties of human performance in human-machine interaction, and the ability to derive transfer function for human operators means they can be directly used in human-machine systems interaction design (Jagacinski & Flach, 2003).

Human interaction with the computing systems is two-way. The user and machine form a closed-loop system, where s/he issues commands through the system's input channels and receives results fed back on output channels. Subsequent input depends on the latest output. The performance is adversely affected when the feedback is subject to delay or lag (Jagacinski & Flach, 2003; MacKenzie & Ware, 1993).

In some genres of interactive systems, which rely heavily on the tracking of hand, head, and/or body motion in a simulated environment, the pretence of reality requires a tight coupling between the user's view and hearing of the environment and the actions, usually hand, head, and body motions, that set the view and hearing. When changes in the environment lag behind input motions, the loss of fidelity is dramatic (MacKenzie & Ware, 1993). In an extension to the previous experiments, we carried out a primary investigation to measure and model the speed, accuracy, and bandwidth of human motor-sensory performance in interactive tasks subject to lag.

Manual control theory suggests that in a simple tracking task the human operator can be modeled via a transfer function that consists of a gain, a lag (an integrator at higher frequencies) and a time delay (Jagacinski & Flach, 2003; Poulton, 1974).

$$Y_h(j\omega) = \frac{Ke^{-j\omega\tau}}{j\omega} \qquad (3)$$

The gain ***K*** is a scaling factor that influences the bandwidth of the control system. The time delay t reflects human reaction time. In simple tracking tasks, the range of the time delay is between 20 to 150 ms, which overlaps with measures of reaction time in response to continuous stimuli. If ***K*** is low, the system will respond very sluggishly, moving only slowly towards the target signal. Conversely, if ***K*** is high then the system is likely to overshoot, requiring adjustment in the opposite direction, which itself may overshoot, leading to oscillation. However, humans adjust their gain to compensate increases or decreases in plant gain. For example, pilots change their behaviour when they switch from Boeing 747 (heavy) to an aerobatic airplane so the total open-loop gain remains constant. So if Y_p represents the plant transfer function and Y_h represents the human transfer function, then:

$$Y_h(j\omega)Y_p(j\omega) = const \qquad (4)$$

The delay τ can also contribute to this behaviour; a high delay makes oscillatory behaviour much more likely (refer to time and delay section discussed earlier). The lag suggests that the human tracker has a low pass characteristic, that is, the human responds to low-frequency components of errors and ignores (or filters out) the high-frequency components of error (MacKenzie & Ware, 1993).

Using the platform in the previous experiment, we did a preliminary investigation to measure and model the accuracy, and bandwidth of human motor-sensory performance in interactive tasks subject to lag. We kept the same format of the second experiment but instead of providing audio feedback to the user's current position, we provided feedback to the user's predicted position, calculated according to equation (3):

$$X_{t+\tau} = X_t + V\tau \qquad (5)$$

The volume of the audio source, which provides feedback about the target's position, is a function of the user's current velocity and position. Here, X_{t+1} and X_t are the user's position at time $t + \tau$ and t or current position and next possible position respectively. τ is the human's time-delay or reflection time, which becomes our "prediction horizon" for the predictive model, and V is the user's speed of motion in the audio space. This has the effect that, as the user moves toward the target, s/he feels him/herself in the position predicted to be reached at time $t + \tau$.

Design

In the second experimental setup, we added a smooth drop-off in the time horizon of the prediction as the target was approached. The falloff began at radius 15 pixels, and once the user was within 5 pixels of the source, the feedback reverted to standard feedback with no predictive element.

In a pilot study, we ran the application for three participants familiar with the Pocket PCs and accelerometer. Neither felt any difference in the two derivative conditions, with and without prediction. In Figures 19(a) and 20(b), we see one of the participant's trajectories when he has moved from the centre to any active target. Providing feedback to the user's future position in the derivative condition has not much changed the user's exploratory behaviour. But users reported a great difference between Doppler with the pre-

Figure 20. The trajectories of 2 subjects in the Doppler and derivative conditions, without and with predictive feedback

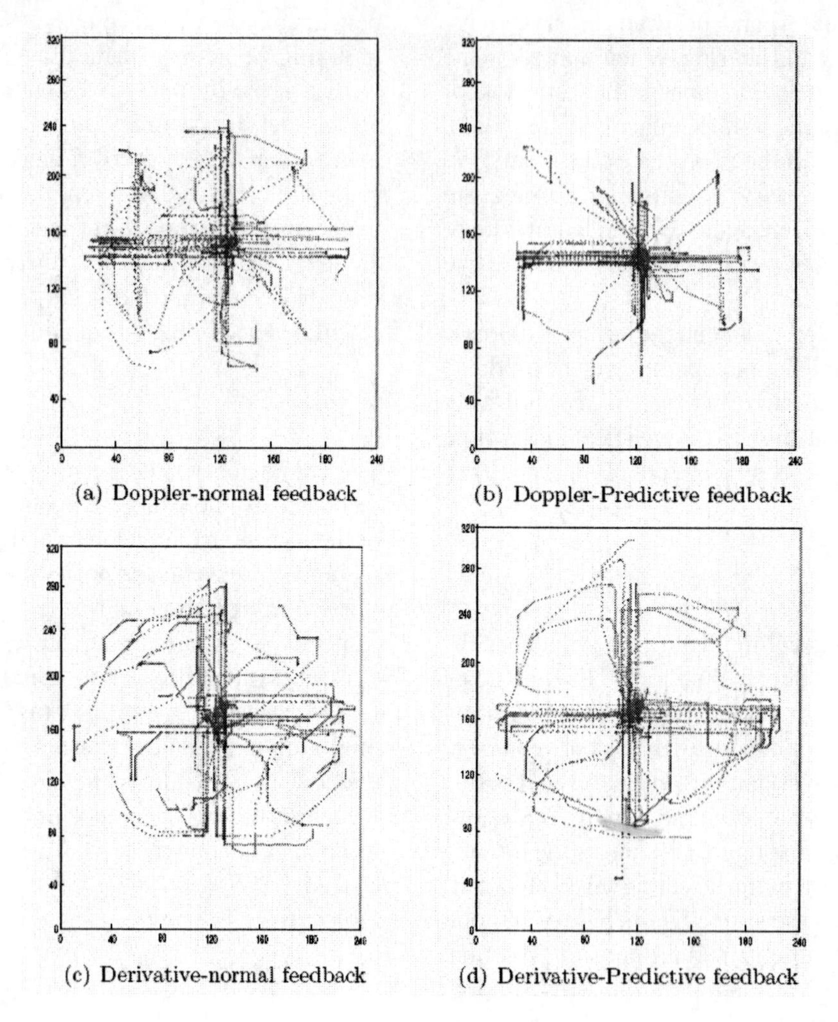

(a) Doppler-normal feedback
(b) Doppler-Predictive feedback
(c) Derivative-normal feedback
(d) Derivative-Predictive feedback

dictive model-based feedback and Doppler with no predictive feedback. They said it felt they were able to acquire the direction of the audio source more quickly in the predictive case, but that it was more difficult to land on the source. In the standard model with no predictive element they felt it was slower to find the direction, but easier to land. Despite their perceptions, the trajectories in Figures 18(a-right) and 20(a) suggest the opposite case, that they performed fewer oscillatory movements around the target in the predictive model, compared to the standard case.

Prediction in the Doppler case allowed the users to converge more rapidly and directly to the target, but it seemed less helpful very close to the target. In the derivative condition, predictive feedback seems to have smoothed the behaviour, but has not improved the initial target localisation. These preliminary explorations suggest that a more detailed investigation of incorporating the predictive element in the feedback system would be of interest.

CONCLUSION AND SUMMARY

This work presents initial experimental results exploring the use of quickened audio displays for localisation and selection based on tilt control of mobile devices. The experiments provided useful exploratory information about how users navigate in such environments, and highlighted some benefits and disadvantages of each of the display options investigated. Users used a search method, which they felt more comfortable with for browsing the space regardless of the sound treatment. Vibration was clearly perceived by users, but led them to spend more time circling around targets.

Average results in the pilot study on the metrics used suggest that participants were more accurate in target selection in the "no Doppler-no vibration" than other conditions. The results do not suggest that the use of Doppler or vibration brought consistently improved accuracy, but some people did very well with Doppler, and most stated that they found the vibration feedback useful. Longer studies might show different use in real-life tasks once users had familiarised themselves with the system.

The main study represented a more focused investigation, with fewer confounding factors. We increased the number of participants, placed all the targets at equal distances from the starting point (centre of the screen), did not include haptic feedback, chose western pop music that was familiar to all users, and allocated more time for allowing users to learn how to use the specific interface, which was new to all. We investigated whether quickening was more useful to users searching for targets in state-space than no quickening audio feedback. We also investigated if there was any advantage of using Doppler feedback over derivative, and if there was an effect of orientation in either Doppler, derivative, or no quickening; therefore, to find out if the results of the first experiment could have been masked by an interaction with the orientation of the targets. It was also found that there was no effect of the angle at which the audio source was located.

In a preliminary investigation to better understand the results and to guide future work, we performed an exploratory experiment with predictive model-based feedback. The model is based on human operator modeling in continuous tracking tasks, and it could take human response delays and lags into account (not considered in this work). Using this predictive model, we could improve the users' performance in the Doppler condition, and reduce their overshoots during landing on the target just by providing audio feedback about the user's predicted position instead of their current position. This suggests further research to investigate the benefits of explicitly incorporating models of human behaviour in the design of feedback methods.

Outlook for Mobile Interface Designers

These results are a useful starting point for further investigation into the types of feedback that are most useful and informative in assisting users of a tilt-controlled mobile device with multimodal feedback. Some of the visualisation tools used will be useful for other designers, but the work also gives an indication of the difficulty of designing experiments that test aspects of low-level perception of multimodal displays, without confounding factors from prior knowledge influencing the results. The experiments also show the need for longitudinal studies. As in early exploration of novel interfaces, much observed behaviour is related to the user exploring the novel interface, and might not be a reliable indicator of typical practiced behaviour. Supporting the design of interaction in mobile devices with multimodal interfaces is a key challenge in mobile HCI. We believe that further development of the model-based prediction techniques we have begun to explore in this chapter will not only give us a better understanding of typical user behaviour, but will provide a promising, scientific basis to support designers in creating more useable systems in a wide range of novel settings, with a range of sensors and displays.

ACKNOWLEDGMENT

The authors gratefully acknowledge the support of IRCSET BRG SC/2003/271 project *Continuous Gestural Interaction with Mobile devices*, HEA project *Body Space*, and SFI grant 00/PI.1/C067, the IST Programme of the European Commission, under PASCAL Network of Excellence, IST 2002-506778. This publication only reflects the views of the authors.

REFERENCES

Birmingham, H. P., & Taylor, F. V. (1954). A design philosophy for man-machine control systems. In *Proceedings of the Institute of Radio Engineers* (pp. 1748-1758).

Blattner, M., Papp, A., & Glinert, E. (1992). Sonic enhancements of two dimensional graphic displays. In *Proceedings of International Conference on Auditory Display, ICAD'92* (pp. 447-470). Santa Fe, NM.

Bootsma, R. J., Fernandez, L., & Mottet, D. (2004). Behind Fitts' law: Kinematic patterns in goal-directed movements. *International Journal of Human-Computer studies, IJHCS, 61*(1), 811-821.

Bregman, A. S. (1990). *Auditory scene analysis: The perceptual organisation of sound.* Cambridge, MA: MIT Press.

Brewster, S. A. (1997). Using non-speech sound to overcome information overload. *Special Issue on Multimedia Displays* (pp. 179-189).

Brewster, S. A. (2002). Overcoming the lack of screen space on mobile computers. In *Proceedings of Personal and Ubiquitous Computing* (pp. 188-205).

Brewster, S. A., & Murray, R. (1998). Presenting dynamic information on mobile computers. *Personal Technologies* (pp. 209-212).

Chion, M. (1994). *Audio-vision.* New York: Columbia University Press.

Cook, P. R. (2002). *Real sound synthesis for interactive applications.* Peters, A K, Limited.

Cook, P. R., & Lakatos, S. (2003). Using Dsp-based parametric physical synthesis models to study human sound perception. *IEEE Workshop, Applications of Signal Processing to Audio and Acoustics* (pp. 75-78).

Crossman, E. R. F. W., & Goodeve, P. J. (1983). Feedback control of hand-movement and Fitts' law. *Quarterly Journal of Experimental Psychology, (Original work presented at the meeting of the Experimental Psychology Society, Oxford, England, July 1963), 35A*, 251-278.

Doherty, G., & Massink, M. (1999). Continuous Interaction and Human Control. In *Proceedings of the XVIII European Annual Conference on Human Decision Making and Manual Control* (pp. 80-96). Loughborough: Group-D Publications.

Dong, L., Watters, C., & Duffy, J. (2005). Comparing two one-handed access methods on a PDA. In *Proceedings of the MobileHCI'05* (pp. 291-295). Salzburg, Austria.

Eslambolchilar, P., Crossan, A., & Murray-Smith, R. (2004). *Model based target sonification on mobile devices.* Paper presented at the Interactive Sonification Workshop, ISON, Bielefeld University, Germany.

Faconti, G., & Massink, M. (2001). Continuous interaction with computers: Issues and requirements. In *Proceedings of Universal Access in HCI, Universal Access in HCI - HCI International 2001* (pp. 301-304), New Orleans.

Fallman, D. (2002a). An interface with weight: Taking interaction by tilt beyond disembodied metaphors. *4th International Symposium on Mobile Human-Computer Interaction* (pp. 291-295).

Fallman, D. (2002b). Wear, point and tilt. In *Proceedings of the Conference on Designing Interactive Systems: Processes, practices, methods, and techniques* (pp. 293-302).

Fitts, P. M. (1954). The information capacity of the human motor system in controlling the amplitude of movement. *Journal of Experimental Psychology, 47*(1), 381-391.

FMOD. (2004). *Music and sound effect system.*

GAPIDraw. (2004). *Graphics API Draw.*

Gaver, W., Smith, R., & O'Shea, T. (1991). Effective sounds in complex systems:The ARKola simulation. In *Proceedings of the ACM Conference on Human Factors in Computing Systems CHI'91* (pp. 471-498).

Gaver, W. W. (1989). *The sonic finder* (Vol. 4). Elsevier Science Publishing.

Gaver, W. W. (1997). *Auditory interfaces* (2nd ed.): *Handbook of human-computer interaction.*

Harrison, B., & Fishkin, K. P. (1998). Squeeze me, hold me, tilt me! An exploration of manipulative user interfaces. In *Proceedings of the ACM Conference on Human Factors in Computing Systems, CHI'98* (pp. 17-24), Los Angeles.

Hermann, T., Hansen, M. H., & Ritter, H. (2000). Principal curve sonification. In *Proceedings of International Conference on Auditory Displays, ICAD'00* (pp. 81-86).

Hermann, T., & Ritter, H. (1999). Listen to your data: Model-based sonification for data analysis. *Intelligent Computing and Multimedia Systems* (pp. 189-194). Baden-Baden, Germany.

Hinckley, K., Pierce, J., Horvitz, E., & Sinclair, M. (2005). Foreground and background interaction with sensor-enhanced mobile devices. *ACM Transactions on Computer-Human Interaction (TOCHI)* (pp. 31-52).

Hoffmann, E. R. (1991). Capture of moving targets: A modification of Fitts' law. *Ergonomics, 34*(2), 211-220.

Holmes, J. (2005). Interacting with an information space using sound: Accuracy and patterns. In *Proceedings of ICAD 05-Eleventh Meeting of the International Conference on Auditory Display* (pp. 69-76). Limerick, Ireland.

Jagacinski, R. J., & Flach, J. M. (2003). *Control theory for humans: Quantitative approaches to modeling performance.* Lawrence Erlbaum Associates, Inc.

Jenison, R. L. (1997). On acoustic information for motion. *Ecological Psychology, 9*(2), 131-151.

Johnson, C., Brewster, S. A., Leplatre, G., & Crease, M. G. (1998). Using non-speech sounds in mobile computing devices. *First Workshop on Human Computer Interaction with Mobile Devices*, Department of Computing Science, University of Glasgow, Glasgow, UK (pp. 26-29).

Kramer, G., Walker, B., Bonebright, T., Cook, P., Flowers, J., Miner, N., et al. (1999). *Sonification report: Status of the field and research agenda.* Report prepared for the National Science Foundation by members of the International Community for Auditory Display.

Langolf, G. D., Chaffin, D. B., & Foulke, J. A. (1976). An investigation of Fitts' Law using a wide range of movement amplitudes. *Journal of Motor Behaviour, 8*, 113-128.

MacKenzie, I. S., & Ware, C. (1993). Lag as a determinant of human performance in interactive systems. In *Proceedings of the ACM Conference on Human Factors in Computing Systems - INTERCHI '93* (pp. 488-493). Amsterdam.

Mackinlay, J. D., Robertson, G. G., & Card, C. K. (1991). The perspective wall: Detail and context smoothly integrated. In *Proceedings of the ACM Conference on Human Factors in Computing Systems, CHI'91* (pp. 173-179). New Orleans, LA.

Oakley, I., Ängeslevä, J., Hughes, S., & O'Modhrain, S. (2004). Tilt and feel: Scrolling with vibrotactile display. In *Proceedings of Eurohaptics 2004*, Munich, Germany.

Partridge, K., Chatterjee, S., Sazawal, V., Borriello, G., & Want, R. (2002). TiltType: Accelerometer-supported text entry for very small devices. *UIST'02: Proceedings of the 15th Annual ACM Symposium on User Interface Software and Technology* (pp. 201-204). Paris.

Patrol, C. A. P.-R. A. (1999). *Mission observer course manual.* Civil Air Patrol.

Poulton, E. C. (1974). *Tracking skill and manual contro.* Academic press.

Rekimoto, J. (1996). Tilting operations for small screen interfaces. *UIST'96: Proceedings of the*

9th Annual ACM Symposium on User Interface Software and Technology (pp. 167-168). Seattle, WA.

Rinott, M. (2004). Sonified interactions with mobile devices. In *Proceedings of the International Workshop on Interactive Sonification*, Bielefeld, Germany.

Sazawal, V., Want, R., & Borriello, G. (2002). The unigesture approach, One-handed text entry for small devices. In *Proceedings of Mobile Human-Computer Interaction - Mobile HCI'02* (pp. 256-270).

Smith, D. R., & Walker, B. N. (2005). Effects of auditory context cues and training on performance of a point estimation sonification task. *Applied Cognitive Psychology, 19*(8), 1065-1087.

Smith, N., & Schmuckler, M. A. (2004). The perception of tonal structure by the differentiation and organization of pitches. *Journal of Experimental Psychology: Human Perception and Performance, 30,* 268-286.

Walker, B. N., & Lindsay, J. (2006). Navigation performance with a virtual auditory display: Effects of beacon sound, capture radius, and practice. *Human Factors, 48*(2), 265-278.

Wigdor, D., & Balakrishnan, R. (2003). TiltText: Using tilt for text input to mobile phones. *UIST'03: Proceedings of the 16th annual ACM symposium on User interface software and technology* (pp. 81-90). Vancouver, Canada.

Williamson, J., & Murray-Smith, R. (2004). Pointing without a pointer. In *Proceedings of the ACM Conference on Human Factors in Computing Systems, CHI'04* (pp. 1407-1410). Vienna, Austria.

Williamson, J., Strachan, S., & Murray-Smith, R. (2006). It's a long way to Monte Carlo: Probabilistic display in GPS navigation. *In Proceedings of Mobile Human-Computer Interaction, Mobile HCI'06* (pp. 89-96). Helsinky, Finland.

KEY TERMS

Continuous Control: A continuous control system measures and adjusts the controlled quantity in continuous-time.

Gestural Interfaces: Interfaces where computers use gestures of the human body, typically hand movements, but in some cases other limbs can be used, for example, head gestures.

Haptic Interfaces: Convey a sense of touch via tactile or force-feedback devices.

Manual Control: A branch of control theory that is used to analyse human and system behaviour when operating in a tightly coupled loop.

Nonspeech Sound: Audio feedback, that does not use human speech. The use of nonspeech sound in interaction has benefits such as the increase of information communicated to the user, the reduction of information received through the visual channel, the performance improvement by sharing information across different sensory modalities.

Prediction Horizon: How far ahead the model predicts the future. When the prediction horizon is well matched to the lag between input and output, the user learns how to control the system more rapidly, and achieves better performance.

Quickened Displays: Displays that show the predicted future system state, rather than the current measured, or estimated state.

Sonically Enhanced Interfaces: Interfaces where sound represent actions or content.

Sonification: The use of nonspeech audio to convey information or perceptualize data.

Sound Localisation: The act of using aural cues to identify the location of specific sound sources.

Chapter XXX
Unobtrusive Movement Interaction for Mobile Devices

Panu Korpipää
Finwe Ltd., Finland

Jukka Linjama
Nokia, Finland

Juha Kela
Finwe Ltd., Finland

Tapani Rantakokko
Finwe Ltd., Finland

ABSTRACT

Gesture control of mobile devices is an emerging user interaction modality. Large-scale deployment has been delayed by two main technical challenges: detecting gestures reliably and power consumption. There have also been user-experience-related challenges, such as indicating the start of a gesture, social acceptance, and feedback on the gesture detection status. This chapter evaluates a solution for the main challenges: an event-based movement interaction modality, tapping, that emphasizes minimal user effort in interacting with a mobile device. The technical feasibility of the interaction method is examined with a smartphone equipped with a sensor interaction cover, utilizing an enabling software framework. The reliability of detecting tapping is evaluated by analyzing a dataset collected with the smartphone prototype. Overall, the results suggest that detecting tapping is reliable enough for practical applications in mobile computing when the interaction is performed in a stationary situation.

INTRODUCTION

The source of innovations in a mobile device user interface lies in combinations of input and output technologies that match the user's needs. In the mobile context, movement sensing, and haptic feedback as its counterpart, offers a new dimension to multimodal interactions. There are use cases where traditional interaction modalities are insufficient, for example, when the device is placed in a pocket or a holster, or if the user is wearing gloves. In these situations the user cannot press or see buttons to interact with the device. Instead, small motion gestures can be used as a limited, but convenient, control modality. The movement of the device can be captured with a 3-axis accelerometer, and the resulting acceleration signal can be used to detect the movement patterns for controlling the device.

One of the main questions in the application of a movement-based interface is how to distinguish gesture movements the user performs from those movements that are produced by various other activities while carrying and using the device. Reliability can be argued to be the most important challenge in developing a mobile device gesture interface. This chapter presents a reliability evaluation of an unobtrusive event-based gesture interface by analyzing a multiuser dataset collected with a smartphone prototype. Another main challenge has been the relatively high power consumption from the continuous measurement of acceleration, which is not acceptable in mobile devices. Novel accelerometers are capable of producing interrupts based on exceeded thresholds; therefore, the detection, initiated by a hardware interrupt, can be implemented as event based and low power. The technical feasibility of event-based tapping detection is examined with a smartphone equipped with a sensor interaction cover, Figure 1, and an enabling software framework. Furthermore, the chapter addresses the issue of flexibly customizing the gesture interface and feedback modalities relevant to aiding the user.

There are various ways of implementing a gesture interface. This chapter focuses on analyzing the tapping interaction, which shows potential as a significant application of accelerometers in future mobile devices. More specifically, the chapter addresses the movement pattern where the user taps the device twice consecutively, which is called a double tap. With an implementation based on abstractions initiated by sensor-driven interrupts, the aim is a low-power, reliable, and customizable user interaction modality.

Gestures can be detected either from a continuous stream or discrete segments of sensor data. In detection from discrete segments, gesture start and end are explicitly marked with a button instead of a continuous flow of device movements. From

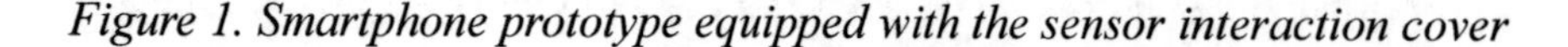

Figure 1. Smartphone prototype equipped with the sensor interaction cover

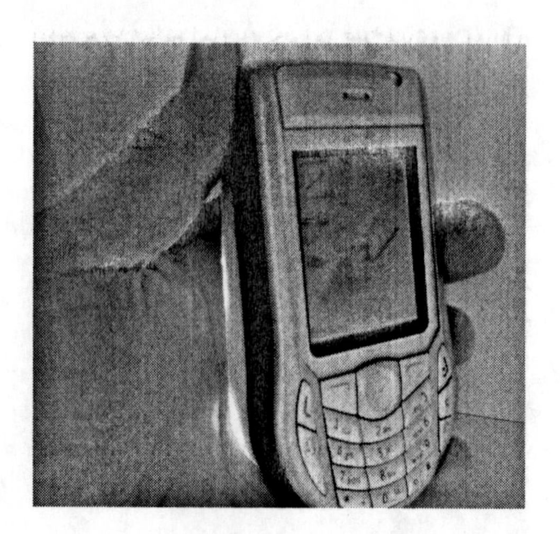

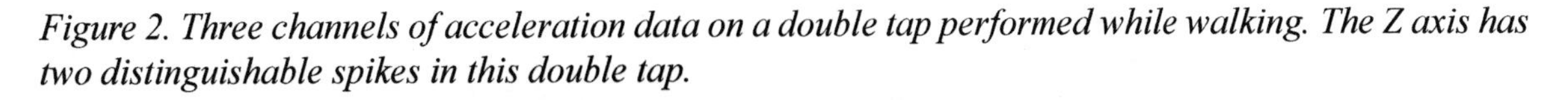

Figure 2. Three channels of acceleration data on a double tap performed while walking. The Z axis has two distinguishable spikes in this double tap.

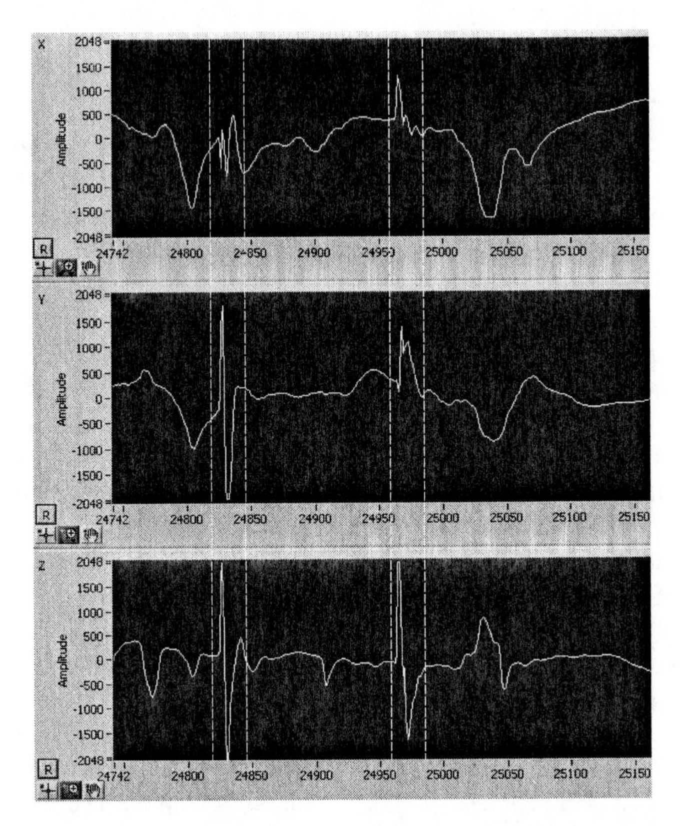

the usability perspective, interaction without explicit marking is preferred, in general, since it requires less attention from the user. However, continuous data streaming and execution of the gesture detection algorithm requires continuous data processing, which normally consumes battery power.

The development in digital acceleration sensor technology enables the integration of programmable interrupt-based solutions that can operate with low current consumption. Such sensors generate interrupts when acceleration on a spatial axis is over or below a set threshold level. Hence, movement detection algorithms, initiated by an exceeded threshold, can be implemented as event based instead of continuously processing a stream of data. The processing load at the mobile device side is similarly reduced since the operating system is woken up less frequently. This development opens up new possibilities for practical application of the technology in mass products such as mobile phones.

The distinguishable form of the tapping pattern, processed after the event threshold, is the basis for the potential reliability of detecting them, even when the detection process is continuously active, Figure 2. By contrast, free-form gesture recognition has a much wider problem setting, requiring a more complex model of the gesture and thus, heavier processing load, making continuous processing much more challenging, especially in mobile devices.

This chapter publishes the first statistical performance evaluation based on a dataset that characterizes the reliability of user-independent tapping interaction in mobile phones. Moreover, the sensitivity of the method to misrecognitions is evaluated with scenarios consisting of various activities. As an introductory topic, applying a smartphone equipped with sensor interaction cover, customization, and feedback of the addressed interaction modality are discussed.

BACKGROUND

In acceleration sensor-based gesture recognition, gestures are detected either from a continuous stream or from discrete segments of sensor data. While this chapter addresses the detection of movement patterns from a continuous stream, there are a lot of studies in the literature on gesture recognition from discrete segments (Feldman, Tapia, Sadi, Maes, & Schmandt, 2005; Mäntyjärvi, Kela, Korpipää, & Kallio, 2004). Specifically, acceleration sensors have been applied in user-trainable and pretrained machine-learning-based gesture recognition systems (Kallio, Kela, Korpipää, & Mäntyjärvi, 2006; Kela, Korpipää, Mäntyjärvi, Kallio, Savino, Jozzo, & Di Marca, 2006). Free-form gesture recognition still has a limitation; it requires an explicit marking of the gesture, for example, with a button, and longer duration gestures to increase the recognition accuracy. Hence the interaction requires more user effort, and gesturing can be socially obtrusive. However, despite the possible obtrusiveness when applied in public places, free-form gestures also have a wide range of potential uses in other settings, such as games, home electronics control, and so forth, where social acceptance does not limit the use of the modality. The social aspect, distinctively important in the mobile usage context, has been addressed by Linjama et al. (Linjama, Häkkilä, & Ronkainen, 2005), Rekimoto (2001), and Ronkainen et al. (Ronkainen, Häkkilä, Kaleva, Colley, & Linjama, 2007). Based on the literature, it can be extrapolated that, when performed with a mobile device such as a phone, smaller gestures are considered more socially acceptable than large ones.

This chapter especially advocates the unobtrusiveness of the interaction; gestures as small and as unnoticeable as possible are preferred, assuming they are more acceptable by the users (Linjama et al., 2005). Examples of possibly useful small-scale gestures include shaking the device, for example, Levin and Yarin (1999), and swinging it from side to side (Sawada, Uta, & Hashimoto, 1999). However, both of these interaction methods can be considered quite noticeable, regardless of scale. Shaking also raises the question of how many repetitions of the shake movement are required until a shake is recognized. A simple accelerometer-based tilting control has been discussed in the literature in many studies over the years, for example, Rekimoto (1996), but also recently, for example, combining tilt and vibrotactile feedback (Oakley, Ängeslevä, Hughes, & O'Modhrain, 2004), scrolling, and switching between landscape and portrait display orientations (Hinckley, Pierce, Horvitz, & Sinclair, 2005). Tilting is another potentially unobtrusive, and very simple to implement, movement-based interaction modality to be applied in carefully selected use cases in mobile computing.

A minimalist extreme in hand gestures is tapping the mobile device, first introduced in Linjama and Kaaresoja (2004). Tapping only requires a small scale of device movement, and can be performed by finger or palm. The technological benefit is that tapping can be relatively straightforwardly captured with a 3-D accelerometer, since the resulting movement pattern has a distinguishable sharp spike form. The detection problem can be narrowed down by applying a small, predefined fixed set of movement patterns: tap events.

The unique usability benefit of the tap interaction is that it is discreet and can be used if the mobile device is located in a pocket or a backpack, since explicit marking is not needed. Furthermore, the user is not required to hold the device or see the keyboard to interact. A good example of a use case where tapping is useful can be found in the Nokia 5500 phone (Nokia, 2006): when a text message arrives, the user has 30 seconds to tap the phone twice and the message will be read aloud to the user. It is useful when the phone is in a pocket or on a belt, or the user is wearing gloves; the message can be read without first taking the phone into the hand and opening the keypad lock. Furthermore, tapping can be used as an additional modality. For instance, phone music player commands, such as play next or previous song, can be controlled by tapping on either side of the phone, which is convenient when the device is worn on a belt or in a pocket. Again, the user does not have to take the phone, open keypad lock, and press a button to perform the control action.

SENSOR INTERACTION COVER

Interrupt-initiated abstracting of movement patterns can be performed using a separate microcontroller, or, ideally, it can be directly integrated in the sensor chip. A sensor interaction test platform was developed to experiment with the interaction concept. The platform consists of a Symbian S60 phone (Nokia 6630) equipped with a sensor and feedback cover attached to the back of the smartphone, Figure 3.

Inside the cover, the hardware includes a 3-D acceleration sensor (STMicroelectronics LIS3LV02DL), a microcontroller (Atmel), an NFC reader, blue LEDs, a buzzer, and a vibra motor, Figure 2. The board is two sided. Tap detection parameters and feedback configuration can be set to the microcontroller from the phone software. The tap detection algorithm and the feedback processing are performed in the cover microcontroller, and the cover transmits recognized tap events to the phone through USB. Thus, the communication between the cover and the phone, as well as power consumption, is minimized.

Figure 3. Sensor interaction cover hardware

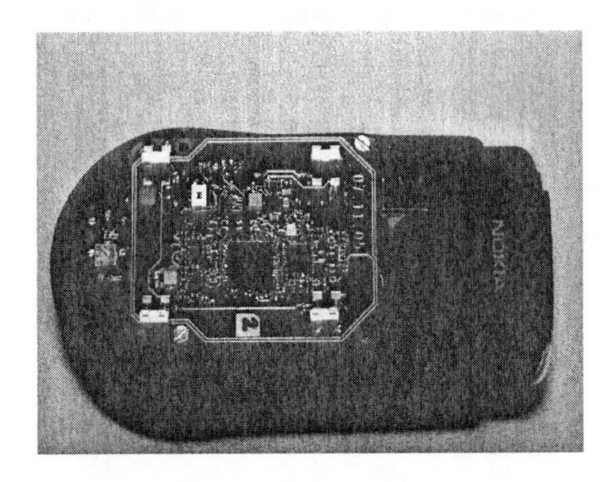

INTERACTION CUSTOMIZATION

Once sensor events are abstracted by the microcontroller and sent to the phone through the USB, they should activate the desired actions in the mobile phone. Flexibly connecting the abstracted sensor events to various application actions requires supporting middleware on the phone side. Instead of connecting an application directly to a device driver, the data is abstracted into a uniform representation applied through Context Framework.

Context Framework (CF) is a blackboard-based software framework for enabling and customizing situation-aware and sensor-based mobile applications (Korpipää 2005; Korpipää, Mäntyjärvi, Kela, Keränen, & Malm 2003). All interaction-related information, including implicit and explicit sensor-based inputs, is treated as context objects within the framework, expressed with a uniform vocabulary. An implemented instantiation of the framework is illustrated in Figure 4 (left-hand side). In this case, the sensor signal abstracting process functionality is on the microcontroller side, illustrated in Figure 4 (right-hand side flow diagram). Sensing, feature extraction, and classification are performed at the cover's microcontroller. Classified movement (context) events are sent over the USB to the phone side, where CF enables controlling any available application action based on the events.

The user can create desired context-action behavior with a mobile phone by creating XML-based rule scripts with the graphical UI of the Customizer. CF handles the background monitoring of context events and the triggering of actions according to the rules. The Application Controller facilitates the application control inference on behalf of the user or application, in other words, provides an inversion of control. The framework completely separates the management of sensor-based context events from application code and the hardware. Hence, by applying CF, no changes need to be made to existing mobile phone applications when they are augmented with sensor-based features.

In the case of tapping input, the events are abstracted into context objects by the sensor cover

Figure 4. Context Framework (CF) architecture example instantiation (left), and pattern recognition flow (right)

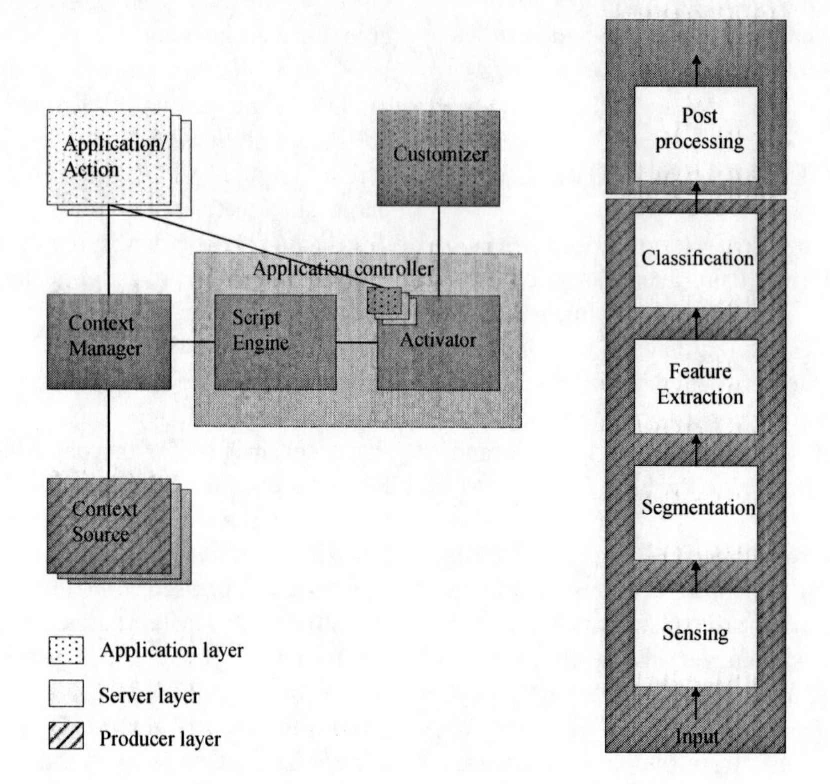

of the phone and delivered to CF. The application developer or the user interface designer can use the Customizer tool to define which application actions are executed by which abstracted sensor events. The definable actions include available feedback modalities, such as tactile, auditory, and visual indications. By creating rules with the Customizer tool, the user can define actions on an operating system level, or for a specific application, by setting a condition part of a rule to include a specific foreground application. For instance, the following accelerometer-based features were defined and executed simply as XML-based rule scripts:

- Playing the next or previous song in music player using double tap
- Activating display illumination using tap
- Unlocking the keypad using double tap

Figure 5 presents a series of screenshots from the Customizer tool, illustrating the definition of a rule that enables the user to unlock the keypad by double tapping the phone.

In Figure 5a, the user selects an action for the rule by navigating through the action type Phone. Keypad and selecting the action value KeypadUnlock. In Figure 5b, the user selects a trigger for the rule by navigating through context type Gesture and selecting context value DoubleTap. The first screenshot in Figure 5c shows the complete rule after the user has selected the elements. After the user selects the option Done, the rule script is generated, and the rule is activated and functional in the context framework. The second screenshot in Figure 5c shows the main rule view with the list of active rules. When the rule conditions are met, the Context Framework automatically performs the action.

Figure 5. Series of screenshots illustrating how to program the phone to open the keypad lock with a double tap.

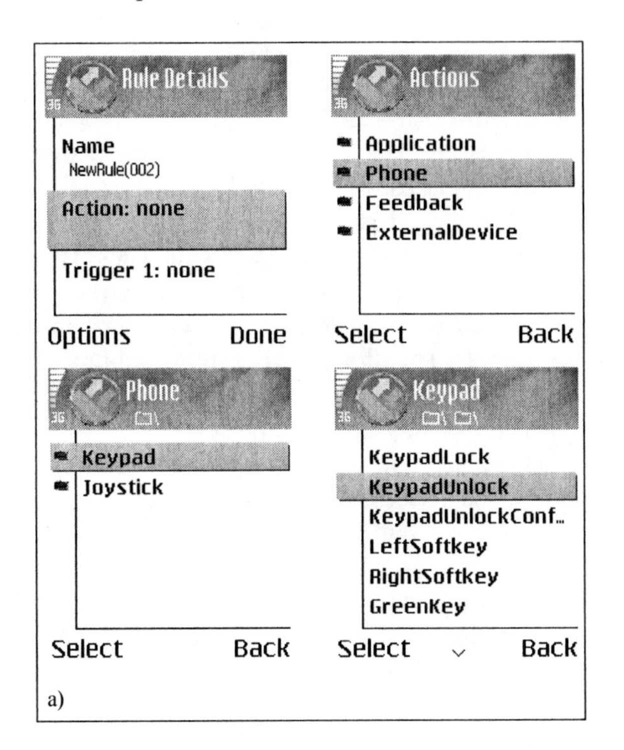

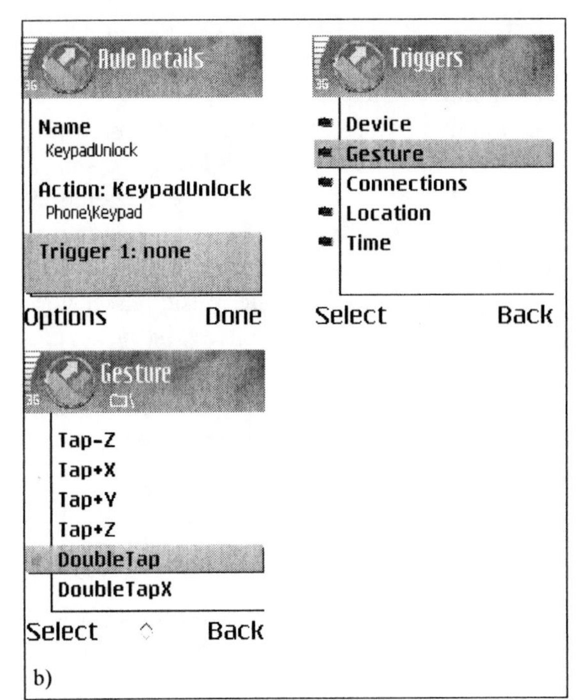

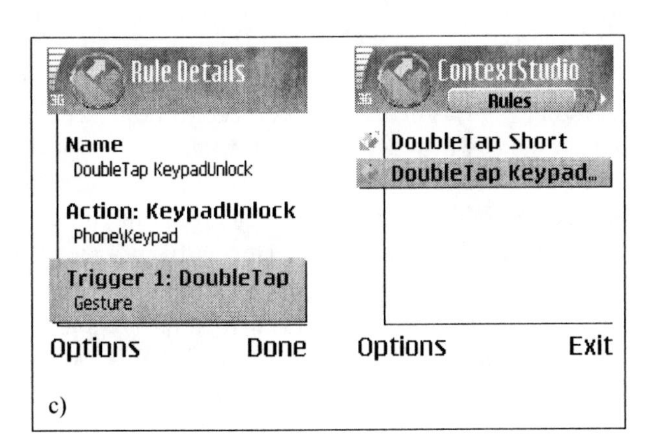

USE CASES AND USABILITY

Evaluating the general usability is an essential aspect in developing tapping interaction, in addition to evaluating the reliability and technical feasibility. As an extensive topic, however, it cannot fit within the scope of this chapter. The purpose of this section is to briefly discuss a few usability-related points as an overview of the experiments studying the usability of the novel interaction modality.

New interaction modalities, like tapping, have certain application areas where they add value, both in terms of utility (usability) and joy (fun of use). The only way of evaluating these aspects is to try the interaction elements in practice, with real hardware and applications. The smartphone sensor interaction cover serves this purpose. It enables the rapid prototyping and iterative development of interaction concepts and demonstrations. User experiences and feedback can be collected during various stages of development, early concept tests, peer evaluations, and end user tests in the lab and in the field.

A number of formal user tests, to be published separately, have been conducted with the smartphone prototype. The tests measure the potential usefulness of tap interaction with a standard Symbian Series 60 phone user interface. For instance, controlling a phone music player with tapping commands, in addition to existing keypad controls, has been studied. The results from the studies indicate that it is very important to maintain consistency in case there are multiple different uses of tap interaction in several applications of the same device. Users may get confused if tapping is used for too many different purposes, such as muting the phone audio in one application and selecting the next or previous music track in another application. Furthermore, using different tap directions, for example, tapping on the device top or the side for activating different controls in different applications, requires delivering specific instructions to the users.

User satisfaction, joy of use, has also been addressed in the usability tests. The enjoyability of the user experience is largely determined by the very details of the interaction; what kind of feedback elements support the user interaction. What is the metaphor behind the observed device behavior that the user learns when using new interaction modalities? All sense modalities in multimodal interaction must be addressed together.

CONTINUOUS DETECTION RELIABILITY: EXPERIMENTS

This study focuses on analyzing the reliability of detecting double taps in various usage situations. The experiments to be presented next aim to answer how accurately double taps can be detected in a general mobile usage setting and how many misrecognitions occur. The results should reflect an essential part of how feasible this interaction method could become, from the reliability viewpoint, when used in mobile phone applications. Detection accuracy is quantitatively analyzed based on acceleration data collected from users performing the interaction, and the results are discussed.

There were 11 users performing the interaction and the scenarios; 7 of the users were male and 4 were female, aged from 25 to 36 years. The subjects were selected randomly from acquaintances of the authors. The subjects were not interviewed and no subjective opinions were collected, only acceleration signals. Therefore, the limited variability of the subjects in the user group was assumed not to bias the results significantly.

There are two categories of use cases for continuous detection of movement events. In the first category, the detection process is initiated by a specific application or a situation, and is active for a certain time. In the other category, the detection process is always active. In the first category, the use cases can be designed such that misrecognitions, false positives, have a minimal effect. In the latter category, false positives usually have a more negative effect since they may result in incorrect operation. In both cases, the sensitivity of detection should yield enough correct recognitions, true positives, to be acceptable for the users.

This section describes the experiments aiming at evaluating how well the tapping interface performs from a statistical point of view, based on collected data. Detecting tapping events is a type of pattern recognition problem (Duda, Hart, & Stork, 2001), although not a very complex one. The aim of the data analysis during the development process was to reach optimal recognition of a double tap pattern, that is, to find detection algorithm parameters that produce a minimal number of false positives while maintaining a high percentage of true positives. The primary goal was to minimize misrecognitions. The algorithm should give the best results as an average when performed by multiple users, not just one specific user. In other words, the aim is to reach optimal user-independent detection accuracy. The experiment involved collecting a dataset on the target patterns performed by several users in controlled stationary conditions. Furthermore, data from several real-world scenarios containing various daily activities was collected to find out how often misrecognitions occurred.

Data Collecting

In order to evaluate the tapping detection reliability statistically, a sufficiently large dataset is required. Dataset size and the variation it contains are in direct relation to the evidence to support generalization. Data was collected in three stages with the sensor cover-equipped smartphone prototypes. The first stage involved exploring a wide set of activities by a user carrying 1-2 prototypes to find out whether there were any specific activities that produced a lot of false positives. The dataset was collected by one user, and the total duration of the activities in the dataset was 5 hours 8 minutes.

The second stage involved having several users perform the target patterns in stationary controlled conditions involving no other activities. This dataset consisted of 11 users performing double-tap patterns. Data was collected in three categories, arranged by the user's skill level and the given advice. The user groups were beginner, people who had never heard about tapping, and advanced, people who knew or were informed about how the tapping user interaction works. There were six users in the beginner group. In the beginner group, the users were only given one piece of advice: to perform the tapping with their hand(s), not by tapping the phone on the table. The second group, five advanced users, were first told to use one hand for tapping and next to use both hands, that is, hold the phone with one hand and tap with the other. Figure 6 shows an example of both ways of tapping interaction.

There were five users, the same ones, in both of the advanced groups. In the three categories, each user performed a double tap 18 times, resulting in total target of 288 repetitions in the dataset. Repetitions were performed in phases of three repetitions and a break, during which the device was put on the table to avoid a routine speed-up and fixation on a certain way of interaction.

The third stage involved having several users perform scenarios involving real-world activities while carrying the prototype in their pocket. The purpose of this dataset was to find the occurrence of false positives during the scenarios, on average over multiple users. There were four to five users in each of the scenarios. The total length of the activity dataset was approximately 54 minutes. The tapping pattern has a sharp spike-form shape, and proper detection requires a relatively high sampling rate. Hence, the total amount of raw data collected for this experiment was approximately 68 megabytes.

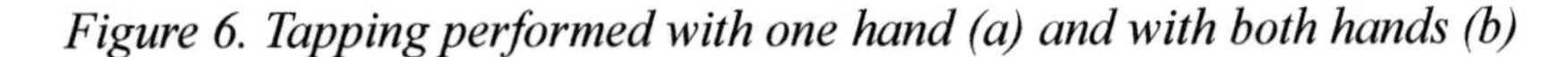
Figure 6. Tapping performed with one hand (a) and with both hands (b)

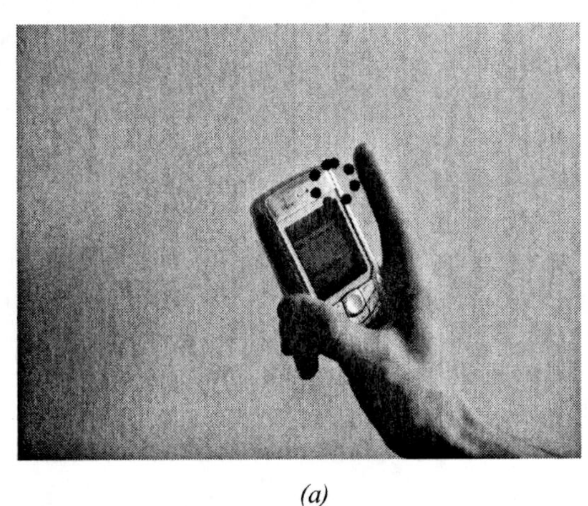
(a)

(b)

EXPERIMENT RESULTS

The collected acceleration data was used to analyze the tapping interaction from multiple aspects. The experiments focused on a specific form of tapping, a double tap. Double tap means performing two consecutive taps within a certain short time span, much like a double click with a mouse. Each aspect of this interaction studied with the collected dataset is described in detail in this section. The experiments produced numerical measurements of the system's tapping detection accuracy. The measurements are briefly introduced here before presenting the results and analyzing them.

The first experiment was an initial pilot test, which was designed to count the number of double-tap patterns detected where they should not exist. In other words, the experiment measures the occurrence of false positives, which can be reported as a number per time unit. For example, the aim could be that there is no more than one false double-tap detection per hour.

False positives can also be represented in relation to how many patterns could be falsely detected from a dataset. The relative number of false positives in a dataset can be given by dividing the occurrence of all detected false positive patterns with all segments of data where there should not be a detected pattern. Here a segment is defined as the maximum time span required to detect one pattern. For example, for double tap pattern the maximum allowed duration is 1.1 seconds. This is due to the algorithm wait time for the second tap to appear after the first one. For instance, in a dataset of 110 seconds, there are 100 segments that could potentially contain a double tap. One false double tap in that dataset would result in one percentage of false positives.

True positive means a correctly detected pattern, for example, a double tap is detected correctly when it is performed by the user. The relative occurrence of true positives can be given by dividing all detected true positive patterns by all actually performed true patterns in a dataset.

Pilot Test

The goal of the pilot test was to explore whether some of the randomly selected ordinary daily activities would produce a high occurrence of false positives. This experiment did not contain any actual double taps performed by the user. The user was assigned to carry one or two prototypes in a pocket during various daily activities, for example, random outdoor activities (cleaning the yard, commuting, driving a car, walking, jogging, biking, cross-country skiing, and roller-skating). The users were free to select which clothes to wear and which pockets to carry the devices in. The tasks were given as, for example, "take the phone with you and go jogging." Table 1 summarizes the results of this test.

There were several activities that did not produce any false positives, such as jogging, various outdoor activities, biking, going for lunch, and roller-skating. The activity that produced the most false positives was cross-country skiing.

Overall, the test indicated that potential problem areas are accidental tapping by hand, ski stick, backpack, and so forth, and when the phone is laying freely on a moving and trembling surface such as a car dashboard. After the pilot test, the detection algorithm and parameters were adjusted to reduce the misrecognitions.

Stationary Conditions

Next, an experiment was performed in controlled stationary conditions. The purpose of the experiment was firstly to gain validation of how well the target patterns are recognized in a stationary situation when there are no external disturbances. Secondly, it is important to know whether there are differences between two groups of users when one has no idea what a tapping interface is and the other has prior knowledge of how to interact with tapping. The results indicate different variations in the first-time use of tapping in terms of gesture signal waveform and the detection accuracy. Thirdly, the results show whether there are major differences between individual users, whether

Table 1. Occurrences of double-tap false positives during random daily activities

Activity	Phone numbers, placement	Duration (min)	False positives
Commuting (dressing, driving, walking, stairs up, stairs down, office)	2, left and right lower jacket pocket	28	2
Travel by car, tarmac road	1, dashboard	70	1
Travel by car, rough gravel road	1, dashboard	20	2
Jogging	2, left and right jacket chest pocket	5	0
Cross-country skiing (walking, changing, skiing, walking, undressing)	2, jacket pocket, backpack	75	7
Outdoor activities (removing snow, walking, putting bike in storage)	2, left and right jacket chest pocket	5	0
Biking (gravel road and tarmac road)	2, jacket pockets	10	0
Going for lunch (stairs down, lunch, walking, stairs up)	1, jeans pocket	25	0
Roller skating	1, loose short trousers front pocket	35	0
Roller skating with sticks	1, pants front pocket	35	1
Total		**5 hours 8 min**	**13**

Table 2. Recognition rate in stationary situation for various user groups in 3-axis detection

User group	Users	True positive %
Beginner	6	55.2
Advanced one hand	5	90.6
Advanced both hands	5	90.2

the interaction is equally assimilated by all users or if there are some individuals that cannot use the method as well. Finally, the interaction by tapping can be performed by using one hand or both hands, and the results indicate which is preferred from the reliability point of view with the evaluated algorithm.

The results can be calculated in two ways: the interaction can be allowed from any of the three axes, or from one selected axis only. In most single application use cases, the direction of tapping is known in advance and can be restricted. For example, music player next and previous commands can only be initiated with a tap on either side of the phone, by utilizing only the x-axis while disregarding the others. Hence, depending on the use case, it is feasible to filter the data from one or two other axes and apply the signal from one axis only. The results are first presented for 3-axis detection, Table 2.

The results show that double taps can be detected fairly well in stationary conditions, except

Table 3. Recognition rate in a stationary situation for each individual user in 3-axis detection

User group	User1	User2	User3	User4	User5	User6	Total
Beginner	94.4	0	77.8	88.9	0	73.3	**55.2**

User group	User1	User2	User3	User4	User5	Total
Advanced one hand	88.2	73.3	100	88.2	100	**90.6**
Advanced both hands	100	50.0	100	100	100	**90.2**
Total	**94.1**	**61.7**	**100**	**94.3**	**100**	**90.4**

in the beginner group. The difference between the beginner and advanced user groups is quite large, which suggests that first-time users may have trouble when starting to apply the method if they are not properly informed. There were also distinct differences between the individual beginner users, Table 3.

The data from the beginner users that produced low accuracies revealed that they performed the taps too lightly. Half of the beginner users chose to perform the tapping with one hand, and half with both hands. One beginner user tapped the top of the device and one the bottom, others from the side. The two beginner users that tapped with one hand had the zero results. The recollection from the actual test situation and data visualization confirm that the two one-hand users having a zero result only touched the device very lightly instead of properly tapping it. In other words, the first-time users' low performance is partly an algorithm sensitivity issue, but most importantly it is due to the lack of information the user has on how to do the tapping in the first place. The results can be improved by modifying the parameters to be more sensitive, but then the false positives tend to increase. The most straightforward way to improve the result is simply to advise the beginner users to tap with the correct intensity. Feedback is one way of giving immediate information to the user.

It must be noted that this experiment produced no information on the learning curve; it simply provides data on how differently first-time users may perform the gesture. There was no feedback or interaction in the test to guide the user on how to improve. In this sense it was a "blind" blank test to examine different users' approaches to performing a double tap, as interpreted from the signal waveform and the resulting detection accuracy. In a normal usage situation, the user would learn that too light taps do not cause the desired operation, and would likely either modify their behavior or abandon the method. In this test the users did not know that they tapped too lightly and thus, could not know how to change their tapping style.

The results in Tables 2 and 3 present the results for a setting where double taps from any direction are allowed. Table 4 presents results where only one predetermined axis signal is applied to detect a double tap. A significant increase in detection accuracy is evident. Furthermore, it is likely to reduce the occurrence of false positives, although it was not tested in this study. In light of

Table 4. Advanced user recognition rate in a stationary situation in 1-axis detection

User group	Users	True positive %
Advanced one hand	5	95.3
Advanced both hands	5	98.8
Total	**5**	**97.0**

the results, it is preferable to restrict the detection axis whenever is possible.

The results indicate that tapping is detected slightly more accurately when performed with one hand in 3-axis detection. In 1-axis detection, the accuracy is slightly better when performed with both hands. However, statistically, a conclusion cannot yet be drawn with this dataset on which way of tapping is more reliable.

False Positives - Multiple Users

The purpose of the experiment with mobile scenarios was to find the occurrence of false positives during the selected common daily activities: walking, walking up stairs, jogging, and roller-skating. Furthermore, the scenarios were performed by multiple users in order to address the issue independent of the user. The scenarios in the experiment were designed to address a usage situation where the phone is in the user's pocket and the user could tap the phone from any direction. The users wore their own clothes and were free to select where to put the phone during the test. No other hard objects were allowed in the same pocket.

The results show that the number of double tap false positives was zero during the total of 54 minutes of activity data. By adjusting the algorithm parameters to more sensitive (which also increased true positives in the stationary test), false positives started to occur. The most false positives occurred on stairs. However, the parameter set that produced zero misrecognitions was generally perceived as sensitive enough, even though there were beginner users who would have benefited from increased sensitivity.

Summary of Results

Overall, the results based on the collected data, Table 5, indicate that detection is reliable enough for practical applications in mobile computing when the user performs the interaction in a stationary situation. Moreover, the number of false positives is low enough for types of mobile applications with at least a restricted scope. The results have significance for commercial applications built on use cases that have a clear usability advantage from the tapping interaction.

Table 5. Overview of the test results

Test	Users	True positive %
Beginner	6	55.2
Advanced one hand 3 axis	5	90.6
Advanced both hands 3 axis	5	90.2
Advanced one hand 1 axis	5	95.3
Advanced both hands 1 axis	5	98.8

The results also show that there is room for improvement. This especially concerns the usability aspect of first-time use. An important question is how to give instruction on using the interface. This experiment took a worst-case scenario where the user was given almost no information, much like when the user does not even read the manual before starting to use the device. In a real learning situation, however, the user may sometimes even look for instructions in the manual, or someone will demonstrate how to use the feature. Thus, the results could be different. Furthermore, unlike in this test, the user would get feedback if the device did not respond to the interaction. Analyzing the learning curve, which is another relevant topic, requires a different experiment setup.

Having zero misrecognitions from four activities performed by four to five users with a total of 54 minutes of data does not yet statistically allow a strong generalization statement, although it is a good result, and shows that practical application is certainly feasible. To gain even wider evaluation, the next phase is to perform longer tests by equipping the users with prototypes for use in their normal daily lives.

FUTURE TRENDS

Although this study did not specifically discuss the user experience side of movement-based interaction, there is one aspect we would like to briefly address when viewing future trends: feedback.

This aspect is still often found insufficient in novel user interfaces. While the presented experiments evaluated the reliability of the double-tap detection, future work includes analyzing the learning curve, the best type of feedback, and its effect on the user experience.

The user experience and learning curve for new interaction modalities can potentially be improved with suitable feedback. For example, if the beginner user makes too light taps in a tutorial mode, the device can indicate this with feedback. In general, feedback gives an indication of the state of the system and guides the users in how to use it. As suggested by O'Modhrain (2004), a key to the design of successful touch and haptic-based mobile applications is in ensuring a good mapping between the tasks, the required sensory cues, and the capabilities of the system on which the application is to be implemented. With the Customizer tool, introduced earlier, developers and user interface designers can easily experiment with different multimodal input and output combinations to find the most suitable and enjoyable solution for their application needs.

Different combinations of the feedback patterns (vibration, LED, sound) available in the interaction cover were implemented in this study. The option of using direct cover feedback in addition to phone vibra in the interaction had the benefit of avoiding possible latencies in feedback generation on the phone side. The vibra feedback was thus precisely adjustable to the desired parameters. Even though experiments on feedback supporting usability were not presented in this chapter, it can be predicted that utilizing minimalist gesture control, together with related haptic feedback elements, has great potential in a mobile device usage and technology context. Haptic content fidelity can be rather low if it is designed to be multimodal; visual and haptic content are applied synchronously to support each other. The interaction and content design are used to promote the adoption of the technology among users.

Continuous detection of small sharp movement events also facilitates forms of gestures other than double tap. As an analogy to mouse control, there is a click and a double click. Obviously, single taps can be utilized for many purposes. However, single taps are more sensitive to various disturbances, such as accidental knocking, dropping, quick swings, turning, and so forth, that can produce a similar sharp pattern to the data and thus, a false positive. Another interesting gesture that feels natural is to swing the device. There are many other possible movement patterns to utilize in the future.

Several research questions remain, such as how to inform the user about the correct intensity of the tapping, and what kind of learning curve the tapping has. Many of the misrecognitions in the beginner group, as well as in the group that used only one hand, were due to too light a touch when tapping the device. In the beginner group, the gestures were even confused with touching in a user's approach. From the detection algorithm point of view, there is a trade-off: the parameters cannot be set too sensitively to avoid increasing the occurrence of false positives. Even though a lighter tap is viewed as more satisfying by some users, this usability increase cannot cost the reliability too much.

Yet another relevant research problem is to examine the recognition accuracy of target patterns during various activities in mobile usage. This study addressed the stationary situation and false positives during scenarios. A relevant question is what happens if the user performs the interaction while jogging, for example, without stopping to do it. Future work includes examining whether and how the continuous gesture interaction algorithms should adapt to the movement situation of the device.

CONCLUSION

Gesture control is increasingly being applied in mobile interaction. Widespread movement interaction application in mobile devices has been delayed by research challenges such as reliably detecting gestures, power consumption, and user experience-related issues such as obtrusiveness and increased effort. This chapter has focused on analyzing and evaluating the reliability of an event-based gesture interaction modality that emphasizes minimal user effort in interacting with a mobile device. The technical feasibility of the interaction modality was examined with an implementation in a smartphone environment. The reliability of continuous detection of sharp movement events produced by the user by lightly tapping the phone was evaluated by analyzing a dataset collected with the prototype.

The results show that for five informed users performing 36 repetitions of double taps in controlled stationary conditions, the target pattern was recognized with 90.4% accuracy for 3-axis detection and 97.0% for 1-axis detection. In four mobile scenarios containing 54 minutes of daily activities, each performed by four to five users carrying the prototype, there were no false positive detections of the pattern. Overall, the results based on a statistical analysis of the collected acceleration data suggested that double-tap detection is reliable enough for practical applications in mobile computing when the user performs the interaction in a stationary situation. Furthermore, it was found that the occurrence of false positives is low enough for application, presuming carefully selected usage situations where possible misrecognitions are not critical. The contribution of this work has significance for commercial utilization.

Several research questions remain to be addressed as future work. These include how to inform the user about the correct intensity of tapping; there were users with too light a touch in the experiments. From the detection algorithm point of view, a balance needs to be found as the parameters cannot be set too sensitive to avoid increasing the occurrence of false positives. Another important research problem is to examine the recognition accuracy of target patterns during various activities in mobile usage. This study addressed the stationary situation and false positives during scenarios.

As to the movement interaction detection performance in general, the trend of development firmly aims toward increased reliability. As a result, the restricted application-specific use cases are likely to be followed by more general platform-level operations, where movement can be used as an additional interaction modality complementary to the existing ones. With emerging commercial utilization, it is easy to see the beginnings of wider adoption of the new interaction modality in mobile computing, while not forgetting that there is still further work to be done.

ACKNOWLEDGMENT

We would like to acknowledge the work of Arto Ylisaukko-oja in the hardware development, Hannu Vasama for designing the cover casing, and other contributors at Finwe, Nokia, and VTT for their kind collaboration.

REFERENCES

Duda, R., Hart, P., & Stork, D. (2001). *Pattern classification* (2nd ed.). John Wiley & Sons.

Feldman, A., Tapia, E. M., Sadi, S., Maes, P., & Schmandt, C. (2005). ReachMedia: On-the-move interaction with everyday objects. In *Proceedings of IEEE International Symposium on Wearable Computers (ISWC'05)* pp. 52-59.

Hinckley, K., Pierce, J., Horvitz, E. & Sinclair, M. (2005). Foreground and background interaction with sensor-enhanced mobile devices. *ACM Transactions on Computer-Human Interaction (TOCHI), 12*(1), 31-52.

Kallio, S., Kela, J., Korpipää, P., & Mäntyjärvi, J. (2006). User independent gesture interaction for small handheld devices. *Special Issue on Intelligent Mobile and Embedded Systems of IJPRAI, 20*(4), 505-524.

Kela, J., Korpipää, P., Mäntyjärvi, J., Kallio, S., Savino, G., Jozzo, L., & Di Marca, S. (2006). Accelerometer based gesture control for a design environment. *Personal and Ubiquitous Computing* (pp. 1-15). Online First Springer.

Korpipää, P. (2005). *Blackboard-based software framework and tool for mobile device context awareness*. Ph.D dissertation. VTT Publications

579. Retrieved from http://www.vtt.fi/inf/pdf/publications/2005/P579.pdf

Korpipää, P., Mäntyjärvi, J., Kela, J., Keränen, H., & Malm E-J. (2003). Managing context information in mobile devices. *IEEE Pervasive Computing Magazine*, *2*(3), 42–51.

Levin, G. & Yarin, P. (1999). Bringing sketching tools to keychain computers with an acceleration-based interface. In *Proceedings of the CHI 98* (pp. 268-269). ACM: New York.

Linjama, J., Häkkilä, J., & Ronkainen, S. (2005, April 3-4). Gesture interfaces for mobile devices—Minimalist approach for haptic interaction. Position paper in *CHI 2005 Workshop "Hands on Haptics."* Portland, Oregon. Retrieved from http://www.dcs.gla.ac.uk/haptic/sub.html

Linjama, J., & Kaaresoja, T. (2004). Novel, minimalist haptic gesture interaction for mobile devices. In *Proceedings of the NordiCHI2004* (pp. 457-458). ACM Press.

Mäntyjärvi, J., Kela, J., Korpipää, P., & Kallio, S. (2004). Enabling fast and effortless customization in accelerometer based gesture interaction. In *Proceedings of the International Conference on Mobile and Ubiquitous Multimedia (MUM)* (pp. 25–31). ACM.

Nokia Corporation. 5500 phone. (2006). Retrieved from http://europe.nokia.com/link?cid=EDITORIAL_8657

Oakley, I., Ängeslevä, J., Hughes, S., & O'Modhrain, S. (2004). Tilt and feel: Scrolling with Vibrotactile Display. In *Proceedings of Eurohaptics* (pp. 316-323).

O'Modhrain, S. (2004). Touch and go - Designing haptic feedback for a hand-held mobile device. *BT Technology Journal*, *22*(4), 139-145.

Rekimoto, J. (1996). Tilting operations for small screen interfaces. In *Proceedings of the 9th Annual ACM Symposium on User Interface Software and Technology* (pp. 167-168).

Rekimoto, J. (2001). GestureWrist and GesturePad: Unobtrusive wearable interaction devices, In *Proceedings of the Fifth International Symposium on Wearable Computers (ISWC)* (pp. 21-27).

Ronkainen, S., Häkkilä, J., Kaleva, S., Colley, A., & Linjama, J. (2007). Tap input as an embedded interaction method for mobile devices. In Proceedings of the First Tangible and Embedded Interaction (pp. 263-270). ACM: New York.

Sawada, H., Uta, S., & Hashimoto, S. (1999). Gesture recognition for human-friendly interface in designer - consumer cooperate design system. In *Proceedings IEEE International Workshop on Robot and Human Interaction* (pp. 400-405). Pisa, Italy.

KEY TERMS

Accelerometer: 3-D accelerometer is a sensor capable of measuring object acceleration along three spatial axes.

Double Tap: Double tap is a form of movement interaction where the user performs two consecutive taps on a mobile device with a finger or palm, each producing a sharp spike waveform in an accelerometer signal measured with a high sampling rate.

Gesture Interaction: Gesture interaction here refers to explicit movements made with a mobile device while holding it in a hand in order to perform any tasks with the device.

False Positive %: False positive percentage is the relative number of falsely detected patterns, given by dividing the occurrence of all detected false positive patterns by all segments of data where a detected pattern in a dataset should not exist.

Pattern Recognition: Pattern recognition is the scientific discipline whose goal is the classification of objects into a number of categories or classes. Objects can be, for example, signal waveforms or any type of measurement that needs to be classified. These objects are here referred to using the generic term "patterns."

Smartphone: A smartphone is an advanced multifunctional mobile phone with a platform open to third-party software.

True Positive %: True positive percentage is the relative number of correctly detected patterns, given by dividing all detected true positive patterns by all actually performed true patterns in a dataset.

Chapter XXXI
EMG for Subtle, Intimate Interfaces

Enrico Costanza
Ecole Polytechnique Fédérale de Lausanne, Switzerland

Samuel A. Inverso
The Australian National University, Australia

Rebecca Allen
UCLA Design | Media Arts, USA

Pattie Maes
MIT Media Lab, USA

ABSTRACT

Mobile interfaces should be designed to enable subtle, discreet, and unobtrusive interaction. Biosignals and, in particular, the electromyographic (EMG) signal, can provide a subtle input modality for mobile interfaces. The EMG signal is generated by a muscle contraction and can be used for volitional control; its greatest potential for mobile interfaces is its ability to sense muscle activity not related to movement. An EMG-based wearable input device, the Intimate Communication Armband, is presented in this chapter to demonstrate this subtle interaction concept. The device detects subtle, motionless gestures from the upper arm. Experimental results show that the gestures are reliably recognized without user or machine training, that the system can be used effectively to control a multimodal interface, and that it is very difficult for observers to guess when a trained user is performing subtle gestures, confirming the subtlety of the proposed interaction.

INTRODUCTION

Mobile communication devices provide ubiquitous connectivity, allowing people to engage in private and personal communication from virtually anywhere. They are often used in public places (offices, libraries, museums, theatres, restaurants) or on public transportation (such as buses and trains), where the user is surrounded by others not involved in the interaction. Using a mobile device in a social context should not cause embarrassment and disruption to the people in the immediate environment. This problem has been reported by social scientists (Fortunati, 2002, Okabe & Ito, 2005), and it is emphasised by the many signs that can be found in public places inviting or ordering people to turn off cell phones. Deactivating these devices is an extreme solution, as it completely annihilates the devices' functions and advantages, and indeed, users are not inclined to do so. Ring-tones' replacement with vibrating alerts in mobile phones constitutes an example of a widespread subtle interface to improve social acceptance, while still allowing access to the device's functionality. Unfortunately, this idea of subtlety and social acceptance has not yet been generalized and is lacking in other parts of the interface design.

Mobile interfaces should be designed to enable subtle, discreet, and unobtrusive interaction. The human-computer interaction (HCI) research community has recently shown increasing interest in the design of mobile and wearable interfaces that are socially acceptable, and that take into account the social context of users. Rekimoto (2001) advocates that, to be accepted in everyday and public situations, wearable input devices should be "as natural and (conceptually) unnoticeable as possible." Lumsden and Brewster (2003) question the social acceptance of speech-based and gesture-based interaction. Marti and Schmandt (2005) address the disruption caused by mobile phone notifications with a subtle notification and vetoing system. The work presented in this chapter extends this research thread, demonstrating how biosignals, and, in particular, the electromyographic (EMG) signal, a biosignal generated by muscular activity, can be used to enable natural and unnoticeable interaction.

This chapter proposes intimate interfaces: discrete interfaces that make interaction with mobile devices private and concealed as much as possible, in order to minimize the disruption of colocated individuals, and let mobile technology gain social acceptance. Even though it has been suggested that making mobile interaction public and evident could help colocated individuals to understand and accept the behaviour of mobile technology users (Hansson, Ljungstrand, & Redström, 2001), users themselves can inform others of their interaction, if they want. What is private can be made public, but not vice-versa.

In a mobile context, users are often involved in a primary activity, ranging from navigation and monitoring of the immediate environment (e.g., waiting for an incoming train) to specific tasks, such as equipment maintenance or field work. Mobile devices are used either for assistance to the primary task–providing access to equipment documentation or data logging–or for *side involvements*, in the sense of collateral activities unrelated but not conflicting with the primary task, as defined by Goffman (1963). Interaction techniques based on EMG signals can provide an extra modality for interaction, one that does not conflict with the primary task.

The EMG signal is generated by muscle contractions and can be used for volitional control. EMG's greatest potential for mobile interfaces is its ability to sense muscle activity not related to movement, allowing the definition of a class of *subtle* or *motionless gestures*.

This chapter covers the design and evaluation of the *Intimate Communication Armband*: a wearable device that detects subtle motionless gestures through EMG signals, and can be used to control existing devices through a Bluetooth interface. The next section provides background about EMG and its applications within human-computer interaction (HCI). Subsequently, the concept of *motionless gestures* and a system to recognize them, the Intimate Communication Armband, are introduced, followed by evaluation through three user studies. The studies assess the basic functionality of subtle gesture recognition, the use of such gestures to control a multimodal interface, and how noticeable the gestures are to bystanders. Finally, suggestions about further work and concluding remarks are presented.

BACKGROUND

The EMG signal is an electrical signal generated by a muscle contraction. Through electromyography, it is possible to sense muscular activity related to movement, such as lifting or folding an arm, and also *isometric* activity: muscular activity that does not produce movement (Tanaka & Knapp, 2002). An example of isometric activity is pushing against a wall; where muscles are activated, but the wall prevents movement; similarly, isometric activity can be produced by flexing the muscles without load, as when "showing off muscles." The sensing of isometric activity has great potential for mobile interfaces, as detailed in the following section.

In the last three decades, biomedical engineering has yielded many effective methods for recording and computer-aided analysis of EMG signals (DeLuca, 1979). This chapter will only consider recording through noninvasive surface electrodes: conductive elements placed on the skin and kept in place, either through adhesive (similar to that commonly found in bandages) or other means (e.g., elastic fabric bands). EMG signals can also be recorded using needle electrodes, introduced through the skin, which produce better signals because they are in close contact with the muscle. While their use can be justified in a medical context, the discomfort that they cause to the user makes them highly impractical for the kind of everyday applications considered in this chapter. Moreover, current integrated circuit technology makes it possible to produce EMG signals from surface electrodes of higher quality than in the past.

Electromyographic (EMG) Signal

The electromyographic (EMG) signal is the result of the superposition of electric voltage generated by each motor unit in a muscle. Being a voltage signal, it is sensed through pairs of differential electrodes, generally located over the muscle of interest, each pair constituting a *channel*. Because surface electrodes record from a large number of motor units at the same time, the resulting EMG signal can be represented as a signal with Gaussian distributed amplitude, typically ranging from 100 μV to about 1 mV (DeLuca, 1979).

Electrodes, Recording, and Applications

Commercial surface electrodes are generally Ag/AgCl plates covered with conductive gel (often solid gel for increased comfort) and attached to the skin with adhesive. The gel is used to improve the electrode to skin interface, lowering the impedance seen from the sensor, and reducing motion artefacts[1]. *Active* or *driven* electrodes are sometimes used to create a feedback control loop between the sensor and the body (Webster, 1992), this method also reduces motion artefacts, eliminating the need for conductive gel: in this case the electrodes are referred to as *dry*. Advances in material technology are producing surface electrodes that are more comfortable for consumer use, for example, electrodes embedded in flexible grids (Lapatki, van Dijk, Jonas, Zwarts, & Stegeman, 2004) or even embedded in fabrics (Paradiso, Loriga, & Taccini, 2004).

The typical biomedical analysis for diagnostic applications involves envelope detection, energy measurement (which relates the signal to physical force), and frequency characterization (DeLuca, 1997). Control applications generally involve signal acquisition from a number of deferential electrodes, feature extraction, and real-time pattern classification. The first examples of EMG-based real-time control systems were for prosthesis control and functional neuromuscular stimulation. Hefftner, Zucchini, and Jaros (1988), for example, report successful results from a system that can recognize two gestures generated from the shoulder and upper arm. The system must be specifically calibrated for each subject, and uses EMG signals from two channels.

Lukowicz, Hanser, Szubski, and Schobersberger (2006) presented a system based on wearable force-sensitive resistors to sense muscle activity. They showed a correlation between the mechanical deformation of the limb (measurable through force sensors placed on an elastic band adherent to the body) and muscle activity, especially fatigue. This approach allows the recording of activity that cannot be obtained through inertial sensors. Unfortunately, no sensing of pure isometric activity is reported. Strachan and Murray-Smith (2004) used accelerometers to measure muscle tremor as a form of isometric muscle activity. The system can detect the gestures of squeezing or holding a PDA in the user's hand, but requires individual calibration.

EMG for Human-Computer Interaction (HCI)

A number of studies have focused on EMG for users with physical disabilities (Coleman, 2001; Guerreiro & Jorge 2006). Putnam and Knapp (1993) developed a reconfigurable system to control generic graphical user interfaces. The system incorporates a continuous control mode where the contraction's amplitude is mapped to a parameter swing (sliders, scrollbars) and a gesture recognition mode that discriminates between two gestures and can be used for discrete selections. Gesture recognition is performed on a dedicated digital signal processing (DSP) board, is based on neural networks, and requires training for each user. Barreto, Scargle, and Adjouadi (1999) propose a system to control a mouse-like point–and-click interface using facial muscles. In addition to amplitude, the EMG signals' spectral features are analysed to increase performance. The system is not reported to require individual calibration for each user, and is implemented on a DSP board.

Other examples of EMG-based HCI include robotic control (Crawford, Miller, Shenoy, & Rao, 2005), unvoiced speech recognition (Manabe, Hiraiwa, & Sugimura, 2003), pointer control (Rosenberg, 1998), affective and emotional state recognition (Benedek & Hazlett, 2005; Healey & Picard, 1998), and a number of musical expression interfaces. For musical expression, the signal is used either in a continuous fashion, for example, with the amplitude being mapped to a variety of sound synthesis parameters, or through gesture recognition. The systems presented in this context are often wearable and allow movement of the performer on stage, yet they are not explicitly designed for the mobile everyday context. Knapp and Lusted (1990) present a generic battery-powered platform to control MIDI systems. Tanaka and Knapp (2002) complement EMG data with inertial sensor information, so that both isometric and isotonic activity can be monitored: muscle tension resulting in no motion and motion with constant muscle tension respectively. Dubost and Tanaka (2002) developed a wearable wireless musical controller supporting preprocessing of EMG signals and output interfacing with different standards (MIDI, RS232, and Ethernet), which requires calibration for every user.

Recent studies focus on the use of EMG for the recognition of an alphabet of discrete gestures. Fistre and Tanaka (2002) propose a system that can recognize six different hand gestures using two EMG channels on the forearm. The device is designed to control consumer electronics and is described as portable. Testing in a mobile context has not been reported. Wheeler and Jorgensen (2003) report the development and successful testing of a neuroelectric joystick and a neuroelectric keypad. Using EMG signals collected from four and eight channels on the forearm, they successfully recognize the movement corresponding to the use of a virtual joystick and virtual numeric keypad. Gestures mimicking the use of physical devices are successfully recognized using hidden Markov models. The system is proposed as an interface for mobile and wearable devices, but an embedded implementation is not reported, nor is testing in a mobile context.

SUBTLE GESTURES AND THE INTIMATE COMMUNICATION ARMBAND

The EMG signal's ability to detect isometric muscle activity (muscle activity not related to movement) allows the definition of a class of "subtle" or "motionless gestures." Motionless gestures are defined as specific, isolated, and volitional muscle contractions that result in little or no visible movement, and that are different from everyday muscle activation patterns. An example of a motionless gesture is a *brief contraction of the upper arm,* a gesture somewhat similar to a brief grasp of an object held in one's hand that can, however, be performed also with free hands. While it might be initially difficult for the reader to imagine such a contraction, minimal feedback about the gesture recognition makes it very easy to learn and to perform reliably, as demonstrated in a user study reported later in this chapter. In fact, the definition of this brief contraction was the result of a user-centred process also described later in the chapter.

Previous studies on the use of EMG for human computer interaction (mobile or not) do not explicitly consider subtlety, leading to a different approach. Tanaka and Knapp (2002) consider it a limitation that EMG cannot distinguish between muscle activity from movement and nonmovement. They remedy this by complementing EMG with inertial sensor (gyros) data in a multimodal fashion. Fistre and Tanaka (2002) and Wheeler and Jorgensen (2003) use EMG for hand-gesture recognition as an alternative to accelerometers or mechanical sensors for movement, but not for subtle gestures.

In addition to the emphasis on subtlety, the approach proposed here is different from other work on EMG (Fistre & Tanaka, 2002, Putnam & Knapp, 1993, Wheeler & Jorgensen, 2003) as it favours avoiding calibration or system training for each user, minimal computational complexity, and robustness against false positives in sacrifice of the variety of gestures recognized.

The Intimate Communication Armband was conceived as a generic input/output peripheral for mobile devices. It is worn on the upper arm invisibly under clothes (Figure 1), senses explicit subtle gestures, and provides localized tactile output. It connects wirelessly via Bluetooth to a phone or PDA, which can sit in the user's pocket or bag. Being a generic i/o device, it emits signals every time a gesture is recognized and accepts signals to activate the tactile display. In this way, complete freedom for a mapping strategy is left to the application designer.

The Intimate Communication Armband does not occupy the user's hands, and does not require hands to operate; hence, it is "hands free." On its own, it can be used for minimal communication and remote awareness: paired armbands can provide a very low bandwidth intimate and unobtrusive communication channel, if one vibrates

Figure 1. The Intimate Communication Armband can be made invisible by hiding it under clothing

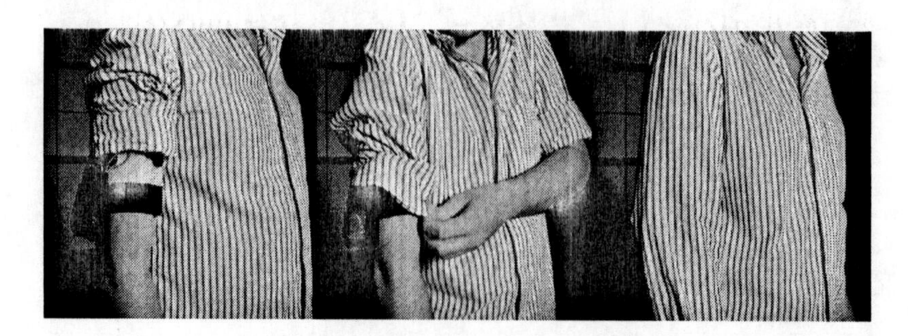

every time the other one detects a motionless gesture. However, the device's greatest potential is realized when combined with a higher-resolution hands-free display to form a closed loop "hands free" system, such as with headphones, loudspeakers, or even high-resolution eyeglass displays. This can be highly advantageous in a number of everyday situations, for example, when the user is carrying objects, as well as for specific domains of applications, such as maintenance in which the users' hands are needed to perform a principal task, and the mobile computing system is used for remote guidance or accessing documentation, for example, an audio guide could be read through headphones and an armband could be used to advance, pause, or rewind the system. A tactile display can be used to give feedback about a subtle gesture being recognized, or it can deliver alerts and notifications.

Hardware

Custom hardware was developed to sense, amplify, and process EMG signals from the upper arm. Commercial EMG amplifiers are generally designed for biomedical applications, where high accuracy and reliability fully justify high prices. Moreover, such equipment is often used in controlled or semicontrolled hospital conditions, so devices are often worn on the belt or the patient's back, and connected through wires to electrodes on other body parts. While this setup allows more flexibility in electrode placement and multiple channel recording, it can be cumbersome to wear in everyday conditions. In contrast, cost for the intimate communication armband is below $100, the most expensive component being the Bluetooth module, which alone accounts for about half. Of course, the accuracy and reliability are not comparable with commercial biomedical devices, yet sufficient for the proposed application.

As detailed in Section 2, the EMG signal is a biopotential in the range of 100 μV to about 1 mV. The general system design for the subtle gesture sensor is illustrated in Figure 2, and it includes:

- Surface electrodes to pick up voltage signals on the body
- A signal amplifier and analog preconditioning stage
- An analog to digital converter
- A digital pattern recognition system
- An interface to applications on a mobile device or PC

The signal preconditioning was performed through analog rather than digital filters to keep the digital processing complexity low. This choice was made based on the detection algorithm's low computational cost, which can run on a low-power 8-bit RISC microcontroller.

Figure 2. Block diagram for the EMG subtle gesture recognition system

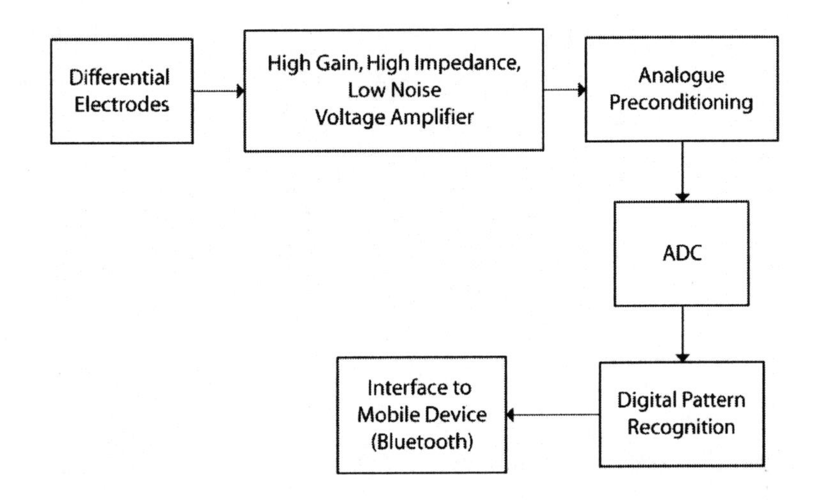

As compared to other sensing methods, EMG has had a number of practical difficulties, due to the need for contact electrodes and their placement (Rekimoto, 2001). However, EMG is worth studying, given its significant advantages in subtlety, and new developments in noncontact electrodes (Trejo, Wheeler, Jorgensen, Rosipal, Clanton, Matthews, et al. 2003) and smart materials.

The system includes two separate circuit boards to minimize interference: one for analog signal amplification and filtering, the other for digital processing and Bluetooth communication. The amplifier design is based on a portable electrocardiogram (ECG) sensor (Company-Bod & Hartmann, 2003). The system uses an integrated instrumentation amplifier in the first stage, with a right-leg driver-feedback stage to reduce noise. The right-leg driver feeds common mode signals back to the source, a design quite common for biosignal amplifiers (Webster, 1992). After the first stage, a first-order high-pass filter at 1.6 Hz is used to eliminate DC components, followed by a second-order Sallen-Key active low-pass Butterworth at 48 Hz with a gain factor of 10, for antialiasing and further noise reduction. A final stage with unity gain is used to offset the signal, centring it with respect to the analog to digital converter (ADC) range. An integrated voltage converter is used to provide +5 V and -5 V supply for the analog stage from a single cell 3.7 V, 130 mAH Li-Po battery. The circuit schematic is illustrated in Figure 3.

An Atmel 8-bit AVR microcontroller, the AT Mega168, is employed for analog to digital conversion, gesture recognition, and to drive the vibrating motor. The motor is driven through pulse width modulation (PWM) to allow fine tuning of the vibration intensity. The BlueGiga WT12 Bluetooth module is used for wireless communication, connected via a serial interface to the microcontroller. Another integrated voltage regulator is used to convert the battery voltage to 3.3 V, as required by the Bluetooth module. The board also includes a C-MOS driver and a protection diode for the vibrating motor, and two LEDs for displaying the microcontroller's status during debugging.

The two boards and the battery are housed in a box of about 3cm x 4cm x 2cm that is inserted into an elastic armband made for a commercial MP3 digital music player, as shown in Figure 4.

Figure 3. Circuit schematic for the EMG amplifier

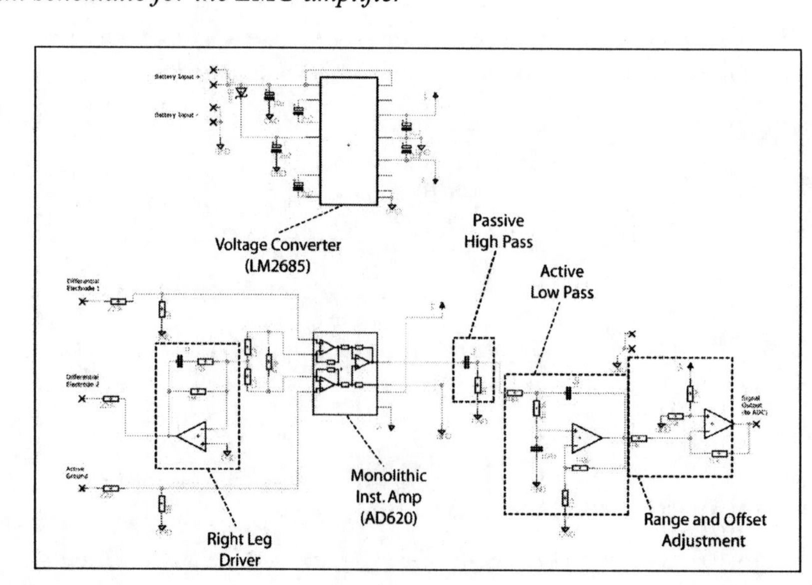

Figure 4. The second generation EMG sensor inside an armband holder for a commercial digital music player. The connector on the left of the photograph is used for recharging the battery and also as a power switch.

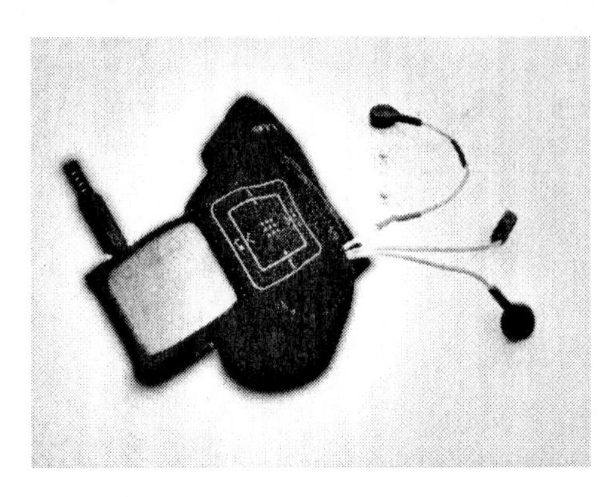

Figure 5. Outline of the design process for the subtle gesture recognition

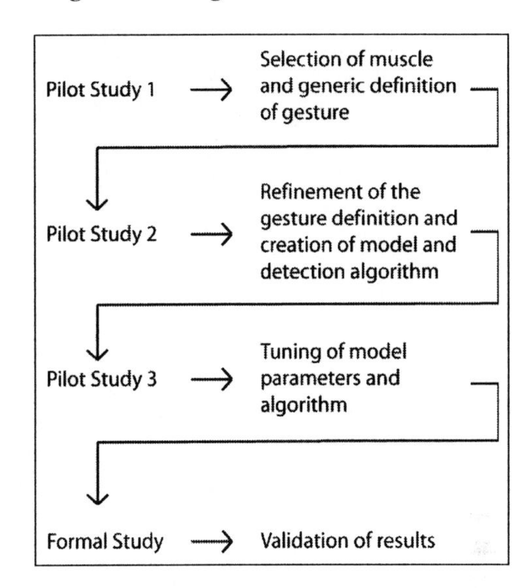

User-Centred Definition of Subtle Gestures

The design of the recognition algorithm and the definition of the gesture were done in parallel to satisfy two requirements: the gesture must be (1) natural for people to perform, and (2) different enough from "normal" muscle activity to avoid misclassification or "false positives." The gesture design was a user-centred iterative process. A number of exploratory, informal user studies were performed to ensure that the system would be natural and easy to use, as summarized in Figure 5.

The process started with a pilot study to select one muscle and subtle isometric contractions that fit the definition of motionless gestures. The test revealed the biceps as the best candidate because it lies superficially, making the signal fairly immune to activity generated by other muscles, and it is well defined, even in nonathletes. The gesture was defined as a brief contraction, such that it could be performed without being noticed, while the arm is unfolded, parallel to the body while the user is standing.

A second informal study was conducted to refine the definition of the subtle gesture and create a model and algorithm for its detection. New subjects participated in the study and were chosen for a variety of muscle volumes. EMG signals were recorded from subjects performing the selected contraction, and compared with the signals generated by other types of muscle activity, such as moving in an indoor space, lifting objects of various weights, and gesticulating while talking.

The subjects were informed about the study's purpose, and the gesture was described to them in a not-detailed way (just as a "brief contraction of the biceps, i.e., the upper arm, that would not be very evident") so that they had some freedom in the way they performed it. This procedure aimed at exploring whether such a definition of "brief contraction" would be consistent across individuals, and to ensure that the gesture definition would be, to a certain extent, natural to perform, rather than defining a gesture a priori, and ask or force the users to learn it.

Subtle Gesture Model

The model resulting from the second study, depicted in Figure 6, is based on the standard deviation of the EMG signal, calculated with a sliding window of duration 0.2 s overlapping for

Figure 6. Model for the subtle gesture (dotted line) and an example gesture recording detected by the algorithm (solid line)

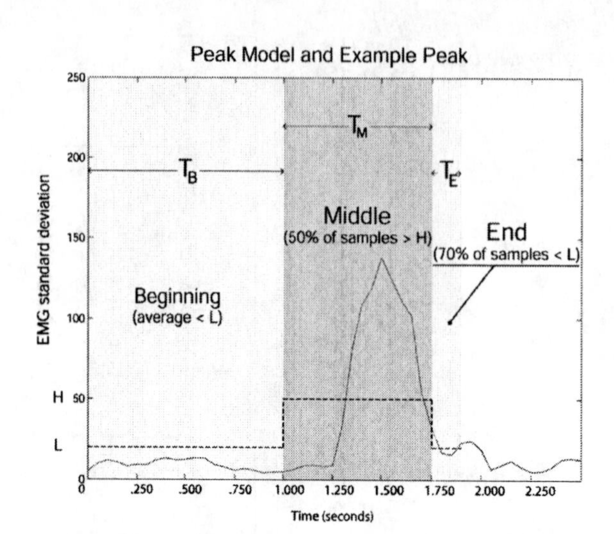

75% of its duration. A mathematical model and a recognition algorithm for the brief contraction were then created heuristically from observation of the data. A brief contraction was observed to correspond to a peak in the signal's standard deviation. Given the noise-like characteristics of the EMG signal (DeLuca, 1979), standard peak-detection techniques could not be employed. Rather, such peaks were modelled as follows: a "beginning" interval of duration T_B of low activity ("silence"), followed by a "middle" interval of high activity of duration T_M and then again, low activity for an "end" interval of duration T_E. High activity and low activity were defined respectively as the signal's standard deviation being above a threshold H and below a threshold L. To allow some tolerance in the model, the condition on the history is imposed on an average of its values; the condition on the middle needs to be satisfied by 50% of the samples, and the condition on the end by 70% of the samples. To increase the resilience to false positives caused by motion artefacts, a zero-crossing counter is included in the detection algorithm to reject low-frequency components.

The model definition is stricter on the contraction's duration than it is on the gesture's intensity. This is because the preliminary study showed that the duration was more consistent than the intensity across users, despite the fact that no specific indication to users was given about either. One disadvantage of this model is it requires a complete gesture before the recognition can take place. The recognition could be made faster by removing the "end condition" for the gesture's closure; however, this would cause an increase in false positives.

The tuning of the model's five parameters required a third informal study. New and returning users were informally asked to test the system. The testing was conducted to stress the system to produce false positives and false negatives. The iterations continued until the number of false positives approached zero and the system recognized contractions performed by any user.

Two Gestures: Long and Short

Once the recognition worked robustly on one gesture, a two-gesture alphabet was explored. The gestures were defined as two short, subtle contractions of different durations. This corresponded to varying the middle interval T_M's duration together with its tolerance. The results obtained at this point were then validated with the first formal user study, described next below.

EVALUATION

Three user studies were performed to validate the design of the EMG-based interaction technique (Costanza, Inverso, & Allen, 2005; Costanza, Inverso, Allen, & Maes, 2007). The novelty of the approach required the evaluation of several different aspects of the interaction, including ease of learning, correct gesture recognition rate, amount of information that can be expressed through subtle gestures, usability of the gestures within a realistic multimodal interface, and noticeability of the interaction to others. Because of the unnoticeable nature of the interaction, it was impossible to define Wizard–of-Oz type studies; therefore, all experiments were conducted using working prototypes of the gesture sensor.

First Study: Learning and Recognition Rate

The first study had three main objectives: (1) to assess whether subjects could learn how to perform the gestures without training, simply by trial and error, receiving minimal feedback; (2) to measure the recognition rate of subtle gestures through the algorithm described; (3) to test whether multiple gestures could be defined on a single muscle.

The experiment was carried out in a simulated mobile scenario: subjects were asked to perform experimental tasks with the device while walking around obstacles in a trafficked walkway in the Media Lab Europe. The setup was similar to one reported by Pirhonen, Brewster, and Holguin (2003), who noted that this mobile context allows us to "take measurements of the usage of the device whilst the users were mobile but was not as formally controlled as a laboratory study, which would lack realism and ecological validity." Subjects were asked to wear the EMG-sensor armband and a pair of wireless headphones so that they could receive auditory cues and feedback while being free to move around. An experimenter applied disposable, solid-gel, self-adhering, Ag/AgCl 9-mm disc surface electromyogram electrodes in three positions around the upper arm of each subject's dominant hand, as illustrated in Figure 7. To ensure signal quality the participant's skin was prepared with an abrasive gel before the electrodes' application.

Participants were 10 adults, 5 women and 5 men, ages 23 to 34, all colleagues from Media Lab Europe, who volunteered to take part in the study. All were naive in that they had not used an EMG-based interface before, with the exception of subject 8, who had taken part in a pilot study.

At the beginning of the experiment, subjects were given written instructions informing them that the study was assessing EMG as a subtle interface for mobile devices, and that the system would recognize brief contractions of the upper arm. The instructions specified that the contraction recognized has a minimum and maximum duration and a minimum strength requirement.

To test how easy the gestures are to learn, subjects were invited to familiarize themselves with

Figure 7. Electrode placement used for the first user study

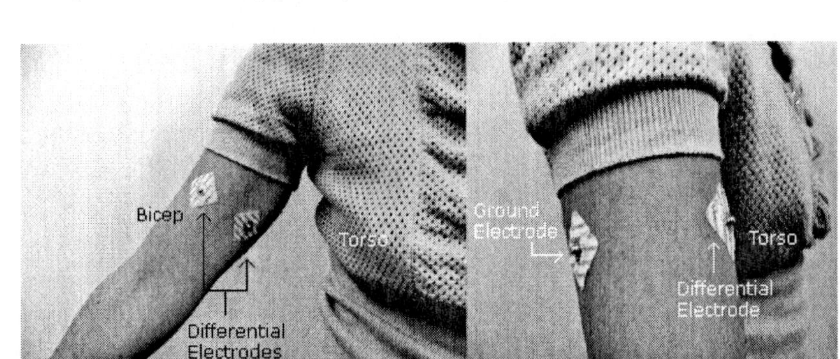

the system until they could comfortably control it. In this phase, participants stood and only heard auditory feedback when the system recognized a contraction. No coaching or further feedback as to the contraction's amplitude or duration was given to the participants; so they were unaware of why the algorithm was or was not recognizing the contraction. They were only aware if the contraction was recognized. If participants did not comfortably control the device within 15 minutes, they were given further feedback by an experimenter who could observe the recorded EMG signals.

After the initial familiarization, participants were asked to engage in the first experimental task: perform a gesture every time they heard an audio cue through the headphones, while walking around the obstacles. The same auditory feedback, as confirmed when a contraction was recognized. Subsequently, subjects were asked to repeat the task three more times, with variations on the gesture duration: in one case, they were asked to always try and perform gestures that were as short (in time) as possible while still being recognized by the system; in another case, to always perform gestures as long as possible (but still recognized by the system); and finally, to perform a mix of "short" and "long" gestures in response to different auditory stimuli. The task with only "long" gestures and the task with only "short" gestures were run in counterbalanced order, so that half of the subjects performed "long" gestures first and the other half "short" gestures first. Each task was preceded by a brief familiarization on "short" or "long" gestures in which again, participants stood and only heard an auditory feedback when the system recognized a contraction. In all cases, the same real-time detection algorithm was used across participants without calibration or modification, and it recognized contractions of duration 0.3 to 0.8 seconds. Therefore, the exact definition of "short" and "long" gestures was left to the individual.

Results and Discussion

The online recognition rates for the four contraction walking tasks were generic 96%, short 97%, long 94%, and mixed 87%. No false positives were detected while online during the first walking task. This accuracy level indicates that EMG-based motionless gesture recognition can be used successfully to control a mobile interface.

In the first familiarization task, participants were able to control the system in an average of 3.75 minutes (SD=2.17), excluding the three participants who reached the 15 minute time limit and required additional feedback. The participants who received feedback (2, 9, and 10), all had the same difficulty that their contractions were too long. They were told, once, to make their contractions shorter, and then they were able to control the system in 11.75, 1.78, and 5.48 minutes, respectively.

Off-line analysis was performed on the data from the short- and long-contraction walking tasks to determine if short and long contractions are separable into two gestures for control. Figure 8 shows the mean and standard deviations for the short and long contraction durations. From the data, a duration boundary of 0.5 seconds was used to create a new recognition algorithm that recognized long and short contractions separately. As with the original algorithm, only the first recognition was counted; any additional recognitions were ignored until the next stimuli. Applying this new short-long detection algorithm to the mixed contraction data resulted in an overall accuracy of 51%, with 55% shorts recognized and 47% longs recognized. The misclassification rate for shorts as longs was 33%, and the misclassification rate for longs as shorts was 11%.

The off-line recognition of short and long contractions using the mixed data set was fairly low. This may have occurred because the online algorithm recognized a small range of contraction durations; therefore, the longs may not have been sufficiently different from the shorts for the participants to accurately produce them. The contraction duration's range was set from pilot studies, which indicated that very long muscle contractions cause most false positives; therefore, a trade-off between reproducibility of long and short contractions and increased false positives can occur if the range is widened.

Figure 8. Mean and standard deviation error bars for long and short contraction durations; closed circles indicate means for short and open circles indicate means for long.

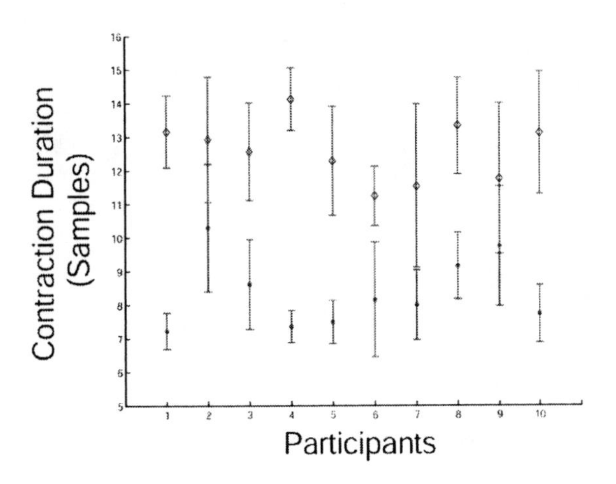

It is important to note the durations of the short and long contractions are subjective because the participants were not given feedback about their actual durations. Therefore, the participants trained themselves on what they considered were long and short contractions. If the participants were given feedback on their contraction durations, they may learn to consistently make different long and short contractions.

After the experiments, some participants stated that they felt longs were more difficult than shorts. In addition, it was noticed that the three participants that required feedback in the first familiarization task became frustrated when they could not make the system recognize their contractions; however, by the end of the experiment they were comfortable using the system.

Second Study: Multimodal Realistic Interaction

Once the first study confirmed the basic functionality of the system, a second study was performed to explore usability in more realistic conditions: subtle gestures were used to select one of four items of an audio-menu. The task required expression of multiple bits of information through subtle gestures, two strategies to achieve this were compared, either by using a time-based interaction or by using two armband devices at the same time. In one condition, defined as *one-arm*, the menu items were iteratively scanned, so that the contraction of just one arm could be used to select the current item. In the other, *two-arm*, condition gestures from one arm were interpreted as "next""and used to advance through the menu items, while gestures from the other arm were used to select the current item.

Two tasks were defined; in each task subjects had to perform a selection in response to a number of cues using one of the conditions described. Wireless headphones were used to display the audio menu, read by a synthetic voice (AT&T, 2004), and deliver audio cues. The cues mimicked incoming phone calls from four callers: each cue consisted of a synthetic voice (the same one used for the menu) announcing "Incoming call from..." followed by the caller's number or name. After each announcement, subjects could access the audio menu and select one item. Subjects were instructed to select a specific menu item in response to each of the four callers. Similar to the first study, participants performed each task while navigating eight-meter laps around obstacles in a regularly trafficked walkway in the MIT Media Lab. Each of the two tasks was preceded by a short familiarization session. All subjects participated in both tasks: within-subjects design and the tasks were performed in fully counterbalanced order.

Figure 9. Electrode placement used in the second user study

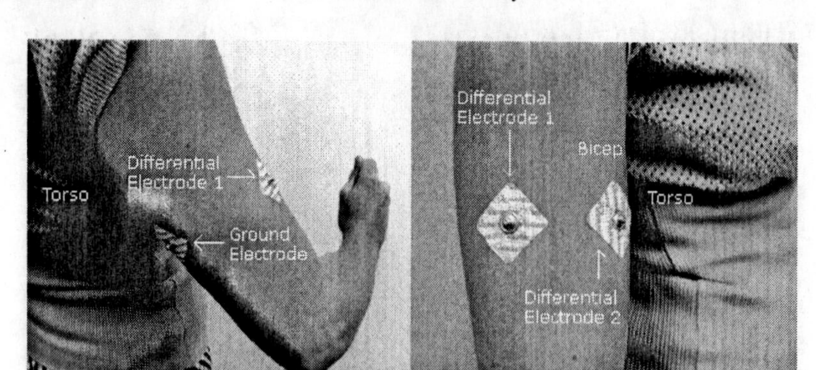

The subject's walking speed during each task was measured as an index for the interface's effectiveness. Petrie, Furner, and Strothotte (1998) pointed out that if a mobile interface has a negative effect on users, it will be reflected in them slowing down while walking. The same measure was later used in other mobile HCI studies (Lumsden & Brewster, 2003; Pirhonen et al. 2003). The subject's preferred walking speed (PWS), that is, the speed at which they walk while not using any mobile device, was measured at the experiment's beginning as a comparison.

Participants were 12 adults; 8 women and 4 men, all volunteers recruited through posters on the MIT campus and university mailing lists. All expressed interest to participate in the study via e-mail, demonstrating a minimum familiarity with computer systems, and they were compensated $10 per hour. All subjects were naive in that they had not used an EMG-based interface before.

Subjects were asked to wear one or two armband devices, depending on the task. Similar to the first experiment, electrodes were placed around each of the subjects' upper arms; however, in this experiment, the participant's skin was not abraded, as an improved amplifier eliminated signal artefacts due to skin creams or lotions. A new electrodes position, illustrated in figure 9, was chosen to avoid the artefacts caused by electrodes pressing against the torso, noticed for some participants in the first study.

Results and Discussion

Overall, subjects performed correct selections of items from the audio menu for 226 of the 235 stimuli presented, corresponding to 96.2% correct selections. Incorrect selections were performed in six cases (2.5%); in all except one of these, an item adjacent to the correct one was selected. In three cases (1.3%) no selection was made. In the two-arms condition, subjects performed correct selections for 120 of the 123 stimuli presented, 97.6% correct; in the same condition, two erroneous selections (1.6%) and only one missed selection (0.8%) occurred. In the one-arm condition, subjects performed correctly for 106 of the 112 stimuli: 94.6%. The number of errors in this condition was four (3.6%) and two misses (1.8%) occurred. Out of the 12 subjects, 7 performed perfectly in both conditions (100% correct selections), while 2 subjects achieved a perfect score on at least one condition. Only five false positives were detected during the entire experiment, but these did not affect the task performance as they happened after a selection was made and before the subsequent stimulus. Additionally, two times subjects reported that an incorrect selection (included in those reported above) was from a false positive.

The high overall accuracy indicates that EMG can, in general, be used successfully in complex and multimodal interfaces. The performance was high in both conditions, demonstrating that the interface bandwidth can be improved either by using controllers on multiple muscles or by using time-based selection strategies. The higher percentage of correct selections in the *two-arm* condition, and the preference expressed by the subjects, suggest that this interaction modality is more efficient than the other one, of course with the extra expense of an additional controller.

A one-way ANOVA showed no significant differences in the subjects' walking speed corresponding to different tasks. Most of the subjects walked slower when operating the interface, however, four subjects walked faster in the *two-arms* conditions than when they were walking without interacting with the device, and three subjects walked faster in the *one-arm* condition than when not operating the interface. These results suggest that controlling an EMG-based interface, with one or two arms, does not involve high workload, nor does it require a high amount of attention (Lumsden & Brewster, 2003; Petrie et al. 1998, Pirhonen et al. 2003). However, further research is required for more conclusive findings.

Eight of the 12 subjects learned to control the device very quickly, and 4 naturally performed the gesture without much arm movement. When asked at the end of the experiment, 7 of 10 subjects expressed a preference for the *two-arms* condition, generally because this provided more control and faster operation; only 2 of 10 subjects preferred the *one-arm* condition, 1 did not express a preference. Most of the subjects spontaneously reported that they enjoyed taking part in the experiment and experienced a novel and unusual way to control a computer interface.

Third Study: Assessing Noticeability

One of the strongest motivations for the use of EMG in the context of mobile HCI is the ability to sense isometric muscular activity, which enables the creation of input interfaces that are subtle, unobtrusive, and unnoticeable by those around the users. An experiment was performed to formally assess how noticeable these gestures are.

The same subjects who took part in the second study were asked to watch a video recording of a trained user activating the interface, and to try and guess when a gesture was performed. The experiment was performed immediately after the completion of the previous one, so all subjects were familiar with the EMG-based interface. The video showed an actor performing subtle gestures with his right upper arm while talking with someone off screen. The recording had no audio and it was divided into three scenes: a medium shot of 135 seconds with the actor wearing long sleeves; a shot of 144 seconds with the same framing and

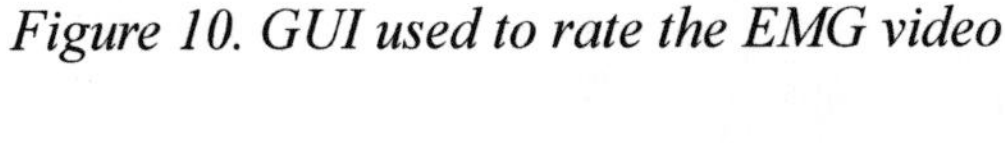
Figure 10. GUI used to rate the EMG video

the actor wearing short sleeves; and a close up of the arm with the electrodes and the armband with duration of 41 seconds. The actor was really activating an armband device, whose recognition output was used as ground truth.

The video was shown on a standard 17" LCD computer display in a video window that measured approximately 8" by 6" at the centre of the screen. Under the video window the sentence "The person in the video is activating the EMG interface" and five buttons labelled "definitively," "very probably," "probably," "possibly," "not sure" were displayed, as illustrated in Figure 10. Subjects were informed about the purpose of the study and instructed to click on one of the five GUI buttons every time they believed the person in the video was performing a gesture recognizable by the EMG-based system.

Figure 11. Results of video rating in the second user study

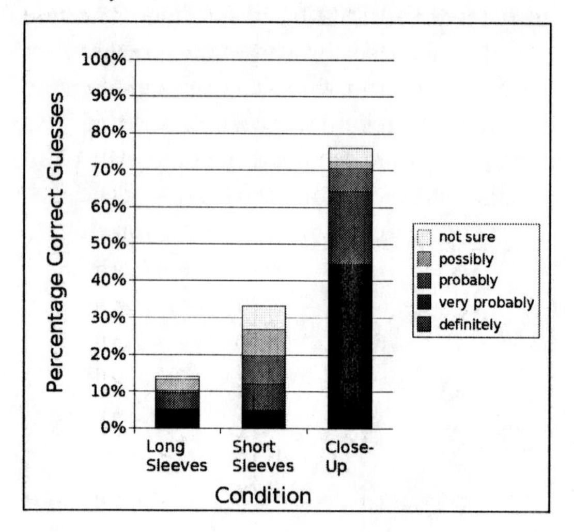

Results and Discussion

Guesses were considered correct if they were within 1.5 seconds of ground truth. For the long sleeves video section, subjects correctly guessed when a contraction was performed for only 13.9% of the attempts (19 correct guesses over 137 attempts). For the short sleeves section, 33.1% of the attempts were correct (47 correct guesses over 142 attempts). Finally, in the close-up case, 75.9% of the attempts were correct (85 over 112 attempts). The subjects' confidences on correct guesses are reported in Table 1. The same data is visualized in Figure 11.

The results show that subjects cannot easily guess when the interface is activated, confirming that the interface is subtle. In the experiment, subjects were told that the person in the video would, at some point, activate the interface; in reality this information would not be available, making the chance of noticing the interface even smaller. Most of the subjects informally reported that they found it difficult to tell when a contraction was performed.

The results can be compared to the probability of a correct uninformed guess, that is, the probability of guessing correctly, assuming that subjects did not look at the video and guessed randomly. This situation can be modelled with the attempts having a uniform random distribution. Considering each "long sleeves" and "short sleeves" sequence separately, and remembering that an attempt is considered correct if it is within 3 seconds of a contraction, a high enough number of attempts evenly spaced in time would give a 100% chance of guessing correctly. The minimum number of attempts for a 100% chance of guessing

Table 1. Video rating results in the third user study

	% overall correct	definitely	very probably	probably	possibly	not sure
Long Sleeves	13.9%	36.84%	31.58%	5.26%	21.05%	5.26%
Short Sleeves	33.1%	14.89%	21.28%	23.40%	21.28%	19.15%
Close-Up	75.9%	58.82%	25.88%	8.24%	2.35%	4.71%

is $N_{100\%} = D_s / D_a$, where D_s is the duration of the sequence and D_a is the uncertainty interval, in this case, 3 seconds. In the "long sleeves" condition, D_s is 135 seconds, so $N_{100\%} = 45$ attempts would give a 100% chance of guessing correctly.

During the experiment, subjects cumulatively attempted to guess 137 times, corresponding to an average of 11.4 attempts per subject, and to an 11.4 / 45 = 25.3% chance of correctly guessing. In the "short sleeves" condition, 142 attempts were made corresponding to an average of 11.8 attempts per subject, over 144 seconds, so $N_{100\%}$ = 144 / 3 = 48 and the uninformed guess chance is 11.8 / 48 = 24.6%.

Therefore, in the long sleeves condition, the subjects guess performance, 13.9%, was much worse than completely random, 25.3%, implying that watching the video did not help guessing, confirming that the contractions are unnoticeable. In the short sleeves case, subjects guessed 8.5 percentage points better than chance; however, overall fairly low. In the close-up condition, subjects guessed correctly most of the time.

General Discussion

The results from the two studies demonstrate that the Intimate Communication Armband can be reliably used in a mobile context. In both the initial and the audio menu experiments, novice subjects learned to use the system very quickly, with little feedback about their performance. Subtle gestures proved to be effective in controlling a multimodal interface even when mobile. Although expressing different subtle gestures with a single arm seems not to be very reliable (at least with the current detection algorithm), subjects did not have problems in using multiple muscles at the same time, nor to use a single muscle to select one of many options presented over time.

The gestures recognized by the armband device are indeed subtle; the last experiment's results showed that it is hard for observers to guess when someone is performing a gesture.

FUTURE TRENDS

Further investigation should explore the use of more advanced analysis techniques for the detection of subtle gestures, such as autoregressive modelling, which has been reported to be successful in some EMG literature (Hefftner et al, 1988). The performance of subtle gestures while users are engaged in tasks that occupy their hands or involve specific movements of their arms should be formally investigated, focussing not only on the physical challenges, but also on the cognitive demands related to performing different motor tasks. To improve the device's comfort, dry electrodes or electrodes embedded in fabric (Paradiso et al., 2004) should be included in the armband design. The potential of localized tactile cues should be explored: tactile stimuli on different parts of the body can convey a large amount of information. Armbands should be paired so that one vibrates when the other recognizes a subtle gesture to form a simple intimate communication system for remote awareness. The device should also be integrated within specific mobile applications, such as browsing audio documentation or navigation guidance. Higher-level evaluation of these applications should analyse how users adopt it for day to day use.

Generally, there are many opportunities to develop biosignal processing techniques that run on embedded devices and the exploration of new application domains, including, for example, games.

CONCLUSION

This chapter has shown that an EMG-based wearable input device, the Intimate Communication Armband, can be effectively employed in a mobile context for subtle and intimate interaction. The device detects *subtle motionless gestures*: explicit muscle contractions resulting in little or no movement. Experimental results show that the gestures are reliably recognized without training, neither for the recognition algorithm nor for users. Subjects were able to reliably control an audio interface using one or two arms while engaged in a walking task without problems. An experiment designed to evaluate the subtlety of the interface revealed that it is very difficult for observers to guess when a trained user is performing subtle gestures. The armband device also includes a tactile display, based on a vibrating motor, and can be made invisible by being worn under clothes.

The design of interfaces and interaction techniques for mobile devices should take into account social acceptance and allow devices to be active but not disruptive. The construction and evaluation of the prototype proposed in this chapter demonstrates that it is possible to realize usable mobile interfaces that are intimate and subtle and therefore, socially acceptable.

ACKNOWLEDGMENT

This research was initiated at Media Lab Europe and continued at MIT Media Lab through the support of the Things That Think (TTT) and the Digital Life (DL) consortia. The authors would like to acknowledge Alberto Perdomo and Juanjo Andres Prado for hardware design and support. Our gratitude goes to Ian Oakley for invaluable suggestions on the user studies design. We are also thankful to Joe Paradiso, Jim Barabas, David Bouchard, Mark Feldmeier, David Merrill and Sajid Sadi at the Media Lab for their insights.

REFERENCES

AT&T. (2004). AT&T natural voices text-to-speech engine. Retrieved from http://www.naturalvoices.att.com/products/tts_data.html

Barreto, A., Scargle, S., & Adjouadi, M. (1999). Hands-off human-computer interfaces for individuals with severe motor disabilities. *HCI (2)* (pp. 970-974).

Benedek, J., & Hazlett, R. (2005). *Incorporating facial emg emotion measures as feedback in the software design process.* Technical report, Microsoft Corporation. Johns Hopkins University School of Medicine.

Coleman, K. (2001). Electromyography based human computer-interface to induce movement in elderly persons with movement impairments. In *WUAUC'01: Proceedings of the 2001 EC/NSF workshop on Universal accessibility of ubiquitous computing* (pp. 75–79). New York, NY: ACM Press.

Company-Bod, E., & Hartmann, E. (2003). Ecg front-end design is simplified with microconverter. *Analog Dialogue, 37*(11).

Costanza, E., Inverso, S. A., & Allen, R. (2005). Toward subtle intimate interfaces for mobile devices using an EMG controller. In *CHI '05: Proceedings of the SIGCHI conference on Human factors in computing systems* (pp. 481-489). New York, NY: ACM Press.

Costanza, E., Inverso, S. A., Allen, R., & Maes, P. (2007). Intimate interfaces in action: Assessing the usability and subtlelty of EMG-based motionless gestures. In *CHI '07: Proceedings of the SIGCHI conference on Human factors in computing systems.* New York, NY: ACM Press.

Crawford, B., Miller, K., Shenoy, P., & Rao, R. (2005). Real-time classification of electromyographic signals for robotic control. In *AAAI* (pp. 523–528).

DeLuca, C. J. (1979). Physiology and mathematics of myoelectric signals. *EEE Transactions on Biomedical Engineering, 26*, 313-325.

DeLuca, C. J. (1997). The use of surface electromyography in biomechanics. *Journal of Applied Biomechanics, 13*(2), 135-163.

Dubost, G., & Tanaka, A. (2002). A wireless, network-based biosensor interface for music. In *Proceedings of International Computer Music Conference (ICMC).*

Fistre, J., & Tanaka, A. (2002). *Real time emg gesture recognition for consumer electronics device control.*. Retrieved from http://www.csl.sony.fr/ atau/gesture/

Fortunati, L. (2002). The mobile phone: Towards new categories and social relations. *Information, Communication & Society, 5*(4), 513-528.

Goffman, E. (1963). *Behavior in public places.* The Free Press.

Guerreiro, T. J. V., & Jorge, J. A. P. (2006). EMG as a daily wearable interface. In *Proceedings of the First International Conference on Computer Graphics Theory and Applications (GRAPP 2006),* Setúbal, Portugal.

Hansson, R., Ljungstrand, P., & Redström, J. (2001). Subtle and public notification cues for mobile devices. In *Proceedings of UbiComp 2001,* Atlanta, Georgia, USA.

Healey, J., & Picard, R. W. (1998). Startlecam: A cybernetic wearable camera. *ISWC* (pp. 42-49).

Hefftner, G., Zucchini, W., & Jaros, G. (1988). The electromyogram (emg) as a control signal for functional neuromuscular stimulation-Part I: Autoregressive modeling as a means of emg signature discrimination. *IEEE Transactions on Biomedical Engineering, 35*(4), 230-237.

Knapp, B., & Lusted, H. (1990). A bioelectric controller for computer music applications. *Computer Music Journal, 14*(1).

Lapatki, B. G., van Dijk, J. P., Jonas, I. E., Zwarts, M. J., & Stegeman, D. F. (2004). A thin, flexible multielectrode grid for high-density surface emg. *J Appl Physiol, 96*(1), 327-336.

Lukowicz, P., Hanser, F., Szubski, C., & Schobersberger, W. (2006). Detecting and interpreting muscle activity with wearable force sensors. In *Proceedings Pervasive 2006.*

Lumsden, J., & Brewster, S. (2003). A paradigm shift: Alternative interaction techniques for use with mobile and wearable devices. In *CASCON '03: Proceedings of the 2003 conference of the Centre for Advanced Studies on Collaborative research* (pp. 197-210). IBM Press.

Manabe, H., Hiraiwa, A., & Sugimura, T. (2003). Unvoiced speech recognition using emg - Mime speech recognition. In *CHI '03 extended abstracts on Human factors in computing systems* (pp. 794–795). New York, NY: ACM Press.

Marti, S., & Schmandt, C. (2005). Giving the caller the finger: Collaborative responsibility for cellphone interruptions. In *CHI Extended Abstracts* (pp. 1633-1636).

Okabe, D., & Ito, M. (2005). Keitai and public transportation. In M. Ito, D. Okabe, & M. Matsuda (Eds.), Personal, portable, pedestrian: Mobile phones in Japanese life. Cambridge: MIT.

Oulasvirta, A., Tamminen, S., Virpi, R., & Kuorelahti J. (2005). Interaction in 4-second bursts: The fragmented nature of attentional resources in mobile HCI. In *CHI '05: Proceedings of the SIGCHI conference on Human factors in computing systems* (pp. 919-928). New York, NY: ACM Press.

Paradiso, G., Loriga, G., & Taccini, N. (2004, August). Wearable health care system for vital signs monitoring - Medicon 2004 conference. In *Proceedings of MEDICON 2004.*

Petrie, H., Furner, S., & Strothotte, T. (1998). Design lifecycles and wearable computers for users with disabilities. In *Proc. First workshop on HCI mobile devices.* Glasgow, UK.

Pirhonen, A., Brewster, S., & Holguin, C. (2003). Gestural and audio metaphors as a means of control for mobile devices. In *CHI '02: Proceedings of the SIGCHI conference on Human factors in computing systems* (pp. 291-298). New York, NY: ACM Press.

Putnam, W., & Knapp, B. (1993, June). The use of the electromyogram in a man-machine interface. In *Proceedings of the Virtual Reality and Persons With Disabilities Conference.*

Rekimoto, J. (2001). Gesturewrist and gesturepad: Unobtrusive wearable interaction devices. In *ISWC '01: Proceedings of the 5th IEEE International Symposium on Wearable Computers* (p. 21). Washington, DC: IEEE Computer Society.

Rosenberg, R. (1998). The biofeedback pointer: Emg control of a two dimensional pointer. In *ISWC* (pp. 162–163).

Strachan, S., & Murray-Smith, R. (2004). Muscle tremor as an input mechanism. In *Proceedings of UIST 2004.* New York, NY: ACM Press.

Tanaka, A., & Knapp, R. B. (2002). Multimodal interaction in music using the electromyogram and relative position sensing. In *Proceedings of New Interfaces for Musical Interaction (NIME).* Medialab Europe.

Trejo, L., Wheeler, K., Jorgensen, C., Rosipal, R., Clanton, S., Matthews, B., Hibbs, A., Matthews, R., & Krupka, M. (2003). Multimodal neuroelectric interface development. *Neural Systems and Rehabilitation Engineering, IEEE Transactions on, 11*(2), 199-203.

Webster, J. G. (1992). *Medical instrumentation: Application and design (*2nd ed.). Houghton Mifflin.

Wheeler, K. R., & Jorgensen, C. C. (2003). Gestures as input: Neuroelectric joysticks and keyboards. *IEEE Pervasive Computing, 2*(2), 56-61.

KEY TERMS

Electrode: Electrically conductive element placed as close as possible to a signal source (the muscle for EMG signals) acting as the interface between the body and the signal recording apparatus

Electromyographic signal or EMG Signal: An electrical voltage signal generated by muscle activity

Intimate Communication Armband: Wearable input/output device capable of detecting subtle gestures from the upper arm and of delivering tactile cues

Intimate Interfaces: Discrete interfaces that make interaction with mobile devices private and concealed as much as possible, in order to minimize the disruption of colocated individuals

Isometric Muscle Activity: Muscular contraction that does not produce movement; that is, the muscle length is constant; for example, pushing against a wall.

Subtle Gesture or Motionless Gesture: Voluntary muscle contractions that result in little or no visible movement—based on isometric muscle activity

Surface Electrode: Conductive metal plates, typically composed of Ag/AgCl metal, placed on the skin surface and kept in place through adhesive or elastic bands. Sometimes conductive gel is used to improve the electrical conduction between the skin and the electrode.

ENDNOTE

1 Impedance is the resistance to current flow. If the impedance between the electrode and skin is high, the muscle's electrical activity will not be conducted through the electrodes properly.

Chapter XXXII
Mobile Camera-Based User Interaction

Tolga Capin
Bilkent University, Turkey

Antonio Haro
D4D Technologies, USA

ABSTRACT

This chapter introduces an approach for user interaction on mobile devices, focusing on camera-enabled mobile phones. A user interacts with an application by moving their device, and the captured camera video is used to estimate phone motion or interact with the real world. We first survey technical issues, recent research results, and then present a prototype implementation and discuss various ways how phone motion can be used for different tasks, such as navigating through large number of media files, and phone motion and shake detection for gaming. The results and discussion may guide interface designers when targeting camera-based user interfaces.

INTRODUCTION

Mobile devices currently support key-modal interfaces through joypad/direction keys and numerical keyboard. On devices with larger form-factors, additional keys provide a better user experience for complex tasks such as navigating through large amounts of content, since keys can be dedicated to specific tasks such as page-up/down and choosing zoom level. Smart phones cannot easily make use of such keys due to limited physical space. Stylus-based interaction with touch-sensitive screens has emerged as an alternative, but it requires two-handed interaction, and has been shown to cause additional attentional overhead in users.

Consequently, alternative interaction techniques are desired. Physical sensors have been added to mobile devices for user interaction, such as accelerometers (Hinckley, Pierce, Sinclair, & Horvitz, 2000), but these can be difficult to integrate into existing consumer-level devices at both the software and hardware level at a low cost. In addition, such sensors are known to have error buildup over time, since some infinitesimal acceleration is always measured.

As another alternative, over the recent years, a number of solutions have been proposed for using the camera as the input device, where incoming video is used to estimate phone motion and to interact with the user's physical environment. These approaches provide a more direct user interaction maximizing the use of the display, minimizing attentional overhead to the user, and permitting one-handed interaction. With camera-based interaction, the user points directly on objects or changes their view by moving the phone. The user is provided with a means of navigating and manipulating individual objects, each of which has a direct display representation. The user applies actions directly to their view or to the objects by selecting them.

There are many application scenarios that could take advantage of camera-based user interaction. For general interaction with the device, the user can be provided a number of camera-based interaction primitives, such as gestures for scrolling and selecting. In games, users can control their viewpoint in the 3-D environment by physically moving their phone around in the real world to look up/down/left/right in the game world. In physical user interfaces, tags can be placed in real world, which can be scanned by the mobile device.

In this chapter, we survey recent research results, survey how camera based UIs can be used for different tasks and applications, and present a prototype implementation.

UNDERSTANDING COMPUTER VISION TECHNOLOGIES

Towards the goal of building applications that support camera-based interaction, a computer vision framework is needed on handheld devices. Computer vision is a large part of camera-based user interaction, and its limitations should be understood for designing interfaces; therefore, we discuss its main issues as the first step in this direction.

The mobile computer vision features required are a subset of the functionality on desktops. There are significant limitations on mobile devices, however:

- Mobile device CPUs have been limited in computing power. Only recently, high-end phones have started to support floating-point units. Hence, the use of floating point computations has to be minimized.
- The optics and image sensor chips of integrated cameras in mobile devices are targeted for imaging and video capture applications and as a result, provide limited quality for image processing tasks. For example, in smart phones, camera calibration is significantly more difficult due to the large amounts of lens distortion present. Such distortion makes the calibration algorithms commonly used in PCs infeasible for smart phones, as the recovered parameters will not be accurate enough for detecting 3-D position.
- Mobile devices do not support modifying camera focus and fixed focus cameras can only be effective in a certain depth range.
- In mobile devices, signal noise is more prevalent. Noise is caused by many factors, for example, bad camera lenses used, electronic noise caused by CCD camera, and "algorithmic" noise introduced by the imaging chain (e.g., white balance correction, exposure, gamma correction, color, shading, geometrical, noise-reduction). Images also have low contrast, varying brightness, and blurred edges.
- Battery power is a major consideration, limiting the type of applications and context that can be supported (e.g., it is not possible to have an always-on interaction scenario).

Correctly interpreting the observed motion of the objects or the global motion of the camera from video requires accurate tracking. To determine the motion direction, various tracking algorithms have been proposed. The solutions proposed in the literature can be divided into two main categories: markerless and marker-based techniques. Markerless tracking-based solutions analyze the video and detect important features, such as edges, corners, or corner-like features; or use motion-flow techniques for the global motion of the camera. Marker-based solutions use a visual tagging system that is based on printed 2-D markers that are placed in the environment and identified by mobile cameras. We describe the two tracking approaches in the next section.

Markerless Tracking

Markerless tracking systems do not assume any presence of a known object or a structure in the capture video. Thus, they provide a more general solution than marker-based systems, described later. Two different approaches are possible to achieve markerless tracking in mobile devices: template matching and optical flow.

- **Template-matching**-based solutions use an image region to track. These regions can be a rectangular block in the video (Figure 1), or arbitrary shapes. Although these systems have reliability issues, speed is their major advantage. The systems can use larger windows to capture more motion, but more processing is needed. Template matching is simple to implement, but requires good features to track.
- **Optical-flow**-based solutions are based on calculating the direction and motion speed of the features in the image, using the velocity field of pixels between two frames. The entire image can be used for tracking, increasing the correctness of the solution. However, the disadvantages of these methods are that the vector field may not be smooth (due to pixel disagreements) and the assumption of constant brightness is not always correct.

Mobile-camera-based markerless tracking has been researched by several groups:

- Rohs (2004) perform tracking based on dividing incoming camera frames into blocks, and then determining how the blocks move given a set of discrete possible translations and rotations.
- Haro et al. (Haro, Mori, Capin, & Wilkinson, 2005) propose a solution based on tracking individual corner-like features observed in the entire incoming camera frames. This allows the tracker to recognize sudden camera movements of arbitrary size, as long as at least some of the features from the previous frame are still visible, at the trade-off of not detecting rotations.
- Hannuksela et al. (Hannuksela, Sangi, & Heikkila, 2005) propose a region-based matching approach, where a sparse set of features are used for motion analysis, to-

Figure 1. Motion simulation. A base image with sliding window locations on the left (a) and a frame with ground truth motion (b) on the right.

(a)

(b)

gether with a Kalman filter-based tracker for estimation. The Kalman tracker has higher motion estimation accuracy, as expected, since the Kalman filter greatly improves the quality of intraframe matching. However, the computational requirements are significantly greater since several matrices must be multiplied and inverted per frame.

- Drab and Artner (2005) present a computationally inexpensive tracking system; however, their system has potential problems with repeating textures, and requires scenes with high dynamic range.

Tagging-Based Systems

Tagging-based tracking has also been researched by several groups. The main principle of these techniques is as follows: first, visual markers, printable with a standard printer, are created and placed in the environment. Then, these markers are detected in the captured video, and the 2-D and 3-D position of each visible marker (relative to camera position) and its rotation (relative to default orientation of the marker) are extracted. Additional information, such as an identification number, can also be detected with these solutions. Each proposed tagging technology has its own advantages and disadvantages:

- CyberCode (Rekimoto & Ayatsuka, 2000) is a visual tagging system based on 2-dimensional barcodes that can be recognized by CMOS and CCD cameras. CyberCodes encode 24 bits of data. In addition to the ID, the system can also compute the 3-D position of the tagged objects. Proposed applications for CyberCodes are augmented reality systems, various direct manipulation techniques involving physical objects, and indoor guidance systems.
- Rohs (2004) provide the Visual Codes system with an address space of 76 bits (83 bits without error detection) and a second guide bar, which allows the recognition of codes at a greater amount of tilting. The algorithm provides the relative x, y, and rotational motion of the phone, representing three degrees of freedom (DOF) input.
- Intelcom (http://www.intelcom.ru) has developed a software development kit for Nokia 7650/3650 for decoding data matrix codes. An example application generates SMS messages from the phone number and text stored in the code.
- Augmented reality and 3-D interaction research on mobile-camera-based tracking systems includes that of Möhring et al. (Möhring, Lessig, & Bimber, 2004), who track a color-coded 3-D marker to estimate 3-D camera pose, after an initial calibration step; and Hachet et al. (Hatchet, Pouderoux, & Guitton, 2005) who use a color-coded target in front of the camera to infer the 3-D DOF position.

Figure 2. Visual code parameters (left) and code coordinate system (right) (Rohs & Zweifel, 2005), with kind permission of Springer Science and Business Media

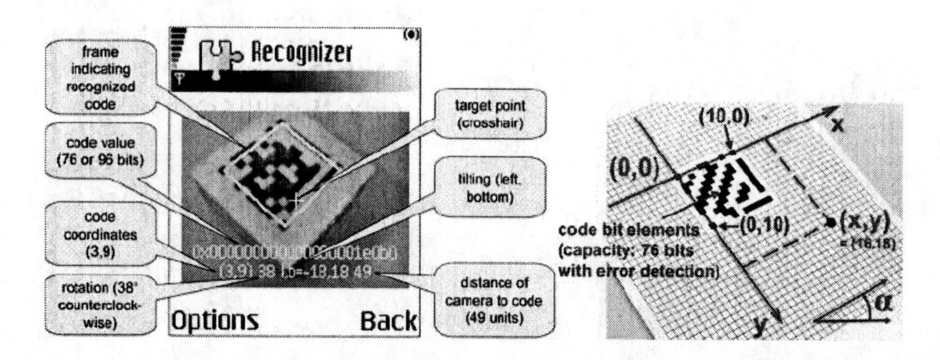

- The popular ARToolkit (Kato & Billinghurst, 1999) also provides visual markers. The toolkit provides optical tracking tools for detecting markers in a live video stream, extracting the 3-D position of the marker (relative to camera position) and its rotation (relative to default orientation of the marker).

The marker-based solutions suffer from the fact that a number of markers should be visible in a frame to detect the motion and orientation of the camera. Furthermore, the required presence of markers in the user's environment limits the range of interaction scenarios.

MAPPING CAMERA MOTION TO USER INTERACTION

As described, there are numerous proposed solutions computer vision technologies for enabling camera-based interfaces, and each solution has its advantages and disadvantages. For designing camera-based interfaces, the use of vision techniques have at least three variations:

- **Camera can be used as a pointing device:** The 1-D, 2-D, and 3-D position and movement of the camera can be acquired by tracking, and these data can used for various pointing tasks.
- **Camera can be used for gestural interaction:** The camera movement can be used as a low-level primitive for high-level gestural interaction techniques. For example, the user can interact with the applications by a set of gestures (e.g., by tilting and shaking the phone).
- **Camera can be used for interacting with the user's environment:** For example, instead of displaying the phone menu on the display of the device, the camera can be used to overlay the user interface in 3-D onto the video of the user's real world.

Next, we will discuss each variation and issues related to their use for interaction.

Using Camera as a Pointing Device

Direct-manipulation interfaces are particularly attractive for mass-market mobile interfaces because users can avoid learning commands and menu sequences, reduce their chance of errors, and keep their attention on the mobile device's small display. However, direct-manipulation interfaces are still difficult to realize on today's devices; currently, only key-modal interfaces are supported through joypad/direction keys and numerical keyboard. Stylus-based interaction is the most popular alternative, but this requires two-handed interaction, and has been shown to cause additional attentional overhead in users.

Using the camera as the pointing device on a mobile device is useful for many direct-manipulation interaction tasks. The range of tasks and their application create a rich set of design alternatives. Camera-based interaction allows for the following pointing tasks (Foley, Wallace, & Chan, 1984):

- **Selection:** Users can choose from a set of items by moving and tilting their phone, instead of pressing the phone's direction keys. This technique can be used for navigating through phone menus, quickly browsing through contacts, or selecting an image in the media library, for example.
- **Positioning:** Users can choose a point in 1-D, 2-D, or 3-D dimensions by moving their phone. Example uses of the position are controlling the mouse pointer on device's display, or selecting a hyperlink in a Web page.
- **Orientation/direction:** Users can choose a direction 1-D, 2-D, or 3-D space. The direction can be used for direction of a motion (e.g., to drag a file from one folder to another), to scroll a document vertically or horizontally, move an input message or a file to trash, or to rotate an image on the display, for example.

- **Path:** Users can rapidly perform a series of positioning and orientation operations. The path may be realized as drawing a curve in a drawing program, simulating handwriting by moving the phone as a stylus, or other direct manipulation tasks.
- **Quantification:** Users can specify a numeric value. The quantification can be used as one-dimensional or two-dimensional selection of integer or real values as parameters, such as the continuous zooming level, while viewing a Web page or a document.
- **Text:** Users can enter, move, or edit text in a two-dimensional space. The pointing device allows one to indicate the location of insertion, modification, or change. Other text-editing tasks, such as formatting a paragraph, can also be realized by moving the camera.

It is ultimately possible to implement a complete direct manipulation interface using the camera as the pointing device, but a way of switching the mode of the input for each of these tasks needs to be available to the user.

Using Camera as a Primitive for Gestural Interaction

In addition to the above low-level pointing tasks, camera can be used for gestural interaction. Gestures have traditionally been defined as a particular movement in front of the camera. However, in mobile interaction, gestures are generally defined by the motion of the camera instead. Each gesture can be defined as a different motion path, which is the output of tracking (Rohs & Zweifel, 2005), or can also include other high-level camera motions, such as shaking of the camera (Haro et al., 2005).

- For shake detection, Haro et al. (2005) determine the magnitude of the physical movement; they use motion history images (MHI) (Davis & Bobick, 1997), which were originally used for performing action and gesture recognition. An additional tracking algorithm provides four directions as application-level events, similar to mouse movement: up, down, left, and right, in the camera plane. The magnitude is also passed as an event where two states are possible: motion magnitude increasing or decreasing.

Figure 3. Possible combination of interaction primitives to build complex gestures for interaction (Rohs & Zweifel, 2005), with kind permission of Springer Science and Business Media

Combination	Interaction cue	Combination	Interaction cue
pointing & rotation	+ highlighted area	rotation & stay	
pointing & tilting	+ highlighted area	rotation & keystroke	
pointing & distance	+ highlighted area	tilting & distance	
pointing & stay	+ highlighted area	tilting & stay	
pointing & keystroke	+ highlighted area	tilting & keystroke	
rotation & tilting		distance & stay	
rotation & distance		distance & keystroke	

- Rohs and Zweifel (2005) propose and evaluate a number of physical gestures that form a basic vocabulary for interaction when using mobile phones. The proposed techniques are based on a visual code system that provides a number of orientation parameters, such as target pointing, rotation, tilting, distance, and relative movement. Their proposed framework defines a set of fundamental physical gestures that form a basic vocabulary for describing interaction when using mobile phones capable of reading visual codes. These interaction primitives can be combined to create more complex and expressive interactions.

Using Camera for Ubiquitous Computing and Augmented Realities

In addition to the tasks mentioned previously, camera input will play a different and more significant role in future mobile interfaces, together with the use of emerging ubiquitous computing and augmented reality paradigms. The interest in ubiquitous (or pervasive) computing has surged in the past few years, thanks to improving mobile processor and sensor technologies. The main characteristic of ubiquitous computing is to break away from the desktop interaction paradigm and move the computational power to the environment surrounding the user. To support this, ubiquitous computing requires user input to move beyond the textual input of keypad and selection from pointing devices, to perceptual interfaces that interact with users and their surroundings.

Augmented reality has emerged as one of the complementary fields that will be the mode of interaction for applications that combine mobile user interfaces with real-world interaction. AR provides a way to overlay computer-generated information (e.g., UI widgets or information) on top of real-world images. Although early AR work has focused on the use of see-through head-mounted displays, recent work has addressed handheld augmented reality, where the images are shown on a mobile device's display:

- Feiner et al. have presented one of the pioneering works in mobile AR interfaces. Their research on their MARS System (Mobile Augmented Reality Systems) began in 1996, and is aimed at exploring AR user interfaces, software, and application scenarios. They have proposed a set of reusable user interface components for mobile augmented reality applications (Feiner, MacIntyre, Höllerer, & Webster, 1997).
- The area of augmented reality on smart phones is still very new, but already there are several toolkit prototypes. The most well-known augmented-reality framework, the AR-toolkit (Kato & Billinghurst, 1999) has been used in a large number of augmented-reality research projects. Work is ongoing by several research groups to provide an efficient implementation of the toolkit available on cell phones and PDAs.
- One of the largest extensions to the AR-toolkit is Studierstube (Schmalstieg, Fuhrmann, Hesina, Szalavari, Encarnação, Gervautz, & Purgathofer, 2002), a framework that aims to make it easier for developers to create collaborative AR applications.

In addition to the frameworks described, a number of companies and developers are creating their own for mixed-reality applications, primarily in gaming. A large number of mixed-reality games for smart phones exist at the moment, ranging from camera-movement tracking to move on-screen crosshairs to simple body part region tracking. The most successful mixed-reality framework and platform for gaming at the moment is Sony's Eyetoy, which consists of a 60fps camera that is attached to a Sony Playstation 2. The processor analyzes incoming video to estimate the player's motion and to segment them from their environment. Eyetoy allows for games to perform game-control input recognition of: moving arms to certain locations to "touch" objects, moving arms up and down, moving the body side to side, and moving the body towards or away from the camera.

In augmented reality, accurate registration between the real and virtual data is essential. In practice, the calibration parameters recovered describing the camera's physical properties are good enough for rendering synthetic objects in augmented-reality applications. In the smart phone domain, camera calibration is significantly more difficult due to the large amounts of lens distortion present. Calibration techniques such as Zhang's accurate error-function minimization approach (1999) have been used successfully.

CAMERA-BASED UI PROTOTYPE

Towards the goal of building interactive applications, we have created a camera-based interaction prototype for smart phones. The prototype is similar in functionality to those on desktops, thus making it easier for computer vision and augmented-reality experts to work on smart phones more easily. Naturally, with smart phones there are hardware restrictions, so not all algorithms are presently possible. However, with a good framework it will be straightforward to extend the library with other algorithms, as well as other layers for content creators who are not vision experts.

While developing this prototype, we have faced a number of design issues. The split of the framework into low-level and high-level components was necessary to allow for expert and nonexpert developers alike to use the library to perform advanced camera input image and video processing. The low-level interface consists of the basic image and linear algebra operations themselves. Necessary geometric image operations, such as rotations, color conversions, image filtering as well as image statistics calculations, are provided. The low level also contains a set of linear algebra classes, including matrices, vectors, and functions such as singular value decomposition (SVD). The high-level interface is more focused on user interaction tasks. For instance, if a developer would like to determine whether the camera is being shaken, they can use library calls in the high-level interface to detect shaking, although the shake detection algorithm will be comprised of numerous calls to functions in the low level.

As described earlier, various tracking algorithms can be used for estimating the movement of the camera. In this chapter, we present a markerless, template-matching-based solution for tracking.

High-Level Algorithm Description

The tracking system was implemented on the Symbian OS. (See Haro et al., 2005 for details.) The process diagram of the tracker is presented in Figure 5. Our tracker uses the current and previous frame captured by the camera for tracking:

- First, the algorithm detects "corner-like" features in the new frame that are matched with the features found in the prior frame. Traditionally in computer vision, features include edges and corners. Edges are usually not significantly temporally coherent (i.e., they might change from one frame to another

Figure 4. Smart phone computer vision framework. This framework is the first step towards building camera-based applications built on a shared framework.

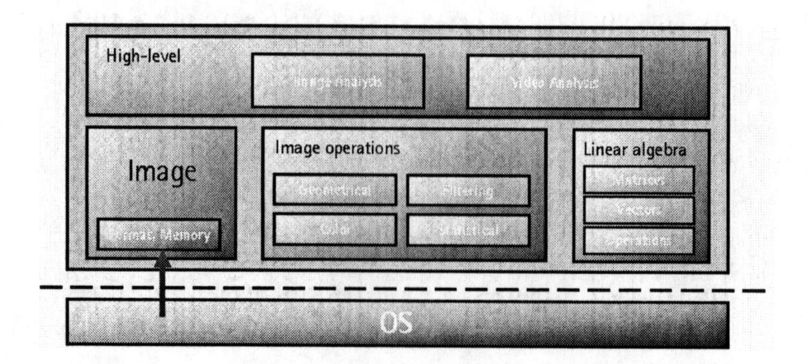

drastically), and corner features are too computationally expensive to find at many image locations while retaining real-time performance. We detect edges on both frames using the well-known Sobel filter. We use a threshold of 50 on both derivatives for each pixel, as this value results in a good number of feature candidates for typical scenes. Feature matching is performed between frames using template matching with 15x15 search windows, which we empirically found to be sufficient for our test hardware.

- Direction estimates are accumulated for a number of frames before a movement direction estimate is made. Direction voting is performed using variables, and the final decision on motion estimation is performed every four frames. This allows several frames to "vote" on the motion, keeping the scrolling from being incorrect due to any errors in other parts of the system.
- The directions of dominant camera motion are computed using the tracking algorithm, but their magnitudes are not known accurately. Camera motion magnitude must be calculated accurately to determine how to adjust the scroll speed in applications that need zoom control. We use motion history images (MHI) (Davis & Bobick, 1997) to estimate camera motion magnitude. Motion histories are encoded in single images such that a single image can be used for simple, robust, and computationally inexpensive gesture recognition. An MHI is computed by performing background subtraction between the current and previous frames. At locations where the pixel values change, the MHI is updated by decrementing by a predefined constant amount. By averaging the intensity values of the MHI, the average camera-motion magnitude is estimated.

Applications

We have implemented several test applications using the proposed prototype to clarify its strengths and limitations. We implemented the tracking algorithm and applications in C++ using the Series 60 second edition feature pack 2 SDK. Our test platform was a Nokia 6630 mobile phone that features an ARM 9 220mhz processor, 10 megabytes of RAM, 176x208 screen resolution, and a 1.3 megapixel camera capable of capturing frames at 15 fps.

Document viewer. Scrolling a document is a commonly difficult task on mobile devices. For instance, Web content designed for desktop computers is vertically much longer since mobile devices have narrower screens. In addition, joystick scrolling is especially difficult when scrolling vertically and horizontally. An alternative is to add an extra hardware button for scrolling. However, an extra button is not a preferable solution for mobile device manufacturers due to the lack of extra physical space on the device, along

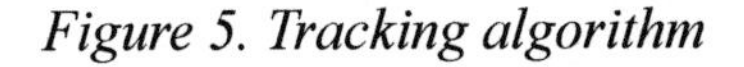

Figure 5. Tracking algorithm

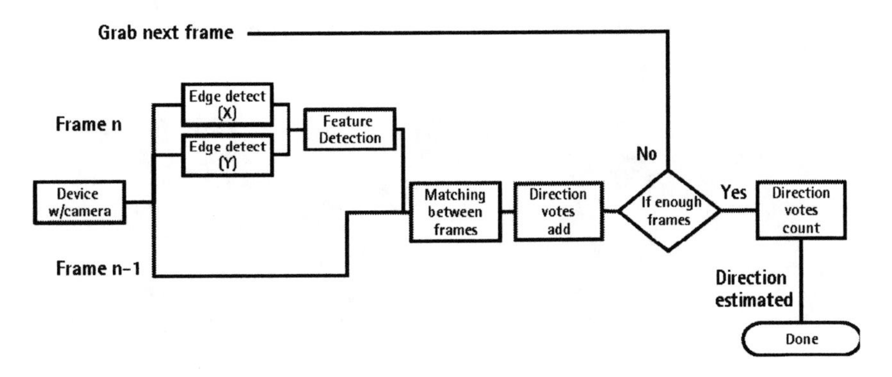

with additional manufacturing costs. In the document viewer prototype application that we implemented (Figure 6(a)), the user can vertically scroll documents by moving the device. Our approach is similar to *AutoZoom* and *GestureZoom* interfaces proposed by Patel et al. (Patel, Marsden Jones & Jones 2004). The scroll speed depends on how fast the user moves the device, which is much more intuitive than changing scrolling speed depending on how long the user presses the joystick, or via menu options and settings. One issue we identified in this application is that at some point, the user has to move the device more than they can reach. For example, if the user is scrolling to the right, at some point they will reach the physical limit of their arm's motion. To address this problem, we use the joystick as a "carriage return" that scrolls the document to the beginning of the next line and allows the user to move their arm back to the left again. After a carriage return, all tracked motion, except movement to the right, is ignored.

Zooming photo browser. As cameras become more widespread on mobile phones and storage size increases, managing photos becomes a more difficult task for the user, as large amounts of information must be viewed with limited input modalities. Current typical photo viewer applications show photo thumbnails as lists, grids, or 3-D carousels. Since image selection and scrolling are done with the joystick, the amount of time a user needs to browse their images is directly related to the number of images that they are browsing. Our photo browser test application (Figure 7) shows thumbnails of the user's photos in a grid layout. The user can scroll in four directions (up, down, left, right, in the camera plane) by physically moving the mobile device. In this case, it is difficult to view all the images, as some zoom control is required when looking for a particular image. If the zoom level is not properly set, it is difficult for a user to select a particular image from the set, as the scrolling will be too fast. To address this problem, we used the adaptive zooming technique introduced by Igarashi and Hinckley (2000). Adaptive zooming, based on the magnitude of the user's physical movement, keeps the scroll speed virtually consistent, allowing the user to browse more thumbnails by only moving the device faster.

3-D Game interaction. Creating an immersive 3-D experience is difficult on mobile devices due to the limited display size. The most immersive experiences are typically created using a combination of large displays reducing peripheral vision as much as possible and/or virtual environment navigation tied to the user's physical motion. In our prototype (Figure 6(b)), we map the user's physical motion to the view-point to create the illusion of a window into a 3-D world. Our renderer loads standard Quake III™ or Quake III Arena™ maps. Textures, light maps, curved surfaces, and lighting calculations are disabled for performance. The rendering is done using the OpenGL ES implementation available in the latest Symbian OS-based Series 60 SDK. Precomputed vertex lighting and fixed point calculations are used to improve performance due to the lack of a floating-point unit on our test hardware. The renderer is able to

Figure 6. (a) Camera-based interaction in a document viewing application. (b) Mapping physical motion to viewing direction creates illusion of a window into an environment.

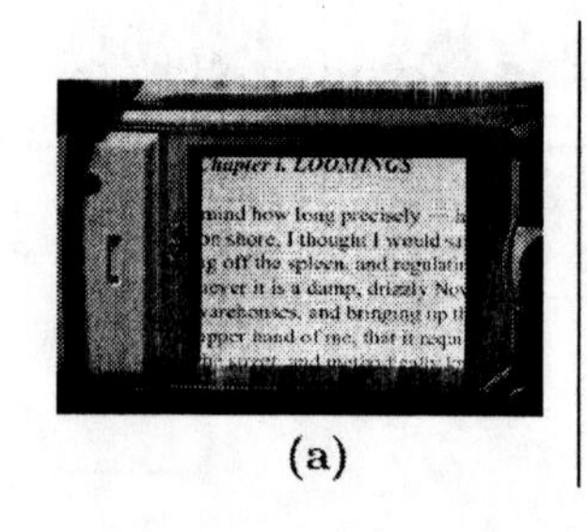

(a)

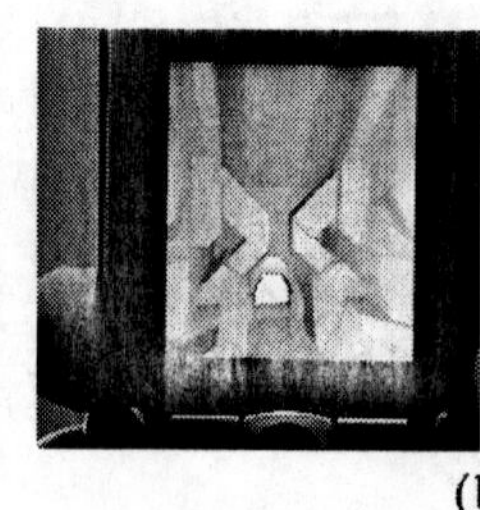

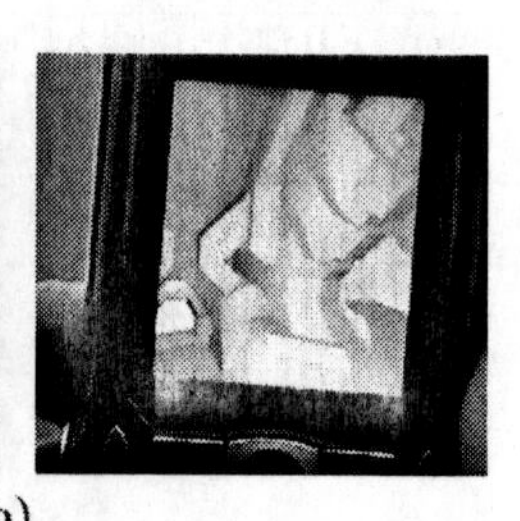

(b)

Figure 7. Picture browser application; the application automatically adjusts the zoom level to help the user browse

realistically render lit virtual environments with several thousand polygons per scene at 3-10 frames per second, depending on the environment that is chosen. Navigation of the virtual environments is performed with a combination of physical motion and keypad presses. Actual movement in the environment is controlled by the keypad. The user looks around in the scene by physically moving the device around their body in the directions that they would like to look. We map the tracked camera motion directions to a trackball as in traditional mouse-based 3-D interaction. The combination of detailed environments, camera-based control, and interactive frame rate create a mobile user experience closer to that using additional hardware or larger displays.

2-D Game interaction. Camera-based user interaction can be used to enhance 2-D games as well as those that are 3-D. Camera motion can be used to add an additional element of interaction in games that require precise movements or very well-timed button presses. We created puzzle and action game prototypes to investigate these ideas using the camera motion and shake-detection algorithms presented. We modified the open source Series 60 port of the "Frozen Bubble" puzzle game (http://fbs60.sourceforge.net/), switching the game control from using the keypad to using the camera (Figure 8(a)). In our version, the user moves their device left and right to aim, and performs sudden shakes to launch their bubble. This has the effect of significantly changing gameplay, as careful arm motions are now required to aim, instead of a number of button presses, which increases the excitement as the game is now more physically based. We created a camera-based action game prototype as well. Using sprites and artwork from Konami's `Track and Field™' game for the Nintendo Entertainment System, a new game (Figure 8(b)) was created. A runner must jump over a never-ending number of approaching hurdles. To jump, the player must time the shaking of their device correctly so that the character does not crash into hurdles. Relying on the camera exclusively

Figure 8. (a) Players move the device left and right to aim, and shake the device to launch a bubble. (b) A jump command is issued by shaking the device at the correct time to avoid tripping on the hurdle.

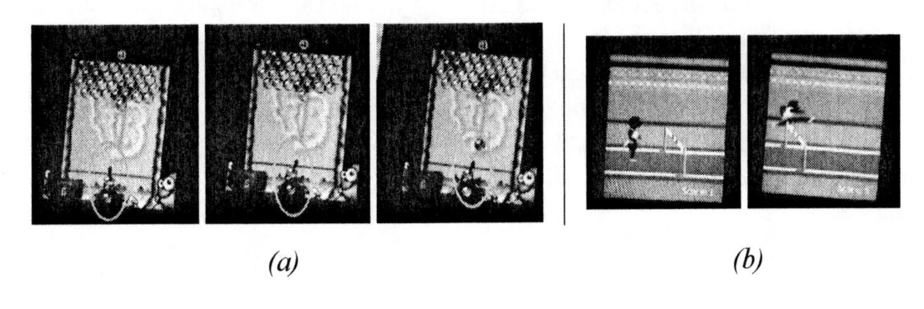

(a) *(b)*

for input results in a game that is very simple to learn and understand but difficult to master, providing a new type of game. Shake detection is performed by thresholding the average intensity of the computed MHI.

DISCUSSION

In order to support intuitive and efficient user interaction, it is important to understand what kind of input is provided by the camera-based interfaces, what type of target task is considered, and what the current limitations are, given the output of the tracking algorithm. The success for a camera-based user interface is achieved when the following criteria are met:

- The speed and accuracy of the used computer vision technique should match the requirements and efficacy of the target task,
- The learning time should be minimal, particularly for gesture-based and augmented-reality interfaces,
- The computational cost and reliability of the used camera input processing techniques should match the target application and user's environment.

The most basic, but potentially most important input, that can be acquired from the tracking algorithm is the two-dimensional movement of the mobile device on a plane parallel to the camera in 3-D. With this type of data, the camera can be used as an input device to capture the device's movement in up/down, left/right directions, as well as its speed in each direction. Mobile camera-based input has restrictions, primarily due to limitations of mobile device hardware. Forward and backward motion cannot be detected with the current generation of mobile phones, so six degree–of-freedom movement is not supported yet. Forward/backward motion is possible to detect; however, this would increase computational demands and reduce the frame rate, impoverishing the user interaction.

Physical movement speed is another challenge for camera-based interaction. The algorithm must perform all of its video analysis in the time between camera frames being captured to support real-time interaction.

Thus, there are implicit limits on the computational complexity of the tracking. In addition, there is a fundamental assumption in tracking algorithms that each frame contains some portion of the prior frame. This assumption is motivated by the observation that users will typically not move their phones erratically when focused on a task. Users usually operate mobile phones with one hand. Mobile phones can also be used anywhere in an office, school, public, home, and so forth. Considering these environments, there are certain interactions that are not appropriate:

- **Precise tasks:** Precise motion is very difficult holding a mobile device with one hand. Interaction should not require operations like "move the device 2.5cm up," or "move the device 34 degrees from the horizontal line." As a result, camera-based interaction will probably be most useful when navigating large amounts of data, or zoom-level-dependent data.
- **Large motion:** This restriction is more serious in some environments, such as in crowded public locations. In such situations, it may be advantageous to provide a "clutch" to turn the tracking on/off. This would emulate the act of lifting a mouse once the edge of a desk is reached in traditional desktop interaction. In our informal testing, we did not provide a clutch; however, in commercial implementations this is a consideration to keep in mind.
- **Extended and/or frequent interaction:** Using single-handed operation, interactions that require extended time and/or frequent movement may fatigue users.

The camera-based input approach works best with coarse selections at different speeds and scales of data. It is critical that visual feedback follows physical motion and that the feedback differs according to motion speed, in order to provide an intuitive user experience. The most typical use case is moving the device to scroll UI content such as a list or a document.

CONCLUSION

In this chapter, we surveyed the technologies underlying mobile camera-based user interaction, and described the concepts and recent research results. We have presented a camera-based toolkit prototype, including design issues that we faced. We demonstrated our approach in several applications using 2-D and 3-D interaction. Initial results suggest that camera-based interaction has a great potential for future user interfaces.

While the proposed tracking solutions are computationally efficient and work well in practice in controlled environments, there are some situations that cannot be handled. Severe lighting differences will cause the template matching to stop working properly. Motion in front of the camera is ambiguous and can affect tracking results, as it is impossible to tell whether the camera is moving or not without other sensors. Shadows may confuse the tracking system, but there are known computer vision techniques for robust tracking in the presence of shadows that will be incorporated into the tracking algorithm once additional processing speed is available.

In the future, we would like to collect user feedback to determine how to improve user interaction further using mobile cameras, using qualitative and controlled experiments. While we applied the camera-based interaction to only viewpoint selections and simple gestures, we would like to investigate its application to more complex gesture-based and augmented-reality interfaces. In the future, we believe that a full camera-based mobile user interface is possible with the described approaches, potentially making the phone keypad unnecessary.

ACKNOWLEDGMENT

This work was partially supported by grants by Nokia Inc. and Bilkent University.

REFERENCES

Davis, J., & Bobick, A. (1997). The representation and recognition of human movement using temporal templates. In *Proceedings of the IEEE Computer Society Conference on Computer Vision and Pattern Recognition (CVPR)* (pp. 928-934).

Drab, S., & Artner, N. (2005). Motion detection as interaction technique for games and applications on mobile devices. In *Extended Abstracts of PERVASIVE: Workshop on Pervasive Mobile Interaction Devices* (pp. 48-51).

Feiner, S., MacIntyre, B., Höllerer, T., & Webster, T. (1997, October 13-14). A touring machine: Prototyping 3-D mobile augmented reality systems for exploring the urban environment. In *Proceedings of the First IEEE International Symposium on Wearable Computers.*

Fishkin, K., Gujar A., Harrison, B., & Want, R. (2000, September). Embodied user interfaces for really direct manipulation. *Communications of the ACM.*

Foley, J. D., Wallace, V. L., & Chan, P. (1984). The human factors of computer graphics interaction techniques. *IEEE Computer Graphics and Applications, 4*(11), 13-48.

Hachet, M., Pouderoux, J., & Guitton, P. (2005, April). A camera-based interface for interaction with mobile handheld computers. In *Proceedings of the 2005 Symposium on interactive 3-D Graphics and Games (Washington, District of Columbia. SI3-D '05)* (pp. 65-72).

Hannuksela, J., Sangi, P., & Heikkila, J. (2005). A vision-based approach for controlling user interfaces of mobile devices. *2005 IEEE Computer Society Conference on Computer Vision and Pattern Recognition.*

Haro, A., Mori, K., Capin, T., & Wilkinson S. (2005). Mobile camera-based user interaction. *Proceedings of ICCV-HCI 2005* (pp. 79-89).

Hinckley K., Pierce, J., Sinclair, M., & Horvitz, E. (2000). Sensing techniques for mobile interaction. *Proceedings of the 13th Annual ACM Symposium on User Interface Software and Technology (UIST 2000)* (pp. 91-100). New York, NY: ACM Press.

Igarashi, T., & Hinckley, K. (2000). Speed-dependent automatic zooming for browsing large documents. In *Proceedings of the ACM symposium on User interface software and technology (UIST 2000)* (pp. 139-148). New York, NY: ACM Press.

Kato, H., & Billinghurst, M. (1999) Marker tracking and HMD calibration for a video-based augmented reality conferencing system. In *Proceedings of the 2nd International Workshop on Augmented Reality* (IWAR 99), San Francisco.

Maes, P., Darrell, T., Blumberg, B., & Pentland A. (1995) The ALIVE system: Full-body interaction with autonomous agents. In *Proceedings Computer Animation* (pp. 11-18, 209).

Möhring, M., Lessig, C., & Bimber, O. (2004). Optical tracking and video see-through AR on consumer cell phones. In *Proceedings of Workshop on Virtual and Augmented Reality of the GI-Fachgruppe AR/VR.* (pp. 193-204).

Patel, D, Marsden, G., Jone,s S., & Jones, M. (2004). An evaluation of techniques for browsing photograph collections on small displays. In *Proceedings Mobile HCI* (pp. 132-143), Glasgow, Scotland.

Rekimoto, J., & Ayatsuka, Y. (2000). CyberCode: Designing augmented reality environments with visual tags. In *Proceedings of DARE 2000.*

Rohs, M. (2004). Real-world interaction with camera-phones. In *Proceedings of 2nd International Symposium on Ubiquitous Computing Systems (UCS 2004)* (pp. 39-48).

Rohs, M., & Zweifel, P. (2005). A conceptual framework for camera phone-based interaction techniques. In *Proceedings of Pervasive 2005,* (pp. 171-189). Schmalstieg, D., Fuhrmann, A., Hesina, G., Szalavari, Zs., Encarnação, L. M., Gervautz, M., & Purgathofer, W. (2002). The Studierstube augmented reality project. *PRESENCE - Teleoperators and Virtual Environments, 11*(1).

Siio, I. (1998). Scroll display: Pointing device for palmtop computers. *Asia Pacific Computer Human Interaction,* 243-248.

Yee, K. P. (2003). Peephole displays: Pen interaction on spatially aware handheld computers. *Proceedings of the SIGCHI conference on Human factors in computing systems* (pp. 1-8).

Zhang, Z. (1999). Flexible camera calibration by viewing a plane from unknown orientations. *International Conference on Computer Vision (ICCV'99)* (pp. 666-673).

KEY TERMS

3-D Interaction: A type of user interaction, where the output is displayed in 3-D (on stereoscopic or 2-D displays), and user input is received through 3-D interaction devices.

Augmented Reality: A field of computer research which deals with the combination of the real world with computer generated data.

Camera-Based User Interaction: A type of interaction, where the user interacts with an application by moving their device, and the captured camera video is used to estimate phone motion or interact with the real world.

Computer Vision: The analysis of image sequences, concerned with computer processing of images from the real world. Computer vision typically requires a combination of low-level image processing and high-level pattern recognition and image understanding to recognize important features in the image.

Gestural Interaction: A type of user interaction, where the user interacts with the computing device through a set of well-defined gestures. Gestures can originate from any bodily motion or state but commonly originate from the face or hand.

Ubiquitous Computing: A computing paradigm, where computation is integrated into the environment, rather than having computers as distinct objects. One of the goals of the field is to embed computation into the environment, and allow everyday objects to be used for interaction.

Zoom Control: Ability to control the zoom level in a user interface or an image or a document.

About the Contributors

Joanna Lumsden is a research officer with the National Research Council of Canada's (NRC) Institute for Information Technology. Prior to joining the NRC, Joanna worked as a research assistant in the Computing Science Department at the University of Glasgow, U.K. where she attained both her undergraduate Software Engineering Honours Degree and her PhD in human computer interaction. Joanna is also an adjunct professor at the University of New Brunswick in Fredericton, where she teaches graduate courses and supervises a number of graduate students. Joanna is the lab manager for the NRC's Mobile Human Computer Interaction Lab—a facility dedicated to investigating mobile interaction design and evaluation.

Matt Jones is a senior lecturer and is helping to set up the Future Interaction Technology Lab at Swansea University. He has worked on mobile interaction issues for the past ten years and has published a large number of articles in this area. He is the co-author of *Mobile Interaction Design*, John Wiley & Sons (2006). He has had many collaborations and interactions with handset and service developers including Orange, Reuters, BT Cellnet, Nokia and Adaptive Info; and has one mobile patent pending. He is an editor of the *International Journal of Personal and Ubiquitous Computing* and on the steering committee for the Mobile Human Computer Interaction conference series. He is married with three mobile, small children; when he's not working he enjoys moving quickly on a bike while listening to music and the occasional podcast.

* * *

Julio Abascal (BSc in physics, Universidad de Navarra, 1978; PhD in informatics, University of the Basque Country-Euskal Herriko Unibertsitatea, UPV-EHU, 1987) is a professor at the Computer Architecture and Technology Department of the UPV-EHU (Spain), where he has been working since 1981. In 1985 he co-founded the Laboratory of Human-Computer Interaction for Special Needs where he has led several R&D projects. His research activity is focused on the application of human-computer interaction methods and techniques to assistive technology, including the design of ubiquitous, adaptive, and accessible user interfaces, and human-robot interfaces for smart wheelchairs. In addition, he leads a research group aiming to develop methods and tools to enhance physical and cognitive accessibility to the web. He has been Spain's representative on the IFIP Technical Committee 13 on "Human-Computer Interaction" since 1991, and is the former and founder chairman (1993) of IFIP WG 13.3 "Human-Computer Interaction and Disability." From 1990 he served as an adviser, reviewer, and evaluator for diverse EU research programmes (TIDE, TAP, IST…), and he is currently a member of the Management Committee of COST 219ter "Accessibility for All to Services and Terminals for Next Generation Networks."

Gregory D. Abowd is an associate professor in the College of Computing and GVU Center at the Georgia Institute of Technology, and director of the Aware Home Research Initiative. His research focuses on an application-driven approach to ubiquitous computing concerning both HCI and software engineering issues. He received a PhD in computation from the University of Oxford. He's a member of the IEEE Computer Society and the ACM.

Thomas Alexander heads the research group on 3D Visualization and Interaction at the FGAN-Research Institute for Communication, Information Processing, and Ergonomics (FKIE) in Wachtberg, Germany. He obtained the Diplom-Ingenieur (MS) and Dr.-Ing. (PhD) degree in 1994 and 2002, respectively. Since 1994 he has been working at FKIE, where he has been responsible for research activities and various projects in the area of digital human models and human-computer interaction (HCI). In 2004-05 he worked as a visiting scientist at the NASA Ames Research Center, CA. Since 1998 he has been German national representative and chairs several international committees on virtual environments.

Over the past three decades, **Rebecca Allen** has been recognized internationally for her ground-breaking work in media art and her research in 3D computer graphics, human motion simulation, artificial life, augmented reality and intimate interfaces. Allen is professor at UCLA Department of Design Media Arts and was founding chair of her department. Previous positions include: senior research scientist and director of the Liminal Devices research group at MIT Media Lab Europe; founding director of the Intel funded research project called Emergence; creative director and 3D visionary for video game company, Virgin Interactive; senior researcher at the world-renowned NYIT Computer Graphics Laboratory and member of the MIT Architecture Machine Group.

Nikolaos Avouris was born in Zakynthos, Greece (1956). He has a degree in electrical engineering from NTUA Greece (1979) and an MSc (1981) and PhD (1983) from the University of Manchester UMIST, UK. He served in various research positions in the UK, Italy, and Greece and then joined the University of Patras, as associate professor (1994-2001) and full professor of software engineering and human-computer interaction (2001-present). He is founder and head of the Human-Computer Interaction Group. His main interests are related to design and evaluation of interactive systems, usability engineering, collaboration technology, context-aware computing systems, and analysis and evaluation of collaborative activities.

Chris Baber received his PhD in human factors of speech technology from Aston University in 1990 before taking up a post at The University of Birmingham, where he is now reader in interactive systems design. His research focuses on human interaction with technology and he is particularly interested in how an understanding of everyday activities can be used to better inform the study of novel technologies.

Cédric Bach is a usability specialist of human virtual environment interactions (HVEI). He conducted his PhD in psychology at the French National Institute for Research in Computer Science and Control (INRIA) until 2004. Currently, he's a postdoctoral researcher in the Research Institute in Computer Science of Toulouse – France (IRIT). His research domain is methods for HVEI ergonomics dedicated to evaluation and user centred design.

Young Kyun Baek has been teaching at Korea National University of Education since 1991. Currently, Dr. Baek's research interest is on designing educational games and simulation, and using mobile devices in education. He has recently published two books entitled *Understanding & Applying Game Based Learning in Classroom* and *Understanding Edutainment* in Korea.

Rafael "Tico" Ballagas is a computer science doctoral candidate and research assistant at RWTH Aachen University in Aachen, Germany. His research interests include post-desktop user interfaces (including mobile phones), physical toolkits, and applying iterative design techniques to ubiquitous computing applications. He developed the iStuff toolkit to rapidly prototype new interactions in ubiquitous computing. He has an MS in electrical engineering from Stanford University and a BS in electrical engineering from Georgia Institute of Technology.

Susanne Bay is a psychologist with a special interest in cognitive ergonomics of small screen and medical devices. The work presented in the current chapter was carried out during her time as a PhD student at the Department of Psychology of RWTH Aachen University, Germany. Currently she is working as a human factors specialist.

Benjamin B. Bederson is an associate professor of computer science at the Human-Computer Interaction Lab at the Institute for Advanced Computer Studies at the University of Maryland, College Park. His work is on mobile device interfaces, information visualization, interaction strategies, digital libraries, and accessibility issues such as voting system usability.

Francesco Bellotti is a researcher at the ELIOS laboratory of the University of Genoa. He is involved in research projects concerning mobile computing, human-computer interaction and multimedia systems in the automotive environment. He received a PhD in electrical engineering from the University of Genoa. Professor Bellotti has authored over 50 papers in international journals and conferences.

Regina Bernhaupt is currently working as assistant professor at the HCI Unit of the ICT&S-Center, working on her habilitation in the area of usability evaluation methods. She holds a masters degree in psychology and in computer science from the Salzburg University. In 2002 she finished her technical dissertation in computer science in the field of intelligent systems (time coded artificial neural networks). She teaches programming courses, user interface techniques and design and human-computer interaction at the Salzburg University and the applied university of Salzburg. She is leading several projects in the area of home entertainment (interactive TV, games, new ways of entertainment) and is responsible for new forms of usability and user experience evaluation in various contexts like mobile interfaces and ambient technologies.

Riccardo Berta is a research consultant at the Department of Biophysical and Electronic Engineering of the University of Genova. He received his degree (MS) in electronic engineering from the University of Genova in 1999, with a thesis on project and development of an optimizing environment for Java bytecode. His current research interests include design, implementation and evaluation of innovative modalities of human-computer interaction for mobile devices. Dr. Berta has authored over 30 papers in international journals and conferences.

Enrico Bertini is a post-doc researcher at University of Fribourg, Switzerland. He holds a PhD in computer engineering from University of Rome "La Sapienza". His research interests span across the whole spectrum of human-computer interaction. In recent years, he has explored issues related to usability in mobile computing, devising novel methodologies for evaluation, and adaptation/personalization techniques. Information Visualization is also one of his main research interests.

Jason Black has a PhD in computer science from Florida State University, an MS degree in computer science from Georgia Institute of Technology, and a BS degree in computer information systems from Florida A&M University. His research interests include mobile computing, collaborative mobile learning (m-learning), user interface design for mobile devices, ambient or ubiquitous computing, and educational technology. He also has a particular interest in developing learning applications for personal digital assistants (PDAs) with a focus outside of math and science, such as reading comprehension. Dr. Black has been as assistant professor at FAMU for seven years.

Ann Blandford is professor of human-computer interaction and director of UCL Interaction Centre. There, she leads a team of 20 academics and researchers studying various aspects of HCI. Her research focuses on seeking a better understanding of interactive systems from a user perspective, and using that understanding to support the design and evaluation of systems. Her research activities include investigating the use of systems in context, and the effects of system design on user experience.

Borja Bonail works at the Laboratory of Human-Computer Interaction for Special Needs of the University of the Basque Country-Euskal Herriko Unibertsitatea, UPV-EHU (Spain) with a Pre-doctoral Fellowship granted by the Basque Government. He holds a BSc in informatics engineering (2004) from the UPV-EHU. His main research areas are the development of methodologies for human-robot interaction in shared-control mode, and the development of intelligent robotics controllers to support such interaction types, both focused on wheelchair users.

Jan Borchers is full professor of computer science and head of the Media Computing Group at RWTH Aachen University. With his research group, he explores the field of human-computer interaction, with a particular interest in new post-desktop user interfaces for smart environments, ubiquitous computing, interactive exhibits, and time-based media such as audio and video. Before joining RWTH, he worked as assistant professor of computer science at Stanford University for two years, and briefly at ETH Zurich. He received his PhD in computer science from Darmstadt University of Technology in 2000, and is a member of ACM, SIGCHI, BayCHI, and GI.

Mike Bradley is senior lecturer in Product Design and Engineering Group at Middlesex University, London, England. He has a BEng in mechanical engineering from Birmingham University, England and an MSc in ergonomics from Loughborough University, England. He was head of human factors at Ford of Europe and his research interests are in interaction design, inclusive design, and human-centred intelligent vehicles.

Stephen Brewster is a professor of human-computer interaction in the Department of Computing Science at the University of Glasgow, UK. Brewster joined Glasgow in 1995, before that working as an EU-funded ERCIM fellow in Finland and Norway. Brewster's work is in the area of multimodal human-computer interaction (HCI), using multiple sensory modalities to make human-computer interaction more effective. He has directed his research towards sound, touch, and smell. His work on sound has focused around the design and use of Earcons, or structured non-speech sounds. His research in this area is now on the use of spatial sound in mobile computing, where the lack of screen space means audio plays an important role. Brewster has also worked on gestural interfaces for mobile devices. One other strand of his work is in haptic (touch-based) interaction, here focusing on applications for visually-impaired people and medical simulation. Brewster is currently an EPSRC Advanced Research Fellow. The aim of the fellowship is to understand how to construct tactile cues for user interfaces, particularly Tactons, or tactile icons as another form of output for mobile devices.

Xavier Briffault is a researcher in the Centre National de la Recherche Scientifique (CNRS) in France. He is a computer scientist with a primary interest in natural language processing. He has been involved in various research projects concerning computer supported collaborative work, knowledge representation, object oriented programming, and multimodal navigation aids for car drivers and blind pedestrians.

Earl Bryenton is a professional engineer and a seasoned veteran of business management with over 50 years of industry experience. He founded his own consulting engineering company in 1983 that later became BRYTECH Inc. Prior to this he managed corporate operations at Bell Northern Research having ascended through the ranks including technical development, systems engineering, and marketing. Earl graduated from the University of New Brunswick with a BSc in engineering and holds a current license as a consulting engineer.

Thorsten Büring is a final year PhD candidate in the Department of Computer Science at the University of Konstanz, Germany. He is a member of the Human-Computer Interaction Group headed by professor Harald Reiterer. His research is supported by the Deutsche Forschungsgemeinschaft and focuses on mobile HCI and information visualization. He holds a MSc in information technology from the University of Glasgow (2003) and a Diploma in electronic business design from the University of Arts, Berlin and the Institute of Electronic Business (2002).

Gary Burnett has 15 years experience in human factors research and development relating to advanced technology within road-based vehicles. He has been a lecturer in human factors and human-computer interaction at the University of Nottingham in the UK since July 2000. Previously, he worked as a research fellow at the HUSAT Research Institute at Loughborough University. His work addresses key safety, usability, and acceptability issues for a number of in-car systems, and he has worked on a number of large-scale collaborative projects within this area (funded by the EU and the UK government). He also acted as a consultant to many of the major car manufacturers and system suppliers (e.g., Honda, Ford/Jaguar, Toyota, Nissan, Alpine). He has published over 50 papers in peer-reviewed journals, conferences, and edited works.

Daniel Cagigas received his PhD degree in computer science from the University of the Basque Country-Euskal Herriko Unibertsitatea, UPV-EHU (Spain) in 2001, for his work on navigation systems for smart wheelchairs. He is now a lecturer at the Department of Computer Architecture and Technology and a member of the 'Robotics and Technical Aids Group' research group, both at the University of Seville (Spain). He lectures on fundamentals of computer science, computer architecture, and parallel architecture systems. His research interests include artificial intelligence, robotics, human-computer interaction, and technologies for rehabilitation.

Mikael Cankar, MSc, is currently a usability specialist at TeliaSonera, where he works to improve the user experience of mobile and Web services. He has recently begun PhD studies and has ambitions to improve the way user research findings are used in product development projects.

Tolga Capin is an assistant professor in computer engineering at Bilkent University, Turkey. He has been involved in the development of mobile graphics and user interface technologies since 2000; most recently contributing to the development of graphics engines as a principal scientist and research manager at Nokia Research Center. His current research interests are mobile graphics platforms, user interaction with graphics and images, and networked graphics platforms. He got his PhD degree in computer science in 1998.

Tiziana Catarci received her PhD in computer science from the University of Rome, where she is currently a full professor. She has published over 100 papers and 10 books in a variety of subjects comprising user interfaces for databases, 2D and 3D data visualization, adaptive interfaces, visual metaphors, usability testing, data quality, cooperative database systems, database integration, and Web access. Dr. Catarci is regularly in the programming committees of the main database and human-computer interaction conferences and is associate editor of *ACM SIGMOD Digital Symposium Collection (DiSC), VLDB Journal, World Wide Web Journal,* and *Journal of Data Semantics.*

Nian-Shing Chen is professor of the Department of Information Management, National Sun Yat-sen University, Taiwan. He is currently the chairman of the MIS Department, NSYSU and adjunct professor of the Griffith Institute for Higher Education, Griffith University, Australia. He has published over 160 research papers in international refereed journals, conference, and book chapters. His research areas include e-learning, knowledge management, and the use and development of online synchronous learning and wireless technologies to enhance learning.

Eleni Christopoulou is a PhD candidate in computer science at the Department of Computer Engineering & Informatics, Patras University, Greece. She received her Engineering Diploma and MSc in computer science and engineering, in 2001 and 2004, respectively, both from the same institution. During 1999-2005 she worked as an R&D Engineer for the Research Academic Computer Technology Institute. During 2004-2005 she taught at the Technological Educational Institute of Patras, Greece. Since 2006 she teaches at the Department of Informatics, Ionian University, Greece. Her current research interests include the use of ontologies and context-awareness in ubiquitous computing systems and semantics in p2p networks and social networks.

Martin Colbert graduated in psychology from the University of Stirling, UK, in 1985. From 1986 to 1998, he was a research associate at the Ergonomics & HCI Unit, University College London (UCL), UK, where he gained his PhD (1995). In 1998, he became senior lecturer at the School of Computing and Information Systems, Kingston University, UK, where he now teaches Human Computer Interaction and Java, and is course leader for MSc Software Engineering. He continues to research and publish in HCI, and conducts contract usability work. Martin likes his family and friends, mountains, bicycles, music and parties, preferably all together.

Jose L. Contreras-Vidal is an associate professor in kinesiology, neuroscience, and bioengineering at the University of Maryland, College Park. His research integrates behavioral, neuroimaging, and computational neuroscience methods to study the neural mechanisms and computational principles underlying human sensory-motor control and learning. His research also includes computer simulations of large-scale, biologically-plausible, neural networks of cognitive-motor systems.

Enrico Costanza is a doctoral student and research assistant at the Laboratoire de design et media at the Ecole Polytechnique Fédérale de Lausanne. He holds an MEng in electronics and communications engineering from the University of York, England, and an MS in media art and science from the Massachusetts Institute of Technology, USA. From 2003 to 2005 he was a research associate at the Media Lab Europe in Dublin, Ireland. Enrico is active in the areas of mobile HCI, visual marker recognition and its HCI applications, and tangible user interfaces.

David Coyle is a PhD candidate with the Department of Computer Science at Trinity College Dublin. His background is interdisciplinary, in areas including electronic engineering, multimedia, computer science, and human computer interaction. His current research is focused on developing user-centred design methodologies for the design and evaluation of adaptable technologies for talk-based mental health interventions. David is currently exploring the design and impact of therapeutic 3D computer games in adolescent mental health interventions.

Murray Crease is a research officer with the National Research Council of Canada based in Fredericton, New Brunswick. He received his BSc (1996) and PhD (2001) in computing science from the University of Glasgow, Scotland. Murray's research interests lie in the general area of human-computer interaction (mobile HCI). Murray is particularly interested in mobile HCI believing that mobile systems should not be limited to mobile devices such as PDAs and mobile phones but should also include ubiquitous computing and alternative interaction techniques such tangible and ambient interaction.

Andrew Crossan graduated with a BSc Hons at the Department of Computing Science at Glasgow University in 1999. He then completed his PhD at the same department before moving to the Hamilton Institute in 2003 where he worked as a research assistant studying continuous control interaction techniques with mobile devices. Since 2005 he has worked as a post-doctoral researcher in the Multimodal Interaction Group (MIG) in the University of Glasgow Computing Science Department. One major focus of his work has been on multimodal interaction, with the main applications being in virtual reality veterinary medical training systems and in developing accessible interfaces for visually impaired people.

Sara Dalzel-Job received an MA in psychology from the University of Glasgow in 2004. She is currently studying for an MSc in informatics at the University of Edinburgh (due to be completed in August 2007), to be followed by a PhD in the same department. Her current interest is human computer interaction, specifically the use of eye tracking technology as a diagnostic tool.

Nelly de Bonnefoy is a PhD student in a joint laboratory involving the IRIT laboratory and EADS. Her work focuses on operators in the context of aeronautical maintenance and the main goal is to "Bring the information they need, when they need it in the most user-friendly and intuitive ways". Three major axes characterise her approach: information filtering, presentation of the useful information, and interaction dynamics for a spatial augmented reality system.

Alessandro De Gloria is full professor of electronics at the University of Genoa. He is the leader of the ELIOS (Electronics for the Information Society) laboratory of the Department of Electronics and Biophysical Engineering at the University of Genoa. His research interests include virtual reality, mobile computing, human-computer interaction, and embedded system design. He has participated and directed several European research projects on these items. Professor De Gloria has authored over 100 papers in international journals and conferences. He is a member of the IEEE.

José Eustáquio Rangel de Queiroz received his electrical engineering degree from the Federal University of Paraíba, Brazil, in 1986 and the DSc degree in electrical engineering from the Federal University of Paraíba (UFPB), Brazil, in 2001. He was a senior engineer at the Brazilian National Institute for Space Research (Instituto Nacional de Pesquisas Espaciais - INPE), from 1986 to 1999, and a lecturer at the Computer Science Department at the Federal University of Campina Grande (UFCG), Brazil, since 2002. His research interests include human-computer interaction, digital image processing and GIS applications.

Li Deng received the bachelor's degree from the University of Science and Technology of China and received the PhD degree from the University of Wisconsin-Madison. In 1989, he joined the Department of Electrical and Computer Engineering, University of Waterloo, Ontario, Canada and became a full professor in 1996. In 1999, he joined Microsoft Research, Redmond, WA as a senior researcher, where he is currently a Principal Researcher. He is also an affiliate professor in the Department of Electrical Engineering at University of Washington, Seattle. He has published over 250 refereed papers in leading international conferences and journals, 12 book chapters, and has given keynotes, tutorials, and lectures worldwide. He has been granted over a dozen US or international patents. He authored two books in speech processing. He is fellow of the IEEE and fellow of the Acoustical Society of America.

Anind Dey is an assistant professor in the human-computer interaction Institute at Carnegie Mellon University. Anind conducts research at the intersection of ubiquitous computing and human-computer interaction. In particular, he has been performing research in context-aware computing for over a decade, looking at issues of end-user control and feedback, application development and tools for building context-aware applications, privacy and information overload. Before joining CMU, Anind was a senior researcher with Intel Research in Berkeley and an adjunct professor at the University of California, Berkeley. He received his PhD in 2000 in computer science from the Georgia Institute of Technology.

Alan Dix is professor in the Department of Computing, Lancaster University, UK. His interests include delays and temporal issues, interaction in mobile and ubiquitous systems, the use of formal methods in HCI, the design of cyberspace and the way information is transforming economics and society, bits of e-learning ... and just about everything.

Gavin Doherty is a lecturer in computer science at Trinity College Dublin and was principal investigator on the HEA "Virtually Healthy" project. Having originally qualified in computer science at TCD he obtained his doctorate in the Human Computer Interaction Group at the University of York. He later worked at CNR Istituto CNUCE, and the Rutherford Appleton Laboratory. His research interests include the application of technology to mental health interventions, and situations where standard approaches to user-centred design are not feasible. He is also interested in the design of visual information representations and user interactions with mobile technology.

Emmanuel Dubois achieved his PhD at the University of Grenoble in 2001 and is now lecturer at the University of Toulouse-IUT Tarbes-France. He is member of the Research Institute in Computer Science of Toulouse-France (IRIT). His main research domain is related to design methods and tools for augmented and mixed reality systems. His approach is based on the modeling of the different facets that form the user's interaction with such systems.

Mark Dunlop is a senior lecturer in the Department of Computing and Information Sciences at the University of Strathclyde in Glasgow, Scotland. He has investigated many aspects of mobile technology including visualisation of complex information, text entry and use of mobile devices to support lectures. He is a member of the international steering committee for MobileHCI and the editorial board of Personal and Ubiquitous Computing, and an associate editor of *Advances in Human Computer Interaction*. Prior to Strathclyde, Mark was a lecturer at The University of Glasgow and a senior researcher at Risø National Laboratory, Denmark.

Parisa Eslambolchilar received the BS degree in computer engineering from the University of Amirkabir (Tehran Polytechnic) in 1998, an MS degree in electrical engineering from the University of Tehran in 2001, and a PhD degree in computer science in 2006. She was a research fellow at the Hamilton Institute, National University of Ireland in Computer Science until 2007 and now works in the Future Interaction Technologies Laboratory at the University of Wales, UK.

Danilo de Sousa Ferreira is a masters student at the Federal University of Campina Grande, currently researching usability evaluation methods for mobile devices. He holds a Bachelor degree in computer science from the same university and has taken part in a fair number of usability evaluations of interactive systems. His main research interest is the usability evaluation of mobile devices.

Bob Fields is a lecturer in the School of Computing Science at Middlesex University, where he is a member of the Interaction Design Centre. His research is concerned with developing understandings of people's interactions with technology and with one another, and using such understandings in the design of new products, systems, and services. This research has investigated collaborative working and human-computer interaction in a number of domains, from aviation and air traffic control to mobile devices and domestic technology.

Georgios Fiotakis was born in Chania, Greece (1977). He holds a degree in electrical and computer engineering from the University of Patras, Greece. He is currently a PhD candidate at the same University pursuing doctoral research in usability engineering. His research interests lie in methods and tools for usability evaluation of mobile and collaborative applications. He is the main designer and developer of the ColAT and ActivityLens usability evaluation tools.

Clifton Forlines is a researcher at Mitsubishi Electric Research Laboratories. His research interests include: the design and evaluation of novel user interfaces, including digital video presentation; collaborative tabletops; multiuser, multidisplay workspaces; and hand-held projectors for augmented reality. Forlines has a masters of Human-Computer Interaction and a masters of Entertainment Technology from Carnegie Mellon University, and is pursuing a PhD in computer science at the University of Toronto.

Peter Fossick is a results-driven professional product designer with expertise in product, new media, and interaction design. He is a full-time academic in the UK. As a designer and as an academic he has worked for sustained periods in Europe, the USA, and SE Asia. He has wide ranging experience in developing high quality innovative, user centric products and services that have received critical acclaim and commercial success. As an academic he has been responsible for developing and leading a wide range of design courses at undergraduate and postgraduate level. As an active researcher he is interested in exploring service and interaction design issues, innovation strategies, e- manufacturing and mass-customisation in product design. He has a BA (Hons) from Manchester Polytechnic and an MSc in CAED from Strathclyde University.

Silvia Gabrielli got a PhD in cognitive sciences from the University of Padova (I) and since 1994 has been working in the field interaction design and usability evaluation of ICT. In recent years, her research focus has been on methodologies for usability and accessibility evaluation of mobile applications, as well as on the design and evaluation of educational environments. Silvia is currently working as a research fellow at HCI Lab (University of Udine, Italy).

Nestor Garay-Vitoria received a BSc degree in informatics (1990), an MSc degree in computer science and technology (1992,) and a PhD in computer science (2000) all from the University of the Basque Country-Euskal Herriko Unibertsitatea, UPV-EHU (Spain). His PhD dissertation received the University's Extraordinary Award in October 2002. Since 1991 he has been working in the Laboratory of Human-Computer Interaction for Special Needs, where he has collaborated on several research projects. These projects are mainly devoted to the alleviation of communication problems for people with severe motor disabilities. From 2004 onwards, he has led several projects related to affective computing. Currently he is an associate professor at the Computer Architecture and Technology Department of the UPV-EHU.

Luis Gardeazabal is an associate professor of the Department of Computer Architecture and Technology of the University of the Basque Country-Euskal Herriko Unibertsitatea, UPV-EHU (Spain). He was a co-founder of the Laboratory of Human-Computer Interaction for Special Needs (1985). He holds a BSc in physics (1981) from the University of Navarra. He also holds a BSc in engineering (1984) and a PhD in informatics for his work on Alternative and Augmentative Communications Systems for People with Disabilities (1999), both from the UPV-EHU. His main research area is the development of methodologies for special interaction devices for communication and mobility aids.

Florence Gaunet is a researcher in the Centre National de la Recherche Scientifique (CNRS), France. She is a cognitive psychologist and has been studying spatial representations in blind people for more than 10 years. She is experienced in human factors for human-computer interface design, and has an interest in spatial wayfinding by blind pedestrians and in the design of navigational aids for visually impaired persons.

David Gibson conducts research at the University of Vermont College of Engineering and Mathematics and is Project Investigator for the National Science Foundation ITEST project "The Global Challenge", a technology-based outreach program for high school students interested in solving global problems through science, technology, engineering, and mathematics. His recent edited book *Games and Simulations in Online Learning* presents research and development frameworks for the field.

Tiong Goh is currently a lecturer with the School of Information Management at Victoria University of Wellington, New Zealand. His research areas include mobile learning, mobile games for education, and mobile database applications development.

Philip Gray is a senior lecturer in the Computing Science Department at the University of Glasgow, Scotland. He has been actively engaged since 1983 in research into models, notations, software architectures and tools for user interfaces. Recently, he has focussed on the description and engineering of interaction techniques for mobile and ubiquitous systems. He is a director of Kelvin Connect Ltd, a small Scottish software company specialising in mobile information systems. He is also vice-chair of of IFIP Working Group 2.7/13.4 (User Interface Engineering) and the UK representative on IFIP Technical Committee 13 (Human Computer Interaction).

Cathal Gurrin is a senior postdoctoral researcher in the School of Computing, Dublin City University. He holds a PhD from DCU (2002) as well as a BSc degree, also from DCU. His PhD was in search engine algorithms and his current research is on developing, implementing, and evaluating efficient techniques for the management of all kinds of digital information including text, image, video, and even genomic data. He has published 44 journal and conference papers and posters, and in addition he serves on many international program committees annually.

Jonna Häkkilä works as a principal scientist at Nokia Research Center, Finland, doing research on mobile user experience. She received her PhD in 2007 from the University of Oulu, Finland. Her PhD research considered usability issues with context-aware mobile applications, and her current research activities include mobile context-awareness, haptic user interfaces, and mobile multimedia.

Dong-Han Ham is senior research fellow in the school of computing science at Middlesex University, United Kingdom. He received his BS in industrial engineering in 1989 from InHa University, and MS and PhD in industrial engineering in 1995 and 2001 respectively from Korea Advanced Institute of Science and Technology (KAIST). He was senior researcher in software & systems engineering at the Electronics and Telecommunications Research Institute (ETRI). His research interests include human-computer interaction joint with intelligent systems, human interaction with automation, software and systems engineering, human-computer interaction, and agent-oriented information systems.

Antonio Haro is a senior research engineer at D4D Technologies. He has worked in the areas of medical graphics, mobile computer vision, image and video based rendering, and augmented reality. He has been active in both computer vision and graphics research, on topics ranging from realistic skin microstructure detail capture and rendering using GPUs to real-time highly robust eye tracking on commodity hardware. He received his doctorate from Georgia Tech in 2003.

Bret Harsham joined MERL in 2001 to pursue interests in speech and multi-modal interfaces and speech-centric devices. Prior to joining MERL, Bret spent four years at Dragon Systems designing and implementing handheld and automotive speech products. Earlier, he was a principal architect of a Firewall and Virtual Private Network product. Bret has a BS in computer science and engineering from MIT.

Lois Wright Hawkes is a professor of computer science, associate dean of the College of Arts & Sciences at Florida State University, and Directory of the High-Performance Computing & Simulation Lab at the FAMU-FSU College of Engineering. Her PhD is from the University of London in electrical engineering (communications). Her research interests include the use of artificial intelligence and technology in education, fault tolerance, and high speed data transfer.

Gillian R. Hayes is a PhD candidate in the College of Computing and GVU Center at the Georgia Institute of Technology. Her research focuses on enabling capture of rich data in informal and unstructured settings and allowing access to that data. She received her BS in computer science and mathematics from Vanderbilt University. She's a member of the IEEE Computer Society and the ACM.

Paul C. Hébert, a critical care physician and clinical researcher, holds the rank of full professor in the Department of Medicine (Critical Care), University of Ottawa, with cross-appointments in the Departments of Epidemiology and Community Medicine, Anesthesiology and Surgery. Currently, he is vice-chair research of the Department of Medicine, University of Ottawa. Dr. Hébert has received several research awards, including "Researcher of the Year Award" from the Ottawa Health Research Institute, and the "Premier's Research Excellence Award" from the Ontario Ministry of Health (2002). With over 180 publications, his current interests include blood transfusion and conservation in critical care units.

Jeongyun Heo is senior researcher of the Mobile UI R&D Lab at LG Electronics Inc. She has received her MS in industrial engineering and BS in chemical engineering in 1991 and 1994, respectively, from the Korea Advanced Institute of Science and Technology (KAIST), where she is preparing her doctoral thesis. Her research has focused on user experience design which could link customer satisfaction to product design, and systematic approaches to achieving this. She is also interested in physical user interaction (PUI), usability issues related with the physical design, the methods of user research to catch the unmet needs considering usage context and culture difference, and usability of usability evaluation for quality improvement.

Steven Herbst has delivered competitively differentiated user experiences in domains ranging from mobile devices, to web applications, to medical equipment for over 15 years. Herbst leads the internationally-based Strategic Design Research team within Motorola's Consumer Experience Design organization. His team is responsible for revealing the underlying, often unarticulated, physical and digital design attributes that influence user satisfaction—and applying this knowledge to new products and platforms. Herbst has held positions at several other companies including Philips, AT&T, Lucent, and Prodigy. Herbst holds several user interface patents, and his research has influenced the designs of dozens of highly successful products.

Rune T. Høegh is a PhD student in the Human Computer Interaction Research Group of the Department of Computer Science at Aalborg University. His research focus is on how to improve the impact of usability activities in the software development process. His PhD project is called "Enhancing the Interplay Between Usability Evaluations and Software Design".

Giovanni Iachello is currently an associate at McKinsey & Co., focusing on IT and media strategy. He earned a PhD in computer science from Georgia Institute of Technology in 2006, with research on privacy and human-computer interaction. Previously, he worked as consultant in a Web services firm. He holds a Laurea from Padua University (Italy) and is member of the ACM, IEEE, IFIP WG9.6/11.7 and SIGCHI.

Samuel A. Inverso is a doctoral candidate in the Human Brain Dynamics Lab under the Visual Sciences Group in the Research School of Biological Sciences at the Australian National University in Canberra, Australia. He received a BS and MS degree in computer science from the Rochester Institute of Technology in Rochester, NY in 2001 and 2004, respectively, and was a research associate at the Media Lab Europe in Dublin, Ireland from 2003 to 2005. Sam's research interests thread computer science and cognitive neuroscience, and include brain-computer interfaces, subtle communication interfaces, and feedback in the cortical hierarchy.

Stephen Johnson graduated with an honours degree in electronic communications engineering from the University of Hull. He joined BT laboratories in 1989 to research statistical speech recognition algorithms and subsequently went on to develop innovative interactive voice demonstration systems. Stephen now works in BT's Mobility Research Centre where he is investigating context awareness as an enabler for mobile applications and services. Stephen is a chartered engineer and a member of the IET.

Gareth Jones is a senior lecturer in the School of Computing at Dublin City University. He obtained a BEng in electrical and electronic engineering in 1989 and a PhD examining the Application of Linguistic Models in Continuous Speech Recognition in 1994, from the University of Bristol. He was a member of the Speech, Vision and Robotics Group, Department of Engineering and Computer Laboratory, University of Cambridge, and lecturer at the University of Exeter. He was a research fellow at the Toshiba R&D Centre, Japan, at the Informedia Project at Carnegie Mellon University, and at the National Institute of Informatics, Japan.

Eija Kaasinen is head of the Media and Mobile Usability research team at VTT, Technical Research Centre of Finland. She has led field studies in several research projects concerning the development of mobile Internet and personal navigation services. Based on the integrated results of those studies, Eija Kaasinen has defined a framework to study user acceptance of mobile services. The framework, the technology acceptance model for nobile services, is currently being applied in several projects. In her recent research activities, Eija Kaasinen has been applying the model to study user acceptance of ubiquitous services.

Anne Kaikkonen is a usability specialist in Nokia Multimedia. She has over 15 years experience in using psychological knowledge in design and development; she has worked in Nokia for 9 years.

Titti Kallio, MA, serves as TeliaSonera's senior expert of design & usability. She has 20 years of experience in the usability area.

Achilles D. Kameas received his engineering diploma (1989) and his PhD in human-computer interaction (1995) from the Department of Computer Engineering and Informatics, Univ. of Patras, Hellas. Since 2003, he has been an assistant professor with the Hellenic Open University, where he teaches software design and engineering. He is also R&D manager with Research Academic Computer Technology Institute (CTI), where he is the head of Research Unit 3 (Applied Information Systems) and the founder of DAISy group (http://daisy.cti.gr). His research interests focus around the design of Ubiquitous Computing and Ambient Intelligence systems, Interaction models and the uses of Ontologies. On these topics he has co-edited 2 books, published over 40 papers, and co-organized several international conferences and workshops. He served as an elected member of the Disappearing Computer steering group and the Convivio network steering group. He is a member of ACM, IEEE CS and Hellenic AI Society.

Anu Kankainen, PhD, is a senior user experience specialist at Idean Research in Espoo, Finland. Anu is a psychologist with 13 years of experience in working together with engineers, designers and marketing people in several international and national projects focusing on mobile technologies.

Amy K. Karlson received her BA and MS in computer science from Johns Hopkins University, and holds an MS in computer science from the University of Maryland, College Park, where she is currently a PhD candidate with the Human-Computer Interaction Lab in the Department of Computer Science. Ms. Karlson's research focuses on developing interfaces and interaction methods to improve information access under the constraints of mobile computing.

Osamu Katai received his BE, ME and PhD degrees in 1969, 1971 and 1979, respectively, from the Faculty of Mechanical and Precision Engineering, Kyoto University. He has been with Kyoto University from 1974. Now, he is a professor at the Dept. of Systems Science, Graduate School of Informatics. From 1980 to 1981, he had been a visiting researcher at National Research Institute for Information Science and Automation (INRIA), France. His current research interests are on the harmonious symbiosis of artificial systems with natural systems including ecological design, ecological interface design, environmental spatial design, community design, etc.

Hiroshi Kawakami received his BE, ME and PhD degrees in 1987, 1989, and 1994, respectively, from the Faculty of Precision Engineering, Kyoto University. From 1989, he was an instructor at the Department of Information Technology, Faculty of Engineering, Okayama University. Since 1998, he has been an associate professor with the Department of Systems Science, Graduate School of Informatics, Kyoto University. His current research interests are on ecological and emergent system design, the co-operative synthesis method and knowledge engineering.

Aki Kekäläinen, BSc, has proudly tackled usability and user experience puzzles offered by a variety of devices and services for 8 years. He is currently working as a usability specialist at TeliaSonera.

Juha Kela is a VP of Finwe Ltd., formerly a research scientist at VTT, and a doctoral student in the University of Oulu's Department of Electrical and Information Engineering. His research interests include multimodal and mobile interaction. He has an MSc in information processing from the University of Oulu.

Julie A. Kientz is a PhD student in the College of Computing and GVU Center at the Georgia Institute of Technology. She is also an assistant director of the Aware Home Research Initiative. Her research focuses on collaborative access to individual experiences. She received her BS in computer science and engineering from the University of Toledo. She's a member of the IEEE Computer Society and the ACM.

Stephen Kimani is currently an academic and research member of Jomo Kenyatta University of Agriculture and Technology (Kenya) and is affiliated with the University of Rome "La Sapienza." He has been a post-doctoral researcher with the University of Rome "La Sapienza" (2004-2006). He holds a PhD in computer engineering (University of Rome "La Sapienza," Italy) and MSc in advanced computing (University of Bristol, UK). His main research interest is in human-computer interaction (HCI). In particular, as HCI relates to areas/aspects such as: user interfaces, usability, accessibility, visualization, visual information access, visual data mining, digital libraries, and ubiquitous computing.

Kinshuk is professor and director of the School of Computing and Information Systems at Athabasca University, Canada. He has been involved in large-scale research projects on adaptive educational environments and has published over 200 research papers in international refereed journals, conferences and book chapters. He is chair of the IEEE Technical Committee on Learning Technology and editor of the SSCI indexed *Journal of Educational Technology & Societ*y (ISSN 1436-4522).

Jesper Kjeldskov is associate professor in Computer Science at Aalborg University, Denmark. Jesper is doing research into "indexical" interaction design for context-aware mobile and pervasive computer systems and exploring new methods and techniques for studying the use of mobile and pervasive computer technologies in laboratory and field settings. He holds a PhD in computer science and engineering and a master of HCI and Sociology from Aalborg University.

Scott Klemmer is an assistant professor of computer science at Stanford University, where he co-directs the Human-Computer Interaction Group. He received a dual BA in art-semiotics and computer science from Brown University in 1999, and an MS and PhD in computer science from UC Berkeley in 2001 and 2004, respectively. His primary research focus is interaction techniques and design tools that enable integrated interactions with physical and digital artifacts and environments. He is a recipient of the UIST 2006 Best Paper Award and the 2006 Microsoft Research New Faculty Fellowship.

Vanja Kljajevic is a cognitive scientist who is currently a guest researcher at NewHeights Software, Ottawa, and a research associate with the Games & Media Group, Human Oriented Technology Lab, Carleton University, Ottawa. Her diverse interests include human-computer interaction, usability, serious games, and language processing in intact and impaired brains.

James F. Knight was awarded his PhD in the ergonomics of Wearable Computers at The University of Birmingham in 2002 and he has published widely on this topic. He currently works as a research fellow on projects funded by the UK Ministry of Defence and the European Union, at the University of Birmingham with interests in the effects of wearing and interacting with wearable computer devices on the human body.

Vassilis Komis was born in Viannos, Crete, Greece (1965). He holds a degree in mathematics from the University of Crete (1987), DEA (1989) and doctoral degrees (1993) in Teaching of Computer Science from the University of Paris 7 - Denis Diderot (Jussieu). He is currently associate professor in the Department of Educational Sciences and Early Childhood Education of the University of Patras. His publications and research interests concern the teaching of computer science, the pupils' representations in the new information technologies and the representations formed during the use of computers in class room, the integration of computers in education, and the conception and development of educational software.

Irina Kondratova is group leader of the People-Centred Technologies Group at the National Research Council of Canada Institute for Information Technology. She is a graduate of, and adjunct professor at, the University of New Brunswick's Engineering Faculty. Dr. Kondratova is a professional engineer with the Association of Professional Engineers for the Province of New Brunswick and has more than 25 years of research, business, and consulting experience. Dr. Kondratova does research work in the area of voice and multimodal technologies that allow the use of speech, along with traditional keyboard or stylus, to interact with computers. These technologies improve usability of mobile devices and allow for better data gathering in the field, or in emergency situations.

Siu Cheung Kong is an associate professor in the Department of Mathematics, Science, Social Sciences and Technology at the Hong Kong Institute of Education. Dr. Kong received a Doctor of Philosophy from the Department of Computer Science of the City University of Hong Kong. Dr. Kong publishes in the fields of information technology in education, information literacy education, collaboration in mobile learning and cognition, and technology in mathematics education. He is currently a member of the Steering Committee on the Strategic Development of Information Technology in Education that was set up by the Government of the Hong Kong SAR.

Panu Korpipää is a CEO of Finwe Ltd., formerly a senior research scientist at VTT. His research interests include enabling situation-aware and sensor-based interaction and applications in mobile computing. He received his doctoral Dr Tech in information processing from the University of Oulu.

Peter Kroeger has had a keen interest in computers and software development since 1985. He graduated from the University of Waterloo with an Honours Bachelor of Mathematics in Computer Science and a minor in Fine Arts (1993). His experience includes nine years at Corel Corporation where he was a project leader producing Web technology features in most of Corel's products. Since Corel, Peter has worked as a software architect and development manager. Currently Peter is a software architect at BRYTECH engaged in medical client-server and human interface technology. Peter founded the Software Professionals Society of Canada in 2003.

Fabrizio Lamberti received his degree in computer engineering and his PhD degree in software engineering from the Politecnico di Torino, Italy, in 2000 and 2005, respectively. He has published a number of technical papers in international journal and conferences in the areas of mobile and distributed computing, wireless networking, image processing and visualization. He has served as a reviewer and program or organization committee member for several conferences. He is member of the editorial advisory board of international journals. He is a member of the IEEE Computer Society.

James Landay is a professor in computer science & engineering at the University of Washington, specializing in HCI. Previously he was director of Intel Research Seattle, a lab exploring ubiquitous computing. His current interests include Automated Usability, Demonstrational Interfaces, Ubicomp, Design Tools, and Web Design. Landay received his BS in EECS from Berkeley in 1990 and his PhD in CS from Carnegie Mellon in 1996. His dissertation was the first to demonstrate sketching in UI design tools. He was also co-founder of NetRaker. In 1997 he joined the faculty in EECS at Berkeley, leaving as an associate professor in 2003.

Hyowon Lee is a postdoctoral researcher in the Centre for Digital Video Processing at Dublin City University where he obtained a PhD in user-interface design issues in keyframe-based video content browsing in 2001. Previously he obtained a BEng in computer science in Soong-Sil University, Korea in 1995, then a MSc in information & library studies at the Robert Gordon University, UK (1996). His area of research is the interaction design for multimedia retrieval systems such as image & video searching/browsing applications and photo & video blogging applications, and the evaluation and user studies involving these applications.

Jaakko Lehikoinen is a research team leader in the Nokia Research Center. He has been with Nokia since 1999 mostly working on and managing user-centered design oriented projects and programs. As a sociologist, he has been focusing on developing user research methodologies in such a way that the results can be utilized in concept creation and design. In addition to privacy, Jaakko's main research interests are social interaction and user experience design.

Rock Leung is a PhD student in the University of British Columbia's Department of Computer Science. He is pursuing research in human-computer interaction, focusing on interface design, mobile computing, and universal usability. His research interests include exploring how to improve computer interfaces and designing innovative assistive technologies for populations with special needs.

Dieter Leyk is professor at the German Sports University Cologne and the head of the Department IV–Military Ergonomics and Exercise Physiology at the Central Institute of the Ferderal Armed Forces Medical Services Koblenz. He studied both medicine and sports sciences and received the MD and PhD degrees. His main research topics are performance diagnostics, exercise physiology, human psychological performance and ergonomics.

Yang Li is a research associate of Computer Science & Engineering at the University of Washington. He was a postdoctoral researcher in EECS at UC Berkeley. He received a PhD in computer science from the Chinese Academy of Sciences in 2002. His research focus is primarily in user interface design tools, activity-based computing, and pen-based user interfaces.

Min Lin is a PhD candidate at UMBC. He received his MS in information systems from UMBC in 2002 and his MS in biology from the University of Science and Technology of China in 1995. Currently, he is a Usability Engineer at Hillcrest Laboratories Inc., where he is responsible for planning, conducting, and analyzing user studies for both hardware and software products. Before joining Hillcrest Labs, he had multiple years of research experience in both cognitive science and human-computer interaction.

Gitte Lindgaard, a full professor in psychology at Carleton University in Ottawa, holds the NSERC/Cognos chair in user-centred design in the Human Oriented Technology Lab (HOTLab) of which she is director. Formerly, as the principal scientist, head of the Human Factors Division at Telstra Research Laboratories, Australia before joining Carleton University, her job was to ensure that all Telstra's systems, services, and products were usable. Her research interests include human decision making, affective- as well as multimodal/multimedia computing, and graph comprehension. She has contributed several hundred peer-reviewed articles and book chapters. She serves on several HCI journal editorial boards.

Jukka Linjama, born in 1959, received his Dr Tech in electrical engineering in 1994, from Helsinki University of technology. His research interests span from acoustics, structural dynamics, and human perception of sound and vibration, to human-computer interaction, haptics and multi-modal interaction. Currently he acts as technology architect at Nokia Technology Platforms, responsible for specifying sensor technology and interaction solutions for future user interfaces in mobile devices.

Yanfang Liu leads the Strategic Design Research team within Motorola's Consumer Experience Design organization. Her team is responsible for design research on physical and digital aspects of mobile devices targeted to Asian markets, including Asian character entry. In her current role, Yanfang has contributed to several innovative character entry solutions that have been successfully delivered to market. Prior to joining the mobile devices business, Yanfang was a member of Motorola Labs' Human Interface Lab. Yanfang received her PhD in applied psychology from the Institute of Psychology at the Chinese Academy of Sciences in 1998.

Bob Longworth has recently completed his Honours degree in computer science at the University of New Brunswick. During his university career he completed five coop work terms three of which were with the National Research Council of Canada (NRC). Bob has interests in pharmaceutical drug design and has developed an interest toward Mobile HCI through his work at the NRC. Bob was also actively involved with his faculty in several ways. He was a peer mentor for three years, president of the Computer Science Association for two years, student representative on the faculty council for two years and was grad class rep.

Rosemarijn Looije received her masters degree in artificial intelligence/man machine interaction in 2006 from the University of Groningen. She is now a research member of the Intelligent Interface group at TNO Defence, Security and Safety in Soesterberg, The Netherlands. Her research interests lie within human-computer interaction, most notably the usability of devices, the dialogue with devices, and the influence of affect on the dialogue.

Saturnino Luz holds a PhD degree in informatics from the University of Edinburgh. Over the past years, he has worked on the development of human-computer interface technology in the areas of mobile systems and computer-supported collaborative work. His research interests also include natural language processing, information visualisation and retrieval, and machine learning. Dr Luz has served in the program committee of several international conferences and the editorial board of an international journal. He has been a member of the Association for Computing Machinery (ACM) since 1994 and contributes regularly to ACM's Computing Reviews.

Pattie Maes is an associate professor in MIT's Program in Media Arts and Sciences. She founded and directs the Media Lab's Ambient Intelligence research group, which focuses on linking the information world into the physical world around us. Previously, she founded and ran the Software Agents group. Prior to joining the Media Lab, Maes was a visiting professor and a research scientist at the MIT Artificial Intelligence Lab. She holds bachelor's and PhD degrees in computer science from the Vrije Universiteit Brussel in Belgium. Her areas of expertise are human-computer interaction and artificial intelligence. Maes is the editor of three books, and is an editorial board member and reviewer for numerous professional journals and conferences. She has received several awards: *Newsweek* magazine named her one of the "100 Americans to watch for" in the year 2000; TIME Digital selected her as a member of the Cyber-Elite, the top 50 technological pioneers of the high-tech world; the World Economic Forum honored her with the title "Global Leader for Tomorrow"; Ars Electronica awarded her the 1995 World Wide Web category prize; and in 2000 she was recognized with the "Lifetime Achievement Award" by the Massachusetts Interactive Media Council.

Massimiliano Margarone is a research consultant at the Department of Biophysical and Electronic Engineering of the University of Genova. He received his masters degree in electronic engineering from the University of Genova, in 1999, with a thesis on asynchronous digital filter development for wireless devices. His current research interests include Web usability, human-computer interaction, and mobile computing.

Masood Masoodian holds a PhD in computer science from the University of Waikato, New Zealand. His research interests include human-computer interaction, interaction design, visualisation, mobile computing and usability. Dr Masoodian has been involved in design, development, and evaluation of numerous graphical user interfaces and visualisation systems for devices ranging from large interactive displays to small handheld devices. He has served as the programme chair, programme committee member, and reviewer for many international conferences and scientific journals in HCI.

Mark Matthews is a PhD candidate with the Department of Computer Science at Trinity College Dublin. His research interests include the design of mobile software for therapeutic activities and developing new approaches to user-centred design for adolescent mental health. He is interested in the design of technology to engage adolescents between therapy sessions and in the use of technology, such as computer games, to develop therapeutic relationships.

Peter Michael Melliar-Smith is a professor in the Department of Electrical and Computer Engineering at the University of California, Santa Barbara. His research interests include distributed computing, network protocols, and fault tolerance. He has published more than 240 conference and journal papers.

Eduardo Mena earned a BS degree in computer science from the University of the Basque Country and a PhD degree in computer science from the University of Zaragoza. He is an associate professor in the Department of Computer Science and Systems Engineering at the University of Zaragoza, Spain. For a year he was a visiting researcher in the Large Scale Distributed Information Systems Laboratory at the University of Georgia. He leads the Distributed Information Systems research group at his university and is a member of the Interoperable Database group at the University of the Basque Country. His research interest areas include interoperable, heterogeneous and distributed information systems, semantic Web, and mobile computing. His main contribution is the OBSERVER project. He has published several papers in international journals, conferences and workshops, including a book on ontology-based query processing. He also has served as a referee for several international journals, and as a program committee member for several international conferences and workshops.

Kristijan Mihalic is a researcher in the HCI & Usability Unit at the University of Salzburg. He received a master's degree in the interdisciplinary field of communication sciences and computer science and holds a doctorate in HCI from the University of Salzburg, Austria. He was co-chair of the MobileHCI 2005 conference, and has co-organized several workshops on context and mobile HCI. Kristijan is involved in several national and international research and industry projects. In addition, he has a sound background in software development and several years of experience in teaching human-computer interaction and telecommunication technologies at national and international institutions. His primary research involves mobile and contextual interfaces, social aspects of mobile systems, and ambient intelligence. He lives on coffee.

John Miller has over 40 years experience in telecommunications and electronics design. He began his career with British Telecom and has held engineering and management positions with a number of companies including Canadian Marconi and Mitel. After completing his PhD in 1988 at Southampton University in England, he returned to Canada to lead a team at the University of Ottawa Heart Institute designing electronics for an artificial heart assist device. He received the 1993 Medforte Innovation Award, American Society of Artificial Organs, for his work on transcutaneous energy transfer. Currently John is VP of product development at BRYTECH Inc.

Nikola Mitrovic earned an MPhil degree in computer science from the University of Zaragoza, Spain. He is currently working toward a PhD degree in the area of intelligent and adaptive user interfaces. Nikola's research interest areas include interoperable, heterogeneous and distributed information systems and mobile computing. He has published several papers in international conferences and workshops and has served as a reviewer for several international conferences.

Michelle Montgomery Masters is a researcher in the Department of Computing and Information Sciences at the University of Strathclyde in Glasgow, Scotland, and is usability director for PyrusMalus™ software design company. She has expertise in designing and conducting usability evaluations on web, desktop and mobile applications. Michelle has a history of research into computer support for education, her current research focus is designing mobile learning systems. Michelle also acts as reviewer for many of the leading HCI publications and conferences. Prior to Strathclyde, Michelle was a researcher at The University of Glasgow.

Louise Elizabeth Moser is a professor in the Department of Electrical and Computer Engineering at the University of California, Santa Barbara. Her research interests span the fields of computer networks, distributed systems, and software engineering. She has published more than 225 conference and journal publications.

Wafaa Abou Moussa is a PhD student at the Research Institute in Computer Science of Toulouse–France (IRIT). Currently he conducts research on distributed 3D interaction and model-driven simulation of mobile mixed systems. He received his electronics engineering diploma from the "Lebanese University" in 2004 and his masters degree in computer science from the "Paul-Sabatier University." France in 2005.

Roderick Murray-Smith received degrees (BEng '90, PhD '94) from the University of Strathclyde, UK. He worked at Daimler-Benz research labs in Berlin from 1990-97, was a visiting researcher in the Dept. of Brain & Cognitive Sciences at MIT in 1994, a research fellow in the Dept. of Mathematical Modeling at the Technical University of Denmark from 1997-99, and since 1999 has been a reader at

Glasgow University in the Department of Computing Science. In 2001 he took up a joint (senior researcher) position at the Hamilton Institute, at NUI Maynooth, Ireland. He heads the Dynamics & Interaction research group, and his interests are in gesture recognition, mobile computing, manual control systems, Gaussian processes, and machine learning.

Bojan Musizza received his BSc in 2003 from the Faculty of Electrical Engineering, University of Ljubljana. Currently he is a PhD student working at the "Jozef Stefan" Institute at the Department of Systems and Control. His research interests are complex oscillatory systems and their interactions. His work mainly focuses on detecting interactions between cardiorespiratory oscillations and brain waves during anaesthesia in humans.

Mark Neerincx is head of the Intelligent Interface group at TNO Defence, Security and Safety, and professor in Man-Machine Interaction at the Delft University of Technology. He has extensive experience in applied and fundamental research. Important results are (1) a cognitive task load model for task allocation and adaptive interfaces, (2) models of human-machine partnership for attuning assistance to the individual user and momentary usage context, (3) cognitive engineering methods and tools, and (4) a diverse set of usability "best practices". He has been involved in the organisation of conferences, workshops and tutorials to discuss and disseminate human factors knowledge.

Shigueo Nomura is presently a JSPS postdoctoral fellow in the laboratory for Theory of Symbiotic Systems, Kyoto University, Japan. He received his BSc, BE, and MSc degrees in 1988, 1992, and 2002, respectively, from ITA at the Aerospace Technical Center, EPUSP at University of São Paulo, and FEELT at Federal University of Uberlândia in Brasil. He won a doctoral scholarship grant from the Japanese Government (Monbukagakusho) in 2002 and received his doctoral degree in 2006 from the Graduate School of Informatics, Kyoto University. His current research interests include human computer interaction, echolocation, pattern recognition, neurocomputing, and morphological image analysis.

Michael O'Grady is an SFI postdoctoral research fellow within the School of Computer Science & Informatics at University College Dublin (UCD). Prior to joining the school, he worked in the commercial software and telecommunications industries. After completing a postdoctoral fellowship in the area of pervasive computing funded by the Irish Research Council for Science and Engineering Technologies (IRCSET), he joined the Adaptive Information Cluster (AIC) group at UCD. His research interests include Ambient Intelligence, mobile multimedia and intelligent systems. He has published some 40 papers in international journals and conferences.

Gregory O'Hare is an SFI investigator and senior lecturer in the School of Computer Science & Informatics at University College Dublin. He was previously a member of faculty at the University of Manchester Institute of Science and Technology (UMIST), UK. He is director of the PRISM (Practice and Research in Intelligent Systems & Media) Laboratory within the School of Computer Science & Informatics. His research focuses upon multi-agent systems (MAS) and mobile & ubiquitous computing. He has published over 200 journal and conference papers in these areas together with two text books.

Kate Oakley completed her doctorate in psychology at Carleton University in 1998. Her dissertation, *Age and Individual Differences in the Realism of Confidence*, focused on decision making and led to a pilot research grant for new investigators from the Alzheimer's Association. Dr. Oakley was subsequently awarded an ORNEC two-year post-doctoral fellow position at the Human Oriented Technology Lab (HotLab), Carleton University, to develop computerized assessment tools in dementia. Currently

NSERC/Cognos research associate at the HOTLab, her research focuses on decision making, and graph comprehension for mobile technology. Kate is also part-time research scientist at the Élisabeth Bruyère Research Institute, Ottawa.

Marianna Obrist graduated in communication and political science at the University of Salzburg. She is a researcher in the HCI & Usability Unit of the ICT&S Center at the University of Salzburg. The focal point of her research lies in human-computer interfaces, user-centered design of interactive services, and in particular the user involvement into the development of new products/systems. She is involved in several research projects concerned with the study of the home environment and the analysis of user requirements for interactive TV. Since September 2006 she is working within the CITIZEN MEDIA project, focusing on co-creation and user experience evaluation in the new user driven media landscape. She was part of the organization team for the MobileHCI 2005 conference. She wrote her PhD on self-motivated adaptation and innovation of interfaces in the context home.

Gustav Öquist is a researcher at the Department of Clinical Neuroscience at Karolinska Institutet, Sweden. He is a computational linguist and his interests include novel concepts and solutions for information access with focus on language, usability, accessibility and mobility. He previously studied at Uppsala University and wrote his PhD thesis about evaluation of readability on mobile devices.

Jeni Paay is an assistant professor at Aalborg University, Denmark, in the Department of Computer Science, where she is a member of the Human-Computer Interaction (HCI) research group. She has completed a multi-discipline PhD in HCI and Architecture from The University of Melbourne, Australia. Her research interests include: pervasive computing in the built environment and the interplay between social interactions and architectural space; indexical interaction design for context-aware mobile computer systems; the representation of context in interface design; and HCI design methods.

Ioanna Papadimitriou was born in Athens, Greece (1978) and holds a degree in early childhood education from the University of Patras (2002), and an MSc in didactics of science (2006). Ioanna is currently a PhD candidate at the same University. Ioanna's research interests include the integration of computers in education and the educational uses of mobile technologies.

Hyungsung Park is a PhD candidate at the Korea National University of Education. His research interests are in using mobile devices for learning, designing educational games, and user interfaces in educational contexts. He has recently published two articles entitled *Development of learning contents in game to support a mobile learning* and *The analysis of the Knowledge Construction Types in Educational on-line Games on the Basis of the Levels of the Self-Regulated Learning* in Korea.

Sanghyun Park is a junior researcher in mobile UI R&D Lab at LG Electronics Inc. Before joining the LG Mobile UI lab in 2005, she also worked in UI Quality assurance Labs at Samsung Electronics Inc. She received an MS in industry engineering from the Korea Advanced Institute of Science and Technology (KAIST) in 2004 and a BS from SungKyunKwan University in 2003. Her research interests center on the user experience design and user research using ethnographic techniques with mobile contexts and industrial environments. She is also interested in physical user interface design and usability evaluation for UI designers to offer valuable user experiences.

Shwetak N. Patel is a PhD student in the College of Computing and GVU Center at the Georgia Institute of Technology. He is also an assistant director of the Aware Home Research Initiative. His research focuses on context-aware mobile phone applications and technology to support ubiquitous computing applications. He received his BS in computer science from the Georgia Institute of Technology. He's a member of the IEEE Computer Society and the ACM.

Antti Pirhonen holds a PhD with a major in educational sciences and minor in information systems. He is a senior researcher in the Department of Computer Science and Information Systems at the University of Jyväskylä Finland. Antti was a visiting researcher at the University of Glasgow's Department of Computing between 2000-2001 and again in 2005. He is the scientific leader of a number of projects funded by the Finnish Funding Agency for Technology and Innovation (Tekes) and the Finnish ICT industry. His research interests concern mobile applications, the use of non-speech audio in user-interfaces, multimodal interaction, and design methodologies.

Frank E. Pollick obtained a PhD in cognitive sciences in 1991 from UC Irvine. From 1991-97 he was a researcher at ATR in Kyoto, Japan. Since 1997 he has been in the Department of Psychology at the University of Glasgow. His research interests include the recognition of human movement.

Bhiksha Raj joined MERL as a staff scientist. He completed his PhD from Carnegie Mellon University (CMU) in May 2000. Dr. Raj works mainly on algorithmic aspects of speech recognition, with special emphasis on improving the robustness of speech recognition systems to environmental noise. Dr. Raj has over seventy five conference and journal publications and is currently writing a book on missing-feature methods for noise-robust speech recognition.

Tapani Rantakokko is a senior SW designer at Finwe Ltd., formerly a research scientist at VTT, and a doctoral student in the University of Oulu's Department of Electrical and Information Engineering. His research interests include mobile user interfaces and pervasive computing. He received his MSc in information processing from the University of Oulu.

Dimitrios Raptis was born in Chalkis, Greece (1980). He holds a degree in electrical and computer engineering from the University of Patras, Greece and is currently a PhD candidate at the same University pursuing doctoral research in the area of mobile usability. His research interests include methods for design of mobile applications, context aware computing, and usability of mobile devices.

Janet Read is a senior lecturer at the University of Central Lancashire in Preston, UK. She is well known for her work on child computer interaction and directs the ChiCI research group at her host institution. One of her areas of interest is in the use of mainstream evaluation methods with children and this research has motivated her interest in Wizard of Oz studies.

Michael Rohs is a senior research scientist with Deutsche Telekom Laboratories at TU Berlin. His primary research interest is in mobile and pervasive user interfaces, with a focus on the integration of physical and virtual aspects of the user's environment. He developed a visual marker system for camera phones and marker-based interaction techniques for passive media as well as electronic displays. He received a PhD in computer science from ETH Zurich, Switzerland, where he was a research assistant at the Institute for Pervasive Computing.

Anxo Cereijo Roibás is senior lecturer at the University of Brighton, visiting lecturer at Westminster University, at the Politecnico di Milano and the National Institute of Design (India). His expertise resides in the user experience in pervasive communication systems. He has been HCI manager at the

Mobile Internet Services Provider, HiuGO SpA, and User Experience Consultant for Vodafone and he has collaborated with the Nokia Research Center. He has coordinated ethnographic research addressing the future use of mobile phones as multimedia tools in collaboration with the Vodafone Group Foundation and the British Royal Academic of Engineering. He has been British Telecom Fellow at the BT IT Mobility Research Centre. Anxo is Events' Chair of the British HCI Group.

Jose A. Royo earned a BS degree in computer science from the University of Zaragoza, Spain, in 2001. In his dissertation he presented the advantages of mobile knowledge driven agents in distributed and heterogeneous systems. He is currently the system analyst of the Electronic Engineering and Communications Department at the University of Zaragoza, Spain. His research interests include: mobile agents; mobile computing; interoperable, heterogeneous and distributed information systems; and semantic Web. He has published several papers in international conferences and workshops, and also has served as a reviewer for some conferences, workshops and journals.

Andrea Sanna graduated in electronic engineering in 1993, and received a PhD degree in computer engineering in 1997, both from Politecnico di Torino, Italy. Currently he has an assistant professor position in the 2nd engineering faculty. He has authored and co-authored several papers in the areas of computer graphics, virtual reality, parallel and distributed computing, scientific visualization and computational geometry. Andrea Sanna is currently involved in several national and international projects concerning grid, peer-to-peer, and distributed technologies. He is a member of ACM and serves as reviewer for a number of international conferences and journals.

Giuseppe Santucci graduated in electrical engineering from the University of Rome "La Sapienza", Rome, Italy, on 1987. Since 1987 he teaches courses in computer science at Italian universities. From 1987 to 1991 he was research assistant at the University of Roma "La Sapienza", where he is now associate professor in the Department of Computer Science. His main research activity concerns both theoretical and practical aspects of visual query language and user interfaces. He is a member of the Steering Committee of the International Workshop on Advanced Visual Interfaces (AVI). He is a member of the Institute of Electrical and Electronics Engineers (IEEE).

Christopher M. Schlick received the Dipl.-Ing. degree in 1992 from Berlin University of Technology, Germany, the Dr.-Ing. and the Habilitation degrees from RWTH Aachen University, Germany, in 1999 and 2004, respectively. He is currently a full professor at the RWTH Aachen Faculty of Mechanical Engineering, where he is the director of the Institute of Industrial Engineering and Ergonomics. He received merits of honor from the German Human Factors Society GfA, RWTH (Borchers insignia 1999) and the Holste Foundation (Holste Price 2004). His research interests include the design and simulation of work and business systems, human-machine systems, and ergonomics.

Bent Schmidt-Nielsen has been developing and testing speech recognition and speech interfaces for automobile applications for the last dozen years. This includes seven years at Dragon Systems and six years at Mitsubishi Electric Research Laboratories. The primary thrust of his work is in the usability of speech interfaces for automobiles. Bent has worked building easy to use computer applications for the last 25 years. However, his academic training is in biology from the University of California at San Diego and the Massachusetts Institute of Technology.

Andrew Sears is a professor and chair of the Information Systems Department at UMBC. Dr. Sears' research explores issues related to human-centered computing with recent projects investigating issues associated with accessibility, mobile computing, and the difficulties IT users experience as a result of their work environments or tasks. He is co-editor of the Human-Computer Interaction Handbook and co-editor-in-chief of the ACM *Transactions on Accessible Computing.* He earned his BS in Computer Science from Rensselaer Polytechnic Institute and his PhD in computer science with an emphasis on human-computer interaction from the University of Maryland, College Park.

Sabine Seymour founded Moondial Inc., which focuses on "the next generation wearables" and the intertwining of aesthetics and function. Sabine introduced the course "Fashionable Technology" at Parsons in New York. She currently curates the exhibition "Cyborgs: Me or Machine" in England. Sabine is writing her PhD dissertation dealing with design and innovation in wearables. She received an MBA from the University of Economics in Vienna and Columbia University in New York and an MPS in Interactive Telecommunications from NYU'S Tisch School of the Arts.

John Sharry is a psychotherapist at the Department of Child and Family Psychiatry, at the Mater Hospital, Dublin, Director of the Parents Plus Charity and was formerly leader of the Therapeutic Technologies research group at Media Lab Europe in Dublin. His specific research interest is exploring the potential of computer technology in the field of psychotherapy with children and adolescents. John is author of nine books including three professional psychotherapy books, *Solution Focused Groupwork* (Sage, 2001), *Becoming a Solution Detective* (BT Press, 2001; Haworth 2003) and *Counselling children adolescents and Families* (Sage, 2004).

Jennifer G. Sheridan is director of BigDog Interactive, a company which creates bespoke interactive installations. Her interest is in digital live art, intersection of human-computer interaction and live art. She directs, designs and develops interactive installations and performance events using sensors and mobile and embedded physical computing technologies to mediate wittingness. She received an MS in HCI from the Georgia Institute of Technology, USA and a PhD in computer science from Lancaster University, UK.

Takayuki Shiose received his BE, ME and PhD degrees in 1996, 1998 and 2003, respectively, from the Faculty of Precision Engineering, Kyoto University. He was engaged as a research fellow with the Japan Society for the Promotion of Science from 1998 to 2000, and as a research associate with the Graduate School of Science and Technology, Kobe University from 2000 to 2002. Also, he was a visiting researcher at ATR Network Informatics Labs from 2002 to 2006. Now, he is an assistant professor at the Graduate School of Informatics, Kyoto University. His current research interests are in assistant systems for a proficient-skill transfer from the viewpoint of ecological psychology.

Katie A. Siek is an assistant professor at the University of Colorado at Boulder in the Department of Computer Science. Her primary research interests are in human computer interaction, health informatics, and ubiquitous computing. More specifically, she is interested in how technology affects medical interventions for underserved populations. Prior to her appointment at Colorado, she completed her PhD and MS at Indiana University, Bloomington in computer science and her undergraduate education in computer science at Eckerd College.

Alexander Sievert is a research associate at the Department of Physiology and Anatomy of the German Sport University Cologne in Cologne, Germany (DSHS). He obtained a Diplom-Sportlehrer (MS) degree and is currently working towards a doctorate in sports sciences. He began working for the DSHS in the area of human physical and psychological performance in 2004. He has also been working as a freelance consultant for local and international companies in the automotive industry and service providers for IT related Education, Training and Development. His main fields of research include work/exercise physiology, and ergonomics as well as human psychological performance.

Mikael B. Skov is associate professor at the Department of Computer Science at Aalborg University, Denmark. His professional interests include various aspects of human-computer interaction (HCI). He primarily conducts research within usability testing and engineering and within design and evaluation of mobile devices. He is involved in research projects on the design and evaluation of awareness-based mobile technologies and is also interested in social aspects of information technology design, evaluation, and use.

Alan Smeaton is a professor of computing and principal investigator in the Adaptive Information Centre at Dublin City University. His background is in text-based information retrieval and his research interests cover all aspects of the management of multimedia information.

Chiwon Song is a junior research fellow at the Mobile Handset R&D Center of LG electronics Company in Seoul, Korea. He received his BS in industrial engineering in 2003 from Ajou University, and MS in industrial engineering in 2003 and 2005, respectively, from Korea Advanced Institute of Science and Technology (KAIST). His research interests include the usability of physical user interfaces (PUI) on mobile devices and the features or areas to extend mobile device user's user experience.

Jan Stage is associate professor at the Department of Computer Science at Aalborg University, Denmark. He conducts research in usability testing and evaluation, design of human-computer interaction, design and evaluation of mobile systems, methods for analysis and design in system development and object-oriented analysis and design. He is currently involved in the following research activities: The USE Project (Usability Evaluation and Software Design: Bridging the Gap), the COST Project MAUSE (Towards the Maturation of Information Technology Usability Evaluation) and a high technology network on mobile systems.

Hanna Stelmaszewska is a PhD student and teaching assistant in the School of Computing Science at Middlesex University, where she is a member of the Interaction Design Centre. Her research interests include applying qualitative methods to investigate users' experience with interactive technology, user interaction and usability of interactive systems, which leads to gaining a better understanding of users' needs and support the design of systems. For her PhD she is evaluating users' experience with camera phones used for social interaction in co-present settings.

Adrian Stoica was born in Braila, Romania (1978). He obtained a degree in economic informatics from the University of Dunărea de Jos of Galaţi, Romania, and is currently a PhD candidate at University of Patras, Greece, pursuing doctoral research in the area of ubiquitous computing. His research interests include the design of middleware for mobile context sensitive computing, interaction design for mobile devices, usability of mobile devices, and distributed user modeling in ubiquitous systems.

Jan Willem Streefkerk has a background in cognitive psychology. Currently he is a PhD student at TNO Defence, Security and Safety and Delft University of Technology in The Netherlands. His research interests focus on the design and evaluation of mobile, adaptive systems and their application in professional, critical domains. Previously, he worked as a lecturer on usability and research methods at The Hague University. He is an active member of CHI Nederland, the Dutch chapter of ACM SIGCHI.

Masanori Sugimoto received BEng, MEng, and Dr Eng degrees from the University of Tokyo, Japan, in 1990, 1992, and 1995, respectively. Currently, he is an associate professor at the Department of Frontier Informatics, Graduate School of Frontier Sciences, University of Tokyo. His research concern is related to human-computer interaction, especially mixed reality, mobile computing, computer supported collaborative work/learning, intelligent systems, and so on. Dr. Sugimoto is a member of ACM, IEEE, IEICE, JSAI, IPSJ, VRSJ, and JCSS.

Khai N. Truong is an assistant professor in the Department of Computer Science at the University of Toronto. His research interests include human-computer interaction and ubiquitous computing. Specifically, he focuses on the design and development of ubiquitous computing applications and then the evaluation of these systems' impact on daily life. He received his PhD in computer science from the Georgia Institute of Technology.

Nikolaos Tselios was born in Athens, Greece (1973). He obtained his Phd (2002) and a degree (1997) in electrical and computer engineering from the University of Patras, Greece. Currently, he is an adjunct lecturer and postgraduate fellow at the University of Patras, Greece. His main interests are related to design and evaluation of software interaction systems (with an emphasis in educational systems), design and evaluation of distance learning systems, cognitive models, context-aware computing systems and artificial intelligence techniques for human computer interaction.

Maria de Fátima Q Vieira Turnell received her Degree in physics from the UFPE in Brazil, in 1975, her masters degree in electrical engineering in 1979 from UFPB, and a PhD in electrical engineering from Bradford University, in the UK, in 1986. She is a senior lecturer at the Electrical Engineering Department of UFCG since 1977. She works in the undergraduate and graduate levels both in the Electrical Engineering Department and the Computer Science Department. Her main research interest is the design and evaluation of human interfaces for automated systems and in particular for mobile devices.

Myra van Esch-Bussemakers holds a masters degree in cognitive science and a PhD in multimodal user interfaces. Her research interests include usability, adaptive mobile user interfaces, multimodal interfaces and innovative user interfaces in complex, high-demand environments. Both as a scientist and as a project manager at TNO Defence, Security and Safety in The Netherlands, she is involved in projects that are more research-oriented as well as projects that are more applied, for instance in developing decision aids for medical patients on the Internet.

William Wong is professor of human-computer interaction, and head of the Interaction Design Centre, Middlesex University, London, UK. Before joining academia, Dr. Wong was an Air Defence Controller, and head of the Systems and Communications Operations Branch, HQ Republic of Singapore Air Force. He joined the University of Otago, New Zealand, in 1992, where he set up and directed the Multimedia Systems Research Laboratory, Department of Information Science, leaving the University as associate professor in 2003. He has conducted research into naturalistic decision making, situation awareness, and representation design in emergency ambulance control in New Zealand, Australia, and the UK; in air traffic control in Europe, exploring 3-D and 4-D representation design, the representation of shared weather information; minimisation of change blindness using visual depth design techniques; and the design of control interfaces for hydro-electricity generation in Australia. Dr Wong gained his PhD in human-computer interaction from the University of Otago, NZ in 1999.

Keiji Yamanaka received his PhD degree in electrical and computer engineering from Nagoya Institute of Technology, Japan, in 1999. He is currently working in the Faculty of Electrical Engineering, Federal University of Uberlândia, Brazil, teaching undergraduate and graduate courses. He has several master and doctoral students under his supervision and his research interests include neuro-computing, artificial intelligence, and pattern recognition.

Nikoletta Yiannoutsou was born in Athens, Greece (1973), and received a degree in psychology (1994) from the University of Athens (NKUA), an MSc in intelligent tutoring systems (1996) from University of Nottingham, and a Phd in educational technology (2005) from the University of Athens (NKUA). Currently, she is an adjunct Lecturer at the University of Peloponnese. Her research interests focus on the role of new representational media in the learning process and on the design of innovative educational activities based on ICT.

Dong Yu joined Microsoft in 1998 and the Microsoft Speech Research Group in 2002. He holds a PhD in computer science from the University of Idaho, an MS degree in computer science from Indiana University / Bloomington, an MS degree in electrical engineering from the Chinese Academy of Sciences, and a BS degree (with honors) in electrical engineering from Zhejiang University (China). His research interests include speech processing, pattern recognition, and computer and network security. He has published dozens of papers in these areas. He is a senior member of IEEE.

Ioannis D. Zaharakis received a BSc in mathematics in 1992 and a PhD in software engineering in 1999, both from the University of Patras, Hellas. He is a member of the Hellenic AI Society and the Hellenic Mathematics Society. During 1993-1999, he was a researcher in the Educational Software Development Lab, University of Patras, where he was involved in the specification, design and implementation of intelligent tutoring systems. He has participated in several EU-funded and national R&D projects. He has authored one textbook for the Hellenic Open University and he is currently teaching in the Technological Educational Institute of Patras. In 2002, he joined CTI/RU3 as a Researcher / R&D Engineer, focusing on the formal specification of component-based ubiquitous computer systems.

Martina Ziefle is a psychologist and professor at RWTH Aachen University, Germany. Her research interests cover a broad field of key issues within human computer interaction, cognitive ergonomics and interface design. A main research topic is concerned with the user-centered interface design of small screen devices and the impact of user diversity for technical devices.

Index

Symbols

3-D visualization 558–575
  GUI design remarks 564
  high-performance 3-D remote visualization 565
    distributed visualization service (DVS) 566
    mobile visualization client 567
  local computation 559
  remote computation 561

A

adaptive
  interfaces in mobile environments 302–317
    abstract user interface adaptation 304
    adaptation to devices 306
      design-time adaptation 306
      mobile agent adaptation 308
      run-time adaptation 307
    adaptation to users 309
    adaptive user interface system (ADUS) 309
  mobile learning management system (AM-LMS) 286–301
    analyzing learning style 296
    structure 295
ambient system (AmS) 369
audio-based memory aid 1031–1048
  personal audio loop (PAL) 1032
    final prototype 1038
    formative evaluations 1033
    making PAL socially and legally acceptable 1044
    making PAL ubiquitous 1043
    making PAL useful 1043
    usefulness of PAL 1036
average ranked list position (ARP) 417

C

camera phones in social contexts 55–68
  barriers to sharing 64
  situated use 58
  social uses 60
cognitive models as usability testing tools 814–829
  architectures 820
    goals, operators, methods, and selection rules (GOMS) 821
  descriptive vs. generative models 822
    atomic component of thought with rational analysis (ACT-R) 823
  ISO quality models 818
collaborative learning 270, 272
  an environment for cognitive engagement 275
    mobile technology supported classroom 275
  cognitive conflict 271
  cognitive elaboration 271
  cognitive tool (CT) 271
  mobile learning 273
  pedagogical design 279
    encouraging reciprocal tutoring 281
collaborative mobile applications
  field study 997–1014
    data analysis 1002
      through ActivityLens 1003
    data collection techniques 999
computer
  -supported collaborative learning (CSCL) 1068
  -supported collaborative work (CSCW) 1068
  -supported intentional learning environments (CSILE) 1068
context 187–204
  -aware mobile interfaces 759–779
    designing 770
    mobile use context 761
    wizard of oz evaluation 770
  for mobile applications? 192
    ontology-based model 194
  mobile context-aware applications 208
    design guidelines 212
    support for interaction design 210
    usability 209
      risks 210
  perils of context-awareness 191
  supporting user interaction 197
  utilisation in mobile applications 190

what is context-awareness? 206
relevance to human-computer interaction (HCI) 207
relevance to mobile HCI 207
what is it? 189

D

disambiguation accuracy (DA) 417
distraction classification 973
three studies 974–978

E

electromyographic (EMG) 524–542
electrodes, recording, and applications 526
for human-computer interaction (HCI) 527
intimate communication armband 528
hardware 529
signal 526
subtle gestures 528
1st study: learning and recognition rate 533
2nd study: multimodal realistic interaction 535
3rd study: assessing noticeability 537
model 531
engineering emergent ecologies 364–385
an example: virtual residence 372
bio-inspired approaches 367
engineering approach 373
AmI spheres and collective behaviour 374
awareness and presence 374
interacting with AmI spheres 374
GAS approach 375
interaction 367
symbiotic AmI spaces 367
ethnography and interface design 3
design sketching 9
informing design 5
interpreting data 4
in the design process 4
prototyping 11
experimental ethno-methods 16–34
experimental prototypes 25
public interactive display 26
results analysis 27
theatre workshops: personas and scenarios 22
to evaluate the user experience with mobile interactive systems 16–34
exploring starfield displays 576–593
fisheye 583
interfaces 584
overvies+detail 580
interfaces 581
smooth-zooming 578
interface 578
extensible user-interface language (XUL) 310
eye movement studies of mobile readability 945–971
cathode ray tube (CRT) 951
evaluation methodology 955
general linear model (GLM) 956
liquid crystal display (LCD) 951
reading on small screens 951
study one: reading on a PDA 957
study three: reading on a mobile phone 963
study two: verifying the results 960
text presentation formats 952
the reading process 947
cognitive processing 948
measuring readability 949
physiological limitations 947
thin-film transistor (TFT) technology 951

F

field laboratory for evaluating in situ 982
close-up video and improved sound 985
increasing battery lifetime 991
minimizing equipment 991
small cameras and video sources 988
flexible organic light emitting diodes (FOLEDs) 179

G

gadgetware architectural style (GAS) 196
generation of GUIs (indirectly) 311
graphical partitioning model(GPM) 274
graphical user interfaces (GUIs) 302

H

heuristic evaluation methods 780–801
appropriating usability heuristics 785
mobile usability issues 786
toward a set of heuristics 787
environment of mobile infrastructure 796
limitations 784
mobile devices, applications, and their context 782
nature of mobile devices 795
strengths 784
human-computer interaction (HCI) 731–744
defining evaluation targets 732
designing an evaluation protocol 739–740

making sense of human activity 736–737
referent models 734
human mobile computing performance 830–846
applying Fitt's law 834
experiment on mobile input performance 835
input time and Fitt's law 839
mobility and HCI 832

I

in-car user-interfaces 218–236
case study: vehicle navigation systems 229
design and evaluation 223
15 second rule 228
field trials 224
keystroke level model (KLM) 228
lane change task 228
peripheral detection task 227
road trials 225
simulator trials 225
human-centered design process 220
environments 223
equipment 222
tasks 221
users 220
types of in-car computing systems 219
individuals with disabilities 609–623
design of assistive technologies 613
instrumented usability analysis 928
case study of walking and tapping 931
example: mobile text entry 930
the Hilbert transform 934
intelligent user interfaces (IUIs) 318–329
artificial intelligence (AI) in mobile computing 322
artificial intelligent (AI) in mobile computing techniques 323
reflections on context 320
device characteristics 321
prevailing environment 321
social situation 322
the intelligent agent paradigm 324
interface definition language (IDL) 369

K

keystrokes per character (KSPC) 417

L

language understanding 469
learning-disabled children 142
method 144
participants' experience with technology 146

M

media services language (MSL) 658
micro-electrical-mechanical systems (MEMS) 160
mobile
applications and mental health 635–656
adaptable systems 646
case study: "mobile mood charting" 649
design of chart 650
design 638
for adolescents 639
for therapists 639
design recommendations 644
multistage prototyping 647
software to support psychotherapy 640
supporting mental health interventions 637
camera-based user interaction 543–557
computer vision technologies 544
markerless tracking 545
tagging-based systems 546
mapping camera motion 547
prototype 550
applications 551–553
high-level algorithm description 550
collaboration in learning environments 1069
collaboration components 1074
paper prototype testing 1070
design for older adults 624–634
meetings/discussions 628
physical interfaces 629
recruiting older target populations 627
virtual interfaces 630
devices as museum guides 256
example of mobile activity design 262
evaluations in a lab environment 910–926
distractions 913
evaluation 1: audio and visual navigational cues 914
evaluation 2: comparison of wearable displays 919
learning 287
environment 288
research trends 289
styles 290
four dimensions 291
index of learning style (ILS) 292
user interface 294
learning in museums 253–269
mixed systems 346
3-D simulation environment 349

ASUR model 350
basic principles 350
extension 351
designing 348
SIMBA 354
element model 355
overall process 354
simulation 358
telephones for rendezvousing 35
a diary study 37
method 38
performance deficits: user experience 43
results 39
design implications 45
model-based sonification 481
doppler effect 483
experiments
one 485–491
two 493–503
human operator modeling 499
quickening 482
multilayered evaluation approach 850
experiment: comparing field and laboratory use of a PDA 851
WebQuest Tool 854
multimodal user interface (MUI) 462
multiplatform e-learning systems 1083
evaluation methodology 1086
overall learner satisfaction score 1090
participants information 1089

N

navigational aid for blind pedestrians 693–710
aids 694
user- and activity-centered approaches 695
activity-centered approach 699
user-centered approach 697
nonspeech audio 676–692
advantages of using our ears 676
benefits 678
ecological psychology approach 678
experimental process 680
sound localization process 679
spatial conceptualization process 684
experiments 686
virtual courses 685
virtual 3-D acoustic space 679

O

one-handed use of mobile devices 86–101
field study 88
thumb movement study 93
design 94
equipment 93
Web survey 90
optical fiber flexible display (OFFD) 178

P

photo management on a mobile device 69–85
designing mobile interface 75
enhancing interaction 77
context-awareness 78
online photoware for sharing and photobBlogging 73
photo browsing techniques 76
stand-alone photoware 73
privacy regulation model 863–876
case study: privacy perception of the PePe system 869
five factors affecting information disclosure 866–868
previous research 865
projected displays for collaboration 594–608
Hotaru (Firefly) 595
intuitive manipulation techniques 599–601
examples 600
of mobile devices 596
user studies 601
experiment 1 602
experiment 2 603
prototyping tools 330–345
building a high fidelity prototype 341
SUEDE 330
topiary 330
with storyboards 332
wizard of oz (WOz) testing 335

Q

question-answer relationships (QAR) 1069

R

radio frequency identification (RFID) technology 657
application fields 660
EuroFlora guide 664
structure of the interface 666
integration of RFID subsystem 662

MADE support 659
location-aware computing 659
mobile applications development environment (MADE) 658
architecture 660

S

smart
garments
applications 184
embedded technologies 177
microprocessors 179
power, radiation, and the environment 180
ergonomics of intelligent clothing 180
aesthetics vs. function 182
cut, connectors, and material 181
wheelchair
adaptability 717
alternative navigation models 724
behaviour-based interaction 725
physical interface 722
structure 712
user interface 713
design constraints 714
what is it? 712
wheelchairs 711–730
speech-based user interfaces (UI) 237
automotive UI design principles 239
recommendations 240
recent automotive spoken UIs 242
speech-in list-out approach (SILO) 245
speech-centric user interface design 461–477
generic MUI architecture 463
modality fusion 470
special considerations for speech modality 465
context-aware language model 469
modality switching 468
resource constrained speech recognition 466
speed-dependent automatic zooming (SDAZ) 589
stroke-based input 426–445
Chinese characters 427
mobile input solutions 428
handwriting recognition 428
pinyin method 428
structure-based methods 429
Motorola iTap™ stroke input method 430

T

technology acceptance model (TAM) 103
for mobile services (TAMM) 106
text entry 408–425
disambiguation 412
evaluation 417
keyboards 409
ambiguous 411
unambiguous 409
stylus-based 414
gesture-based input 416
handwriting 415
on-screen keyboards 414
tourist digital assistant (TDA) 658
transgenerational designs 122–141
assessments 126
implications for design 135
independent and dependent variables 124
learnability effects 131
menu navigation performance 130

U

ubiquitous mobile input 386–407
design space of input devices 387
orient 394
positioning tasks 388
continuous direct interactions 390
continuous indirect interactions 388
discrete direct interactions 391
discrete indirect interactions 391
positioning techniques 392
spatial layout of design space 401
text 399
UI design in a closed environment 1015
competing technologies 1019
paticipatory design 1023
patient monitoring unit (PMU) 1017, 1025
physiological monitoring 1018
strategic user needs analysis (SUNA) 1020
steps 1021
usage context 1017
user-centred design (UCD) 1019
unobtrusive movement interaction 507–523
continuous detection reliability: experiments 514
customization 511
sensor interaction cover 511
use cases and usability 513
usability
evaluation methods (UEMs) 745–758
case study: towards a real world lab 752
current UEM framework 747
cultural probes 750
for mobile applications 746
factors of mobile phones 877–896

case studies 890–892
developing a framework 881
hierarchical model of impact factors 883
phones and tasks 879
user acceptance of mobile services 102–121
applicability of earlier approaches 105
design implications 110
perceived ease of adoption 116
perceived ease of use 112

V

validity laboratory test results 897–909
challenges of mobility 899
suggestions for field testing 904
logistics 905
usability testing 900
principles 900
visualising meeting recordings on small screens 1052
meeting browser evaluation test (BET) 1057
voice-enabled user interfaces 446–460
the prototype 448
managed applications 450–454
program manager 454
graphical user interface (GUI) 454
underlying speech technology 449
speech recognition 449
speech synthesis 450

W

W3C device independence activities 1082
wearable computers 158–175
computer response to physical activity 164
emotional impact 168
finding and retrieving information 166
human factors 158–175
form-factor and physical attachment 160
navigation and wayfFinding 165
perceptual impacts 163
physical effects 161
reducing size and separating components 162
supporting memory 165
wizard of oz for evaluating 802–813
in the development lifecycle 804
method 803
studies for mobile technology 805
variability 806
wozzing 806
cautions 810

Z

zoomable user interface (ZUI) 577